Essentials of 3D Biofabrication and Translation

Essentials of 3D Biofabrication and Translation

Edited by

Anthony Atala, MD and James J. Yoo, MD, PhD
Wake Forest Institute for Regenerative Medicine,
Wake Forest School of Medicine,
Winston-Salem, NC, USA

AMSTERDAM • BOSTON • HEIDELBERG • LONDON • NEW YORK • OXFORD • PARIS
SAN DIEGO • SAN FRANCISCO • SINGAPORE • SYDNEY • TOKYO

Academic Press is an Imprint of Elsevier

Academic Press is an imprint of Elsevier
125, London Wall, EC2Y 5AS, UK
525 B Street, Suite 1800, San Diego, CA 92101-4495, USA
225 Wyman Street, Waltham, MA 02451, USA
The Boulevard, Langford Lane, Kidlington, Oxford OX5 1GB, UK

British Library Cataloguing-in-Publication Data
A catalogue record for this book is available from the British Library

Library of Congress Cataloging-in-Publication Data
A catalog record for this book is available from the Library of Congress

ISBN: 978-0-12-800972-7

For information on all Academic Press publications
visit our website at http://store.elsevier.com/

Publisher: Mica Haley
Acquisition Editor: Mica Haley
Editorial Project Manager: Lisa Eppich
Production Project Manager: Julia Haynes
Designer: Matt Limbert

Typeset by Thomson Digital

Printed and bound in USA

Dedication

This book is dedicated to Katherine,
Christopher, and Zachary

Anthony Atala

This book is dedicated to Yook-IL, Kyung Whan,
Kyung Jin, and Kyung Min

James J. Yoo

Contents

18 Bioprinting of Cartilage

Kuilin Lai and Tao Xu

19 Biofabrication of Vascular Networks

James B. Hoying and Stuart K. Williams

20 Bioprinting of Blood Vessels

Anthony J. Melchiorri and John P. Fisher

21 Bioprinting of Cardiac Tissues

Daniel Y.C. Cheung, Bin Duan, and Jonathan T. Butcher

22 Bioprinting of Skin

Julie Marco, Anthony Atala, and James J. Yoo

23 Bioprinting of Nerve

Christopher Owens, Francoise Marga, and Gabor Forgacs

24 Bioprinting: An Industrial Perspective

Kristina Roskos, Ingrid Stuiver, Steve Pentoney, and Sharon Presnell

List of Contributors

Kenichi Arai
Graduate School of Science and Engineering for Research, University of Toyama, Toyama, Japan

Anthony Atala
Wake Forest Institute for Regenerative Medicine, Wake Forest School of Medicine, Medical Center Boulevard, Winston-Salem, NC, USA

Rashid Bashir
Department of Bioengineering; Department of Electrical and Computer Engineering; Micro and Nanotechnology Laboratory, University of Illinois at Urbana-Champaign, Urbana, IL, USA

Danielle Beski
Materialise USA, LLC, Plymouth, MI, USA

Jonathan T. Butcher
Department of Biomedical Engineering, Cornell University, Ithaca, NY, USA

Hyung-Gi Byun
Division of Electronics, Information and Communication Engineering, Kangwon National University, Korea

James K. Carrow
Department of Biomedical Engineering, Texas A&M University, College Station, TX, USA

Sylvain Catros
Tissue Bioengineering, University Bordeaux Segalen, Bordeaux, France

Daniel Y.C. Cheung
Department of Biomedical Engineering, Cornell University, Ithaca, NY, USA

Dong-Woo Cho
Department of Mechanical Engineering, Pohang University of Science and Technology (POSTECH), Pohang, Kyungbuk, South Korea

David Dean
Department of Plastic Surgery, The Ohio State University, Columbus, OH, USA

Aurora De Acutis
Research Center E. Piaggio, University of Pisa, Pisa, Italy

Carmelo De Maria
Research Center E. Piaggio; Department of Ingegneria dell'Informazione, University of Pisa, Pisa, Italy

Bin Duan
Department of Biomedical Engineering, Cornell University, Ithaca, NY, USA

Tom Dufour
Materialise N.V., Leuven, Belgium

John P. Fisher
Fischell Department of Bioengineering, University of Maryland, College Park, MD, USA

Colleen L. Flanagan
Department of Biomedical Engineering, University of Michigan, Ann Arbor, MI, USA

Gabor Forgacs
Department of Physics and Astronomy, University of Missouri; Modern Meadow Inc. Missouri Innovation Center; Department of Biomedical Engineering, University of Missouri, Columbia, MO, USA

Jean-Christophe Fricain
Tissue Bioengineering, University Bordeaux Segalen, Bordeaux, France

Akhilesh K. Gaharwar
Department of Biomedical Engineering; Department of Materials Science and Engineering, Texas A&M University, College Station, TX, USA

Frederik Gelaude
Mobelife N.V., Leuven, Belgium

Glenn E. Green
Department of Otolaryngology – Head and Neck Surgery, Division of Pediatric Otolaryngology, University of Michigan, Ann Arbor, MI, USA

Fabien Guillemot
Tissue Bioengineering, University Bordeaux Segalen, Bordeaux, France

Scott J. Hollister
Department of Biomedical Engineering; Department of Mechanical Engineering; Department of Surgery, University of Michigan, Ann Arbor, MI, USA

James B. Hoying
Department of Physiology, Cardiovascular Innovation Institute, University of Louisville, Louisville, KY, USA

Jeung Soo Huh
Department of Materials Science and Metallurgy, Kyungpook National University, Korea

Ashok Ilankovan
Materialise SDN. BHD., Selangor, Malaysia

Shintaroh Iwanaga
Institute of Industrial Science, University of Tokyo, Tokyo, Japan

Manish K. Jaiswal
Department of Biomedical Engineering, Texas A&M University, College Station, TX, USA

Jinah Jang
Division of Integrative Biosciences and Biotechnology, Pohang University of Science and Technology (POSTECH), Pohang, Kyungbuk, South Korea

Hyun-Wook Kang
Biomedical Engineering, School of Life Sciences, Ulsan National Institute for Science and Technology, Ulsa, Korea

Carlos Kengla
Wake Forest Institute for Regenerative Medicine, Wake Forest School of Medicine; School of Biomedical Engineering and Sciences, Wake Forest University Virginia Tech, Winston-Salem, NC, USA

Punyavee Kerativitayanan
Department of Biomedical Engineering, Texas A&M University, College Station, TX, USA

Virginie Keriquel
Tissue Bioengineering, University Bordeaux Segalen, Bordeaux, France

Maryna Kvasnytsia
Materialise N.V., Leuven, Belgium

Joseph M. Labuz
Department of Biomedical Engineering, College of Engineering, University of Michigan, Ann Arbor, MI, USA

Kuilin Lai
Medprin Regenerative Technologies Co. Ltd; School of Materials Science and Engineering, South China University of Technology, Guangzhou, Guangdong Province, China

Michael Larsen
Department of Plastic Surgery, The Ohio State University, Columbus, OH, USA

Hui Chong Lau
Department of Biomedical Science, Joint Institute for Regenerative Medicine, Kyungpook National University, Korea

Mike Lawrenchuk
Materialise USA, LLC, Plymouth, MI, USA

Jin Woo Lee
Department of Molecular Medicine, Graduate School of Medicine, Gachon University, Incheon City, South Korea

Sang Jin Lee
Wake Forest Institute for Regenerative Medicine; School of Biomedical Engineering and Sciences, Wake Forest School of Medicine, Winston-Salem, NC, USA

Brendan M. Leung
Department of Biomedical Engineering, College of Engineering, University of Michigan, Ann Arbor, MI, USA

Grace J. Lim
Department of Biomedical Science, Joint Institute for Regenerative Medicine, Kyungpook National University, Korea

Giriraj Lokhande
Department of Biomedical Engineering, Texas A&M University, College Station, TX, USA

Ihor Lukyanenko
Materialise N.V., Leuven, Belgium

Julie Marco
Wake Forest Institute for Regenerative Medicine, Wake Forest School of Medicine, Medical Center Boulevard, Winston-Salem, NC, USA

Francoise Marga
Department of Physics and Astronomy, University of Missouri; Modern Meadow Inc. Missouri Innovation Center, Columbia, MO, USA

Anthony J. Melchiorri
Fischell Department of Bioengineering, University of Maryland, College Park, MD, USA

Tyler K. Merceron
Wake Forest Institute for Regenerative Medicine, Wake Forest School of Medicine, Winston-Salem, NC, USA; Vanderbilt University School of Medicine, Vanderbilt University, Nashville, TN, USA

Michael Miller
Department of Plastic Surgery, The Ohio State University,
Columbus, OH, USA

Mariam Mir
Materialise N.V., Leuven, Belgium

Ruchi Mishra
Department of Plastic Surgery, The Ohio State University,
Columbus, OH, USA

Christopher Moraes
Department of Biomedical Engineering, College of
Engineering, University of Michigan, Ann Arbor,
MI, USA

Lorenzo Moroni
Department of Complex Tissue Regeneration (CTR),
Institute for Technology-Inspired Regenerative
Medicine (MERLN), Maastricht University, Maastricht,
The Netherlands

Robert J. Morrison
Department of Otolaryngology – Head and Neck Surgery,
Division of Pediatric Otolaryngology, University of
Michigan, Ann Arbor, MI, USA

Carlos Mota
Department of Complex Tissue Regeneration (CTR),
Institute for Technology-Inspired Regenerative Medicine
(MERLN), Maastricht University, Maastricht, The
Netherlands

Sean V. Murphy
Wake Forest Institute for Regenerative Medicine,
Wake Forest School of Medicine, Winston-Salem,
NC, USA

Makoto Nakamura
Graduate School of Science and Engineering for Research,
University of Toyama, Toyama, Japan

Hassan Nasser
Department of Otolaryngology – Head and Neck Surgery,
Division of Pediatric Otolaryngology, University of
Michigan, Ann Arbor, MI, USA

Lars Neumann
Materialise N.V., Leuven, Belgium

Anthony Nguyen
Materialise USA, LLC, Plymouth, MI, USA

Christopher Owens
Department of Physics and Astronomy, University of
Missouri, Columbia, MO, USA

Falguni Pati
Department of Mechanical Engineering, Pohang
University of Science and Technology (POSTECH),
Pohang, Kyungbuk, South Korea

Steve Pentoney
Organovo, Inc, San Diego, CA, USA

Sharon Presnell
Organovo, Inc, San Diego, CA, USA

Ritu Raman
Department of Mechanical Science and Engineering;
Micro and Nanotechnology Laboratory, University of
Illinois at Urbana-Champaign, Urbana, IL, USA

Kristina Roskos
Organovo, Inc., San Diego, CA, USA

Emilie Sauvage
Materialise N.V., Leuven, Belgium

Young-Joon Seol
Wake Forest Institute for Regenerative Medicine,
Wake Forest School of Medicine, Winston-Salem,
NC, USA

Aleksander Skardal
Wake Forest Institute for Regenerative Medicine, Wake
Forest School of Medicine, Winston-Salem, NC, USA

Ana Soares
Mobelife N.V., Leuven, Belgium

Ingrid Stuiver
Organovo, Inc., San Diego, CA, USA

Shuichi Takayama
Department of Biomedical Engineering; Macromolecular
Science and Engineering Center, College of Engineering;
Biointerfaces Institute, University of Michigan,
Ann Arbor, MI, USA

Katrien Vanderperren
Materialise N.V., Leuven, Belgium

Dieter Vangeneugden
Materialise N.V., Leuven, Belgium

Giovanni Vozzi
Research Center E. Piaggio; Department of Ingegneria
dell'Informazione, University of Pisa, Pisa, Italy

Matthew B. Wheeler
Institute for Genomic Biology and Department of Animal
Sciences, University of Illinois, Urbana-Champaign, IL,
USA

Stuart K. Williams
Department of Physiology, Cardiovascular Innovation
Institute, University of Louisville, Louisville, KY, USA

Tao Xu
Department of Mechanical Engineering, Bio-
manfuactuirng Center, Tsinghua University, Beijing;
Medprin Regenerative Technologies Co. Ltd, Guangzhou,
Guangdong Province, China

James J. Yoo
Wake Forest Institute for Regenerative Medicine,
Wake Forest School of Medicine, Medical Center
Boulevard, Winston-Salem, NC, USA

David A. Zopf
Department of Otolaryngology – Head and Neck Surgery,
Division of Pediatric Otolaryngology, University of
Michigan, Ann Arbor, MI, USA

Preface

Biofabrication, also known as additive manufacturing, is driving major innovations in various fields, including engineering, manufacturing, and medicine. This transformative technology is enabling rapid construction of 3D structures with complex geometries using a diverse range of synthetic and natural materials. Recent breakthroughs have enabled 3D printing of cells, biocompatible materials, and supporting components into complex 3D functional living tissues. This developing field, commonly referred to 3D bioprinting, promises to revolutionize the field of medicine addressing the dire need for tissues and organs suitable for transplantation.

While the overall philosophy of 3D bioprinting is similar to that currently applied for nonbiological 3D printing, there are many additional complexities, such as the choice of materials, cell types, growth/differentiation factors, and the technological challenges that occur due to the sensitivities of these living, biological structures. Addressing these new complexities requires the integration of cutting edge technologies from the fields of engineering, biomaterials, cell biology, physics, computer sciences, and medicine. Current advances have facilitated the generation of transplantable tissue structures including multilayered skin, vascular and nerve grafts, tracheas, muscle, bone, cartilage, and heart tissues. Other applications include using high throughput 3D bioprinted tissue models for research and drug discovery.

This book brings together leaders in the field to define this important area of development. It provides a comprehensive overview of the applications of 3D bioprinting technologies, and covers the essentials of 3D biofabrication for translation, including hardware setup, software, design, bioink, and applications. This book aims to provide critical information and insights on the current approaches and future prospect of 3D biofabrication. We hope that this book will serve as a guide and resource to assist researchers in interfacing between the biological (cells, supportive factors, etc.), the materials (hydrogels, polymers, bioinks, etc.), the devices and the support technologies (lasers, jets, nozzles, software, etc.) used for the fabrication process.

We would like to thank the authors who contributed to the making of this book as well as the publishing team at Elsevier.

Anthony Atala
James J. Yoo

Chapter 1

Bioprinting Essentials of Cell and Protein Viability

Aleksander Skardal

Wake Forest Institute for Regenerative Medicine, Wake Forest School of Medicine, Winston-Salem, NC, USA

Chapter Outline

ABSTRACT

Bioprinting has emerged in recent years as an attractive method for engineering of 3D tissues and organs in the laboratory, which can subsequently be implemented in a number of regenerative medicine applications. Currently, the primary goals of bioprinting are to (1) create complete replacements for damaged tissues in patients and (2) rapidly fabricate small-sized human-based tissue models or organoids for high-throughput diagnostics, pathology modeling, and drug development. Regardless of which of these end applications are targeted, successful biofabrication using bioprinting technology relies on a set of four essential characteristics that must be optimized. In this chapter, we will discuss and evaluate integration of cell sourcing, biomaterial support, bioprinting device compatibility, and postfabrication tissue support, and how these characteristics are mandatory considerations for practical realization of viable and functional tissue engineered organ and organoid structures.

Keywords: bioprinting; biomaterials; biocompatibility; viability; biofabrication; stability; bioink; cells; hydrogel

1 AN INTRODUCTION TO BIOPRINTING

Bioprinting has emerged as a flexible tool in regenerative medicine with potential in a variety of applications. Bioprinting is a relatively new field within biotechnology that can be described as a robotic additive biofabrication that has the potential to build or pattern viable organ-like or tissue structures [1]. In general, bioprinting uses a computer-controlled 3D printing device to accurately deposit cells and biomaterials into precise geometries with the goal being the creation of anatomically correct biological structures. Generally, bioprinting devices have the ability to print cell aggregates, cells encapsulated in hydrogels or viscous fluids, or cell-seeded microcarriers – all of which can be referred to as "bioink" – as well as cell-free polymers that provide mechanical structure or a more hospitable environment [2,3]. Biologically inspired computer-assisted designs can be used to guide the placement of specific types of cells and materials into precise, planned geometries that mimic the architecture of actual tissue construction, which can subsequently be matured into functional tissue constructs or organs [4–6].

Essentials of 3D Biofabrication and Translation. http://dx.doi.org/10.1016/B978-0-12-800972-7.00001-3

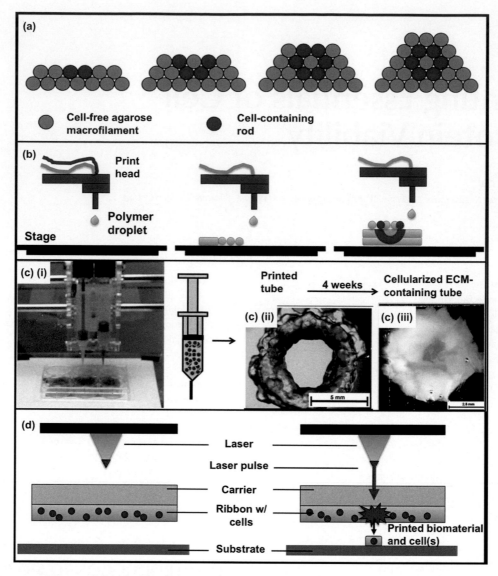

FIGURE 1.1   Common printing modalities for bioprinting applications. (a) Scaffold-free printing; (b) inkjet droplet printing; (c) extrusion printing; and (d) laser-induced forward transfer (LIFT)-based printing. (c) (i) A syringe-based extrusion printer; (c) (ii) Extruded cell-hydrogel tubes that (c) (iii) mature over time into cellularized ECM-containing tubes.

To date, a complete fully functional organ has not been printed, but it remains the primary long-term goal of bioprinting. However, small-scale "organoids" are currently being implemented in a number of applications, including pathology modeling, drug development, and toxicology screening.

1.1   Printing Modalities

A number of bioprinting approaches have been recently explored, encompassing the use of inkjet-like printers, extrusion devices, and laser-assisted devices. To better understand the essential components required for successful printing of viable tissue and organ structures, here we will describe some of the most common printing modalities currently being employed (Fig. 1.1). Further discussion of these technologies and their applications will be discussed in later chapters.

1.1.1   Inkjet Printing

Inkjet printing, also referred to as drop-by-drop bioprinting, is one bioprinting approach that is being explored for creating 3D biological structures, and is closely related to technologies used for cell patterning. Where basic cell patterning creates a 2D pattern comprised of cells on a surface, by incorporating a hydrogel or other cell-friendly biomaterial, 3D cellularized structures can be fabricated drop by drop [7,8]. These types of bioprinters often use a cartridge-based delivery system mounted on an *XYZ* plotting device. The cartridge system is similar to that used in traditional inkjet printing such that cells and biomaterial components can be loaded into individual cartridges for computer-controlled deposition. Examples of this implementation for 3D construct fabrication include collagen-encapsulated smooth muscle cells that were printed in droplet form to create muscular patches [9], and the use of alginate and fibrin gel droplets for creating structures such as cellular fiber and multilayered cell sheets [10]. We have shown that this approach to bioprinting is also effective for skin printing to aid wound healing, which we will further describe later in this chapter [11]. In terms of 3D fabrication, the drop-by-drop approach relies on being able to quickly polymerize or stabilize the printed material in place, so that subsequent droplets can be added to the growing structure. Polymerization rates of the printed materials are a direct product of the various cross-linking chemistries innate to materials used, and are an essential consideration for successful inkjet printing. The requirement for a fast-gelling material places a limitation on the types of materials that can successfully be applied in this manner. Additionally, as the printable droplets are typically small volumes, scaling up to fabricate a large organ structure might be difficult. On the other hand, the small droplet volumes support high-resolution printing of intricate structures.

1.1.2 Scaffold-Free Printing

"Scaffold-free" bioprinting technology is based on the principles of tissue liquidity and tissue fusion. In this approach, aggregates or rods comprised solely of cells are printed in geometric patterns or shaped and allowed to fuse over time to form larger constructs [12]. Typically, multiple layers of aggregates or rods are printed, and after fusing, singular 3D structures remain. In a pioneering work, this approach was used to build branched vascular structures [13], and more recently nerve grafts [14]. It should be noted that this type of bioprinting does actually rely on biomaterials, and as such the "scaffold-free" term could be taken as a misnomer of sorts. In most cases, the cell aggregates or cell rods are either printed into a biopolymer or additively stacked using space-holding biomaterials to preserve the appropriate structures during the tissue fusion and maturation process. Generally, these space-holders are eventually removed when the construct is sufficiently fused and possesses the mechanical properties to support itself. The strength of this method lies within its high cellularity, which allows for rapid fusion between discretely printed pieces. Unfortunately, preparation times, printing speeds, and building material volumes currently limit its scalability. Nevertheless, this method has been explored extensively and is the basis for commercially available technologies. Additionally, we have explored a hydrogel-based approach that mimicked this technique, by printing hydrogel and cell-hydrogel rods to create tubular constructs *in vitro* [15].

1.1.3 Extrusion-Based Deposition

Extrusion-based deposition, generally from syringe-like equipment, is an additional approach for 3D bioprinting that relies on the mechanical and temporal properties of the polymer materials being printed. In this modality, the properties of the printed polymer or hydrogel are used to facilitate extrusion through a syringe tip, commonly driven by pneumatic pressure or mechanical pistons controlled by a computer. The reliance on the material properties for printing means that the material must be soft or fluid enough to facilitate extrusion through the small diameter tip or nozzle, but must also support itself mechanically after deposition. One common approach is employing melt-curable polymers such as polycaprolactone, which when heated can be deformed and printed at relatively high resolutions, and cool down to a solid material. These materials can be used to build rather large and intricate structures, capable of mimicking physiological structures. However, melt-cure printing cannot be done with cells. Cells must be seeded at a later time, and as such other cell-friendly materials, such as hydrogels, must be incorporated. Printing with hydrogels via extrusion techniques can be difficult when working with materials that rely on time for gelation to occur. Mistiming the deposition process can result in either a structure that collapses because cross-linking has not occurred quickly enough, or conversely, clogging of the bioprinting device as a result of polymerization that was too fast. However, numerous studies have implemented novel cross-linking chemistries and methods to facilitate spatial and temporal control over materials that can contain encapsulated cells for printing cell-containing 3D structures [16–18].

1.1.4 Laser-Induced Forward Transfer

Laser-induced forward transfer (LIFT)-based bioprinting is a recently introduced method that has been adopted from other fields by researchers pursuing bioprinting [19,20]. LIFT technology was initially developed for high-resolution patterning of metals for use in areas such as computer chip fabrication. More recently it has been employed to create micropattern peptides, DNA, and cells. LIFT technology is comprised of a laser beam that is pulsed at desired time lengths and a donor material "ribbon" comprised of the printable material. This is supported on a transport layer, such as gold or titanium, which absorbs the laser energy and transfers it to the ribbon. When the laser pulses on the ribbon, the focused energy generates an incredibly small, high-pressure bubble that propels a droplet of the donor material onto a collecting substrate and stage. By either moving the stage or the laser in relation to the ribbon, material can be patterned on the collecting substrate [21–23]. In the case of LIFT-based bioprinting, the ribbon may be comprised of a biopolymer or protein, and can contain cells within. In this scenario the laser pulse-driven ribbon droplets contain cells, which are then deposited in a pattern on the substrate to create cellular structures and patterns. The lack of a nozzle in LIFT is a departure from other printing modalities, and does away with the need to prevent clogging issues. This results in increased flexibility in the printing materials, as long as they can be sufficiently transferred by the energy supplied by the laser. Studies have shown little to no negative effects on cell viability [24–26] and the ability to print nearly a single cell per droplet [27], positioning LIFT as a bioprinted modality with much potential in the future.

The noted high resolution of LIFT is directed by a number of variables, such as the laser itself, the material properties of the printable material, the relative hydrophilic/hydrophobic nature of the substrate material to the printable material, cell density, and the distance between the ribbon and the substrate [28]. Thus, there are also some challenges that need to be overcome. The high resolution and subsequently small printing volume per laser pulse requires fast gelation kinetics of the printable material and a fast moving stage for fabrication. In current LIFT methods, preparation times of the ribbon, especially when containing cells and thus cell-friendly biomaterials, can be time consuming. Furthermore, to create structures of size, multiple ribbons are often employed, requiring reloading during the printing process.

1.2 Essential Considerations

Regardless of the specific modality employed for bioprinting, there are four general variables that need to be thoroughly considered to successfully bioprint viable and functional tissue constructs. First is the cellular component comprising the living portion of the construct, which can be comprised of one cell type, but more often than not requires a complex but elegant interplay between multiple cell types to achieve significant tissue function. Second is the inclusion of supportive biomaterials, generally in the form of proteins and polymers, that (1) facilitate the deposition method by mechanical means and (2) provide support and protection to the cells during and after the tissue construct fabrication process.

These biomaterials can encompass the physical environment inside of which the cells will reside, as well as the biochemical signals cells need to function as they would in the body. Third is the actual bioprinting or biofabrication device itself and the associated method by which fabrication is performed, be it inkjet, extrusion printing, or another modality. The way that these first three variables are integrated plays the largest role in successful bioprinting. However, equally important for the end application is a fourth consideration, namely, postfabrication maintenance and maturation. This step often employs the use of bioreactors, which either condition the tissue constructs in some manner, priming them for their end application, or in some cases, the operation of the bioreactor is the end application such as in *in vitro* applications like toxicology screening. It is these four aspects of bioprinting just described that we will discuss throughout this chapter, focusing within these areas on the essential requirements needed to biofabricate viable and functional tissue constructs. Last, we will briefly discuss some of the most promising end applications in which biofabricated tissues are being implemented (Fig. 1.2).

2 CELL SOURCING

The choice of the cellular component in bioprinting is often the first and foremost consideration to enter the minds of most researchers. This makes sense, as in almost all cases, researchers have a target tissue in mind for biofabrication. Naturally, the first building block a tissue engineer would consider would be the cells that would give the target tissue its primary function. Beyond these initial thoughts, the choice of a single cell type, multiple cell types, and source of cells becomes far more complicated. With all of the advances in biology and cell culture over

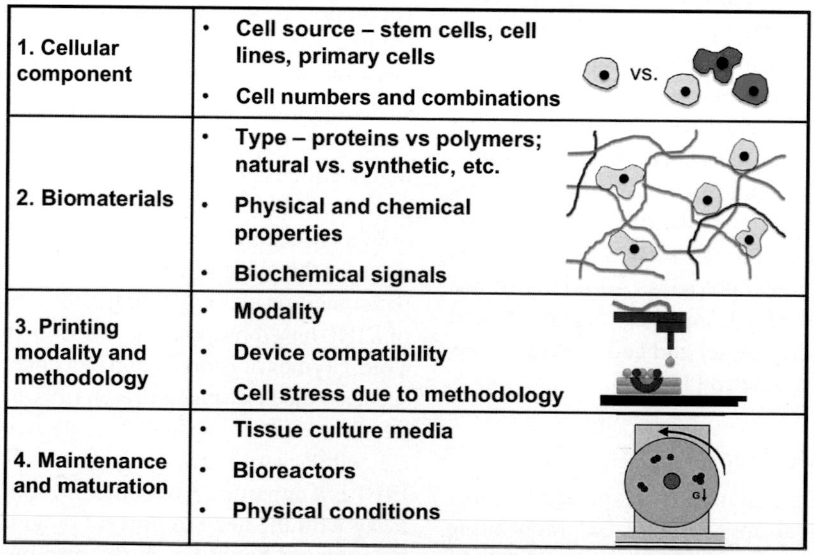

1. Cellular component	• Cell source – stem cells, cell lines, primary cells • Cell numbers and combinations	
2. Biomaterials	• Type – proteins vs polymers; natural vs. synthetic, etc. • Physical and chemical properties • Biochemical signals	
3. Printing modality and methodology	• Modality • Device compatibility • Cell stress due to methodology	
4. Maintenance and maturation	• Tissue culture media • Bioreactors • Physical conditions	

FIGURE 1.2   Considerations essential to successfully bioprint viable and functional tissue and organ constructs.

the years, we now have numerous options for cell sources at our disposal, such as established cell lines, primary cell lines taken from actual animal or human tissues, and stem cells and cells derived from those stem cells that have been differentiated into downstream functional cells. Furthermore, a substantial number of these cell sources are now commercially available to most laboratories. Even more of these cell sources have been further developed and customized in individual laboratories. Together, this comprises a massive catalog of cell types that can be considered for implementation when bioprinting tissue and organ constructs. Yet, some cell types are rare or difficult to expand in culture, and as such can prove to be a limiting factor in bioprinting applications.

2.1 Cell Lines

Cell lines consist of transformed cell populations with the ability to divide indefinitely. This is typically due to immortalization in the laboratory or because the cell line is derived from a tumorigenic source from a patient or animal. Cell lines can be invaluable in research and have led to numerous important discoveries throughout medicine. They are generally robust in nature, requiring relatively simple conditions and tissue culture. As such, these kinds of cells are optimal for proof of concept work such as development and initial implementation of bioprinting devices and techniques. However, while cell lines do retain some of the normal functionality of the cell types they are derived from, this functionality is often significantly diminished. Therefore, to construct truly functional bioprinted constructs that can accurately model or replace human tissues, other cell sources need to be considered.

2.2 Primary Cells

Primary cells specifically refer to cells that are isolated or harvested directly from living tissue or organs. With the exception of stem cell and progenitor populations, the majority of primary cells are terminally differentiated cells with a highly specific type and high level of function. These cells are perhaps the optimal cell type for creating tissue constructs or organs with respect to functionality. For example, primary liver hepatocytes are vastly superior to hepatocyte-derived cell lines, such as HEPG2 and HEPG2 C3A, in terms of secretion of substances such as albumin and urea, as well as the ability to metabolize drugs and toxins. However, as terminally differentiated cells, many primary cells possess limited to no proliferative capacity and are sensitive and fragile in *in vitro* environments such as standard 2D tissue cultures. As such they are often only available in limited quantities and require customized, supportive environments that mimic their *in vivo* environments to maximize viability and function [29].

2.3 Stem Cells and Stem Cell-Derived Cells

Stem cells are often defined by their unique ability to self-renew and produce progeny that can differentiate into specific functional cells. Stem cells are usually broken down into four principal categories. Embryonic stem cells (ESCs) are derived from the inner cell mass of an embryo and are considered pluripotent – the ability to differentiate into all three germ layers. Second, adult somatic stem cells exist in most tissues in the body and are generally constricted to differentiation into only the cell types that reside in that tissue. Third, fetal stem cells (FSCs), which can be obtained from fetal environments, such as the placenta, umbilical cord, and amniotic fluid, are often considered to have plasticity somewhere between ESCs and adult stem cells. Last, artificially generated, induced pluripotent stem (iPS) cells result from reprogramming terminally differentiated cells. Each of these stem cell types has advantages and disadvantages in tissue engineering applications such as bioprinting. For example, ESCs are attractive because of their ability to differentiate into potentially any other cell type. However, there are ethical concerns associated with procurement of ESCs, and upon transplantation, they often form teratomas, thereby drastically limiting their clinical use. Adult stem cells, such bone marrow-derived mesenchymal cells (BM-MSCs), are already being used clinically and are working their way through the regulatory pathways for various applications [30,31], but their use is often restricted by limited proliferative capacity *in vitro* and a limit on differentiation potential placed on them by already being partially differentiated. iPS cells are less well characterized, but have been shown to sometimes form teratomas after transplantation [32]. Our laboratory has extensively explored the use of FSCs, particularly those isolated from the amniotic fluid. However, they can be isolated from additional sources, such as the amniotic membrane, placenta, umbilical cord, and Wharton's jelly [33]. Unlike ESCs, FSCs carry with them no ethical concerns because their isolation does not put the developing fetus at risk. Early rise of cells in the extraembryonic environment during pregnancy may give them a higher degree of plasticity than adult stem cells, such as BM-MSCs or adipose-derived stem cells [34]. Importantly, like MSCs, amniotic fluid-derived stem (AFS) cells, and likely other FSCs, secrete a wide range of trophic factors that are immunomodulatory and promote regeneration [35–37].

As described earlier, differentiation potential among stem cell types varies greatly. However, differentiated cells from stem cell sources are a potentially optimal source for biofabrication applications. First, differentiated cells may be able to more sufficiently mimic the level of function of the corresponding primary cells than cell lines can. That being said, the ability to adequately differentiate stem cells into cells that can completely replace primary cell function is still difficult in many cases. Second, since some stem cell varieties possess heightened proliferative capacity, they theoretically can be expanded to an extraordinary number prior to differentiation, thereby providing a nearly limitless source of relatively highly functioning cells.

3 BIOMATERIALS AND BIOINKS

Today the term *biomaterials* encompasses a vast range of materials, including technologies that did not exist even a decade ago. Biomaterials range from cell supportive soft hydrogels, to stiff metal or ceramic implants; from nanoparticles and quantum dots for drug delivery and imaging, to complex functioning medical devices such as left ventricular assist devices and artificial hearts. As proficiency in material science and biology continues to expand, so does the number classifications of biomaterial types [38,39].

In the context of bioprinting, biomaterials currently are limited to two primary categories. The first category is that of curable polymers that result in mechanically robust (i.e., stiff or durable) materials that provide structure to printed constructs. The second category is that of soft biomaterials such as hydrogels, generally with a high water content, inside of which cells are capable of residing. In this section, we provide an overview of these types of materials, together with examples, after which we will describe the considerations required for the implementation of these materials to achieve long-term cellular viability and function in bioprinting.

3.1 Curable Polymers for Structural Support

Three-dimensional printing technologies stemmed from applications requiring fabrication of structures comprised of metals and thermoplastics that employed melting and curing in the fabrication steps. Many of these approaches involved high temperatures or toxic organic solvents or cross-linking agents, rendering them incompatible with living cells and biological materials such as proteins and cellular signaling molecules. However, due to the robust mechanical properties associated with such materials once cured, they have been implemented extensively as the foundation of devices as well as the structural components of biological constructs.

Polycaprolactone (PCL) is an example of a synthetic polymer that is commonly employed in bioprinting as a scaffolding component. PCL is a polyester-based material that due to its ability to be biodegraded by the body and its relatively low melting temperature of 60°C is commonly used as a structural printing component. PCL is well established for long term implantable devices and constructs, but other than non specific binding of cells to hydrophobic PCL, it lacks natural motifs that provide specific binding sites for cells that facilitate tissue integration. Because of this, PCL is often combined with other functionalized materials or naturally derived materials, such as the hydrogels described later, to create more complex hybrid structures. Indeed this integrative bioprinting approach is being implemented in our laboratory to create numerous types of tissue constructs and organs [40].

Other melt-curable polymers, such as polystyrene, can be used in printing applications; however, the high temperatures associated with reaching a flowing melt state that can be extruded has limited their applications beyond structural components.

3.2 Cell-Supportive Soft Materials and Hydrogels

Here we focus specifically on hydrogel biomaterials and their implementation in regenerative medicine applications such as cell therapy and tissue engineering. With the exception of bone and teeth, hydrogels allow for mimicry of the range of elastic modulus (E') values associated with the soft tissues of the body. Furthermore, processing techniques to generate sol–gel transitions can be designed to be noncytotoxic, allowing simple encapsulation of cells. This is important as there is an increasing movement from 2D to 3D cell and tissue culture in the fields of tissue engineering, regenerative medicine, and tissue modeling [41]. Hydrogels that support encapsulation procedures are infinitely more efficient for 3D uses than rigid scaffold seeding approaches of the past. While not every example discussed in this chapter reflects 3D use, the majority do, and it is important to note that in general, 3D applications in regenerative medicine provide cellular environments more like those in the body.

The majority of hydrogel biomaterials typically fall within one of two major categories: (1) synthetic hydrogels, which are completely synthesized in the laboratory, or (2) naturally derived hydrogels, which are purified from natural sources and often further modified in the laboratory. Common examples of synthetic hydrogels include polyethylene glycol (PEG)-based materials, such as PEG diacrylate (PEGDA), as well as polyacrylamide (PAAm)-based gels. Examples of naturally derived materials that are commonly used to generate hydrogels include collagen, hyaluonic acid, alginate, and fibrin. In general, synthetic materials allow for more fine-tuned control over molecular weight numbers and distributions, as well as cross-linking densities, allowing for precise modulation of specific mechanical properties such as E' or stiffness. Natural hydrogels, on the other hand, often have an innate bioactivity that aids with cell and tissue integration and biocompatibility. In this chapter we briefly discuss the use of some common synthetic hydrogel biomaterials, but primarily focus on naturally derived hydrogel biomaterials, as they are more efficient at mimicking the biological nature of the native ECM.

3.2.1 Synthetic Polymer Hydrogels

A variety of synthetic materials have been implemented as hydrogels for applications in regenerative medicine. Synthetic polymers are advantageous for one primary reason – as indicated earlier, they allow for precise control over their chemical and physical properties. Scientists can maintain precise chemical control over molecular weight, functional groups, and hydrophobicity/hydrophilicity at a monomer level. As a result, cross-linking rates and mechanical properties can be precisely controlled. PEG and polyacrylamide are examples of commonly used synthetic polymers in biomedical applications. PEG, which is perhaps most common, has long been used as medical device coatings to control host immune responses or appended to drug constructs to reduce degradation *in vivo*. It can also be manipulated to form a variety of hydrogels for cell culture and stem cell differentiation. PEG is often chemically modified with acrylate groups to create a photopolymerizable PEGDA in which cells can quickly be encapsulated.

The same features that allow such precise control over the chemical and mechanical properties also translate into an inherent drawback. Since synthetic polymer chains typically do not contain natural attachment sites that can interact with cells, all biological activity must be artificially preprogrammed into the material. PEG requires chemical immobilization of cell adhesion motifs in order to support cell adherence. Alternatively, many hydrogels derived from natural polymers and peptides retain some, if not all, of their original biological activity.

3.2.2 Collagen

Collagen is one of the most frequently used natural materials for a cell substrate, since it is the most abundant component of the ECM in most tissues [42]. Isolation and purification processes are well established, particularly for collagen type I, so using collagens as surface coatings and gels for cell culture has become an industry-wide practice. Inherent in the collagen structure are important arginine–glycine–aspartic acid (RGD) amino acid sequences that allow cells to adhere and proliferate via integrin–RGD binding. However, in normal tissue and ECM, collagen is but one of many components. Collagen biomaterial matrices are indeed useful and have yielded many important biological advances, but ~100% collagen matrices may limit cell migration and locomotion due to strong cell attachment. The lack of other common ECM components such as elastin, fibrinogen, laminin, and glycosaminoglycans (GAGs), may result in biological signaling that can induce unanticipated cellular changes. Furthermore, collagen fibers and gels primarily contain hydrophobic peptide motifs. As such, when used as implants or cell delivery agents, collagen gels can exclude water and contract, potentially resulting in decreased function, decreased diffusion of nutrients and gases, and cell death. Despite this limitation, collagen is still used extensively in tissue culture. However, its future application might be improved with the development of new hybrid biomaterials consisting of combinations of collagen and other ECM components with superior properties.

3.2.3 Hyaluronic Acid

Hyaluronic acid (HA), or hyaluronan, is a versatile nonsulfated GAG consisting of repeating disaccharide units and is present in tissues as a major constituent of the ECM that has shown great potential in regenerative medicine [43,44]. Unmodified HA has been used clinically for over three decades [45], in applications such as the treatment of damaged joints [46,47]. More recently, HA has been commonly chemically modified to become a more useful and robust biomaterial that can be cross-linked or loaded with other functional molecules [48].

HA hydrogels are often implemented by photocrosslinking methacrylate groups appended to the HA chains that can undergo free radical polymerization when exposed to ultraviolet (UV) irradiation to form soft hydrogels, referred to here as MA-HA hydrogels. Photocrosslinkable MA-HA hydrogels have been used in many settings, from cutaneous and corneal wound healing [49] to prototype vessel structure bioprinting [16]. Thiol-modification of HA also yields a material by which hydrogels can be formed through Michael-type addition cross-linking. Like the MA-HA variety of HA, thiol-modified HA, particularly a thiolated carboxymethyl HA (CMHA-S), has been implemented in many applications in regenerative medicine such as wound healing [50], tumor modeling [51], and bioprinting of cellularized structures [17]. One example that addressed this problem was the combination of HA with dextran to form a semi-interpenetrating network-based gel that had the appropriate rheological properties for printing [18]. To address these issues we investigated various cross-linking techniques using HA-based hydrogels. First, we discovered that gold nanoparticles (AuNPs) could serve as thiophilic cross-linking agents when paired with thiolated HA and gelatin solutions. The gold–thiol interactions resulted in a hydrogel that gelled slowly, increasing in elastic modulus over the course of 96 h. This slow reaction produced a large window during which the material was extrudable for bioprinting (at about 24 h of cross-linking). We printed cellularized tubular structures that after layer-by-layer deposition, fused in culture during the next several days. After 4 weeks in culture the constructs had become opaque with proliferating cells and cell-secreted ECM as they remodeled the construct. This cross-linking strategy was also reversible, allowing us to use cell-free AuNP gels as structural supports and space holders that could be washed away by interrupting the gold–thiol bonds, resulting a flexible system for building constructs [17]. We also explored the use of photocrosslinkable methacrylated HA and gelatin for continuous bioprinting deposition. This cross-linking strategy allowed an initial partial gelation step which left the gel in a soft and extrudable, but structurally sound state during which cellularized tubular constructs were fabricated. After layer-by-layer deposition, the individual segments were fused and stiffened with a secondary photocrosslinking step. Like the previous example, these constructs were remodeled as the cells proliferated and deposited ECM [16]. More recently we have been further exploring HA hydrogels mixed with tissue-specific growth factors and ECM components [29] and multiple cross-linking chemistries to created flexible bioinks for bioprinting tissue organoids of varying stiffnesses.

3.2.4 Gelatin

Gelatin is a mixture of peptide sequences derived from collagen that has undergone partial hydrolysis. This degraded product can then be dissolved in aqueous solutions more easily than collagen, while still maintaining the ability to form simple gels when solutions are brought to low temperatures through hydrophobic cross-linking. However, the gelation/melt temperature of gelatin solutions/gels typically lies between 30°C and 35°C, which limits its use in this gel form to applications that are below physiological temperatures. Due to this limitation, gelatin often requires additional chemical modification, alternative cross-linking techniques, or combination with other proteins or polymers for implementation in living systems such as skin.

Gelatin-fibrinogen cross-linked with glutaraldehyde has been used in *in vitro* studies with dermal fibroblasts to develop dermal matrices to be used in wound repair. These matrices showed characteristics, such as collagen production, cellular infiltration, and eventual biodegradation, all of which are important in wound healing treatments [52]. Other *in vitro* studies explored electrospun gelatin-polycaprolactone nanofibers as scaffolds for human dermal fibroblasts, keratinocytes, and mesenchymal stem cells. These scaffolds and gelatin-only nanospun scaffolds were then implemented *in vivo*, resulting in accelerated wound closure and epithelialization compared to gauze treatments [53]. Gelatin was combined with its precursor, collagen, to form a sponge from which loaded bFGF could be released sustainably. In a pressure-induced diabetic ulcer mouse model, bFGF-loaded scaffolds achieved accelerated dermis-like tissue formation, wound closure, and new blood vessel formation in comparison to control scaffolds loaded with saline [54]. Similar scaffolds were also loaded with a platelet-derived lysate containing TGF-ß1, PDGF, VEGF, and bFGF, which also accelerated dermis-like tissue formation [55]. Gelatin-based materials have also been used for cell delivery to wounds. In one study, gelatin-polyethylene glycol matrices were used to encapsulate MSCs through thiol-ene cross-linking and applied to full thickness wounds in rats. This combinatorial treatment decreased the overall immune response, reducing immune cell infiltration and foreign giant cell formation, while accelerating wound closure, re-epithelialization, and neovascularization [56].

3.2.5 Alginate

Alginate is a natural polysaccharide that is derived from algae or seaweed. It is common in regenerative medicine applications due to the ease at which it can form a hydrogel through an almost instantaneous sodium–calcium ion exchange reaction. This has made alginate the material of choice for microencapsulation of cells, in which easily available and inexpensive alginic sodium salt, or sodium alginate, which is unmodified, is quickly cross-linked into calcium alginate hydrogel microspheres [57]. These constructs have been extensively used for creating hydrogel capsules containing trapped liver cells or pancreatic islets [58]. However, without chemical modification, alginate, like PEG, is mostly inert, and use for cell and tissue culture is limited without incorporating cell-adherent motifs. Additionally, the reagents commonly used for creating cell-laden hydrogel microspheres, such as $CaCl_2$, the cross-linking reagent, as well as sodium citrate and ethylenediaminetetraacetic acid (EDTA), commonly used chelators, can have a detrimental effect on cell viability during the encapsulation process [59]. However, due to the ease with which alginate gels can be formed, it remains a popular and effective choice as a material in applications requiring cell encapsulation.

3.2.6 Fibrin

Another natural-sourced material for generating hydrogels is fibrin, which has been implemented for culture of various tissues types. Fibrin is comprised of fibrinogen monomers that are joined by thrombin-mediated cleavage cross-linking. In the body, it has an important role in blood clotting, wound healing, and tumor growth. In a concentrated glue-like form, it has been used clinically as a hemostatic agent and sealant in surgery. More recently, less concentrated fibrin gels have been used as a scaffold for regenerative medicine due to its quick cross-linking rates and robust mechanical properties [60]. In the context of bioprinting, our laboratory has used a fibrin–collagen blend to bioprint hydrogels containing stem cells over full thickness wounds to accelerate skin regeneration (Table 1.1) [37].

3.3 Biomaterial Features

3.3.1 Modulation of Mechanical Properties

It has been shown that the mechanical properties of a substrate or environment directly influence cell behavior, particularly differentiation of stem cells. Several stem cell types show a preference to select particular lineages and differentiate toward a given tissue type when cultured in a material or on a substrate that mimics the stiffness, or elastic modulus, of that tissue. In one seminal example explored by Engler et al., MSCs were cultured on collagen-coated polyacrylimide substrates of 0.1–1, 8–17, and 25–40 kPa, and the cells were predisposed toward neural, muscle, and osteogenic characteristics, respectively [61]. Such stiffness-directed behavior has also been shown by using softer

TABLE 1.1 Commonly Used Materials for Bioprinting and Associated Characteristics

Hydrogel	Modification	Cross-Linking Method	Cross-Linking Speed	Pros	Cons	Common Applications
Collagen	–	Hydrophobic bonding	0.5–1 h	• Naturally cell adherent • Major component of native ECM	• Slow gelation • Can be associated with tissue fibrosis	• Implants • Cell encapsulation • Substrate coating • Dermal substitutes
Fibrin	–	Thrombin-catalyzed fibrin polymerization	Seconds	• Fast gelation • Cell adherent	• Difficult to control geometry due to fast gelation	• Cell delivery • Cell encapsulation • Surgical glue
Hyaluronic acid	Thiolated	Thiol group cross-link	15–30 min	• Commercially available in kit form w/ gelatin for cell adherence • Mechanical properties can be modulated by cross-linker geometry and MW	• Generally low mechanical properties	• Cell encapsulation • Cell delivery • Wound healing
Hyaluronic acid	Thiolated	UV photopolymeration (thiol-ene)	Seconds	• Easily controllable fast gelation • Mechanical properties can be modulated by cross-linker geometry and MW	• Generally low mechanical properties	• Cell delivery • Bioprinting • Wound healing
Hyaluronic acid	Methacrylated	UV photopolymeration	Minutes	• Gelation speed modulated by UV intensity	• Low mechanical properties	• Cell encapsulation
Gelatin	–	Temperature-based hydrophobic bonding	Minutes to hours	• Naturally cell adherent	• Unstable	• Cell encapsulation
Gelatin	–	Glutaraldehyde	Hours	• Naturally cell adherent • Stable after cross-linking	• Cross-linking must be accomplished prior to addition of cells	• Scaffolds and films for cell seeding
Sodium alginate	–	$CaCl_2$ ion exchange	Seconds	• Easy to create gel microspheres	• Difficult to control geometry due to fast gelation • Not cell-adherent without modification • $CaCl_2$ can induce toxicity w/ prolonged exposure to cells	• Cell encapsulation in microspheres
Silk (fibroin)	–	Hydrophobic formation of semicrystalline structures	Minutes to hours	• Stable • Strong mechanical properties	• Slow degradation limits use to more permanent implantations	• *In vivo* implantation
PEG	Acrylated PEGDA	UV photopolymerization	Minutes	• Easy to control mechanical properties by changing MW	• Not cell-adherent without modification	• Cell encapsulation • Cell delivery
Polycaprolactone	–	Melt-cure	–	• Robust mechanical properties	• Requires high temperature • No cell encapsulation	• Structural support • 3D scaffolding for cell seeding • Drug delivery

alginate hydrogels to increase neural stem cell differentiation shown by increased levels of neural markers [62]. Likewise, muscle stem cells were shown to self-renew on lower stiffness PEG-laminin hydrogels of 12 kPa, but not on stiff plastic surfaces [63]. The fact that stiffness plays an important role in stem cell lineage specification highlights the importance of taking into consideration the mechanical properties of 3D biofabrication environments. Beyond just differentiation, our laboratory recently demonstrated that elastic modulus also plays a role in determining the therapeutic trophic factor secretion levels of amniotic fluid-derived stem cells [64]. This result is significant, as stem cell, progenitor cell, and even adult subpopulations within tissues are responsible for secreting trophic factors, which can improve viability and function within tissues and biofabricated tissue constructs. Consideration of the appropriate mechanical properties to facilitate this, is therefore important.

Fortunately, polymeric materials, such as hydrogels, can support a range of elastic moduli, matching the mechanical properties of most tissues, with the exception of tooth and bone. In most cases, hydrogels are formed by cross-links between polymer chains, although some are formed between proteins or peptides. These cross-links can be covalent (e.g., photocrosslinked HA or PEGDA [16,65,66]), ionic (e.g., alginate [67]), or physical interactions (e.g., unmodified collagen or gelatin [68]). Manipulation of material elastic modulus E' can be achieved in several common ways, with varying levels of difficulty depending on the material in question. Perhaps the simplest method to increase or decrease E' is to increase or decrease the concentration of polymer or protein within a given volume. Simply put, the more dense a material is, the stiffer it will be in most cases. However, in the case of hydrogel materials, there is often a point at which the polymer or protein can no longer be dissolved into solution, placing an upper limit on using concentration to increase stiffness. Conversely, there is also a point at which the concentration of material can be too low for a solid gel to form. E' can also be manipulated by altering the molecular weight (MW) of individual polymer or protein chains comprising the bulk material. Since polymer solutions' and hydrogels' material properties typically arise from the organization of the polymer or protein chains within the material, MW changes can cause significant changes in their physical characteristics. Increasing chain MW can cause changes in two ways. First, larger changes cause more "drag" during random chain movement, a phenomenon studied in polymer science, as they move past one another, thus increasing the effective stiffness and strength of a material. Second, if the change in MW is between a set number of cross-links, the overall hydrogel network will expand with larger pores, or gaps, in the network. This decreases E' as there is more room in between chains in which water can reside. Likewise, decreasing chain MW between

cross-links tightens the polymer network, thereby increasing E'. More recently, the geometry of cross-linking molecules has been explored as a method for manipulating E'. For example, if one transitions from a linear cross-linking molecule, such as when PEGDA is used to cross-link thiolated HA chains [48,69], to a multiarmed PEG cross-linker of comparable MW, the network tightens up and pore sizes decrease, resulting in substantial increases in E' as we have described in a previous work [15].

3.3.2 Biochemical Signal Loading

Another essential consideration in bioprinting or biofabricating tissue and organ constructs is what biochemical signals need to be available to the cells. Some of these signals can be provided directly to the cells by the biomaterial, such as in the case of integrin binding to peptide motifs on collagen or gelatin, or CD44 binding to HA. However, there exists a vast quantity of soluble factors that can be integral in supporting cell and tissue function, which may need to be incorporated in bioprinting studies. These factors (growth factors (GFs), cytokines, chemokines, etc.) can simply be added to the cell culture media administered to the bioprinted constructs after fabrication is complete, but access to cells within the construct is then limited by diffusion kinetics, which decrease as one moves from the outside to the inside of a 3D structure.

A more elegant and effective method to provide biochemical signaling to cells is to incorporate release mechanisms into the biomaterial employed. This can be achieved in several ways, each of which is highly dependent on the material being used. First, drugs or GFs can be encapsulated within biomaterials upon formation. In the case of most hydrogels, these drugs or GFs will diffuse out of the constructs quickly in a burst release. However, in melt-curable materials such as PCL or PLGA, release kinetics are governed by the slow degradation of the scaffolds, resulting in a sustained release over time [70,71]. Second, bioactive molecules that have an affinity for soluble signaling molecules can be tethered to polymer chains within hydrogels that provide sustained release. The most common of such applications is the use of heparin or heparan sulfate pendant chains, which are covalently bound to the chains of the hydrogel network. Heparin and heparan sulfate naturally bind to a variety of GFs. When this method is employed in biofabricated tissue constructs or implanted hydrogel constructs, extended GF release kinetics can be achieved, resulting in increased biological function compared to simple burst releases without heparin/heparan sulfate. A common example of this approach includes loading of VEGF and FGF to increase angiogenesis and maturation of blood vessels [72–75]. In our laboratory, we are currently employing this method within bioprintable HA-based bioinks that contain GFs derived from tissue sources, such as liver [29] and heart, to bioprint highly functional tissue organoids for use in drug and toxicology screening.

4 INTEGRATION WITH BIOFABRICATION DEVICES

With the tissue construct building blocks chosen – cells, biomaterials, biochemical signals – for the given target tissue construct to be fabricated, an appropriate device is necessary for the physical fabrication steps. Depending on the type of device or printing modality, as discussed earlier, specific concerns arise based on deposition methodology.

4.1 Pressure and Shear Stress

In the case of extrusion-based printing, driving the physical printing process is by pneumatic pressure or mechanical force. Both of these methods result in pressure being translated to the material being extruded. In the case of melt-curable polymers, the pressure required to perform efficient extrusion can be quite high, exceeding 60 psi even. The interplay between nozzle size and driving pressure determines the build-up of pressure that the material experiences. Importantly, when cells are being printed within biocompatible hydrogels, they also experience pressure from the device, which can significantly impact cell viability. For example, in one study, the effects of dispensing pressures on cell viability and death were evaluated. This work demonstrated that when tissue constructs were printed at 40 psi viability decreased by nearly 40% in comparison to tissue constructs printed at 5 psi [76]. Similarly, shear stress placed on cells as they move through the printing device – for example, against the walls of a syringe or syringe needle tip – can impact cells as well. The same study described earlier found that shear stress as an effect of nozzle size had less of an effect on viability than overall pressure did [76]. However, shear stress is directly related to the speed at which the cells and surrounding material move through the bioprinter. Therefore, one must weigh the value of fast printing versus maintenance of cell viability.

4.2 Temperature

Temperature can come into play during bioprinting in several ways. First and foremost, as described earlier, some biomaterials require elevated temperatures in order to be printed. These conditions are almost always incompatible with printing methodologies in which cells are to be printed also. Because of this, when cells are printed within materials such as hydrogels, care should be taken to maintain physiological temperature for the duration of the printing. Alternatively, if temperatures must change, the magnitude and duration of temperature changes should be minimized. Indeed, in our experience, we observed that if we bioprinted primary hepatocytes at ambient room temperature during a 30–60 min protocol, viability was not as high as we had hoped. By incorporating environmental controls into the bioprinter and maintaining the bioprinter stage and cell-bioink reservoirs at 37 °C, we were able to significantly improve cell viability within the bioprinted tissue constructs and increase the functional output of the constructs.

4.3 Nutrient Availability

During the time period in which the printing process occurs, the cells that will be printed are often encapsulated in a material or are suspended in hydrogel precursors that will be cross-linked during or after deposition. During this time, it is important that the potential requirement of nutrients and oxygen that a cell may need is considered. The sensitivity of the cells to stresses and the duration of the printing procedure are the determining factors here. For example, if the printing procedure is short and the cell component is comprised of robust cell lines, cell viability may not be impacted by the printing. However, if the print time (including preparation) is long, and the cells in question are fragile and sensitive primary cells, then without supplying an extra nutritional component to the procedure, viability may decrease severely. In recent work under these latter conditions we have observed that by incorporating both tissue-derived growth factors, as described earlier, and cell culture media into the bioprintable materials, we can significantly improve the viability of primary cells within bioprinted constructs.

5 MAINTENANCE AND MATURATION OF CONSTRUCTS

After biofabrication is finalized and a structure containing viable cells is complete, there often remains significant work to be done to ensure that the tissue or organ construct not only lives for an extended period of time, but also is able to function as intended. In this postfabrication stage, the cells and tissue must be maintained in such a manner that the cells form the appropriate connections with one another for communication, mobilize if necessary, have the opportunity to reorganize or secrete their own matrix components, and in some cases can be conditioned physically so that they can function as they would in the body.

5.1 Cellular Self-Organization

Provided that cells are in an appropriate environment, they possess the ability to use the surrounding matrix to move around and interact with one another. Over time cells reach an equilibrium state between cell–matrix adhesions, such as integrins, and cell–cell adhesions, such as tight junctions and cadherins. This equilibrium state is variable depending on the cell type or types, as well as the matrix. For example, in many cases epithelial lineages favor cell–cell adhesions, expressing increased ZO-1 or occluding (tight junction markers) and increased E-cadherin and Ep-CAM [77,78]. Conversely, cells of mesenchymal phenotypes often have an increased expression level of cell–matrix adhesions versus cell–cell adhesions, resulting in a more mobile-capable phenotype [79]. These preferential interactions between like cells or between certain types of lineages supports the ability of cell populations to spontaneously reorganize within a 3D environment, a property known as tissue liquidity [80,81]. This spontaneous self-organization happens during development *in vivo*, but has been recapitulated in numerous *in vitro* applications. For example, cell spheroids can be placed into geometric architectures, which over time fuse together into seamless structures. This has been performed to make rings, tubes, and branches vasculature [13,82]. In a more complex example, when a cell spheroid comprised of mixed smooth muscle and endothelial cell spheroids is bioprinted into a hydrogel together, the cells naturally self-organize into a new architecture in which endothelial cells form a lumen-like structure inside a smooth muscle-based layer. If multiple uniluminal spheroids are then placed adjacently, they can fuse together, and reorganize into one larger multilayered luminal spheroid [83]. By manipulating the host biomaterial environment composition, self-organization can be controlled. Migration of cells in 3D can be expedited or minimized depending on the ratios of matrix components such as collagens and glycosaminoglycans [84]. This external manipulation can also be used to spatially organize uniluminal spheroids and harness their fusion to create tubular structures, rather than fusion into one larger spheroid [82].

Cells also have the ability to reorganize their surrounding matrix. In particular, cells of mesenchymal phenotype, such as fibroblasts, can break up or modify the matrix material used in the biofabrication stages, and secrete their own matrix materials during the maturation process. This will often provide increased strength to the construct. In fact, we observed this behavior in several studies. Using different bioprinting techniques we fabricated tubular vessel-like prototype tissue constructs using hydrogels and either 3T3 fibroblasts or HEPG2 cells. In both studies, the bioprinted constructs were maintained in culture for several weeks after which the constructs had transitioned from translucent to completely opaque; an effect of both cell growth and cell secretion of additional extracellular matrix proteins. This secretion activity was verified by stains for collagen as well as cytoplasmic procollagen, the internal precursor to collagen fibers [16,17]. Supplying a supportive environment that allows the cellular components of biofabricated constructs to freely reorganize themselves and their environment is ultimately important for long term development of mature functional tissues.

5.2 Mechanical Stimulation

The phenomena involved in maturation that were discussed earlier can sometimes be accelerated using techniques such as mechanical conditioning. Periodic stretching, pulsing, or compression of a tissue construct that mimics the physical forces that its corresponding tissue or organ experiences *in vivo* can increase strength and flexibility, as well as increase matrix reorganization and maturation of the construct [85]. This concept has been explored extensively with biofabricated blood vessels and skeletal muscle constructs, typically employing perfusion bioreactors and tensile conditioning bioreactors [86]. While mechanical properties can be modulated through the biomaterial composition [87], often additional conditioning is required. It is well documented that pulsatile flow through blood vessel constructs increases the production of collagen within the construct walls. This increases the elastic moduli of the tissue [88,89], which in turn increases the pressure of flow that the construct can withstand. In the case of skeletal muscle, material manipulation such as fiber alignment can induce muscle cell organization, but to achieve a functional contracting tissue, cyclic mechanical preconditioning is often required. Applying unidirectional tensile loading to muscle constructs aids in achieving cellular alignment, muscle fiber formation, and increases the force that the constructs can generate during contraction [90]. These improvements in function due to preconditioning are crucial for applications where fabricated constructs are implanted *in vivo* and are expected to integrate with surrounding tissue and function appropriately [91]. Other examples exist beyond vascular and muscular structures, and for those constructs, the appropriate type of bioreactor (perfusion, tensile, compression, rotating, air–liquid interface, etc.) must be chosen based on the particular mechanical forces that correspond to the type of tissue being matured.

6 CONCLUSIONS

Bioprinting can be implemented in a variety of regenerative research directions and end applications. The ultimate goal is to create organ structures that can act as complete replacements of diseased or damaged organs in patients. This goal is becoming realized by tissue engineering in some areas, such as bladder and ureter replacements [92–94]. However, in the case of more complex solid organs, there remains much work to be accomplished before fully functional replacement organs can be realized. That being said, substantial advancements are being made on a daily basis in laboratories around the world. Tissue engineering approaches, such as bioprinting, can be used immediately in several other arenas including creation of niche environments for maintaining or expanding primary cells and stem cells, and fabrication of small-scale, high-throughput models for pathology modeling, pathway exploration, and drug and toxicology screening.

Tissues in the body occupy a wide range of stiffnesses. For example, the elastic modulus of brain tissue is in the range of 0.5–2 kPa, muscle is near 10–15 kPa, while that of bone is more than 50 kPa. Unfortunately, standard procedures call for culturing stem cells (as well as other cells) on simple surfaces comprised on plastic that exceed 100,000 kPa [95]. Synthesis of biologically useful materials with softer mechanical properties may provide tailorable microenvironments to act as cell niches, allowing scientists to employ favorable 3D environments that increase the potential for physiological and clinical applications [96]. These kinds of environments have the potential to act as tools for maintenance and expansion of stem cells and primary cells, while preserving their function more effectively than standard tissue culture conditions. These cells can then be implemented within other clinical and research-based applications.

A key aspect for accurately modeling pathologies and behavior of tissues, and thus creating better screening platforms, is the use of 3D systems since they allow cells to grow, differentiate, and interact with each other and the surrounding matrix, representing more *in vivo*-like conditions [97,98]. Bioprinting is a tool that can efficiently fabricate these 3D systems. Although useful, 2D culture environments are not natural. For example, cancer drug testing using 2D cultures have shown some levels of success with candidate drugs; however, when doses are scaled appropriately to patient levels, they are often ineffective. Conversely, 3D tumor models show an appropriate increased resistance to drugs, serving as better testing platforms [99]. This is likely due to the fact that *in vivo* tumor environments are 3D and much more complex in cell arrangement and tissue architecture. Furthermore, the diffusion properties of the drug are vastly different in 3D tissue in comparison to 2D monolayers. In terms of disease modeling, it has been shown in numerous 3D organoid types that organotypic 3D tissue models respond to bacterial and viral infections in ways that reflect the natural infection process *in vivo* – as evidenced by differences in tissue pathology, adherence and invasion, apoptosis, and host biosignature profiles (proteomics and cytokine/chemokine profiling) [100–107]. Simply put, the closer to actual *in vivo* tissue a model system is, the more dependable and accurate it is in drug and toxicology screening.

These are just a few examples of how bioprinting can be used in end applications. More detailed descriptions and many other examples of biofabrication will be discussed in subsequent chapters. The emergence of bioprinting has created a set of tools that can be used to create living structures and customized environments that have the potential to change the way medicine is practiced. Regardless of which of these end applications are realized, it is important to consider the essentials of bioprinting technology that must be optimized, which will have an integral impact on the practical realization of the successful application of viable and functional tissue-engineered constructs and entire organ structures in clinical and research settings.

GLOSSARY

Adult stem cell One of many types of stem cells that exist in most tissues in the body that can be multipotent or unipotent.

Alginate A natural polysaccharide that is derived from algae or seaweed.

Amniotic fluid-derived stem cell A fetal stem cell isolated from the amniotic fluid.

Bioink A term used to describe materials that are bioprinted. Bioinks can describe materials containing cells or materials that are cell-free.

Biofabrication Construction of 3D biological structures. Bioprinting is one form of biofabrication.

Biomaterial A material that interacts with a biological system.

Bioreactor A device that supplies appropriate environmental cues (physical forces, nutrients, oxygen) in order to maintain or mature cells, cellular tissues, or engineered tissue constructs.

Collagen A triple-helix protein that is the most abundant ECM protein in the body.

Cross-link A linkage between polymer or peptide chains – generally chemical, physical, or ionic – that results in the formation of a mesh-like network.

Cytotoxic Detrimental to cell viability.

Elastic modulus A value representing the elasticity of a given material.

Embryonic stem cell A pluripotent stem cell isolated from the blastocyst of an embryo.

Extracellular matrix The highly complex physical network of proteins and polymers in the body that act as structural support, or scaffolding, for the cells.

Fetal stem cell A multipotent stem cell isolated from the fetal environment (placenta, umbilical cord, amniotic fluid).

Fibrin Fibrinogen monomers that have been joined by thrombin-mediated cleavage cross-linking.

Gelatin A mixture of peptide sequences derived from collagen that has undergone partial hydrolysis.

Gelation The transition of a material from a liquid to a hydrogel, often due to the physical or chemical bond formation.

HEPG2/HEPG2 C3A Transformed human-derived liver cell lines.

Hyaluronic acid A nonsulfated glycosaminoglycan consisting of repeating disaccharide units that is present in tissues as a major constituent of the ECM.

Hydrogel A network of hydrophilic polymer chains comprising a gel-like material with a high water content.

Hydrophilic Having a strong affinity to water.

Hydrophobic Having a tendency to repel water.

In vitro Referring to scientific work performed using cells or molecules outside their normal animal or human environment.

In vivo Referring to scientific work performed in an animal or human.

Induced pluripotent stem cell An artificially generated stem cell, created by reprogramming other, usually adult, cells.

Mesenchymal stem cell An adult stem cell able to differentiate into a variety of mesenchymal lineages.

Monomer A single unit repeated to form a polymer.

Organoid A small-sized (smaller than an analogous organ) tissue construct comprised of cells and often a supporting material or scaffold.

Peptide A short, multiamino-acid sequence which when combined with others, results in a protein.

Polyacrylamide A type of synthetic polymer.

Polycaprolactone A type of synthetic polymer.

Polyethylene glycol A simple hydrophilic synthetic polymer

Polyethylene glycol diacrylate A PEG chain modified to have diacrylate functional groups on both ends for cross-linking.

Polymer A large molecule comprised of many repeating subunits.

Polystyrene A type of synthetic polymer.

ABBREVIATIONS

2D	Two-dimensional
3D	Three-dimensional
AFS	Amniotic fluid-derived stem (cell)
bFGF	Basic fibroblast growth factor
BM-MSC	Bone marrow-derived mesenchymal stem cell
CD44	A cell surface receptor
DNA	Deoxyribonucleic acid
E'	Elastic modulus
ECM	Extracellular matrix
ESC	Embryonic stem cell
FSC	Fetal stem cell
GF	Growth factor
iPS	Induced pluripotent stem (cell)
LIFT	Laser-induced forward transfer
MW	Molecular weight
PAAm	Polyacrylamide
PCL	Polycaprolactone
PDGF	Platelet-derived growth factor
PEG	Polyethylene glycol
PEGDA	Polyethylene glycol diacrylate
PSI	Pressure per square inch
VEGF	Vascular endothelial growth factor

REFERENCES

[1] Visconti RP, Kasyanov V, Gentile C, Zhang J, Markwald RR, Mironov V. Towards organ printing: engineering an intra-organ branched vascular tree. Expert Opin Biol Ther 2010;10:409–20.

[2] Fedorovich NE, Alblas J, de Wijn JR, Hennink WE, Verbout AJ, Dhert WJ. Hydrogels as extracellular matrices for skeletal tissue engineering: state-of-the-art and novel application in organ printing. Tissue Eng 2007;13:1905–25.

[3] Mironov V, Boland T, Trusk T, Forgacs G, Markwald RR. Organ printing: computer-aided jet-based 3D tissue engineering. Trends Biotechnol 2003;21:157–61.

[4] Boland T, Mironov V, Gutowska A, Roth EA, Markwald RR. Cell and organ printing 2: fusion of cell aggregates in three-dimensional gels. Anat Rec A Discov Mol Cell Evol Biol 2003;272:497–502.

[5] Mironov V, Kasyanov V, Drake C, Markwald RR. Organ printing: promises and challenges. Regen Med 2008;3:93–103.

[6] Derby B. Printing and prototyping of tissues and scaffolds. Science 2012;338:921–6.

[7] Catros S, Fricain JC, Guillotin B, Pippenger B, Bareille R, Remy M, et al. Laser-assisted bioprinting for creating on-demand patterns of human osteoprogenitor cells and nano-hydroxyapatite. Biofabrication 2011;3:025001.

[8] Guillotin B, Guillemot F. Cell patterning technologies for organotypic tissue fabrication. Trends Biotechnol 2011;29:183–90.

[9] Moon S, Hasan SK, Song YS, Xu F, Keles HO, Manzur F, et al. Layer by layer three-dimensional tissue epitaxy by cell-laden hydrogel droplets. Tissue Eng Part C Methods 2010;16:157–66.

[10] Nakamura M, Iwanaga S, Henmi C, Arai K, Nishiyama Y. Biomatrices and biomaterials for future developments of bioprinting and biofabrication. Biofabrication 2010;2:014110.

[11] Skardal A, Mack D, Kapetanovic E, Atala A, Jackson JD, Yoo JJ, et al. Bioprinted amniotic fluid-derived stem cells accelerate healing of large skin wounds. Stem Cells Transl Med 2012;1(11):792–802.

[12] Jakab K, Norotte C, Damon B, Marga F, Neagu A, Besch-Williford CL, et al. Tissue engineering by self-assembly of cells printed into topologically defined structures. Tissue Eng Part A 2008;14:413–21.

[13] Norotte C, Marga FS, Niklason LE, Forgacs G. Scaffold-free vascular tissue engineering using bioprinting. Biomaterials 2009; 30:5910–7.

[14] Marga F, Jakab K, Khatiwala C, Shepherd B, Dorfman S, Hubbard B, et al. Toward engineering functional organ modules by additive manufacturing. Biofabrication 2012;4:022001.

[15] Skardal A, Zhang J, Prestwich GD. Bioprinting vessel-like constructs using hyaluronan hydrogels crosslinked with tetrahedral polyethylene glycol tetracrylates. Biomaterials 2010;31:6173–81.

[16] Skardal A, Zhang J, McCoard L, Xu X, Oottamasathien S, Prestwich GD. Photocrosslinkable hyaluronan-gelatin hydrogels for two-step bioprinting. Tissue Eng Part A 2010;16:2675–85.

[17] Skardal A, Zhang J, McCoard L, Oottamasathien S, Prestwich GD. Dynamically crosslinked gold nanoparticle – hyaluronan hydrogels. Adv Mater 2010;22:4736–40.

[18] Pescosolido L, Schuurman W, Malda J, Matricardi P, Alhaique F, Coviello T, et al. Hyaluronic acid and dextran-based semi-IPN hydrogels as biomaterials for bioprinting. Biomacromolecules 2011;12:1831–8.

[19] Bohandy J, Kim B, Adrian F. Metal deposition from a supported metal film using an excimer laser. J Appl Phys 1986;60:1538.

[20] Barron JA, Ringeisen BR, Kim H, Spargo BJ, Chrisey DB. Application of laser printing to mammalian cells. Thin Solid Films 2004;453:383–7.

[21] Chrisey DB. Materials processing: the power of direct writing. Science 2000;289:879–81.

[22] Colina M, Serra P, Fernandez-Pradas JM, Sevilla L, Morenza JL. DNA deposition through laser induced forward transfer. Biosens Bioelectron 2005;20:1638–42.

[23] Dinca V, Kasotakis E, Catherine J, Mourka A, Ranella A, Ovsianikov A, et al. Directed three-dimensional patterning of self-assembled peptide fibrils. Nano Lett 2008;8:538–43.

[24] Hopp B, Smausz T, Kresz N, Barna N, Bor Z, Kolozsvari L, et al. Survival and proliferative ability of various living cell types after laser-induced forward transfer. Tissue Eng 2005;11:1817–23.

[25] Gruene M, Deiwick A, Koch L, Schlie S, Unger C, Hofmann N, et al. Laser printing of stem cells for biofabrication of scaffold-free autologous grafts. Tissue Eng Part C Methods 2010;17(1):79–87.

[26] Koch L, Kuhn S, Sorg H, Gruene M, Schlie S, Gaebel R, et al. Laser printing of skin cells and human stem cells. Tissue Eng Part C Methods 2010;16:847–54.

[27] Guillotin B, Souquet A, Catros S, Duocastella M, Pippenger B, Bellance S, et al. Laser assisted bioprinting of engineered tissue with high cell density and microscale organization. Biomaterials 2010;31:7250–6.

[28] Guillemot F, Souquet A, Catros S, Guillotin B. Laser-assisted cell printing: principle, physical parameters versus cell fate and perspectives in tissue engineering. Nanomedicine (Lond) 2010;5:507–15.

[29] Skardal A, Smith L, Bharadwaj S, Atala A, Soker S, Zhang Y. Tissue specific synthetic ECM hydrogels for 3-D *in vitro* maintenance of hepatocyte function. Biomaterials 2012;33:4565–75.

[30] Keating A. Mesenchymal stromal cells: new directions. Cell Stem Cell 2012;10:709–16.

[31] Mulder GD, Lee DK, Jeppesen NS. Comprehensive review of the clinical application of autologous mesenchymal stem cells in the treatment of chronic wounds and diabetic bone healing. Int Wound J 2012.

[32] Fu X, Xu Y. Challenges to the clinical application of pluripotent stem cells: towards genomic and functional stability. Genome Med 2012;4:55.

[33] Abdulrazzak H, Moschidou D, Jones G, Guillot PV. Biological characteristics of stem cells from foetal, cord blood and extraembryonic tissues. J R Soc Interface 2010;7(Suppl. 6):S689–706.

[34] De Coppi P, Bartsch G Jr, Siddiqui MM, Xu T, Santos CC, Perin L, et al. Isolation of amniotic stem cell lines with potential for therapy. Nat Biotechnol 2007;25:100–6.

[35] Moorefield EC, McKee EE, Solchaga L, Orlando G, Yoo JJ, Walker S, et al. Cloned, CD117 selected human amniotic fluid stem cells are capable of modulating the immune response. PLoS ONE 2011; 6:e26535.

[36] Caplan AI. Adult mesenchymal stem cells for tissue engineering versus regenerative medicine. J Cell Physiol 2007;213.341–7.

[37] Skardal A, Mack D, Kapetanovic E, Atala A, Jackson JD, Yoo J, et al. Bioprinted amniotic fluid-derived stem cells accelerate healing of large skin wounds. Stem Cells Transl Med 2012;1:792–802.

[38] Williams D. The continuing evolution of biomaterials. Biomaterials 2011;32:1–2.

[39] Williams DF. On the nature of biomaterials. Biomaterials 2009; 30:5897–909.

[40] Xu T, Binder KW, Albanna MZ, Dice D, Zhao W, Yoo JJ, et al. Hybrid printing of mechanically and biologically improved constructs for cartilage tissue engineering applications. Biofabrication 2013;5. 015001.

[41] Prestwich GD. Evaluating drug efficacy and toxicology in three dimensions: using synthetic extracellular matrices in drug discovery. Acc Chem Res 2008;41:139–48.

[42] Hesse E, Hefferan TE, Tarara JE, Haasper C, Meller R, Krettek C, et al. Collagen type I hydrogel allows migration, proliferation, and osteogenic differentiation of rat bone marrow stromal cells. J Biomed Mater Res A 2010;94:442–9.

[43] Allison DD, Grande-Allen KJ. Review. Hyaluronan: a powerful tissue engineering tool. Tissue Eng 2006;12:2131–40.

[44] Knudson CB, Knudson W. Cartilage proteoglycans. Semin Cell Dev Biol 2001;12:69–78.

[45] Kuo JW. Practical aspects of hyaluronan based medical products. Boca Raton: CRC/Taylor & Francis; 2006.

[46] Galus R, Antiszko M, Wlodarski P. Clinical applications of hyaluronic acid. Pol Merkur Lekarski 2006;20:606–8.

[47] Schiavinato A, Finesso M, Cortivo R, Abatangelo G. Comparison of the effects of intra-articular injections of Hyaluronan and its chemically cross-linked derivative (Hylan G-F20) in normal rabbit knee joints. Clin Exp Rheumatol 2002;20:445–54.

[48] Prestwich GD, Kuo JW. Chemically-modified HA for therapy and regenerative medicine. Curr Pharm Biotechnol 2008;9:242–5.

[49] Miki D, Dastgheib K, Kim T, Pfister-Serres A, Smeds KA, Inoue M, et al. A photopolymerized sealant for corneal lacerations. Cornea 2002;21:393–9.

[50] Kirker KR, Luo Y, Morris SE, Shelby J, Prestwich GD. Glycosaminoglycan hydrogels as supplemental wound dressings for donor sites. J Burn Care Rehabil 2004;25:276–86.

[51] Liu Y, Shu XZ, Prestwich GD. Tumor engineering: orthotopic cancer models in mice using cell-loaded, injectable, cross-linked hyaluronan-derived hydrogels. Tissue Eng 2007;13:1091–101.

[52] Dainiak MB, Allan IU, Savina IN, Cornelio L, James ES, James SL, et al. Gelatin-fibrinogen cryogel dermal matrices for wound repair: preparation, optimisation and *in vitro* study. Biomaterials 2010;31:67–76.

[53] Dubsky M, Kubinova S, Sirc J, Voska L, Zajicek R, Zajicova A, et al. Nanofibers prepared by needleless electrospinning technology as scaffolds for wound healing. J Mater Sci Mater Med 2012;23:931–41.

[54] Kanda N, Morimoto N, Ayvazyan AA, Takemoto S, Kawai K, Nakamura Y, et al. Evaluation of a novel collagen-gelatin scaffold for achieving the sustained release of basic fibroblast growth factor in a diabetic mouse model. J Tissue Eng Regen Med 2012;8(1):29–40.

[55] Ito R, Morimoto N, Pham LH, Taira T, Kawai K, Suzuki S. Efficacy of the controlled release of concentrated platelet lysate from a collagen/gelatin scaffold for dermis-like tissue regeneration. Tissue Eng Part A 2013;19:1398–405.

[56] Xu K, Cantu DA, Fu Y, Kim J, Zheng X, Hematti P, et al. Thiol-ene Michael-type formation of gelatin/poly(ethylene glycol) biomatrices for three-dimensional mesenchymal stromal/stem cell administration to cutaneous wounds. Acta Biomater 2013;9:8802–14.

[57] Santos E, Zarate J, Orive G, Hernandez RM, Pedraz JL. Biomaterials in cell microencapsulation. Adv Exp Med Biol 2010;670:5–21.

[58] Opara EC, Mirmalek-Sani SH, Khanna O, Moya ML, Brey EM. Design of a bioartificial pancreas(+). J Investig Med 2010;58:831–7.

[59] Cohen J, Zaleski KL, Nourissat G, Julien TP, Randolph MA, Yaremchuk MJ. Survival of porcine mesenchymal stem cells over the alginate recovered cellular method. J Biomed Mater Res A 2011;96:93–9.

[60] Ahmed TA, Dare EV, Hincke M. Fibrin: a versatile scaffold for tissue engineering applications. Tissue Eng Part B Rev 2008;14:199–215.

[61] Engler AJ, Sen S, Sweeney HL, Discher DE. Matrix elasticity directs stem cell lineage specification. Cell 2006;126:677–89.

[62] Banerjee A, Arha M, Choudhary S, Ashton RS, Bhatia SR, Schaffer DV, et al. The influence of hydrogel modulus on the proliferation and differentiation of encapsulated neural stem cells. Biomaterials 2009;30:4695–9.

[63] Gilbert PM, Havenstrite KL, Magnusson KE, Sacco A, Leonardi NA, Kraft P, et al. Substrate elasticity regulates skeletal muscle stem cell self-renewal in culture. Science 2010;329(5995):1078–81.

[64] Skardal A, Mack D, Atala A, Soker S. Substrate elasticity controls cell proliferation, surface marker expression and motile phenotype in amniotic fluid-derived stem cells. J Mech Behav Biomed Mater 2013;17:307–16.

[65] Musumeci G, Loreto C, Carnazza ML, Strehin I, Elisseeff J. OA cartilage derived chondrocytes encapsulated in poly(ethylene glycol) diacrylate (PEGDA) for the evaluation of cartilage

restoration and apoptosis in an *in vitro* model. Histol Histopathol 2011;26:1265–78.

[66] Fairbanks BD, Schwartz MP, Bowman CN, Anseth KS. Photoiniti-ated polymerization of PEG-diacrylate with lithium phenyl-2,4,6-trimethylbenzoylphosphinate: polymerization rate and cytocompat-ibility. Biomaterials 2009;30:6702–7.

[67] Augst AD, Kong HJ, Mooney DJ. Alginate hydrogels as biomateri-als. Macromol Biosci 2006;6:623–33.

[68] Chen MY, Sun YL, Zhao C, Zobitz ME, An KN, Moran SL, et al. Substrate adhesion affects contraction and mechanical properties of fibroblast populated collagen lattices. J Biomed Mater Res B Appl Biomater 2008;84:218–23.

[69] Serban MA, Prestwich GD. Making modular extracellular matrices: solutions for the puzzle. Methods 2008;45:93–8.

[70] de Araujo TM, Teixeira Z, Barbosa-Sampaio HC, Rezende LF, Boschero AC, Duran N, et al. Insulin-loaded poly(epsilon-caprolactone) nanoparticles: efficient, sustained and safe insulin delivery system. J Biomed Nanotechnol 2013;9:1098–106.

[71] Yang J, Zeng Y, Zhang C, Chen YX, Yang Z, Li Y, et al. The pre-vention of restenosis *in vivo* with a VEGF gene and paclitaxel co-eluting stent. Biomaterials 2013;34:1635–43.

[72] Elia R, Fuegy PW, VanDelden A, Firpo MA, Prestwich GD, Peattie RA. Stimulation of *in vivo* angiogenesis by in situ crosslinked, dual growth factor-loaded, glycosaminoglycan hydrogels. Biomaterials 2010;31:4630–8.

[73] Peattie RA, Nayate AP, Firpo MA, Shelby J, Fisher RJ, Prest-wich GD. Stimulation of *in vivo* angiogenesis by cytokine-loaded hyaluronic acid hydrogel implants. Biomaterials 2004;25:2789–98.

[74] Peattie RA, Rieke ER, Hewett EM, Fisher RJ, Shu XZ, Prest-wich GD. Dual growth factor-induced angiogenesis *in vivo* using hyaluronan hydrogel implants. Biomaterials 2006;27:1868–75.

[75] Pike DB, Cai S, Pomraning KR, Firpo MA, Fisher RJ, Shu XZ, et al. Heparin-regulated release of growth factors *in vitro* and angio-genic response *in vivo* to implanted hyaluronan hydrogels contain-ing VEGF and bFGF. Biomaterials 2006;27:5242–51.

[76] Nair K, Gandhi M, Khalil S, Yan KC, Marcolongo M, Barbee K, et al. Characterization of cell viability during bioprinting processes. Biotechnol J 2009;4:1168–77.

[77] Balzar M, Prins FA, Bakker HA, Fleuren GJ, Warnaar SO, Litvinov SV. The structural analysis of adhesions mediated by Ep-CAM. Exp Cell Res 1999;246:108–21.

[78] Howarth AG, Stevenson BR. Molecular environment of ZO-1 in epithelial and non-epithelial cells. Cell Motil Cytoskel 1995;31:323–32.

[79] Scanlon CS, Van Tubergen EA, Inglehart RC, D'Silva NJ. Biomark-ers of epithelial-mesenchymal transition in squamous cell carcino-ma. J Dental Res 2013;92:114–21.

[80] Jakab K, Damon B, Marga F, Doaga O, Mironov V, Kosztin I, et al. Relating cell and tissue mechanics: implications and applications. Dev Dyn 2008;237:2438–49.

[81] Jakab K, Neagu A, Mironov V, Markwald RR, Forgacs G. Engineer-ing biological structures of prescribed shape using self-assembling multicellular systems. Proc Natl Acad Sci USA 2004;101:2864–9.

[82] Mironov V, Visconti RP, Kasyanov V, Forgacs G, Drake CJ, Mark-wald RR. Organ printing: tissue spheroids as building blocks. Bio-materials 2009;30:2164–74.

[83] Fleming PA, Argraves WS, Gentile C, Neagu A, Forgacs G, Drake CJ. Fusion of uniluminal vascular spheroids: a model for assembly of blood vessels. Dev Dyn 2010;239:398–406.

[84] Mironov V, Prestwich G, Forgacs G. Bioprinting living structures. J Mater Chem 2007;17:2054–60.

[85] Goldstein AS, Christ G. Functional tissue engineering requires bio-reactor strategies. Tissue Eng Part A 2009;15:739–40.

[86] Mironov V, Kasyanov V, McAllister K, Oliver S, Sistino J, Mark-wald R. Perfusion bioreactor for vascular tissue engineering with capacities for longitudinal stretch. J Craniofac Surg 2003;14:340–7.

[87] Lee SJ, Liu J, Oh SH, Soker S, Atala A, Yoo JJ. Development of a composite vascular scaffolding system that withstands physiologi-cal vascular conditions. Biomaterials 2008;29:2891–8.

[88] Hahn MS, McHale MK, Wang E, Schmedlen RH, West JL. Physi-ologic pulsatile flow bioreactor conditioning of poly(ethylene glycol)-based tissue engineered vascular grafts. Ann Biomed Eng 2007;35:190–200.

[89] Rashid ST, Fuller B, Hamilton G, Seifalian AM. Tissue engineering of a hybrid bypass graft for coronary and lower limb bypass surgery. FASEB J 2008.

[90] Moon du G, Christ G, Stitzel JD, Atala A, Yoo JJ. Cyclic mechani-cal preconditioning improves engineered muscle contraction. Tis-sue Eng Part A 2008;14:473–82.

[91] Machingal MA, Corona BT, Walters TJ, Kesireddy V, Koval CN, Dannahower A, et al. A tissue-engineered muscle repair construct for functional restoration of an irrecoverable muscle injury in a mu-rine model. Tissue Eng Part A 2011;17:2291–303.

[92] Atala A. Tissue engineering for bladder substitution. World J Urol 2000;18:364–70.

[93] Atala A. Bladder regeneration by tissue engineering. BJU Int 2001;88:765–70.

[94] Raya-Rivera A, Esquiliano DR, Yoo JJ, Lopez-Bayghen E, Soker S, Atala A. Tissue-engineered autologous urethras for pa-tients who need reconstruction: an observational study. Lancet 2011;377:1175–82.

[95] Vanderhooft JL, Alcoutlabi M, Magda JJ, Prestwich GD. Rheologi-cal properties of cross-linked hyaluronan-gelatin hydrogels for tis-sue engineering. Macromol Biosci 2009;9:20–8.

[96] Nuttelman CR, Rice MA, Rydholm AE, Salinas CN, Shah DN, An-seth KS. Macromolecular monomers for the synthesis of hydrogel niches and their application in cell encapsulation and tissue engi-neering. Prog Polym Sci 2008;33:167–79.

[97] Lavik E, Langer R. Tissue engineering: current state and perspec-tives. Appl Microbiol Biotechnol 2004;65:1–8.

[98] Ifkovits JL, Burdick JA. Review: photopolymerizable and degrad-able biomaterials for tissue engineering applications. Tissue Eng 2007;13:2369–85.

[99] Ho WJ, Pham EA, Kim JW, Ng CW, Kim JH, Kamei DT, et al. In-corporation of multicellular spheroids into 3-D polymeric scaffolds provides an improved tumor model for screening anticancer drugs. Cancer Sci 2010;101:2637–43.

[100] Yamada KM, Cukierman E. Modeling tissue morphogenesis and cancer in 3D. Cell 2007;130:601–10.

[101] Griffith LG, Swartz MA. Capturing complex 3D tissue physiology *in vitro*. Nat Rev Mol Cell Biol 2006;7:211–24.

[102] Kabelitz D, Medzhitov R. Innate immunity – cross-talk with adap-tive immunity through pattern recognition receptors and cytokines. Curr Opin Immunol 2007;19:1–3.

[103] Straub TM, Honer zu Bentrup K, Orosz-Coghlan P, Dohnalkova A, Mayer BK, Bartholomew RA, et al. *In vitro* cell culture infectivity assay for human noroviruses. Emerg Infect Dis 2007;13:396–403.

[104] Nickerson CA, Ott CM. A new dimension in modeling infectious disease. ASM News 2004;70:169–75.

[105] Carterson AJ, Honer zu Bentrup K, Ott CM, Clarke MS, Pierson DL, Vanderburg CR, et al. A549 lung epithelial cells grown as three-dimensional aggregates: alternative tissue culture model for *Pseudomonas aeruginosa* pathogenesis. Infect Immun 2005;73:1129–40.

[106] Honer zu Bentrup K, Ramamurthy R, Ott CM, Emami K, Nelman-Gonzalez M, Wilson JW, et al. Three-dimensional organotypic models of human colonic epithelium to study the early stages of enteric salmonellosis. Microbes Infect 2006;8:1813–25.

[107] LaMarca HL, Ott CM, Honer Zu Bentrup K, Leblanc CL, Pierson DL, Nelson AB, et al. Three-dimensional growth of extravillous cytotrophoblasts promotes differentiation and invasion. Placenta 2005;26:709–20.

Chapter 2

Software for Biofabrication

Danielle Beski*, Tom Dufour**, Frederik Gelaude†, Ashok Ilankovan‡, Maryna Kvasnytsia**, Mike Lawrenchuk*, Ihor Lukyanenko**, Mariam Mir**, Lars Neumann**, Anthony Nguyen*, Ana Soares†, Emilie Sauvage**, Katrien Vanderperren**, and Dieter Vangeneugden**

*Materialise USA, LLC, Plymouth, MI, USA; **Materialise N.V., Leuven, Belgium; †Mobelife N.V., Leuven, Belgium; ‡Materialise SDN. BHD., Selangor, Malaysia

Chapter Outline

ABSTRACT

Software is the key to 3D printing. Central to almost all medical printing applications is the creation of a digital representation of a patient's anatomy from 3D image data obtained from a medical scan of the patient. The quality and usability of such a model depends crucially on software. When proceeding from the design phase to the actual printing, the model is prepared and optimized for the material, the printing process, and the printer model of choice. Dedicated software ensures a final high-quality physical model. While homemade solutions currently prevail in bioprinting, we expect standardized and medical-grade solutions developed in "conventional" 3D printing to enter the field of bioprinting as it progresses from research into real-life applications. In this chapter, we describe state-of-the-art 3D printing software to go from medical images to the printed object and show how bioprinting benefits from conventional 3D printing.

Keywords: bioprinting; software; 3D printing; medical; mimics; 3-matic; magics; implants; segmentation; orthopedics

1 INTRODUCTION

Medical 3D printing is a powerful and flexible solution for the medical restoration of the human body. From patient-specific

Essentials of 3D Biofabrication and Translation. http://dx.doi.org/10.1016/B978-0-12-800972-7.00002-5

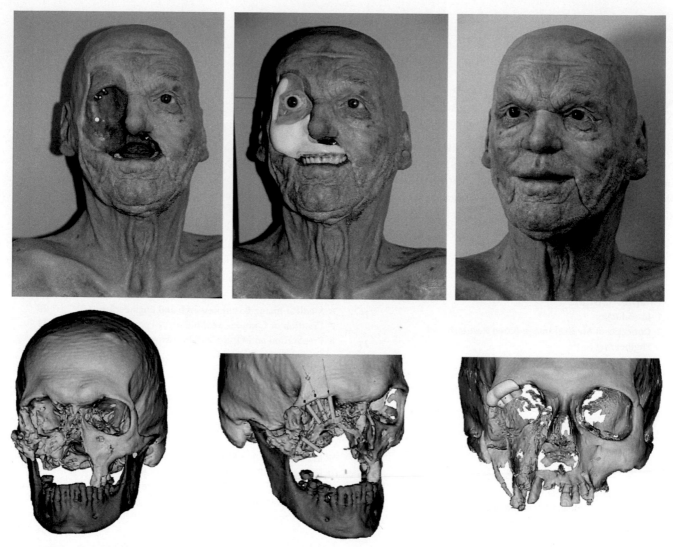

FIGURE 2.1   Top: Facial reconstruction by Jan De Cubber, CCE Zaventem. Bottom: Intervention planning with the Mimics Innovation Suite, Materialise N.V. The appearance of the model was changed in order to protect the identity of the patient. *Images courtesy of Jan De Cubber, CCE Zaventem, and Materialise NV, Belgium.*

implants to the restoration of a face and its aesthetics, human skin with natural pigmentation, bone replacements, or the replication of complex organs like the liver or kidney, 3D printing is already a valuable tool. Medical 3D printing creates patient-specific solutions that perfectly mimic the physical shape and can potentially restore the original functionality of a body part.

Central to the process of medical 3D printing is the software that enables the workflow from a patient's medical images to an accurate 3D model. This model is the starting point for design or analysis operations and for 3D printing and device manufacturing. Its quality is the decisive factor as to whether a medical solution will benefit a patient. While the field of 3D bioprinting is relatively young, full software solutions already exist in the medical 3D printing market.

An impressive example displaying the capabilities of current medical and 3D printing software is the facial

reconstruction by Jan De Cubber, CCE Zaventem, presented in Fig. 2.1. Recurring bouts of cancer forced doctors to remove a large area of soft and hard tissue from this patient's face that included his right eye, cheek bone, and upper jaw. The resulting deformation had severe functional and aesthetic consequences for this patient.

The successful treatment had to address two reconstruction challenges: (1) the missing bone tissue and (2) the missing soft tissue. The new bone tissue would have to provide the shape and physical strength to fix dentures and give the patient the ability to once again chew food. The new soft tissue would restore the natural aesthetics.

Biomedical engineering software has been developed with this type of application in mind: to create a digital 3D representation of a body part like the face, perform computer-aided design operations directly on the anatomy, and turn the design into a physical device by 3D printing.

Model creation and design operations are performed digitally and are largely independent of the materials used later in the printing process. Whether the desired solution requires metals, plastics, or biomaterials, researchers, engineers, or clinicians follow the same workflow in the design of implants and prosthesis.

The material properties and the printing hardware are relevant when the design is ready for printing. Also, the existing software corrects and optimizes the design for a specific printing process and technology and ensures a high-quality end product.

In the model depicted in Fig. 2.1, De Cubber visibly achieved an incredibly realistic reconstruction of the patient's original face. The original aesthetics were successfully restored, including the shape and texture of the face. The functional properties were also fully restored.

In the process, De Cubber and clinical engineers at Materialise created an accurate digital 3D model of the patient's head from CT images. As all design operations were performed on this model, its quality was crucial for the outcome. The missing bone was reconstructed by mirroring the part of the model that corresponded to the healthy side of the patient's face. Printed using titanium, the replacement bone served as the base for the dentures and then the shape of the face was restored in a similar way. Due to the complex natural texture of skin, its natural appearance was modelled with silicone. The implants provided the required support and shape.

Current software and processes now enable researchers, engineers, and clinicians to achieve such realistic results and create 3D printed customized devices and implants. 3D bioprinting promises to expand these possibilities further. The term "bioprinting" commonly denotes the 3D printing of biocompatible, biodegradable, or "living" materials. While "conventional" 3D printing allows the creation of bio-inert devices in metals and plastics, bioprinting aims at designing functional human tissue that integrates and interacts with the human body.

This means that for the reconstruction depicted in Fig. 2.1 bioprinting will – eventually – provide the clinicians and engineers with functional skin and vascularized bone tissue to reconstruct the patient's anatomy in every respect.

2 CONCEPTS OF MEDICAL IMAGE-BASED RESEARCH AND ENGINEERING

Computer-aided design (CAD) operations are a central tool for biomedical researchers and engineers. As these operations are performed on, and with, a patient's actual anatomy, a high-quality and three-dimensional digital model of the patient's anatomy is a precondition. Motivated by the needs of physicians and radiologists for accurately transferring image-based planning to the operation theater, the design and fabrication of patient-specific guides and implants have become important applications. Medical image-based engineering has also become a powerful tool in research.

This section introduces the overall concepts of medical image-based research and engineering. As illustrated in Fig. 2.2, we explain the concepts used within a typical

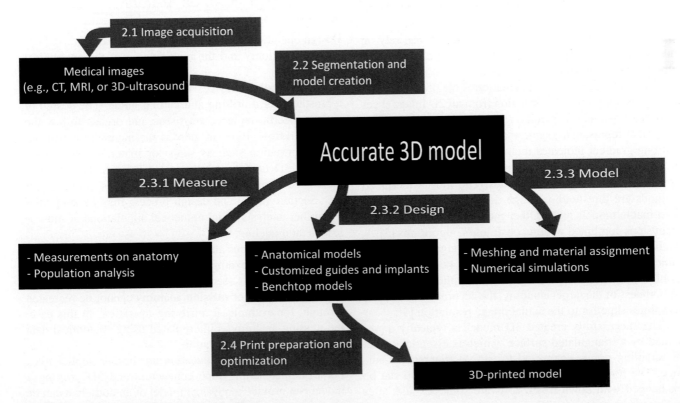

FIGURE 2.2   Sketch of typical biomedical workflows.

workflow from the acquisition of patient images through the creation of an accurate digital 3D model to design operations and the final 3D printing of the model.

2.1 The Acquisition of Patient Images

The acquisition of medical images of a patient is the very first step in medical image-based research and engineering. The quality of the medical images directly impacts the accuracy for further steps along the workflow. The choice of an appropriate imaging protocol is essential, and recent developments in medical imaging techniques have been oriented toward higher resolution and sensitivity. Often, the accurate reconstruction of patient anatomy requires a different protocol than the one used for pure diagnostic purposes.

Standard imaging modalities include magnetic resonance imaging (MRI), computed tomography (CT), and 3D ultrasound. Most modalities acquire the data as a set of successive, two-dimensional (2D) image slices. Within these slices, information is stored in the form of gray values that correspond to the measured type of signal. As the signal depends on the material properties of the bodily tissue, certain material properties and spatial information may be extracted.

2.2 The Creation of a 3D Model from Patient Images

Image segmentation refers to the process of converting the 2D gray value images to a high-quality 3D model. As all subsequent steps, from planning a surgical intervention to custom device design and numerical simulations, rely on the quality of this 3D model, it is therefore clear that accurate image segmentation is crucial.

On a technical level, segmentation is the process of collecting and combining information from all 2D image slices to form a 3D model. Typically, regions of comparable intensity values represent the same anatomical structure, whereas a strong gradient indicates their boundaries. A wide range of segmentation algorithms cater to different segmentation needs and imaging modalities. The most widespread methods are thresholding, edge detection techniques, and deformable models [1,2]. Other methods include enhancement steps combined with reconstruction algorithms [3].

Image resolution and segmentation algorithms determine the smallest feature size that can be reconstructed from the images. Depending on the application and the smoothness of the target anatomy, the accuracy can be up to five times superior to the actual image resolution [4].

The successfully created 3D model is typically represented by a triangulated surface, which is essentially a set of sampling points connected to form a triangulated surface. The format is a standard in 3D printing and can be exchanged with downstream software solutions and, after postprocessing, sent for 3D printing.

2.3 From an Accurate 3D Model to New Results

The high-quality 3D model created by image segmentation is the starting point for all further engineering operations.

2.3.1 Measure – Measurements and Population Analysis

The 3D model permits accurate measures of parameters such as lengths, angles, diameters, volumes, and many more. So-called center lines map vascular and pulmonary systems and automatically provide information on the branching structure.

The statistical comparison of several 3D models, for example of the femur anatomy, is commonly referred to as population analysis [4a,4b]. This analysis allows the assessment of the differences between subpopulations like Asian versus European subpopulations for female versus male subpopulations. Population analysis is particularly useful in two cases:

1. Based on population analysis results, standard implants can be designed to optimally cover a specific population.
2. Cancer or accidents can impact or destroy a patient's anatomy. Statistical analysis may help to reconstruct the anatomy by comparison with a fitting instantiation of an appropriate population model.

2.3.2 Design – CAD on Anatomy and Restoration of Anatomical Defects

Design operations encompass the design of customized devices on anatomy and the modification of the anatomical model.

Traditionally, "device" meant a medical implant like a prosthesis or a drilling and cutting guide. The design of patient-specific replacement tissue and organs follow the same workflow. Relevant here is the incorporation of patient information such as shape or material composition from the 3D model into the design process. This ensures that the customized device perfectly fits the specific patient (see Section 10.4). The design process may be supported by further analysis using numerical simulations or musculoskeletal modeling.

Design operations also allow reconstructing missing anatomical data, for example, as in the case shown in Fig. 2.1. Population analysis (see Section 2.3.1) can be a valuable tool if a defective or missing anatomy cannot be recreated through, for example, a mirroring operation. In this case, the missing anatomy is interpolated using anatomical data from a comparable population.

Device testing and education are further applications. For example, adding pipe connectors and 3D printing a model can provide a physical model of an aorta that can be connected to a circuit for actual fluid experiments.

2.3.3 Model – Preparation for Numerical Simulations

Numerical simulation techniques, such as finite element analysis or computational fluid analysis, have gained increasing relevance in research and development. Simulations allow a noninvasive insight into the human body or developing better implants by predicting their functionality in different scenarios. The 3D model defines the boundaries of the computational domain.

Computation time is one central limitation to the simulations. As medical 3D models are derived from medical images, the model contains a certain level of roughness and small features. If not compensated for, these can lead to a large number of mesh points and drive up the cost. The necessarily limited image quality can also lead to holes in the surface of the 3D model. These holes can make meshing impossible or, like a blood vessel with holes in its boundary, lead to incomplete or simply wrong simulation domains.

Dedicated preprocessing operations like smoothing or wrapping prepare the 3D model for numerical simulations and help to control the simulation time and cost. These and additional dedicated operations help to improve the mesh quality, which is crucial as the accuracy of the numerical simulations depends strongly on the mesh.

After remeshing, the mesh should be ready for numerical simulations in a dedicated numerical solver, whether it involves fluid, solid, or multiphysics problems.

The measure, design, and model steps are illustrated in Fig. 2.3.

2.4 Preparing and 3D Printing a Model

3D printing, also referred to as additive manufacturing, refers to a vast number of different technologies with the same goal: to turn a 3D computer model into a physical object. 3D printing uniquely permits the realization of highly customized physical objects like patient-specific implants in a cost-effective manner.

Printing technologies differ in their requirements and constraints in terms of the available materials and the properties of the printed object. Prior to printing, every design has to be optimized for a specific printing technology.

The wall thickness is an example of a parameter that typically needs correction depending on the printing technology. Walls too thin cannot be printed or are mechanically unstable. In general, and depending on the prior step in the workflow, the triangulated surface of the model often requires processing.

Creating support structures is another highly relevant task when printing in liquid or living materials. Support structures prevent the collapsing of overhangs and enable complex structures like vascular networks by connecting them with the build platform. Number, shape, and position of these supports are set automatically by dedicated software for additive manufacturing. The quality of the final

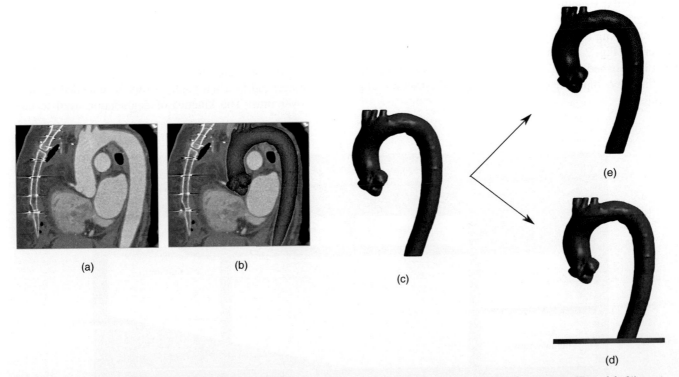

(a)   (b)   (c)   (d)   (e)

FIGURE 2.3   The figure illustrates a cardiovascular workflow. (a) Original medical image, (b) segmentation step to retrieve the 3D model of the aorta, (c) surface smoothing, (d) surface and volume meshing, and (e) design of a base plate.

product may also depend on how the 3D model is positioned within the printing volume of the specific printer.

From this brief summary, it is evident that the tools available today to physicians and engineers in the biomedical field enable noninvasive retrieval of patient-specific geometries and rapid creation of corresponding computer models. These can be further used in additive manufacturing, thus enabling medical staff to test and assess operational procedures. The same computer models are used for numerical simulations that provide an alternative to *in vivo* measurements and are increasingly used to improve our understanding of certain diseases or to design prosthetic implants. Much of the technology described here is also applicable in the area of bioprinting, which will be defined within the next section.

3 "BIOPRINTING" MEANS MANY THINGS IN 3D PRINTING

Regenerative medicine is the term covering the broad field of science around the process of creating living, functional tissues to repair or replace tissue or organ function lost due to age, disease, damage, or congenital defects. As part of regenerative medicine, tissue engineering is the domain that focuses on combining scaffolds, cells and biologically active molecules. This process has the aim of creating functional human, tissue which can range from skin and cartilage to more complex forms of tissues with vascular structures, to eventually, maybe, fully functioning organs.

Bioprinting is mostly used within the field of tissue engineering (Fig. 2.4). No clear definition of bioprinting has been established so far, and so the definitions range from printing biocompatible materials to the more strict printing of cells.

3.1 Scaffold Printing

Scaffolds are used to promote tissue growth. A known orthopedic application is a scaffold structure on bone interfacing locations to enhance bone ingrowth. These scaffolds most often consist of a bio-inert material like titanium.

The main approach for tissue engineering involves the seeding of a scaffold, in this case a biocompatible/bioresorbable 3D construct, with donor cells. This scaffold should mimic the extracellular matrix (ECM) properties like mechanical structure, support, and behavior to stimulate donor cell proliferation.

There are multiple ways to manufacture a scaffold. Among this variety of methods is chemical and gas foaming, solvent casting, particulate leaching, phase separation, and melt molding. New ways of producing scaffolds have presented themselves with the rise of additive manufacturing (AM). Among these AM technologies are stereolithography, selective laser sintering, fused deposition modeling, 3D printing, and 3D plotting [5–7].

In contrast to the conventional techniques, additive manufacturing brings more control to both internal and external scaffold geometry. Parameters like pore size, total porosity, and pore connectivity play a role in the mass transport of oxygen and nutrients and thus the tissue ingrowth. At the same time, this will, together with the choice of material, determine the stiffness of the whole construct. Optimal parameters depend on the tissue being engineered. Optimal pore sizes range from 100 μm to 500 μm. AM techniques are capable of reaching the needed accuracy if proper care is given to optimization of the processing of the materials. AM techniques that need to remove residual raw materials like powder or resins might hit constraints with inner microstructures of said dimensions [8,9].

The scaffold is a temporary structure intended to degrade over time. The kinetics of degradation need to be

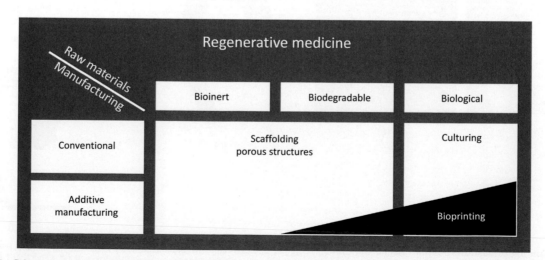

FIGURE 2.4 Schematic representation of where bioprinting is situated in the field of regenerative medicine and tissue engineering.

linked to the kinetics of formation of the new biological ECM. A range of materials for printing scaffolds have been investigated, including a variety of ceramic, metallic, polymeric, and composite materials. Among these materials, PCL, TCP, PLLA, and PLGA have shown potential [5,10].

3.2 Cell Printing

The introduction of additive manufacturing has the potential to overcome the constraints of insufficient and/or nonuniform cell seeding. This problem is often associated with the postfabrication cell seeding of scaffolds. More control over cell density and distribution will result in more viable tissue engineered constructs.

The incorporation of printing living cells adds to the complexity of the printing solution. The additive manufacturing machine needs to handle multiple materials: at least one scaffold material and one cell mix. The cell mix would be a hydrogel carrying the biological material. Several groups reported on printing cell constructs with only hydrogel. These constructs have shown difficulties in becoming end applications because of the mechanical weaknesses.

The actual dispensing of the cell mix demands control over various parameters so as to not compromise cell viability. Flow rate, temperature, nozzle tip size, pressure, layer thickness, and others need fine-tuning and control during the manufacturing process to enable sufficient diffusion of oxygen to keep the cells alive [11].

The microcontrollers needed to handle the printing of multiple materials with high demand parameter control on a microscale may be simultaneous and will have to be high-performance devices. Furthermore, bench-top manufacturing applications like the BioCell Printing device incorporate sterilization and a bioreactor, both of which add to the complexity. Nevertheless, it has been shown that these complex devices can produce cell-laden constructs with high cell viability. This is promising with respect to simple tissue solutions, but more research needs to be done in order to have fully functioning bioprinted constructs [12].

3.3 An Additive Manufacturing Definition of "Bioprinting"

The integration of "conventional" additive manufacturing in the medical world is now a fact. Patient-specific approaches have been established in multiple fields, and the technology allows pushing innovative areas like regenerative medicine and tissue engineering.

The term "bioprinting" ranges from using additive manufacturing for creating biocompatible devices to the strict printing of cells. Bioresorbable materials play an important role in scaffolds for tissue engineering and in non-tissue-engineered solutions like temporary stents. Restricting the definition of bioprinting to cell printing would not include printed applications that provide temporary function to the biological environment, where, after absorption, improved functioning of the treated tissues is achieved. Hence, a more suitable definition of bioprinting could be the manufacturing, by means of AM, of a device that results in a fully integrated and functionally restored biological environment.

4 BIOPRINTING AND CONVENTIONAL 3D PRINTING REQUIRE SIMILAR WORKFLOWS AND TOOLS

Bioprinting is a biomedical variant of additive manufacturing or 3D printing [13]. Even though bioprinting has many methods and workflows used in conventional 3D printing, the requirements specific to tissue engineering result in important differences.

As mentioned in the Chapter 1 and shown in Fig. 2.4, central in bioprinting are the two methods of scaffold printing and cell printing. The latter application dictates the materials and material classes of choice that will be employed in the printing. Relevant material classes include bio-inert, biodegradable, and biological or "living" materials. The properties of these material classes influence the entire fabrication process from the preferred software workflow to the printing technology.

Bioprinting consists of three consecutive steps: preprocessing, processing, and postprocessing [14].

4.1 Preprocessing

Medical images are the starting point for medical 3D printing. The major focus of conventional medical 3D printing is the design of implants and guides on anatomy. In contrast, bioprinting aims at regenerating the anatomy.

Studies highlight that the natural geometry of an organ is an important factor in tissue durability and organ function [15]. For instance, replicating the natural geometries of the aortic root, cups, and sinus wall is crucial for efficient hemodynamics and coronary flow [15]. An accurate anatomy modeling based on medical images is a very important phase in preprocessing within a bioprinting workflow.

For bio-inert materials, the design process can be considered as a separate phase prior to printing. For fully biological materials like living cells; however, the design development must incorporate the postprinting dynamics of living matter [14]. These dynamics are often required to fully develop the intended design.

Once the design is defined, dedicated software tools prepare the design prior to 3D printing and correct potential defects that may affect the printing quality. This step is similar for conventional 3D printing and bioprinting. It is parameterized with the printing technology and the printing material of choice.

TABLE 2.1 Summary of the Most Relevant Additive Manufacturing or 3D Printing Technologies

Technique	Resolution (μm)	Advantages	Disadvantages	Applications
SLS	400	High porosity, good mechanical properties, high accuracy, broad range of materials available	High temperature during process, feature resolution depends on laser beam diameter, porosity uncontrolled	Bone, cartilage
FDM	250–700	Good mechanical properties	High temperature during processing, available materials restricted due to the necessity of a molten phase	Bone, adipose, cartilage
3D plotting/direct ink writing	1000	Mild condition of process allows printing/plotting of drugs and biomolecules (proteins and living cells)	Limited resolution, slow processing, unfavorable mechanical properties	Bone, cartilage
SLA	70–250	Good mechanical properties, easy to remove support materials, high resolution	Limited choice of materials due to required photosensitivity and biocompatibility, exposure of material to laser	Bone, heart valves

4.2 Processing

Printing technologies and processes are a topic on their own. Since manufacturing methods define the design freedom of the researcher or engineer, we briefly summarize the major techniques in Table 2.1 with regard to their spatial resolution, advantages, and limitations [5,7,16].

In conventional 3D printing, software empowers the printing process in many parameters such as the optimization of the print quality, the utilization of the printing volume, and the occupancy of the printer itself. The economic success of conventional 3D printing has resulted in the foundation of large printing facilities that require control software to organize a large number of printers efficiently.

While conventional 3D printing imposes high requirements on the process and environmental stability, as well as parameters such as temperature or humidity, bioprinting demands even higher process control to ensure the survival of living matter. In addition, the complex internal structures of organs or bones translate into additional design requirements around chemical compositions, mechanical properties, degradation profiles, nutrient transports, and cellular (self-) organization. For instance, the strength degradation kinetics of porous scaffolds is highly affected by pore size, geometry, and orientation with respect to loading direction [17]. Surface properties, such as chemistry, surface charge, and topology, also influence hydrophilicity and, in turn, cell–material interactions for bone tissues' ingrowth.

4.3 Postprocessing

Biological or "living" materials continue to evolve after the printing process. Bioreactors have been designed to support this reorganization by providing a chemical and mechanical environment typical for native tissue [14]. This ongoing reorganization must be accounted for in the initial design process.

Quality control and process improvement require a comparison between the printed object and the initial design and optimization of the printing parameters [17]. Medical image-based engineering software often provides the tools to analyze and compare different 3D models. The researcher or engineer can simultaneously use different 2D and 3D imaging modalities in the analysis. Solutions dedicated to cell image analysis exist to estimate the validity of the fabricated tissue [18].

Bioprinting benefits from techniques and software solutions developed for conventional 3D printing technology. However, it requires adjustments in workflow to cater to new applications and material classes. The successful translation of bioprinting technologies into clinical applications will depend on the entire complex of printing technologies, materials, software, and their seamless integration. Software assists in the design process, prepares the design for the printing technology and material of choice, and is a tool for quality control.

The following section provides an overview of software solutions that may be used for bioprinting applications.

5 A SOFTWARE REVIEW

Listed in Fig. 2.5 are a number of software tools that are currently used in medical image-based research and engineering. These tools can also contribute to typical workflow in bioprinting as presented in the previous sections. This section is not intended to be a comprehensive overview of every software solution available, but focusses on the most relevant products only. In the interest of conciseness, common CAD packages, such as SolidWorks, CATIA, and Rhino, are not included in this review section.

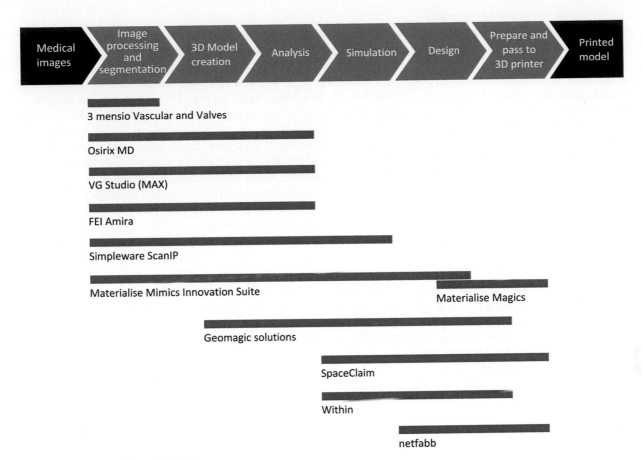

FIGURE 2.5 Overview on biomedical software.

3mensio focuses, since 2007, on specific workflows for both endovascular aortic aneurysm repair and more recently on transcatheter aortic valve implantation. *3mensio Vascular* allows importation of CT images, segment vessels, and visual inspection of calcification levels and the performing of measurements, whereas *3mensio Valves* focuses on segmentation of the ascending aorta.

OsiriX MD is an image-processing software dedicated to DICOM images produced by imaging equipment. It allows one to view, reslice, and stitch image sets and render 3D volumes and surfaces from the imported images.

Founded in 1997, Volume Graphics GmBH offers the following software products that can be used within the bioprinting workflow: *VGStudio* MAX and VGStudio. These products are dedicated to visualize and analyze CT data (e.g., looking at wall thickness or porosity).

Amira, from the company FEI, offers the ability to manipulate images through digital filters, visualization, and the editing of volumes. These volumes can be further analyzed with the available geometrical measurement tools of distances, angles, surfaces, and volumes.

ScanIP, from the company Simpleware founded in 2000, offers the ability to import image data, with which the user can visualize and perform segmentation and volume rendering. The images can be converted into computational models for CAD, finite element analysis, and computational fluid dynamics (CFD).

Founded in 1990, Materialise offers two software tools to cover the whole bioprinting workflow: The Mimics® Innovation Suite (MIS) and Magics.

The Mimics Innovation Suite, in turn, combines two applications: Mimics and 3-matic®. Mimics can be used for the manipulation and segmentation of (DICOM) images coming from MRI, CT, µ-CT, 3D Ultrasound, and other image modalities, from which 3D models of the various anatomies can be created. It offers a wide variety of measurement, analysis, and simulation tools in both 2D and 3D. Three-matic combines CAD tools with preprocessing or meshing capabilities. The software is geared toward the design of patient-specific implants or surgical guides but can be used to perform 3D measurements or engineering analyses and prepare anatomical data or implants for finite element analysis (FEA) or CFD as well.

Magics allows one to visualize, edit, and repair 3D models in preparation for additive manufacturing or 3D printing. In addition, users can perform hollow or Boolean operations, label, apply textures, perform cuts, and prepare the placing of each part on the build platform to ensure an optimal 3D build.

The *Geomagic* Solutions group within 3D Systems integrates combined activities from four smaller companies previously active in the 3D modeling field: Rapidform, Geomagic, Sensable, and Alibre. Their solutions can be roughly divided into three categories: scanning and 3D imaging software, 3D inspection software, and 3D design tools. Geomagic Studio (and to a lesser extent also Geomagic Wrap) allows transforming 3D scanned data into surface, polygon, or native CAD models. To model your objects further in a more manufacturing-oriented way, Geomagic Freeform and Freeform Plus can be used. Measurements and analyses can be performed with Geomagic Verify and Geomagic Control.

SpaceClaim Corporation has existed since 2005 but has recently been integrated into the technical software company ANSYS. Their main product, SpaceClaim Engineer, focuses on flexible and hands-on 3D design and links directly to FEA and CAD packages and additive manufacturing.

Within is an engineering design software and consulting company, focusing on optimizing the design of objects with a variable density lattice structure in preparation for additive manufacturing. Besides their flagship software, Within Enhance, they also offer Within Medical, allowing implant designers to create porous implants.

Netfabb is a software tool, allowing one to visualize, edit, and repair 3D models in preparation for additive manufacturing. The availability of functions depends on the version of netfabb Studio which comes only in a basic version but also as netfabb Private and netfabb Professional. The latter can be extended with a tool to create complex lightweight structures (netfabb Selective Space Structures software) or the possibility to import CAD files or automatically nest your objects for 3D printing.

In the next three sections we explain how the functionalities these tools incorporate are used in medical image-based research and engineering workflows.

6 MEDICAL IMAGE-BASED RESEARCH AND ENGINEERING

Bioprinting holds the promise of advancing implantable medical devices that can restore the body's anatomy and physiology so that they closely resemble normal conditions. As printers and materials progress toward more feasible solutions, it is important to note that the software infrastructure required for bioprinting has progressed greatly through parallel applications in the development of custom titanium implants and biocompatible surgical guides in the orthopedic and craniomaxillofacial fields and highly detailed medical models in the cardiovascular field. Looking at some workflows, this chapter reviews the software technologies behind the Mimics Innovation Suite that are available to advance bioprinting applications. Please notice that many of the concepts introduced in Section 2 will return here in an actually implemented form.

3D printing has played an invaluable role in the medical setting by opening opportunities to try different surgical approaches, providing medical practitioners a tool set to improve surgical outcome, limiting intra-operative planning, and preparing medical devices with anatomically relevant fittings. Software has been crucial for these opportunities. The MIS, shown in Fig. 2.6, allows for the accurate reconstruction of a patient's anatomy from medical images, detailed analysis of the anatomical condition, and virtual planning of a surgery. In the same suite, viewers can see design- and patient-specific implants and instrumentation, personalization of the fixation of implants, and conduct virtual tests to determine how the implants perform in the body over time and how they restore the musculoskeletal kinematics of the body.

Segmentation is the process of generating highly detailed and accurate 3D reconstructions of the anatomy based on medical images. The reconstruction of the heart in Fig. 2.7 showcases the level of detail that can be achieved – if the appropriate images are acquired. In these images, a cardiac CT protocol is used to distinguish the lumen from the soft tissue with contrast and acquires the images at a resolution of 0.6 mm. The segmentation and reconstruction of the lumen of the left compartment of the heart is shown on the left. With the preparation of the model, the myocardium of the heart can be reconstructed, as shown in the center. Cutting open the model, a surgeon can observe the papillary muscles. For bioprinting, a highly accurate model of the heart can be used as the basis to 3D print a collagen network of the heart for cells to be seeded.

In another example, an accurate reconstruction of the bones from medical images allows an assessment of a specific condition. In the left image in Fig. 2.8, the anatomical reconstruction shows a patient who suffers from lateral compartment arthritis. Aside from an anterior cruciate ligament reconstruction and lateral meniscectomy, the patient required an open wedge osteotomy to reposition his right femur to a normal anatomical alignment to improve his mobility. The improper alignment can be seen by comparing the right and left femurs. Furthermore, the ailment can be more concretely assessed by identifying the anatomical landmarks in 3D and measuring the angulation of the two femurs. The images show that the right femur has a 14° lateral angulation compared to the left. To restore the normal anatomical angulation of the right leg, an open-wedge osteotomy can be performed. These reconstructed models can be used to discuss with the surgeon how to approach the surgery. In the image, we show an open wedge realignment of the bone if the surgeon's plan is to rotate the shaft of the femur open 18° around a defined axis 7 mm away from the medial surface at a point between the condyle and the shaft.

Virtual planning of a surgery offers the doctor an opportunity to explore various surgical approaches and then by using measurements to analyze the anatomical outcome of each approach. This provides the doctor with an additional toolset to achieve the perfect surgery, and discard surgical

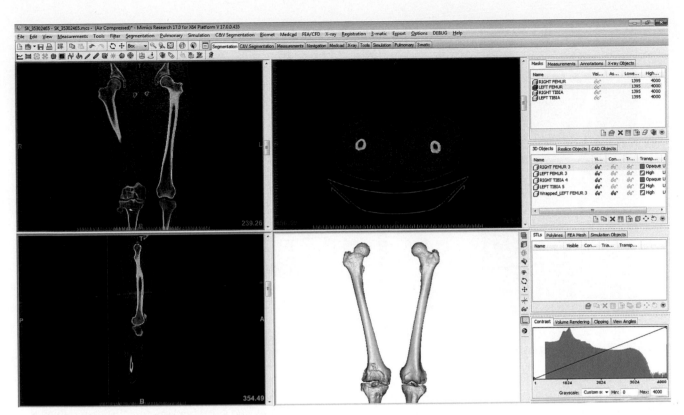

FIGURE 2.6   Medical images showing the lower body are processed with the Mimics Innovation Suite.

approaches that may ultimately result in a suboptimal result or require the doctor to make changes intraoperatively.

As one example of bioprinting, the 3D models of the planned postoperative situation can be used to create a bioresorbable scaffold to ensure optimal bone regrowth. The design features of MIS offer a variety of solutions for the design of such an implant. A wedge can be easily designed to fit cover the open portion, as shown on the right in Fig. 2.8.

Where a tumor has been osteotomized from a bone and does not require a realignment, the software can be used to mirror the anatomy from the good side as a basis to design an implant with realistic geometry, as shown in Fig. 2.9. The image on the left shows the left side of the pelvis mirrored and aligned (in green) onto the right side (in natural color). The second image from the left shows the implant that is created with the accurate geometry of the mirrored left pelvis. In the absence of a reflected model, MIS includes tools

FIGURE 2.7   The heart model is created from medical images and shows the exemplary quality that a segmentation process can reach.

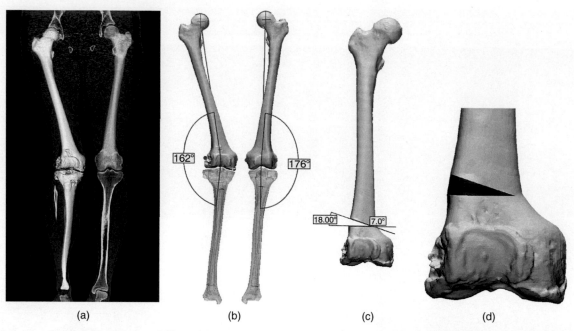

(a) (b) (c) (d)

FIGURE 2.8 Overlay of medical image and 3D model. (a) Lower leg, (b) angle measurements, (c) intervention planning, and (d) scaffold fitting.

that can reconstruct the anatomy based on the existing portion of the anatomy. The two images on the right in Fig. 2.9 show a reconstruction of a 10-cm portion of the femur shaft.

Virtual testing with FEA is valuable to analyze the performance of each unique implant. The MIS bridges the gap between anatomical models and FEA packages by including a solution for mesh generation. The software can be used to generate a tetrahedral mesh with optimal triangulation for use in FEA packages. Tetrahedral meshes are used due to their superiority in recreating complex structures with a high degree of accuracy, and by automatically using adaptive meshing MIS ensures geometric complexity is maintained while reducing mesh density where required. Furthermore,

one is able to generate an assembly mesh that combines anatomy, bioscaffold, and any other structure that may influence the virtual testing. In Fig. 2.10, the previously designed scaffold has been meshed as an assembly of the osteotomized bone. Aside from the automated adaptive meshing, a higher density mesh has been intentionally created at the interface between the bone and scaffold. The left image shows the original triangulation, the center image the optimized triangulation, and the right image the tetrahedral mesh.

Note that CFD is often used in the case of cardiovascular or respiratory devices to study the influence of the flow of blood or air. The mesh generation tools mentioned can also be used for CFD applications.

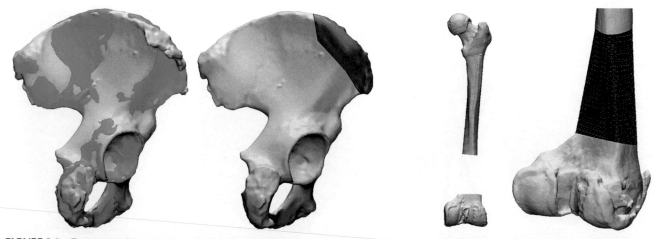

FIGURE 2.9 Reconstruction operations include the comparison and mirroring of healthy and pathological anatomy (left two images), and the interpolation of defect anatomy with a scaffold structure (right two images).

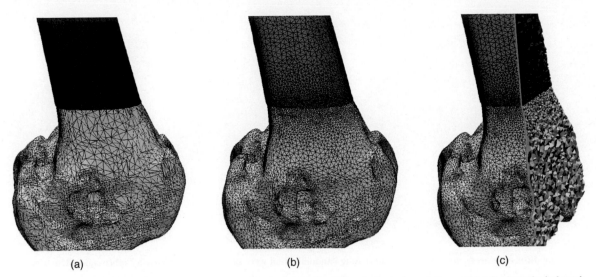

(a) (b) (c)

FIGURE 2.10 **Preparation for numerical simulations.** (a) original triangulation, (b) optimized triangulation, and (c) final tetrahedral mesh.

As software designed specifically with biomedical professionals in research and development in mind, the MIS offers a tool set dedicated and optimized for biomedical applications. Where traditional CAD packages are designed for technical designs, anatomy has complex and freeform geometries. The software contains tools that capitalize on the existing anatomical information that we can extrapolate from medical images and solutions for creating anatomical designs that meet the biomedical aspirations. Bioprinting applications can benefit from the solutions available in the software, as they are further developed to stay ahead of the fast-changing 3D printing landscape.

7 CREATION OF COMPLEX SCAFFOLDING OR POROUS STRUCTURES

As introduced in Section 2.4, scaffolds, or porous structures, are highly relevant tools in regenerative medicine. From a software perspective, their creation, handling, and printing are complex and demanding tasks.

Porous structures are highly interconnected, web-like objects that can be regular or random in shape. Several examples are presented in Fig. 2.11. Their creation requires dedicated algorithms that build such a structure from a combination of designed unit cells and periodicity or randomization parameters. Porous structures allow modifying the mechanical properties of the complete design by their internal structure. In orthopedics, porous structures also shorten the healing process by supporting bone ingrowth.

The complexity then leads to high memory requirements, such that objects of realistic sizes and complexities cannot be processed and printed by the 3D printing systems. Also here, dedicated software like Materialise's Magics and Build processors is required to preprocess the design for printing.

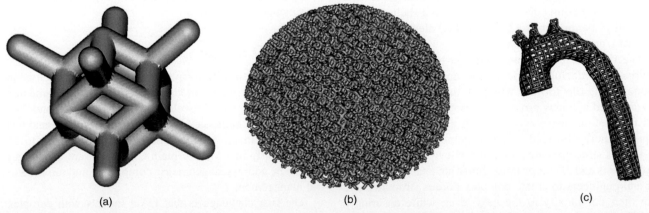

(a) (b) (c)

FIGURE 2.11 A unit cell (left) is the starting point in the design of complex porous structures (middle). The lightweight structures software also allows creating shell structures that surround a particular volume (right).

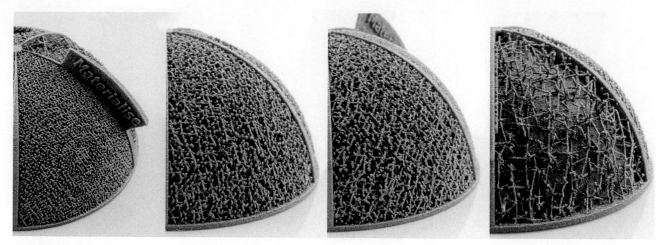

FIGURE 2.12 Four different porous structures with increasing randomization were design in Materialise's Lightweight Structures software and 3D printed.

The Lightweight Structures software allows the creation of foam-like porous structures within a defined volume. The starting point is a so-called cell-element that is multiplied to fill a specific volume. The software contains a library of standard cell elements; however, the researcher or engineer has the freedom to create new cell elements in Materialise's 3-matic STL software or in any other CAD package. Control parameters include the size of a cell-element as well as its orientation, strut thickness, and porosity level.

Randomization of these elements as shown in Fig. 2.12 allows the creation of an organic and natural profile that imitates the porous structure of bone. The level of randomization is chosen by the user. Reproducibility is ensured through the possibility to recreate the same porous structure within different volumes.

While in orthopedic applications, for example, the osteotomy in Section 7, the porous structure fills out the volume, other applications require surrounding a volume with a shell structure. Figure 2.11 shows such an example. The software also permits the creation of such a shell structure and control parameters such as the shell element size, the thickness, and its orientation.

The finalized designs are then sent to a 3D printer for additive manufacturing. Generally, designs are exported through STL file format, which is the industry standard and accepted by many programs. However, highly complex and memory-intensive objects, such as porous structures, require a more advanced data management.

Porous structures are especially relevant in biomedical applications and 3D bioprinting. Software enables researchers and engineers to create complex porous structures and take their complexity to the next level, while ensuring a smooth printing process. The initial preparation of a design for 3D printing is the focus of the next section.

8   PREPARATION AND OPTIMIZATION – FROM THE FINAL DESIGN TO THE PRINTED OBJECT

In Sections 6 and 7, we showed the workflow from medical images through the computer-aided design phase to the finalized device design, which could be an implant in the form of a porous structure for example. While the design phase focuses on CAD operations, the next phase is oriented at realizing the design through additive manufacturing or 3D printing.

Ideally, an AM machine should be able to print a part with a simple click of a button. Unfortunately, a series of actions are required to build a part. The work flow will vary depending upon the AM technology utilized, but it generally starts with a CAD model that is converted into an STL file, edited, placed on a machine platform, build instructions set, built, and finally postprocessed for application (Fig. 2.13). Our objective is to focus on the software aspect of the industry and will highlight this aspect throughout each of the aforementioned stages.

8.1   CAD Design

The AM process will always start from CAD, which is an invaluable tool where modifications can be implemented in product design and development. CAD-based candidate designs can be built via AM processes to visualize and test prior to manufacturing. In order for a CAD model to be built properly in AM, two main CAD software challenges need to be addressed: geometric complexity and multimaterial management.

The first challenge is that CAD models with complex external geometry may contain regions difficult for AM machines to build. For example, a bone tissue scaffold that mimics

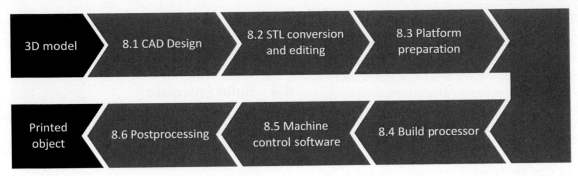

FIGURE 2.13   Workflow from a 3D model to the printed object.

bone morphology would contain thin-walled structures. Due to technology-dependent limitations, wall-thickness analysis must be performed to ensure the thickness of a wall will not be less than the resolution of an AM machine's print capability. AM software (Fig. 2.14) provides a functionality that analyzes the wall thickness of a part and creates a color gradient to identify troubled areas.

The second challenge is that CAD models for a multimaterial part do not contain material information but only geometric information. The reason for this is that CAD software has not been developed to specify a part's material composition. Fused deposition modeling (FDM) is an extrusion-based process. The Fortus machine, developed by Stratasys (Edina, USA), utilizes two nozzles on one head and is capable of printing discrete multiple material parts. Since multiple material information is not provided by CAD software, a user must pause the build using Stratasys' Insight software and manually change the material cartridge.

The limitation is also evident when trying to determine the material composition of multiple material parts produced by Objet, an original equipment manufacturer (OEM). The Objet Connex machine utilizes two different photopolymers and has heads that contain over 1500 individual nozzles for rapid deposition. The Connex machine can print about 25 different materials by mixing different ratios of photocurable resin and curing agent (Fig. 2.14). Additional software tools have been created to allow the user to designate and assign material attributes to different regions of an STL file.

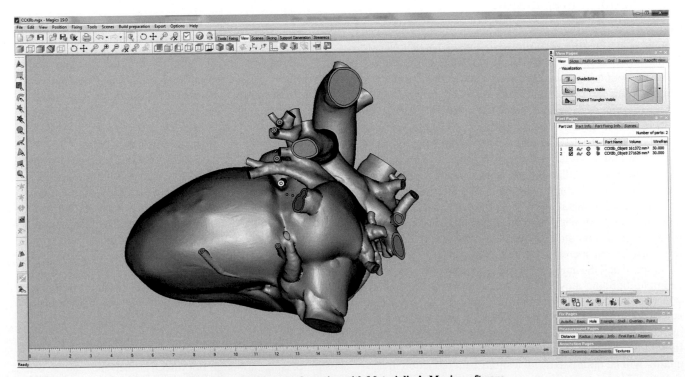

FIGURE 2.14   A human heart is prepared for additive manufacturing with Materialise's Magics software.

8.2   STL Conversion and Editing

The STL file is a generic file format that is the standard input file for most AM machines. The term STL was coined by 3D Systems and was derived from its commercial technology called stereolithography. Due to technical innovations in the AM industry, novel processes differing from stereolithography have been developed and are now readily utilized. Thus, an ASTM committee renamed the file to Standard Tessellation Language. An STL file consists of only three vertices stored in a counter-clockwise fashion with a unit normal that defines the outward face of a triangle. In order to produce a fully enclosed solid or "watertight" model, each triangle edge must be shared with the edge of one other adjacent triangle and their normals must be facing the proper direction. Euler's characteristic equation for a solid state is as follows:

$$\text{Number of faces} + \text{Number of vertices} - \text{Number of edges} = 2 \times \text{Number of shells}$$

where a shell is defined as a group of triangles properly connected to one another and their normals facing the proper direction.

Unfortunately, some CAD software will not produce a watertight model due to gaps in surfaces and/or solids during design while others create errors during STL conversion. STL repair and editing software resolves these issues by performing automatic repair as well as manual repair functionalities for more complex CAD models.

8.3   Platform Preparation

Quality, speed, and supports (for AM technologies that need supports) are the primary criteria to consider when placing a part on a machine platform. AM machines build in the X and Y direction significantly faster than in the Z direction. Supports are structures that have different functionalities with different AM technologies. For example, parts built with stereolithography require supports to anchor the part to the platform so that the part will not float away or move during the build. Metal parts need supports to be anchored to the platform as well, although not to prevent unnecessary translation but to prevent warping due to thermal stresses.

Supports for metal parts also have a critical function as heat sinks to provide thermal dissipation. When a CAD model is placed on an AM machine platform, part orientation must be considered to optimize the quality of a part and build speed and the amount of required support structure. For example, building a cylindrical rod lengthwise horizontally will maximize build speed but sacrifice part quality (build layers are visible on the part and require extra finishing and increase the amount of support structure needed to successfully build the part). By building the part vertically, the part quality will be optimized with a significant reduction in necessary support structure at the cost of increased build time.

Essential to any AM software is an orientation optimizer as well as an automatic support generator with manual editing capabilities.

8.4   Build Processor

As mentioned earlier, STL files only contain geometrical part information and thus build instructions, such as material composition, slicing strategy, and hatching strategies, must be provided to successfully print a part.

For example, SLM is an OEM that focuses on the selective laser melting technology that has a software called Build Processor. The software application allows a user to adjust 166 different parameters. By doing so, the user is able to configure the machine for each part on the platform. The parameters will vary, depending upon the capabilities and limitation of the AM technology utilized and thus every technology will require the OEM to produce its own software to assign attributes to external geometries provided by STL files. With every OEM producing its own supplementary software, the utilization of multiple AM machines would result in an accumulation of software as well as a need to learn how to utilize each one. Like 2DP before 3DP, each 2DP OEM has its own print processor (drivers) for each of its 2D printers installed on personal computers that are consolidated into a library.

Similarly, AM software, like Materialise's Build Processor, can communicate with a multitude of AM machines and consolidate the machines for many AM OEMs into a single library.

8.5   Machine Control Software

Each AM technology and each AM system within each technology must define the build behavior of its machine. The machine control software (MCS) manages, commands, directs, and regulates the behavior of an AM machine to build parts layer by layer and consists of two components: (1) the framework and (2) the logic or behavior.

The framework is the hardware, the machine components, and can be compared to a house with walls, doors, electrical switches, and roof all well defined. The logic or behavior defines how the machine operates (e.g., warm-up phase, platform position, cool-down phase, exposure times) and can be likened to the arrangement of the furniture and the installation locations of the lamps according to the owners' (OEM) preference. As hardware technology evolves, the machine control software has to be flexible, adaptable, and reusable as the machine components are upgraded.

8.6   Postprocessing

Postprocessing, although not directly software related, is a critical stage in the AM process. Once a build job has been completed, the parts must be removed from the platform. The removal process utilized will depend upon the AM technology. Parts built via stereolithography may simply be pulled off the platform, while parts built via metal technologies will require a bandsaw or electrical discharge machining (EDM) to remove. After removal, the part will require additional treatment, depending upon the application. Painting, priming, plating, and infiltration are examples of some of the treatments that could be laborious and require experienced manual manipulation.

Current software solutions provide flexibility to modify the supports as well as orient the part to reduce the amount of postprocessing required.

Today, nearly every biofabrication research lab has built its own bioprinter and thus has developed its own software, which is similar to the situation today with AM OEMs. Thus, there is a need to standardize and centralize all the various OEM software solutions into a single accessible library. Each OEM produces specialized machines, thus a demand exists for multiple machines to cover the gamut of production applications. A machine management system was created to standardize and network AM machines similar to the way 2D printers are managed on personal computers. As the number of bioprinter OEMs begins to increase, bioprinting software will benefit by following its predecessors in standardizing all OEM software. With an integrated software workflow and a standardized bioprinting software library in place, biofabrication will be one step closer to standardized mass production of living tissue.

9 MANAGEMENT OF A MULTIPLE-PRINTER FACILITY

The success of 3D printing and the global demand for AM manufactured parts brought about the inception and proliferation of industrial-scale 3D printing service bureaus. To meet the demands of all production processes and expand their product portfolios, AM service bureaus acquire and operate 3D printers from a variety of manufacturers, printer generations, and printing technologies.

The integration of different 3D printers into a single manufacturing facility will inevitably create new challenges, chief among them: How to use printing resources with maximal efficiency? How can an array of orders be tracked, especially if different orders are combined and filled by a single machine? How are design revisions stored, accessed, and traced? Quality control is essential to medical device fabrication and is dependent upon machine and printing processes. Automation will become an absolute necessity as a maturing market forces companies to decrease costs.

While industrial-scale production is not yet prevalent in bioprinting, technological progress, and new market generations will bring about extensive bioprinting facilities and thus the need for automation and control software systems.

10 EXAMPLES OF BIOPRINTING APPLICATIONS BENEFITING FROM ADDITIVE MANUFACTURING SOFTWARE

Four exemplary cases show how researchers and engineers today use 3D printing software solutions in 3D bioprinting. In the first case, researchers intend to replicate a patient's aorta using cell printing. Resorbable materials are used in the second case for a degradable implant. A different approach employing scaffolds and porous structures is presented in the third case. The last case shows how far tissue properties can be incorporated into the design of an implant to optimize functionality.

10.1 Toward Replicating a Patient's Aorta

Cardiovascular diseases remain the leading cause of deaths worldwide and often require vascular reconstruction. Autografts and blood vessel transplantation are the most common treatments in cardiovascular diseases. However, autologous veins are not available in 40% of cases as result of trauma or underlying diseases [19]. The problem is further amplified by chronicity of the diseases resulting in a group of patients who will require reoperation.

The fabrication of blood vessels remains a very challenging field of research. Ideal tissue-engineered blood vessels have to fulfil a range of biological, mechanical, and commercial requirements and should function immediately upon implantation [19].

The traditional approach in tissue engineering included seeding cells into synthetic, biological, or composite scaffolds or porous structures. The construction of entirely biomimetic blood vessels has not yet been achieved. The main challenges that remain are the poor mechanical properties of materials, materials degradation, and poor cell–material interaction [20].

Bahattin Koç et al. from Sabanci University [21] investigated scaffold-free soft tissue engineering techniques. The group developed computer-aided algorithms to model and 3D bioprint multicellular cylindrical aggregates and their support structures with biomaterials.

For the optimal organ functionality, it is crucial to reproduce the original vessel shape and branching system. The researchers from Sabanci University have presented a proof of concept study on bioprinting of macrovascular structures based on patient-specific anatomy [22]. As shown in Fig. 2.15, they created an accurate 3D model of abdominal human aorta based on MRI data using the Mimics Innovation Suite for image segmentation. However, the vessel wall thickness cannot be derived from images due to limitations in imaging technology. The researchers assigned literature wall thickness values to the 3D aorta model with the design environment of the Mimics Innovation Suite.

One of the challenges of scaffold-free engineering is the requirement to provide support for cell layers to preserve

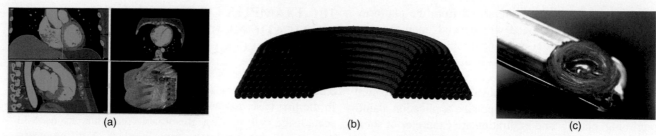

FIGURE 2.15 (a) Creating a heart model with the Mimics Innovation Suite. (b) Cross-section of 3D self-supporting model with cells (red) and supports structure (blue). (c) 3D bioprinted tissue constructs after three days of incubation. *Images courtesy of Bahattin Koç, Sabanci University.*

the required topology. Researches defined two approaches for supports generation: "cake" support and "zig-zag" support. In the case of "cake" supports, the cylindrical cell and support rods are printed in between the preceding layer's support structure. In case of "zig-zag" supports, the layers are formed by crossing vertical and horizontal support rods at successive layers. The "zig-zag" approach is more suitable for complex organ geometries.

The Novogen MMX Bioprinter was used for hybrid 3D printing with support biomaterials. The researchers created path-planning scripts in Rhino3D with Rhinoscript language for cell aggregates and their support structure. Thus, the designed anatomical model was used to create bioprinting instructions. The vessel segments (about 3.5 mm long) with support structures were built using fibroblast culture and Novogel.

As Koç emphasizes, preprocessing, such as converting medical images to bioprinter instructions, is an important problem. The topology of the bioprinting process (where and how to print cell and biomaterials) should be optimized under all the biology/cell-related and process-related constraints. The study illustrated that the combination of medical imaging, CAD systems, and path planning algorithms, with 3D printing technologies in one pipeline made one more step toward the construction of functional blood vessels. As the next step toward biomimetic vessel generation, researchers see the simultaneous printing of the endothelium and muscle vessel cells to reproduce all vessel layers.

10.2 Creating Patient-Specific Implants with Resorbable Materials

Scott Hollister and Gleen Green from the University of Michigan used a similar software workflow to work on a different material class, when they collaborated to treat Kaiba Gionfriddo, a patient with severe tracheobronchomalacia [23,24]. Tracheobronchomalacia is a rare condition that is characterized by weak cartilage in the trachea. This condition can result in difficulty in breathing and, in severe cases, can cause a patient's bronchi to collapse without warning [25]. Kaiba Gionfriddo was one of approximately 200 babies born every year with this most severe version of tracheobronchomalacia [23].

When other treatments for Kaiba proved unsuccessful, Dr Hollister and Dr Green received an emergency-use exemption from the FDA to create a patient-specific splint in a novel bioresorbable material called polycaprolactone. The intent was that the stent would support the bronchus locally to prevent collapse and eventually resorb, leaving an intact remodeled bronchus [24].

The splint was designed directly from a CT scan of the patient. The Mimics Innovation Suite was used to create a 3D CAD model of the trachea/bronchi, and from there was used to custom design the splint (generated using a custom written MATLAB program) to fit to the bronchi (Fig. 2.16). This design was exported and sent to a customized EOS 3D printer as an STL file. Here, the device was built through additive manufacturing by laser sintering layers of

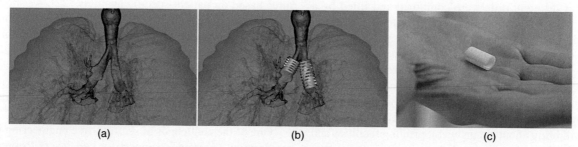

FIGURE 2.16 (a and b) CT rendering of the trachea/lungs and the simulated intervention for a second case. (c) 3D-printed splint built of the polycaprolactone material for the case described in the text [23].

polycaprolactone (PCL) powder using techniques that Dr Hollister and colleagues had previously developed at the University of Michigan [26–28]. The end result was a completely novel implantable device: a fully bioresorbable splint, custom-made specifically to the anatomy of a single patient.

On 2 February 2012 the splint was placed and sewn around Kaiba's bronchi to provide support and aid in the proper growth of the airway. The result of the surgery was immediate and dramatic: according to Dr Green the patient's lungs "began going up and down right away…we knew he was going to be OK [24]." The patient was fully off of ventilator support after 21 days and, since then, the patient has not experienced any trouble breathing [23].

Due to the nature of the PCL material and design of the device, it is expected that the splint will be fully resorbed by the body after three years [24]. PCL proved to be a good choice for a polymer in this case, and it has been applied by Dr Hollister and his lab in other preclinical models. For example, it has been shown that PCL is effective in promoting tissue growth in temporomandibular applications [28], as well as other bone and soft tissue applications [26,27,29].

This case is an exceptional example of the possibilities that exist today with both tissue engineering and with medicine in general. In collaborative fashion, Dr Hollister and Dr Green used advances in CT image-based design, 3D printing, and innovative bioresorbable polymers to provide a novel treatment for a patient for whom no other option was presented. This is the promise that tissue engineering holds for the future: "personalized" medicine that can truly save lives.

10.3 Designing Scaffolds and Porous Structures for Medical Applications

Biological structures like organs or bones consist of several types of cells arranged in a specific spatial shape. Engineering an organ requires the replication of its biological composition and the internal structure simultaneously. A different approach to address this challenge is taken in this case by using a scaffold or porous structure [30]. A porous structure is usually a complex physical construct into which cells are seeded to grow and eventually fill the complete structure. Porous structures are the initial building blocks in the engineering of fully functional organs.

The choice of a particular porous structure is motivated by its intended behavior and researchers are faced with the challenge of creating an ideal scaffold. The overall shape reflects the geometrical fitting into a patient's anatomy. The internal structure determines the structural performance and integrity (both internally as well as externally). On a cellular level, the pore size and shape of the structure are key factors in the in-growth process of cells [31] and thus the success of the particular device. In addition to these characteristics, the material and technology being used for printing plays a key role in the success of the scaffold.

The laboratories of Dr Wei Sun at Drexel University and Tsinghua University address this challenge by developing optimized scaffolds as shown in Fig. 2.17a [32]. A complex example created by Materialise with its Lightweight Structures software is presented in Fig. 2.17b. Effective software, on the one hand, enables researchers to design highly complex shapes. Hardware and printing processes, on the other hand, are the key to translating these designs into physical structures.

At Dr Sun's laboratory, molding, soft lithography, bioprinting, and photo-patterning are just a few of the methods that his team has explored in designing scaffolds [33,34]. Wei also developed a novel approach to using the in-house developed precision extrusion deposition (PED) system with an assisted cooling (AC) device.

The PED allows for the fabrication of a scaffold through the means of liquefying and extruding a polymer through its tip in a 3D orientation. The highly precise movements of this device make it possible to use different computer-aided design programs to create complex internal and external architectures that more closely mimic native tissue.

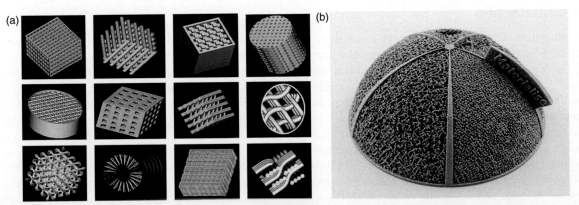

FIGURE 2.17 (a) Various computer-generated scaffolds. (b) Randomized and printed porous structure created by Materialise NV with its Lightweight Structures software. *Figure 2.17a is reprinted with permission by Sun et al. [32].*

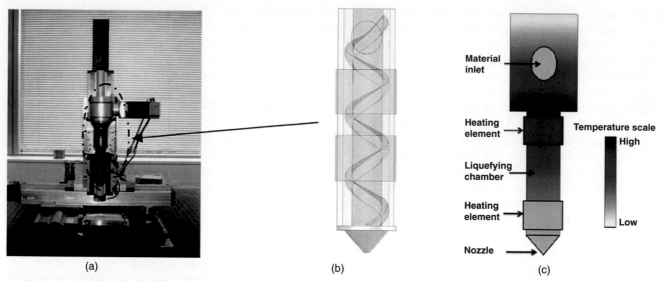

FIGURE 2.18 **The precision extrusion deposition (PED) system with an assisted cooling (AC) device.** (a) Front view of the PED, (b) schematic view of the material delivery chamber, (c) temperature gradient schematic view of the material delivery chamber where the top heat element has a higher working temperature in comparison to the lower heating element. *Reprinted with permission by IOP from Ref. [33].*

The PED is made up of several parts, including a material delivery system with two heating components that control the temperature of the polymer material being extruded. Figure 2.18 depicts the PED device and its delivery system in more detail. In one particular study, the team used a proprietary polymer known as Polyglycolide (melting point of 180 °C) to test the fabrication of three scaffolds with varying degrees of porosity. The study aimed to show that each scaffold developed with the PED can exhibit a different porosity (small, medium, or large pore size).

As the design and fabrication of highly complex scaffolds continues to improve, we will move closer to engineering a fully functional organ. Although hurdles still exist at this point, Dr Sun and his team are continuing their research and efforts in the design and fabrication of an optimized scaffold structure.

10.4 Implant Design with Optimal Biological Interaction

Total hip arthroplasty (THA) is an orthopedic surgical procedure for the treatment of traumatic or degenerative conditions of the hip joint [35–38] such as osteoarthritis and femoral neck fractures. The prevalence of THA is increasing due to the increasing age of the population and an increasing number of patients under 60 years receiving THA [39,40]. The procedure may fail due to a number of different complications often necessitating revision, that is, replacement surgery of the implant. On average, around 15% of patients receiving THA require acetabular revision surgery [41]. For some patients, especially after multiple revisions, this results in defects with extensive bone loss and/or bone deformation that drastically complicate conventional THA

approaches. Due to the increasing prevalence of THA and consequently multiple hip revisions, the need for complex hip surgery is likely to increase.

To deal with the challenges in complex THA surgery, a patient-specific approach is most suitable. Mobelife provides orthopedic surgeons with such custom acetabular implants that are developed based on CT data by combining state-of-the-art image processing tools and FEA. The process, presented in Fig. 2.19, begins with the creation of a 3D model of the patient's pelvic bone based on CT images. Mobelife employs Mimics medical imaging software from Materialise, enabling the quick creation of the bone anatomy surface from the CT images. The defect bone anatomy is then virtually reconstructed to mimic the healthy situation [42,43].

Based on this 3D reconstruction, Mobelife uses in-house software and Mimics software from Materialise to design an acetabular implant to fit the patient's bone defect [42]. The precise orientation of the newly created hip implant is anatomically analyzed [44]. In cases of dramatic pelvic bone loss, implants may extend onto the major bones of the pelvis for fixation. These patient-specific implants comprise a porous augment interfacing the implant and bone [35]. The distance between the bone and the porous structure is minimized and its contact surface is maximized due to the optimal fit, making bone ingrowth easier and possible over a larger surface area. Concomitantly, screw placement is planned preoperatively, taking into account innovative parameters such as the quality and thickness of the patient's bone.

Once the design phase is finalized, FEA of the patient-specific implant is performed using the Abaqus software environment to analyze mechanical integrity of the implant

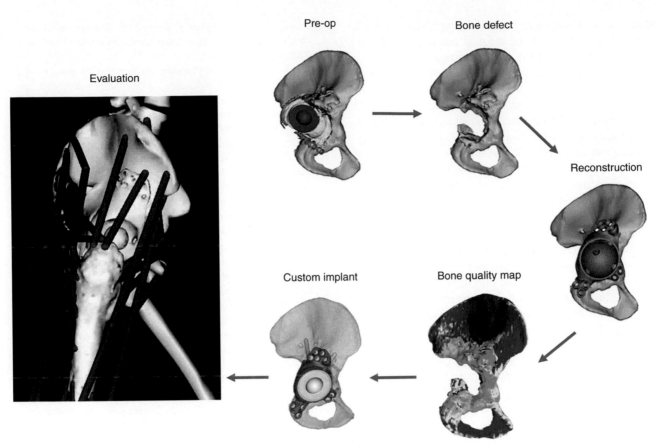

FIGURE 2.19 Schematic representation of Mobelife design process from a 50-year-old female patient with pelvic dislocation (pre-op). Large pelvis dissociation (bone defect) is reconstructed by a patient-specific implant with a porous augment (reconstruction). Optimal screw positioning (custom implant) is based on the patient bone quality (bone quality map). An analysis of the implant integrity is then performed with a patient-specific numerical model comprising muscle attachments and trajectories.

and its interaction with the surrounding bone. The location of muscle attachment regions on the bony structures and the interconnection trajectories of the muscles are included into the FEA. Specific forces acting on the pelvic bone and on the implant are determined based on patient weight and muscle activation [45-47]. The thickness and properties of both cortical and trabecular bone are automatically calculated from the patient CT data and included into the FEA. Finally, material properties of the titanium implant components are assigned.

After the evaluation of the implant-bone interfaces, Mobelife implants are 3D printed, allowing the implant's shape to be perfectly fitted to the patient's bone. Furthermore, custom drill guides are 3D printed in plastic to aid in accurate drilling and screw fixation, and plastic models of both the implant and bone are provided to surgeons to confirm the implant's fit and aid in preoperative planning.

11 CONCLUSIONS

Bioprinting has started to expand the field of medical 3D printing toward new medical applications. Crucial to the suc-

cess of medical bioprinting will be software that effectively and accurately supports a biomedical bioprinting workflow.

The commercial success of "conventional" medical 3D printing has led to the existence of a variety of software tools designed to help researchers, engineers, and clinicians in their daily tasks. These software tools support typical biomedical workflows that usually start with the acquisition of patient images and the conversion of these images into a high-quality digital 3D model. This 3D model is the starting point for all further operations and is crucial to the overall process quality, for example, the design of a patient-specific implant. The creation of the physical device by additive manufacturing, commonly called 3D printing, requires dedicated software to preprocess the design and optimize it for the printing process of choice.

Bioprinting essentially follows the same biomedical workflow as "conventional" 3D printing. As we have shown, researchers, engineers, and clinicians can reuse software solutions already in the market and reduce their development costs.

The success of "conventional" medical 3D printing and related biomedical engineering resulted in the on-going standardization of software, hardware, and processes. High regulatory requirements in the medical industry and the increasing interest of regulatory bodies in medical 3D printing emphasize the relevance of standardization. While bioprinting today relies heavily on "homemade" solutions for the bioprinter and software written by individual researchers, we expect standardization to be become an important topic in the near future. Software developed in "conventional" medical 3D printing is often already standardized and may offer ready-to-use and tested solutions. The development of software standards will further facilitate interaction with different bioprinting hardware and enable mass production of bioprinted devices.

Software will continue to play a crucial role in bioprinting that it already has in "conventional" 3D printing. Bioprinting researchers and engineers will benefit from the extensive research and development that has already gone into "conventional" 3D printing and thus shorten their development times, improve time-to-market, save resources, and allow cost-effectiveness. In other words, software should allow bioprinting to cross the gap between research and commercial applications allowing it to shift from a research-driven vision into a viable and active application area through which it eventually should touch and improve the lives of millions of patients.

ABBREVIATIONS

μ-CT	Microcomputed tomography
AC	Assisted cooling
AM	Additive manufacturing
CAD	Computer-aided design
CFD	Computational fluid dynamics
CT	Computed tomography
ECM	Extracellular matrix
EDM	Electrical discharge machining
FDA	Food and Drug Administration
FDM	Fused deposition molding
FEA	Finite element analysis
MCS	Machine control software
MIS	Mimics® Innovation Suite
MRI	Magnetic resonance imaging
OEM	Original equipment manufacturer
PED	Precision extrusion deposition
STL	Stereolithography/standard tessellation language
THA	Total hip arthroplasty

REFERENCES

[1] Ma Z, Tavares J, Jorge R, Mascarenhas T. A review of algorithms for medical image segmentation and their application to the female pelvic cavity. Comput Methods Biomech Biomed Engin 2010;13(2):235–46.

[2] Pham D, Xu C, Prince J. Current methods in medical image segmentation. Annu Rev Biomed Eng 2000;2:315–37.

[3] Lorensen W, Cline H. Marching cubes: a high resolution 3D surface construction algorithm. Comput Graph (ACM) 1987;21(4):163–9.

[4] Gelaude F, Vander Sloten J, Lauwers B. Accuracy assessment of CT-based outer surface femur meshes. Comput Aided Surg 2008;13(4):188–99.

[4a] Engelborghs K, Kaimal V. Finding the best fit. Orthopedic Design Technol 2012;8:56–9.

[4b] Mahfouz M, ElHak E, Fatah A, Smith Bowers L, Scuderi G. Three-dimensional morphology of the knee reveals ethnic differences. Clin Orthop Relat Res 2012;470:172–85.

[5] Bose S, Vahabzadeh S, Bandyopadhyay A. Bone tissue engineering using 3D printing. Mater Today 2013;16(12):496–504.

[6] Sachlos E, Czernuszka JT. Making tissue engineering work. Review: the application of solid freeform fabrication technology to the production of tissue engineering scaffolds. Eur Cell Mater 2003;5:29–39.

[7] Zhu N, Chen X. Biofabrication of tissue scaffolds. In: Pignatello R, editor. Advances in biomaterials science and biomedical applications. InTech; 2013. p. 315–28.

[8] Peltona SM, Melchels FP, Grijpma DW, Kellomäki M. A review of rapid prototyping techniques for tissue. Ann Med 2008;40(4):268–80.

[9] Melchels FPW, Domingos MAN, Klein TJ, Malda J, Bartolo PJ, Hutmacher D. Additive manufacturing of tissues and organs. Prog Polym Sci 2012;37(31):1079–104.

[10] Kontakis GM, Pagkalos JE, Tosounidis TI, Melissas J, Katonis P. Bioabsorbable materials in orthopaedics. Acta Orthop Belg 2007;73(2):159–69.

[11] Pati F, Shim JH, Lee JS, Cho DW. 3D printing of cell-laden constructs for heterogeneous tissue generation. Manuf Lett 2013;1(1):49–53.

[12] Bartolo P, Domingos M, Gloria A, Ciurana J. BioCell printing: integrated automated assembly system for tissue engineering constructs. Manuf Technol 2011;60(1):271–4.

[13] Bartolo PJ, Domingos M, Patricio T, Cometa S, Mironov V. Biofabrication strategies for tissue engineering. In: Rui P, Jorge P, editors. Advances on modeling in tissue engineering. Springer; 2011. p. 137–76.

[14] Rezende RA, Selishchev SV, Kasyanov VA, da Silva JVL, Mironov VA. An organ fabrication line: enabling technology for organ printing. Part I: From Biocad to Biofabricators of Spheroids. Biomed Eng 2013;47(3):116–20.

[15] Hockaday LA, Kang KH, Colangelo NW, Cheung PY, Duan B, Malone E, et al. Rapic 3D printing of anatomically accurate and mechanically heterogeneous aortic valve hydrogel scaffolds. Biofabrication 2012;4(3). 035005.

[16] Wüst S, Müller R, Hofmann S. Controlled positioning of cells in biomaterials – approaches towards 3D tissue printing. J Funct Biomatter 2011;2(3):119–54.

[17] http://www.organovo.com.

[18] http://www.cellprofiler.org.

[19] Pallua N, Suscheck CV. Tissue engineering. Berlin, Heidelberg: Springer; 2011.

[20] Nemeno-Guanzon JG, Lee S, Berg JR, Jo YH, Yeo JE, Nam BM, et al. Trends in tissue engineering for blood vessels. J Biomed Biotechnol 2012;956345.

[21] http://www.sabanciuniv.edu/en.

[22] Kucukgul C, Ozler B, Karakas HE, Gozuacik D, Koc B. 3D hybrid bioprinting of macrovascular structures. Procedia Eng 2013;59:183–93.

[23] Zopf DA, Hollister SJ, Nelson ME, Ohye RG, Green GE. Bioresorbable three-dimensional printed airway splint. N Engl J Med 2013;368:2043–5.

[24] University of Michigan Health System, http://www.uofmhealth.org/news/archive/201305/baby%E2%80%99s-life-saved-groundbreaking-3d-printed-device.

[25] NIH, http://rarediseases.info.nih.gov/gard/7791/tracheobronchomalacia/resources/1.

[26] Williams JM, Adewunmi A, Schek RM, Flanagan CL, Krebsbach PH, Feinberg SE, et al. Bone tissue engineering using polycaprolactone scaffolds fabricated via selective laser sintering. Biomaterials 2005;26:4817–27.

[27] Partee B, Hollister SJ, Das S. Selective laser sintering process optimization for layered manufacturing of CAPA 6501 polycaprolactone bone tissue engineering scaffolds. J Manuf Sci Eng J Manuf Sci Eng 2006;128:531–40.

[28] Smith M, Flanagan CL, Kemppainen JM, Sack J, Chung H, Das S, et al. Computed tomography-based tissue engineered scaffolds in craniomaxillofacial surgery. Int J Med Robotics Comput Assist Surg 2007;3:207–16.

[29] Zopf DA, Mitsak AG, Flanagan CL, Wheeler MB, Green GE, Hollister SJ. Computer-aided designed, 3-dimensionally printed porous tissue bioscaffolds for craniofacial soft tissue reconstruction. Otolaryngol Head Neck Surg 2015;152(1):57–62.

[30] Agrawal CM, Ray RB. Biodegradable polymeric scaffolds for musculoskeletal tissue engineering. J Biomed Mater Res A 2001;55(2):141.

[31] Whang K, Healy KE, Elenz DR, Nam EK, Tsai DC, Thomas CH, et al. Engineering bone regeneration with bioabsorbable scaffolds with novel micro architecture. Tissue Eng 1999;5:35.

[32] Sun W, Starly B, Darling A, Gomez C. Computer-aided tissue engineering: application to biomimetic modelling and design of tissue scaffolds. Biotechnol Appl Biochem 2004;39:49–58.

[33] Hamid Q, Snyder J, Wang CY, Timmer M, Hammer J, Sun W, et al. Fabrication of three-dimensional scaffolds using the precision extrusion deposition with an integrated assisted cooling. Biofabrication 2011;3. 034109.

[34] Huang G, Zhou L, Zhang Q, Chen Y, Sun W, Xu F, et al. Microfluidic hydrogels for tissue engineering. Biofabrication 2011;3. 012001.

[35] Chang RW, Pellisier JM, Hazen GB. A cost-effectiveness analysis to total hip arthroplasty for osteoarthritis of the hip. JAMA J Am Med Assoc. 1996;11:858–65.

[36] Cram P, Lu X, Kaboli PJ, Vaughan-Sarrazin MS, Cai X, Wolf BR, et al. Clinical characteristics and outcomes of Medicare patients undergoing total hip arthroplasty 1991–2008. JAMA J Am Med Assoc 2011;305(15):1560–7.

[37] Faulkner A, Kennedy LG, Baxter K, Donovan J, Wilkinson M, Bevan G. Effectiveness of hip prostheses in primary total hip replacement: a critical review of evidence and an economic model. Health Technol Assess Winch Engl 1998;2(6):1–133.

[38] Katz JN, Phillip CB, Baron JA, Fossel AH, Mahomed NN, Barrett J, et al. Association of hospital and surgeon volume of total hip replacement with functional status and satisfaction three years following surgery. Arthritis Rheum 2003;48(2):560–8.

[39] Jones CA, Voaklander DC, Johnston DW, Suarez-Almazor ME. The effect of age on pain, function, and quality of life after total hip and knee arthroplasty. Arch Intern Med 2001;161(3):454–60.

[40] OECD. Hip and knee replacement. Health at a Glance 2011.

[41] Lie SA, Havelin LI, Furnes ON, Engesaeter LB, Vollset SE. Failure rates for 4762 revision total hip arthroplasties in the Norwegian Arthroplasty Register. J Bone Joint Surg Br 2004;86(4):504–9.

[42] Gelaude F, Clijmans T, Bross PL, Lauwers B, Vander Sloten J. Computer-aided planning of reconstructive surgery of the innominate bone: automated correction proposals. Comput Aided Surg Off J Int Soc Comput Aided Surg 2007;12(5):286–94.

[43] Vanden Berghe P, Demol J, Vander Sloten J. Automatic reconstruction of large acetabular bone defects using statistical shape models. Presented at CORS2013; 2013 Venice.

[44] Gelaude F, Clijmans T, Delport H. Quantitative computerized assessment of the degree of acetabular bone deficiency: total radial acetabular bone loss (TrABL). Adv Orthop 2011;e494382.

[45] Gelaude F, Broos P, Mulier M, Vandenbroucke B, Kruth J-P, Lauwers B, et al. Treatment of massive acetabular defects with excessive bone loss: from automated computer-bases reconstruction proposal to biomechanically justified defect-filling Triflange Cuf implant. Presented at the International Society for Computer Assisted Orthopaedic Surgery. Heidelberg, Germany; 2007.

[46] Bartels W, Demol J, Gelaude F, Jonkers I, Vander Sloten J. Computer tomography-based joint locations affect calculation of joint moments during gait when compared to scaling approaches. Comput Methods Biomech Biomed Engin 2013;18(11):1238–1251.

[47] Delp SL, Anderson FC, Arnold AS, Loan P, Habib A, John CT, et al. OpenSim: open-source software to create and analyze dynamic simulations of movement. IEEE Trans Biomed Eng 2007;54(11):1940–50.

Chapter 3

Design and Quality Control for Translating 3D-Printed Scaffolds

Scott J. Hollister*,**,†, Colleen L. Flanagan*, David A. Zopf‡, Robert J. Morrison‡, Hassan Nasser‡,
Matthew B. Wheeler§, and Glenn E Green‡

*Department of Biomedical Engineering, University of Michigan, Ann Arbor, MI, USA; **Department of Mechanical Engineering, University of Michigan, Ann Arbor, MI, USA; †Department of Surgery, University of Michigan, Ann Arbor, MI, USA; ‡Department of Otolaryngology – Head and Neck Surgery, Division of Pediatric Otolaryngology, University of Michigan, Ann Arbor, MI, USA; §Institute for Genomic Biology and Department of Animal Sciences, University of Illinois, Urbana-Champaign, IL, USA*

ABSTRACT

3D printing has become widely utilized for regenerative medicine research due to its ability to fabricate patient-specific scaffolds with well-controlled porous architecture and the capability of printing cells in 3D configurations. These characteristics, combined with the unique capability of producing implants and scaffolds for small, specific patient populations not feasible with other manufacturing techniques, have generated significant interest in using 3D printing to transplant implants and scaffolds for clinical use. However, like any clinically translated device, 3D printed scaffolds are subject to design control and quality control requirements to achieve regulatory approval. 3D printing, with its ability to make patient-specific and custom scaffolds; however, brings a number of challenges for design and quality control. In this chapter, we present an example design and quality control for a laser-sintered 3D-printed, resorbable polycaprolactone splint to treat tracheobronchalmalacia (TBM). This splint has been used clinically to save three children with life threatening TBM. We specifically describe a design control for the PCL resorbable splint, detailing design requirements for surgical, mechanical, and biomaterial needs of the splint. Design control entails not only the design requirements for the device, but also the tests to verify that the final device meets the design inputs (design verification) and the preclinical and clinical tests to verify that the fabricated device meets the clinical requirements (design validation). Finally, since design verification and validation are dependent on the laser sintering fabrication process, we detail the parameters of the laser sintering process as well as the methods used to assess devices made using the laser sintering process.

Keywords: splint design; laser sintering; quality control; 3D printing; design verification; design validation

1 INTRODUCTION

Tracheobronchomalacia (TBM) is characterized by congenital or acquired deficiency of supporting tracheal cartilage, which may result in airway collapse, respiratory difficulties, acute life-threatening events, or death. Estimated incidence of congenital tracheomalacia is 1 in 2100 newborn infants, when isolated, and 1 in 2200 when tracheomalacia as one of multiple airway lesions are included [1]. While mild cases often resolve by 24 months with conservative measures, more severe TBM necessitates intervention beginning with positive pressure ventilatory support and, if necessary, tracheostomy

Essentials of 3D Biofabrication and Translation. http://dx.doi.org/10.1016/B978-0-12-800972-7.00003-7

and further ventilator support. Poor clinical response necessitates move invasive intervention with several options including aortopexy, tracheal stenting, or tracheoplasty. A Cochrane review reports that currently available interventions are associated with high rates of failure and complications [2].

The need for mechanical support of severely malacic trachea and mainstem bronchi has long been recognized. Horvath et al. [3] noted that 5 of 26 patients required reoperation following vascular suspension, including two patients who died of persistent TBM. The question is whether to provide mechanical support from within the trachea (stenting) or external to the trachea (splinting). Stenting has been associated with numerous complications including stent migration and granuloma formation [4–6], leading to the need for stent removal. There was additional concern about controlling mechanical properties of the stent, since stents that are too stiff cause tracheal erosion and granulation tissue, while stents that are too compliant tend to migrate [7].

External splinting has also previously been utilized clinically for TBM, although primarily with permanent materials like polypropylene (marlex) [8–11] or polytetrafluoroethylene [12,13] (PTFE = Teflon). However, the presence of a permanent material can lead to complications including sterile serous fluid collection [12] and tracheal erosion [12], which was attributed to the splint being too stiff.

Bugmann et al. [14] treated TBM with a polydioxanone degradable plate fashioned in the operating room (OR) as a splint. A successful outcome was achieved, although there was an indentation in the trachea due to the rapid polydioxanone degradation. Furthermore, these splints were ad hoc devices with no reproducibility in design or mechanical behavior, and thus not developed under the Food and Drug Administration (FDA) design control guidelines. The desire to provide temporary mechanical support without long-term complications has led to further investigations of degradable materials including poly-lactic-co-glycolic acid (PLGA) [15–17]. The PLGA had a relatively short degradation time (4 months, not the 2 years necessary to resolve symptoms [17]) and there was no capability to vary or control the stent design to systematically vary mechanical properties. In general, previous splints were not produced under rigorous design or manufacturing controls, nor were their degradation properties sufficient to resolve the TBM. Furthermore, these devices, especially the splints, were largely developed ad hoc without specific biomaterial, surgery and mechanical designs requirements that could be tested in reference to clinical outcomes.

Results from previous studies suggest that there is still a significant clinical need to develop a rigorously defined and designed device to treat TBM. Although clinical data are limited, recent studies suggest that external stabilization of collapsed airways may provide superior results to internal stenting for children. However, previous clinical studies have all utilized permanent materials like ePTFE and polypropylene in children. Clinical outcomes show that children with mild to moderate forms of TBM outgrow the symptoms, suggesting that if the airway can be protected and allowed to grow the TBM will resolve. The use of permanent materials, then, may raise issues and complications as the patient ages. The ideal intervention would be an external splint that provides initial mechanical protection and patency, but then degrades and provides a smooth transition that allows gradual increased mechanical stimulation of the airway such that the airway will remodel to normal tissue. Such a device, however, clearly requires significant thought concerning its design to have the right mechanical stiffness, and biomaterial and degradation characteristics to achieve this complex transition.

The goal of this chapter is to outline a step-by-step process to define design requirements for a splint to treat TBM, design the splint, fabricate the splint, verify how closely the splint meets design requirements, and evaluate how verified splints affect clinical outcomes. In addition, this chapter will describe how specific 3D printing (specifically laser sintering) parameters affect splint performance and the performance of 3D-printed resorbable devices in general. No such study can be exhaustive, but the hope is that this chapter will provide a roadmap for those wanting to clinically translate 3D-printed resorbable devices. Finally, we will discuss areas needing further development and characterization for 3D-printed resorbable devices, using the PCL splint as a model.

2 SPLINT DESIGN CONTROL

Design control is a rigorous process covering the definition of design requirements denoted as design inputs, design, design review, verification that the final device meets design inputs, and validation that the design as postulated mitigates the clinical condition. Thorough descriptions and discussions of design control may be found on the FDA website [18] as well as in independent publications [19].

2.1 Design Control Overview

The first step in design control is to define the clinical problem, in this case TBM, as well as the target patient population. TBM is described in Section 3.1, and its incidence in the US based on the Centers for Disease Control and Prevention estimates (4,130,665) of live births is approximately 1800 new cases per year.

The next step in design control is to outline the clinical objectives and goals, which are often described in qualitative terms. The obvious ultimate clinical goal in TBM is to create and maintain airway patency, eventually leading to remodeling and development of a mechanically competent airway that can withstand vascular compression and respiratory pressures. This ultimate qualitative clinical goal can be further refined into specific requirements that if achieved should satisfy the clinical goal. For the splint, Zopf et al. [20,21] defined these specific clinical requirements (CR) as follows:

CR1. The splint should provide radial compressive mechanical support to keep the trachea/bronchus open and patent.

CR2. The splint should provide this radial mechanical support for a period of 24–30 months to allow tracheal or bronchial remodeling and development.

CR3. The splint should allow transverse and bending displacement, not interfering with cervical motion.

CR4. The splint should allow growth and expansion of the tracheobronchial complex during this 24–30 month period.

CR5. The splint should not cause adverse tissue reaction or remodeling.

CR6. It is desirable that a second surgical procedure should be avoided to remove the splint; the splint should therefore be bioresorbable.

CR7. The splint should not interfere with the mucociliary architecture of the tracheal or bronchial lumen; it should therefore be placed externally.

CR8. Surgical placement of the splint and attachment of the trachea or bronchus into the splint should be straightforward.

These qualitative requirements may be broadly classified into mechanical/mass transport (CR1–CR4), biomaterial (CR4–CR5), and surgical (CR6–CR8) requirements. The *tracheobronchial splint design hypothesis* is that scaffolds meeting design requirements one to eight will successfully mitigate TBM and reverse respiratory distress. There are immediately two questions that arise from these qualitative design requirements and clinical design hypothesis. First, how do we know if a given splint meets these design requirements? Second, how do we know that a splint that does meet these design requirements actually mitigates or successfully treats TBM? These two questions, namely if the fabricated device meets the design input and if so, does the device mitigate the clinical conditions, are universal for any medical device for any clinical application.

To answer the first question of whether a device meets its requirements, we must be able to convert our *qualitative requirements* into *quantitative design inputs* that can be objectively tested. Design inputs are actually the first step in design control, and incorporate the process for definition of the clinical problem to qualitative design requirements and ultimately the quantitative design inputs. When we have quantitative design inputs, for example, governing the biocompatibility, mechanical, mass transport, and surgical performance of the device, we can then devise tests, noted appropriately as *design outputs*, to measure device performance in all these areas. Design verification is the process of comparing the design outputs to the design inputs. If the design outputs match the design inputs within predefined limits, then we can confidently answer the first question that the splint (or any other device of interest) conforms to the design requirements.

The CR and subsequent design requirements are typically generated based on a postulate or clinical design hypothesis concerning how a device will mitigate or treat a clinical condition. Such hypotheses are the basis for the second question of whether a device that meets CRs and design input actually mitigates the clinical condition. Obviously, we cannot cleanly answer this question if we do not have a device that meets design inputs, as these design inputs are generated based on the clinical design hypothesis. Just as with design verification, we must have objective tests that we can perform with the verified device to test how the parameters of this device affect clinical outcomes. The tests will include computational modeling and bench tests, but ultimately the required tests will include a preclinical animal model for an FDA class 2 device and a preclinical animal model and human clinical trials for an FDA class 3 device. The process of testing how a verified device performs in all these tests up through human clinical trials is termed *design validation*.

2.2 Quantitative Interpretation of Splint CRs – Design Inputs

Once we have defined the splint CRs (CR1–CR8), we need to develop quantitative interpretations of these CR for design inputs. The most straightforward of the CR are CR7 and CR8 governing *surgical* requirements. CR8 requires the use of resorbable materials to avoid a second surgery. Although the materials are resorbable, they should last at least 1½ to 2 years to allow sufficient airway growth to overcome collapse upon exhalation. One resorbable material whose degradation profile fulfills this requirement is ε-polycaprolactone (PCL). CR7 requires that surgical placement of the splint be straightforward. Given that the airway must be suspended within the splint to keep it open and patent, this requirement means that the surgeon must be able to place sutures through the airway wall at periodically spaced points along the malacic segment, and attach these sutures to the splint. Thus, this requirement boils down to a geometric one. The splint must have periodically placed suture holes throughout its structure and large enough to accommodate a suturing needle of approximately 1 mm diameter. Furthermore, the holes should be rounded within the splint wall to avoid elevated stress concentrations, which may lead to splint fracture. Finally, the splint should be designed to be surgically placed exterior to the airway. This again is a geometric requirement for the splint to have an ellipsoidal or circular cross-section with a defined opening angle such that it can be placed around the airway.

The predominant biomaterial/biocompatibility requirement is CR5. CR5 requires that there is no adverse tissue reaction to the splint material, and that as the splint degrades that the degradation products also do not cause an adverse tissue response. The accepted biocompatibility tests for device materials in general are the ISO 10993 "Biological Evaluation of Medical Devices." This standard has 18 parts, but the ones most relevant to the splint are 10993-3 "Genotoxicity," 10993-5 "*In Vitro* Cytotoxicity," 10993-6 "Local Effects Implantation," 10993-11 "Systemic Toxicity," and 10993-13 "Identification and quantification of degradation products from polymeric devices." These standards cover both *in vitro* cell tests and *in vivo* implantation tests to look at cell and tissue response to the chosen biomaterial.

CR1–CR4 are the predominant mechanical requirements for the splint to maintain patency of an initially collapsed airway, allow airway growth over time, and to provide a transition that allows increasing mechanical stimulation of the airway for tissue remodeling. Taken together, these mechanical requirements dictate a splint design that is stiff in compression to protect the airway as it is suspended by sutures in the splint, but relatively compliant in transverse opening to accommodate airway growth. In addition, a slowly resorbing material is desirable in which splint stiffness gradually declines over time to allow gradually increasing airway deformation, thereby mechanically stimulating tissue remodeling and development.

A major difficulty is that forces on the airway, especially in cases of vascular compression present in TBM, are largely unknown. Respiration pressures have been reported by Costantino et al. [22] and Bagnoli et al. [23] where maximum inhalation pressures of +30 cm H_2O (0.003 MPa) and maximum exhalation pressures of −60 cm H_2O (−0.006 MPa). For arterial pressure Khanafer et al. [24] reported a maximum pressure of 240 mm Hg (0.03 MPa). We assume a basic linear relationship between geometric stiffness of the splint as follows:

$$K^s u^s = f^{\text{respiration/arterial}} \tag{3.1}$$

where K^s is the geometric stiffness of the splint, u^s is the radial displacement of the splint, and $f^{\text{respiration/arterial}}$ the force transmitted to the splint from respiration and arterial compression. The force transmitted through the splint will be equal to the pressure (respiration or arterial) multiplied by the external area of the splint to which force is applied. The external area of the splint can be approximated by the splint length L^s times the splint circumference. The splint circumference is equal to 2π minus the opening angle OA^s times the maximum radius of the splint, OR^s. The transmitted force may then be written as:

$$f^{\text{respiration/arterial}} = (P^{\text{respiration/respiration}} \text{N} / \text{mm}^2) (L^s \text{ mm})(2\pi OR^s - OA^s OR^s) \tag{3.2}$$

where $P^{\text{arterial/respiration}}$ is the arterial and respiration pressure applied to the splint. We can further make an estimate on the

limit of splint airway displacement that may be accepted to maintain patency as a fraction of the internal splint diameter:

$$u^s = x * \text{ID}^s \tag{3.3}$$

where u^s is the splint displacement, ID^s is the initial splint inner diameter, and x is the allowable displacement as a percentage of the splint inner diameter. Combining Eqs (3.1)–(3.3), we can derive a design criterion for splint geometric stiffness under both compression:

$$K^s = \frac{(\text{sf})\left(P^{\text{arterial/respiration}} \text{N} / \text{mm}^2\right)(L^s \text{ mm})(2\pi OR^s - OA^s OR^s \text{mm})}{(x)(\text{ID}^s \text{ mm})} \tag{3.4}$$

where K^s is the splint geometric stiffness, $P^{\text{respiration/arterial}}$ is the applied respiration and arterial pressure, L^s is the splint length, OR^s is the outer radius of the splint (distance from the splint center to the outer radius of the outside of the splint), OA^s is the opening angle of the splint, ID^s is splint inner diameter, x is the percentage of the inner diameter allowed to displace, and sf denotes a safety factor introduced to offset assumptions and uncertainty made in the calculation. We typically use a safety factor of 4. A sample calculation for a 10 mm long splint with an 8 mm inner diameter, 2.5 mm wall thickness giving an outer radius of 6.5 mm, an opening angle of 90° or $\pi/2$, a safety factor of 4, x of 10%, and arterial pressure of 0.03, we have:

$$K^s = \frac{(4)\left(0.03 \text{ N} / \text{mm}^2\right)(10 \text{ mm})\left(2\pi 6.5 - (\pi / 2)6.5 \text{ mm}\right)}{(0.1)(8 \text{ mm})}$$
$$= 45.9 \text{ N/mm}$$

Thus, a 10 mm long, 8 mm inner diameter splint with a 2.5 mm wall thickness and 90° opening angle should have at least a 45.9 N/mm geometric compression stiffness to prevent displacement greater than 10% of inner diameter. Longer splints require greater geometric stiffness.

The other important mechanical requirement is that the splint allows airway growth. We can still utilize Eq. (3.4) to estimate an opening geometric stiffness for the splint; in other words, the geometric stiffness the splint must allow it to open during growth. Data on tracheal growth in children from Wright et al. [25] indicates maximal tracheal growth over 3 years of 20%. For growth, we do not need to impose a safety factor since failure to resist growth loads will not collapse the airway as in compression. Therefore, assuming a requirement of 20% opening of the splint to accommodate growth and assuming that pressure due to growth must be equal to arterial pressure of the airway otherwise growth will be retarded, we can calculate an opening stiffness that allows growth as follows:

$$K^s_{\text{growth}} = \frac{\left(0.03 \text{ N} / \text{mm}^2\right)(10 \text{ mm})\left(2\pi 6.5 - (\pi / 2)6.5 \text{ mm}\right)}{(0.2)(8 \text{ mm})}$$
$$= 5.75 \text{ N/mm}$$

With these calculations and the biocompatability, we now have quantitative design inputs for the splint. If we denote the quantitative design inputs for clinical design requirements as clinical requirement design input (CRDI) we can write these as follows:

CRDI1. *Mechanical*: The splint should provide this radial mechanical support for a period of 24–30 months to allow tracheal or bronchial remodeling and development.

$$K^s_{compression} \geq \frac{1.2\,\text{N}\,/\,\text{mm}^2(3\pi\,/\,2)\,\text{OR}^s\text{mm}\,L^s\,\text{mm}}{\text{ID}^s\,\text{mm}}$$

CRDI2. *Mechanical*: The splint should allow transverse and bending displacement, not interfering with cervical motion.

CRDI3. *Mechanical*: The splint should allow growth and expansion of the tracheobronchial complex during this 24–30 month period.

$$K^s_{opening} \leq \frac{0.03\,\text{N}\,/\,\text{mm}^2(3\pi\,/\,2)\,\text{OR}^s\,\text{mm}\,L^s\,\text{mm}}{\text{ID}^s\,\text{mm}}$$

CRDI4. *Biomaterial*: The splint should not cause adverse tissue reaction or remodeling.

Satisfy Select ISO 10993 Biocompatibility requirements; tissue reaction preclinical model

CRDI5. *Biomaterial*: It is desirable that a second surgical procedure should be avoided to remove the splint; the splint should therefore be bioresorbable.

Splint is made from slowly resorbing polymer PCL

CRDI6. Surgical: The splint should not interfere with the mucociliary architecture of the tracheal or bronchial lumen; it should therefore be placed externally.

Splint designed with opening angle and varying inner diameter to go around airway

CRDI7. Surgical: Surgical placement of the splint and attachment of the trachea or bronchus into the splint should be straightforward.

Splint wall has periodically placed suture holes to allow airway suspension

2.3 Design Outputs

Design outputs are tests performed to measure the design inputs. These tests must be performed under standard operating procedures (SOPs) using calibrated equipment. Design outputs specify not only the tests used to evaluate the design inputs, but also the acceptable mean and variation in the test result to accept that the design input has been met.

The mechanical design outputs are mechanical tests performed to determine the compressive, bending, and opening stiffness of the splint. Splint compression testing is carried out by having each splint situated between two flat plates (2″ wide) attached to the testing machine. The plates are then advanced toward each other until a force of 20 N is

measured. The geometric stiffness is calculated as the slope of the load displacement curve.

Splint bending tests are performed to determine the bending geometric stiffness of the splint. These tests follow much of the ASTM F2606-08 "Three Point Bending of Balloon Expandable Vascular Stents and Stent Systems," except for the fact that the tracheobronchial splint has a higher diameter-to-length ratio than vascular stents, is an open cylinder as opposed to vascular stents, which are complete cylinders, and a maximum three-point bending load of 50 N is applied.

The final test is an opening test in which an opening load is applied through hooks to the opening edge of the splints (Fig. 3.1). The splint is positioned within the grasping hooks at a resting force of less than 0.05 N and caliber measurement of less than 1 mm resting displacement. The testing plates are then separated at a constant rate and resistance force of the splint was measured until the splint fails.

Thus, the design outputs for the mechanical design inputs are compression, three-point bending, and opening tests on fabricated splints, poststerilization. The compression stiffness must be greater than the value given in CRDI1 and the opening stiffness must be less than the stiffness given in CRDI3.

The design outputs for CRDI4-5 are biomaterial outputs as governed by ISO 10933, specifically 10993-3 "Genotoxicity," 10993-5 "*In Vitro* Cytotoxicity," 10993-6 "Local Effects Implantation," 10993-11 "Systemic Toxicity," and 10993-13 "Identification and quantification of degradation products from polymeric devices." In addition to the ISO 10993 tests, tissue response and resorption in a large preclinical animal model will be the ultimate arbiter of biocompatibility for the PCL splint.

Design outputs for CRDI6-7, although for surgical implementation, are all determined by geometry and dimensional tolerancing. CRDI7 requires the splint to be surgically placed externally. This requires that the splint must have an opening to be placed around the airway and that the splint inner diameter is large enough to accommodate the airway. The splint opening and inner diameter can be measured manually using calipers or by nondestructive imaging using, for example, microcomputed tomography (micro-CT).

2.4 Design Implementation

Obviously, to meet the design inputs from Section 2.2, and subsequently test the finalized device using design outputs from Section 2.3, we need to generate a splint design with design variables that may be adjusted to achieve design inputs. The splint design we developed is a bellowed, open cylinder design with periodically spaced suture holes. A custom MATLAB™ (The Mathworks, Natick, MA; www.mathworks.com) code was written to automatically

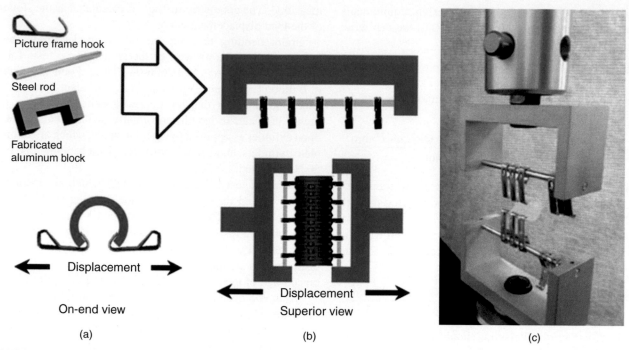

FIGURE 3.1   Schematic and actual opening hook test of splint. (a) Schematic view of the components and end view of the displacement. (b) Superior view of opening test. (c) Actual opening test being performed on the splint.

generate .tif files of image slices on the basis of the input of 12 design variables, listed as follows:

1. Inner "*a*" splint diameter — #*a* in mm
2. Inner "*b*" splint diameter — #*b* in mm
3. Maximum wall thickness — wt in mm
4. Splint length — sl in mm
5. Bellow period length — bp in mm
6. Bellow height as % reduction in weight — bh from 0 (no bellow) to <100%
7. Wedge angle to cut from cylinder — wa (degrees) typically 90
8. Strut size — Distance between suture holes typically 2 mm
9. Hole size — Size of suture holes, typically 2 mm
10. Spiral angle — Total angle through which splint opening rotates (degrees)
11. Starting point of spiral — Point in mm along splint length that spiral starts
12. Ending point of spiral — Point in mm along splint length that spiral ends

Figures 3.2–3.4 illustrate the 12 design variables for the splint.

The range and acceptable variations in the splint design variables are shown in Table 3.1.

The design shown in Figs 3.1–3.3 has significant geometric complexity, especially if a spiral angle is incorporated. In addition to geometric complexity, there is a need to adjust the scaffold for each individual patient due to differences in the malacic segment length, patient airway diameter, and the number of scaffolds needed to treat one or multiple malacic segments. These design requirements, in addition to the small pediatric patient population, make it technically and economically difficult to make the splint using traditional manufacturing techniques like injection molding. Therefore, splint fabrication is best done through 3D printing technology, for which we use laser sintering, described in Section 3.

3D printing requires a .STL file (surface representation by triangles). The splint .STL representation is generated by importing image data from the MATLAB splint design program into MIMICS™ software by Materialise (Leuven,

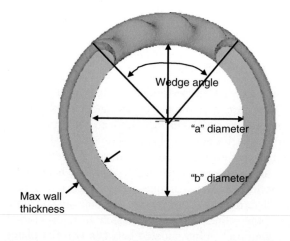

FIGURE 3.2   Illustration of splint design parameters inner diameter "a," inner diameter "b," wall thickness, and opening angle.

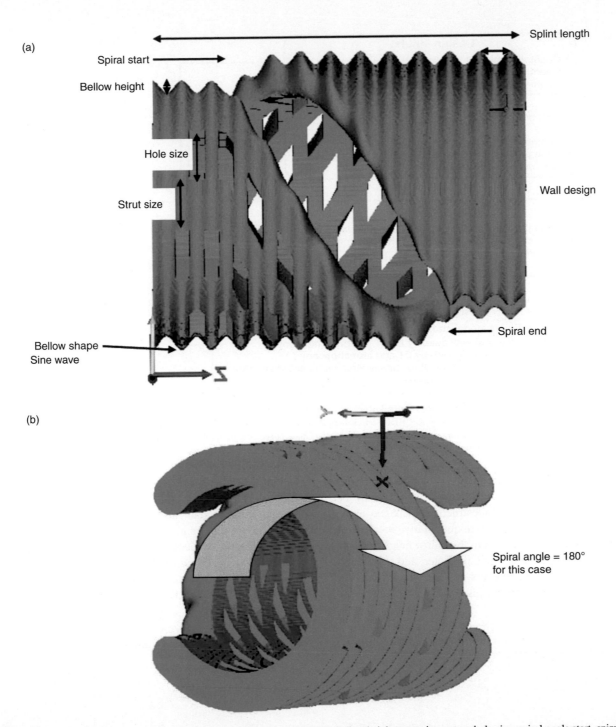

FIGURE 3.3   (a) Splint design parameters splint length, bellow period length, bellow height, strut size, suture hole size, spiral angle start, spiral angle end, and bellow shape. (b) Illustration of spiral angle for splint.

Belgium; www.materialise.com) and subsequently generating the isosurface representation.

2.5   Design Verification

A critical step toward finalizing a design for clinical use is to ensure that the final manufactured device meets the design inputs. This step in design control is known as *design verification*. If a device in reality is not what we designed, and it performs differently than expected in preclinical or clinical testing, we cannot know if this is due to our clinical hypotheses being off base or the device not being designed to the design inputs. Design verification is the process of ensuring the device meets design inputs, and thereby eliminating this uncertainty if the device does not perform as postulated in preclinical and clinical testing.

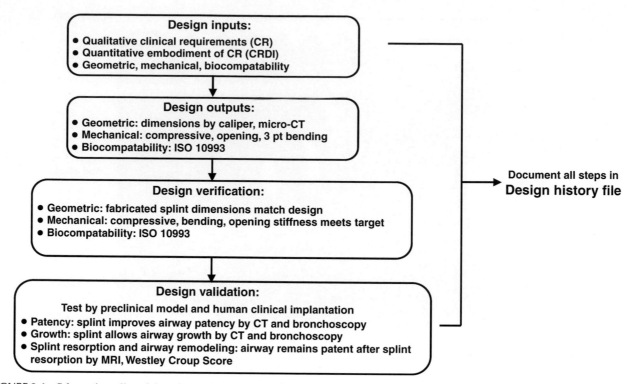

FIGURE 3.4 **Schematic outline of the splint design control process.** The process begins with design input in which qualitative clinical requirements (CRs) are defined and interpreted as Quantitatively as CRDIs. Specific Design Outputs are then defined, which are tests performed on the fabricated device. Design Verification involves determining if the design outputs meet the Quantitative Design Inputs (CRDI) to within specified bounds. Finally, in Design Validation the verified device is tested to determine if it mitigates the clinical condition as hypothesized.

Splint design is verified through geometric and mechanical evaluation. For geometric evaluation, splints may be measured with calipers or nondestructively imaged using micro-CT. Micro-CT allows a more quantitative measure of inner diameter, wall thickness, length, bellow height, bellow period, opening angle, suture hole diameter, and strut thickness. These geometric parameters are measured and compared to the mean inputs and accepted variations (Table 3.1).

The splint is geometrically verified if all geometric design variables are within the variation about the mean.

The splint is mechanically verified by performing compression, three-point bending, and opening (Fig. 3.1) tests on fabricated splints. Geometric stiffness is measured and compared to values given in CRDI1–CRDI3. If the fabricated, sterilized splint satisfies the geometric stiffness requirements of CRDI1–CRDI3, then the splint is verified for

TABLE 3.1 Range and Acceptable Variation in Geometric Design Parameters for the Tracheal Splint

Design Parameter	Inner a Diameter	Inner b Diameter	Max. Wall Thickness	Splint Length	Bellow Period	Bellow Height (% Reduction of Max. Thickness	Opening Wedge Angle	Suture Hole Thickness (open pore wall design)	Strut Thickness	Spiral Angle	Spiral Angle Start–Finish
Range	5–15 mm (depends on patient anatomy)	5–15 mm (depends on patient anatomy)	2–3 mm	10–30 mm (depends on patient defect anatomy)	1–4 mm	0–40% max. thickness	60°–120°	2–3 mm	2–3 mm	0°–180°	0-splint length
Acceptable variation	± 0.5 mm	± 0.5 mm	± 15% of range	± 5% of range	± 15% of range	± 15% of range	± 5°	± 15% of range	± 15% of range	± 5°	± 0.5 mm

mechanical performance. After geometric and mechanical verification, the splint can be evaluated in preclinical models and clinical trials.

2.6 Design Validation

Developing a medical device or regenerative medicine platform is inherently based on a clinical hypothesis or postulate as to how the design of the device will affect clinical outcomes, specifically how the device will mitigate or treat a clinical condition. Before obtaining regulatory approval and subsequent clinical use of the device, the developer, and the FDA must be satisfied that the device treats or mitigates the clinical conditions as originally hypothesized. These clinical hypotheses are tested in a series of bench, preclinical animal model, and clinical trials in a process known as *design validation*. All the steps in design control including establishing design inputs, design outputs, and design verification lead to the battery of tests for clinical design validation. Specifically for the preclinical model and clinical trial, outcome measures must be established that will allow testing of the clinical design hypothesis for design validation.

For the splint, the key clinical outcomes are that the airway patency is restored, the splint does not cause adverse tissue remodeling, the splint allows airway growth, and finally, that normal respiration is maintained after splint resorption. As with design verification, we need specific clinical tests that will provide quantitative evaluation of these outcomes. Clinically, the Westley Croup Scale is used to assess respiratory distress before and after splint placement. Hydraulic diameter from CT scans can be used to evaluate airway patency, while MRI is used *in vivo* to assess when the splint is completely resorbed and therefore if respiration is maintained after splint resorption.

Preclinical testing of the splint has been performed in a large animal model, the Yorkshire Pig [20]. In this first model, tracheal collapse was initiated by surgically resecting five cartilage rings in the trachea. A total of six pigs had the procedure, with three pigs receiving no treatment (control) and three pigs receiving the splint. Pigs were assessed using the Westley Croup Score. The difficulty with creating animal models of TBM, however, is that these models do not replicate the etiology of the human condition. Indeed, the five-cartilage ring resection model is an extremely severe model. Therefore, a second model in the pig was pursued specifically to assess the effect of the splint on tracheal growth and long-term tissue reaction by placing splints on intact trachea. In this model, CT scans were taken at 4 and 8 months post-surgery. Digital models of the tracheal lumen were then created using MIMICS (Materialise, Leuven, Belgium, www.materialise.com). The hydraulic diameters were then measured along the tracheal lumen from the 4 and 8 week CT scans.

Clinical data on the splint was obtained in three patients, ages 3, 16, and 5 months at the time of surgery, in which splints were placed on an emergency basis to treat life-threatening TBM. Emergency clearance was received from the FDA to place in the splints, in conjunction with Institutional Review Board from the University of Michigan. The first child received a left bronchial splint in February 2012 at 3 months old and this case was reported in May 2013 with 1 year follow up [21]. The second child received bilateral bronchial splints at 16 months of age in January 2014 [26]. The third child received a splint for left bronchialmalacia at 5 months of age in March 2014. All children continue to do well, and have been followed using CT scans and bronchoscopy to assess bronchial patency, as well as MRI to determine if the splint is still present. The CT, MRI, and bronchoscopy evaluations along with regular physician examinations are clinical outcomes used to assess device performance.

Design control is implemented to ensure that every step of the device engineering process can be traced. This is especially important if adverse events occur using the device, as a rigorously documented design control increases the probability of pinpointing the cause of the adverse event. Documented design control further enhances the capability of the device manufacturer to reproducibly produce quality devices. Design control documentation is denoted as the design history file (DHF). The DHF is essentially a guide that explains how to produce a device from the initial clinical goals to design verification and validation. For the splint, the outline of the design control process is illustrated in Fig. 3.4.

3 LASER SINTERING PCL SPLINTS

The fundamental step to successfully implementing device design control is manufacturing or fabricating a device or scaffold that fulfills the design requirements. If this task cannot be achieved, the device will never be clinically translated. Thus, being able to reproducibly manufacture a scaffold/device through a controlled manufacturing process and documenting the quality of this process is the lynchpin of clinical translation. The manufacturing process for the tracheal splint is laser sintering, a 3D printing (also known as additive manufacturing) process. This process, however, is not just about the machine that produces the splint, but also entails the design, material procurement, material processing, postprocessing, standard operating procedures, and quality measures that ensure the most reproducible and quality manufacturing process possible.

The integrated design and manufacturing process used to create the tracheal splint is shown in Fig. 3.5.

3.1 Laser Sintering Parameters

The first step in the process is obtaining PCL and hydroxyapatite (HA), which is used as an agent to help powder flow when swept by the recoated arm across the build platform. The University of Michigan purchases PCL from Polysciences

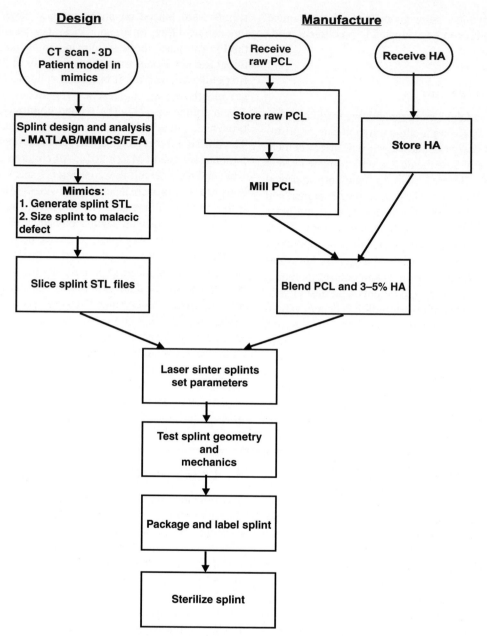

FIGURE 3.5   Outline of the integrated design and fabrication process used to manufacture PCL splints by laser sintering.

(www.polysciences.com) and HA from Plasma Biotal (www.plasma-biotal.com). The percentage of HA in addition to beta Tri-calcium phosphate (b-TCP) reported in the literature is between 0% and 20% [27–42] and for the splint the range is 3–5%. Following purchase, the raw materials in most cases must be processed via milling to obtain a powder size range that works best with the laser sintering process. Smaller particle sizes enable fabrication of finer feature sizes; however, too small particle sizes may inhibit spreading of the powder over the build platform due to increased particle electrostatic interactions resulting from increased particle surface area. There are a number of vendors who perform milling, including Jet Pulverizer (www.jetpulverizer.com),

Fraunhaufer Institute (www.umsicht.fraunhofer.de/en.ht), and Evonik (www.north-america.evonik.com) [27–42]. The University of Michigan utilizes Jet Pulverizer to mill under Good Manufacturing Processes. Once the PCL is milled, it is mixed with 3–5% by volume HA, which is within the range established by other investigators [27–42].

After material processing, the material is now ready to be loaded into the laser sintering system to fabricate splints. The University of Michigan utilizes an EOS P 100 laser sintering system (www.eos.info/en) to fabricate PCL scaffolds. Laser sintering utilizes a laser to sinter, or melt, powdered materials together on a layer-by-layer basis to form a 3D structure. There are a number of parameters

TABLE 3.2 Laser Sintering Manufacturing Parameters for Building PCL Implants and Scaffolds

Manufacturing Parameter	Literature Reported Ranges	Splint Ranges
Laser power	1–9 W	4 W
Laser scan speed	900–3800 mm/s	1000–1500 mm/s
Laser scan spacing	0.07–0.2 mm	0.15–0.2 mm
Fabrication bed temperature	38–56°C	50–56°C
Particle size	0.04–0.125 mm	0.04–0.06 mm

Middle column gives ranges reported in the literature for laser sintering PCL scaffolds and implants. Right column gives ranges for laser sintering splints.

controlling the laser and its interaction with the powdered material, as well as powder characteristics and environmental conditions of the powder, which affect the fabrication quality of the implant or scaffold. The laser parameters include laser power, laser scanning speed, laser scanning spacing, laser scanning pattern, and laser beam offset that affect fabrication quality. Powder characteristics include the material melting temperature, material age, and the material particle size distribution. Environmental conditions include the storage conditions of the powder, most critically the temperature and humidity. All of these parameters will affect fabrication quality, and as part of quality system, ranges must be established for these manufacturing parameters. A number of investigators have reported these manufacturing parameters for PCL laser sintering. Table 3.2 presents the range of these laser-sintering parameters established in the literature [27–42]. The parameter values we used for manufacturing the splint were determined based on these ranges, specifically using parameters we established in Partee et al. [31].

3.2 Laser Sintering Quality Control

Quality control for manufacturing, like design verification, requires a measure of quality for the fabricated implant or scaffold, in relation to a specific manufacturing parameter. This presents a conundrum, as 3D printing technologies like laser sintering allow an almost unlimited range of implants and scaffolds to be manufactured. However, this broad range in implant geometries makes it difficult to establish a range of laser sintering parameters (Table 3.2) that gives the best manufactured devices, since different implant/scaffold geometry and porous architecture may require different laser sintering parameters. Therefore, a middle ground is to build a standard geometry that can be compared across multiple builds, denoted as a coupon, in addition to the specific implant(s) and/or scaffold(s) that will be used from the build. Building a coupon that can be tested over time or for different machine and material parameters provides a consistent standard for verifying the laser sintering process. An example showing the effect of laser scan speed on the geometry (cylindrical diameter) and modulus of a cylindrical coupon is shown in Fig. 3.6.

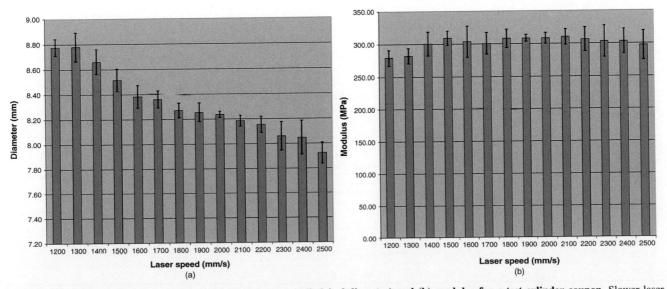

FIGURE 3.6   Effect of laser scan speed on the (a) geometry (cylindrical diameter) and (b) modulus for a test cylinder coupon. Slower laser scanning speed equates with more laser heat on specific powder, causing thermal growth of a given structure.

Perhaps the most rigorous study of laser sintering parameter effects was performed by Partee et al. [31]. They examined the effect of a wide range of laser sintering parameters including laser power, laser scan speed, bed temperature, and laser scan spacing on three outcomes: (1) the ease of part removal from the build, (2) the fabricated part dimensions compared to design dimensions, and (3) the density of the part, with 100% dense sintering being desired. Using these metrics in a design of experiments (DOE), Partee et al. suggested optimal parameters of 4.1 W laser power, 1079.5 mm/s laser scan speed, 0.152 mm scan spacing, and 46°C bed temperature for laser scanning PCL scaffolds and implants.

It is clear from the literature and our own work that fabrication quality in terms of how closely the fabricated implant or scaffold matches the original design, and thus the ability to verify an implant or scaffold, is significantly dependent on laser sintering parameters and material powder characteristics.

The conundrum for custom devices like the splint, however, is that the geometry will vary with different patients and different splints, that is, for splints with different lengths, inner diameters, wall thickness, suture holes, and so on. Verifying splint fabrication requires that the splint meet both geometric design targets (specified inner diameter, length, wall thickness, suture hole size, bellow period, and bellow height) as well as mechanical design targets (compression stiffness, opening stiffness, and potentially fatigue limits, discussed at the end of the chapter). Failure to meet these design targets could result from a breakdown in the design process, or more likely, variations in material characteristics or laser sintering parameters for the specific build. However, if the fabricated patient-specific device differs from any previously built, it will be difficult to compare its fabrication conditions. Thus, standard test coupons with consistent geometry and known mechanical properties should be included in every build to provide a standard with which to compare builds performed at different times.

4 SPLINT DESIGN VERIFICATION

As noted in Section 2.5, splint design verification requires that fabricated splints meet the quantitative CRDI 1–7. For the geometric CRDI 6 and 7, in addition to the geometric design variables and ranges given in Table 3.1, caliber measures and micro-CT analysis are performed on sample splints from a build.

For the mechanical design requirements, compression and opening tests are performed on sample splints from a build. The results are compared to thresholds from CRDI1–3. Compression stiffness tests (Table 3.3) and opening stiffness tests (Table 3.4) have been performed on splints implanted in three children under emergency use and in a preclinical pig model.

Results from both geometric and mechanical tests, especially for the splints implanted in human patients, demonstrated that fabricated and sterilized splints met geometric and mechanical CR design inputs.

5 DESIGN VALIDATION – PRECLINICAL MODEL RESULTS

The development of the CRDI is based on a design hypothesis that if the splint meets specified biocompatibility, geometric, and mechanical requirements it will perform well clinically. Just as there are design outputs corresponding to design inputs to verify device/implants design, specific clinical outcomes must be defined to assess if the device/implants

TABLE 3.3 Compressive Stiffness Measures from Mechanical Tests on Splints Used in Preclinical Models and in Human Clinical Patients

Application	Wall Thickness, Length, Bellow Height	Testing Mode (Compression or Opening)	Result (N/mm)	CRDI Value (N/mm)
Preclinical	2 mm; 25 mm, 30%	Compression	17.8 ± 0.6	90.8
Preclinical	2 mm; 25 mm, 0%	Compression	32.0 ± 4.2	90.8
Preclinical	3 mm; 25 mm, 0%	Compression	162.4 ± 48.2	100.9
Preclinical	4 mm; 25 mm, 30%	Compression	158.1 ± 5.8	111.0
Preclinical	4 mm; 25 mm, 0%	Compression	187.8 ± 34.8	111.0
Preclinical	6 mm; 25 mm, 0%	Compression	449.9 ± 7.8	131.2
Patient 1	7 mm; 35 mm, 30%	Compression	129.0 ± 11.9	169.5
Patient 2, splint 1	2.5 mm; 11 mm, 30%	Compression	72.2 ± 14.6	50.5
Patient 2, splint 2	2.5 mm; 23 mm, 30%	Compression	195.8 ± 16.2	105.6

Results are compared to CRDIs for the splint dimensions. Compressive results should be *greater* than the CRDI value.

TABLE 3.4 Opening Stiffness Measures from Mechanical Tests on Splints Used in Preclinical Models and in Human Clinical Patients

Application	Wall Thickness, Length, Bellow Height	Testing Mode (Compression or Opening)	Result (N/mm)	CRDI Value (N/mm)
Preclinical	2 mm; 25 mm, 30%	Opening	0.38 ± 0.02	2.25
Preclinical	2 mm; 25 mm, 0%	Opening	0.80 ± 0.02	2.25
Preclinical	3 mm; 25 mm, 0%	Opening	1.97 ± 0.12	2.50
Preclinical	4 mm; 25 mm, 30%	Opening	3.39 ± 0.10	2.75
Preclinical	4 mm; 25 mm, 0%	Opening	4.31 ± 0.15	2.75
Patient 1	2.5 mm; 35 mm, 30%	Opening	2.77 ± 0.26	4.20
Patient 2, splint 1	2.5 mm; 11 mm, 30%	Opening	1.43 ± 0.12	1.25
Patient 2, splint 2	2.5 mm; 23 mm, 30%	Opening	2.43 ± 0.15	2.62

Results are compared to CRDIs for the splint dimensions. Opening stiffness measures should be *less* than CRDI values.

performs well clinically. For the splint to mitigate TBM it must create airway patency, allow airway growth, and allow airway remodeling to create a mechanically sufficient airway as the splint resorbs. Airway patency and airway growth are determined by CT scans from which digital models are made to measure hydraulic diameter, as well as bronchoscopy. Airway remodeling and sufficiency following splint resorption requires, of course, that the degradation status of the splint be tracked *in vivo*. Splint presence *in vivo* is determined by MRI. Finally, the Westley Croup Score is used to quantify the severity of respiratory distress. Thus, quantitative assessment of clinical outcome in the preclinical model and human use is determined by CT scans and bronchoscopy to determine airway patency and growth, MRI scans to determine the status of splint resorption and if the airway patency is thus maintained following airway resorption, and finally the Westley Croup Score to assess respiratory distress.

We have tested airway splints in a preclinical large animal model, the Yorkshire pig, by placing splints in a surgically created TBM model as well as on intact trachea. We have used both conditions as follows: (1) although a surgically created model can replicate the physical manifestations of TBM, it does so through a mechanism that is very different than the human condition, and (2) the difficulty of creating TBM by surgically resecting cartilage rings can easily allow perforation of the airway lumen, leading to infection.

We presented a study [22] investigating if the splint can mitigate severe, life-threatening TBM in a Yorkshire pig preclinical model. Six pigs had surgically created TBM through the resection of five contiguous tracheal cartilage rings. Three pigs served as control with no intervention following cartilage ring resection. Three pigs had splints implanted following cartilage ring resection. Clinical

outcomes were assessed using the Westley Croup Score. All three control pigs had to be euthanized within 24 h of surgically created TBM due to loss of consciousness and respiratory distress. The three pigs with implanted splints had moderate (1 pig) to normal (2 pigs) Westley Croup Scores and survived significantly longer than the control group. All pigs in the splint group died 4–7 days post-surgery, however, due to infections from rents in the airway lumen that allowed bacteria to pass into the airway tissues. These results indicated that the surgically created TBM model was limited as a model for the human condition, but also demonstrated the splints' capability to mitigate airway collapse and associated respiratory distress.

To delineate whether any complications in the pig TBM model were due to splint material, and to assess long-term outcomes with the splint, a second preclinical experiment was performed in which the splint was placed on the intact trachea. This procedure eliminated the possibility of lumen mucosal rents and subsequent infection, while allowing assessment of airway growth over time. Two pigs with splints placed on intact trachea were followed for 8 months, with CT scans taken at 4 and 8 months. Digital models were created of the tracheal lumen from the CT scans and hydraulic diameter measures were taken along the trachea (Fig. 3.7).

The tracheal lumen hydraulic diameter grew between 20% and 22% for the two pigs between 4 and 8 months for splints with an opening stiffness of 0.43 ± .05 N/mm, which is less than CRDI 3. Furthermore, there was no evidence of adverse tissue reaction or infection on the intact trachea. The intact trachea more closely represents the human condition as the cartilage rings are malformed in the human condition, but no perforation exists of the lumen. Thus, the preclinical model validates that splint opening stiffness less than CRDI3 allows airway growth, and that the splint mitigates TBM.

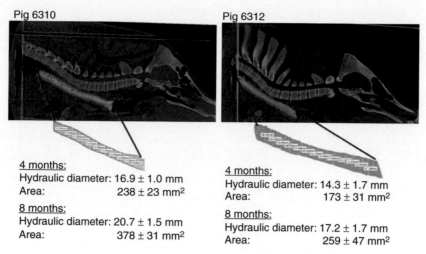

Pig 6310

4 months:
Hydraulic diameter: 16.9 ± 1.0 mm
Area: 238 ± 23 mm²

8 months:
Hydraulic diameter: 20.7 ± 1.5 mm
Area: 378 ± 31 mm²

Pig 6312

4 months:
Hydraulic diameter: 14.3 ± 1.7 mm
Area: 173 ± 31 mm²

8 months:
Hydraulic diameter: 17.2 ± 1.7 mm
Area: 259 ± 47 mm²

FIGURE 3.7 **Measurements of tracheal lumen hydraulic diameter taken from digital models generated from CT scans of Yorkshire pigs with splints on intact trachea.** Results show lumen growth in hydraulic diameter of 20–22% in 4 months, which is consistent with normal growth.

6 DESIGN VALIDATION – CLINICAL RESULTS

Laser-sintered PCL splints have been implanted in three children to mitigate life-threatening TBM under emergency clearance from the University of Michigan Institutional Review Board (IRB) and the FDA. The first child received a splint for left mainstem bronchomalacia at 3 months of age in February 2012 [21]. The second child received bilateral splints for left and right mainstem bronchomalacia at 16 months of age in January 2014. The third child received a splint for left mainstem bronchomalacia at 5 months of age in March 2014. As noted in Tables 3.3 and 3.4, splints for children 1 and 2 met compression and opening stiffness CRDI, except that the splint used in patient 1 had 23% less compressive stiffness than CRDI 1 and the shorter of the two splints used in patient 2 had 14% greater opening stiffness than CRDI 3.

Results from all three children show that airway patency was increased in all three children after splint placement, as demonstrated by hydraulic diameter measurements pre-op and at 1 month post-op, all of which were statistically signif-

icantly greater at 1 month (Table 3.5). In addition, the treated left mainstem bronchus of patient 1 continued to grow as measured at the same rate as the untreated right bronchus.

Clinical results, although not in a designed clinical study, also validate that splints satisfying mechanical design criteria had positive outcomes in terms of creating airway patency and allowing airway growth. In addition, MR imaging can detect the presence of the splint *in vivo*, making it possible to determine when the splint has completely resorbed in a patient. Since the current longest duration patient is out 32 months, less than the projected 36 month PCL resorption, a final scan has not been taken to determine if the splint has completely resorbed and that the airway remains patent upon resorption.

7 THE FUTURE OF 3D LASER-SINTERED PCL DEVICES

Laser-sintered PCL tracheobronchial splints have had initial clinical success, saving the lives of three children [21,26]. Initial results in preclinical animal models have also been

TABLE 3.5 Hydraulic Diameter Measurements on Patient Digital Airway Models Created from Patient CT Scans

Patient/Implant Site	Pre-Op (mm)	1 Month Post-Op (mm)	6 Months Post-Op (mm)	12 Months Post-Op (mm)	30 Months Post-Op (mm)
Patient 1 left bronchus treated	1.4 ± 1.5	2.9 ± 0.2	3.8 ± 0.4	3.5 ± 0.5	4.0 ± 0.9
Patient 1 right bronchus control	3.2 ± 0.4	2.9 ± 0.4	3.6 ± 0.5	3.9 ± 0.5	4.5 ± 1.0
Patient 2 left bronchus treated	2.3 ± 1.1	4.8 ± 0.5			
Patient 2 right bronchus treated	4.0 ± 0.6	5.4 ± 0.6			
Patient 3 left bronchus treated	1.9 ± 1.1	4.0 ± 0.1			

There is a statistically significant increase in patency from pre-op to 1 month post-op for all treated bronchi. In addition, the treated left bronchus in patient 1 continues to grow at a pace akin to the untreated control right bronchus.

encouraging. Furthermore, the integration of patient-specific image-based design with laser sintering of PCL provides an extensive capability to engineer devices for a wide variety of clinical applications. In fact, we have published preclinical work in a variety of clinical applications including spine fusion [43], craniofacial bone and joint reconstruction [44], periodontal reconstruction [45], and craniofacial soft tissue reconstruction [46].

The question of course is how to transition these devices from preclinical models to clinical use. The starting point is design control, as design control is the basis for regulatory approval. This means laying out specific clinical goals at the beginning of a project, and developing quantitative design inputs that can be tested using specific design outputs to verify the device, and validate that the specific clinical design hypotheses embodied in the device design parameters actually treat the clinical condition, as per Fig. 3.4.

There is significant difficulty, however, for relevant organizations to implement design control for small patient populations like pediatrics or for rare conditions. It is not economically feasible for device companies to commit large amounts of resources when economic returns will not cover investments. Academic medical centers obviously treat both pediatric and rare condition populations, but implementation of design control often does not fit within the framework of the basic research environment and limited personnel resources of these academic centers.

3D printing, however, does provide the means to address the needs of pediatric and rare disorder populations, namely providing the ability to customize devices and produce low volumes of devices in an economically feasible manner. This chapter sought to provide a blueprint for implementing the design control of a sample bioresorbable device, the tracheobronchial splint, engineered using patient-specific image-based design with 3D printing via laser sintering of PCL. Although a blueprint is presented with results for design inputs, design outputs, design verification, and design validation, it is far from a complete design history file in which the design control process is documented. Complete ISO 10993 biocompatibility testing needs to be completed, as does a complete clinical trial.

Future considerations for design verification, design validation, and quality control for 3D printed resorbable scaffolds should also account for the effect of 3D printing parameters on both fatigue and degradation behavior. It is highly likely that the effect of 3D printing parameters seen on static mechanical properties (see Fig. 3.6) will also affect fatigue and degradation properties. The fatigue, degradation, and the coupling between these behaviors is a critical area for future study, especially for 3D-printed devices. It will especially be important to characterize the relationship between 3D printing parameters, device quality in terms of production variation, material voids, anisotropy, and so on, and fatigue/degradation.

In conclusion, implementing rigorous design and quality control will be critical to future translations of 3D-printed devices to clinical use. 3D-printed devices present new challenges and paradoxes for quality and design control. On the one hand, 3D-printed devices must meet the same standards as devices produced using other manufacturing technologies. On the other hand, 3D printing will allow a much broader range of new and custom devices for niche and small patient markets that could not be produced previously. These small volumes, and the fact that such devices will be made by academic medical centers traditionally geared more toward basic research, make the enhanced focus and funding of 3D-printed device development and quality control crucial to continued clinical translation and progress.

GLOSSARY

3D printing A general set of manufacturing technologies that build three dimensional (3D) structures on a layer by layer basis with a support mechanism

TBM Softening or collapse of the trachea and/or bronchus

PCL A resorbable polyester biopolymer

Design control A framework defined in the United States Code of Federal Regulations (CFR) that is implemented to ensure rigorous and proper design of medical devices

Design inputs Design requirements that a medical device should meet that are developed to ensure that the medical device diagnoses, treats, and/or mitigates a specific clinical condition

Design outputs Tests and characterizations with associated target values that are applied to a medical device to determine how closely the device meets design inputs

Design verification A portion of design control requiring that the final medical device fulfill design inputs and demonstrated by design outputs

Design validation A portion of design control specifying that the final verified device mitigate or address the specified clinical condition for which the device is designed. Essentially, that the final design meets the patient/end user needs for which the design inputs were originally developed

Clinical requirements A set of principles, often qualitative, the guide the characteristics a medical device should possess to address a clinical need

CRDI A quantitative interpretation or embodiment of a CR for which a definitive test can be performed and a measurement made. This measurement is a design output

ABBREVIATIONS

CR	Clinical requirement
CRDI	Clinical requirement design input
FDA	Food and Drug Administration
HA	Hydroxyapatite
ISO	International Standards Organization
PCL	Polycaprolactone
TBM	Tracheobronchalmalacia

REFERENCES

[1] Boogaard R, Huijsmans SH, Pijnenburg MW, Tiddens HA, de Jongste JC, Merkus PJ. Tracheomalacia and bronchomalacia in children: incidence and patient characteristics. Chest 2005;128(5):3391–7.

[2] Masters IB, Chang AB. Interventions for primary (intrinsic) tracheomalacia in children. Cochrane Database Syst Rev 2005;4(4). CD005304.

[3] Horvath P, Hucin B, Hruda J, Sulc J, Brezovský P, Tuma S, et al. Intermediate to late results of surgical relief of vascular tracheobronchial compression. Eur J Cardiothorac Surg 1992;6:366–71.

[4] Bugmann P, Beghetti M, Berner M, Habre W, Le Coultre C, Hanquinet S. 11-year follow-up of two Gianturco Z stents placed in the airways of an infant: a prospective analysis of stent damage. J Vasc Interv Radiol 2003;14:818–9.

[5] Geller KA, Wells WJ, Koempel JA, St John MA. Use of the Palmaz stent in the treatment of severe tracheomalacia. Ann Otol Rhinol Laryngol 2004;113:641–7.

[6] Fayoux P, Sfeir R. Management of severe tracheomalacia. J Pediatr Gastroenterol Nutr 2011;52(Suppl. 1):S33–4.

[7] Vondrys D, Elliot MJ, McLaren CA, Noctor C, Roebuck DJ. First experience with biodegradable airway stents in children. Ann Thorac Surg 2011;92:1870–4.

[8] Gangadharan SP, Bakhos CT, Majid A, Kent MS, Michaud G, Ernst A, et al. Technical aspects and outcomes of tracheobronchoplasty for severe tracheobronchomalacia. Ann Thorac Surg 2011;91(5): 1574–80. discussion 1580,1581.

[9] Fitzgerald PG, Walton JM. Intratracheal granuloma formation: a late complication of Marlex mesh splinting for tracheomalacia. J Pediatr Surg 1996;31:1568–9.

[10] Filler RM, Buck JR, Bahoric A, Steward DJ. Treatment of segmental tracheomalacia and bronchomalacia by implantation of an airway splint. J Pediatr Surg 1982;17:597–603.

[11] Vinograd I, Filler RM, Bahoric A. Long-term functional results of prosthetic airway splinting in tracheomalacia and bronchomalacia. J Pediatr Surg 1987;22:38–41.

[12] Ley S, Loukanov T, Ley-Zaporozhan J, Springer W, Sebening C, Sommerburg O, et al. Long-term outcome after external tracheal stabilization due to congenital tracheal instability. Ann Thorac Surg 2010;89:918–25.

[13] Takazawa S, Uchida H, Kawashima H, Tanaka Y, Masuko T, Deie K, et al. External stabilization for severe tracheobronchomalacia using separated ring-reinforced ePTFE grafts is effective and safe on a long-term basis. Pediatr Surg Int 2013;29:1165–9.

[14] Bugmann P, Rimensberger PC, Kalangos A, Barazzone C, Beghetti M, Lang FJW. Extratracheal biodegradable splint to treat life-threatening tracheomalacia. Ann Thorac Surg 2004;78:1446–8.

[15] Robey TC, Valimaa T, Murphy HS, Törmälä P, Mooney DJ, Weatherly RA. Use of internal bioabsorbable PLGA "finger-type" stents in a rabbit tracheal reconstruction model. Arch Otolaryngol Head Neck Surg 2000;126:985–91.

[16] Sewall GK, Warner T, Connor NP, Hartig GK. Comparison of resorbable poly-L-lactic acid-polyglycolic acid and internal Palmaz stents for the surgical correction of severe tracheomalacia. Ann Otol Rhinol Laryngol 2003;112:515–21.

[17] Hartig GK, Warner T, Connor NP, Thielman MJ. Evaluation of poly-L-lactic acid and polyglycolic acid resorbable stents for repair of tracheomalacia in a porcine model. Ann Otol Rhinol Laryngol 2001;110:11.

[18] Design Control Guidance for Medical Device Manufacturers, http://www.fda.gov/medicaldevices/deviceregulationandguidance/guidancedocuments/ucm070627.htm. [accessed 11.03.97]

[19] Teixeira MB, Bradley R. Design controls for the medical device industry. New York: Marcel Dekker, Inc; 2003.

[20] Zopf DA, Flanagan CL, Wheeler M, Hollister SJ, Green GE. Treatment of severe porcine tracheomalacia with a 3-dimensionally printed, bioresorbable, external airway splint. JAMA Otolaryngol Head Neck Surg 2014;140:66–71.

[21] Zopf DA, Hollister SJ, Nelson ME, Ohye RG, Green GE. Bioresorbable airway splint created with a three-dimensional printer. N Engl J Med 2013;368:2043–5.

[22] Costantino ML, Bagnoli P, Dini G, Fiore GB, Soncini M, Acocella F, et al. A numerical and experimental study of compliance and collapsibility of preterm lamb tracheae. J Biomech 2004;37:1837–47.

[23] Bagnoli P, Acocella F, Di Giancamillo MD, Fumero R, Costantino ML. Finite element analysis of the mechanical behavior of preterm lamb tracheal bifurcation during total liquid ventilation. J Biomech 2013;46:462–9.

[24] Khanafer K, Duprey A, Zainal M, Schlicht M, Williams D, Berguer R. Determination of the elastic modulus of ascending thoracic aortic aneurysm at different ranges of pressure using uniaxial tensile testing. J Thorac Cardiovasc Surg 2011;142:682–6.

[25] Wright A, Ardran GM, Stell PM. Does tracheostomy in children retard the growth of trachea or larynx? Clin Otolaryngol 1981;6:91–6.

[26] http://www.npr.org/blogs/health/2014/03/17/289042381/doctors-use-3-d-printing-to-help-a-baby-breathe.

[27] Ciardelli G, Chiono V, Vozzi G, Pracella M, Ahluwalia A, Barbani N, et al. Blends of poly-(epsilon-caprolactone) and polysaccharides in tissue engineering applications. Biomacromolecules 2005;6:1961–76.

[28] Partee B, Hollister SJ, Das S. Selective laser sintering of polycaprolactone bone tissue engineering scaffolds. Material research symposium proceedings, vol. 845, 2005.

[29] Williams JM, Adewunmi A, Schek RM, Flanagan CL, Krebsbach PH, Feinberg SE, et al. Bone tissue engineering using polycaprolactone scaffolds fabricated via selective laser sintering. Biomaterials 2005;26:4817–27.

[30] Tan KH, Chua CK, Leong KF, Cheah CM, Gui WS, Tan WS, et al. Selective laser sintering of biocompatible polymers for applications in tissue engineering. Biomed Mater Eng 2005;15:113–24.

[31] Partee B, Hollister SJ, Das S. Selective laser sintering process optimization for layered manufacturing of CAPA 6501 Polycaprolactone bone tissue engineering scaffolds. ASME J Manuf Sci Eng 2006;128:531–40.

[32] Wiria FE, Leong KF, Chua CK, Liu Y. Poly-epsilon-caprolactone/hydroxyapatite for tissue engineering scaffold fabrication via selective laser sintering. Acta Biomater 2007;3:1–12.

[33] Leong KF, Wiria FE, Chua CK, Li SH. Characterization of a poly-epsilon-caprolactone polymeric drug delivery device built by selective laser sintering. Biomed Mater Eng 2007;17:147–57.

[34] Huang H, Oizumi S, Kojima N, Niino T, Sakai Y. Avidin–biotin binding-based cell seeding and perfusion culture of liver-derived cells in a porous scaffold with a three-dimensional interconnected flow-channel network. Biomaterials 2007;28:3815–23.

[35] Yeong WY, Sudarmadji N, Yu HY, Chua CK, Leong KF, Venkatraman SS, et al. Porous polycaprolactone scaffold for cardiac tissue engineering fabricated by selective laser sintering. Acta Biomater 2010;6:2028–34.

[36] Eshraghi S, Das S. Mechanical and microstructural properties of polycaprolactone scaffolds with one-dimensional, two-dimensional, and

three-dimensional orthogonally oriented porous architectures produced by selective laser sintering. Acta Biomater 2010;6:2467–76.

[37] Eosoly S, Brabazon D, Lohfeld S, Looney L. Selective laser sintering of hydroxyapatite/poly-e-caprolactone scaffolds. Acta Biomater 2010;6:2511–7.

[38] Sudarmadji N, Tan JY, Leong KF, Chua CK, Loh YT. Investigation of the mechanical properties and porosity relationships in selective laser-sintered polyhedral for functionally graded scaffolds. Acta Biomater 2011;7:530–7.

[39] Eshraghi S, Das S. Micromechanical finite-element modeling and experimental characterization of the compressive mechanical properties of polycaprolactone-hydroxyapatite composite scaffolds prepared by selective laser sintering for bone tissue engineering. Acta Biomater 2012;8:3138–43.

[40] Lohfeld S, Cahill S, Barron V, McHugh P, Durselen L, Kreja L, et al. Fabrication, mechanical and *in vivo* performance of polycaprolactone/tricalcium phosphate composite scaffolds. Acta Biomater 2012;8:3446–56.

[41] Eosoly S, Vrana NE, Lohfeld S, Hindie M, Looney L. Interaction of cell culture with composition effects on the mechanical properties of polycaprolactone-hydroxyapatite scaffolds fabricated via selected laser sintering (SLS). Mat Sci Eng C 2012;32:2250–7.

[42] Doyle H, Lohfield S, McHugh P. Predicting the elastic properties of selective laser sintered PCL/beta-TCP bone scaffold materials using computational modeling. Ann Biomed Eng 2014;42:661–77.

[43] Kang H, Hollister SJ, LaMarca F, Park P, Lin CY. Porous biodegradable lumbar interbody fusion cage design and fabrication using integrated global-local topology optimization with laser sintering. ASME J Biomech Eng 2013;135:101013–8.

[44] Smith MH, Flanagan CL, Kemppainen JM, Sack JA, Chung H, Das S, et al. Computer tomography-based tissue-engineered scaffolds in craniomaxillofacial surgery. Int J Med Robotics Comput Assist Surg 2007;3:207–16.

[45] Park CH, Rios HF, Jin Q, Sugai JV, Padial-Molina M, Taut AD, et al. Tissue engineering bone-ligament complexes using fiber-guiding scaffolds. Biomaterials 2012;33:137–45.

[46] Zopf DA, Mitsak AG, Flanagan CL, Wheeler M, Green GE, Hollister SJ. Computer aided-designed, 3-dimensionally printed porous tissue bioscaffolds for craniofacial soft tissue reconstruction. Otolaryngology – Head and Neck Surg 2015;152(1):57–62.

Chapter 4

Inkjet Bioprinting

Shintaroh Iwanaga*, Kenichi Arai**, and Makoto Nakamura**

*Institute of Industrial Science, University of Tokyo, Tokyo, Japan; **Graduate School of Science and Engineering for Research, University of Toyama, Toyama, Japan

Chapter Outline

ABSTRACT

Printing technology has recently been applied in the fabrication of three-dimensional (3D) structures and also in medical applications. Drop-on-demand printing that is one of the inkjet printing technologies, enables the printing of complex and precise sections of living tissues or organs on the culture substrates utilizing cells and/or biomaterials as bioinks. This technique has the potential to accelerate the study of tissue biofabrication thanks to the development of bioinks and biopapers that are suitable for inkjet bioprinting and remarkable progress in 3D printing technology. Additionally, the technology of printing biomaterials is drawn upon drug delivery systems, which have also become an essential technique in tissue biofabrication in recent years. This work presents the research trends in inkjet printing as a bioapplicable technology, especially in the fields of tissue engineering and drug delivery systems.

Keywords: inkjet printing; 3D bioprinting; biofabrication; tissue engineering; regenerative medicine; drug delivery system

1 INTRODUCTION

Computer and device technologies have changed the concept of manufacturing during the last 20–30 years, and have brought great innovations to the industrial field and human lives. Now, they are also bringing another great innovation. This innovation is occurring in the field of biomedical engineering and tissue and organ engineering. "Biofabrication" is one of the emerging research topics in these fields [1–4]. The term "biofabrication" directly represents the procedures for manufacturing various biological products. In the researches on biofabrication, all sorts of promising technologies are aggressively introduced and applied, such as computer technology, printing technology, rapid prototyping and digital fabrication technologies, and micro-nano-machining and robotics technologies. All of these technologies are applied and developed focusing only on manufacturing biological products artificially. Considering the marked progress in manufacturing in the industrial field by the emergence of computer-assisted machines, biofabrication also has great potential to change the present concept of tissue engineering, and to bring great advancements in the medicine and life science fields. Especially, printing technologies are receiving a lot of attention as one of the useful systems for biofabrication.

The remarkable development of printing techniques enables us to print photos taken with a digital camera with high resolution at home. Inkjet is one of the printing technologies for printing something on paper by ejecting ink droplets. Variations of inkjet printing systems are generally divided into two types: a continuous inkjet printing system and an on-demand inkjet printing one [5]. Continuous inkjet printers use inks that excel in fast drying, and they are good at printing something on printing objects with low

Essentials of 3D Biofabrication and Translation. http://dx.doi.org/10.1016/B978-0-12-800972-7.00004-9

moisture adsorption or on objects with a curved surface. However, the inkjet nozzles are easily clogged because of the quick-drying property of inks; continuous inkjet printers need to keep ejecting inks at all times to preventing clogging. Moreover, as excess amounts of ejected ink droplets would be collected in a reservoir tank, the ink droplets are required to be charged to change their track. On the other hand, on-demand printing systems can eject ink droplets as much as needed, and does not require ejecting ink continuously similar to continuous inkjet printing systems. On-demand printing can be classified into three types: (1) the bulb inkjet system, where ink droplets are pressed and ejected by opening and closing a bulb [6], (2) the piezoelectric inkjet system, where ink droplets are ejected by vibration of the piezoelectric element by voltage impression [7], and (3) the bubble inkjet system, where ink droplets are ejected with the pressure of bubbles by heating ink locally [8]. The physical values of inks, such as surface tension or viscosity, would need to be changed in each manner, and regulating the penetration of ink to the paper enables an on-demand inkjet printing system to print in high definition. In this way, the efficiency of inkjet printing depends largely on the inks and the object for printing, which is mainly paper. Three-dimensional (3D) printing has recently become possible. In the case of normal two-dimensional printing, printing is done by ejecting ink droplets onto the paper. For three-dimensional printing, forming three-dimensional architectures is done by laminating printed layers where the shaping liquid is ejected to the stereoscopic shaping powder, or where ejected curable ink is cured by heat or ultraviolet (UV) irradiation [9,10]. In this situation, many researchers have tried to apply inkjet printing to the tissue-engineering field or the pharmaceutical field using biomaterials and cells as bioinks [11–13]. The on-demand printing system is especially suitable for bioprinting a little amount of protein or cells because of its property of on-demand printing. Above all, since inkjet printing enables contactless printing, nozzle tips are not contaminated by getting in contact with the printing object. This is of great advantage when applied to cell culturing or fabricating implantable grafts that must avoid contamination. Moreover, because the inkjet printing system enables ejecting various kinds of inks by controlling the ejected points correctly, anyone can print complex photos or figures of two-dimensional living body tissue structures, such as the sections of tissues or organs, if they can prepare the computer and printing data. Three-dimensional bioprinting is also possible by laminating the printed tissue layers. Nowadays, the advance in technologies in medicine easily enables the preparation of tomographic images of internal organs of the body noninvasively. Therefore, it is easy to imagine that the age of order-made biofabrication is coming, where individual tissues or organs are printed three-dimensionally by inkjet printing technology.

A lot of researchers have studied and applied the inkjet bioprinting technique as one of the tools in the research on tissue engineering; however, it was very difficult to eject the living cells during the earliest days of inkjet bioprinting research [14]. Bioprinting of biomaterials that did not include cells was easily performed and the preparation of two-dimensional patterning surfaces or scaffolds for cell culturing was achieved. Nevertheless, printing the bioinks that included cells was a process of trial and error. Some of the reasons why printing the living cells is difficult are the property of fast drying of inks and the high hygroscopicity of printing papers. As previously described, the high-resolution printing of inkjet printing technology depends on the quick drying of inks and the moisture adsorption of the printing objects. However, with a very few exceptions, such as keratinocyte, cells are generally vulnerable to dryness. Since dryness causes cell death, prevention of drying and maintaining wet conditions are required when printing cells. That is, the selection or development of novel bioinks that have the opposite property to the previous ones is required. Now, various types of bioinks have been developed to solve these problems, and inkjet bioprinting technology has been advancing as an elemental technology for the fabrication of three-dimensional tissues step by step. In this chapter, we introduce how the inkjet printing technology has been utilized to the tissue engineering or pharmaceutical field.

2 THE ADVENT OF NEW ERA – BIOFABRICATION OF TISSUES AND ORGANS

Since the dawn of history, the development of medicine has been as long and as remarkable. When ill, we can cure a disease using some drugs or by surgical treatment. The development of artificial organs enabled us to complement the lost biofunction of patients, and that facilitated the improvement of patients' quality of life (QOL). Many researchers have tried to create living artificial organs or tissues made from cells in order to serve more and more patients suffering from congenital or acquired illnesses. Takahashi et al. carried out a clinical trial of transplanting a retina derived from a patient's iPS cells in 2014 [15], and thus it is expected that tissue engineering would become a more important technology for a lot of patients. The new era for biofabrication is coming right now.

2.1 Why Is It Required to Fabricate Biological Tissues and Organs Artificially?

As mentioned earlier, biofabrication is aimed at manufacturing biological products. The purposes of producing such artificially engineered tissues and organs are intrinsically

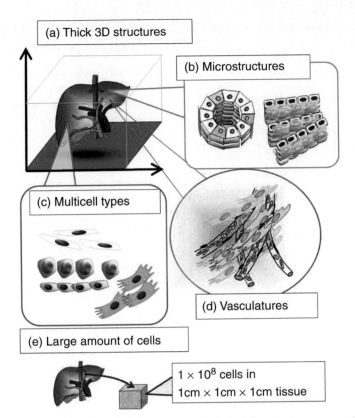

FIGURE 4.1 **Schematic drawing of the characteristics of functional organs.** In the high functional organs, a variety of cells are arranged at the right places and the functional structures are formed at a micro level. The tissues or organs as an integration of functionable microstructures fulfill not the sum of their functions but the higher functions. There is no doubt that the vascular networks spread over the whole tissues contribute to maintaining the function of centimeter-scaled large tissues or organs. These highly functionable organs are constructed from approximately 100 million cells per cubic centimeter, and thus it would be almost impossible to fabricate the functional tissues or organs without applying the biofabrication technology using computers and machines.

for their medical application to treat several intractable diseases and organ failures. Owing to the great progress in organ transplantation, many patients suffering from serious organ failure have been treated and saved by replacing their dysfunctional organs with healthy organs. The clinical effectiveness of organ replacement has already been proven without doubt; however, it has become very difficult to obtain healthy organs, which are indispensable for this therapy. Because of the absolute lack of donor organs, millions of patients are dying during the waiting period every year. In addition, ethical and religious problems in receiving organs from other people are also serious and still controversial. When encountering some serious problems, human beings develop the science and technologies to solve these problems. The researches on tissue engineering and regenerative medicine are also among these actions [16]. Therefore, we should keep in mind such essential mission when we attempt to manufacture tissues and organs.

In addition, another possibility has been raised by the developments in some tissue engineering technologies. It is their application to the effective biological tissue models, which serve as experimental models for exploratory researches on cellular physiology and disease pathogenesis, as well as test tools for effect and toxicity examinations for drug discovery and drug screening. It is predicted that new therapeutics will be developed through the researches using engineered tissues [17]. Therefore, these activities toward the development of life science also have precious missions.

2.2 Histological Difficulties in Fabricating Tissues

However, despite the big hopes and the great efforts of many researchers, success has been limited in a few simple tissues such as skin, cornea, and cartilages. All of these are with simple structures and simple components. Meanwhile, based on the histological considerations of the important organs, such as the heart, liver, and kidney, the tissues of such organs have the following major characteristics as shown in Fig. 4.1 [4,14]:

1. Thick 3D architectures
2. Characteristic microstructures for respective organs

3. Heterogeneous structures composed of multiple types of cells and extra-cellular matrices
4. Capillary rich tissues
5. Tissues composed of large amount of cells.

However, conversely, those histological characteristics become rather difficulties in manufacturing tissues. Such thick and complicated tissues have not been achieved yet, and even 100 μm thickness is still a big hurdle. How can we overcome this?

2.3 Computer-Aided Tissue Engineering

The solution determined is the concept that the specialized fabrication machines for biofabrication should be developed and used; that is, we need a new technology on computer-aided tissue engineering and biofabrication. Considering these histological difficulties, the following technologies are considered necessary:

1. 3D fabrication or 3D deposition technologies
2. Microscaled cell manipulation technologies
3. Fabrication technologies of heterogeneous structures
4. Construction of perfusion structures
5. Feasible technologies for manipulation of large amount of cells.

Then, it was considered that handling and arranging individual cells directly onto the targeted 3D spatial positions with the corresponding resolution of the histological microstructures is needed. However, living cells, which are the most important materials for biofabrication, have never been used as materials for fabrication or manufacturing in the usual industrial field. The sizes of individual cells are only 10–30 μm in diameter. Besides, they are soft and wet, and must be treated in wet physiological environments. In addition, the biofabrication researchers must manipulate enormous amounts of cells during fabrication, reasonably at least within a few hours or so. It is said that a 1 cm cube of biological tissues is composed of about 100 million cells. Therefore, more than a few hundred million cells should be handled, individually. It is impossible to do so even for the most skilled technician in the world using their own hands. But we suppose that it will become possible by developing and using some high-performance machines instead of manual procedures.

For these reasons, many researchers started to explore effective technologies to develop a novel computer-aided manufacturing (CAM) machine together with computer-aided design (CAD) [4,18]. To achieve this, ink-jet bioprinting technology is now expected to be applied to biofabrication as the microcell handling technology and the 3D fabrication technology by layer-by-layer (LBL) procedure.

3 INKJET BIOPRINTING TECHNOLOGY FOR TISSUE ENGINEERING

There are various types of inks for bioprinting, specifically, bioinks: a solution suspended with cells, biomaterials from which scaffolds or drug carriers derive, and a solution that contains proteins or deoxyribonucleic acid (DNAs). That all materials are for cell culturing or implantable grafts is quite certain. As many biomaterials are used for bioinks, they are roughly classified into two categories: (1) materials that are printed with cells together and (2) materials whose cells are seeded and cultured after ejecting materials. In order to distinguish these two cases, we call the former "direct printing" and the latter "indirect printing" [19]. To begin with, we have the reason why we think of cells as the main point, and that is the issue of drying cells. The details of each feature are described further.

3.1 Indirect Printing

In the case of indirect printing, the quick-drying inks, which are commercial inks for the inkjet printer, or the inks that are dissolved in organic solvent, could be used as bioinks because cells would be seeded after printing and drying the bioinks. For example, it is possible to prepare the patterning surfaces of cell adhesive areas and nonadhesive areas by ejecting cell nonadhesive materials as bioinks onto the culturing substrates [20–26]. In contrast, it is easy to prepare the patterning surfaces by ejecting cell adhesive materials onto the low cell adhesive substrates such as a petri dish [19].

Kim et al. printed various patterns of poly(lactic-co-glycolic acid) (PLGA) on a polystyrene (PS) substrate for stem cell patterning (Fig. 4.2) [27]. They used a mixture of PLGA and N,N-dimethylformamide as bioinks, and evaluated the relationship between the concentration of polymer solutions and the viscosity. Patterned surfaces were designed by commercial software, Adobe Photoshop CS. Although cells were not perfectly patterned onto the PLGA printed surfaces, they succeeded in printing the synthetic polymer onto the plastic substrates and prepared the cell patterning surfaces utilizing the inkjet printing system.

The authors also succeeded in preparing the various patterning surfaces by ejecting bioinks where 2-methacryloyloxyethyl phosphorylcholine polymer or zwitterion polymer was dissolved in organic solvents. Also, the patterning with little bleeding onto culture substrates was possible by changing ethanol into dimethyl sulfoxide (DMSO) as the dissolving organic solvent (Fig. 4.3). Of course, these bioinks would not have any effect on cell viability because cells were seeded after organic solvents were volatilized. These results showed that we can easily prepare the patterned surfaces with high resolution by changing the property of inks. Moreover, the merit of inkjet bioprinting is

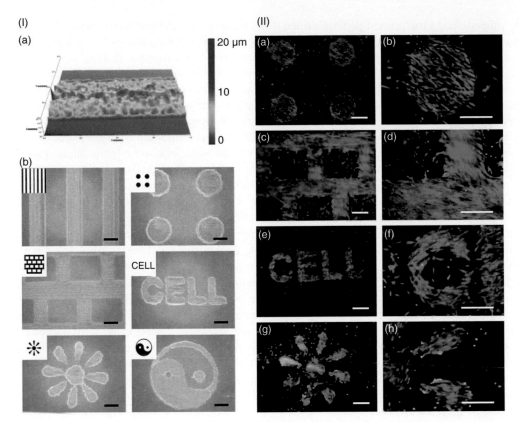

FIGURE 4.2 The two-dimensional patterning of PLGA onto the culture substrates. (I) Optical micrographs of inkjet printed PLGA patterns. The inserts represent the patterns designed using Adobe Photoshop CS software. (a) Surface profile of line pattern, (b) microscopic images of various patterns. Scale bars represent 500 μm. (II) Fluorescence microscope images of hASCs on PLGA-patterned PS substrates after 5 days of culture. (a and b) Dot pattern, (c and d) brick pattern, (e and f) "CELL" letter pattern, and (g and h) flower pattern. White bars represent 500 μm. *(Figure reprinted with permission from Kim et al. [27].)*

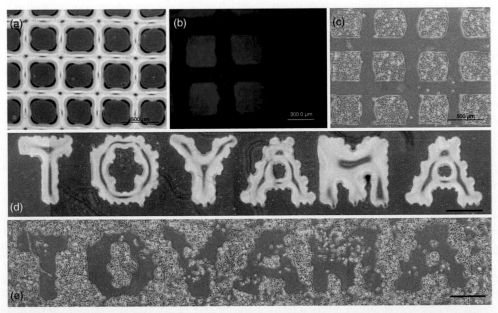

FIGURE 4.3 The patterning by printing non-cell-adhesive polymer (MPC polymer) dissolved in DMSO as bioink (data given by Iwanaga et al. from University of Toyama). (a–c) Lattice patterning printed on the polystyrene film. FITC-labeled BSA adsorbed onto the nonprinted surfaces, whereas MPC printed surfaces inhibited the adsorption of protein. Cells were also adhered and proliferated onto the nonprinted regions. The lattice patterning was printed using the bioprinter of the raster printing method. (d and e) Printed letter of "TOYAMA" and cell patterning using the bioprinter of the vector printing method.

that we can quite simply design the printed patterns on the computer. It is possible to print the cell patterning surfaces as positive type or negative type by changing the combination of inks and culture substrate. However, it enables us to alter the cell-adhesive regions by inverting the printed areas on computational design using the same bioinks. It is also possible to prepare the patterned surfaces with a concentration gradient on two-dimensional surfaces by gradational printing of inks, and these patterned surfaces are useful for eliminating the effect of the concentration gradient of some materials to cell migration [28–30].

The plasmid DNA or protein printed surfaces could be used for drug or gene delivery to cultured cells [31–33]. Normally, protein or gene delivery to cells would be carried out using virus vectors or cesium phosphate particles, or these materials would be transferred to cells by compulsion through the nanometer-scale holes that were made on the cell membranes temporally by electroporation. Gene-transferred cells are required to be selected from whole cells for culturing in any case. On the other hand, if you prepare the culture substrates printed with proteins or genes and culture the cells onto the patterned surfaces, you can easily find protein- or gene-transferred cells cultured onto the patterned regions. In the normal method, there are some problems that the loss of proteins or genes will be significant because these materials are required to be dispersed into the culture medium, or that cells may be hurt by electrostimulation. However, in the case of culturing cells onto the patterned surfaces prepared by inkjet printing, cells would be effectively transferred with very small amounts of proteins or genes. Thus, the inkjet bioprinting technology of proteins or genes is greatly useful for materials delivery system into cells.

Bioinks are also ejected onto the cultured cell layers as printed substrates. Matsusaki et al. reported the LBL alternating immersion of fibronectin and gelatin for fabricating 3D tissue structures, and they used an inkjet bioprinting system that can perform multicolor printing for this LBL method (Fig. 4.4) [34]. Because the inkjet bioprinting system is under computer control, the desired inks are printed exactly on the desired positions, and the amount of samples to be used as ink may be little. Therefore, the inkjet bioprinting system could be utilized as a prominent technology to solve the problems of reproducibility and cost performance. It is expected that many biologists may apply the inkjet bioprinting system for their biological studies in the near future.

Although we have mainly demonstrated the two-dimensional indirect printing, lamination of the layers printed on two dimensions enables the inkjet printing technology to expand into three-dimensional bioprinting, and it is possible to fabricate the scaffolds in arbitrary shape for cell culturing. The three-dimensional scaffolds were fabricated using polylactic acid, polycaprolactone (PCL),

or a copolymer derived from these materials, which have high biocompatibility and good biodegradability. The main method for fabricating the scaffold is using the mold and the precursor materials of the scaffold. We can fabricate the free-shaped scaffolds to some extent with this molding method; however, the articles are limited to the shapes that can be removed from the mold and it is difficult to create a scaffold of the complex and entangled structures. Furthermore, if you desire to architect a hierarchical scaffold made from various kinds of materials, it is extremely hard to create the ones that you desire using the molding method. For the inkjet bioprinting method, we can prepare a huge variety of scaffolds in shape because a mold is not necessary. Also, the property of inkjet's multicolor printing enables us to fabricate the scaffolds that are locally formed with different ingredients without any difficulty. Thus, it is possible to print three-dimensional scaffolds where cells may be controlled for their proliferation, differentiation, or migration by arranging proteins or growth factors on the desired positions in the scaffolds. Chung et al. have succeeded in fabricating artificial bones with high biocompatibility by ejecting a solution of sodium chondroitin sulfate as binder onto calcium phosphate powder (Fig. 4.5) [35,36]. They calculated and estimated the optimum shapes of artificial bone grafts that would be implanted to bone defect parts using the patients' computed tomography (CT) data, and printed the bone grafts based on the CAD data. This is truly the made-to-order bioprinting technology based on individual patient's data. Although the adapted defect parts are limited to parts that do not carry load, such as the face or the head, they have already carried out several clinical trials and have succeeded in verifying the effectiveness of their artificial bones.

There are numerous studies on fabrication of the scaffolds made from not only solid materials but also soft matters like hydrogel, which contains plenty of water. The hydrogel scaffolds work as a protection against drying cells in the case of direct printing, which is detailed later, and would be one of the indispensable bioink materials in bioprinting technology. We can point out several examples for bioprinting of soft matter: photo-crosslinkable materials, such as polyethylene glycol (PEG), hyaluronic acid, or gelatin, that are modified with methacrylates [37–42]; gelling agents that have a temperature responsibility, such as block polymer based on poly (N-isopropyl acrylamide) [43]; or materials that can form gels by mixing with two other liquid materials [44]. Especially, alginate hydrogel is utilized as a bioink for three-dimensional bioprinting quite frequently because alginate has a high biocompatibility and easily forms hydrogel in the presence of calcium ion. Boland et al. prepared scaffolds made from the mixture of alginate and gelation by an inkjet bioprinting system (Fig. 4.6) [45]. They were successful in the three-dimensional bioprinting of macroscopic hydrogel

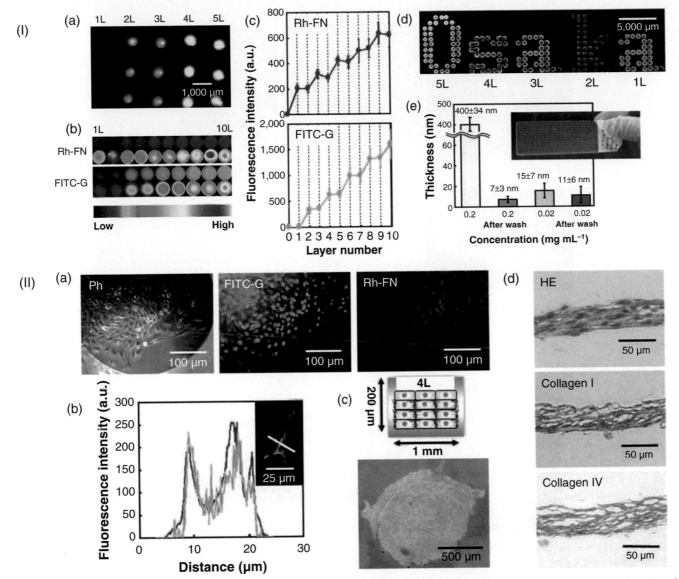

FIGURE 4.4 **Printing protein bioinks onto the cell layers for laminating cell sheets by alternating immersion.** (I) (a, b, d) Fluorescence merged image of layered spots prepared by the LBL printing of fibronectin (FN, labeled with FITC) and gelatin (G, labeled with rhodamine). (c) Dependence of the layer number of the FN-G nanofilm spots on the fluorescence intensity. (e) Thickness of the 10 layers of FN-G nanofilm spots by changing the ejected FN concentrations. (II) (a) Phase contrast and fluorescence microscopic images of cells after the LBL printing of Rh-FN and FITC-G. (b) Fluorescence intensity of FITC-G (green) and Rh-FN (red) on protein-printed cell by confocal laser microscopy scanning. (c) Four-layered cells cultured in 50-well plates. (d) Hematoxylin and eosin (HE), collagen I, and collagen IV staining histological images of four-layered cells. (*Figure reprinted with permission from Matsusaki et al. [34].*)

scaffolds by ejecting calcium chloride solution as the bio-ink into the mixed solution of alginate and gelatin. On the other hand, we succeeded in fabricating various shapes of alginate hydrogel using sodium alginate solution as the bioink [46,47]. We were able to fabricate alginate hydrogel beads, two-dimensional hydrogel sheets, and three-dimensional tubular structures by controlling the motion of the nozzle head. Moreover, we can create more complex structures, such as pyramid structures and honeycomb structures (Fig. 4.7).

3.2 Direct Printing

Previously, we demonstrated the various applications of the indirect printing method. Although there are some limitations for the ink materials as for the viscosity, almost any type of material could be used as bioink in the indirect printing method. However, since cells would be printed with bioink materials, the physical and chemical properties of bioinks are extremely limited. First, the solvent of bioinks needs to be water, and pH, osmolality, and ion intensity of

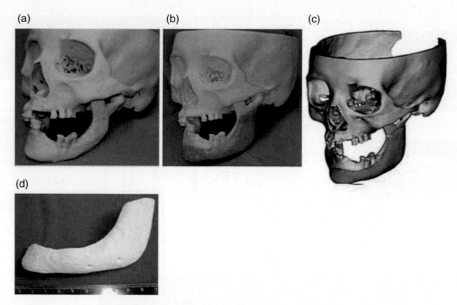

FIGURE 4.5   The custom-made printing of artificial bones using the patients' CT data. (a) Preoperative 3D plaster model showing the deformity of the transplanted autograft in the left lower jaw. (b) The design of an artificial bone created on the 3D plaster model with special radiopaque wax (pink) by the surgical operator. (c) Extraction of the CAD data of the created artificial bone (red) based on CT image. (d) Macroscopic image of the inkjet-printed custom-made artificial bone (IPCAB). *(Figure reprinted with permission from Saijo et al. [35,36].)*

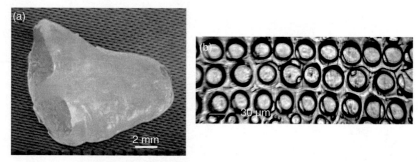

FIGURE 4.6   The three-dimensional printed hydrogel scaffold. Macroscopic and microscopic views of printed alginate structures. (a) Gross view of a branched chambered structure. (b) Light micrograph of the structure, top view 40×. *(Figure reprinted with permission from Boland et al. [45].)*

the solution must be adjusted to be the same as the physiological condition. As mentioned earlier, it is required to prevent printed cells from drying and spreading after printing on the substrate. Furthermore, the effect of physical stress when ejecting toward the cells is a big problem.

Saunders et al. evaluated the effect of waveform amplitudes and rise times of an inkjet printer to the cell viability using human sarcoma cell line, HT-1080 [48]. They confirmed that over 90% of cells were alive after printing with various ranges of waveform amplitudes and rise times. In evaluating cell proliferation potency by Alamar Blue assay, although some cell populations showed a lower proliferation potency compared with the control, which is pipetted cell population, many ejected cells were alive and proliferated for 96 h. Also, the authors confirmed that the physical stress of the ejection from the nozzle head did not affect the cell viability by utilizing various kinds of cells such as cancer cells, normal cells, and primary cells.

After confirming that the cell viability is mostly influenced by the process of ejection from the inkjet nozzle, a lot of research on direct printing has been reported in earnest. However, there was a basic problem that we need to print cells as cells would be prevented from drying and bleeding. Although the authors had also printed cells suspended in culture medium onto the culture substrates, the printed cells were immediately dried because the ejected droplets have ultramicro volumes (approximately 100 pL; pL means 10^{-12} L). Since cells were ejected into the medium of the buffer to prevent cells from drying, it was difficult to keep the patterned shapes on account of dispersion of cells in the solutions. Thus, through creative thinking, our group was able to resolve the problem by fabricating tissue structures using some hydrogel materials. The hydrogel enabled us to prevent cells from drying and bleeding, and to fabricate the three-dimensional structures at the same time that cells were arranged within the gel. Nowadays, the hydrogel is

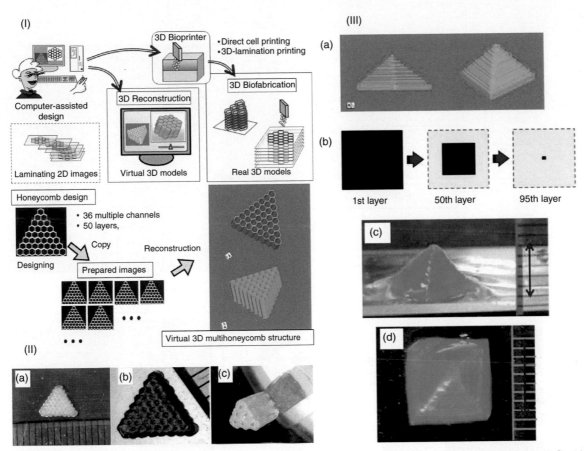

FIGURE 4.7 **Computer-aided design, manufacturing, and engineering enables the fabrication of the arbitrary gel constructs for biofabrication.** (I) Conception of computer-aided tissue fabrication. (II) Multihoneycomb structures made from alginate hydrogel. (III) (a) 3D computer model of a pyramid structure reconstructed from 95 different bitmap images. (b) The fabrication process with multiple images. (c, d) Fabricated three-dimensional pyramid structure, about 9 × 9 mm in size and about 5 mm in height. *(Figure reprinted with permission from Arai et al. [47].)*

commonly used as bioinks in direct printing. The material that is most frequently used for direct printing is alginate hydrogel, which has been mentioned in Section 3.1. We have also fabricated tissue-like structures with cells using alginate bioinks (Fig. 4.8) [47,49,50]. In our group, we have fabricated gel fibers, two-dimensional gel sheets, and three-dimensional tubular structures by drawing using the raster method. The multicolor printing technology enables us to arrange various types of cells at the same time, and we succeeded in fabricating the vessel-like structures by printing a double-layered tube whose outer layer included smooth muscle cells and inner layer included endothelial cells. Moreover, we can fabricate the arbitrarily shaped tissue structures by drawing using the vector method toward the fabrication of more complex structures. Alginate hydrogel contains a large amount of water and has a very short gelling time, and thereby, it has become possible to deter drying cells and bleeding cell patterning. Furthermore, because alginate forms gels onto the printed surfaces, it works not only as a bioink but also as a biopaper in some manner and enables us to construct the three-dimensional structures by laminating the alginate biopapers. To top it

off, alginate hydrogel has a high biocompatibility and cells do not suffer any damage when alginate forms the gel. The important point in the case of direct printing is to print cells under the conditions of how to reduce the cellular cytotoxicity. Alginate hydrogel is one of the excellent materials that make it possible to solve this problem; however, there is one major issue when using alginate hydrogel – its poor cell adhesiveness. Many types of cells attach to the scaffolds or substrates and grow. It is well known that an adherent cell, with the exception of floating cells such as some blood cells, die out step by step if they do not attach to something. Accordingly, reports on three-dimensional bioprinting utilizing a variety of unique hydrogels are now coming in.

Tao et al. fabricated tissue-like structures consisting of multiple kinds of cells by printing a calcium chloride solution suspended with cells as bioink into the mixture of an alginate and collagen solution (Fig. 4.9) [51]. Since alginate hydrogel has poor cell adhesiveness and collagen takes a lot of time to form a gel, it is hard to fabricate the tissue structures using only alginate or collagen materials. Thus, they succeeded in constructing three-dimensional tissue-like

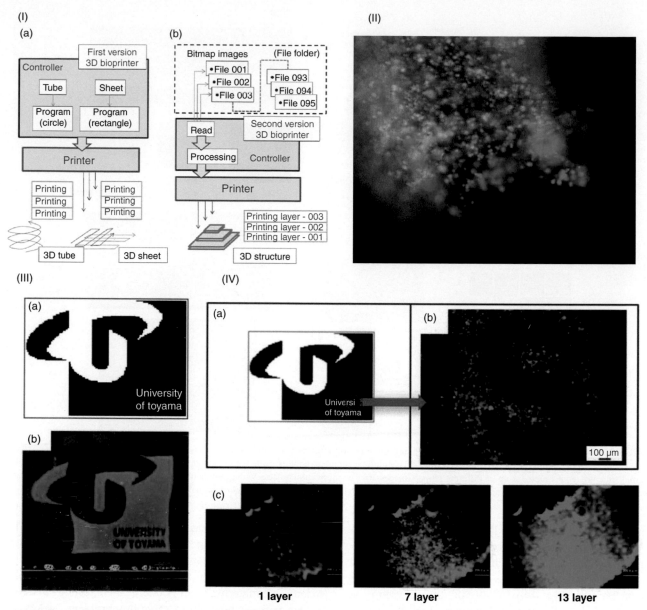

FIGURE 4.8 **The tissue-like structures consisting of alginate hydrogel and cells.** (I) Comparison of the 3D bioprinters with the raster method or the vector method. Schematic diagram of the processes used to fabricate 3D structures. (a) Printing is performed through a prepared program for each product with the raster type bioprinter, and (b) performed by reading serial bitmap files with the vector type bioprinter. (II) Vessel-like structure fabricated with the raster type bioprinter. The outer layer included primary human aortic smooth muscle cells (red), and the inner layer included primary human umbilical vein endothelial cells (green). (III) (a and b) Printed logo mark of University of Toyama by laminating three layers of alginate hydrogel sheet with the vector type bioprinter. (IV) Printed gel structures encapsulated living cells. (a) The logo image of the University of Toyama. (b) Fluorescent image of "S" character in the fabricated logo that included living cells. Cells were stained with calcein AM (green) and propidium iodide (red). (c) Z-stacked images of the fabricated pyramid structure that included living cells. Cells were stained with calcein AM (green) and propidium iodide (red). These images were obtained with a focus on the top, middle, and bottom sections of the fabricated pyramid structure. *(Figure reprinted with permission from Arai et al. [47,50].)*

structures where cells could adhere and proliferate inside by mixing these two gelling materials.

Cui et al. reported the fabrication of cartilaginous tissue made from the hydrogel of photocrosslinkable PEG (Fig. 4.10) [52]. Alginate is the physical hydrogel and it forms a gel by crosslinking with Ca ion. On the other hand, PEG diacrylate (PEGDA) is the chemical hydrogel that forms a gel by UV irradiation. Polymerization initiators turn into free radicals according to the UV irradiation, and generated radicals attach to PEGDA. Next PEGDAs turn into free radicals and crosslink with each other. Then, PEGDA forms a hydrogel. In this way, there is a risk that cells would be hurt when using the chemical hydrogel, whereas the physical hydrogel can form a gel under relatively mild conditions

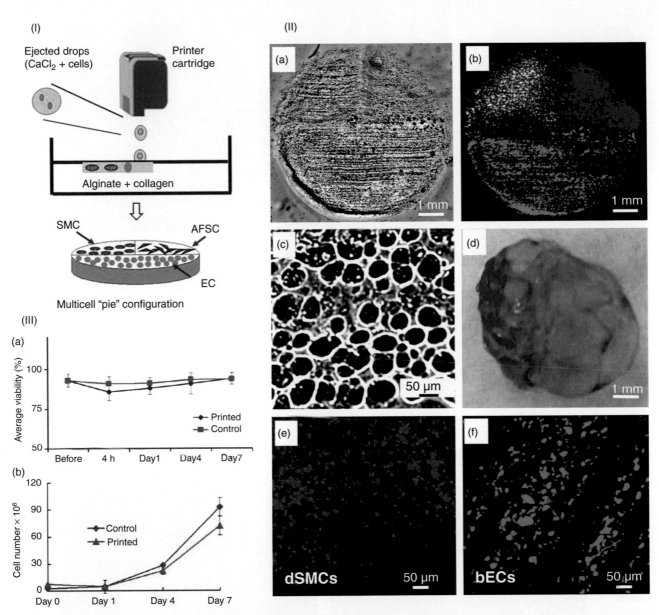

FIGURE 4.9   The tissue-like structure with multiple cell types fabricated using hybrid bioinks. (I) Schematic drawing of the proposed bioprinting method to fabricate multicell heterogeneous tissue constructs. (II) (a) Light and (b) fluorescence microscopic top views of a complete 3D multicell "pie" construct before implantation. The cells that appear green are bovine endothelial cells (bECs) labeled with PKH 26 dyes; the cells that appear blue are human amniotic fluid-derived stem cells (hAFSCs) tagged with CMHC dyes; the cells that appear red are canine smooth muscle cells (dSMCs) labeled with PKH 67 dyes. (c) Microscopic image of the microstructure of the printed "pie" scaffold. (d) Gross view of the retrieved "pie" 2 weeks postimplantation. (e and f) Fluorescence images of dSMCs and bECs within the "pie" implant, respectively. (III) (a) Viability of dSMCs within the printed 3D constructs was evaluated by using live/dead assay and compared with the nonprinted samples (the control). (b) Proliferation of bECs within the printed 3D constructs were analyzed by using mitochondrial metabolic activity (MTT) assay and compared with the nonprinted samples (the control). *(Figure reprinted with permission from Xu et al. [51].)*

to cells. However, they succeeded in finding out the mild reaction condition by examining the time or power of UV irradiation, and fabricating the tissue-like structures. They confirmed that chondrocytes proliferated within the PEGDA hydrogel by changing the combination of growth factors. The merits of utilizing these photocrosslinkable hydrogel are as follows: one is that various types of biomaterials could

be easily modified and used as precursors of photocrosslinkable hydrogel, for example, hyaluronic acid, gelatin, and chitosan. The other is that it is easy to prepare a hydrogel with high mechanical strength compared with the physical hydrogel. It has been possible to utilize these chemical hydrogels as a bioink for the fabrication of tissues or organs, such as cartilage, which require very high strength.

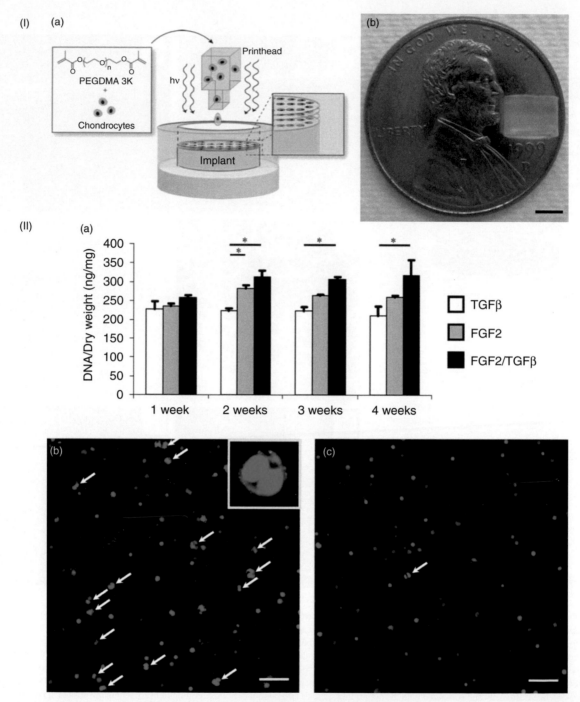

FIGURE 4.10 **Fabrication of the three-dimensional printed gel utilizing chemical hydrogel.** (I) Bioprinted cell-laden constructs for cartilage tissue engineering. (a) Schematic of bioprinting with simultaneous photopolymerization process. (b) A printed PEG hydrogel construct with 4 mm in diameter and 4 mm in height using layer-by-layer assembly. Scale bar: 2 mm. (II) Human chondrocyte proliferation within printed 3D cell-laden PEG hydrogel constructs. (a) Cell proliferation within PEG hydrogel treated with TGF-b1 (white bars), FGF-2 (grey bars), and FGF-2/TGF-b1 (black bars) for the first week, then TGF-b1 for all groups. (b and c) Confocal images of Calcein AM stained cells treated with FGF-2/TGF-b1 and TGF-b1 only after 2 weeks in culture. The dividing cells were marked with arrows. Asterisks indicate statistical significance between assigned groups. Scale bars, 100 μm. *(Figure reprinted with permission from Cui et al. [52].)*

Pati et al. reported on three-dimensional bioprinting using unique bioinks derived from living organs (Fig. 4.11) [53]. They used extracellular matrixes originated from cartilage, heart, and adipose tissues. First, these tissues or organs were processed for decellularization, and the frameworks of matrixes were obtained. The decellularized extracellular matrixes were treated chemically and enzymatically, and then they obtained the novel bioinks named extra

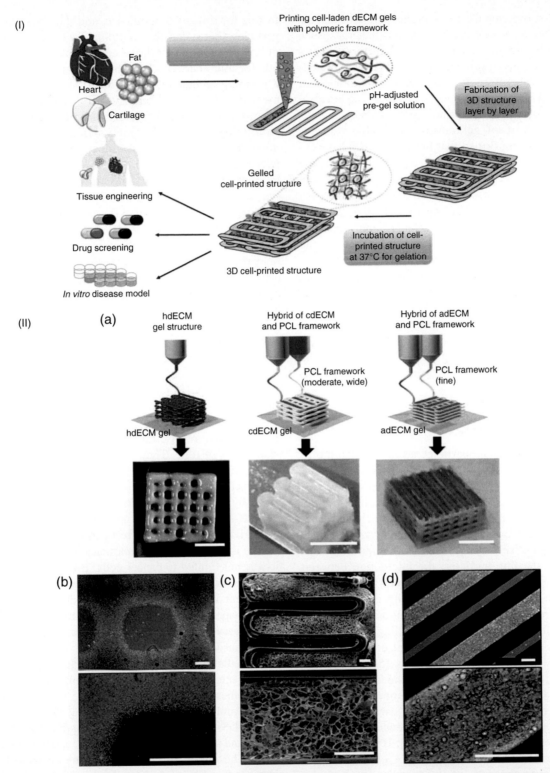

FIGURE 4.11 Bioprinting of tissue-like structures using unique bioinks derived from extracellular matrixes of decellularized tissues. (I) Schematic elucidating the tissue-printing process using dECM bioink. Respective tissues were decellularized after harvesting with a combination of physical, chemical, and enzymatic processes, solubilized in acidic condition, and adjusted to physiological pH. (II) Printing process of particular tissue constructs with dECM bioink. (a) Heart tissue construct was printed with only heart dECM (hdECM). Cartilage and adipose tissues were printed with cartilage dECM (cdECM) and adipose dECM (adECM), respectively, and in combination with PCL framework (scale bar, 5 mm). (b) Representative microscopic images of hdECM construct (scale bar, 400 mm). (c) SEM images of hybrid structure of cdECM with PCL framework (scale bar, 400 μm) and (d) microscopic images of cell-printed structure of adECM with PCL framework (scale bar, 400 μm). *(Figure reprinted with permission from Pati et al. [53].)*

cellular matrix. In living tissues or organs, the cells that are forming the organs would produce various materials, such as collagen, glycosaminoglycan, or growth factors, and create a microenvironment suitable for the subsistence of cells by themselves. Therefore, the decellularized extracellular matrix derived from heart tissue, for instance, is quite ideal biomaterials for culturing cardiomyocyte. These bioinks have the property of sol–gel transformation similar to collagen, and it was confirmed that these materials have excellent cell adhesiveness and proliferation. However, the speed of forming a gel was not so fast, and it was possible to print these bioinks with cells for fabricating the two-dimensional construction, whereas it was hard to laminate into three dimensions. Thus, they printed and fabricated the solid frame by ejecting PCL bioink, then printed the bioink of the decellularized extracellular matrix and cells between the PCL frames. They succeeded in fabricating three-dimensional tissue-like structures by laminating the PCL frames.

As shown here, a lot of researchers have tried to fabricate three-dimensional tissues approximate to the vital tissues by being creative in the preparation of bioinks or in the fashion of laminating the printed layers.

4 INKJET BIOPRINTING TECHNOLOGY FOR PHARMACEUTICAL APPLICATIONS

Functional micro- or nanoparticles are paid attention to in the field of medical and biological engineering, such as medicine or drug development, and not used only in industrial applications [54–57]. In the medical field, some drugs are supported by these fine particles, and they will be used for releasing drugs *in vitro* or *in vivo*, which is called the drug delivery system (DDS). To support the object drugs on the fine particles, the interaction between them is one of the quite important factors. In general, drugs will be supported on carriers with covalent bond [58–60], hydrogen bond, electrostatic interaction, or hydrophobic interaction [61–69]. Controlling drug release from carriers will be directly dependent on the force of interaction between drugs and carriers [58,68]. Therefore, controlling interactions is one of the key factors for drug release. On another front, drugs supported on carriers would be released from a surface of carriers by diffusion. Thus, speed of release will be dependent on the size of particles to a great extent [65,70,71]. If a particle has a perfect spherical shape, as the volume of the particles is proportional to the cube of the particle diameter, the surface area is proportional to the square of the particle diameter, and thus the specific surface area is inversely proportional to the particle diameter, the smaller the size of particles become, the quicker the speed of drug release from carriers would be [65,72]. Moreover, a supporting amount of drugs within the carrier is one of the important factors for time of drug release, and thus, the porosity of particles, that is density, would have a big role in the time of drug release [73]. A large number of fabrication methods for microparticles have been reported. For example, a grinding technique with a mill [74–76], spray drying method [77–79], and emulsification method [62,80,81] are quite famous for fabricating microparticles. Inkjet bioprinting is also increasingly being utilized to prepare microparticles for pharmaceutical application. The inkjet printing system enables us to eject the ink droplets in large quantities, rapidly and uniformly. Thus, some research groups have focused on this property of the inkjet printing system to be applied for fabricating various kinds of monodisperse microparticles.

In pharmaceutical applications especially the DDS field, the size and shape of the carrier particles are quite important factors because these microparticles are involved in the release kinetics of loaded drugs. Lee et al. manufactured drug-loaded polymer microparticles with arbitrary shapes using the inkjet printing system (Fig. 4.12) [82]. They printed four types of microparticles (circle, grid, honeycomb, and ring) using a mixture of PLGA, paclitaxel (PTX), and *N,N*-dimethylacetamide as bioinks. Their results showed that PTX-loaded PLGA microparticles with different geometries exhibited different drug release rates mainly due to the different surface areas. Moreover, they confirmed that the cumulative release ratio of PTX is proportional to the square root of time. They also tested the effectiveness of their microparticles as a DDS carrier for cancer cells. They confirmed that the viability of HeLa cancer cells was reduced by adding the PTX-loaded microparticles, whereas the PTX-free microparticles did not induce the cytotoxicity to the HeLa cells during the test period.

The application of microparticles in the pharmaceutical field is not limited to cancer cells. Boehm and colleagues fabricated miconazole-coated microneedles for use in treating fungal infections [83]. Microneedles are small-scale lancet-shaped structures that may be utilized for transdermal delivery of pharmacologic agents and vaccines as well as for transdermal biosensing [84–88]. They fabricated the microneedle structures by dynamic mask microstereolithography or micromolding, and used the inkjet printing system to load miconazole into microneedles. The inkjet printing system would also be useful in creating drug delivery systems (DDSs) with poorly soluble pharmacologic agents, on-demand individualized medications, and low-dose medication. The authors succeeded in loading miconazole to the microneedles by using piezoelectric inkjet printing, and exhibited antifungal activity of the drug-loaded microneedles against *Candida albicans*. Since it is possible to print various ink materials on the desired position exactly, we can prepare the drug carriers where pharmacologic functions were added to the existing microstructures using the inkjet printing system.

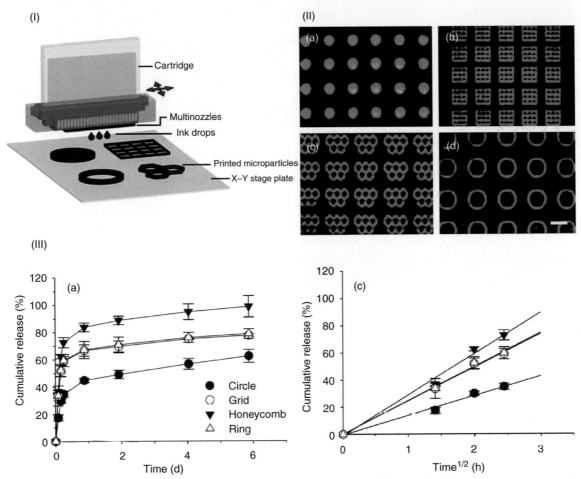

FIGURE 4.12 The fabrication of microparticles with arbitrary geometries utilizing the inkjet printing system. (I) Schematic representation of the piezoelectric inkjet printing system for fabrication of PTX-loaded PLGA microparticles. (II) Fluorescence micrographs of PTX-loaded PLGA microparticles with different geometries. (a) Circle, (b) grid, (c) honeycomb, and (d) ring. Scale bar represents 500 μm. (III) Release profiles in cumulative percentage of PTX released from PLGA microparticles with four different geometries (closed circle: circle, open circle: grid, closed inverted triangle: honeycomb, and open inverted triangle: ring) in PBS containing 0.1% (v/v) Tween 80 at 37°C for (a) 6 d and (c) the cumulative release (%) as a function of the square root of time. Data are shown as means ± standard deviations ($n = 3$). *(Figure reprinted with permission from Lee et al. [82].)*

Researchers have also tried to apply the inkjet bioprinting system in the pharmaceutical field. Our group fabricated uniform microparticles by combining the drying process with the inkjet printing system, and the novel concept of a tandem drug delivery system (T-DDS) using micro- and nanoparticles was suggested (Fig. 4.13) [89]. When preparing drug-loaded micro- or nanocarriers, the drugs are normally loaded to the carriers by immersing them in the solutions in which the drugs are dissolved in. However, a large amount of drugs would be lost with this method. Then, we paid attention to the drying process for fabricating particles. The drying process would be expected to enable microparticles to include nonvolatile drugs inside without loss of drugs. Moreover, if the nanoparticle-supported drugs would be loaded to microparticles, T-DDS would be expected to be used for medical and biological applications, the concept of which

is the double release of drugs from nano- and microparticles. We prepared alginate microparticles loaded with different sizes of nanoparticles (25, 100, and 250 nm). First, we evaluated the relationship between the alginate concentration and the size of the alginate microparticles. Microparticle volume increased linearly with the increase in alginate concentration. And we can manufacture uniform-size-controlled alginate hydrogel microparticles by immersing dried particles into a calcium chloride solution. Furthermore, the release ratio of nanoparticles could be controlled by changing the size of nanoparticles. We confirmed that the size of alginate hydrogel microparticles cannot be controlled by changing the alginate concentration without passing through the drying process. It is possible to manufacture the microparticles with various kinds of materials, and the sub-microparticles also can be fabricated with our method. Inkjet bioprinting has great

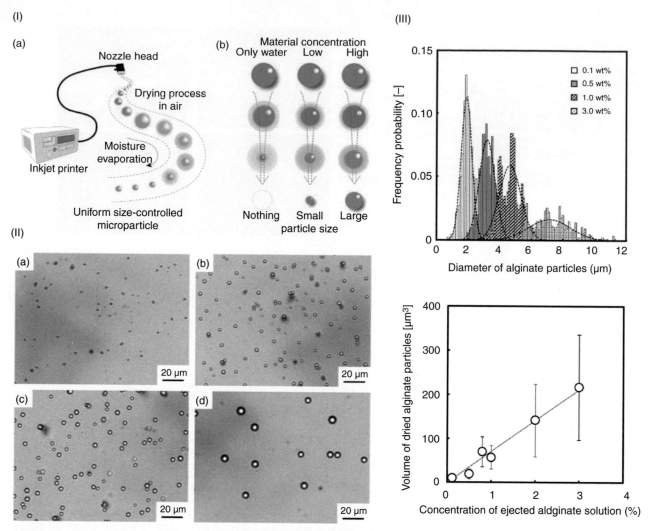

FIGURE 4.13 **The manufacture of microparticles utilizing inkjet printer and dry process.** (I) Concept of fabricating uniform-size-controlled microparticles with inkjet printer. (a) Microparticles are easily fabricated by evaporating moisture in air, and (b) size of particles could be controlled with changing materials concentration. (II) Photographs of dried alginate microparticles fabricated with concentration of (a) 0.1, (b) 0.5, (c) 1.0, and (d) 3.0% alginate solutions. Size-controlled spherical microparticles were easily manufactured by only changing the concentration of alginate solution. (III) (top) A distribution graph of diameter of dried alginate microparticles prepared with concentration of 0.1, 0.5, 1.0, and 3.0% alginate solutions. Diameters of approximately 1000 microparticles were analyzed, respectively. Dash lines show each fitted normal distribution curve. (bottom) Concentration of ejected sodium alginate solution affected size of prepared microparticles. Concentration was proportional to volume of microparticles. *(Figure reprinted with permission from Iwanaga et al. [89].)*

potential using the novel DDS concept or the fabrication process in the pharmaceutical field.

5 CONCLUSIONS

In this chapter, we introduced the inkjet bioprinting approach in the tissue engineering and pharmaceutical field. It has been shown that the inkjet printing technique is very useful not only for industrial application but also as a tool for the biology and medical fields. Other than inkjet technology, there are also several approaches in biofabrication such as laser cell printing, manipulation of cellular blocks,

laminating cell sheets, and transfer printing. However, we think that they are still in the beginning stage as inkjet 3D biofabrication.

There is one approach in conventional tissue engineering in which tissues are incubated *in vivo* after scaffolds have been implanted. This approach is called "*in situ* tissue engineering." Here, the concept of "in-factory tissue engineering" is proposed. In order to produce high-grade tissues and organs, it is essential for them to pass through several deliberate processes. Biofabrication and bioprocessing are important processes in this concept. Owing to the advancement of biofabrication, we feel the door of "in-factory tissue engineering" has just opened.

In the unexplored research field, we will encounter unknown problems and difficulties. However, we believe that we can overcome these difficulties by developing science and technology aggressively. We hope that human beings can succeed in producing high-grade bioproducts, such as physiologically functional tissues and organs that meet the requirements for medical use, in the future.

GLOSSARY

Biofabrication The technologies to reconstruct living tissues and organs, or the technologies to produce bioengineered tools and materials such as scaffolds, DNA and protein arrays, and cell-containing biodevices such as organ-on-a-chip.

Bioink Materials for bioprinting systems, for example, cells, proteins, DNAs, drugs, natural materials and synthetic materials, etc.

Bioprinting One of the biofabrication technologies by printing or dispensing cells and/or some biomaterials for preparing 2D or 3D structures.

Computer-aided tissue engineering The technologies of computer-controlled biofabrication to be specialized in the tissue-engineering field. Bioprinting is one of the technologies on computer-aided tissue engineering.

In factory tissue engineering The concept of tissue engineering in which the tissue engineered products are produced via a series of many processes, such as pretreating cells, preparing materials, three-dimensional construction, three-dimensional assembly, cultivation, scaling-up, organization, evaluating, checking, finalizing, and packaging etc. And all the processes are advanced with automated machineries, just like several industrial products are produced via many processes in the factory.

Inkjet printing Printing system that prints objects by ejecting ink droplets via the nozzle head, and the system is divided into continuous printing system and on-demand printing system.

ABBREVIATIONS

CAD	Computer-aided design
CAM	Computer-aided manufacturing
CT	Computed tomography
DNA	Deoxyribonucleic acid
DMSO	Dimethyl sulfoxide
DDS	Drug delivery system
LBL	Layer-by-layer
PTX	Paclitaxel
PCL	Polycaprolactone
PEG	Polyethylene glycol
PEGDA	Polyethylene glycol diacrylate
PLGA	Poly(lactic-coglycolic acid)
T-DDS	Tandem drug delivery system
3D	Three-dimensional
UV	Ultraviolet

REFERENCES

[1] Sun W. Welcome to biofabrication. Biofabrication 2009;1. 010201.

[2] Mironov V, Trusk T, Kasyanov V, Little S, Swaja R, Markwald R. Biofabrication: a 21st century manufacturing paradigm. Biofabrication 2009;1:022001.

[3] Guillemot F, Mironov V, Nakamura M. Bioprinting is coming of age: report from the International Conference on Bioprinting and Biofabrication in Bordeaux (3B'09). Biofabrication 2010;2:010201.

[4] Nakamura M. Reconstruction of biological three-dimensional tissues: bioprinting and biofabrication using inkjet technology. In: Ringeisen BR, Spargo BJ, Wu P, editors. Cell and Organ Printing. Springer; 2010. p. 23–34.

[5] Weng B, Shepherd RL, Crowley K, Killard AJ, Wallace GG. Printing conducting polymers. Analyst 2010;135:2779–89.

[6] Olkkonen J, Leppäniemi J, Mattila T, Eiroma K. Sintering of inkjet printed silver tracks with boiling salt water. J Mater Chem C 2014;2:3577–82.

[7] Tekin E, Smith PJ, Schubert US. Inkjet printing as a deposition and patterning tool for polymers and inorganic particles. Soft Matter 2008;4:703–13.

[8] Gonzalez-Macia L, Morrin A, Smyth MR, Killard AJ. Advanced printing and deposition methodologies for the fabrication of biosensors and biodevices. Analyst 2010;135:845–67.

[9] Jones N. Science in three dimensions: the print revolution, three-dimensional printers are opening up new worlds to research. Nature 2012;487:22–3.

[10] Zheng Y, He Z-Z, Yang J, Liu J. Personal electronics printing via tapping mode composite liquid metal ink delivery and adhesion mechanism. Sci Rep 2014;4:4588.

[11] Murphy SV, Atala A. 3D bioprinting of tissues and organs. Nat Biotechnol 2014;32:773–85.

[12] Durmus NG, Tasoglu S, Demirci U. Bioprinting: functional droplet networks. Nat Mater 2013;12:478–9.

[13] Wüst S, Müller R, Hofmann S. Controlled positioning of cells in biomaterials – approaches towards 3D tissue printing. J Funct Biomater 2011;2:119–54.

[14] Nakamura M, Iwanaga S, Henmi C, Arai K, Nishiyama Y. Biomatrices and biomaterials for future developments of bioprinting and biofabrication. Biofabrication 2010;2. 014110.

[15] Reardon S, Cyranoski D. Japan stem-cell trial stirs envy. Nature 2014;513:287–8.

[16] Langer R, Vacanti JP. Tissue engineering. Science 1993;260: 920–6.

[17] Griffith LG, Naughton G. Tissue engineering – current challenges and expanding opportunities. Science 2002;295:1009–14.

[18] Nakamura M, Kobayashi A, Takagi F, Watanabe A, Hiruma Y, Ohuchi K, et al. Biocompatible inkjet printing technique for designed seeding of individual living cells. Tissue Eng 2005;11: 1658–66.

[19] Nakamura M, Henmi C, Sasaki K, Iwanaga S. Indirect cell printing for fabrication of engineered capillary vessels. In: Proc. International Conference on Biofabrication. Philadelphia, USA; 2010.

[20] Roth EA, Xu T, Das M, Gregory C, Hickman JJ, Boland T. Inkjet printing for high-throughput cell patterning. Biomaterials 2004;25:3707–15.

[21] Ilkhanizadeh S, Teixeira AI, Hermanson O. Inkjet printing of macromolecules on hydrogels to steer neural stem cell differentiation. Biomaterials 2007;28:3936–43.

[22] Phillippi JA, Miller E, Weiss L, Huard J, Waggoner A, Campbell P. Microenvironments engineered by inkjet bioprinting spatially direct adult stem cells toward muscle- and bone-like subpopulations. Stem Cells 2008;26:127–34.

[23] Wilson WC Jr, Boland T. Cell and organ printing 1: protein and cell printers. The Anatomical Record Part A 2003;272:491–6.

[24] Klebe RJ. Cytoscribing: a method for micropositioning cells and the construction of two- and three-dimensional synthetic tissues. Exp Cell Res 1988;179:362–73.

[25] Sanjana NE, Fuller SB. A fast flexible ink-jet printing method for patterning dissociated neurons in culture. J Neurosci Meth 2004;136:151–63.

[26] Watanabe K, Miyazaki T, Matsuda R. Growth factor array fabrication using a color ink jet printer. Zoolog Sci 2003;20:429–34.

[27] Kim JD, Choi JS, Kim BS, Choi YC, Cho YW. Piezoelectric ink-jet printing of polymers: stem cell patterning on polymer substrates. Polymer 2010;51:2147–54.

[28] Cai K, Dong H, Chen C, Yang L, Jandt KD, Deng L. Inkjet printing of laminin gradient to investigate endothelial cellular alignment. Colloids Surf B 2009;72:230–5.

[29] Khatiwala C, Law R, Shepherd B, Dorfman S, Csete M. 3D cell bioprinting for regenerative medicine research and therapies. Gene Ther Regul 2012;7:1230004.

[30] Miller ED, Phillippi JA, Fisher GW, Campbell PG, Walker LM, Weiss LE. Inkjet printing of growth factor concentration gradients and combinatorial arrays immobilized on biologically-relevant substrates. Com Chem High T Scr 2009;12:604–18.

[31] Goldmann T, Gonzalez JS. DNA-printing: utilization of a standard inkjet printer for the transfer of nucleic acids to solid supports. J Biochem Biophys Methods 2000;42:105–10.

[32] Xu T, Rohozinski J, Zhao W, Moorefield EC, Atala A, Yoo JJ. Inkjet-mediated gene transfection into living cells combined with targeted delivery. Tissue Eng Part A 2009;15:95–101.

[33] Kato K, Umezawa K, Miyake M, Miyake J, Nagamune T. Transfection microarray of nonadherent cells on an oleyl poly(ethylene glycol) ether-modified glass slide. BioTechniques 2004;37:444–52.

[34] Matsusaki M, Sakaue K, Kadowaki K, Akashi M. Three-dimensional human tissue chips fabricated by rapid and automatic inkjet cell printing. Adv Healthcare Mater 2013;2:534–9.

[35] Saijo H, Igawa K, Kanno Y, Mori Y, Kondo K, Shimizu K, et al. Maxillofacial reconstruction using custom-made artificial bones fabricated by inkjet printing technology. J Artif Organs 2009;12:200–5.

[36] Saijo H, Kanno Y, Mori Y, Suzuki S, Ohkubo K, Chikazu D, et al. A novel method for designing and fabricating custom-made artificial bones. Int J Oral Maxillofac Surg 2011;40:955–60.

[37] Hynes WF, Doty NJ, Zarembinski TI, Schwartz MP, Toepke MW, Murphy WL, et al. Micropatterning of 3D microenvironments for living biosensor applications. Biosensors (Basel) 2014;4:28–44.

[38] Biase MD, Saunders RE, Tirelli N, Derby B. Inkjet printing and cell seeding thermoreversible photocurable gel structures. Soft Matter 2011;7:2639–46.

[39] Lin H, Zhang D, Alexander PG, Yang G, Tan J, Cheng AW, et al. Application of visible light-based projection stereolithography for live cell-scaffold fabrication with designed architecture. Biomaterials 2013;34:331–9.

[40] Skardal A, Zhang J, McCoard L, Xu X, Oottamasathien S, Prestwich GD. Photocrosslinkable hyaluronan-gelatin hydrogels for two-step bioprinting. Tissue Eng Part A 2010;16:2675–85.

[41] Bertassoni LE, Cardoso JC, Manoharan V, Cristino AL, Bhise NS, Araujo WA, et al. Direct-write bioprinting of cell-laden methacrylated gelatin hydrogels. Biofabrication 2014;6:024105.

[42] Hockaday LA, Kang KH, Colangelo NW, Cheung PYC, Duan B, Malone E, et al. Rapid 3D printing of anatomically accurate and mechanically heterogeneous aortic valve hydrogel scaffolds. Biofabrication 2012;4:035005.

[43] Iwami K, Noda T, Ishida K, Morishima K, Nakamura M, Umeda N. Bio rapid prototyping by extruding/aspirating/refilling thermoreversible hydrogel. Biofabrication 2010;2:014108.

[44] Iwanaga S, Arai K, Nakamura M. Novel approach for fabricating 3-D tissues by printing biomatters. In: Proc. PACIFICHEM 2010. Hawaii, USA. 2010.

[45] Boland T, Tao X, Damon BJ, Manley B, Kesari P, Jalota S, et al. Drop-on-demand printing of cells and materials for designer tissue constructs. Mat Sci Eng C 2007;27:372–6.

[46] Nishiyama Y, Nakamura M, Henmi C, Yamaguchi K, Mochizuki S, Nakagawa H, et al. Development of a three-dimensional bioprinter: construction of cell supporting structures using hydrogel and state-of-the-art inkjet technology. J Biomech Eng 2009;131:035001.

[47] Arai K, Iwanaga S, Toda H, Genci C, Nishiyama Y, Nakamura M. Three-dimensional inkjet biofabrication based on designed images. Biofabrication 2011;3:034113.

[48] Saunders RE, Gough JE, Derby B. Delivery of human fibroblast cells by piezoelectric drop-on-demand inkjet printing. Biomaterials 2008;29:193–203.

[49] Henmi C, Nakamura M, Nishiyama Y, Yamaguchi K, Mochizuki S, Takiura K, et al. New approaches for tissue engineering: three dimensional cell patterning using inkjet technology. Inflamm Regen 2008;28:36–40.

[50] Calvert P. Printing cells. Science 2007;318:208–9.

[51] Xu T, Zhao W, Zhu J-M, Albanna MZ, Yoo JJ, Atala A. Complex heterogeneous tissue constructs containing multiple cell types prepared by inkjet printing technology. Biomaterials 2013;34:130–9.

[52] Cui X, Breitenkamp K, Lotz M, D'Lima D. Synergistic action of fibroblast growth factor-2 and transforming growth factor-beta1 enhances bioprinted human neocartilage formation. Biotech Bioeng 2012;109:2357–68.

[53] Pati F, Jang J, Ha DH, Kim SW, Rhie JW, Shim JH, et al. Printing three-dimensional tissue analogues with decellularized extracellular matrix bioink. Nat Commun 2014;5:3935.

[54] Delie F. Evaluation of nano- and microparticle uptake by the gastrointestinal tract. Adv Drug Deliv Rev 1998;34:221–33.

[55] Wischke C, Schwendeman SP. Principles of encapsulating hydrophobic drugs in PLA/PLGA microparticles. Int J Pharm 2008;364:298–327.

[56] Hernández RM, Orive G, Murua A, Pedraz JL. Microcapsules and microcarriers for *in situ* cell delivery. Adv Drug Deliv Rev 2010;62:711–30.

[57] Serda RE, Godin B, Blanco E, Chiappini C, Ferrari M. Multi-stage delivery nano-particle systems for therapeutic applications. Biochim Biophys Acta 2011;1810:317–29.

[58] Wang X, Wu G, Lu C, Zhao W, Wang Y, Fan Y, et al. A novel delivery system of doxorubicin with high load and pH-responsive release from the nanoparticles of poly (α,β-aspartic acid) derivative. Eur J Pharm Sci 2012;47:256–64.

[59] Kato KS, Ishikura K, Oshima Y, Tada M, Suzuki T, Watabe AI, et al. Evaluation of intracellular trafficking and clearance from HeLa cells of doxorubicin-bound block copolymers. Int J Pharm 2012;423:401–9.

[60] Nakamura J, Nakajima N, Matsumura K, Hyon SH. *In vivo* cancer targeting of water-soluble taxol by folic acid immobilization. J Nanomed Nanotechnol 2011;2:1000106.

[61] Licciardi M, Stefano MD, Craparo EF, Amato G, Fontana G, Cavallaro G, et al. PHEA-graft-polybutylmethacrylate copolymer microparticles for delivery of hydrophobic drugs. Int J Pharm 2012;433:16–24.

[62] Lee WL, Seh YC, Widjaja E, Chong HC, Tan NS, Loo SCJ. Fabrication and drug release study of double-layered microparticles of various sizes. J Pharm Sci 2012;101:2787–97.

[63] Chung JE, Yokoyama M, Okano T. Inner core segment design for drug delivery control of thermo-responsive polymeric micelles. J Controlled Release 2000;65:93–103.

[64] Shimizu T, Kishida T, Hasegawa U, Ueda Y, Imanishi J, Yamagishi H, et al. Nanogel DDS enables sustained release of IL-12 for tumor immunotherapy. Biochem Biophys Res Commun 2008;367:330–5.

[65] Siepmann J, Faisant N, Akiki J, Richard J, Benoit JP. Effect of the size of biodegradable microparticles on drug release: experiment and theory. J Controlled Release 2004;96:123–34.

[66] Nakamura K, Nara E, Akiyama Y. Development of an oral sustained release drug delivery system utilizing pH-dependent swelling of carboxyvinyl polymer. J Controlled Release 2006;111:309–15.

[67] Suedee R, Jantarat C, Lindner W, Viernstein H, Songkro S, Srichana T. Development of a pH-responsive drug delivery system for enantioselective-controlled delivery of racemic drugs. J Controlled Release 2010;142:122–31.

[68] Lassalle V, Ferreira ML. PLGA based drug delivery systems (DDS) for the sustained release of insulin: insight into the protein/polyester interactions and the insulin release behavior. J Chem Technol Biotechnol 2010;85:1588–96.

[69] Doane T, Burda C. Nanoparticle mediated non-covalent drug delivery. Adv Drug Deliv Rev 2013;65:607–21.

[70] Golomb G, Fisher P, Rahamim E. The relationship between drug release rate, particle size and swelling of silicone matrices. J Control Release 1990;12:121–32.

[71] Chorny M, Fishbein I, Danenberg HD, Golomb G. Lipophilic drug loaded nanospheres prepared by nanoprecipitation: effect of formulation variables on size, drug recovery and release kinetics. J Controlled Release 2002;83:389–400.

[72] Berkland C, King M, Cox A, Kim KK, Pack DW. Precise control of PLG microsphere size provides enhanced control of drug release rate. J Controlled Release 2002;82:137–47.

[73] Andersson J, Rosenholm J, Areva S, Lindén M. Influences of material characteristics on ibuprofen drug loading and release profiles from ordered micro- and mesoporous silica matrices. Chem Mater 2004;16:4160–7.

[74] Thibert R, Akbarieh M, Tawashi R. Morphic features variation of solid particles after size reduction: sonification compared to jet mill grinding. Int J Pharm 1988;47:171–7.

[75] Kürti L, Kukovecz Á, Kozma G, Ambrus R, Deli MA, Révész PS. Study of the parameters influencing the co-grinding process for the production of meloxicam nanoparticles. Powder Technol 2011;212:210–7.

[76] Hamidi HA, Edwards AA, Mohammad MA, Nokhodchi A. Glucosamine HCl as a new carrier for improved dissolution behaviour: effect of grinding. Colloid Surf B 2010;81:96–109.

[77] Schafroth N, Arpagaus C, Jadhav UY, Makne S, Douroumis D. Nano and microparticle engineering of water insoluble drugs using a novel spray-drying process. Colloid Surf B 2012;90:8–15.

[78] Broichsitter MB, Schweiger C, Schmehl T, Gessler T, Seeger W, Kissel T. Characterization of novel spray-dried polymeric particles for controlled pulmonary drug delivery. J Controlled Release 2012;158:329–35.

[79] Fu N, Zhou Z, Jones TB, Tan TTY, Wu WD, Lin SX, et al. Production of monodisperse epigallocatechin gallate (EGCG) microparticles by spray drying for high antioxidant activity retention. Int J Pharm 2011;413:155–66.

[80] Alipour S, Montaseri H, Tafaghodi M. Preparation and characterization of biodegradable paclitaxel loaded alginate microparticles for pulmonary delivery. Colloid Surf B 2010;81:521–9.

[81] Reis CP, Neufeld RJ, Vilela S, Ribeiro AJ, Veiga F. Review and current status of emulsion/dispersion technology using an internal gelation process for the design of alginate particles. J Microencapsul 2006;23:245–57.

[82] Lee BK, Yun YH, Choi JS, Choi YC, Kim JD, Cho YW. Fabrication of drug-loaded polymer microparticles with arbitrary geometries using a piezoelectric inkjet printing system. Int J Pharm 2012;427:305–10.

[83] Boehm RD, Miller PR, Daniels J, Stafslien S, Narayan RJ. Inkjet printing for pharmaceutical applications. Mater Today 2014;17:247–52.

[84] Gittard SD, Ovsianikov A, Monteiro-Riviere NA, Lusk J, Morel P, Minghetti P, et al. Fabrication of polymer microneedles using a two-photon polymerization and micromolding process. J Diabetes Sci Technol 2009;3:304–11.

[85] Gittard SD, Ovsianikov A, Chichkov BN, Doraiswamy A, Narayan RJ. Two-photon polymerization of microneedles for transdermal drug delivery. Exp Opin Drug Deliv 2010;4:513–33.

[86] Miller PR, Gittard SD, Edwards TL, Lopez DM, Xiao X, Wheeler DR, et al. Integrated carbon fiber electrodes within hollow polymer microneedles for transdermal electrochemical sensing. Biomicrofluidics 2011;5:013415.

[87] Windmiller JR, Zhou N, Chuang M-C, Valdés-Ramírez G, Santhosh P, Miller PR, et al. Microneedle array-based carbon paste amperometric sensors and biosensors. Analyst 2011;136:1846–51.

[88] Miller PR, Skoog SA, Edwards TL, Lopez DM, Wheeler DR, Arango DC, et al. Multiplexed microneedle-based biosensor array for characterization of metabolic acidosis. Talanta 2012;88:739–42.

[89] Iwanaga S, Saito N, Sanae H, Nakamura M. Facile fabrication of uniform size-controlled microparticles and potentiality for tandem drug delivery system of micro/nanoparticles. Colloids Surf B 2013;109:301–6.

Chapter 5

In Vivo and *In Situ* Biofabrication by Laser-Assisted Bioprinting

Sylvain Catros, Virginie Keriquel, Jean-Christophe Fricain, and Fabien Guillemot
Tissue Bioengineering, University Bordeaux Segalen, Bordeaux, France

Chapter Outline

ABSTRACT

Bioprinting represents a broad spectrum of methods and devices dedicated to micron-scale organization of biological elements. In general, it aims to manipulate cells and biomaterials for the microfabrication of spatially organized three-dimensional structures. These emerging methods can easily be applied to tissue engineering for the fabrication of composite tissues or preclinical models for tissue regeneration. In the perspective of computer-assisted medical interventions, we present here a novel approach of bioprinting consisting in the layer-by-layer deposit of a biomaterial *in vivo* and *in situ*. First, the concept and the critical steps of this approach are detailed, then preliminary results of laser-assisted bioprinting (LAB) of nanohydroxyapatite (nHA) in mice calvarial defects are exposed. A custom workstation dedicated to LAB has been designed with a specific mouse holder. The innocuousness of the procedure on the animals was evidenced, then 30 stacked layers of nHA were successfully printed in mice calvarial defects. Based on histological decalcified sections and microcomputed tomography quantifications, we have observed bone regeneration after 4 weeks in the target area. Although heterogeneous, these preliminary results demonstrate that *in vivo* and *in situ* bioprinting is possible. Bioprinting may prove to be helpful in the future for medical robotics and computer-assisted medical interventions.

Keywords: biofabrication; laser-assisted bioprinting; *in vivo*; tissue engineering

1 MERGING COMPUTER-ASSISTED SURGERY AND BIOFABRICATION

The development of computer-assisted surgery (CAS) emerged after the convergence of medicine, biomaterials science, informatics, and robotics [1–3]. CAS allows the surgeon to plan, simulate, and execute mini-invasive medical interventions accurately and safely.

In parallel, tissue engineering was proposed to face the limitations of standard clinical methods for tissue repair, which are mainly due to the limited amount of autografts and to the poor integration of natural or synthetic materials. The principle of this approach is to apply the methods of engineering and life sciences toward the fundamental understanding and development of biological substitutes to restore, maintain, and improve human tissue functions [4]. The schematic sequence of tissue engineering comprises the design of a scaffold, in which cells are seeded and cultured in a controlled environment, before the implantation of the mature material.

In this concept, biofabrication has emerged as an important technological progress in tissue engineering because it allows tissue engineers to precisely control histoarchitecture and thus to guide cell assembly and tissue morphogenesis [5,6]. Among biofabrication technologies, bioprinting methods are appealing because they allow the generation of three-dimensional (3D) complex structures. It has been thus defined as "the use of computer-aided transfer processes for patterning and assembling living and nonliving materials with a prescribed 2D or 3D organization in order to produce bioengineered structures serving in regenerative medicine, pharmacokinetic, and basic cell biology studies" [7,8]. Bioprinting represents a conceptual shift because it aims to organize the individual components of a tissue during its fabrication through the layer-by-layer deposit of biological materials products, and before its maturation.

Inkjet printers have successfully been applied to pattern biological assemblies [9–11]. Pressure-operated mechanical extruders were used to handle living cells and cell aggregates [12]. At the same time, laser-assisted printing technologies have emerged as alternative methods for assembling and micropatterning biomaterials and cells. Laser-assisted bioprinting (LAB) of biological materials and living cells is based on the laser-induced forward-transfer technique, in which a pulsed laser is used to induce the transfer of material from a source film [13,14]. Under suitable irradiation conditions, and for liquids presenting a wide range of rheologies, the material can be deposited in the form of well-defined circular droplets with a high degree of spatial resolution [15–17].

After the first conceptualization of the potential use of bioprinting technologies for computed assisted surgeries [18], our goal was to establish a relevant model for *in vivo* and *in situ* biofabrication.

The aim of this article is to introduce bioprinting technology for *in vivo* and *in situ* tissue reconstruction in a preclinical model of bone regeneration. In this objective, we first create a custom bioprinting system, and then we elaborate a dedicated process for *in vivo* and *in situ* printing.

Finally, we present some preliminary results dealing with *in vivo* and *in situ* printing of nanohydroxyapatite (nHA) into mice calvarial defects.

2 CUSTOMIZED BIOPRINTING SYSTEM FOR *IN VIVO* AND *IN SITU* INTERVENTIONS

A workstation dedicated to LAB has been elaborated, as described in Ref. [19] (Fig. 5.1). This set-up was composed of a solid Nd:YAG crystal laser (Navigator I, Newport Spectra Physics, $\lambda = 1064$ nm, $\tau = 30$ ns, $f = 1$–100 kHz, $P = 7$ W) and a 5-axes positioning system. The laser beam was driven over the ribbon surface using a high-speed scanning system (up to 2000 mm/s) composed of two galvanometric mirrors (SCANgine 14, ScanLab) and a large field optical F-theta lens (S4LFT, Sill Optics, France) ($F = 58$ mm). The ribbon was a quartz blade, which was coated with a thin absorbing layer of titanium (60 nm), deposited with a high vacuum titanium coater. Then, 30 μL of bioink were homogeneously spread on the ribbon surface with a "doctor blade" device (Film Applicator 3570, Elcometer, France) to obtain a 30-μm thick layer. Droplets ejection from the substrate occurred after the interaction of the laser beam with the sacrificial layer (titanium coating) and the creation of a vapor bubble that induced the formation of a jet of bioink (Fig. 5.2). More details on bioinks, ribbon preparation, and droplet ejection mechanism by LAB are presented in Refs [19,20].

Focal settings on the ribbon and substrate positioning in three dimensions were carried out with a charge-coupled device (CCD) camera through the optical scanning system. Substrate positioning, carousel driving, video observation,

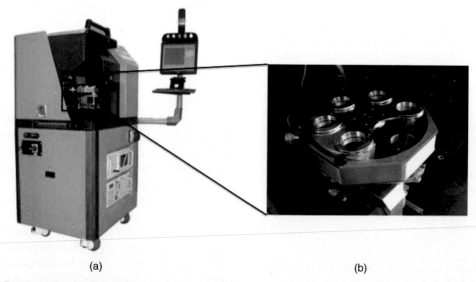

(a)   (b)

FIGURE 5.1   LAB Set-up. (a) General view of the set-up for LAB experiment, and (b) magnified picture of the carousel holder with a loading capacity of five different ribbons.

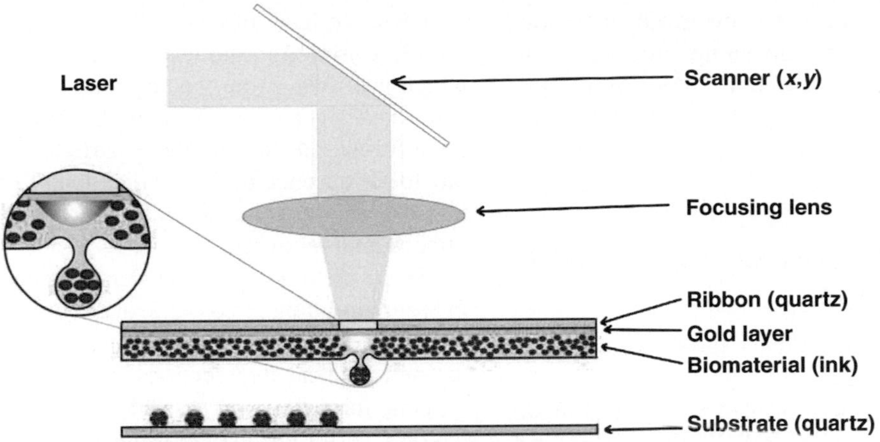

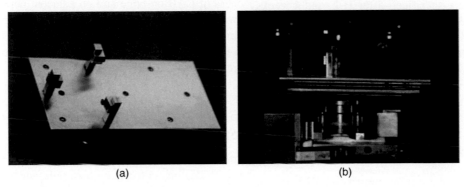

Laser

Scanner (*x*,*y*)

Focusing lens

Ribbon (quartz)
Gold layer
Biomaterial (ink)

Substrate (quartz)

FIGURE 5.2   Laser-assisted bioprinting principle. *(Figure reprinted with permission from Catros et al. [35].)*

(a)   (b)

FIGURE 5.3   Mouse holder for *in vivo* and *in situ* laser-assisted bioprinting.

and pattern designs were computer-driven using a dedicated software developed by Novalase S.A. (Canéjan, France).

For *in vivo* and *in situ* experiments, a specific mouse holder (Fig. 5.3), inspired from a stereotactic headholder for rodents, was designed and adapted to the LAB set-up. It allowed inserting a living mouse inside the printing device, instead of the conventional quartz receiving substrate for LAB. The target area for bioprinting was visualized thanks to the CCD camera [21].

3   PROCEDURE FOR *IN VIVO* AND *IN SITU* BIOPRINTING INTO BONE CALVARIAL DEFECTS

The bone calvarial defects were prepared (Fig. 5.4), then the mouse was placed into its holder, and introduced inside the workstation equipped with (*x*, *y*, *z*) motorized translation stages.

The calvarial defects were visualized and focused by translating mouse holder according to the *z*-axis, thanks to the video system (CCD camera in live mode). Dura mater surface was recorded into the software as the physical position of the substrate following *z*-axis. The center of the

FIGURE 5.4   Bilateral calvarial defects in mouse calvaria.

right defect was then targeted, inducing the translation of the mouse holder according to (*x*, *y*) axes. This final position was recorded as the origin of the printed pattern.

The pattern was computed in the software and coupled with laser parameters (power, frequency), the scanning speed and the printing gap distance. Three-dimensional printing was performed by reproducing this pattern several times to stack the printed material layer-by-layer. Between each layer, the holder was shifted automatically downward following *z*-axis, at a distance depending on it. At the end of the process, the holder was lowered at a suitable distance to avoid any contact between the mouse and the ribbon. The mouse was removed from its holder, sutured, and given conventional postoperative care.

4 PROOF OF CONCEPT OF *IN VIVO* AND *IN SITU* BIOPRINTING

The detailed experiment that validated this concept was previously published in Ref. [22].

Briefly, a nHA bioink was printed *in vivo* and *in situ* in bone calvarial defects of 30 OF-1 mice. Every contralateral defect was left empty. In order to obtain a 3D structure, 30 layers of nHA were stacked inside each defect. The animals were sacrificed after 1, 2, and 4 weeks and the samples were processed for Micro Scanner analysis and decalcified histological sections (stained with hematein–eosin–safran).

The pattern was a 3 mm diameter disk of nHA bioink printed with specific parameters: the laser energy was 12 µJ per pulse (with a 40-µm spot size), the frequency was 5 kHz, the scanner speed was 200 mm/s and the printing gap was 1500 µm. This pattern was stacked 30 times to complete 3D printing, and the mouse holder was lowered 20 µm between each layer.

The detailed results of our pilot study using nHA printed in mice calvarial defects are presented in Ref. [22].

The first objective of this work was to evaluate the feasibility of the method and how the animals could withstand this experiment. Twenty-nine out of 30 animals recovered the *in vivo* and *in situ* bioprinting process in calvarial defects, without any neurological disorders. None of them showed signs of infection or malignant disorders, as evidenced by micromagnetic resonance imaging, microscanner, and histology follow-up. This absence of side effects of the *in vivo* and *in situ* bioprinting in mice using LAB validated the feasibility of our concept, and the main hypothesis of the study.

The second objective of this study was to evaluate bone regeneration of the critical sized calvarial defects treated by 30 layers of nHA printed by LAB. Macroscopic observations done at the end of the printing process revealed the deposit of nHA inside the target areas. Histological observations of decalcified sections were done at three time points and they revealed after 1 week the presence of nHA in close contact to the dura mater, after 1 month the presence of immature bone in the defects, and after 3 months mature bone tissue in the defects (Fig. 5.5). These observations were confirmed by microcomputed tomography quantifications that showed a tendency of a higher bone surface in test than in contralateral control sites (empty defects). However, in some

samples, we have observed by histology a migration of the nHA printed in the test side toward the control side, which was away from the target area. This could explain some heterogeneous effect of the printing observed on bone formation. Indeed, these particles may have moved due to the absence of immobilization of printed materials into the recipient site and/or pressures applied on the skin after animal recovery.

These preliminary results demonstrate that *in vivo* and *in situ* bioprinting is possible, and may find helpful future applications in the field of medical robotics and computer-assisted medical interventions. Despite inconstant effects on bone formation, *in vivo* and *in situ* bioprinting of nHA has been performed in mice using a custom LAB set-up. We have established the proof of concept of *in vivo* and *in situ* bioprinting. First, we have demonstrated the innocuousness of LAB on bone and on adjacent neural tissues. nHA was successfully printed in the targeted area and resulted in the regeneration of bone in the test sites. However, some of the printed material migrated and led to heterogeneous results in some samples.

5 FROM *IN VITRO* TO *IN VIVO* AND *IN SITU* BIOFABRICATION

In the following section, we describe how *in vitro* experiments dealing with cell, biomaterials, and three-dimensional printing could be translated in the future to *in vivo* and *in situ* biofabrication.

5.1 *In Vitro* Bioprinting of Biomaterials and Human Cells

Numerous studies have shown successful LAB of several prokaryotic and eukaryotic cells, which is comprehensively reviewed in Ref. [23]. Indeed, no alteration of cell proliferation and differentiation or DNA damage has been reported using LAB, as compared to conventional cell seeding. Considering human primary cells, the following types have been printed by LAB: human umbilical vein endothelial cells and human umbilical vein smooth muscle cells [24,25], human mesenchymal stem cells [25–27], adipose-tissue derived stem cells, and endothelial colony-forming cells [28], as well as human bone-marrow derived osteoprogenitors [29].

It is noteworthy that these cells have been systematically embedded in specific bioinks: culture medium alone [24], in combination with sodium alginate [30], thrombin [17], combination of hyaluronic acid and fibrinogen [28], or a combination of blood plasma and sodium alginate [26,31]. Culture medium supplemented with sodium alginate, or hydrogels like Collagen type I and Matrigel™ have been used as well.

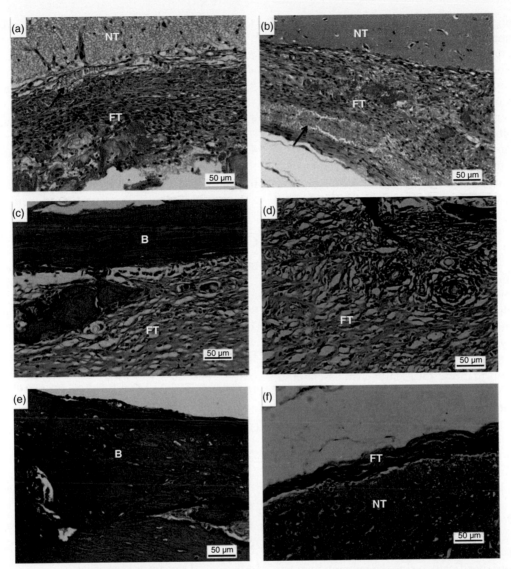

FIGURE 5.5   Decalcified histology of mouse calvaria after 1 week (a, b), 1 month (c, d), and 3 months (e, f) of healing. (a) Test site, 1 week: laser-printed nHA (arrow) in close contact with the brain surface. (b) Control site, 1 week: some nHA (arrow) is seen at a distance from the brain surface. (c) Test site, 1 month: mature and immature bones are observed. (d) Control site, 1 month: fibrous tissues are present in the defect. (e) Test site, 3 months: mature bone tissue repaired the entire defect in this case. (f) Control site, 3 months: no bone tissue is present in the center of the defect. B, bone; NT, nervous tissues; FT, fibrous tissues. *(Figure reprinted with permission from Keriquel et al. [22].)*

5.2   Three-Dimensional Printing Strategies

Three-dimensional printing of tissue constructs *in vitro* remains a challenging task. Tissue engineering applications including hybrid structures for tissue repair demand the management of perfusion and histological complexity. Such applications require a sophisticated construct with defined properties throughout its entire volume [6,7]. At first sight LAB seems not well suited for building large volume (cm³ size) tissue structures, since the characteristic droplet volume is in the order of 1 pL [19]. Some materials can be used in a layer-by-layer approach to provide volume and or biochemical properties that the bioink may not supply, to stabilize the pattern of the printed cells

and support the construct in its whole. The layer-by-layer approach has been done using hydrogels [26] or biopapers [32].

5.3   Future of *In Vivo* and *In Situ* Bioprinting and Biofabrication

In our pilot study of *in vivo* and *in situ* printing, the calvarial defect model in mice was chosen because it was critical to avoid any contact of the LAB printing device (ribbon surface) with adjacent tissues, thus we have finally selected a printing site in the superficial tissues. In subsequent similar studies, superficial organs as the skin of nude mice [33] or entire microorganisms [34] have been selected. Even if several clinical applications exist for the regeneration of superficial organs, future developments of the printing set-ups and methods are needed for the *in vivo* and *in situ* regeneration of internal organs. We have chosen a nHA bioink to print in the calvarial defect model because it is the main mineral component of bone. It is also completely biocompatible and resorbable in the nonsintered state. The composition of the solution and the printing parameters were adapted for printing by LAB [35]. Moreover, it was possible to follow the fate of the material during time on histological sections and we have observed bone regeneration in the area where nHA was printed. In general, for *in vivo* printing, the nature and the rheological properties of the bioink should be adapted to the tissue and to the bioprinting method, respectively.

Concerning the migration of patterns after completion of the printing process, it is a common complication observed *in vitro* that we have confirmed here *in vivo*. The first way to limit the migration of patterns after printing is to spread a hydrogel layer on the receiving substrate before printing: this layer increases printed cell survival by providing a soft mattress [30] and is also essential to keep the pattern shape by blocking the cells after their penetration inside the gel. It is also possible to maintain cell position and printed pattern shape by using biopapers stacked above the pattern [32]. Future protocols that will be developed for *in vivo* and *in situ* bioprinting should include a system to maintain the shape of printed materials among time and to limit their migration out of the target area.

The integration of bioprinting technology in current medical robots will improve their spatial resolution, and the volume of material or drug deposited could be highly reduced [17]. More specifically, using the integration of laser technologies in these devices for assisted surgery would allow combining other laser-based methods for treating patients like tissue ablation or polymerization of implanted polymeric materials.

Interestingly, the results described here have already been reproduced for the regeneration of different tissues, using relevant biological materials and dedicated bioprinting technologies. Syringe extruders were used to print fibroblast in the injured skin of nude mice [33] and two-photon polymerization was used to immobilize microorganisms inside a scaffold made of a photopolymerizable hydrogel [34].

6 CONCLUSIONS

The technical procedure described here could be reproduced with several bioprinting set-ups like inkjet printing or other dispensing methods [6]. In this regard, the recurrent steps are as follows:

1. To provide a bioprinting device adapted to deposit a pattern of specific biological material.
2. To select a living substrate that includes a tissue target accessible for bioprinting.
3. To orientate this target with respect to the bioprinting device.
4. To image the substrate and to reveal the target features (size, shape, type of tissue, etc.).
5. To generate the bioprinting pattern corresponding to these features.
6. And finally to deposit the biological material onto the target, according to the determined pattern.

GLOSSARY

Biofabrication Technologies developed for the manufacturing of complex living or nonliving bioengineered microarrays, scaffolds, tissue models or substitutes.

Bioprinting Group of computer-controlled technologies allowing micron-scale organization of biological elements, based on previously designed patterns.

Tissue Engineering Applications of methods and principles of life sciences, material sciences and physics for the design and fabrication of experimental models in basic research in biology and/or for the production of customized tissue substitute for regenerative medicine.

Computer Assisted Surgery Application of computer-aided design and computer assisted manufacturing (CAD-CAM) in surgery. It involves (1) the planning of surgical interventions; (2) the training of surgeons; (3) supporting the surgeon to perform accurate interventions; (4) guiding the surgery based on predefined surgical procedure planning.

ABBREVIATIONS

LAB	Laser-assisted bioprinting
nHA	Nanohydroxyapatite
CAS	Computer-assisted surgery
2D	Two-dimensional
3D	Three-dimensional
CCD	Charge-coupled device

ACKNOWLEDGMENTS

No competing financial interests exist in this work. The authors would like to thank Région Aquitaine for funding this study.

REFERENCES

[1] Nof SY. Springer handbook of automation. Berlin: Springer; 2009.
[2] Kaladji A, Lucas A, Cardon A, Haigron P. Computer-aided surgery: concepts and applications in vascular surgery. Perspect Vasc Surg Endovasc Ther 2012;24(1):23–7.
[3] De Almeida EO, Pellizzer EP, Goiatto MC, Margonar R, Rocha EP, Freitas AC Jr, et al. Computer-guided surgery in implantology: review of basic concepts. J Craniofac Surg 2010;21(6):1917–21.

[4] Langer R, Vacanti J. Tissue engineering. Science 1993;260(5110): 920–6.

[5] Hutmacher DW, Sittinger M, Risbud MV. Scaffold-based tissue engineering: rationale for computer-aided design and solid free-form fabrication systems. Trends Biotechnol 2004;22(7):354–62.

[6] Guillotin B, Guillemot F. Cell patterning technologies for organotypic tissue fabrication. Trends Biotechnol 2011;29(4):183–90.

[7] Mironov V, Visconti RP, Kasyanov V, Forgacs G, Drake CJ, Markwald RR. Organ printing: tissue spheroids as building blocks. Biomaterials 2009;30(12):2164–74.

[8] Guillemot F, Mironov V, Nakamura M. Bioprinting is coming of age: report from the International Conference on Bioprinting and Biofabrication in Bordeaux (3B'09). Biofabrication 2010;2(1):010201.

[9] Boland T, Xu T, Damon B, Cui X. Application of inkjet printing to tissue engineering. Biotechnol J 2006;1(9):910–7.

[10] Nakamura M, Kobayashi A, Takagi F, Watanabe A, Hiruma Y, Ohuchi K, et al. Biocompatible inkjet printing technique for designed seeding of individual living cells. Tissue Eng 2005;11(11–12):1658–66.

[11] Saunders RE, Gough JE, Derby B. Delivery of human fibroblast cells by piezoelectric drop-on-demand inkjet printing. Biomaterials 2008;29(2):193–203.

[12] Jakab K, Norotte C, Marga F, Murphy K, Vunjak-Novakovic G, Forgacs G. Tissue engineering by self-assembly and bio-printing of living cells. Biofabrication 2010;2(2):022001.

[13] Young D, Auyeung RCY, Piqué A, Chrisey DB, Dlott DD. Plume and jetting regimes in a laser based forward transfer process as observed by time-resolved optical microscopy. Appl Surf Sci 2002;197–198:181–7.

[14] Bohandy J, Kim BF, Adrian FJ. Metal deposition from a supported metal film using an excimer laser. J Appl Phys 1986;60(4):1538–9.

[15] Barron JA, Wu P, Ladouceur HD, Ringeisen BR. Biological laser printing: a novel technique for creating heterogeneous 3-dimensional cell patterns. Biomed Microdevices 2004;6(2):139–47.

[16] Barron JA, Krizman DB, Ringeisen BR. Laser printing of single cells: statistical analysis, cell viability, and stress. Ann Biomed Eng 2005;33(2):121–30.

[17] Guillotin B, Souquet A, Catros S, Duocastella M, Pippenger B, Bellance S, et al. Laser assisted bioprinting of engineered tissue with high cell density and microscale organization. Biomaterials 2010;31(28):7250–6.

[18] Campbell PG, Weiss LE. Tissue engineering with the aid of inkjet printers. Expert Opin Biol Ther 2007;7(8):1123–7.

[19] Guillemot F, Souquet A, Catros S, Guillotin B, Lopez J, Faucon M, et al. High-throughput laser printing of cells and biomaterials for tissue engineering. Acta Biomater 2010;6(7):2494–500.

[20] Guillemot F, Souquet A, Catros S, Guillotin B. Laser-assisted cell printing: principle, physical parameters versus cell fate and perspectives in tissue engineering. Nanomedicine 2010;5(3):507–15.

[21] Guillemot F, Catros S, Keriquel V, Fricain J-C. Bioprinting Station, Assembly Comprising Such Bioprinting Station and Bioprinting Method. WO/2011/107599; 2011.

[22] Keriquel V, Guillemot F, Arnault I, Guillotin B, Miraux S, Amédée J, et al. *In vivo* bioprinting for computer- and robotic-assisted medical intervention: preliminary study in mice. Biofabrication 2010;2(1):014101.

[23] Schiele NR, Corr DT, Huang Y, Raof NA, Xie Y, Chrisey DB. Laser-based direct-write techniques for cell printing. Biofabrication 2010;2(3):032001.

[24] Wu PK, Ringeisen BR. Development of human umbilical vein endothelial cell (HUVEC) and human umbilical vein smooth muscle cell (HUVSMC) branch/stem structures on hydrogel layers via biological laser printing (BioLP). Biofabrication 2010;2(1): 014111.

[25] Gaebel R, Ma N, Liu J, Guan J, Koch L, Klopsch C, et al. Patterning human stem cells and endothelial cells with laser printing for cardiac regeneration. Biomaterials 2011;32:9218–30.

[26] Koch L, Kuhn S, Sorg H, Gruene M, Schlie S, Gaebel R, et al. Laser printing of skin cells and human stem cells. Tissue Eng Part C Methods 2010;16(5):847–54.

[27] Gruene M, Deiwick A, Koch L, Schlie S, Unger C, Hofmann N, et al. Laser printing of stem cells for biofabrication of scaffold-free autologous grafts. Tissue Eng Part C Methods 2010;17(1):79–87.

[28] Gruene M, Pflaum M, Hess C, Diamantouros S, Schlie S, Deiwick A, et al. Laser printing of three-dimensional multicellular arrays for studies of cell–cell and cell–environment interactions. Tissue Eng Part C Methods 2011;17(10):973–82.

[29] Catros S, Fricain J-C, Guillotin B, Pippenger B, Bareille R, Remy M, et al. Laser-assisted bioprinting for creating on-demand patterns of human osteoprogenitor cells and nano-hydroxyapatite. Biofabrication 2011;3(2):025001.

[30] Catros S, Guillotin B, Bacáková M, Fricain J-C, Guillemot F. Effect of laser energy, substrate film thickness and bioink viscosity on viability of endothelial cells printed by laser-assisted bioprinting. Appl Surf Sci 2011;257(12):5142–7.

[31] Gruene M, Pflaum M, Deiwick A, Koch L, Schlie S, Unger C, et al. Adipogenic differentiation of laser-printed 3D tissue grafts consisting of human adipose-derived stem cells. Biofabrication 2011;3(1):015005.

[32] Catros S, Guillemot F, Nandakumar A, Ziane S, Moroni L, Habibovic P, et al. Layer-by-layer tissue microfabrication supports cell proliferation *in vitro* and *in vivo*. Tissue Eng Part C Methods 2012;18(1): 62–70.

[33] Binder K, Zhao W, Dice D, Atala A, Yoo J. In situ bioprinting of the skin for burns. Seoul: Termis WC; 2009.

[34] Torgersen J, Baudrimont A, Pucher N, Stadlmann K, Cicha K, Heller C, et al. *In vivo* writing using two-photon polymerization. Proceedings of LPM2010 – The 11th International Symposium on Laser Precision Microfabrication; 2010.

[35] Catros S, Fricain J-C, Guillotin B, Pippenger B, Bareille R, Remy M, et al. Laser-assisted bioprinting for creating on-demand patterns of human osteoprogenitor cells and nano-hydroxyapatite. Biofabrication 2011;3:025001.

Chapter 6

Stereolithographic 3D Bioprinting for Biomedical Applications

Ritu Raman*,‡ and **Rashid Bashir****,†,‡

*Department of Mechanical Science and Engineering, University of Illinois at Urbana-Champaign, Urbana, IL, USA; **Department of Bioengineering, University of Illinois at Urbana-Champaign, Urbana, IL, USA; †Department of Electrical and Computer Engineering, University of Illinois at Urbana-Champaign, Urbana, IL, USA; ‡Micro and Nanotechnology Laboratory, University of Illinois at Urbana-Champaign, Urbana, IL, USA

Chapter Outline

Essentials of 3D Biofabrication and Translation. http://dx.doi.org/10.1016/B978-0-12-800972-7.00006-2

ABSTRACT

In recent years, stereolithographic fabrication has advanced greatly in the quality, resolution, and accuracy of manufactured parts. The concurrent development of photocurable resins that are biocompatible, biodegradable, and bioactive has enabled a vast array of biomedical and translation medical applications of stereolithography-based fabrication technologies. Stereolithographic techniques have been readily integrated with medical imaging technologies in order to improve disease diagnosis, preoperative planning, quality and morphology of prosthetics and implants, and functional success of complex surgeries. Furthermore, stereolithography has established itself as one of the primary enabling tools that will be useful for regenerative medicine applications in the coming years. As a whole, the versatility in design, scale, resolution, and broad applicability of stereolithographic technologies render them the ideal enabling technology for biomedical and translational medical applications.

Keywords: stereolithography; prosthetics; implants; tissue engineering; regenerative medicine

1 INTRODUCTION

Stereolithographic 3D printing is a solid freeform additive layer manufacturing technology that was pioneered by the 3D Systems manufacturing company in 1986. Originally intended for use in rapid prototyping for manufacturing sectors, such as the automotive and aeronautic industries, stereolithography revolutionized this field by eliminating the need for inefficient and expensive methods of manufacture. As the first commercially available and most popular form of solid freeform fabrication technology, stereolithography has undergone decades of further developments in efficiency and accuracy, yielding surface finish quality comparable to traditional machine milling and rendering it one of the most commercially viable additive manufacturing technologies available at present. The many advancements in this field and the advantages associated with this versatile manufacturing technology thus encourage its widespread use and adaptation to a variety of industry sectors, most interestingly applications in biomedical and translational research [1].

Stereolithography demonstrates greater versatility in scale of fabricated parts (submicron to decimeter) and has the highest fabrication accuracy and resolution as compared to other additive layer manufacturing technologies [2]. Alternative additive manufacturing apparatus, such as selective laser sintering (SLS), sheet lamination (LOM), or adhesion bonding (3DP) machines, are restricted to fabricating parts with less complex internal geometries because they are limited by unused material being trapped inside internal holes. Fused deposition modeling (FDM) apparatus, like SLA, are not limited by this, but do have high heat effects on the raw material used for fabrication [3]. The cost of fabricating parts via stereolithography is comparable to other additive manufacturing apparatus and is counterbalanced by the versatility of design, high resolution, and superior quality of parts fabricated via SLA [4].

The main limitation facing the advancement of stereolithography for translational biomedical applications is the scarcity of suitable biocompatible and biodegradable photopolymerizable liquid polymer resins. As shall be seen in this chapter, this limitation is being daily reduced by the development of a vast array of stereolithographic resin materials including families of natural and synthetic polymers, ceramics, composites, hydrogels, and even living cells. These resins have been altered and tuned to target a myriad array of biomedical applications.

This chapter outlines the use of stereolithographic manufacturing processes to fabricate a variety of models and structures that have translational applications in biomedical engineering. Section 2 introduces the physics and materials chemistry governing the stereolithographic fabrication process, with special emphasis placed on the design challenges pertinent to biofabrication applications. Section 3 elaborates on applications of stereolithographic fabrication in surgical procedures, prosthesis, and implants, introducing concepts that promise to improve upon currently existing clinical practices that are in wide use and operation. Section 4 covers the use of stereolithography to fabricate biomaterial scaffolds and tissue substitutes for novel and rapidly evolving applications in tissue engineering and regenerative medicine. A discussion of the current challenges present in the field of stereolithography, as well as the anticipated advancements in this field, is presented in Section 5. A presentation of stereolithographic biofabrication in the broader context of translational 3D biofabrication and a discussion of future research trends concludes this chapter.

2 THE STEREOLITHOGRAPHIC PROCESS

Stereolithographic systems rely on the process of photopolymerization, or light-initiated polymerization, to fabricate 3D structures. In general, stereolithographic processes can be separated into two broad categories: single-photon and multiphoton methods, which differ in the method of light excitation and absorption that trigger the polymerization process. Single-photon methods can be further divided into: (1) Visible radiation systems that employ light in the visible wavelength range; (2) "Conventional" stereolithography systems that employ ultraviolet (UV) radiation; (3) IR stereolithography systems that employ infrared (IR) radiation; (4) Stereo-thermal lithography systems that combine UV and IR radiation to initiate polymerization [5]. All these single-photon polymerization processes can be implemented using direct-laser writing, physical mask projection, or digital mask projection machines, as shall be described in later sections. Of the diverse array of stereolithographic systems just listed,

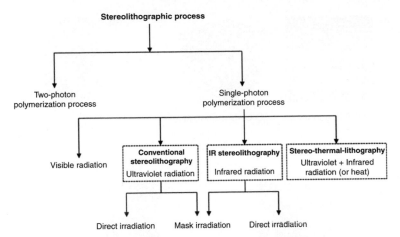

FIGURE 6.1   Broad categorization of the forms of stereolithographic processes into single-photon and multiphoton methods. Single-photon methods, which are the most commonly employed for translational biomedical applications, can be implemented using several forms of excitation radiation [5].

conventional single-photon stereolithography apparatus (SLA) are by far the most popular (Fig. 6.1).

Conventional SLA machines utilize the energy from an UV light source to drive the conversion of UV-irradiation-sensitive liquid oligomers into cross-linked solid/gel-like polymeric networks. These machines, as well as SLA machines based on visible light irradiation, possess the advantage of providing precise spatial and temporal control of reaction kinetics as they are governed purely by the ease of light manipulation. Recent advances in developing controlled light sources, such as lasers, have thus greatly enhanced the advantages of using such systems. Conventional stereolithography systems have likewise seen the greatest use and adaption in biomedical and translational research, as they have the ability to drive polymerization under physiologic conditions with minimal heat production and damage to the raw material. This is of special relevance in 3D fabrication of structures containing encapsulated living cells, as discussed in Section 4.

As with most additive layer manufacturing technologies, the first step in fabricating a 3D structure by stereolithography involves creating a digital model of the part, using computer-aided design software (CAD). Lately, advanced digital scanners have also been used to convert complex structures into virtual 3D models. This is of special interest in the context of biomedical and translational applications, as many of the scanning technologies that are used in this field have their basis in clinical imaging technologies such as magnetic resonance imaging (MRI) and computed tomography (CT). Complex structures, such as those found inside the human body, can thus be imaged and readily converted into 3D digital models for manufacturing patient-specific models, implants, and tissue-engineered replacements for damaged tissue.

In order to be prepared for 3D printing, all digital models must first be converted to a standard tessellation language file format (STL) that represents the surface geometry of the 3D model as a series of interconnecting tessellated triangles. Specialized software is then used to virtually slice the model into sequential layers of specified thickness, often determined by the user-specified part size, required resolution, and desired accuracy of the final part. The resulting information is then sent to the stereolithography apparatus, which builds the 3D model layer-by-layer sequentially from the bottom up. Schematics of various forms of SLA machines are further described in Section 2.1 (Figs 6.2 and 6.3).

2.1 Stereolithographic Fabrication Apparatus

2.1.1 Single-Photon Stereolithography

Single-photon stereolithographic fabrication processes are so termed because the process of photoinitiator excitation in this process is driven by the absorption of a single photon. Conventional UV light-based stereolithography falls under this category of photopolymerization, as does visible light-based stereolithography. The two most basic and widely adopted apparatus for single-photon photolithography in the context of biomedical applications are direct laser writing and mask-based UV light-based stereolithography.

A direct laser writing stereolithographic apparatus uses a high-energy laser to trace lines across the resin surface, serially polymerizing (i.e., "rasterizing") two-dimensional cross-sections of a three-dimensional design. Sequential polymerization of these two-dimensional cross-sections layer by layer from the bottom up, with the aid of a computer-controlled stage, drives formation of three-dimensional structures [5].

Similarly, mask-based stereolithography filters a high-energy light source through a patterned physical or digital mask, allowing for curing of an entire two-dimensional

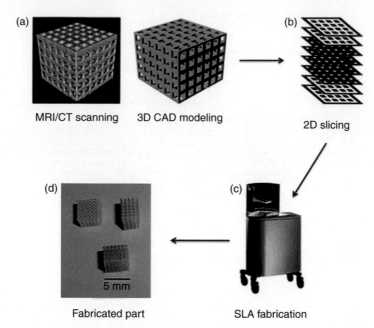

MRI/CT scanning   3D CAD modeling   2D slicing

5 mm

Fabricated part   SLA fabrication

FIGURE 6.2 Process flow for fabricating 3D parts via stereolithography. (a) Scanning to create a 3D digital image of a design or creating a 3D solid model via computer-aided design software (CAD); (b) digitally slicing the 3D model layer-by-layer into 2D sections; (c) fabricating the 3D model layer-by-layer using a stereolithographic apparatus; (d) the final fabricated part with the desired feature sizes, scale, resolution, and surface finish [1].

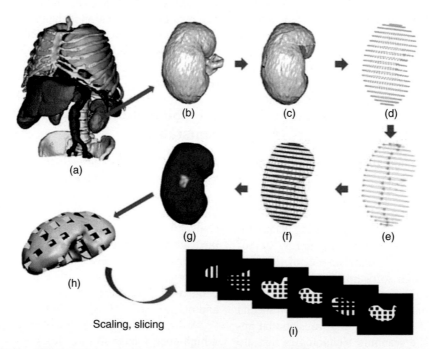

Scaling, slicing

FIGURE 6.3 Digital scanning of biological tissues with complex 3D structures is a multistep process. (a) CT imaging data are gathered; (b) point cloud data are extracted from scan using reverse engineering software; (c) point cloud data are "cleaned" to remove imaging noise/defects; (d) point cloud is sliced into 2D cross-sections; (e) point cloud layers are converted into spline curves; (f) a lofted surface is created from the splines; (g) a solid geometry model of the biological tissue is created; (h) the solid model is converted into a porous scaffold for applications in tissue engineering and regenerative medicine; (i) the porous model is scaled sliced into 2D layers and sent to an SLA machine for fabrication [36].

cross-section within a single exposure, rather than serially tracing lines as with a laser-based apparatus. This process is thus considered higher-throughput than laser-based processes, as complete layers of resin can be polymerized within a single light exposure, thereby reducing build-time. It is of importance to remember, however, that resolution of feature sizes obtainable via mask-based projection stereolithography systems is inversely proportional to the size/scale of

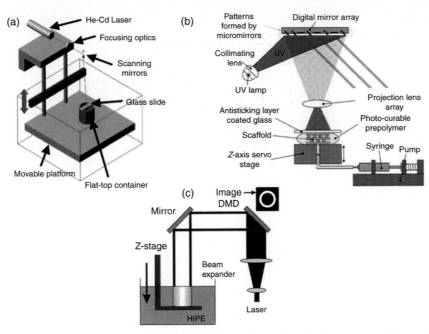

FIGURE 6.4 Schematic of (a) laser-based SLA that directly "writes" patterns in photopolymerizable resin by serially tracing 2D cross-sections of 3D designs [41]; (b) mask-based SLA that polymerizes 2D cross-sections in a single projection to create 3D designs layer-by-layer [49]; (c) schematic of SLA that combines a laser light source and digital micromirror mask device to fabricate large-scale 3D parts with complex and high-resolution features [6].

the final manufactured part. Hence, fabricating large parts with very small feature sizes can be a time-consuming process, as it will require rasterizing several 2D projections to completely polymerize an entire 2D cross-section of the 3D model. Novel SLA that employ a combination of a laser light source and mask-based projection system have been developed to target this mismatch between part size, resolution, and fabrication time, but these apparatus are yet to be widely employed [6].

Physical masks used in this type of mask-based projection photolithography are manufactured via the microfabrication approaches that were initially developed to target applications in semiconductor manufacturing. Greater design flexibility is enabled by the use of digital masks that can display a vast array of different patterns. One of the most commonly used digital masks used in this type of stereolithography is inspired by the digital light processing technology in projectors, projection-based television sets, digital signs, and digital cinema projection. The digital mask, or digital micromirror device (DMD), is comprised of an array of millions of microscopic mirrors that can be precisely and independently rotated into an "on" state or an "off" state, generating a pixelated-pattern image. Patterning and projection of light through this digital mask enables the photopolymerization of a stereolithographic resin in precisely defined patterns [7,8].

Both laser-based and mask-based SLA rely heavily on computer-controlled building stages that move the 2D polymerized cross-sections by a precisely defined amount to ensure adherence of sequential layers to one another. Understanding of the cure depth relationship of the resin in use, as well as precise calibration of the light energy of the apparatus, is thus of great importance in manufacturing 3D parts via single-photon stereolithography (Fig. 6.4).

2.1.2 Multiphoton Stereolithography

Two-photon stereolithographic fabrication processes represent the simplest case of multiphoton absorption, and involve the sequential or simultaneous absorption of two relatively low-intensity photons in order to excite a photosensitive resin to a high-energy radical state. This method of excitation depends quadratically on the incident light intensity [5], as opposed to the linear relationship for single-photon stereolithography, allowing for extremely rapid fabrication in three dimensions with submicron resolution (Fig. 6.5).

In two-photon initiated stereolithography, which was first demonstrated by Kawata and coworkers in 1997 [9], femtosecond light pulses from a near-IR laser are focused into a 3D volume/vat of liquid resin. The polymerization process is initiated in the precisely defined focal volume of the laser, limiting interaction with the resin through which the laser passes to reach the focal volume. Shifting the focal plane of the laser enables the fabrication of complex three-dimensional structures with high-resolution feature sizes [10]. Similarly, three-photon approaches to photopolymerization of 3D structures with submicron resolution have also been demonstrated [11]. There is great

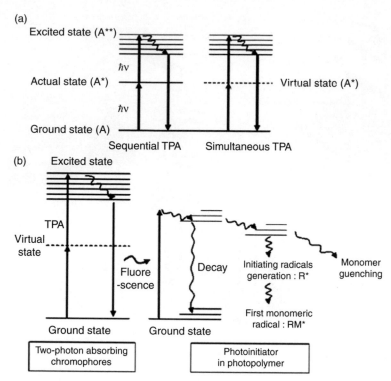

FIGURE 6.5 **Multiphoton stereolithographic polymerization involves the absorption of multiple photons in order to excite a photosensitive resin to a high-energy radical state.** (a) For two-photon polymerization, this high-energy radical state can be accomplished via sequential or simultaneous absorption of two photons. The energy of incident photons is represented by $h\nu$ (the product of Planck's constants divided by 2π and angular frequency of incident light, respectively). A and A* denote energy levels; (b) schematic diagrams of multiphoton-initiated polymerization showing valence electrons in a photoinitiator excited to a high-energy state and transforming by: (1) decaying back to photoinitiator with emission of light, (2) generating an excited state quenching by oxygen, or (3) yielding an initiator species for polymerization. All three competing processes are required for efficient photopolymerization [10].

potential for using multiphoton fabrication approaches to manufacture three-dimensional scaffolds with complex internal microarchitectures for tissue engineering applications [12]. Multiphoton methods for patterning living cells encapsulated within photosensitive resins have also been demonstrated [13]. It is to be noted that photoinitiators used for conventional stereolithography are often unsuitable for applications in two-photon initiated stereolithography (see Section 2.2.1), as these chemicals demonstrate reduced photosensitivity to light in the near-IR wavelength range [5] (Fig. 6.6).

2.1.3 Interference Stereolithography

A novel form of photolithography known as interference lithography (alternatively known as holography) is based on creating an interference pattern between multiple coherent light waves in order to create a pattern of high intensity and low intensity fringes of light [14]. Photosensitive resins exposed to this interference-derived pattern of light are thus polymerized in regions of high intensity fringes. This method can be used to create patterns with nanoscale resolution and offers the additional advantage of more rapid polymerization than attainable via conventional stereolithography [15]. However, because interference lithography

is restricted in the number and type of patterns that can be created via light interference, it is only of interest in biomedical applications that require repetitive structures [16]. For example, tissue engineering of cancellous/trabecular bone requires the formation of an ossified "spongy" scaffold with a repetitive porous structure. This type of scaffold, and other scaffolds that require similar repetitive porous structures, could be rapidly and accurately fabricated via an interference lithography fabrication apparatus (Fig. 6.7).

2.2 Stereolithographic Resins

A conventional stereolithography single-photon photofabrication apparatus uses the energy from UV light to selectively polymerize, or "cure" photosensitive liquids called resins to form solid or gel-like structures. Photosensitive resins, which are the raw materials used in conventional stereolithography, are composed primarily of polymerizable oligomers, or "prepolymers," and a radical photoinitiator. The polymerization process of these resins can be simply described in three steps: initiation, propagation, and termination. During *initiation*, components of the resin are irradiated by light and form reactive radical species. These radicals then *propagate* between polymerizable oligomers

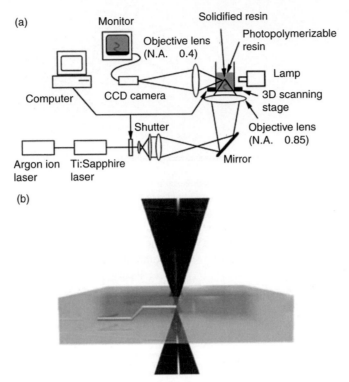

FIGURE 6.6   Multiphoton stereolithographic apparatus. (a) Optical system for two-photon laser-based photolithographic fabrication approach [9]; (b) polymerization occurs via tracing of femtosecond laser through 3D vat of photosensitive resin. Shifting the plane of laser focus shifts the plane of polymerization, enabling high-resolution fabrication of complex 3D structures [13].

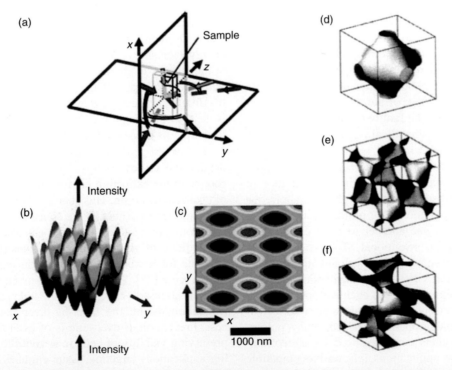

FIGURE 6.7   Interference stereolithography relies on creating interference patterns between multiple coherent light waves in order to create patterns of high and low intensity light fringes. (a) Beam and sample geometry for creating interference pattern. Arrows of the same color represent coherent beams and arrows of different colors represent mutually incoherent (i.e., orthogonally polarized) beams; (b) intensity pattern generated by laser beam interference; (c) color map of intensity pattern projected onto the x–y plane; (d) simple cubic P structure intensity map for interference stereolithographic fabrication; (e) diamond-like structure with FCC translational symmetry for interference stereolithographic fabrication; (f) gyroid-like structure with BCC translational symmetry for interference stereolithographic fabrication. (Figure 6.7a–c reprinted with permission from Tondiglia et al. [14]; Figure 6.7d–f reprinted with permission from Ullal et al. [16]).

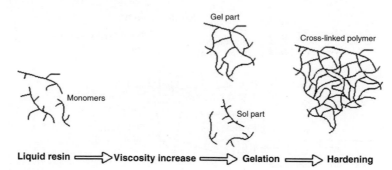

FIGURE 6.8 Schematic of cure mechanism of polymerizable oligomer. Resin is composed of liquid photosensitive resin that forms reactive radical species when irradiated with light. These radicals begin to propagate via cross-linking to form gel-like networks. Following termination of the polymerization process, a solid cross-linked network is formed [5].

to form cross-linked networks. The formation of covalent bonds in these networks *terminates* the polymerization process (Fig. 6.8).

Radical photoinitiators are a crucial component of stereolithographic resins. They belong to a family of reactive chemical compounds that decompose to form high-energy free radicals, or molecules with unpaired valence electrons, when exposed to light. In single-photon stereolithography processes, a single photon from the irradiating light source is sufficient to excite the photoinitiator into this high-energy state. In multiphoton stereolithography processes, the simultaneous or sequential absorption of two relatively low-intensity photons is required to excite a photosensitive resin to a high-energy radical state. In both types of processes, photoinitiators help to initiate the stereolithographic polymerization process by forming the reactive radical species that drive polymerization of oligomers in the resin material. The choice of photoinitiator determines the rate of polymerization, or curing, of the resin. As several photoinitiators are cytotoxic, the family of photoinitiators suitable for applications in biomedical engineering is limited, as discussed in Section 2.2.1.

The primary component of a stereolithographic resin is the polymerizable oligomer. Following the initiation of the polymerization process by radical photoinitiators, the functional groups of polymerizable oligomers form active radicals that react with one another to form cross-linked polymerized networks. Conventional SLA processes rely heavily upon fabrication with acrylate and methacrylate oligomers, epoxide, and vinyl ether-based resins that cure rapidly upon irradiation and whose chemical formulas can readily be modified to obtain materials with a variety of geometric and material properties. Indeed, many polymerizable oligomer systems currently used for stereolithographic biofabrication applications rely on biocompatible oligomers containing acrylate and methacrylate functional groups that form radicals in response to photoexcitation. A detailed description of the composition and material properties of the most commonly used stereolithographic resins for biomedical applications is presented in Section 2.2.2.

The process of curing photopolymerizable resins via conventional stereolithography is very well understood and is easily characterized via an adapted form of the Beer–Lambert equation, which is a relationship between the intensity of a light source and the exponential decay it experiences as it passes through an absorbing medium. This adapted equation, termed the cure-depth equation (Eq. 6.1), demonstrates the relationship between the amount of UV radiation that a liquid resin is exposed to and the depth to which the resin is cured as an effect of radiation penetration:

$$C_d = D_p \ln\left[\frac{E_{max}}{E_c}\right] \tag{6.1}$$

Where C_d represents the maximum cure depth of the resin, D_p represents the penetration depth at which UV energy intensity is reduced to $1/e$ of its maximum value at the surface of the resin (e is the mathematical constant that forms the base of the natural logarithm), E_{max} is the maximum UV energy intensity at the resin surface, and E_c is the critical energy intensity required to cure the resin by triggering its transition into a solid phase.

Resins are generally characterized by their penetration depths and critical energy. The penetration depth corresponds to the extinction coefficient, or molar absorptivity, of the irradiated resin. This characteristic value can be tuned and modified by adjusting the concentration of photoinitiator solution present in the resin. It can also be modulated by the addition of visible-light or UV-absorbing dyes, which compete for irradiation absorption during the polymerization process [17]. Values of D_p are tuned to prevent "overcure," or excessive exposure to UV light caused by repeated irradiation during the layer-by-layer manufacturing process. Prevention of over-cure is of essential importance in preserving viability of cells in stereolithographic bioprinting applications targeting tissue engineering, as excessive UV exposure has a detrimental effect on cellular viability (see Section 4). The value of the critical energy for a resin, E_c, is likewise dependent on the concentration of photoinitiator present in the solution, as well as other dissolved gases (e.g., oxygen) or liquids (e.g., colored dyes) that could compete for the absorbed light energy. The energy dose of the irradiating light source at the surface of the resin, corresponding to E_{max}, is regulated by modulating the scanning speed (for laser-based systems) or specifying the exposure time (for projection mask-based systems).

Precise calibration and modulation of the light energy dose and composition of the photosensitive resin therefore provides a simple and reliable method of controlling the cure depth. Plots of cure depth vs. energy dose for a given combination of stereolithographic apparatus and resin are termed "working curves." These working curves are used to select the conditions that provide cure depths that ensure adhesion to the fabrication support platform as well as to adjacent layers of the 3D fabricated part. Reduced cure depths provide more accurate control of thickness and provide higher quality finishes to manufactured parts. However, decreased cure depths correspond to increase number of layers and hence increased build-time.

For applications in conventional manufacturing sectors, parts manufactured via stereolithography are often subjected to a postcuring process, which involves uniform exposure to UV irradiation that helps improve the mechanical properties of the fabricated parts. This step is often not required for translational biomedical applications, as many such applications do not require extremely high mechanical strengths, but various postfabrication steps used to clean and sterilize fabricated parts have been employed, as discussed later in this chapter.

2.2.1 Radical Photoinitiators for Stereolithographic Biofabrication

Different photoinitiators are distinguished by the wavelength range in which they present high-energy absorption, and are thus readily described by their unique absorption spectra. The choice of excitation light source wavelength and photoinitiator composition are thus inextricably linked to one another. A general trend in excitation wavelength is that the shorter the wavelength of UV light, the better the resolution of fabricated feature sizes. However, this decrease in excitation wavelength is paired with lower penetration depth into the sample and hence increased build-time [18].

A variety of photoinitiators have been developed to suit a range of photopolymerization applications, but most photoinitiators are cytotoxic (i.e., cause cell death), which renders them inappropriate for use in biomedical applications. For such translational medical applications, the wavelength criterion poses a further challenge for adapting stereolithographic technology, as it is essential in such cases to limit the oxidative damage to biological components caused by exposure to high-energy irradiation. These stringent limitations severely constrain the size of the relevant family of photoinitiators that can be used in biomedical applications. However, by characterizing the cytotoxicity of relevant

photoinitiators and incorporating postprocessing steps, such as repeated "washes" and sterilizations, researchers have succeeded in fabricating 3D structures that can safely integrate with biological tissues and organs. The success of finding a range of photoinitiators that match the design criteria required for biomedical applications has demonstrated feasibility of using stereolithography for a broad range of applications including patient-specific prosthetics, implants, and scaffolds for tissue engineering.

Tissue engineering applications pose the most stringent restrictions on the photoinitiator compositions that can be incorporated in stereolithographic resins, as they require living cells to come into direct and prolonged contact with these reactive compounds. Furthermore, the viability of living cells exposed to high-intensity irradiation is very sensitive to the wavelength of the light source – cells are more viable upon irradiation of wavelengths in the UV A (400–315 nm) range than in the UV B (315–280 nm) or UV C (280–100 nm) range. Thus, photoinitiators with absorption spectra in the UV A range are most suitable for applications involving resins containing living cells. It is to be noted that the absorption spectra of photoinitiators can be modified by adjusting the concentration of photoinitiator present in the photopolymer resin, as increasing the photoinitiator concentration ensures that irradiation from a broader range of target wavelengths can be utilized.

The most commonly used commercially available photoinitiator for applications in tissue engineering by conventional stereolithography is 2-hydroxy-1-[4-hydroxyethoxy) phenyl]-2-methyl-1-propanone, also known as Irgacure 2959. The cytotoxicity profile of this photoinitiator has been very well characterized by several studies. Elisseeff and coworkers published an extensive comparison of the cytotoxicity profile of Irgacure 2959 and other commonly used UV light sensitive photoinitiators on a broad range of mammalian cell types and species [19]. This study demonstrated minimal cytotoxicity of Irgacure 2959, as compared to other UV irradiation sensitive photoinitiators. Furthermore, this study demonstrated that resin cytotoxicity is highly sensitive to the concentration of photoinitiator, as expected, and that different cell types have varying degrees of sensitivity to identical concentrations of a given photoinitiator. The advantageous properties of Irgacure 2959 as demonstrated by this and other studies have, therefore, contributed to its popular use in conventional stereolithography-based systems to manufacture 3D structures with encapsulated cells (Fig. 6.9).

Advances in photoinitiator chemistry have been further driven by Anseth and coworkers, who have developed a photoinitiator, lithium phenyl-2,4,6-trimethylbenzoylphosphinate (LAP) that demonstrates increased sensitivity to UV light and is also sensitive to wavelengths in the visible light range [20]. By enabling the use of reduced photoinitiator concentrations and lowered light intensities, LAP has demonstrably superior capacity to preserve the viability of

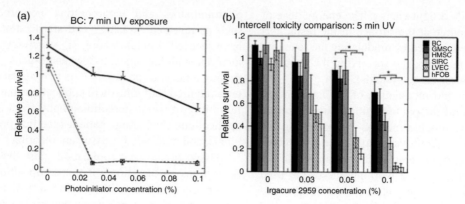

FIGURE 6.9 **The relatively low cytotoxicity of Irgacure 2959, as compared to other commonly used photoinitiators, renders it favorable for use in encapsulating cells via stereolithography.** (a) Relative survival of BC cell line after 7 min of UV exposure in resin containing Irgacure 2959, HPK, or Irgacure 651 photoinitiator; (b) relative survival of six cell types (BC, gMSC, hMSC, SIRC, LVEC, hFOB) after 5 min of UV exposure in resin containing variable concentrations of Irgacure 2959 [19].

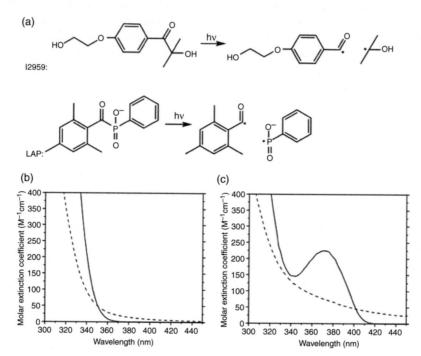

FIGURE 6.10 **Irgacure 2959 and LAP are two of the most successfully used photoinitiators for biocompatible stereolithography with resins containing living cells.** (a) Schematic showing photon absorption leading to cleavage of Irgacure 2959 and LAP into radical reactive species; (b) molar absorptivities of Irgacure 2959 (-) and cleavage products (--); (c) molar absorptivities of LAP (-) and cleavage products (--) [20].

cells encapsulated in stereolithographic resins. As this field continues to grow and applications of stereolithography in regenerative medicine become more popular, it is likely that further research will target the discovery of photoinitiator compounds with advantageous properties similar to or perhaps greater than those demonstrated by Irgacure 2959 and LAP (Fig. 6.10).

2.2.2 Biocompatible Polymerizable Oligomers for Stereolithographic Biofabrication

Stereolithography systems for conventional manufacturing applications have, as mentioned in an earlier section of

this chapter, primarily used reactive acrylate/methacrylate, epoxide, or vinyl ether-based resins, or some hybrid composition thereof [5]. As stereolithographic fabrication technology expands to target translational medical applications, the number of suitable resins decreases slightly, but is not as restricting as the requirements posed on photoinitiators discussed in Section 2.2.1. This is because most cross-linked polymers are not toxic in themselves, though the unreacted monomers and residues of reactive photoinitiator residues may have cytotoxic effects. In many cases, post-processing steps such as washes with water and/or alcoholic solutions, sterilization with UV light irradiation, or treatment with supercritical carbon dioxide (which introduces microporosity into fabricated structures in addition to removing toxic residues [21]) have proven sufficient to ensure biomedical applicability of a resin composition.

It is to be noted, however, that there are still many quality criteria that must be met in order for a stereolithographic resin to be considered suitable and desirable for biofabrication. Photosensitive resins for biomedical applications must, in addition to efficiently absorbing and curing in response to light irradiation, be biocompatible. That is, they must integrate with the target biological host without having a toxic effect on the living system. Biomedical applications also often require materials that are biodegradable (controllably disintegrate in response to host environmental cues leaving only nontoxic byproducts) and bioactive (actively promote beneficial and regenerative effects in the host biological system).

2.2.2.1 High-Strength Resins

For biomedical applications that require high mechanical strength, such as load-bearing implants, polymer-ceramic composite materials combine strong mechanical properties with demonstrable biocompatibility [22]. Formulation of these ceramic-based resin materials is primarily accomplished by mixing ceramic particles, most often alumina or hydroxyapatite, with a liquid photopolymer [23–25]. It is to be noted that such powder-infused resin compositions are often extremely viscous, and must hence be mixed with diluents as described previously. Furthermore, the ceramic particles in these resins must be significantly smaller than the resolved feature sizes and layer thickness within the fabricated part in order to maintain fabrication quality/accuracy [26].

2.2.2.2 Elastomeric Resins

An important constraint on relevant stereolithographic resins is posed by the mechanical properties that are required of biomedical devices. Most conventional SLA resins are glass-like/brittle because they are composed primarily of low molecular weight monomers that react to form rigid cross-linked networks. By contrast, biomedical applications require materials that are more elastomeric (i.e., compliant/flexible) in nature. Resins formulated for these applications thus comprise high molecular weight macromers with relatively low glass transition temperatures. Since liquid suspensions of such macromers are often very viscous, fabrication with such resins often requires mixing with either nonreactive or reactive diluents, such as N-methylpyrrolidone, N-vinyl-2-pyrrolidone (NVP), diethyl fumarate (DEF), or water, to reduce resin viscosity and increase fabrication resolution and accuracy [1]. These diluents have demonstrably increased the hydrophilicity and biocompatibility of SLA-fabricated structures for a variety

of resin compositions [27,28]. It is to be noted that nonreactive diluents, such as N-methylpyrrolidone, are more applicable when the macromer contains more reactive groups such as acrylates or methacrylates.

Historically, stereolithographic systems have made extensive use of (meth)acrylate-based and epoxy-based resins for fabrication. Epoxy-based resins are generally cytotoxic and hence unsuitable for translational medical applications. On the other hand, modified polyether (meth)acrylate-based systems, such as those developed by Emons and coworkers, have been proven to be biocompatible via in vitro tests on mouse fibroblasts [29]. Furthermore, these biocompatible resins form elastomeric polymers that are more flexible than conventional SLA-fabricated materials, displaying a broad range of tunable material properties (hardness, stiffness, etc.) that can be tailored to suit various biomedical applications (Fig. 6.11).

2.2.2.3 Biodegradable Resins

Often, in vivo applications of SLA fabricated biomedical devices require that the fabricated structures biodegrade in a precisely controlled and nontoxic manner, in order to encourage the regeneration of host tissue at the site of disease or trauma-induced damage. As a consequence, the field of research targeting formulations of biodegradable photosensitive resins has been very active in recent years.

Matsuda et al. initiated early developments in this field by developing a photosensitive biodegradable oligomer through ring-opening polyaddition of trimethylene carbonate (TMC) with ε-caprolactone [30,31]. Degradation behaviors of these polymers induced by hydrolytic surface erosion were measured in vivo via subcutaneous implantation of SLA-fabricated microstructures in rats [32]. Similar work on TMC-based biodegradable polymers by Cho and coworkers demonstrated the use of such materials to create scaffolds for regeneration of cartilage and bone [33].

In early demonstrations by Dean and coworkers, SLA systems were successfully used to manufacture biocompatible and biodegradable structures with poly(propylene fumarate) (PPF) based photosensitive resins mixed with DEF diluent [34]. Further characterization and in vitro testing of such PPF/DEF biodegradable systems have promising implications for a diverse array of tissue engineering applications [28,35,36].

Grijpma and coworkers have developed SLA-compatible poly(D,L-lactide) (PDLLA) biodegradable resins mixed with NVP diluent, creating hydrophilic polymerized networks with good cell-adhesive properties as demonstrated by in vitro tests with mouse preosteoblasts [27]. Further developments by this group eliminated the need for such reactive diluents, which are nonbiodegradable, by replacing them with the nonreactive diluent ethyl lactate and further improving the biodegradability of SLA-fabricated structures [37].

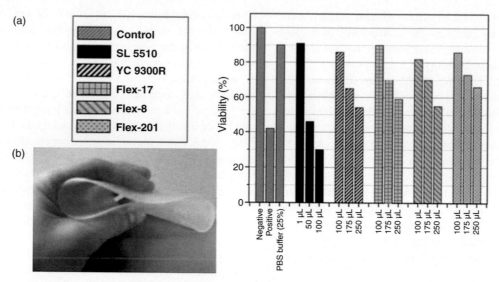

FIGURE 6.11 **Elastomeric resins for stereolithography that allow for the fabrication of compliant/flexible parts are of especial relevance in biomedical and translational applications.** (a) Cytocompatibility of elastomeric resins developed by Emons and coworkers. SL 5510 is an acrylate/epoxy-based formulation, YC 9300 R is an acrylate-based formulation, and the Flex materials are polyether(meth)-acrylate-based formulations; (b) a flexible breathing mask fabricated using the novel biocompatible Flex material [29].

2.2.2.4 Hydrogel Resins

The resins presented in this section have thus far been compatible for applications that involve fabrication of biomedical devices/structures that come into contact with biological systems after postprocessing steps. However, next-generation regenerative medicine applications that fall under the category of "tissue engineering" herald the need for stereolithographic resins that contain living cells. The most promising of photosensitive polymeric materials that support this requirement are a broad class of hydrophilic and biocompatible materials known as "hydrogels."

Cross-linked polymeric hydrogel networks are highly porous and absorbent, facilitating the diffusion of biochemical signals and essential nutrients through the polymeric network. Their structural and functional properties, such as network porosity and hydrophilic swelling behavior, can be readily tailored via chemical modifications to the monomers present in the resin. Their mechanical properties, which fall in the range of 1–100 kPa, are more suited to applications targeting engineering of soft tissues such as those present in the human body. Furthermore, hydrogels also have the ability to respond to changes in temperature, pH, illumination, or other physical and chemical stimuli. This is particularly advantageous for biomedical applications that rely on real-time response behavior such as dynamic medical implants or bioactuators that generate force in response to external stimuli. For these reasons, systems incorporating hydrogel polymers have seen widespread use in translational medical applications in recent years [38,39].

In a pioneering study in this field, Boland and coworkers presented the first demonstration of encapsulating live cells within a photopolymerizable hydrogel via stereolithography. A hydrogel prepolymer solution comprising a mixture of poly(ethylene oxide) and poly(ethylene glycol) dimethacrylate (PEGDMA), reminiscent of the (meth)acrylate-based polymer chemistry discussed earlier in this chapter, was mixed with Irgacure 2959 photoinitiator and Chinese hamster ovary (CHO) cells and polymerized using a direct-laser writing single-photon stereolithography machine [40]. Viability of cells cultured *in vitro* within these polymerized hydrogel constructs was assessed via live/dead assay, proving that SLA-based technologies could be used to fabricate high-density elastomeric tissue-like constructs.

Wicker and coworkers expanded upon this early work, which demonstrated fabrication of relatively simple geometry constructs, to viably encapsulate living cells in PEGDMA hydrogels with more complex 3D geometries via SLA [41]. The hydrogel prepolymer formulation in this case used a combination of two photoinitiators, Irgacure 2959 and 2-hydroxy-2-methyl-1-phenyl-1-propanone. The soluble fraction and swelling ratios of fabricated hydrogels were characterized, providing an understanding of the underlying mechanisms that govern the geometric properties of fabricated parts as a result of polymer resin chemistry (Fig. 6.12).

2.2.2.5 Bioactive Resins

These and other early studies with PEG-based hydrogels, while promising, were still limited by the fact that cells do not naturally form strong adhesive interactions with networks of synthetic polymers. To target this, the study by Wicker and coworkers explored the effect of covalently incorporating the cell-adhesive ligand RGDS (Arg–Gly–Asp–Ser tetrapeptide) within the fabricated hydrogel matrix

Well	Initiator concentration (µL)	Live cells (%)	Total no. of live cells
1	0	91.58	$8.62 \times 10^4 \pm 2.23 \times 10^4$
2	50	64.54	$6.07 \times 10^4 \pm 5.11 \times 10^4$
3	100	50.26	$4.74 \times 10^4 \pm 2.47 \times 10^4$
4	150	28.32	$2.67 \times 10^4 \pm 1.82 \times 10^4$
5	200	41.71	$3.93 \times 10^4 \pm 3.17 \times 10^4$
6	250	11.22	$1.06 \times 10^4 \pm 0 \times 10^4$

FIGURE 6.12 **The porous and absorbent nature of polymeric hydrogel resins make them the ideal biocompatible resin for stereolithographic fabrication with living cells.** (a) A complex 3D design of a chess rook immediately after fabrication: (i) following deformation via mechanical compression, (ii) and after drying, (iii) (scale bar 5 mm) [41]; (b) relative viability of CHO cells encapsulated in elastic hydrogel matrices via stereolithography. Increasing photoinitiator concentration leads to decreased viability [40].

to provide cell-adhesive attachment sites throughout the polymerized structure. Bashir and coworkers expanded upon this work by assessing the effect of RGDS on the long-term viability of cells encapsulated within PEG-based hydrogels of varying molecular weights [42]. Addition of these adhesive peptide sequences showed significant increases in cell viability, proliferation, and spreading as compared to control samples, demonstrating that the incorporation of bioactive moieties within polymeric networks could remove this limitation from SLA-fabricated hydrogel structures. West and coworkers expanded this work by using two-photon laser scanning photolithography to precisely dictate the placement of RGDS peptides within 3D fabricated acrylate-modified PEG hydrogel architectures [43] (Fig. 6.13).

Others have similarly investigated the effects of incorporating different types of bioactive compounds within polymeric matrices. Farsari et al. utilized a multiphoton photopolymerization apparatus to immobilize biotin linked to fluorescently labeled streptavidin to 3D fabricated structures, demonstrating the ability to bind proteins to structures via SLA [11]. Similar work by Woodbury and coworkers utilizing a scanning laser-based SLA in conjunction with methacrylate-based polymer resins to selectively graft peptides onto 3D fabricated microstructures and study postsource decay sequencing of peptides [44]. Bashir and coworkers have demonstrated spatially selective and controllable patterning of proteins on hydrogel polymers by combining a microcontact printing approach with a direct-laser writing SLA setup, showing that cell patterning and alignment in 3D structures can be readily controlled by such a process [45]. Roy and coworkers have also demonstrated the use of SLA to spatiotemporally pattern scaffolds with extracellular matrix components, such as heparan sulfate,

and polymeric microparticles containing heparan-binding growth factors, such as basic fibroblast growth factor-2 [46]. By demonstrating controlled predesigned spatiotemporal distribution of multiple bioactive factors within SLA-fabricated polymeric matrices, these and other studies have set the stage for engineering complex tissue-engineered structures through multilineage differentiation of a single encapsulated population of stem cells.

An alternative approach to separately incorporating bioactive moieties into polymerized matrices is to modify the chemistry of oligomeric monomers used in photosensitive resins. Liska and coworkers used a stereolithographic process to polymerize various methacrylate-based gelatin derivatives, demonstrating the ability to fabricate arbitrary cellular structures with a range of cell types [47]. Other modified methacrylated PEG- and gelatin-based systems incorporating oligopeptides into hydrogel networks have shown that such polymeric scaffolds significantly enhance cell adhesion and growth as compared to nonmodified materials [48]. Chen and coworkers have made extensive use of similar gelatin methacrylate (GelMA) hydrogel materials as a photosensitive resin for high-resolution 3D patterning using a projection SLA, demonstrating the biological functionality of such manufactured constructs [49].

Methacrylate modification of naturally occurring polysaccharides, such as hyaluronic acid [50] and chitosan [51], has likewise been used in photosensitive resin compositions for applications in stereolithographic biofabrication targeting tissue repair and regeneration. Recent studies by Schober and coworkers have even demonstrated multiphoton laser-based photofabrication of unmodified native polymers, such as collagen and fibrinogen, and liquids, such as natural human blood and fetal calf serum [52].

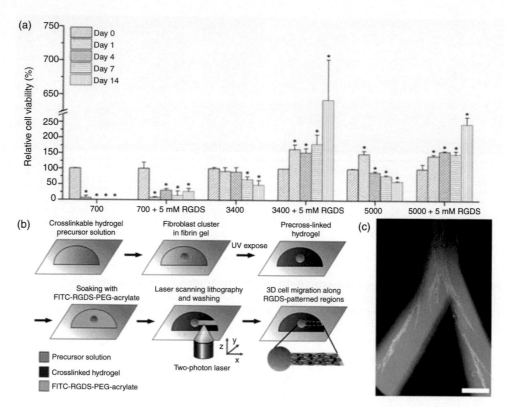

FIGURE 6.13 **Incorporating bioactive moieties, such as the cell-adhesive ligand RGDS, in stereolithographic resins can improve the viability and proliferation of cells encapsulated within the resins.** (a) Relative viability of cells encapsulated in PEG-based hydrogels of varying molecule molecular weight with/without incorporated RGDS ligands [42]; (b) selective patterning of RGDS peptides within 3D PEG-based hydrogels via two-photon stereolithography [43]; (c) confocal microscope image of human dermal fibroblasts cells (red and blue) undergoing migration within the RGDS-patterned region (green) of a PEG hydrogel (scale bar = 100 μm) [43].

2.2.3 Multimaterial Stereolithographic Fabrication

While all the polymers described in this section are useful for a variety of biomedical applications, true versatility of the stereolithographic process can only be attained by multimaterial fabrication of complex 3D structures. This type of multimaterial SLA fabrication can be accomplished by selectively polymerizing portions of layers with a single resin type, washing away unpolymerized resin, and then adding and polymerizing a different resin, as demonstrated by Wicker and coworkers [8]. Bashir and coworkers incorporated a similar multistep system to fabricate 3D cell-encapsulating multimaterial hydrogel structures using a laser-based apparatus [42]. This set the stage for using multimaterial SLA processes to coculture systems of cells, such as neurons and skeletal muscle, to study intercellular interactions and establish a basis for engineering complex tissues and organs containing multiple encapsulated cell types [53] (Figs 6.14 and 6.15).

2.2.4 Novel Resin Systems

As stereolithography has grown to become a fairly well established commercial process, the number of photosensitive resins that rapidly solidify in response to light-excitation has correspondingly increased. The morphological, mechanical, and material properties of these polymerizable resins cover a broad range and are broadly relevant for a variety of translational applications in the field of biomedical engineering. However, the number of photopolymerizable resins for stereolithography that are compatible with applications in bionanotechnology are still limited. A few of the most commonly used families of resins in this field have been elaborated upon in this section, and detailed information regarding photosensitive noncytotoxic resins are listed in more extensive reviews [54,55], but the rapidly changing nature of this field ensures that the variety of available resins will continue to increase and diversify in the coming years.

2.3 Applications of Stereolithography in Biomedical Engineering

The vast array of biomedical and translational applications of stereolithographic fabrication fall into a series of broad categories: Manufacturing of (1) data visualization and surgical planning tools to aid clinicians; (2) individualized prosthetics; (3) customized implants and surgical tools;

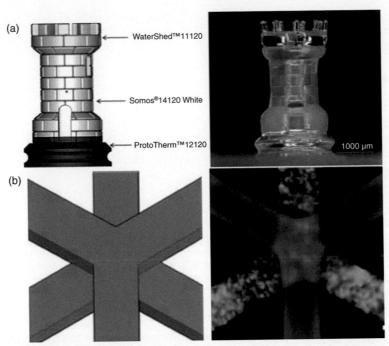

FIGURE 6.14 **Multimaterial stereolithography enables the fabrication of complex 3D structures with varying mechanical, material, and functional properties.** (a) Chess rook design fabricated using three different photosensitive resins [8]; (b) NIH/3T3 cells tagged with either green or red dye encapsulated in different distinct layers using stereolithography, showing viability of multimaterial approach for resins containing living cells [42].

(4) biomaterial scaffolds for tissue engineering; (5) high-density cellular constructs for tissue engineering. These applications are further elaborated upon in Sections 3 and 4.

3 APPLICATIONS OF STEREOLITHOGRAPHY IN SURGICAL PROCEDURES, PROSTHESES, AND IMPLANTS

The vast majority of currently used pre- and postoperative surgical procedures, prostheses, and implants could benefit greatly from the integration of patient-specific models and customized design parameters. While this individualized treatment model would be infeasible using standardized manufacturing techniques, the relative ease of generating accurate digital models and rapidly fabricating prototypes via medical imaging and stereolithography renders patient-specific treatment feasible. Use of stereolithographic manufacturing to manufacture preoperative planning models, external prosthetics, implantable devices, and surgical guides and tools are discussed in this section.

3.1 Stereolithographic Fabrication of Preoperative Visualization and Planning Tools to Aid Clinicians

Structural models and analysis have been extensively used in the fields of architecture and civil engineering for decades,

motivating the complex analysis of clinically relevant 3D models via a similar approach. Advances in clinical imaging of 3D structures via CT and MRI have greatly advanced the practicing clinician's ability to easily visualize exterior surfaces and some internal structures. Manufacturing of patient-specific models and three-dimensional representations of clinical data via stereolithography promises to carry these advances further by building physical representations of a patient's anatomy. These models can be tuned to represent both hard and soft tissues and colored to distinguish different internal structures or pathologies, such as tumors, thereby augmenting the complexity of patient-specific models and predicting/controlling the effectiveness and aesthetic result of surgery [56]. Tools and models manufactured via stereolithography promise to aid clinicians in gaining a complex spatial understanding of patient anatomy, studying pathologies, making accurate diagnoses, and preoperative planning.

3.1.1 Patient-Specific Models

Stoker and coworkers demonstrated early efforts in the construction of patient-specific anatomical models by using an SLA to manufacture plastic models of the internal anatomy within a closed skull, as imaged by CT [57]. Providing a 3D physical representation of the internal anatomy of the skull has obvious applications in both physician and patient education, and can further aid as a model for surgical planning. Further developments in this field by

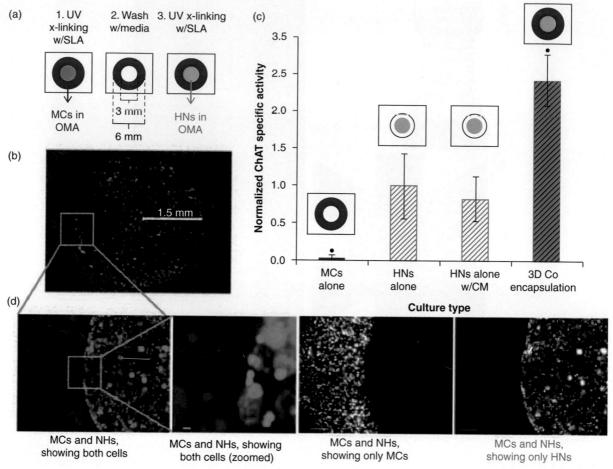

FIGURE 6.15 Multimaterial stereolithography can enable fabrication of 3D cell coculture systems to study intercellular interactions and establish a basis for engineering complex and functional 3D tissues. (a) Schematic of stereolithographic fabrication of resin containing either hippocampus neurons or skeletal muscle myoblast cells; (b) neurons and myoblasts encapsulated within single 3D structure via stereolithography; (c) bar graph showing increase in choline acetyltransferase specific activity for cells encapsulated in 3D multimaterial coculture system; (d) magnified fluorescence images of patterned multimaterial structure [53].

Wittenberg and coworkers demonstrated the use of stereolithography in maxillofacial operation planning by checking the "fit" of a human donor source for cranioplasty prior to surgical intervention [58]. This surgical planning step enabled by stereolithography was shown to demonstrable reduce operative risks and treatment time and improved postoperative results.

3.1.2 Preoperative Planning Tools

More recently, the efficacy of using CT/SLA-fabricated templates for oral implant surgeries have shown that such preoperative planning tools have a higher likelihood of implant survival and significantly lower deviation from planned implant positions [59]. Indeed, studies have shown that such stereolithographic surgical templates, in addition to allowing precise translation of surgical treatment plans into practice, also offer significant benefits over traditional procedures in more complex cases [59].

Implants that must function in high load-bearing environments, such as total hip replacements, require even more careful preoperative planning to ensure proper surgical placement. De Momi et al. utilized an SLA-fabricated preoperative planning model to visualize the physiological movement of the hip joint during daily life activities and generate data on patient gait analysis [60]. Understanding the load-bearing capacity of this implant in the context of the specific patient model provided surgeons with an accurate guided approach to performing the actual implantation procedure (Fig. 6.16).

3.1.3 Data Visualization

Stereolithography has also found use as an aid to generate physical representations of clinical data that help clinicians visualize patient-specific scenarios. Maurer and coworkers introduced this concept by manufacturing polyacrylic hard copies of echocardiographic patient data in order to enhance

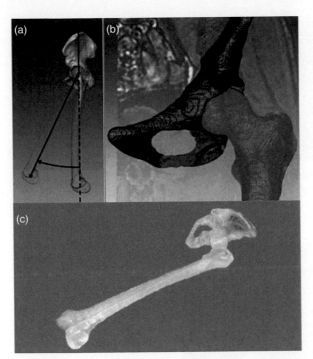

FIGURE 6.16 Preoperative planning tools manufactured via stereolithography can provide surgeons with a guided approach to performing complex surgical procedures. (a) Simulations used to perform patient gait analysis; (b) combined MRI and CAD modeling to produce model of required implant; (c) stereolithographic reconstruction of patient-specific model that can be used by surgeons for practicing placement in preoperative planning [60].

surgeon's spatial perception of the anatomy and pathology of the heart [61]. By providing anatomically correct models of mitral valve anatomy and pathology in patients, this 3D representation of data acquired via echocardiography helped improve the efficiency of diagnoses as well as preoperative planning for treatment of cardiovascular pathologies.

These and other studies first introduced stereolithographic fabrication into the field of medicine, by providing a customizable planning and educational tool for clinicians as well as patients. SLA-fabricated physical models have been used to fabricate complex external and internal structures within the human body and three-dimensional representations of clinical data sets. By providing clinicians with test-platforms and patient-specific models for preoperative planning, stereolithography has the potential to greatly impact the efficiency and accuracy of the diagnosis and treatment of a diverse set of pathologies.

3.2 Stereolithographic Fabrication of Individualized Prosthetics

Building upon this early work using stereolithography to fabricate patient-specific models and guide surgical planning, researchers began to use SLA processes to generate anatomically accurate 3D models that could be used

directly as prosthetics. These individualized stereolithographic prosthetics can be customized to target a range of external patient defects [62].

3.2.1 Cosmetic Prosthetics

Initial studies in this field demonstrated much success in using stereolithography to fabricate high-quality patient specific hearing aids, setting the stage for mass customization of prosthetic devices [63]. Wilkinson and coworkers carried these developments further by utilizing a stereolithographic fabrication process to manufacture a customized whole-ear prosthetic from a wax resin [64]. In this study, information gathered from a digitized magnetic resonance image of a patient's ear was used to create a mirror-image 3D model of the missing contralateral ear. The digitized data were then converted into a 3D mold with a cavity corresponding to the shape of the missing ear and fabricated from a photopolymerizable silicone resin via stereolithography. A wax resin was then injected into the silicone mold to generate a prosthetic ear of the precise dimensions and morphology corresponding to the patient. Other auricular prostheses and personalized maxillofacial prostheses manufactured via mirror-image conversion of digital models have seen similar success as applications of stereolithographic fabrication [65]. As SLA technology has continued to evolve and the number of available resins has increased, direct mold-less fabrication of customized prosthetic devices has also been enabled (Fig. 6.17).

3.2.2 Load-Bearing Prosthetics

The manufacturing of more complex prosthetics for functional and weight bearing limbs, as opposed to mainly cosmetic enhancements, has likewise been enabled by new developments in chemically modified SLA resins with strong mechanical properties (i.e., hardness, stiffness, corrosion resistance, etc.). Preliminary studies in this field have used SLA technologies to test preliminary fitting of external prosthetics on patient-specific "test sockets." These sockets serve the purpose of testing the weight-distribution of the prosthetic to ensure that soft tissue is compressed only in pressure-tolerant areas. Ensuring a good fit through such test socket procedures is essential to the successful integration of the prosthetic with the patient. Stereolithographic fabrication of test-sockets for high load-bearing structures, such as transtibial prosthetics, has been successfully demonstrated [66] (Fig. 6.18).

3.3 Stereolithographic Fabrication of Customized Implants and Surgical Guides

The advances in prosthetics enabled by stereolithography illustrated in the previous section lead naturally to

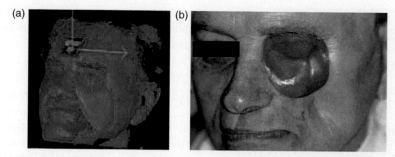

FIGURE 6.17 Cosmetic maxillofacial prosthetics manufactured via stereolithography can be readily customized and tailored to individual patients and different states of motion (e.g., mouth opening). (a) Digital scan and reconstruction of patient's face provides a model for rapid prototyping of cosmetic prosthetic; (b) prosthetic fabricated with the aid of stereolithography can be placed at the sight of damage and finished using conventional techniques [65].

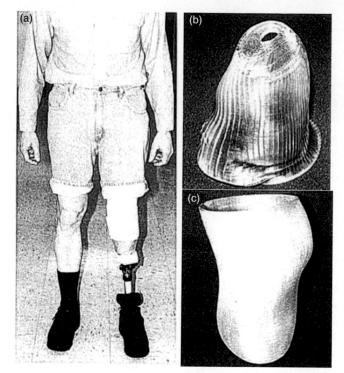

FIGURE 6.18 Manufacturing and testing test sockets for load-bearing prosthetics is critical to the long-term success of the prosthetic devices. (a) Transtibial amputee testing a socket fabricated via stereolithography for fit and functionality; (b, c) magnified view of two potential patient-specific test sockets for transtibial implant [66].

the use of stereolithography to manufacture customized patient-specific implants. These customized biocompatible implants, which must permanently integrate with a host biological system, promise to greatly improve the prospect for invasive surgical procedures targeting synthetic replacements of native tissue and organs.

Since implants, unlike external prosthetics, are in direct contact with native and regenerated tissue *in vivo*, it is critical to ensure the long-term biocompatibility of implant materials fabricated via stereolithography. A study by Popov et al. demonstrated that removal of potentially toxic residues of photosensitive resins (monomers, low molecular weight oligomers, etc.) could be accomplished via treatment with supercritical carbon dioxide, significantly

enhancing the biocompatibility of the fabricated structures when implanted *in vivo* [21]. Other sterilization procedures using UV radiation or washes with alcoholic solutions have been similarly proven successful, as discussed further in Section 4.

3.3.1 Cosmetic Implants

Drawing upon the early advances in using stereolithography to fabricate external maxillofacial prosthetics, there have been many successful attempts to manufacture patient-specific implants for cranioplasty manufactured via stereolithography. Some approaches used CT data and SLA apparatus to design molds that were then used

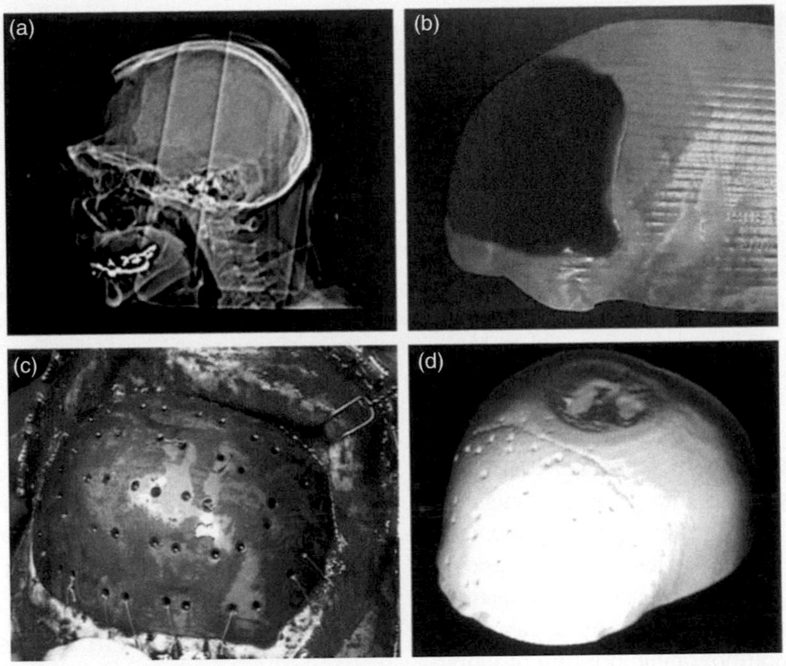

FIGURE 6.19 Cosmetic maxillofacial prosthetics can be customized to patient-specific defects and readily fabricated via stereolithography. (a) Digital scan of patient's large bifrontal defect; (b) model of patient skull fabricated via stereolithography and used as mold for implant generation; (c) patient-specific implant; (d) postoperative CT scan of implant showing excellent fit and integration with patient host [68].

to fabricate thermally polymerized custom implants from acrylic materials [67], or carbon-fiber reinforced polymeric composites [68]. More direct SLA-fabrication approaches have been demonstrated by Day and coworkers, who used a digital model obtained via contiguous helical CT scans to directly fabricate a patient-specific mandible via stereolithography to fit a large surgical defect obtained as a result of a tumor [69]. Fitting of the manufactured implant into the defect site provided aesthetically appealing results with good fit and integration of the fabricated mandible into the surrounding bone (Fig. 6.19).

3.3.2 Load-Bearing Implants

Customized design of load-bearing implants has been enabled by the creation of SLA-fabricated replica models of individualized patients. Early studies in this field aimed at individualizing total knee replacements for patients were shown to successfully return weight-bearing capacity to damaged limbs in both animal and human models [70,71]. Such customized implantation procedures have great value for all patient models, but especially in complicated cases where more individualized morphologies must be taken into account. As with the prosthetic devices mentioned earlier, further developments in creating SLA resins with strong mechanical properties (such as the polymer-ceramic composites detailed in Section 2) promise to result in the direct fabrication of patient-specific load-bearing implants.

3.3.3 Surgical Guides

SLA technologies have also been used to manufacture surgical guides that ensure proper placement of implants within patients. These guides are critical to the operative procedure, and are conventionally employed to direct implant drilling and placement systems. Customized surgical guides manufactured via SLA can greatly augment surgical outcome by improving the ultimate functional performance and design aesthetics of the implanted device. Rosenstiel and coworkers have demonstrated the reliability of surgical guides manufactured via stereolithography, showing that a significantly lower mean angular deviation was obtained for SLA surgically guided implants as compared to implants placed without SLA surgical guides [72]. These and other similar studies have demonstrated the ability to rapidly and accurately manufacture patient-specific surgical guides via stereolithography, thereby ensuring a high quality of pre-surgical planning and surgical execution on a customized case-by-case basis (Fig. 6.20).

4 APPLICATIONS OF STEREOLITHOGRAPHY IN TISSUE ENGINEERING AND REGENERATIVE MEDICINE

The field of tissue engineering is centered on the aim of creating complex biological substitutes for native tissue using a combination of cells and instructive biomaterials. This

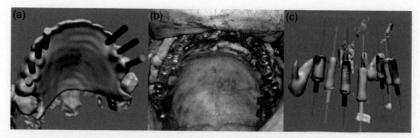

FIGURE 6.20 **Patient-specific surgical guides fabricated via stereolithography can greatly improve the ultimate functional performance and design aesthetics of an implanted medical device.** (a) CT scans and CAD modeling used to generate digital models of treatment planning guides; (b) implants placed using surgical guides fabricated via stereolithography; (c) demonstration of good match between planned implants (various colors) and actual placed implants (gray) [72].

emerging field promises to revolutionize modern medicine by providing an alternative to tissue and organ transplantation, and perhaps eliminating the need for implants and transplants altogether by aiding the rapid development and testing of new drugs and therapies [73]. The use of stereolithographic fabrication to fabricate biocompatible scaffolds for tissue engineering, followed by advances in this field pioneered by stereolithographic printing of living cells, is described in the following sections. As previously discussed in Section 2.2, advances in creating biocompatible, biodegradable, and bioactive photosensitive resins have enabled many of the advances in this field. These resins are tailored to prevent inflammatory response upon implantation, controllably degrade into nontoxic byproducts that can achieve good renal clearance, and actively encourage the formation of new regenerated tissue [74,75].

4.1 Stereolithographic Fabrication Strategies for Tissue Engineering and Regenerative Medicine

4.1.1 "Top–Down" Fabrication

The conventional approach to tissue engineering has been the fabrication of porous biocompatible and biodegradable scaffold structures as platforms for cell seeding. In the early stages of these studies, scaffolds were often generated using conventional microfabrication approaches for patterning natural and synthetic materials [76]. Researchers have even explored approaches in which cadaveric tissue is decellularized and repopulated with living cells to yield functional tissue and whole organs [77,78]. More recently, direct fabrication of biocompatible scaffolds via SLA processes has seen much success in a variety of tissue engineering applications (see Sections 4.2–4.5)

Scaffold-based approaches provide the ability to place cells within precise morphological designs while maintaining the cell–cell and cell–matrix interactions observed *in vivo*. However, these approaches come with the challenges of regulating the distribution of cells within the scaffolds, as well as the inability to engineer adequate vascular networks that can provide supplies of nutrients to large-scale

engineered tissues. These challenges render scaffold-based "top–down" approaches undesirable for applications in tissue engineering of complex tissues and organs, as they are unable to provide truly precise spatial control of cells and biochemical signals in three dimensions [79].

4.1.2 "Bottom–Up" Fabrication

The rise of "bottom–up" approaches in biofabrication, triggered by the advent of highly absorbent biocompatible stereolithographic resins, provides and attractive alternative approach to manufacturing complex biological substitutes for native tissue and organs. In this approach, engineered tissues can be directly assembled layer by layer from the bottom up by using SLA resins containing encapsulated cells, allowing for complex three-dimensional control over the morphological and functional properties of engineered tissue.

The single-photon and multiphoton stereolithographic systems presented earlier in this chapter are ideal candidates for targeting applications in bottom–up tissue engineering. These systems have demonstrable high-resolution patterning capabilities and have been proven to viably encapsulate cells in polymeric networks while preserving viability and functionality in long-term studies by minimizing heat production and damage to raw material during fabrication.

4.1.3 Macro- and Microscale Architecture

High-resolution fabrication of three-dimensional porous structures with intricate internal geometries is a necessary requirement of top–down and bottom–up methods of manufacturing viable tissue-engineered constructs. Recent studies on the effect of microscale porous internal architectures of scaffolds on the repair and regeneration rate of damaged tissue have demonstrated the need for creating 3D structures with controllable morphologies on the macro- and microscales. This level of microenvironmental control has the further advantage of facilitating architecture for vascular networks and conduits, setting the stage for developing large-scale engineered tissues and organs in the future. High-resolution spatiotemporal patterning of bioactive moieties can also drive advances in multilineage differentiation of a single population of encapsulated patient stem cells, creating truly customized replacement tissues. Hence, while many of the pioneering studies in SLA-based tissue engineering utilized the direct laser writing strategies and conventional UV irradiation driven polymerization, the rise of digital mask projection micro-SLA promise to further improve our ability to fabricate biodegradable and biocompatible architectures with single micron resolution [36]. Two-photon methods improve upon this even more by establishing biocompatible fabrication mechanisms with submicron resolution [12]. The fundamental challenge underlying all these fabrication strategies is based on negotiating the balance between fabricating macroscale structures while incorporating microscale features.

The use of SLA fabrication strategies to target regenerative medicine applications in tissue-engineered bone, cartilage, and cardiovascular tissue/networks are presented in the following sections. SLA technologies that target a range of other tissue/organ systems, such as liver, connective tissues, and neural conduits, have also been demonstrated, proving the readily customizable nature of this fabrication process.

4.2 Stereolithographic Fabrication for Bone Tissue Engineering

Tissue-engineered substitutes for bone are motivated by clinical need: skeletal defects, caused by disease or traumatic fracture, are increasingly common in a growing and ageing population and require treatment with bone grafts in order to retain functional output of the skeletal system [80]. Traditionally, autografts and allografts have been used to replace damaged tissue. However, these approaches come with the limitations of available supply, as well as an inability to directly match the patient-specific morphological and functional design requirements of the replacement bone to the native tissue that surrounds it [81].

4.2.1 Biocompatible Materials for Bone Tissue Engineering

Synthetic alternatives to bone grafts fabricated by traditional and SLA-based processes have been proven to have biocompatible material properties that render them useful in implant applications. Inspired by this early work in ceramic implants, Chu et al. demonstrated the use of a SLA to polymerize UV-curable polymer-ceramic composite scaffolds that were biocompatible and suitable for use as cell-seeding platforms [25]. These hydroxyapatite implants were implanted *in vivo* in Yucatan minipig models and shown to support the regeneration of bone tissue, as dictated by the internal architecture of the engineered scaffold. Similarly, work with vinyl ester resin based bone regeneration scaffolds fabricated via SLA and implanted *in vivo* in a rabbit model have shown active bone ingrowth and regeneration into the defect site [17] (Fig. 6.21).

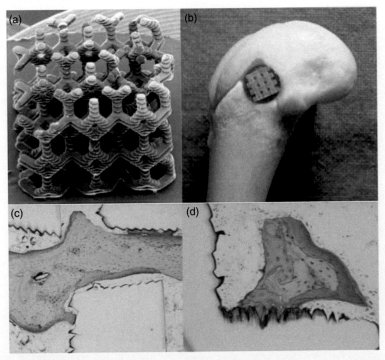

FIGURE 6.21 Biocompatible scaffolds for bone tissue engineering manufactured via stereolithography. (a) SEM image of SLA-fabricated scaffold manufactured with a biocompatible vinyl ester resin; (b) placement of SLA-fabricated scaffold *in situ*; (c) magnified view of fabricated scaffold 8 weeks after *in vivo* implantation showing newly formed bone with excellent host system integration; (d) bone apposition along serrated surface shows enhanced proliferation and growth in response to microscale scaffold architecture [82].

4.2.2 Biodegradable Materials for Bone Tissue Engineering

Synthetic structural materials often exhibit poor integration with the surrounding tissue; however, as a result of the mismatch between their morphological and mechanical properties and those of surrounding bone. Furthermore, as they are not bioresorbable, they do not represent a truly biointegrated solution to the skeletal defect. As a result, the field of regenerative bone engineering can only be advanced by developing complex 3D substitutes for bone tissue that can mimic the load-bearing mechanical properties and the morphological architecture observed *in vivo*, while allowing for biodegradation and active biointegration with the host/patient over time. Stereolithography is ideally suited to the task of creating such complex three-dimensional biodegradable substitutes for damaged bone tissue, and studies that make use of this technology have demonstrated significant advances in this field.

Early work in SLA fabrication using biodegradable resins for bone regeneration made extensive use of PPF resins diluted with DEF, as previously described in Section 2. Complex 3D scaffolds with intricate internal architectures were fabricated using this resin, with promising implications for bone ingrowth in implantation applications [28,34].

Grijpma and coworkers furthered this field by using stereolithography to manufacture scaffolds for bone tissue engineering using a novel PDLLA resin [37]. While other biodegradable macromers used in stereolithographic fabrication of bone tissue engineering scaffolds, such as TCM or PPF require mixing with a reactive diluent such as DEF to obtain appropriate viscosity [28], PDLLA is not so restricted. This ensured that a greater percentage of the final fabricated scaffold would biodegrade and resorb into engineered bone upon implantation of the scaffold into the biological system, promoting better biointegration with the host/patient. Furthermore, PDLLA was proven to be much stiffer than its polymeric counterparts, demonstrating an elastic modulus (a measure of material stiffness) of 3 GPa, which approaches the properties of bone tissue *in vivo* (3–30 GPa). Attachment and proliferation of mouse preosteoblasts cultured at physiologic conditions within these scaffolds demonstrated the viability of this approach for translational applications in bone tissue engineering (Fig. 6.22).

SLA-fabricated bone regeneration scaffolds have also been fabricated by chemically modifying natural polymers, such as the biocompatible and biodegradable polysaccharide chitosan. Wen and coworkers have shown that subcutaneous *in vivo* implantation of these natural material-based scaffolds in a rat model supported osteoconductivity and regeneration [51].

4.2.3 Microscale Architecture in Bone Tissue Engineering

Digital mask-based photolithographic methods have further improved on the precise fabrication of spatially patterned

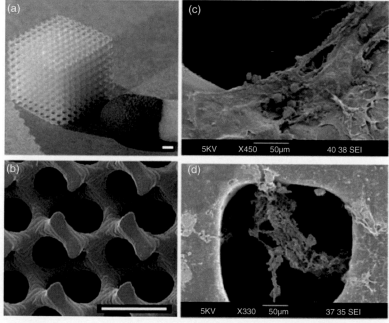

FIGURE 6.22 Biodegradable scaffolds for bone tissue engineering support the ingrowth and regeneration of bone. (a) SLA-fabricated scaffold manufactured using biodegradable resin; (b) SEM image of fabricated structure showing complex gyroid architecture; (c) 3D cell culture in SLA-fabricated scaffold showing cell adhesion, proliferation, and spreading on the scaffold surface 1 week postseeding; (d) cell migration and invasion in 3D into internal pores 4 weeks postseeding. (*Figure 6.22a,b reprinted with permission from Melchels et al. [37]; Figure 6.22c,d reprinted with permission from Lee et al. [35].*).

scaffolds for bone tissue engineering. Roy and coworkers have recently employed a DMD-based projection SLA to precisely distribute bioactive factors within a three-dimensional poly(ethylene glycol) diacrylate (PEGDA) scaffold with complex internal architecture [83]. The advantageous functional properties of these bioactive scaffolds were proven to drive osteogenic differentiation of marrow-derived stem cells *in vitro*, as indicated by postseeding mineralization of engineered tissue constructs.

Two-photon laser-based polymerization has also been used to target applications in bone tissue engineering by Ovsianikov et al. [84]. The photosensitive resin used in this study was a methacrylamide-modified gelatin derived from native collagen, and demonstrated chemical properties that mimicked the cellular microenvironment *in vivo*. The biodegradation properties of this material in response to collagenase digestion were quantified and proven to be tunable – a desirable property for any tissue engineered scaffold. Porcine mesenchymal stem cells seeded within these scaffolds demonstrated improved adhesion and proliferation behaviors. Osteogenic stimulants in the culture medium drove differentiation of these stem cells into the osteogenic line, with osteogenic cell byproduct calcium phosphate deposition observed in the scaffold.

4.2.4 Multifunctional Components for Bone Tissue Engineering

In addition to applications that target recreating the structure of cancellous and cortical bone, researchers have used stereolithography to fabricate the connective tissues found in bone marrow. Roy and coworkers used a DMD-based projection stereolithography system to manufacture scaffolds with intricate pore geometries, such as 3D honeycomb-like structures of interconnecting hexagons [7,83]. These 3D printed scaffolds then underwent surface modification to be functionalized with the ECM protein, fibronectin, and were sterilized prior to seeding with D1 cells, a murine bone marrow progenitor cell line. Successful seeding, attachment, and proliferation of D1 cells within these scaffolds was observed and quantified.

SLA-based fabrication technologies have, therefore, had significant impact on the bone regeneration techniques developed in recent years. By providing a mechanism of fabricating custom-fit and patient-specific biodegradable implants to support the regeneration of mineralized tissue, stereolithography is arguably one of the primary enabling technologies promoting advancements in this field.

4.3 Stereolithographic Fabrication for Cartilage Tissue Engineering

Many of the pioneering advances in tissue engineering have targeted applications in tissue engineering of cartilage,

since this type of tissue is often relatively homogeneous and largely avascular, and hence relatively easy to fabricate. The ability to regenerate load-bearing and articular cartilaginous tissues can address many ongoing challenges in replacing tissues damaged by degenerative diseases such as progressive arthritis, aging, or traumatic injury [85].

4.3.1 Cosmetic Tissue-Engineered Cartilage

Reminiscent of the external maxillofacial prosthetics discussed earlier in this chapter, Naumann et al. used a combination of CT and UV laser-based stereolithographic fabrication to fabricate three-dimensional bioresorbable scaffolds for tissue engineering of an auricle from the hyaluronic acid derivative Hyaff 11 [86]. As hyaluronan is an important component of native cartilage, this resin was chosen and tailored to suit the particular application targeted by this study. Histomorphology and immunohistochemistry performed on Hyaff 11 scaffolds seeded with living chondrocytes demonstrated the homogeneous expression of cartilage-specific collagen type II within the engineered tissue after 4 weeks of *in vitro* culture at physiologic conditions (Fig. 6.23).

4.3.2 Load-Bearing Tissue-Engineered Cartilage

Other resins used for tissue engineering of cartilage have been chosen for the close match between their material properties, such as water uptake and swelling depth, and mechanical properties, such as stiffness, and those of native cartilage tissue. For applications in tissue engineering of load-bearing cartilage, such as that found in intervertebral discs, match of the material and mechanical properties of regenerated and native tissue is of utmost importance in functional output. Cho and coworkers targeted this challenge by manufacturing biodegradable scaffolds from TMC-based oligomers, whose material and mechanical properties match those of native cartilage [33]. Chondrocytes extracted from articular cartilage and seeded within these scaffolds were cultured at physiologic conditions and supplemented with transforming growth factor (TGF) and insulin-like growth factor. Viability of cells over 2 weeks was determined by a quantitative measure of the mitochondrial metabolic activity of adhered cells (3-(4,5-dimethylthiazol-2-yl)-2,5-diphenyltetrazolium bromide assay), showing strong adhesion and activity of seeded chondrocytes.

4.3.3 "Bottom–Up" Fabrication of Tissue-Engineered Cartilage

Ellisseeff and coworkers improved upon these scaffold-based fabrication techniques by incorporating a suspension

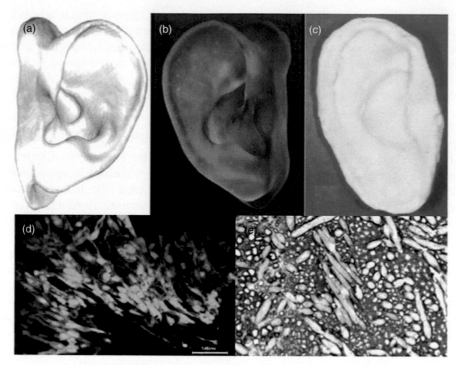

FIGURE 6.23 Cosmetic tissue engineered cartilage. (a) CAD model of contralateral (i.e., undamaged) ear; (b) Stereolithographic fabrication of mirror image design; (c) bioresorbable hyoluronan-based scaffold fabricated with the aid of stereolithography; (d) confocal microscopy of scaffold after 4 weeks *in vitro* culture showing homogeneous distribution of vital chondrocytes (green) distributed throughout scaffold, with relatively few avital cells (red); (e) histomorphology and immunohistochemistry of chondrocyte constructs after 4 weeks *in vitro* culture showing homogenous expression of cartilage-specific collagen type II [86].

of embryonic stem cells that had formed embryoid bodies (EB) into a photosensitive PEG hydrogel-based prepolymer solution [87]. A simple UV light and a physical mask-based SLA was used to encapsulate EBs into a three-dimensional architecture and cultured *in vitro* in chondrogenic medium containing (TGF)- β1, and bone morphogenic protein-2. Extensive characterization of regenerated tissue fabricated in this manner via gene expression and protein analysis as well as histological analysis, suggest that EBs encapsulated in these hydrogels via stereolithography up-regulated cartilage-relevant markers and induced a chondrocytic phenotype. The basophilic exctracellular matrix deposited in the engineered 3D environment was characteristic of neocartilage, suggesting that this apparatus is a promising approach to engineering complex functional replacements for native cartilage.

4.4 Stereolithographic Fabrication for Cardiac and Vascular Tissue Engineering

The increasing prevalence of cardiovascular disease, which is the primary cause of death in the United States and other developing countries, motivates the development of tissue-engineered replacements for tissues in the cardiovascular system. Moreover, advancements pertaining to the regeneration of interconnected vascular networks are broadly applicable to all regenerative medicine applications. Vascularization of engineered tissues, which is the key challenge facing researchers aiming to create large-scale replacements for native tissue and organs, must be addressed before regenerative medicine technologies are adapted for widespread clinical use [88].

4.4.1 2D Cardiac Tissue Engineering

As cardiac tissue is inherently an actuator in the body, advances in tissue engineering of cardiac muscle powered actuators have many applications in restoring pulsatile/beating function to damaged or ischemic cardiac tissue. Bashir and coworkers fabricated protein surface functionalized hydrogel cantilever-like substrates via a laser-based stereolithographic 3D printing apparatus and seeded primary cardiomyocytes derived from neonatal rats upon these 2D scaffolds [89]. The cardiomyocytes formed a connected cell sheet attached to the substrate and were able to drive actuation of the hydrogel substrate via contraction of the engineered tissue (Fig. 6.24).

4.4.2 3D Cardiac Tissue Engineering

Sodian et al. explored 3D scaffolds for cardiac tissue engineering by fabricating stereolithographic plastic models for thermoplastic modeling of two elastomers, poly-4-hydroxybutyrate and polyhydroxyoctanoate, in the shape

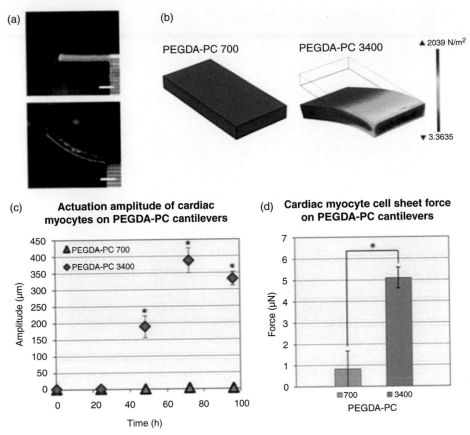

FIGURE 6.24 **2D engineered cardiac tissue manufactured with the aid of stereolithography.** (a) Primary neonatal cardiomyocytes seeded on PEG-based cantilevers of high stiffness (top) and low stiffness (bottom) fabricated via stereolithography; (b) computer aided simulation showing the distribution of stresses in the prototyped cantilevers as a result of contraction of the engineered cell sheet; (c) actuation amplitude of cardiac myocytes on PEG-based cantilevers; (d) force exerted by the engineered cardiac myocyte cell sheet [89].

of a trileaflet heart valve scaffold [90]. These valve scaffolds were placed inside a bioreactor and subjected to pulsatile flow, and demonstrated synchronous opening and closing behaviors similar to those seen *in vitro*. This technique, rendered feasible by stereolithography, demonstrates the ability to accurately reconstruct physiological valve design and bypassing the need for a human allograft.

Butcher and coworkers demonstrated further improvements on this technology by combining stereolithographic fabrication and bioplotting technologies to generate 3D cardiac tissue scaffolds for manufacturing heterogeneous aortic valves. PEGDA hydrogels were supplemented with alginate and cured by an UV LED crosslinking module, demonstrating a broadly tunable range of mechanical properties for printed valve scaffolds. Porcine aortic valve interstitial cells seeded within these scaffolds and cultured *in vitro* at physiologic conditions for 21 days showed cell viability and spreading on this biocompatible fabricated scaffold (Fig. 6.25).

4.4.3 Vascular Tissue Engineering

Kong and coworkers demonstrated the use of a direct laser-writing apparatus to fabricate "Living" microvascular stamps that could controllably pattern functional neovessels when tested *in ovo* on the chick chorioallantoic membrane [92]. These SLA-fabricated stamps, comprised of fibroblasts encapsulated in PEG-based hydrogels that were stimulated to secrete the angiogenic molecule vascular endothelial growth factor (VEGF), proved to demonstrate orchestrated control over the placement and geometry of neovasculature.

Work by Khademhosseini and coworkers extended this work by fabricating very-high-resolution three-dimensional porous scaffold architectures with intricate geometries using a projection stereolithography system [49]. The mechanical properties of these 3D scaffolds were tuned and regulated by varying the chemical composition of the GelMA resins used as photosensitive prepolymers in this study. The complex interconnected pore structure rendered feasibly via stereolithographic fabrication led to the formation of a high cell-density network of seeded human umbilical vein endothelial cells (HUVECs) within the scaffold. Immunohistochemistry of these tissue-engineered substitutes showed that seeded cells maintained their endothelial phenotype over time and were well distributed throughout the scaffold during the observed culture period, thereby demonstrating a promising method of engineering 3D

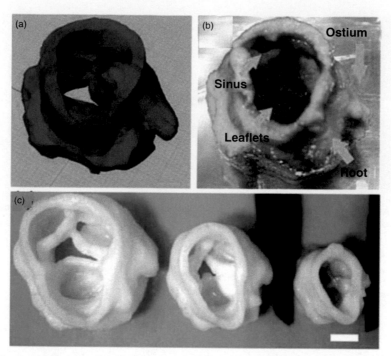

FIGURE 6.25   3D tissue-engineered aortic valve. (a) Porcine aortic valve model; (b) printed aortic valve model formed from PEG-based resins. Root (blue) formed from stiffer hydrogel and leaflets formed from less stiff hydrogel; (c) aortic valve scaffolds printed at different scales (Inner diameter = 22, 17, and 12 mm) for shape fidelity and resolution analysis [91].

vascular networks within SLA-fabricated tissue constructs (Fig. 6.26).

4.5   Stereolithographic Fabrication for Other Tissue Engineering Applications

SLA-based technologies are broadly applicable in a vast array of tissue engineering applications. The techniques used for fabrication are the same across tissue types, but the types of chemically modified resins and 3D macro- and microscale architectures have been modified to suit the specific application in question. A couple notable examples that use next-generation multiphoton SLA processes are mentioned in this section.

4.5.1   Liver Tissue Engineering

Wan and coworkers fabricated three-dimensional microstructure scaffolds for tissue engineering of liver using a two-photon laser scanning photolithography technique [93]. In this study, a commercially available photocurable polymer manufactured by 3D systems (the company that pioneered stereolithographic fabrication) was selectively polymerized using laser pulses. Sterilized scaffolds were then functionalized with collagen and seeded with primary rat hepatocytes. To assess for liver-specific function of the engineered tissue, the culture medium was assayed for albumin and urea secretion and demonstrated that the cells had

received adequate nutrient transport within the fabricated scaffolds and were able to preserve their functionality.

4.5.2   Neural Tissue Engineering

Claeyssens and coworkers used a novel multiphoton polymerization approach to fabricate polylactide-based scaffolds for neural tissue engineering [94]. The photosensitive polylactide resin (PLA) was cured using femtosecond laser pulses of IR irradiation, with a maximum resolution of 800 nm achieved for scaffold feature sizes. Neuroblastoma cells were cultured on these structured PLA scaffolds, demonstrating cell viability and proliferation on scaffolds to provide proof of the biocompatibility of these structures (Fig. 6.27).

The examples listed in this section are just a few of the many studies that serve as strong demonstrations of the potential of stereolithographic fabrication technologies to suit a wide variety of applications in tissue engineering and regenerative medicine.

5   CONCLUSIONS

5.1   Current Challenges in Stereolithographic Fabrication

5.1.1   Multimaterial Fabrication

One of the primary challenges facing widespread adaptation of stereolithography for a vast array of translational

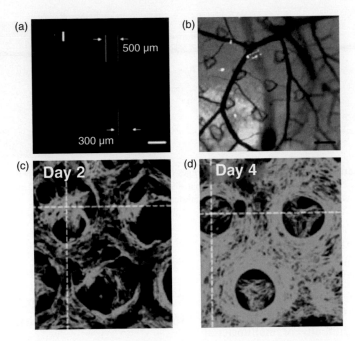

FIGURE 6.26 **Tissue engineered vasculature is required to create large-scale replacements for any tissue/organ system.** (a) Hydrogel patch containing VEGF-secreting fibroblasts fabricated via stereolithography; (b) patterned formation of neovasculature in response to *in ovo* incubation of the hydrogel on the chick chorioallantoic membrane; (c, d) adhesion and spreading of HUVECs on GelMA scaffold fabricated via stereolithography shown Day 2 and Day 4 postseeding. *(Figure 6.26a,b reprinted with permission from Jeong et al. [92]; Figure 6.26c,d reprinted with permission from Gauvin et al. [49].).*

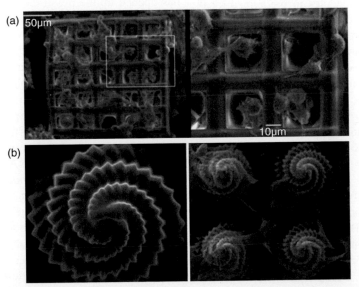

FIGURE 6.27 **Neural tissue engineering.** (a) 3D microscaffolds manufactured via multiphoton stereolithography using a PLA resin. Neuroblastoma cells cultured within these scaffolds demonstrate excellent adhesion and spreading 5 days postseeding; (b) complex microscale architectures, such as seashells, can be printed using stereolithography and used to control the directional spreading of seeded cells [94].

biomedical applications is the difficulty of fabricating multimaterial 3D structures. Wicker and coworkers demonstrated wash/refill steps that could be employed to fabricate 3D multimaterial structures with a projection micro-SLA [8]. Bashir and coworkers extended this work to encapsulate multiple cell types in 3D patterned hydrogels using a single photon laser-based SLA [42]. This approach, while efficacious, causes a significant increase in part build-times. Build time can be reduced by employing automated syringe-pump systems for washing/filling resins. Arcaute et al. recently proposed an alternative method for fabrication using multiple resins by employing the use of an array of "minivats," which allow for selective spatially controlled variation of material composition in 3D architectures [95].

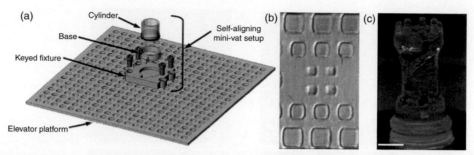

FIGURE 6.28 Multimaterial fabrication is one of the primary challenges facing stereolithography. (a) Schematic of self-aligning minivat setup proposed by Arcaute et al.; (b) spatially controlled variation of resin within a single printed layer; (c) complex 3D chess rook structure with fluorescent internal staircase manufactured using this multimaterial stereolithographic fabrication approach [95].

Further advancements of this and other multimaterial strategies promise to remove the limitation from stereolithographic biofabrication (Fig. 6.28).

5.1.2 Microscale Control of Architecture

The limitations posed by the difficulty of multimaterial fabrication are mainly contingent on the resultant inability to create precise predefined spatial patterns of mechanical, material, and biochemical properties within 3D structures. For tissue engineering applications, the processes currently in common use do not allow for accurate placement of single cells or bioactive molecules in predefined locations. Rather, cells and molecules are mixed within resins, assumed to be uniformly mixed, and then selectively polymerized prior to washing and refilling steps. Advances in various single-cell and molecule manipulation technologies may address some of the main concerns regarding this limitation of stereolithographic biofabrication.

Timp and coworkers addressed this challenge by using optical tweezers to selectively manipulate single cells into precisely defined 3D spatial positions within a prepolymer solution prior to encapsulation via photopolymerization [96]. This approach, while effective, would prove to be extremely time consuming for the fabrication of large 3D structures containing millions of cells and biomolecules. Bashir and coworkers have demonstrated a higher throughput approach based on the principle of dielectrophoresis (DEP). In this process, a set of electrodes is incorporated on the build platform of a single photon SLA, allowing for selective and simultaneous patterning of large arrays of cells prior to photopolymerization [97] (Fig. 6.29).

Pioneering work in this field has recently been published by Deforest and Anseth, who have demonstrated hydrogels with tunable range of properties (mechanical, material, biochemical, etc.) that can be spatiotemporally controlled with high 3D resolution [98]. This spatiotemporal control is accomplished via a novel cytocompatible "click-based" chemistry that uses wavelength-specific photochemical reactions to dynamically conjugate or cleave bioactive moieties to 3D cell-culture systems *in vitro* (Fig. 6.30).

5.2 New Developments in Stereolithography for Biomedical Applications

Stereolithographic technologies have enabled the rise of reverse-engineering native tissue for applications in regenerative medicine, but this is not the only field of research opened up by work in this area. By giving researchers the ability to build systems of cells, stereolithography has opened up the possibility of designing novel systems that harness the innate dynamic abilities of cells to self-organize and respond to environmental cues. This idea of forward-engineering integrated cellular systems, or "biological machines," with multiple functionalities using stereolithography as an enabling tool has many potential applications.

Bashir and coworkers have recently designed a biological machine, or "biobot," that utilizes the autonomous and synchronous contraction of engineered cardiac tissue to power locomotion of an SLA-fabricated soft robotic device [99]. Further studies in this field that focus on creating biological machines that can accomplish such objectives of robotics as sensing, storage, and processing of signals, and a resultant response (such as actuation) have many potential applications. Biological machines that dynamically respond to environmental cues can target a diverse set of translational medical applications including drug-delivery, noninvasive surgery, dynamic implants, and biocompatible microelectronics. These machines and other studies that build upon them will demonstrate the power using SLA fabrication and cells as building blocks to engineer the machines and systems of the future.

5.3 Stereolithographic 3D Bioprinting for Biomedical Applications

The future of modern medicine is undoubtedly rooted in the core philosophy of customized/personalized healthcare.

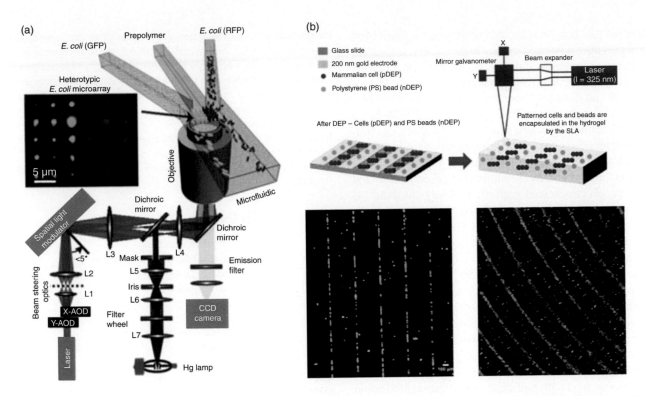

FIGURE 6.29 **High-resolution patterning of single cells within a complex 3D structure is a major challenge facing stereolithography.** (a) Optical tweezers used to precisely manipulate arrays of fluorescently labeled *Escherichia coli* before encapsulation within a photosensitive polymer [96]; (b) DEP used as a high-throughput approach to manipulating fluorescently labeled cells and beads into precisely defined spatial patterns prior to encapsulation in a photosensitive polymer [97].

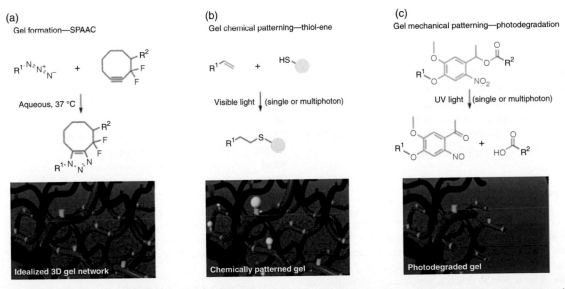

FIGURE 6.30 **High-resolution spatiotemporal control of mechanical, material, and biochemical properties within a 3D structure is the next-generation requirement for stereolithographic fabrication.** (a) Formation of a 3D hydrogel; (b) single or multiphoton visible light used to chemically pattern bioactive moieties within a 3D hydrogel matrix; (c) single or multiphoton UV light used to selectively degrade portions of the polymerized gel [98].

As we learn more about the underlying design principles and molecular mechanisms governing biological systems, we can develop precisely individualized and targeted cures that address patient-specific needs. This motivates the need for developing an enabling technology that allows us to fabricate complex 3D structures that are biocompatible, biodegradable, and bioactive. Such biofabricated structures can readily integrate with a host biological system to perform a specified task, such as targeting diseased tissues and promoting tissue regeneration.

Over the past few decades, stereolithographic fabrication technologies have advanced greatly in the quality, resolution, and accuracy of manufactured parts. Recent developments in creating photocurable resins that are biocompatible, biodegradable, and bioactive have enabled a vast array of biomedical and translation medical applications of SLA-based fabrication technologies. In a progressive manner that mimics the progression of the medical field itself, SLA machines have been used to manufacture: (1) data visualization and surgical planning tools to aid clinicians; (2) individualized prosthetics; (3) customized implants and surgical tools; (4) biomaterial scaffolds for tissue engineering; and (5) high-density cellular constructs for tissue engineering. In each of these broad types of applications, stereolithographic techniques have been readily integrated with medical imaging technologies (MRI, CT) in order to improve disease diagnosis, preoperative planning, quality and morphology of prosthetics and implants, and functional success of complex surgeries. Furthermore, SLA has established itself as one of the primary enabling tools that will be useful for regenerative medicine applications in the coming years.

Limitations of stereolithographic systems, such as the difficulties involved with multimaterial construct fabrication and high-resolution spatiotemporal control over the placement of specific moieties within complex 3D structures, are being addressed by advances in SLA technologies (projection stereolithography, multiphoton methods, automated resin dispersal systems, arrays of multimaterial "minivats," etc.) as well as advances in novel resin chemistries (dynamic photoconjugation and photocleavage of moieties within 3D structures). New biodegradable and bioactive resins continue to be developed as this technology achieves even broader use in the field of biomedical engineering.

As a whole, the versatility in design, scale, resolution, and broad applicability of stereolithographic technologies render them the ideal enabling technology for biomedical and translational medical applications.

GLOSSARY

Bioactive A term used to describe a substance or material that has a biological effect on a living cell, tissue, or organism.

Biocompatible A term used to describe a substance or material that is not harmful to living cells, tissues, or organisms.

Biodegradable A term used to describe a substance or material that can be decomposed by a living cell, tissue, or organism.

Cytotoxic A term used to describe a substance that is toxic to living cells.

Elastomer A material that is compliant or flexible, that is, characterized by a low Young's Modulus and relatively high failure strain.

Hydrogel Term describing a broad class of hydrophilic materials that are highly absorbent and have mechanical properties similar to native tissue.

Photoinitiator A reactive chemical compound that decomposes to form high-energy free radicals, or molecules with unpaired valence electrons, when exposed to light.

Photosensitive Prepolymer (Photopolymer) A solution composed of a polymer and a photoinitiator that can be excited into a radical state through exposure to light and driven to cure into a solid or gel-like structure.

Resin General descriptive term for a photosensitive prepolymer that is used for fabrication using a SLA.

Stereolithography A subset of additive manufacturing technology that relies on using light to cure a photosensitive prepolymer.

ABBREVIATIONS

BMP	Bone morphogenic protein
CAD	Computer-aided design
CHO	Chinese hamster ovary
CT	Computed tomography
DEF	Diethyl fumarate
DEP	Dielectrophoresis
DMD	Digital micromirror device
ECM	Extracellular matrix
EB	Embryoid body
GelMA	Gelatin methacrylate
gMSC	Gingival mesenchymal stem cells
hFOB	Human fetal osteoblastic cells
hMSC	Human mesenchymal stem cells
HUVEC	Human umbilical vein endothelial cells
IR	Infrared
LAP	Lithium phenyl-2,4,6-trimethylbenzoylphosphinate
LVEC	Large vessel endothelial cells
MRI	Magnetic resonance imaging
NIH/3T3	Mouse fibroblast cell line
NVP	*N*-Vinyl-2-pyrrolidone
PDLLA	Poly(D,L-lactide)
PEG	Poly(ethylene glycol)
PEGDMA	Poly(ethylene glycol) dimethacrylate
PLA	Polylactide
PPF	Poly(propylene fumarate)
RGDS	Arg–Gly–Asp–Ser tetrapeptide
SEM	Scanning electron microscopy
SIRC	Statens seruminstitut rabbit corneal cells
SLA	Stereolithography apparatus
TGF	Transforming growth factor
TMC	Trimethylene carbonate
UV	Ultraviolet
VEGF	Vascular endothelial growth factor

ACKNOWLEDGMENTS

We thank our funding sources: National Science Foundation (NSF) Science and Technology Center (STC) Emergent Behavior of Integrated Cellular Systems (EBICS) Grant CBET-0939511, the National Science Foundation (NSF) Grant 0965918 IGERT: Training the Next Generation of Researchers in Cellular and Molecular Mechanics and Bio-Nanotechnology, and National Science Foundation (NSF) Graduate Research Fellowship Program (GRFP) Grant DGE-1144245.

REFERENCES

[1] Melchels FPW, Feijen J, Grijpma DW. A review on stereolithography and its applications in biomedical engineering. Biomaterials 2010;31:6121–30.

[2] Pham D, Gault R. A comparison of rapid prototyping technologies. Int J Mach Tools Manuf 1998;38:1257–87.

[3] Yang S, Leong K-F, Du Z, Chua CK. The design of scaffolds for use in tissue engineering. Part II. Rapid prototyping techniques. Tissue Eng 2002;8:1–11.

[4] Wendel B, Rietzel D, Kühnlein F, Feulner R, Hülder G, Schmachtenberg E. Additive processing of polymers. Macromol Mater Eng 2008;293:799–809.

[5] Bartolo PJ. Stereolithographic Processes. In: Bártolo PJ, editor. Stereolithography: materials, processes and applications. Boston, MA: Springer US; 2011. p. 1–36.

[6] Johnson DW, Sherborne C, Didsbury MP, Pateman C, Cameron NR, Claeyssens F. Macrostructuring of emulsion-templated porous polymers by 3D laser patterning. Adv Mater 2013;25:3178–3181.

[7] Han L-H, Mapili G, Chen S, Roy K. Projection microfabrication of three-dimensional scaffolds for tissue engineering. J Manuf Sci Eng 2008;130. 021005.

[8] Choi J-W, MacDonald E, Wicker R. Multi-material microstereolithography. Int J Adv Manuf Technol 2009;49:543–51.

[9] Maruo S, Nakamura O, Kawata S. Three-dimensional microfabrication with two-photon-absorbed photopolymerization. Opt Lett 1997;22:132–4.

[10] Lee KS, Kim RH, Yang DY, Park SH. Advances in 3D nano/microfabrication using two-photon initiated polymerization. Prog Polym Sci 2008;33:631–81.

[11] Farsari M, Filippidis G, Drakakis TS, Sambani K, Georgiou S, Papadakis G, et al. Three-dimensional biomolecule patterning. Appl Surf Sci 2007;253:8115–8.

[12] Weiß T, Hildebrand G, Schade G, Liefeith K. Two-photon polymerization for microfabrication of three-dimensional scaffolds for tissue engineering application. Eng Life Sci 2009;9:384–90.

[13] Ovsianikov A, Gruene M, Pflaum M, Koch L, Maiorana F, Wilhelmi M, et al. Laser printing of cells into 3D scaffolds. Biofabrication 2010;2. 014104.

[14] Tondiglia VP, Natarajan LV, Sutherland RL, Tomlin D, Bunning TJ. Holographic formation of electro-optical polymer-liquid crystal photonic crystal. Adv Mater 2002;14:187–91.

[15] Moon JH, Ford J, Yang S. Fabricating three-dimensional polymeric photonic structures by multi-beam interference lithography. Polym Adv Technol 2006;17:83–93.

[16] Ullal CK, Maldovan M, Thomas EL, Chen G, Han Y-J, Yang S. Photonic crystals through holographic lithography: simple cubic, diamond-like, and gyroid-like structures. Appl Phys Lett 2004;84:5434.

[17] Heller C, Schwentenwein M, Russmueller G, Varga F, Stampfl J, Liska R. Vinyl esters: low cytotoxicity monomers for the fabrication of biocompatible 3D scaffolds by lithography based additive manufacturing. J Polym Sci Part A1 Polym Chem 2009;47:6941–54.

[18] Davis FJ, Mitchell GR. Polymeric materials for rapid manufacturing. Stereolithography: materials, processes and applications. New York, NY: Springer; 2011. p. 113–39.

[19] Williams CG, Malik AN, Kim TK, Manson PN, Elisseeff JH. Variable cytocompatibility of six cell lines with photoinitiators used for polymerizing hydrogels and cell encapsulation. Biomaterials 2005;26:1211–8.

[20] Fairbanks BD, Schwartz MP, Bowman CN, Anseth KS. Photoinitiated polymerization of PEG-diacrylate with lithium phenyl-2,4,6-trimethylbenzoylphosphinate: polymerization rate and cytocompatibility. Biomaterials 2009;30:6702–7.

[21] Popov VK, Evseev AV, Ivanov AL, Roginski VV, Volozhin AI, Howdle SM. Laser stereolithography and supercritical fluid processing for custom-designed implant fabrication. J Mater Sci Mater Med 2004;15:123–8.

[22] Chu GT-M, Brady GA, Miao W, Halloran JW, Hollister SJ, Brei D. Ceramic SFF by direct and indirect stereolithography. In: Materials Research Society Symposium; 1999: 119–123.

[23] Provin C, Monneret S, Chimie D De. Complex ceramic-polymer composite microparts made by microstereolithography. In: Proceedings of SPIE;2001: 535–542.

[24] Licciulli A, Esposito Corcione C, Greco A, Amicarelli V, Maffezzoli A. Laser stereolithography of ZrO_2 toughened Al_2O_3. J Eur Ceram Soc 2005;25:1581–9.

[25] Chu TMG, Orton DG, Hollister SJ, Feinberg SE, Halloran JW. Mechanical and *in vivo* performance of hydroxyapatite implants with controlled architectures. Biomaterials 2002;23:1283–93.

[26] Hinczewski C, Corbel S, Chartie T. Ceramic suspensions suitable for stereolithography. J Eur Ceram Soc 1998;18:583–90.

[27] Jansen J, Melchels FPW, Grijpma DW, Feijen J. Fumaric acid monoethyl ester-functionalized poly (D,L-lactide)/N-vinyl-2-pyrrolidone resins for the preparation of tissue engineering scaffolds by stereolithography. Biomacromolecules 2009;10:214–20.

[28] Lee K-W, Wang S, Fox BC, Ritman EL, Yaszemski MJ, Lu L. Poly(propylene fumarate) bone tissue engineering scaffold fabrication using stereolithography: effects of resin formulations and laser parameters. Biomacromolecules 2007;8:1077–84.

[29] Bens A, Seitz H, Bermes G, Emons M. Non-toxic flexible photopolymers for medical stereolithography technology. Rapid Prototyp J 2007;13:38–47.

[30] Matsuda T, Mizutani M, Arnold SC. Molecular design of photocurable liquid biodegradable copolymers. 1. Synthesis and photocuring characteristics. Macromolecules 2000;33:795–800.

[31] Matsuda T, Mizutani M. Molecular design of photocurable liquid biodegradable copolymers. 2. Synthesis of coumarin-derivatized oligo(methacrylate)s and photocuring. Macromolecules 2000;33:791–4.

[32] Matsuda T, Mizutani M. Liquid acrylate-endcapped biodegradable poly(epsilon-caprolactone-co-trimethylene carbonate). II. Computer-aided stereolithographic microarchitectural surface photoconstructs. J Biomed Mater Res 2002;62:395–403.

[33] Lee SJ, Kang HW, Park JK, Rhie JW, Hahn SK, Cho DW. Application of microstereolithography in the development of three-dimensional cartilage regeneration scaffolds. Biomed Microdevices 2008;10:233–41.

[34] Cooke MN, Fisher JP, Dean D, Rimnac C, Mikos AG. Use of stereolithography to manufacture critical-sized 3D biodegradable scaffolds for bone ingrowth. J Biomed Mater Res B Appl Biomater 2002;64:65–9.

[35] Lee JW, Lan PX, Kim B, Lim G, Cho D-W. 3D scaffold fabrication with PPF/DEF using micro-stereolithography. Microelectron Eng 2007;84:1702–5.

[36] Choi JW, Wicker R, Lee SH, Choi KH, Ha CS, Chung I. Fabrication of 3D biocompatible/biodegradable micro-scaffolds using dynamic mask projection microstereolithography. J Mater Process Technol 2009;209:5494–503.

[37] Melchels FPW, Feijen J, Grijpma DW. A poly(D,L-lactide) resin for the preparation of tissue engineering scaffolds by stereolithography. Biomaterials 2009;30:3801–9.

[38] Stampfl J, Liska R. Polymerizable hydrogels for rapid prototyping: chemistry, photolithography, and mechanical properties. In: Bártolo PJ, editor. Stereolithography: materials, processes and applications. Boston, MA: Springer US; 2011. p.161–182.

[39] Hinkley JA, Morgret LD, Gehrke SH. Tensile properties of two responsive hydrogels. Polymer 2004;45:8837–43.

[40] Dhariwala B, Hunt E, Boland T. Rapid prototyping of tissue-engineering constructs, using photopolymerizable hydrogels and stereolithography. Tissue Eng 2004;10:1316–22.

[41] Arcaute K, Mann BK, Wicker RB. Stereolithography of three-dimensional bioactive poly(ethylene glycol) constructs with encapsulated cells. Ann Biomed Eng 2006;34:1429–41.

[42] Chan V, Zorlutuna P, Jeong JH, Kong H, Bashir R. Three-dimensional photopatterning of hydrogels using stereolithography for long-term cell encapsulation. Lab Chip 2010;10:2062–70.

[43] Lee SH, Moon JJ, West JL. Three-dimensional micropatterning of bioactive hydrogels via two-photon laser scanning photolithography for guided 3D cell migration. Biomaterials 2008;29:2962–8.

[44] Northen TR, Brune DC, Woodbury NW. Synthesis and characterization of peptide grafted porous polymer microstructures. Biomacromolecules 2006;7:750–4.

[45] Chan V, Collens MB, Jeong JH, Park K, Kong H, Bashir R. Directed cell growth and alignment on protein-patterned 3D hydrogels with stereolithography. Virtual Phys Prototyp 2012;7:219–28.

[46] Mapili G, Lu Y, Chen S, Roy K. Laser-layered microfabrication of spatially patterned functionalized tissue-engineering scaffolds. J Biomed Mater Res B Appl Biomater 2005;75:414–24.

[47] Schuster M, Turecek C, Weigel G, Saf R, Stampfl J, Varga F, et al. Gelatin-based photopolymers for bone replacement materials. J Polym Sci Part A1 Polym Chem 2009;47:7078–89.

[48] Zimmerman J, Bittner K, Stark B, Mulhaupt R. Novel hydrogels as supports for *in vitro* cell growth: poly(ethylene glycol)- and gelatine-based (meth)acrylamidopeptide macromonomers. Biomaterials 2002;23:2127–34.

[49] Gauvin R, Chen YC, Lee JW, Soman P, Zorlutuna P, Nichol JW, et al. Microfabrication of complex porous tissue engineering scaffolds using 3D projection stereolithography. Biomaterials 2012;33:3824–34.

[50] Smeds KA, Pfister-Serres A, Hatchell DL, Grinstaff MW. Synthesis of a novel polysaccharide hydrogel. J Macromol Sci Part A Pure Appl Chem 1999;36:981–9.

[51] Qiu Y, Zhang N, Kang Q, An Y, Wen X. Chemically modified light-curable chitosans with enhanced potential for bone tissue repair. J Biomed Mater Res A 2009;89:772–9.

[52] Gebinoga M, Katzmann J, Fernekorn U, Hampl J, Weise F, Klett M, et al. Multiphoton structuring of native polymers: a case study for structuring natural proteins. Eng Life Sci 2013;13:368–375.

[53] Zorlutuna P, Jeong JH, Kong H, Bashir R. Stereolithography-based hydrogel microenvironments to examine cellular interactions. Adv Funct Mater 2011;21:3642–51.

[54] Schuster M, Turecek C, Kaiser B, Stampfl J, Liska R, Varga F. Evaluation of biocompatible photopolymers I: photoreactivity and mechanical properties of reactive diluents. J Macromol Sci Part A 2007;44:547–57.

[55] Schuster M, Turecek C, Mateos A, Stampfl J, Liska R, Varga F. Evaluation of biocompatible photopolymers II: further reactive diluents. Monatshefte für Chemie – Chem Mon 2007;138:261–8.

[56] Seitz H, Tille C, Irsen S, Bermes G, Sader R, Zeilhofer H-F. Rapid prototyping models for surgical planning with hard and soft tissue representation. Int Congr Ser 2004;1268:567–72.

[57] Mankovich NJ, Cheeseman AM, Stoker NG. The display of three-dimensional anatomy with stereolithographic models. J Digit Imaging 1990;3:200–3.

[58] Bill JS, Reuther JF, Dittmann W, Kübler N, Meier JL, Pistner H, et al. Stereolithography in oral and maxillofacial operation planning. Int J Oral Maxillofac Surg 1995;24:98–103.

[59] Sarment DP, Al-Shammari K, Kazor CE. Stereolithographic surgical templates for placement of dental implants in complex cases. Int J Periodontics Restorative Dent 2003;23:287–95.

[60] De Momi E, Pavan E, Motyl B, Bandera C, Frigo C. Hip joint anatomy virtual and stereolithographic reconstruction for preoperative planning of total hip replacement. Int Congr Ser 2005;1281:708–12.

[61] Binder TM, Moertl D, Mundigler G, Rehak G, Franke M, Delle-Karth G, et al. Stereolithographic biomodeling to create tangible hard copies of cardiac structures from echocardiographic data: *in vitro* and *in vivo* validation. J Am Coll Cardiol 2000;35:230–7.

[62] Goiato MC, Santos MR, Pesqueira AA, Moreno A, dos Santos DM, Haddad MF. Prototyping for surgical and prosthetic treatment. J Craniofac Surg 2011;22:914–7.

[63] El-Siblani A. Advantages of utilizing DMD based rapid manufacturing systems in mass customization applications. Proc SPIE 2010;7596:75960G1–175960G.

[64] Coward TJ, Watson RM, Wilkinson IC. Fabrication of a wax ear by rapid process modeling using stereolithography. Int J Prosthodont 1999;12:20–7.

[65] Runte C, Dirksen D, Deleré H, Thomas C, Runte B, Meyer U, et al. Optical data acquisition for computer-assisted design of facial prostheses. Int J Prosthodont 2002;15:129–32.

[66] Freeman D, Wontorcik L. Stereolithography and prosthetic test socket manufacture: a cost/benefit analysis. J Prosthetics Orthot 1998;10:17–20.

[67] D'Urso PS, Earwaker WJ, Barkert TM, Redmond MJ, Thompson RG, Effeney DJ, et al. Custom cranioplasty using stereolithography and acrylic. Br J Plast Surg 2000;53:200–4.

[68] Wurm G, Tomancok B, Holl K, Trenkler J. Prospective study on cranioplasty with individual carbon fiber reinforced polymer (CFRP) implants produced by means of stereolithography. Surg Neurol 2004;62:510–21.

[69] Morris CL, Barber RF, Day R. Orofacial prosthesis design and fabrication using stereolithography. Aust Dent J 2000;45:250–3.

[70] Liska WD, Marcellin-Little DJ, Eskelinen EV, Sidebotham CG, Harrysson OLA, Hielm-Bjorkman AK. Custom total knee replacement in a dog with femoral condylar bone loss. Vet Surg 2007;36:293–301.

[71] Minns RJ, Bibb R, Banks R, Sutton RA. The use of a reconstructed three-dimensional solid model from CT to aid the surgical management of a total knee arthroplasty: a case study. Med Eng Phys 2003;25:523–6.

[72] Ozan O, Turkyilmaz I, Ersoy AE, McGlumphy EA, Rosenstiel SF. Clinical accuracy of 3 different types of computed tomography-derived stereolithographic surgical guides in implant placement. J Oral Maxillofac Surg 2009;67:394–401.

[73] Griffith LG, Naughton G. Tissue engineering–current challenges and expanding opportunities. Science 2002;295:1009–14.

[74] Yamaoka T, Tabata Y, Ikada Y. Distribution and tissue uptake of poly(ethylene glycol) with different molecular weights after intravenous administration to mice. J Pharm Sci 1994;83:1–6.

[75] He S, Timmer M, Yaszemski M, Yasko A, Engel P, Mikos A. Synthesis of biodegradable poly(propylene fumarate) networks with poly(propylene fumarate)–diacrylate macromers as crosslinking agents and characterization of their degradation products. Polymer (Guildf) 2001;42:1251–60.

[76] Zhang H, Hutmacher DW, Chollet F, Poo AN, Burdet E. Microrobotics and MEMS-based fabrication techniques for scaffold-based tissue engineering. Macromol Biosci 2005;5:477–89.

[77] Ott HC, Matthiesen TS, Goh S-K, Black LD, Kren SM, Netoff TI, et al. Perfusion-decellularized matrix: using nature's platform to engineer a bioartificial heart. Nat Med 2008;14:213–21.

[78] Song JJ, Guyette JP, Gilpin SE, Gonzalez G, Vacanti JP, Ott HC. Regeneration and experimental orthotopic transplantation of a bioengineered kidney. Nat Med 2013;19:646–651.

[79] Andersson H, van den Berg A. Microfabrication and microfluidics for tissue engineering: state of the art and future opportunities. Lab Chip 2004;4:98–103.

[80] Salgado AJ, Coutinho OP, Reis RL. Bone tissue engineering: state of the art and future trends. Macromol Biosci 2004;4:743–65.

[81] Healy KE, Guldberg RE. Bone tissue engineering. J Musculoskelet Neuronal Interact 2007;7:328–30.

[82] Heller C, Schwentenwein M, Russmueller G, Varga F, Stampfl J, Liska R. Vinyl esters: low cytotoxicity monomers for the fabrication of biocompatible 3D scaffolds by lithography based additive manufacturing. J Polym Sci Part A-1: Polym Chem 2009;47: 6941–54.

[83] Lu Y, Mapili G, Suhali G, Chen S, Roy K. A digital micro-mirror device-based system for the microfabrication of complex, spatially patterned tissue engineering scaffolds. J Biomed Mater Res A 2006;77:396–405.

[84] Ovsianikov A, Deiwick A, Van Vlierberghe S, Dubruel P, Möller L, Dräger G, et al. Laser fabrication of three-dimensional CAD scaffolds from photosensitive gelatin for applications in tissue engineering. Biomacromolecules 2011;12:851–8.

[85] Schüller-Ravoo S, Teixeira SM, Feijen J, Grijpma DW, Poot AA. Flexible and elastic scaffolds for cartilage tissue engineering prepared by stereolithography using poly(trimethylene carbonate)-based resins. Macromol Biosci 2013;13:1711–9.

[86] Naumann A, Aigner J, Staudenmaier R, Seemann M, Bruening R, Englmeier KH, et al. Clinical aspects and strategy for biomaterial engineering of an auricle based on three-dimensional stereolithography. Eur Arch Otorhinolaryngol 2003;260:568–75.

[87] Hwang NS, Kim MS, Sampattavanich S, Baek JH, Zhang Z, Elisseeff J. Effects of three-dimensional culture and growth factors on the chondrogenic differentiation of murine embryonic stem cells. Stem Cells 2006;24:284–91.

[88] Novosel EC, Kleinhans C, Kluger PJ. Vascularization is the key challenge in tissue engineering. Adv Drug Deliv Rev 2011;63:300–11.

[89] Chan V, Jeong JH, Bajaj P, Collens M, Saif T, Kong H, et al. Multimaterial bio-fabrication of hydrogel cantilevers and actuators with stereolithography. Lab Chip 2012;12:88–98.

[90] Sodian R, Loebe M, Hein A, Martin DP, Hoerstrup SP, Potapov EV, et al. Application of stereolithography for scaffold fabrication for tissue engineered heart valves. ASAIO J 2002;48:12–6.

[91] Hockaday LA, Kang KH, Colangelo NW, Cheung PY, Duan B, Malone E, et al. Rapid 3D printing of anatomically accurate and mechanically heterogeneous aortic valve hydrogel scaffolds. Biofabrication 2012;4:035005.

[92] Jeong JH, Chan V, Cha C, Zorlutuna P, Dyck C, Hsia KJ, et al. "Living" microvascular stamp for patterning of functional neovessels; orchestrated control of matrix property and geometry. Adv Mater 2012;24:58–63. 1.

[93] Hsieh TM, Ng CWB, Narayanan K, Wan ACA, Ying JY. Three-dimensional microstructured tissue scaffolds fabricated by two-photon laser scanning photolithography. Biomaterials 2010;31: 7648–52.

[94] Melissinaki V, Gill AA, Ortega I, Vamvakaki M, Ranella A, Haycock JW, et al. Direct laser writing of 3D scaffolds for neural tissue engineering applications. Biofabrication 2011;3:045005.

[95] Arcaute K, Mann B, Wicker R. Stereolithography of spatially controlled multi-material bioactive poly(ethylene glycol) scaffolds. Acta Biomater 2010;6:1047–54.

[96] Mirsaidov U, Scrimgeour J, Timp W, Beck K, Mir M, Matsudaira P, et al. Live cell lithography: using optical tweezers to create synthetic tissue. Lab Chip 2008;8:2174–81.

[97] Bajaj P, Marchwiany D, Duarte C, Bashir R. Patterned three-dimensional encapsulation of embryonic stem cells using dielectrophoresis and stereolithography. Adv Healthc Mater 2012;2:450–458.

[98] Deforest CA, Anseth KS. Cytocompatible click-based hydrogels with dynamically tunable properties through orthogonal photoconjugation and photocleavage reactions. Nat Chem 2011;3:925–31.

[99] Chan V, Park K, Collens MB, Kong H, Saif TA, Bashir R. Development of miniaturized walking biological machines. Sci Rep 2012;2:1–8.

Chapter 7

Extrusion Bioprinting

Falguni Pati*, Jinah Jang**, Jin Woo Lee†, and Dong-Woo Cho*

*Department of Mechanical Engineering, Pohang University of Science and Technology (POSTECH), Pohang, Kyungbuk, South Korea; **Division of Integrative Biosciences and Biotechnology, Pohang University of Science and Technology (POSTECH), Pohang, Kyungbuk, South Korea; †Department of Molecular Medicine, Graduate School of Medicine, Gachon University, Incheon City, South Korea

Chapter Outline

ABSTRACT

Extrusion bioprinting has exciting prospects in fabricating organized tissue constructs by simultaneous dispensing of cells and matrix materials to repair or replace damaged or diseased tissues and organs. The exhilarating capability of this technique is to generate constructs with spatial variations of cells along multiple axes with high geometric complexity. A variety of bioprinters based on extrusion principle have been developed around the world and many are even commercialized. These computer-controlled technologies to design and fabricate tissues will enhance our understanding of the governing factors of tissue formation and function. Moreover, it will provide a valuable tool for several emerging technologies including tissue engineering. In this chapter, we discuss the rationale of extrusion-based bioprinting for fabrication of tissue constructs. Current strategies of building tissue constructs are discussed; in particular, their merits and demerits and the requirements to move the current concepts to practical application are highlighted.

Keywords: biofabrication; extrusion-based printing; synthetic polymer; hydrogel; bioink; 3D construct; hybrid structure; tissue regeneration

1 INTRODUCTION

Bioprinting is defined as the automated computer-aided deposition of living cells, extracellular matrix (ECM) components, and biochemical factors at a specified position with adequate numbers and right combinations for the development of three-dimensional (3D) tissue constructs [1,2]. This technique offers reproducible control over the placement and distribution of cells or bioactive factors within the scaffold by layer-by-layer depositions of cells and matrix materials. Despite the initial challenges, bioprinting has experienced a rapid growth in comparison to other emerging technologies. The growing interest in bioprinting is primarily because of its innumerable merits over other existing technologies. Among these merits are the following [1–4]: (1) straightforward operation; (2) rapid and inexpensive method to generate geometrically well-defined scaffolds; (3) enables the use of polymers or ceramics and other bioactive molecules; (4) capable of manipulating cells via a computer-controlled fabrication process; (5) capable of high-throughput creation of models

Essentials of 3D Biofabrication and Translation. http://dx.doi.org/10.1016/B978-0-12-800972-7.00007-4

of spatially and temporally well-controlled complex constructs; (6) provides 3D complexity by multilayer printing; and (7) computer-aided design (CAD) combined with medical imaging techniques to make anatomically shaped implants. The fundamental principle of bioprinting is that the fabrication of structure can be directed by simultaneous placement of cells and matrix materials, rather than by fabricating a solid support structure alone [5]. Even though it is still at an early stage of concept development and proof-of-principle experiments, bioprinting has the potential to deliver clinical solutions on the longer term [6]. With bioprinting, anatomically shaped implants can be made using CAD combined with medical imaging techniques [7]. This method can open up new avenues to scalable and reproducible mass production of tissue precursors and solve the problem of tissue/organ shortage [1]. In recent years, there has been an increasing consideration to apply this technology for various applications in biology and medicine. The development of diverse bioprinting technologies also enables us to include highly sensitive stem cells. This can be witnessed by the utilization of various kinds of stem cells such as human bone marrow stem cells (BMSCs), adipose-derived stem cells (ASCs), and even highly sensitive embryonic stem cells (ESCs) [8–10]. Moreover, the fate of stem cells can also be directed by controlling the microenvironments with spatially engineered gradients of immobilized macromolecules through bioprinting [11–13].

Based on the working principles, bioprinting systems can primarily be classified as: (1) laser based, (2) inkjet based, or (3) extrusion based. Extrusion-based bioprinting technique is focused on the printing of living cells or bioactive molecules by extruding continuous filaments made of biomaterials. Honestly, this process is a practical combination of a fluid-dispensing system and an automated three-axis robotic system for extrusion and printing, respectively [14]. Biomaterial is generally dispensed by a pressure-assisted system, either pneumatic or piston-driven during printing, which results in precise deposition of encapsulated cells or bioactive molecules in the form of cylindrical filaments for producing desired 3D structures. Extrusion-based bioprinting was believed to provide reasonably better structural integrity due to continuous deposition of filaments. This technology lays the foundation for cell patterning and is most effective for rapid fabrication of 3D porous cellular structures [15].

However, this technique has some limitations such as shear-stress-induced cell deformation and limited material selection. Potential apoptotic effects during and after bioprinting have been considered as the foremost concern by the potential future end-users of this technology. This is primarily because of the pressure drop associated with extrusion-based systems as the process requires extruding cells or materials through a micronozzle. However, this issue can be resolved to a large extent by optimization of the process parameters such as biomaterial concentration, nozzle pressure, nozzle diameter, and loaded cell density [16]. Furthermore, when hydrogels are used they should be solidified or gelled rapidly subsequent to extrusion and thus avoid deformation of the printed structure. Cellular viability and long-term functionality post-printing with the hydrogels are known to influence the future tissue building process.

In case of extrusion-based dispensing, printing fidelity generally increases with increasing viscosity, as material with higher viscosities and instant gelation (cross-linking) would facilitate the maintenance of the filamentous shape after deposition [17]. But high viscosity, on the other hand, can cause clogging of the nozzle tip and should be optimized based on the diameter of the nozzle tip [15,18]. Furthermore, highly concentrated materials would provide a restrictive environment for mass transport, cell proliferation, migration, and ECM accumulation [19]. Unfortunately, restricted biomaterials choice and limited resolution hinder the widespread application of extrusion-based systems. In this chapter, we describe the principle and methods of several extrusion-based bioprinting systems. We also discuss various synthetic polymers as well as hydrogels used for extrusion bioprinting along with their processing conditions. Recent achievements along with the challenges and future directions were discussed.

2 EXTRUSION-BASED BIOPRINTING SYSTEM

Extrusion-based bioprinting systems rely on dispensing of larger polymer or hydrogel strands through a micronozzle and positioning them via computer-controlled motion either of the printing heads or collecting stage. For this method, polymer or hydrogel is generally loaded in metallic or plastic syringes and dispensed via either pneumatic, piston-driven, or screw-driven force on a building platform (Fig. 7.1). The resolution that can be achieved with extrusion-based printing is in the order of 200 μm, which is considerably low compared to laser- or inkjet-based systems. On the contrary, fabrication speed using this technique is significantly higher and anatomically shaped constructs can be generated [20]. The dispensing of materials is based on pneumatic, piston-driven, or screw-driven force and each has its own merits and demerits. Pneumatic systems are generally associated with a delay in dispensing due to the compressed gas; however, they work better for highly viscous molten polymer. On the other hand, the piston-driven deposition generally offers more direct control over the flow of the hydrogel from the nozzle. Conversely, screw-based systems provide more spatial control and are valuable for the dispensing of hydrogels with higher viscosities [21,22]. Importantly, cells can be incorporated within the hydrogels and be deposited with high viability using pneumatic and piston-driven systems. One concern using the screw-driven

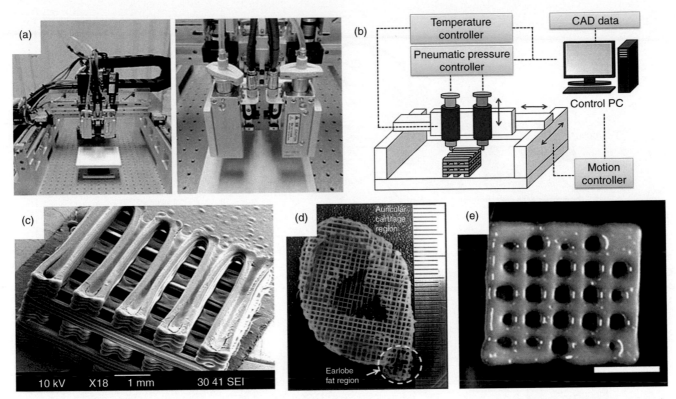

FIGURE 7.1 Images of in-house developed 3D printer and printed structures. (a) A front view of multihead dispensing system (MHDS) and (b) its schematic diagram. (c) A SEM image of the fabricated hybrid scaffold, (d) an image of cell-printed heterogeneous ear tissue, and (e) a printed decellularized heart ECM construct with cells (scale bar, 5 mm) (Unpublished data). *(Figure reprinted with permission from Refs. [6,39,120,141].)*

extrusion system is that it can generate larger pressure drops at the nozzle, which can be detrimental for the encapsulated cells. However, this problem can be resolved by specifically designing the screw for bioprinting rather than using off-the-shelf screws designed for other applications. Taken together, extrusion bioprinting can be regarded as the most promising as this technology allows the fabrication of organized constructs of clinically relevant sizes within a realistic time frame [5].

2.1 Scaffold Fabrication Based on Extrusion Principle

2.1.1 Fused Deposition Modeling

Fused deposition modeling (FDM) is one of the 3D printing technologies commonly used in bioprinting. This technology was developed by Crump in the late 1980s and was successfully commercialized in 1990 [23]. The traditional FDM system consists of a z-direction controllable head-heated liquefier attached to a carriage moving in the horizontal x–y plane. The main function of the liquefier assembly is to heat and extrude the filament materials through a nozzle directly onto the building platform following a preprogrammed path, essentially based on computer-aided manufacturing (CAM). Subsequent to fabrication of

one layer, the x–y plane of the machine moves up one step in the z direction to deposit the next layer. The final product is produced by a layer-by-layer process. By changing the direction of material deposition for successively deposited layers and the spacing between the filaments, scaffolds with highly controllable pore geometry and complete pore interconnectivity can be manufactured.

In comparison to other 3D printing techniques, the FDM method does not require any solvent and it offers several advantages, such as ease and flexibility in handling and processing of materials. The extrusion of filament type materials also allows for continuous production without the need for replacing feedstock. However, the FDM technique suffers from the need for preformed fibers of uniform size and material properties to feed through the rollers and nozzle. In addition, only a limited number of biodegradable materials are being used like poly-caprolactone (PCL), polylactic acid (PLA), and poly lactic-co-glycolic acid (PLGA). Therefore, several modified FDM processes were developed and evaluated to overcome the limitations of the traditional FDM process for tissue scaffold fabrication [2–13,24–27]. The efficacy of the commercialized FDM technique for scaffold design and fabrication was demonstrated by various groups (Fig. 7.2) [24,25]. Scaffolds with regular geometrical honeycomb pores were manufactured with pore/channel sizes ranging from 160 μm to 700 μm and with porosities of

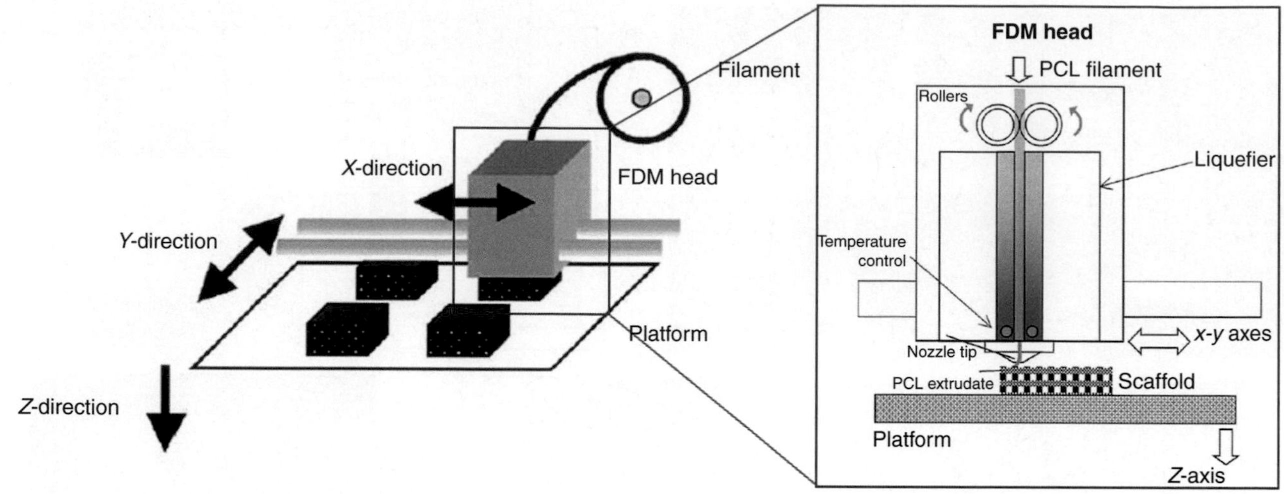

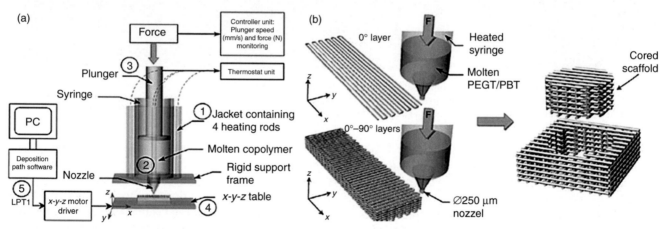

FIGURE 7.2   Schematic diagram of the FDM extrusion and deposition process. The nozzle tip could be changed into different sizes for different road width and thickness. The layers of roads were fused together upon solidification to form a 3D structure. *(Figure reprinted with permission from Zein et al. [24].)*

FIGURE 7.3   (a) The fiber deposition device and (b) the 3D deposition process where 250 μm PEGT/PBT fibers are successively laid down in a computer-controlled pattern (0°– 90° orientation shown). Scaffolds are subsequently cored from the deposited bulk material. *(Figure reprinted with permission from Woodfield et al. [28].)*

48–77%. The mechanical properties of these scaffolds were found to be generally dependent on porosity, regardless of the lay-down pattern and channel size. These results are in agreement with theoretical concepts on the structure–property relationships of porous solids.

Wang et al. developed and characterized a fiber deposition technique for producing 3D PEG-terephthalate (PEGT)-poly(butylene terephthalate) (PBT) block copolymer scaffolds with a 100% interconnecting pore network (Fig. 7.3) [28]. They used the plunger to squeeze the molten copolymer and four heating rods were installed to melt the polymer. Here, the printing head is stationary and the x–y–z table is in motion for the fabrication of scaffolds. By varying the PEGT/PBT composition, porosity, and pore geometry, 3D-deposited scaffolds were produced with an equilibrium modulus and a dynamic stiffness ranging from 0.05 MPa to 2.5 MPa and 0.16 MPa to 4.33 MPa, respectively.

Our group developed the precision extruding deposition (PED) system (Fig. 7.4) [29]. In contrast to the conventional FDM process, which requires the use of precursor filaments, the PED process directly extrudes scaffolding materials from a granulated or pellet form without filament preparation and freeform deposits according to the designed microscale features. The typical pore size of fabricated PCL scaffold ranges from 200 μm to 300 μm, which is near the optimal size suggested for bone tissue scaffold applications. The compressive modulus of scaffolds ranges from 150 MPa to 200 MPa. The preliminary results of biological experiments demonstrate the biocompatibility of the PED process and PCL materials.

Our group also developed a multihead deposition system (MHDS) having four printing heads and capable of dispensing four different polymers or gels or both simultaneously. They demonstrated the adhesion and proliferation of human

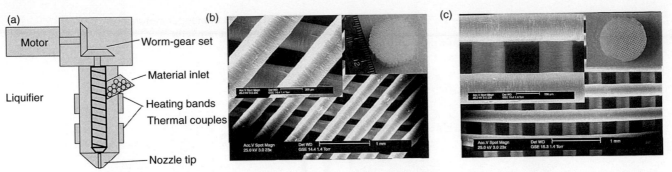

FIGURE 7.4 (a) Schematic of material miniextruder. (b) SEM image of the scaffold with 0°/120° layout pattern. (c) SEM image of the scaffold with 0°/90° layout pattern. *(Figure reprinted with permission from Wang et al. [26].)*

BMSCs on scaffolds made of various biodegradable materials by MHDS (Fig. 7.1) [29,30]. The MHDS is considered to be superior to other commercialized systems based on the FDM process as it can conveniently and quickly fabricate scaffolds from various materials. Furthermore, being a MHDS, a structure combination with synthetic polymer and hydrogel, that is, a hybrid structure, can be fabricated. The MHDS enables the fabrication of 3D tissue scaffolds with a resolution of several tens of microns. In addition, multiple printing heads installed in the MHDS permit rapid fabrication and manufacture of composite scaffolds with various biomaterial compositions. PCL, PLGA, blended PCL/PLGA, and blended PCL/PLGA/TCP scaffolds, which have the same inner/outer architecture and interconnected porosity, were effectively manufactured.

2.1.2 Solution-Based Deposition System

A thermal process to fabricate polymeric scaffolds using FDM cannot be utilized to encapsulate bioactive factors, such as bone morphogenetic proteins (BMPs), within the scaffolds due to the thermal protein denaturation. Therefore, a solution-based deposition system that can fabricate the scaffold below or around physiological temperature (≤37°C) has been investigated. Generally, the polymeric biomaterials are dissolved in organic solvent. When the dissolved polymer solution is deposited by the use of a syringe in 3D, a fusing module of a thermoplastic polymer, which causes the protein denaturation, is not needed for deposition. Vozzi et al. developed a scaffold fabrication system using the pressure-assisted microsyringe, which is based on the use of a microsyringe that utilizes a computer-controlled, three-axis micropositioner and allows control of motor speeds and position (Fig. 7.5) [31,32]. A PLGA solution was deposited with great control from the needle of a syringe by the application of a constant pressure of 20–300 mm Hg. They fabricated PLGA scaffolds with feature sizes of approximately 10–30 μm.

Our group developed a polymer solution dispensing MHDS to fabricate the HA-PLGA scaffold, which is conjugated with BMP-2 (Fig. 7.6a) [33]. They used dichloromethane and chloroform to dissolve the biopolymer. To accelerate the evaporation, heated air blower/air knife was installed at the side of the scaffold fabrication plane (Fig. 7.6c). Also, a suction system was installed to prevent the release of harmful solvents to the outside (Fig. 7.6b).

2.1.3 Direct-Write Electrospinning

One common nanofibrous scaffold fabrication technique is electrospinning, which was originally proposed in 1934 [34]. It is a process that utilizes a large electric field between a polymer solution dispensing micronozzle and a collection plate to induce a stable jet to be ejected from the solution reservoir and travel toward the collecting plate [35]. Solution conditions (pH, concentration, solvent), device conditions (the distance between the tip and plate, strength of the electric field, nozzle dimensions), and collection methods (plate vs. rotating mandrel, speed of collection) affect fiber diameter and fibers from micro- to nanosize can be fabricated. The result of this process is a nonwoven fiber mat that is randomly oriented, because the whipping motion of electrospun nanofibers makes the geometry difficult to control. To solve this alignment issue, many studies have been progressed and direct-write electrospinning (DWES) was developed. DWES can write a pattern of single nanofiber or nanofibers. Therefore, it can be used to fabricate the 3D scaffold with well-controlled geometry.

Jeong's group adopted focusing and scanning functionality on DWES to fabricate a patterned nanofibrous mat with regular pores (resulting from patterns) in microscale and ECM-like random deposition of nanofiber, which can be used as a layer of stacked nanofibrous 3D scaffold (Fig. 7.7) [36]. As a result, nanofibrous 3D scaffolds with well-defined pores and geometry can be fabricated. Bellan and Craighead proposed a fascinating method using a secondary electrode for focusing and jet-steering (Fig. 7.8) [37]. They used electric fields to confine and steer an electrospun polymer jet for controlled deposition of functional materials and used an electrode between the electrospinning tip and grounded sample to suppress the chaotic whipping

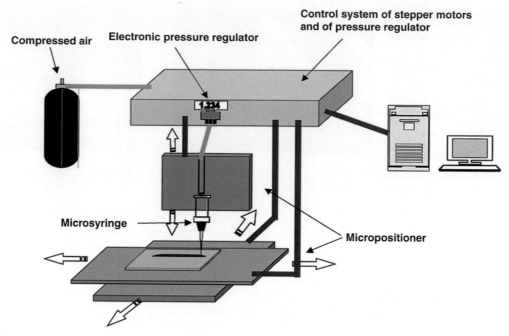

Compressed air

Electronic pressure regulator

Control system of stepper motors and of pressure regulator

1.234

Microsyringe

Micropositioner

FIGURE 7.5 **Schematic elucidating the microsyringe method.** *(Figure reprinted with permission from Vozzi et al. [32].)*

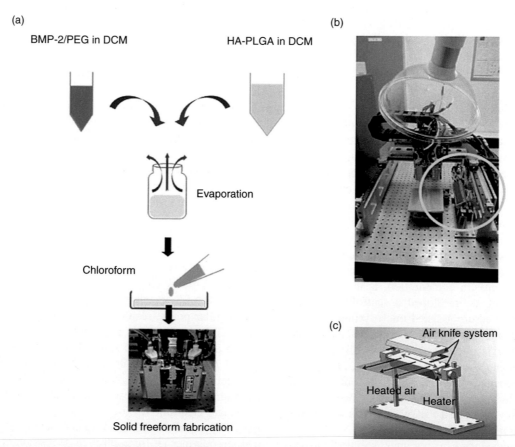

(a)

BMP-2/PEG in DCM

HA-PLGA in DCM

(b)

Evaporation

Chloroform

(c)

Air knife system

Heated air

Heater

Solid freeform fabrication

FIGURE 7.6 (a) Schematic illustration of the preparation of feeding solution for 3D printing of TE scaffolds using an MHDS. (b) Photograph of the solution-based MHDS with heated air blower/air knife (green circle) and suction systems. (c) Schematic representation for the heated-air-blower/air-knife system in the MHDS. *(Figure reprinted with permission from Park et al. [33].)*

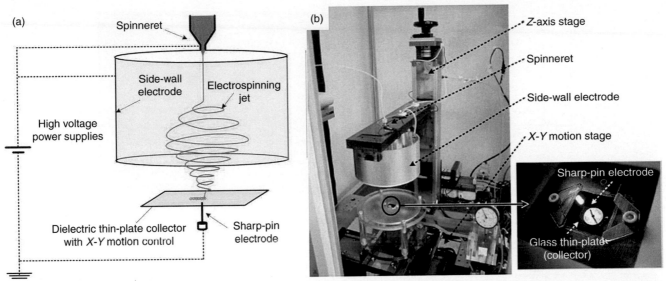

FIGURE 7.7 (a) Schematic of the direct-write electrospinning apparatus. (b) Actual experimental setup. A borosilicate glass plate with a thickness of about 100 μm was used as the collector. *(Figure reprinted with permission from Lee et al. [36].)*

mode, thereby focusing the characteristic spot size of the deposited fibers to a smaller diameter. Dalton et al. demonstrated that simple nanofibrous patterns with line widths as small as 500 μm can be fabricated via melt electrospinning with a larger tip-to-collector distance and lower speed, using a plate collector with two-dimensional (2D) motion (Fig. 7.9) [38]. The electrospun fibers collected in focused spots was used in the patterning and drawing of a cell adhesive scaffold.

2.2 Cell Printing technology Based on Extrusion Principle

The cell printing technique is essentially a route for fabricating cell-laden construct. This technique demands the use of an aqueous gel of hydrogel as living cells are usually encapsulated within the materials. The material for embedding cells in this process is called "bioink" and it should meet several physicochemical requirements. The suitable medium selection for fabricating cell-laden structure requires investigation of various materials, which have an appropriate yield stress, viscosity, and functionality for successfully building an engineered tissue. The hydrothermal controlled environment also needs to be fostered around the bioprinting system so as to prevent the fabricating structure from desiccation. In addition, the optimized bioprinting process and conditions, including a temperature, feed rate, dispensing speed, and nozzle diameter, have to be developed according to the types of bioink.

The hydrogel is widely used as a bioink in bioprinting processes due to its similarity to the native ECM. It also can provide the embedded cells with the proper biochemical and physical stimuli to guide cellular processes. The highly hydrated network structure permits the exchange of nutrients, wastes, and gases, and it allows easy formulation of flowable inks containing cells. Hydrogel precursor solutions are typically extruded by a 3D printer through tapered conical needles to minimize thixotropic behavior, and the applied pressure is in the range of 30–400 kPa [16,39] in the case of a pneumatic pressure. After fabricating a hydrogel-based cell-laden structure, the structure needs to have an adequate stability and mechanical properties, which is done by proper cross-linking methods for *in vitro* culture and *in vivo* implantation. In addition, the swelling or contraction characteristics of the bioink must be considered so that the deformation of the final construct can be prevented by the use of similar swelling behavior.

2.2.1 Hydrogels for Cell Printing

The hydrogels are classified into several types with respect to their structural, chemical, and biological characteristics and used both as a construction material and as a cell delivery vehicle [40–42]. Hydrogels are polymeric networks that absorb large amounts of water without solubilizing and preserving their characteristic 3D structure. This can be attributed to the large number of physical or chemical links present in between the polymer chains. Depending on the nature of the network, they can be distinguished as either physical or chemical gel. In the case of physical hydrogels, the network formation is reversible. In contrast, the chemical hydrogels are formed by irreversible, covalent cross-links. There also exist hydrogels with a combination of physical and chemical networks, for example, gelatin modified with methacrylamide groups [43]. The biocompatibility of hydrogels is determined by their hydrophilicity, which make them attractive as drug and cell carriers, and specifically

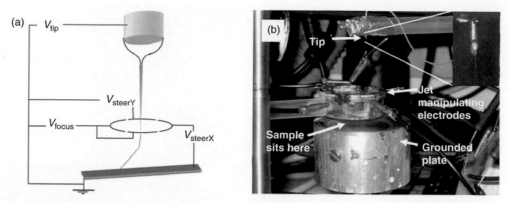

FIGURE 7.8 (a) Schematic of controlled electrospinning system not to scale, and (b) picture of controlled electrospinning setup. *(Figure reprinted with permission from Bellan and Craighead [37].)*

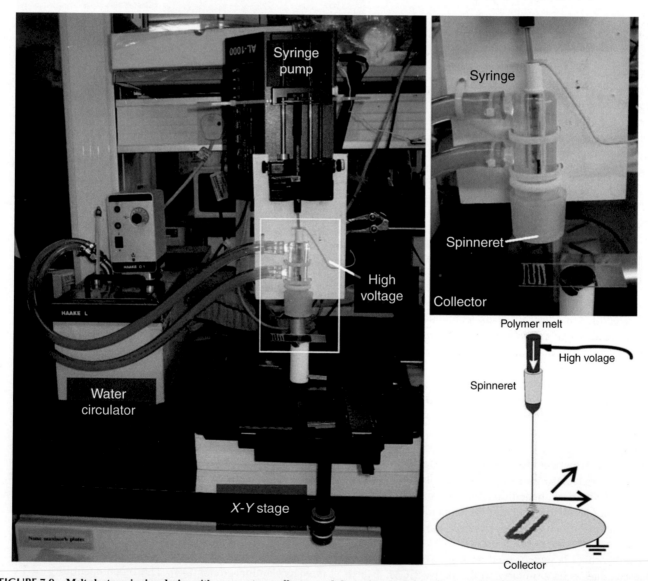

FIGURE 7.9 **Melt electrospinning device with an *x*–*y* stage collector, and the syringe and the collector enlarged.** As proposed with the inset schematic, buckling and bending occur close to the collector, and fibers are collected in a focused patch. The movement of the *x*–*y* stage and collector results in the controlled collection of the fibers. *(Figure reprinted with permission from Dalton et al. [38].)*

for the design and fabrication of tissue constructs [44]. The cell encapsulation strategies have frequently been applied to fabricate cell-laden constructs [45–48]. A wide variety of cells, including fibroblasts, adipocytes, chondrocytes, hepatocytes, smooth muscle cells, neuronal cells, and even stem cells, have been viably encapsulated within hydrogels [49]. The hydrogels are especially attractive for repairing and regenerating soft tissue due to their feeble mechanical properties [50,51].

Based on their origin, hydrogels are usually classified as either naturally derived or synthetic. Naturally derived gels are often ECM constituents and thus generally support good cellular attachment and function. However, they suffer from some intrinsic problems such as limited processability, batch-to-batch variation, and possibility of disease transfer. Synthetically derived hydrogels bear none of these disadvantages as their production is well standardized, but they often lack biofunctionality. Apart from these two types of hydrogels, hybrid gels having natural and synthetic components are gaining popularity in bioprinting due to their tailorable mechanical properties, swelling behavior, and degradation kinetics properties. For example, the methacrylate chemistry was effectively utilized for the functionalization of naturally derived hydrogels, such as gelatin, hyaluronic acid (HA), and dextran, to enable (photo-initiated) cross-linking in combination with robotic dispensing [52,53]. This functionalization with methacrylate or methacrylamide groups is versatile and can be extended to more naturally derived hydrogels, including alginate [54]. The chemical modification of naturally derived hydrogels allows preservation of the typical characteristics of both the polymers, like intrinsic biofunctionality of natural hydrogel with the tunability of synthetic hydrogel. Furthermore, by introducing the chemical cross-links at controlled densities, post-printing distortion of fabricated shapes can be prevented. On the other hand, to introduce the biofunctionality in synthetic gels, they are increasingly being functionalized with biologically active components such as cell-adhesive peptides, covalently bound growth factors, heparan sulfate, and protease-cleavable cross-links [55].

From the bioprinting point of view, the main challenge of employing hydrogels is shaping them in predesigned geometries. During bioprinting of 3D tissue constructs, a hydrogel precursor solution with encapsulated cells is generally operated to create a defined, designed shape that is subsequently stabilized by gelation. Higher viscosity of the hydrogel usually overcomes surface-tension-driven droplet formation, which enables drawing of thin strands of material and eventually well-defined shapes, and also prevents cells from settling during the fabrication process. A relatively quick gelation is thus required to retain the shape of the fabricated structure. This gelation typically occurs through cross-linking reaction, initiated either by light (photosensitive), by a chemical, by hydrophobic or complexation interactions (ionic), or by a thermal transition (thermosensitive) [56–59]. All of these approaches utilize sol–gel reactions based on photo/thermo/chemical gelation phenomena, and have shown promising outcomes for utilizing hydrogels to construct 3D structures. However, this cross-linking reaction and the overall bioprinting process obviously should not compromise cell viability. Adequate mechanical strength is another important requirement to retain the designed and fabricated shape. Most bioprinting processes require gels with superior mechanical properties than that of gels used for casting and molding. 3D structures with interconnected porosity can only be accurately and reproducibly prepared when the elastic modulus and gel strength are sufficiently high.

Hydrogels can provide the embedded cells with a native tissue mimicking 3D environment and thus are being encouragingly used for fabrication of cell-laden constructs. The hydrogels are required to meet the challenges associated with cell encapsulation and tissue development. Most of the hydrogels used for cell encapsulation are chemically cross-linked. Covalent cross-linking was mostly used to increase the stability of those hydrogels. However, these cross-linking create firmer 3D networks of polymer chains with meshes that often are in the orders of magnitude smaller than cells. This imposes a huge opposing effect on the mobility of encapsulated cells; predominantly cell migration. Cell migration is an essential process for tissue remodeling and regeneration. As a result of this the proliferation of the encapsulated cells is also completely arrested due to the space crunch [55]. The hydrogel should allow space for these processes to occur for cell proliferation, remodeling, and vascularization within the engineered tissues. However, this process can be reversed if suitable degradation sites are incorporated into hydrogels, allowing for cell-mediated matrix degradation and permit cell migration and proliferation [19,60]. A number of mechanisms are available to incorporate the degradation sites within the hydrogel like hydrolytically driven [61], or even light-driven by incorporation of photo-degradable linkers [62]. Tailored microgeometry like inclusion of macroporosity in the hydrogel construct can facilitate vascularization and thus designed branched vascular networks were increasingly becoming an essential component of a manufactured tissue [63]. However, commonly there lies a tissue-specific approach for engineering the hydrogels for certain applications. For example, in cartilage tissue engineering (TE), cell proliferation and migration are not always welcome at the initial stage. Often, a highly dense cell population is encapsulated for the sole aim of achieving high matrix production. The beneficial effect of using high densities of cells and their associated ECM has been demonstrated for cartilage tissue repair [64]. Nevertheless, the mesh size is still an important parameter as it influences the diffusion of nutrients and cellular secretion like proteins and glycosaminoglycans throughout the gel [65].

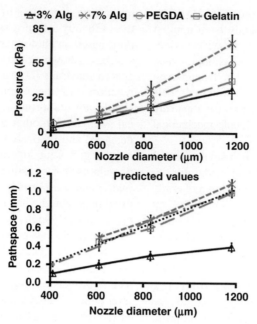

FIGURE 7.10 Predicted optimal printing parameters for varying nozzle diameters and hydrogels. Criteria for optimal parameter were defined as: length accuracy above 90%, resolution index 1, and layer thickness 75% of the nozzle diameter. (a) Pressure versus diameter for each material. (b) Actual pathspace versus diameter for each material. Error bars represent effective range. Black dotted line represents predicted optimal path space values determined from modeling. *(Figure reprinted with permission from Kang et al. [67].)*

A particular challenge for the application of cell-laden hydrogels in extrusion bioprinting is to find a suitable concentration that can meet the processing conditions for accurate printing and also support cell viability and function. This is a challenging task as often these criteria impose opposing requirements. Printing fidelity generally increases with polymer concentrations and cross-link density, whereas high polymer concentration hinders cell migration and proliferation and subsequent ECM formation. Thus, a bioprinting window exists for each polymer that can be used for both printing and cell encapsulation; however, often this is a very small processing window. This bioprinting window generally differs from one hydrogel system to another as the polymer concentration and cross-linking density influence the processability. Researchers have now taken a leap toward optimizing gel parameters and processing conditions in a systematic and quantitative ways after several years of predominantly proof-of-principle studies demonstrating the (bio) printability of a gel [66]. The development of new polymers specifically for bioprinting of cell-laden constructs may help overcome the limitations of current gels and expand the bioprinting window.

2.2.2 Process Parameters

The cell-laden structure is fabricated using the patient's computed tomography (CT)-derived CAD so that it can perfectly fit the native tissue or organ structure. However, it has to satisfy various physical properties such as complexity, high resolution, and shape stability. The resolution of

the extrusion bioprinting can be determined by changing the concentration of materials or the condition of the bioprinter, including the temperature, feed rate, dispensing speed, and nozzle diameter (Fig. 7.10) [67]. The parametric study with different process parameters needs to be conducted to analyze and verify cell injury. The high pressure and small nozzle diameters are the best example to increase the damage to the cells. The size of the droplet, which can be generated by applying the force to the bioink-laden syringe, and the speed of the printer head (feed rate) determine the line width of the construct [68]. Although, use of low pressures and bigger sized nozzle is favorable for cell viability after printing, it can result in a structure with low shape fidelity or limited resolution. Hence, it is important to take into consideration the advantages and drawbacks of using various process parameters and select the conditions for the fabrication of the tissue construct.

2.2.3 Rheology of Hydrogel

Rheology, the study of the flow of material in response to an applied force, is highly relevant to bioprinting. Newtonian fluids can be characterized by disclosing a constant viscosity for a varying shear rate; whereas, the viscosity of non-Newtonian fluids change with the shear rate. Most of the hydrogel is generally classified into the non-Newtonian fluids. Furthermore, it presents a shear thinning behavior in the shear rate due to the shear-induced reorganization of the polymer chains. The analysis of viscoelastic behaviors of the hydrogel can be carried out using a rheometer, and two different rheological measurements, which are rotational and oscillatory tests, are widely used to characterize the hydrogel.

3D printing of clinical-sized tissue, which is considered a size larger than 10 mm, with high anatomical precision has been primarily demonstrated for bone and cartilage. On the other hand, the soft tissues, such as the heart, liver, and lung, have been fabricated up to 5 mm tall or less because these are dominated by the geometric complexity or non-self-supporting geometries [69]. The viscosity is the resistance of a fluid to flow upon application of stress, and the viscosity of a polymer solution, such as a pregel, is determined by the polymer concentration and molecular weight. A high viscosity usually prevents the collapse of deposited structures during printing and it directly influences shape fidelity postprinting. However, the high number of investigations does not take rheology into account when developing or evaluating hydrogels for bioprinting. On the other hand, investigation for standardizing the range of viscosity for bioprinting is very helpful for successful tissue reconstruction.

The bioink should have specific material properties that not only satisfy the fabrication parameters to print a structure but also meet the biological needs to maintain cell viability and functionality. A high shear stress can be generated when high viscous bioink is used and can cause shear-induced cell death. However, the harmfulness of the high shear stress is still controversial. It is because when shear stress is directly applied to the cells in a 2D environment, it influences the cell attachment, whereas the embedded cells in the hydrogel can be protected from the immediate stress. For example, chondrocytes under a 1–10 Pa of shear stress start to change in cell morphology and metabolic activity [70], and hamster kidney cells are detached from the substrate under a 34 Pa shear stress [71]. On the other hand, the printed endothelial cells have been shown to stay alive for shear stress levels up to 1150 kPa [72]. This level of shear stress is at least two to three orders of magnitude higher than typical values for detaching cells from a surface or influencing cell morphology and metabolism. The apoptotic and necrotic cells can also be generated by the process-induced mechanical perturbations damage to cell membrane integrity. Moreover, most of the studies have been focused on verifying the short-term exposure effects; the effects of the high shear stress on cell function over longer periods must be investigated.

2.2.4 Cross-Linking Mechanisms of Hydrogels

The bioink, after dispensing through the bioprinter, is usually stabilized via several cross-linking methods. Hydrogel polymerization can occur via physical, chemical, or combinatorial processes. Physical cross-linking mechanism is a nonchemical reversible interaction and this is the most prominent cross-linking method used for the bioprinting process. Ionic cross-linking is a reversible cross-linking mechanism and it is applied to form the alginate or chitosan hydrogel by adding di/trivalent cations. Effective conditions for ionic cross-linking are relatively well established so that it can be directly used to the bioprinting system. Thermosensitive polymers form a gel by physical interactions, including entanglements of polymer chain, hydrogen bridges, or hydrophobic interactions between the polymer units. The printed thermosensitive polymers rapidly form a gel by changing the temperature. Those physical cross-linking mechanisms generally provide weak mechanical properties; thus, the modification of hydrogel with photosensitive sequences that result in covalent cross-linking is the popular approach in bioprinting. Chemical cross-linking can lead to the formation of a hydrogel by connection of gel precursors through newly formed covalent bonds. It is usually achieved by mixing two low-viscosity solutions with gel precursors, such as monomers and an initiator, or UV-light exposure. Cross-linking kinetics of hydrogel precursors has a very stringent condition due to its low viscosity. One possibility to exploit chemical cross-linking for biofabrication is the use of a reactive mixing head (Fig. 7.11).

3 BIOFABRICATION STRATEGIES

3.1 Scaffold-Free vs Scaffold-Based Tissue Manufacturing Approaches

From the TE perspective, two major distinct premises are being heavily explored regarding the aspect of cell migration, proliferation, differentiation, and organization behavior [3]. Among them, the first approach is based on the presupposition that a 3D biomaterial scaffold serves as an ECM mimetic and acts as a necessary cell guide and supporting template for the seeded or embedded cells [73,74]. On the contrary, the second approach considers that cells have an extensive potential to self-organize through cell–cell interactions and is described as "scaffold-free TE" [3]. Basically, the former theory capitalizes on the role of a supporting structure as a cell guide and minimizes the potency to self-assembly of cells, whereas the latter theory takes full advantage of the self-assembly process of cells.

The foundation of using cells or aggregates of cells as building blocks for the manufacture of tissue constructs without any preformed structure is that aggregates of cells can fuse through cell–cell and cell–ECM interactions and transform into larger structures, similar to embryonic development [75]. This process is advantageous to direct tissue formation as it is believed that aggregates of thousands of cells (also referred to as tissue spheroids or embryoid bodies) should be used for tissue manufacturing instead of suspended single cells. Self-assembly of closely placed cell aggregation was perceived by the tissue fusion process

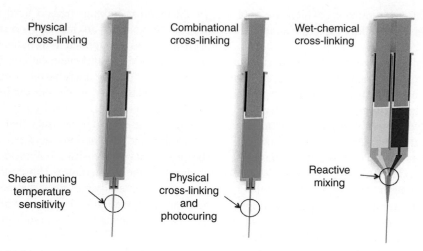

FIGURE 7.11 **Graphical illustration of physical, combinational, and wet-chemical cross-linking mechanisms for extrusion-based biofabrication.** *(Figure reprinted with permission from Malda et al. [17].)*

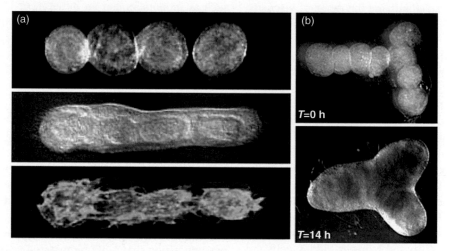

FIGURE 7.12 **Bioprinting of segments of intraorgan branched vascular tree using unilumenal vascular tissue spheroids.** (a) Sequential steps of tissue fusion of vascular tissue spheroids placed in collagen type 1 hydrogel. (b) Fabrication branched vascular segments from unilumenal vascular tissue spheroids in collagen type 1 hydrogel (before and after tissue fusion process). *(Figure reprinted with permission from Mironov et al. [75].)*

into macrotissue constructs in a permissive environment (Fig. 7.12) [75]. An example of this approach is the preparation of vascular grafts from cell aggregates (Fig. 7.13) [63]. On the other hand, Nakamura et al. believed that cells should be positioned with the highest resolution possible for fabrication of biological tissues because they are not a random mixture of cells, but rather the definite microstructures were built by specific arrangement of cells and synthesized ECM [76]. Construction of 3D architecture and precise control of the inner composition is essential for mimicking the functions and behavior of cells *in vitro* [76].

Cell-sheet technology has shown a promising option; micropatterned coculture of fibroblasts and endothelial cells were exercised to generate prevascularized tissue from stacks of cell sheets [77]. However, the challenge lies in achieving the missing third dimension to produce anatomically relevant tissue constructs. By combining the cell sheet technology with dispensing techniques the missing third dimension can be attained through deposition of structured hydrogels onto and in-between cell sheets.

3.2 Bioprinted Scaffolds

3.2.1 Scaffolds Produced By Fused Deposition Modeling

FDM has been used to fabricate a number of scaffolds with highly interconnecting and controllable pore structure, mainly for bone TE applications using synthetic polymers such as PCL and PLGA and their composites with ceramics [78]. The remainder of this paragraph describes the representative applications using FDM. Hutmacher et al. investigated the mechanical properties and responses of cells cultured on PCL scaffolds designed and fabricated using commercialized FDM (Stratasys Inc., Eden Prairie,

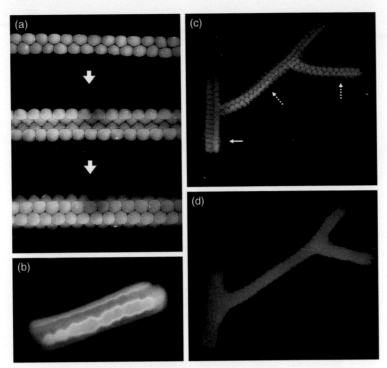

FIGURE 7.13 **Fusion patterns of multicellular spheroids assembled into tubular structures.** (a) Assembled HSF spheroids. (b) Fusion pattern after 7 days of a tube assembled from fluorescently labeled red and green sequences of CHO spheroids. (c) Branched structure built of 300 mm HSF spheroids with branches of 1.2 mm (solid arrow) and 0.9 mm (broken arrows). (d) The fused branched construct after 6 days of deposition. *(Figure reprinted with permission from Norotte et al. [63].)*

MN) [25]. Two types of PCL scaffolds with an area of 32.0 (length) × 25.5 (width) × 13.5 mm (height) and 61% porosity were manufactured using different patterns (0°/60°/120° and 0°/72°/144°/36°/108°), to give a honeycomb-like pattern of triangular and polygonal pores, respectively. They performed the *in vitro* studies over a period of several weeks to show the biocompatibility of PCL scaffolds using human fibroblasts and periosteal cell culture systems. This result showed that FDM allows for the design and fabrication of highly reproducible biodegradable 3D scaffolds with a fully interconnected pore network (Fig. 7.14).

Cao et al. demonstrated successful *in vitro* coculturing of osteoblasts and chondrocytes on PCL scaffolds for more than 50 days using a commercialized FDM apparatus [79]. Rectangular-shaped, honeycomb-like scaffolds were fabricated with a three-angle lay-down pattern (0°/60°/120°). The porosity ranged from 60% to 65%, and the pore size ranged from 300 μm to 580 μm. The 10 × 10 × 3.2 mm PCL scaffold was partitioned vertically into two halves with a gap between them. One-half (the bone compartment) of the partitioned scaffold was designated for BMSC seeding, and the other half (the cartilage compartment) was designated for chondrocyte seeding. It was reported that osteoblasts and chondrocytes produced a rich ECM in their respective scaffold compartments. At the interface region, a mixture of cell types was observed. Therefore, it was demonstrated that 3D porous PCL scaffolds produced by FDM

are biocompatible, as evidenced by extensive cell adhesion and proliferation. Furthermore, they are suitable for *in vitro* osteochondral constructs in TE.

Sun and coworkers [80–82] developed a multinozzle bioprinting system with the capability to simultaneously deposit cells and multiple biomaterials. They performed a rheological study and cell viability assay for estimating mechanical-stress-induced cell damage during the printing process [83]. The cell viability was shown to be influenced by material flow rate, material concentration, dispensing pressure, and nozzle geometry. This investigation can serve as a guideline for future studies and optimization of the deposition system.

Woodfield et al. studied the formation of articular cartilage using a developed 3D PEGT-PBT block copolymer scaffolds [28]. They used bovine chondrocytes to evaluate the cellular responses on the 3D-printed scaffolds. Cartilage tissue formation was observed following dynamic culture *in vitro* and subcutaneous implantation in nude mice, as demonstrated by the presence of articular cartilage ECM constituents, GAG, and type II collagen throughout the interconnected interior pore volume (Fig. 7.15). They also found similar results when expanded human articular chondrocytes were used and homogenous distribution of viable cells after 5 days of dynamic seeding was observed.

Our group applied the scaffold fabricated by MHDS to bone tissue regeneration [84,85]. By the use of

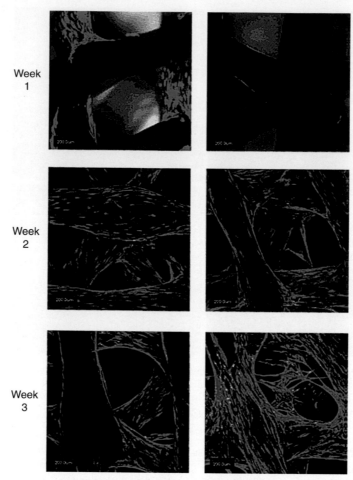

Week
1

Week
2

Week
3

FIGURE 7.14   Confocal micrographs revealed a netlike proliferation pattern on the PCL columns and bars. Cell/scaffold constructs were prepared for CLM by the staining of viable cells green with FDA and dead cells red via PI. In the 1st week, a denser cell network was qualitatively observed in the three-angle architecture, whereas at week 3 the five-angle scaffolds showed more cells. Both matrix architectures presented a low rate of apoptosis starting only at week 3 (left row, 0/60/120° lay-down pattern, and right row, 0/72/144/36/108° lay-down pattern). *(Figure reprinted with permission from Hutmacher et al. [25].)*

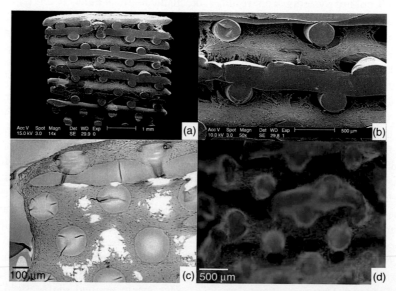

FIGURE 7.15   (a and b) SEM, (c) safranin-O, and (d) live/dead sections showing attachment, proliferation and high percentage of live (green) expanded human articular chondrocytes throughout interconnecting pores on 3D-deposited 300/55/45 scaffolds following 5 days dynamic seeding. *(Figure reprinted with permission from Woodfield et al. [28].)*

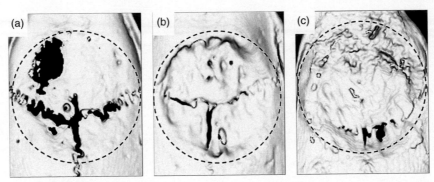

FIGURE 7.16 μ-CT images of regenerated bones on (a) the control, (b) HA-PLGA scaffold, and (c) HA-PLGA/PEG/BMP-2 scaffold after implantation for 4 weeks. Dashed circular lines represent the bone-defect area. *(Figure reprinted with permission from Park et al. [33].)*

multideposition heads, we manufactured blended scaffolds with various biomaterial compositions such as PCL/PLGA and PCL/PLGA/TCP scaffolds. The mechanical testing and cell proliferation results show that the blended PCL/PLGA/TCP scaffold was superior to other scaffolds. *In vivo* bone regeneration in rat calvaria also showed the superiority of PCL/PLGA/TCP scaffold. And they observed the encouraged osteogenic differentiation of seeded ASCs in PCL/PLGA/β-TCP scaffold. PCL/PLGA/β-TCP thin membrane for guided bone regeneration was developed and they showed a better bone formation performance and structure stability of this membrane than in a commercial collagen membrane *in vivo* animal study.

3.2.2 Scaffolds Produced By Solution-Based Deposition System

A solution-based deposition technology that can use biomolecules, such as growth factors, directly showed a promising possibility for bone tissue regeneration [33]. Tissue-engineering scaffolds of HA-PLGA conjugates encapsulating BMP-2/PEG complex developed by our group showed the successful *in vitro* osteogenic differentiation and *in vivo* bone formation ability. *In vitro* assessment of optical cell density and the gene-expression levels of osterix, osteocalcin, and alkaline phosphatase (ALP) in MC3T3-E1 preosteoblast cells confirmed the more efficient cell growth on HA-PLGA/PEG/BMP-2 scaffold than on HA-PLGA scaffold. Furthermore, after implantation into the calvarial bone defects in SD rats for 4 weeks, micro-CT and histological analyses with Masson's trichrome and H&E staining revealed active bone regeneration on the HA-PLGA/PEG/BMP-2 scaffold (Fig. 7.16). Namely, the BMP-2 released from the scaffold was thought to contribute to enhanced bone regeneration.

3.2.3 Scaffolds Produced By Direct-Write Electrospinning

Although the electrospun scaffolds provide favorable cellular interaction, cellular migration into the 3D electrospun scaffolds has been limited. This is because of their inherent small pore sizes and limited control over their porosity. To solve this shortcoming, researchers came up with some novel and emerging approaches to produce advanced electrospun 3D scaffolds, namely, DWES for tissue regeneration applications [86]. Jeong and coworkers fabricated a 3D nanofibrous scaffold with regular pores using a DWES nanofibrous pattern mat followed by stacking of the mats onto a 3D structure [36,87]. They compared the cellular interaction on the structure produced by DWES, conventional electrospinning, and salt leaching method (Fig. 7.17). Scanning electron microscopy (SEM) images and confocal images of F-actin and nuclei expression showed the superiority of both DWES and conventional electrospinning techniques related to the attachment and spreading of MIH3T3 cells. The stained cross-sectional images exhibited the better cell migration at the DWES scaffold to the inner space than the conventional electrospinning scaffold. It is expected that these scaffolds with controlled geometry will eventually improve cellular migration into the core and support 3D tissue formation.

Dalton et al. demonstrated the melt electrospinning of PCL in a direct writing mode onto a rotating cylinder [88]. This allowed the design and fabrication of tubes using 20 μm diameter fibers with controllable micropatterns and mechanical properties. The produced structure supports cell colonization and was noncytotoxic in nature when evaluated by culturing primary human osteoblasts (hOBs) and mesothelial cells as well as mouse osteoblasts (mOBs) (Fig. 7.18). The *in vitro* cell culture results showed the potential of these scaffolds for TE.

3.3 Bioprinting with Cell-Laden Hydrogels

Hydrogels are very popular for cell encapsulation; however, only a handful of hydrogels are being employed in bioprinting successfully. Among the various hydrogels, alginate, which is derived from brown seaweed, is the most widely used bioink in the current bioprinting technology (Fig. 7.19a), and it is commonly cross-linked by divalent cations (Ca^{2+}, Ba^{2+}). The ideal conditions for using alginate hydrogel in 3D printing have been investigated [72],

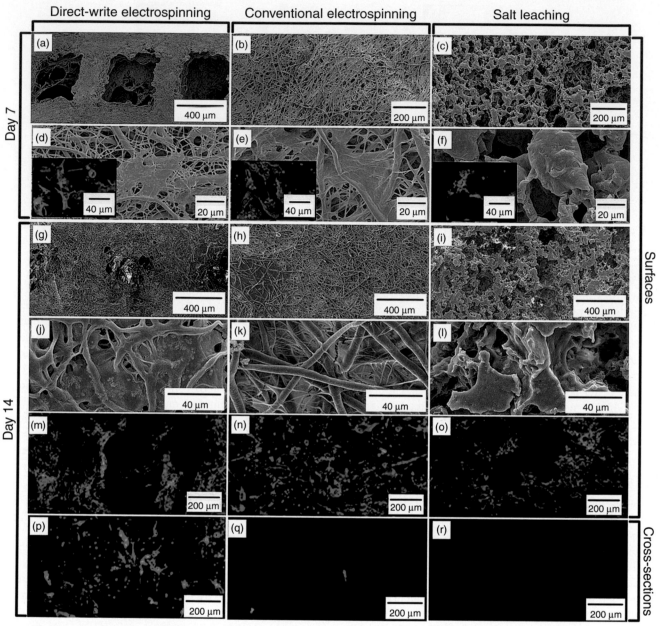

FIGURE 7.17 **SEM and F-actin images of three types of scaffolds.** (a–f) After 7 days, and (g–o) after 14 days at the surfaces and (p–r) at the cross-sections. *(Figure reprinted with permission from Lee et al. [87].)*

and the widely used concentration of alginate is between 1% and 10% (w/v) [72,89–91]. Average viability of cells encapsulated in the alginate gel is above 70%; however, the low adhesive characteristics of the alginate gel may hamper to stretch the cell and stimulate the endogenous repair after implantation. In this context, RGD modified alginate matrix (alginate-RGD) [92], alginate/fibrin [93], and alginate/gelatin [94] have been studied according to the targeted site (Fig. 7.20). Agarose displays a temperature-sensitive solubility in water and has found widespread applications in TE [95]. However, for application in bioprinting, the elevated temperature needed to

dispense alginate is the major concern. Alginate can be formed as gels or beads or filaments; this allows cells that have been cultured in monolayers to be redifferentiated [89,96]. Chitosan is a semicrystalline polymer with a high degree of biocompatibility *in vivo* [97]. A temperature-sensitive carrier and an injectable material can be developed using chitosan. It forms gels at body temperature and has the ability to deliver and interact with growth factors and adhesion proteins [98]. However, chitosan is soluble in acidic condition and becomes a gel at physiological pH. Hence, chemical modification is required to render solubility at physiological pH [99].

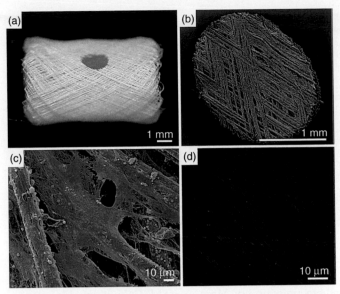

FIGURE 7.18   **Culture of mOBs on melt electrospun PCL fibers taken from a tube with a winding angle of 30.** mOBs initially showed good attachment and then proliferated over a culture period of 4 weeks. (a) After 4 weeks of culture a 2 mm biopsy punch was used to harvest the specimen shown in Fig. 7.18b–d. SEM showed that the mOBs formed a mineralized ECM not only on the scaffold fibers but also inside and across the pore architecture (b and c). (d) CLSM revealed by using DAPI/Phalloidin staining alignment of mOBs along the fiber axis. *(Figure reprinted with permission from Brown et al. [88].)*

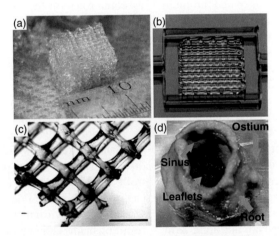

FIGURE 7.19   **Examples of bioprinted structures using various bioinks.** (a) A porous 3D alginate scaffold was fabricated and cross-linked by applying divalent cations. (b) Lutrol F127 printed in the form of a 3D vascular network within silicon ink border. (c) A construct of UV-light cross-linked gelMA with HA was fabricated in four layers. (d) Porcine aortic valve model was printed, where root (blue) was formed with 700 MW PEG-DA hydrogel while the leaflets (red) were formed with 700/8000 MW PEG-DA hydrogels. *(Figure reprinted with permission from Refs [69,72,105,106].)*

The protein-based hydrogels, including collagen [39], gelatin [100], Matrigel [101], and decellularized extracellular matrix (dECM) [6], enable to provide an abundance of chemical signals to the cells resulting in high viability and proliferation rates. It can also induce the formation of neo-tissues and differentiation of the cells into the specific lineage of interest. Collagen, an important component of mammalian ECM, provides cellular recognition for regulating cell attachment and proliferation [102]. Collagen can be used for bioprinting by utilizing its thermo-responsive gelation properties. Gelatin is a water-soluble protein obtained by denaturation of collagen and it forms thermoreversible hydrogels below their upper critical solution temperature of 25–35°C. Gelatin is frequently used as a basement polymer to enhance tissue regeneration or printability by blending or conjugating the adjuvant materials such as alginate [103], chitosan [104], and methacrylamide [105]. Gelatin-methacrylamide (gelMA) is an example resulting from the introducing of unsaturated photopolymerizable methacrylamide groups to the gelatin, allowing the matrix to be covalently cross-linked by UV light after printing (Fig. 7.19c). It can be selected for bioprinting due to its low-cost,

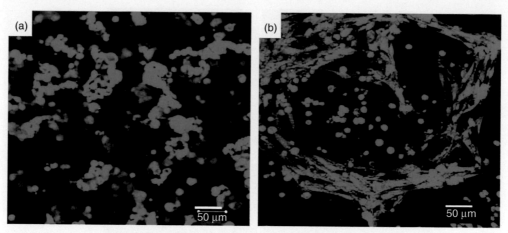

FIGURE 7.20 Bioprinting of functionalized bioink. (a) Confocal analysis of live/death assay of printed hCMPCs in unmodified and (b) RGD-modified alginate scaffold after 2 weeks in culture. *(Figure reprinted with permission with permission from Gaetani et al. [92].)*

abundance, ease of processing, and biocompatibility. Moreover, degree of methacrylation, and shear stress and elastic modulus of gelMA can be systemically tuned. The widely used concentration of gelMA is around 10–20% (w/v) [105,106]. Fibrin plays an important role in wound healing and thereby prevents blood loss during injury. Fibrin glue is readily formed from the mixture of fibrinogen and thrombin and allows them to solidify. It is often used as a carrier for cells and in conjunction with other scaffold materials [107].

Hyaluronan (HA) is an anionic polysaccharide that is used as a carrier for cells to regenerate various tissues. It can be made as an injectable and used to fill irregularly shaped defects and implanted with minimal invasion. HA can be used as matrix material for bioprinting as it is not antigenic and elicits no inflammatory or foreign body reaction [108]. dECM has been known as an excellent material having the potential to represent the complexity of natural ECM and providing an optimized microenvironment to the encapsulated cells. dECM hydrogel was utilized as a bioink for a bioprinting process and it showed the capability to provide crucial cues for cells engraftment, survival, and long-term function [6].

Our group has utilized alginate hydrogel formed by extrusion of a gel precursor to produce a construct by extrusion-based method [91]. Alginate hydrogel is known to be formed by an ion-bonded reaction based on egg-box-shaped bonding structures with polysaccharide molecules of alginate and calcium ion. We have observed better chondrogenic function when human inferior turbinate-tissue derived mesenchymal stromal cells (hTMSCs) were encapsulated in alginate gel and printed into tissue constructs (Fig. 7.21). We have developed a silk-gelatin bioink for printing 3D tissue constructs [109]. We used two innocuous cross-linking methods, that is, sonication-induced β-sheet crystallization and tyrosinase-induced covalent cross-linking. The developed silk-gelatin bioink supported long-term viability and functions. Multilineage differentiation of the encapsulated

hTMSCs within silk-gelatin bioink was also observed (Fig. 7.22). However, these materials cannot represent the complexity of the natural ECM and thus are unable to reconstitute the function typical to native tissue. dECM has been shown to facilitate the constructive remodeling of many different tissues. There is growing evidence to support the essential roles for the structural and functional characteristics of dECM [110]. We have shown by decellularization and solubilization that dECM can be applied as a bioink for bioprinting [6]. The dECM bioink supports better cellular proliferation and function than the other printable materials like collagen and alginate. We have also observed particular tissue formation and tissue-specific gene expression within a definite dECM bioink.

A major disadvantage of natural hydrogels is their low mechanical strength, which makes handling and application difficult. The application of synthetic polymers, such as Lutrol F127 [111] and poly(ethylene glycol) diacrylate [112], can offer high shape fidelity; however, these materials provide the inert environment without active binding sites often resulting in low cell viability. For the functionalization of the existing hydrogel, various approaches have been developed by adding the functional compound like peptide sequences and biomolecules or blending the multiple hydrogels. Lutrol F127 is the photopolymerizable thermosensitive hydrogels based on a triblock copolymer of polyethylene-oxide and polypropylene-oxide modified with photo-cross-linkable methacrylamide groups (Fig. 7.19b). The mechanical properties and degradation profile can be tuned by changing the photoinitiator like an Irgacure 2959. The used concentration of Lutrol F127 for bioprinting is from 25% to 40% (w/v), but it shows relatively lower cell viability than that of the native hydrogel. Hyaluronic acid methacrylate (HA-MA) [52], and the mixture of poly(ethylene glycol) diacrylate and alginate [112] are representative examples of the blended or conjugated to enhance cell functionality and printability for bioprinting processes.

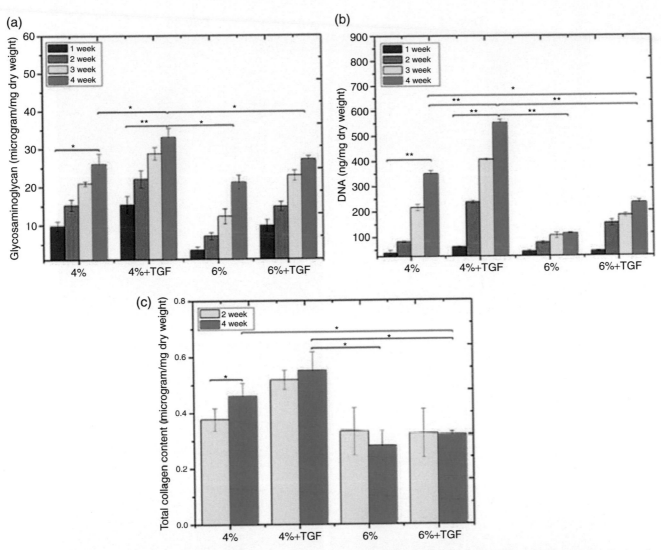

FIGURE 7.21 (a) Glycosaminoglycans content (mg/mg dry weight). (b) DNA content (ng/mg dry weight). (c) Total collagen content (mg/mg dry weight) analysis from different chondrocyte cell-seeded PCL–alginate constructs, namely: 4% alginate – TGFb (4%); 4% alginate + TGFb (4% + TGF); 6% alginate – TGFb (6%); 6% alginate + TGFb (6% + TGF). The PCL–alginate constructs were cultured *in vitro* for 1, 2, 3, and 4 weeks for the determination of GAGs and DNA content, while for the total collagen content the experiments were conducted for 2 and 4 weeks. The values represents mean SD; $n = 3$, *$p < 0.05$, **$p < 0.001$. *(Figure reprinted with permission from Kundu et al. [91].)*

Photo-cross-linkable gels have also been bioprinted utilizing their prompt solidification upon exposure to light. For instance, methacrylated dextran was bioprinted by mixing with high-molecular weight HA to increase the viscosity [53]. The high viscosity enables printing of a porous structure that can be fixed subsequently by photo-cross-linking. It is expected that the development of more hydrogels tailored for specific bioprinting techniques will greatly increase the potential of bioprinting.

Hybrid gels are also being used in bioprinting productively. One of the few examples is a thermosensitive block copolymer based on PEG–PPO–PEG copolymer, which allows dispensing of a cell suspension at ambient temperature and subsequently solidifies upon incubating at 37°C. However, the major concern is that the gel does not support

cell viability in culture for a long term as most of the cells die within a few days despite remaining viable during the printing process. Furthermore, the printed thermogel slowly dissolves into the culture media [113]. To improve the cytocompatibility, the terminal hydroxyl units of PEG–PPO–PEG were functionalized with a peptide linker followed by a methacrylate group that resulted in increased viability over 3 weeks. Moreover, it also makes the gel biodegradable [111]. The same group also designed a synthetic gel based on an ABA block copolymer composed of poly(*N*-(2-hydroxypropyl)methacrylamide lactate) A-blocks and hydrophilic poly(ethylene glycol) B blocks of a molecular weight of 10 kDa. This gel allows for thermal gelation as well as UV-initiated chemical cross-linking [114]. The hydrophobic A-blocks induce lower critical solution

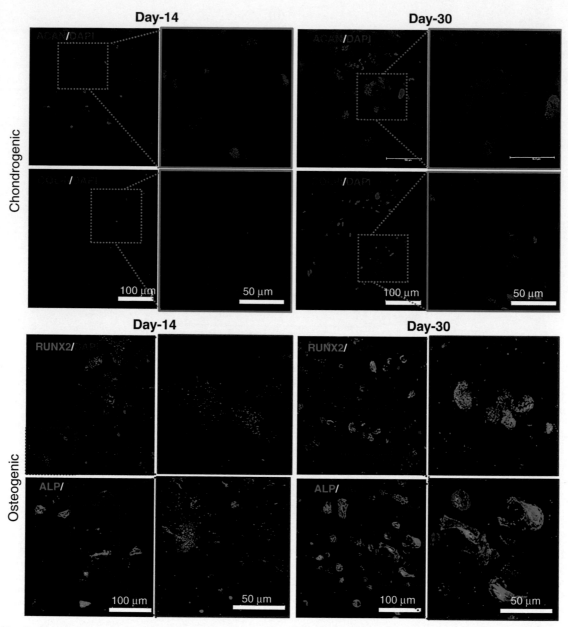

FIGURE 7.22 **Confocal images for assessing the chondrogenesis and osteogenesis of htMSC laden tyrosinase induced cross-linked silk-gelatin constructs.** For chondrogenesis, aggrecan (ACAN) and collagen II (COL-II) expression and for osteogenesis, runt related transcription factor (RUNX2) and ALP expression at day 14 and 30. *(Figure reprinted with permission from Das et al. [109].)*

temperature behavior and additionally it also allows for photopolymerization as the A-blocks partly derivatized with methacrylate groups. When utilized for bioprinting, this gel results in a structure with increased strength and shape stability.

3.4 Convergent Bioprinting Strategies

3.4.1 Bioprinting with Sacrificial Materials

For the fabrication of a personalized bioprinted structure, the 3D scan of the part of interest of the human body can

be converted to CAM models of complex anatomical structures [15]. The CAM model can be printed using the extrusion-based bioprinters. However, in many occasions such models consist of overhang geometries due to internal cavities, or due to the outer contour of complex anatomical structures. Though the number of overhang geometries can be minimized by smart rotation of the 3D design, the remaining overhang geometries need to be temporarily supported during bioprinting, as it demands the deposition of material at times above an empty cavity. This is done by printing with sacrificial materials to support the overhangs temporarily during printing. Subsequent to printing,

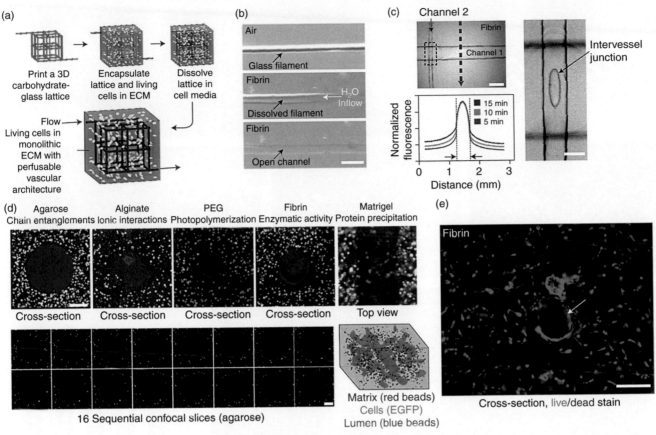

FIGURE 7.23 **Monolithic tissue construct containing patterned vascular architectures and living cells.** (a) Schematic overview. An open, interconnected, self-supporting carbohydrate-glass lattice is printed to serve as the sacrificial element for the casting of 3D vascular architectures. The lattice is encapsulated in ECM along with living cells. The lattice is dissolved in minutes in cell media without damage to nearby cells. The process yields a monolithic tissue construct with a vascular architecture that matches the original lattice. (b) A single carbohydrate-glass fiber (200 m in diameter, top) is encapsulated in a fibrin gel. Following ECM crosslinking, the gel and filament are immersed in aqueous solution and the dissolved carbohydrates are flowed out of the resulting channel (middle). Removal of the filament yields an open perfusable channel in the fibrin gel (bottom, scale bar, 500 m). (c) A fibrin gel with patterned interconnected channels of different diameters supports convective and diffusive transport of a fluorescent dextran injected into the channel network (upper left, phase contrast, scale bar, 500 m). Line plot of normalized fluorescence across the gel and channel (blue arrow) shows a sinusoidal profile in the channel (between dotted black lines) characteristic of a cylinder and temporal diffusion from the channel into the bulk gel. Enlargement of the dotted box region shows an oval intervessel junction between the two perpendicular channels (right, scale bar, 100 m). (d) Cells constitutively expressing enhanced green fluorescent protein (EGFP) were encapsulated (5×10^6 m/L) in a variety of ECM materials and then imaged with confocal microscopy to visualize the matrix (red beads), cells (10T1/2, green) and the perfusable vascular lumen (blue beads). They are also shown schematically (bottom right). The materials have varied crosslinking mechanisms (annotated above the images) but were all able to be patterned with vascular channels. Scale bars, 200 m. (e) Representative cross-section image of unlabeled HUVEC (1×10^6 m/L) and 10T1/2 (1×10^6 m/L) cocultures (not expressing EGFP) encapsulated uniformly in the interstitial space of a fibrin gel (10 mg m/L) with perfusable networks after 2 days in culture were stained with a fluorescent live/dead assay (green, Calcein AM; red, Ethidium Homodimer). Cells survive and spread near open cylindrical channels (highlighted with white arrow). Scale bar, 200 μm. *(Figure reprinted with permission from Miller et al. [116].)*

the sacrificial material can be washed away from the target structure [115,116]. For application as sacrificial component, the material should fulfil a certain criteria. The material should be printable at lower temperature if coprinted along with cell-laden hydrogel. Furthermore, the material should be soluble in aqueous solution for its safe and effective removal. Most importantly, as the sacrificial materials are a component of cell-laden hydrogel constructs, the material should be cytocompatible. The sacrificial materials have been effectively employed for creating channel networks within either cell-laden hydrogel constructs or only

hydrogels, either by casting [117] or by combining printing and casting [43,118]. Miller et al. [116] printed a vascular network from carbohydrate glass, a solution of sucrose, glucose, and dextran, through extrusion-based bioprinting. A hydrogel encapsulating cells of interest was subsequently cast and allowed to cross-link around this network. Later, the construct was placed in culture medium to eliminate the printed carbohydrate glass without any cytotoxic effect on the encapsulated cells (Fig. 7.23). The strong point of this approach is that it provides exceptional control over the shape of the internal vascular network as the carbohydrate

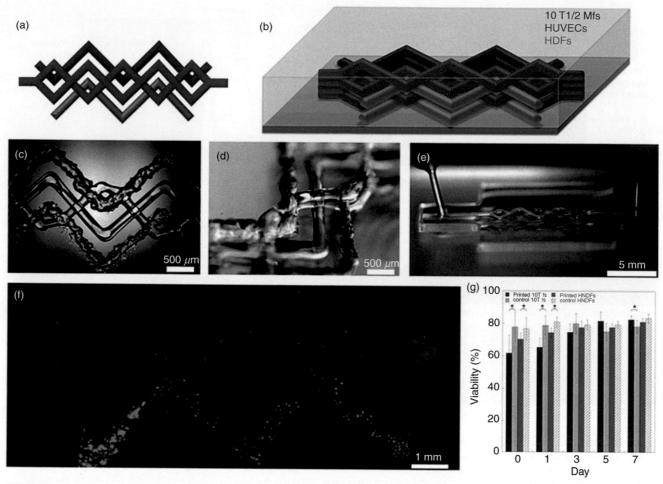

FIGURE 7.24 (a and b) Schematic views of the top–down and side views of a heterogeneous engineered tissue construct, in which blue, red, and green filaments correspond to printed 10T1/2 fibroblast-laden gelMA, fugitive, and GFP HNDF-laden gelMA, inks, respectively. The gray shaded region corresponds to pure gelMA matrix that encapsulates the 3D printed tissue construct. Note: The red filaments are evacuated to create open microchannels, which are endothelialized with RFP HUVECs. (c) Bright-field microscopy image of the 3D-printed tissue construct, which is overlayed with the green fluorescent channel. (d) Image showing the spanning and out-of-plane nature of the 3D printed construct. (e) Image acquired during fugitive ink evacuation. (f) Composite image (top view) of the 3D-printed tissue construct acquired using three fluorescent channels: 10T1/2 fibroblasts (blue), HNDFs (green), HUVECs (red). (g) Cell-viability assay results of printed 10T ½ fibroblast-laden and HNDF-laden gelMA features compared to a control sample (200–300 μm thick) of identical composition. The asterisks indicate differences with $p < 0.05$ obtained from Student's t-test. *(Figure reprinted with permission from Kolesky et al. [106].)*

glass was printed with relative ease and adeptness. However, as the process of making hydrogel construct is essentially based on casting, this reduced the control over the architecture of the surrounding hydrogel construct. The limitation of this approach is that the sacrificial carbohydrate glass filaments were printed at temperatures above 100°C and thus it cannot be coprinted with cell-laden hydrogels. As mentioned before, for fabrication of multicellular tissue constructs, the control deposition of different cell types or bioactive substances at a predesigned position is essential. If sacrificial material is coprinted along with the cell-laden hydrogel in a bottom–up bioprinting approach, the generation of complex tissue constructs can be realized. Kolesky et al. [106] bioprinted vascularized, heterogeneous cell-laden tissue constructs by coprinting fugitive Pluronic

F127 for the vasculature and gelMA as a bulk matrix and cell carrier (Fig. 7.24). The fugitive Pluronic F127 can be removed from the structure subsequent to cross-linking of gelMA by reducing the temperature and applying suction. This highly scalable platform allows one to produce engineered tissue constructs in which vasculature and multiple cell types are programmably placed within extracellular matrices.

Alternatively, a thermoplastic polymer (e.g., polyethylene glycol (PEG)) can be codeposited as a sacrificial component that forms a stable interface with the hydrogel construct [119,120]. However, the removal process of such thermoplastic polymers should not be detrimental to the embedded cells. In most of the cases, the printed thermoplastic structure serves as a mold and only supports

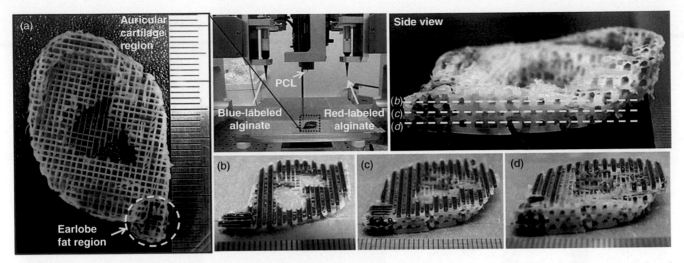

FIGURE 7.25 **Acellular printed structure using 3D bioprinting technology with the sacrificial layer process, auricular cartilage region (red) and lobe fat region (blue).** (a) Fabricated ear-shaped structure with dual hydrogel type. (b–d) Images of printed structure having dual hydrogel-type and PCL framework. *(Figure reprinted with permission from Lee et al. [120].)*

the outer contours [121]. On the contrary, codeposition of two stable hydrogels allows simultaneous fabrication of complex constructs with temporary support of internal cavities [119,122]. However, the target component needs to be cross-linked selectively in order to dissolve the sacrificial component [121].

Our group printed composite tissue with a complex shape like ear with PEG as the sacrificial material [120]. The ear has a complex shape and composition and that is difficult to fabricate using traditional methods. We used 3D printing technology including a sacrificial layer process to regenerate the auricular cartilage and fat tissue (Fig. 7.25). The main part was printed with polycaprolactone (PCL) and cell-laden hydrogel. At the same time, PEG was used as a sacrificial layer to support the overhangs. After complete fabrication, PEG can be easily removed in aqueous solutions without any deleterious effect on the cell viability.

3.4.2 Bioprinting in Combination with Thermoplastic Polymers

Although hydrogels were mostly used for fabrication of constructs through bioprinting, they generally resulted in constructs with lower stiffness than their target tissue, especially when used in the musculoskeletal system [105,123]. For such application, a stiff and coherent hydrogel construct will be required in the human body. One option is to do preculturing cells in these constructs to increase the stiffness due to the deposition of a specific tissue matrix [124]. Nonetheless, this process is not economically and practically possible as it demands high cell concentrations and a substantial preculturing period. Another way of increasing the stiffness of the hydrogel is by increasing the

hydrogel cross-link density. Unfortunately, this retards new tissue formation as it restricts the diffusion of nutrients and waste products through the highly cross-linked hydrogel system [72,124–126].

The superior way to achieve the favorable biological and mechanical properties is the reinforcement of hydrogels as they have been reinforced by use of double networks (DN) [125] and interpenetrating polymer networks [127], as well as by incorporation of nanoparticles [128], nanotubes [128,129], or electrospun fibers [130,131]. Interestingly, without changing the cross-link density of the hydrogel, the mechanical property can be effectively increased, which also allows for adequate tissue formation. However, most of these approaches requires casting or an additional crosslinking reaction and not compatible with the bioprinting. Therefore, researchers around the world recently develop multiple-tool biofabrication processes. One such example is the reinforcement of hydrogel constructs by codeposited thermoplastic polymer fibers [39,119,131–133]. Specifically, this has been achieved by combining hydrogel and PCL in extrusion bioprinting [39,132,133] and by combining electrospinning techniques with inkjet printing [131] or laser-induced forward transfer printing [134]. Here are the dual advantages of these methods like the opportunity to process the hydrogels at low and the shape strength of the overall construct are secured by the thermoplastic polymer network. Furthermore, these processes repeatedly resulted in the fabrication of complex-shaped tissue constructs [119]. Moreover, the mechanical properties of the developed construct can be modulated to match the target tissue better by adjusting the thermoplastic polymer network [132,133].

Our group has developed a hybrid scaffold with synthetic biomaterials and hydrogel through MHDS [39]. The

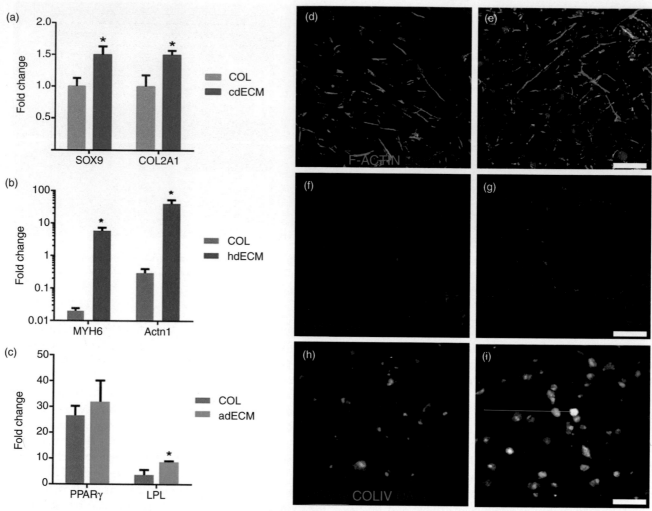

FIGURE 7.26 Evaluation of differentiation into tissue-specific lineages and structural maturation of cells. Comparative gene expression analysis for (a) chondrogenic (SOX9 and COl2A1), (b) cardiogenic (myosin heavy chain and Actn1) and (c) adipogenic (PPARγ and LPL) in COL and particular dECM (cdECM or hdECM or adECM) at day 14. Immunofluorescence images showing chondrogenic differentiation of htMSCs in (d) COL and (e) cdECM constructs showing COL type II staining (COLII, red), cell nuclei (DAPI, blue) and F-actin (green) (scale bar, 50 mm). Structural maturation of myoblasts in (f) COL and (g) hdECM construct showing Myh7 (red) and cell nuclei (DAPI, blue) (scale bar, 200 mm). Adipogenic differentiation of hASCs in (h) COL and (i) adECM constructs showing PPARg (red), COL IV (COL IV, green) and cell nuclei (DAPI, blue) (scale bar, 50 mm). All experiments were performed in triplicate. Error bars represent s.d. (*$p < 0.05$). *(Figure reprinted with permission from Pati et al. [6].)*

PCL framework was printed and the collagen was intentionally infused into the space between the lines of PCL. The cellular efficacy of the hybrid scaffold was validated using rat primary hepatocytes and a mouse preosteoblast MC3T3-E1 cell line. In addition, the collagen hydrogel, which encapsulates cells, was dispensed and the viability of the cells observed. We demonstrated superior effects of the hybrid scaffold on cell adhesion and proliferation and showed the high viability of dispensed cells. We also produced a cell-printed construct based on the mentioned hybrid structure fabrication method using dECM from cartilage and heart tissue and PCL framework through an in-house developed multihead tissue/organ building system [6]. We observed better tissue-specific gene expression and function of the printed tissue constructs (Fig. 7.26).

Electrospinning techniques can become handy as it is capable of producing fibers with a higher resolution [131,134] compared to extrusion-based printing, and can mimic the structural features of natural ECM. However, the traditional electrospinning techniques are not able to control fiber deposition, and the limited pore size of the resulting random meshes limits cell migration [135]. Some melt electrospinning writing techniques [136] address both these limitations [137], since fibers can be deposited with high spatial resolution and orientation. Combining this technique with hydrogel deposition approaches will allow for the generation of reinforced hydrogel constructs with high control over the intricate spatial organization, although grafting between fibers and the hydrogel needs to be addressed in order to biofabricate truly integrated constructs.

Jeong's group has produced a construct with mechanically enhanced hydrogel by reinforcement with the electrospun nanofiber and for load-bearing application [138]. The PCL/alginate multimaterial electrospun 3D composite structure is a wonderful example for the evaluation of this phenomena. Alginate solution is continuously fed during the electrospinning of PCL nanofiber and the ethanol was concurrently coated to the electrospun PCL nanofiber by using a dual-nozzle system. The ethanol coating technique allows to successfully consolidate the hydrophilic and hydrophobic material into one structure.

4 FUTURE DIRECTIONS

Bioprinting is a vibrant research area in which few successful proof of principle studies have been done in the last decade. However, it holds great promise toward successful tissue/organ regeneration. Another exciting feature of this technique is that it can be translated into *in situ*, *in vivo* tissue printing for healing of wounds. Furthermore, it can be used in the development of models of tissues or organs for evaluating the efficacy of pharmaceutical drugs or biomolecules more realistically. Moreover, tissue/tumor models can be generated and that can help the studies move from basic science to clinicians. Within the next few years, extrusion bioprinting is expected to advance to meet the needs of specific applications. The use of cell printers was predicted to be as common as the current use of microscopes for biological academic, clinical, and industrial laboratories [14]. This technology really holds the promise of meeting the crisis of tissue on demand.

Until now, most published results have been based on the laboratory-scale demonstrating the deposition of cells encapsulated in various hydrogels. Among these, only a few studies have investigated process parameters either for predictions or optimization strategies in a systematic way. Thus, studies targeted toward understanding the relationship between process parameters with the structure and functions of the printed constructs are urgently needed. Moreover, modern fabrication schemes rely on mathematical modelling and computer simulations for optimizing the process design and predictions [139,140]. Using computer simulations, the constructs can be predicted and hence optimized before printing. However, this approach needs more attention for the precise use of new biological processes and 3D tissues.

Most importantly, a single bioprinting technology cannot fulfil all the requirements and certainly the convergence of two or more bioprinting technologies can realize the functional tissue or organ building. We have seen a few attempts based on the convergent approaches in the last few years. However, in the coming days the tissue manufacturing approaches will definitely consider the utilization of more than one bioprinting technology. The real potential of the computer-controlled tissue fabrication process can only be harnessed by doing so.

5 CONCLUSIONS

Extrusion bioprinting enables the production of cell-laden constructs in a computer-controlled manner, thereby bypassing costly and poorly controlled postfabrication manual cell seeding. Although big steps have been taken since the development of this technique, the technology is still in its infancy. It is now critical to address key issues in novel material development, construct design (including vascularization of the construct), and utilization of patient-specific cells and suitable materials in compliance with good manufacturing practices. With the combined effort of researchers from polymer chemistry, mechatronics, computer engineering, information technology, biology, and medicine, extrusion bioprinting techniques can evolve into a technology platform that allows users to create tissue-engineered constructs economically in the coming years and lower the gap between demand and supply of tissues or organs.

GLOSSARY

Bioprinting 3D printing of biomaterials and cells via a computer-controlled layer-by-layer process.

Fused deposition modeling (FDM) Technology for printing fused thermoplastic materials through a nozzle directly onto the build platform following a preprogrammed path.

Direct-write electrospinning (DWES) An electrospinning process that utilizes a large electric field between a polymer solution dispensing micronozzle and a collection plate to induce a stable jet to be ejected from the solution reservoir and travel toward the collecting plate.

Cell printing 3D printing of cells via encapsulating them in an appropriate vehicle called a bioink into 3D-structured tissue.

Bioink Biomaterials solution used for 3D printing that acts as vehicle for loading cells and biologicals by providing a suitable environment to them.

Thermoplastics Polymeric materials that melt at certain temperature range and can go back to original solid phase upon cooling, ideal material for extrusion-based 3D printing.

ABBREVIATIONS

2D	Two-dimensional
3D	Three-dimensional
ALP	Alkaline phosphatase
ASCs	Adipose-derived stem cells
BMPs	Bone morphogenetic proteins
BMSCs	Bone marrow stem cells
CAD	Computer-aided design
CAM	Computer-aided manufacturing
dECM	Decellularized extracellular matrix
DWES	Direct-write electrospinning
ECM	Extracellular matrix
FDM	Fused deposition modeling
gelMA	Gelatin-methacrylamide
hTMSCs	Human inferior turbinate-tissue derived mesenchymal stromal cells

HA	Hyaluronic acid
HA-MA	Hyaluronic acid methacrylate
MHDS	Multihead deposition system
mOBs	Mouse osteoblasts
PBT	Poly(butylene terephthalate)
PCL	Poly-caprolactone
PED	Precision extruding deposition
PEG	Polyethylene glycol
PEGT	Poly(ethylene glycol) terephthalate
PLGA	Poly lactic-co-glycolic acid
PPO	Polypropylene oxide
SEM	Scanning electron microscope
TCP	Tricalcium phosphate
TE	Tissue engineering

ACKNOWLEDGMENT

This work was supported by the National Research Foundation of Korea (NRF) grant funded by the Korea government (MSIP) (No. 2010-0018294).

REFERENCES

[1] Mironov V, Kasyanov V, Drake C, Markwald RR. Organ printing: promises and challenges. Regen Med 2008;3(1):93–103.

[2] Moon S, Hasan SK, Song YS, et al. Layer by layer three-dimensional tissue epitaxy by cell-laden hydrogel droplets. Tissue Eng Part C Methods 2010;16(1):157–66.

[3] Mironov V, Boland T, Trusk T, Forgacs G, Markwald RR. Organ printing: computer-aided jet-based 3D tissue engineering. Trends Biotechnol 2003;21(4):157–61.

[4] Hamid Q, Snyder J, Wang C, Timmer M, Hammer J, Guceri S, et al. Fabrication of three-dimensional scaffolds using precision extrusion deposition with an assisted cooling device. Biofabrication 2011;3(3). 034109.

[5] Derby B. Printing and prototyping of tissues and scaffolds. Science 2012;338(6109):921–6.

[6] Pati F, Jang J, Ha D-H, et al. Printing three dimensional tissue analogues with decellularized extracellular matrix bioink. *Nature Communications* 2014;5:3935.

[7] Ballyns JJ, Bonassar LJ. Image-guided tissue engineering. J Cell Mol Med 2009;13:1428–36.

[8] Gruene M, Pflaum M, Hess C, Diamantouros S, Schlie S, Deiwick A, et al. Laser printing of three-dimensional multicellular arrays for studies of cell–cell and cell–environment interactions. Tissue Eng Part C Methods 2011;17(10):973–82.

[9] Gruene M, Deiwick A, Koch L, Schlie S, Unger C, Hofmann N, et al. Laser printing of stem cells for biofabrication of scaffold-free autologous grafts. Tissue Eng Part C Methods 2011;17(1):79–87.

[10] Koch L, Kuhn S, Sorg H, Gruene M, Schlie S, Gaebel R, et al. Laser printing of skin cells and human stem cells. Tissue Eng Part C Methods 2010;16(5):847–54.

[11] Ker ED, Chu B, Phillippi JA, Gharaibeh B, Huard J, Weiss LE, et al. Engineering spatial control of multiple differentiation fates within a stem cell population. Biomaterials 2011;32(13):3413–22.

[12] Ker ED, Nain AS, Weiss LE, Wang J, Suhan J, Amon CH, et al. Bioprinting of growth factors onto aligned sub-micron fibrous scaffolds for simultaneous control of cell differentiation and alignment. Biomaterials 2011;32(32):8097–107.

[13] Phillippi JA, Miller E, Weiss L, Huard J, Waggoner A, Campbell P. Microenvironments engineered by inkjet bioprinting spatially direct adult stem cells toward muscle- and bone-like subpopulations. Stem Cells 2008;26(1):127–34.

[14] Mironov V. Printing technology to produce living tissue. Expert Opin Biol Ther 2003;3(5):701–4.

[15] Melchels FPW, Domingos MAN, Klein TJ, Malda J, Bartolo PJ, Hutmacher DW. Additive manufacturing of tissues and organs. Prog Polym Sci 2012;37:1079–104.

[16] Nair K, Gandhi M, Khalil S, Yan KC, Marcolongo M, Barbee K, et al. Characterization of cell viability during bioprinting processes. Biotechnol J 2009;4:1168–77.

[17] Malda J, Visser J, Melchels FP, Jüngst T, Hennink WE, Dhert WJ, et al. 25th Anniversary article: engineering hydrogels for biofabrication. Adv Mater 2013;25(36):5011–28.

[18] Lee J, Sato M, Kim H, Mochida J. Transplantation of scaffold-free spheroids composed of synovium-derived cells and chondrocytes for the treatment of cartilage defects of the knee. Eur Cells Mater 2011;22:275–90.

[19] Nicodemus GD, Bryant SJ. Cell encapsulation in biodegradable hydrogels for tissue engineering applications. Tissue Eng Part B Rev 2008;14:149–65.

[20] Francoise M, Karoly J, Chirag K, Shepherd B, Dorfman S, Hubbard B, et al. Toward engineering functional organ modules by additive manufacturing. Biofabrication 2012;4(2):022001.

[21] Maher PS, Keatch RP, Donnelly K, Mackay RE, Paxton JZ. Construction of 3D biological matrices using rapid prototyping technology. Rapid Prototyping J 2009;15(3):204–10.

[22] Duarte Campos DF, Blaeser A, Weber M, Jäkel J, Neuss S, Jahnen-Dechent W, et al. Three-dimensional printing of stem cell-laden hydrogels submerged in a hydrophobic high-density fluid. Biofabrication 2013;5:015003.

[23] Crump SS, Inventor Stratasys Inc., Minneapolis, MN, assignee. Apparatus and method for creating three-dimensional objects. US patent US 5,121,329. 1992.

[24] Zein I, Hutmacher DW, Tan KC, Teoh SH. Fused deposition modeling of novel scaffold architectures for tissue engineering applications. Biomaterials 2002;23(4):1169–85.

[25] Hutmacher DW, Schantz T, Zein I, Ng KW, Teoh SH, Tan KC. Mechanical properties and cell cultural response of polycaprolactone scaffolds designed and fabricated via fused deposition modeling. J Biomed Mater Res 2001;55(2):203–16.

[26] Wang F, Shor L, Darling A, Khalil S, Sun W, Güçeri S, et al. Precision extruding deposition and characterization of cellular poly-ε-caprolactone tissue scaffolds. Rapid Prototyping J 2004;10(1):42–9.

[27] Jakab K, Norotte C, Marga F, Murphy K, Vunjak-Novakovic G, Forgacs G. Tissue engineering by self-assembly and bio-printing of living cells. Biofabrication 2010;2(2):022001.

[28] Woodfield TBF, Malda J, de Wijn J, Péters F, Riesle J, van Blitterswijk CA. Design of porous scaffolds for cartilage tissue engineering using a three-dimensional fiber-deposition technique. Biomaterials 2004;25(18):4149–61.

[29] Kim JY, Park EK, Kim SY, Shin JW, Cho DW. Fabrication of a SFF-based three-dimensional scaffold using a precision deposition system in tissue engineering. J Micromech Microeng 2008;18(5):055027.

[30] Kim JY, Yoon JJ, Park EY, Kim DS, Kim SY, Cho DW. Cell adhesion and proliferation evaluation of SFF-based biodegradable

scaffolds fabricated using a multi-head deposition system. Biofabrication 2009;1:015002.

[31] Vozzi G, Previti A, Rossi DD, Ahluwalla A. Microsyringe-based deposition of two-dimensional and three-dimensional polymer scaffolds with a well-defined geometry for application to tissue engineering. Tissue Eng 2002;8(6):1089–98.

[32] Vozzi G, Flaim C, Ahluwalia A, Bhatia S. Fabrication of PLGA scaffolds using soft lithography and microsyringe deposition. Biomaterials 2003;24(14):2533–40.

[33] Park JK, Shim J-H, Kang KS, Yeom J, Jung HS, Kim JY, et al. Solid free-form fabrication of tissue-engineering scaffolds with a poly(lactic-co-glycolic acid) grafted hyaluronic acid conjugate encapsulating an intact bone morphogenetic protein–2/poly(ethylene glycol) complex. Adv Funct Mater 2011;21(15):2906–12.

[34] Huang Z-M, Zhang YZ, Kotaki M, Ramakrishna S. A review on polymer nanofibers by electrospinning and their applications in nanocomposites. Composites Sci Technol 2003;63(15):2223–53.

[35] Boland ED, Matthews JA, Pawlowski KJ, Simpson DG, Wnek GE, Bowlin GL. Electrospinning collagen and elastin: preliminary vascular tissue engineering. Front Biosci 2004;9:1422–32.

[36] Lee J, Lee SY, Jang J, Jeong YH, Cho DW. Fabrication of patterned nanofibrous mats using a direct-write electrospinning. Langmuir 2012;28(18):7267–75.

[37] Bellan LM, Craighead HG. Control of an electrospinning jet using electric focusing and jet-steering fields. J Vac Sci Technol B 2006;24(6):3179–83.

[38] Dalton PD, Joergensen NT, Groll J, Moeller M. Patterned melt electrospun substrates for tissue engineering. Biomed Mater 2008;3:034109.

[39] Shim JH, Kim JY, Park M, Park J, Cho DW. Development of a hybrid scaffold with synthetic biomaterials and hydrogel using solid freeform fabrication technology. Biofabrication 2011;3(3):034102.

[40] Awad HA, Wickham MQ, Leddy HA, Gimble JM, Guilak F. Chondrogenic differentiation of adipose-derived adult stem cells in agarose, alginate, and gelatin scaffolds. Biomaterials 2004;25(16):3211–22.

[41] Benoit DSW, Schwartz MP, Durney AR, Anseth KS. Small functional groups for controlled differentiation of hydrogel-encapsulated human mesenchymal stem cells. Nat Mater 2008;7:816–23.

[42] Elisseeff J, Anseth K, Sims D, McIntosh W, Randolph M, Langer R. Transdermal photopolymerization for minimally invasive implantation. PNAS 1999;96:3104–7.

[43] Van Vlierberghe S, Dubruel P, Lippens E, Masschaele B, Van Hoorebeke L, Cornelissen M, et al. Toward modulating the architecture of hydrogel scaffolds: curtains versus channels. J Mater Sci Mater Med 2008;19(4):1459–66.

[44] Patterson J, Martino MM, Hubbell JA. Biomimetic materials in tissue engineering. Mater Today 2010;13:14–22.

[45] Li SJ, Xiong Z, Wang XH, Yan YN, Liu HX, Zhang RJ. Direct fabrication of a hybrid cell/hydrogel construct by a double-nozzle assembling technology. J Bioact Compat Pol 2009;24:249–65.

[46] Li SJ, Yan YN, Xiong Z, Weng CY, Zhang RJ, Wang XH. Gradient hydrogel construct based on an improved cell assembling system. J Bioact Compat Pol 2009;24:84–99.

[47] Lee W, Debasitis JC, Lee VK, Lee JH, Fischer K, Edminster K, et al. Multi-layered culture of human skin fibroblasts and keratinocytes through three-dimensional freeform fabrication. Biomaterials 2009;30(8):1587–95.

[48] Smith CM, Stone AL, Parkhill RL, Stewart RL, Simpkins MW, Kachurin AM, et al. Three-dimensional bioassembly tool for generating viable tissue-engineered constructs. Tissue Eng 2004;10:1566–76.

[49] Lutolf MP, Gilbert PM, Blau HM. Designing materials to direct stem-cell fate. Nature 2009;462:433–41.

[50] Tsang VL, Chen AA, Cho LM, Jadin KD, Sah RL, DeLong S, et al. Fabrication of 3D hepatic tissues by additive photopatterning of cellular hydrogels. FASEB J 2007;21(3):790–801.

[51] Matsusaki M, Yoshida H, Akashi M. The construction of 3D-engineered tissues composed of cells and extracellular matrices by hydrogel template approach. Biomaterials 2007;28(17):2729–37.

[52] Skardal A, Zhang J, McCoard L, Xu X, Oottamasathien S, Prestwich GD. Photocrosslinkable hyaluronan-gelatin hydrogels for two-step bioprinting. Tissue Eng Part A 2010;16:2675–85.

[53] Pescosolido L, Schuurman W, Malda J, Matricardi P, Alhaique F, Coviello T, et al. Hyaluronic acid and dextran based semi-IPN hydrogels as biomaterials for bioprinting. Biomacromolecules 2011;12:1831–8.

[54] Möller L, Krause A, Dahlmann J, Gruh I, Kirschning A, Dräger G. Preparation and evaluation of hydrogel-composites from methacrylated hyaluronic acid, alginate, and gelatin for tissue engineering. Int J Artif Organs 2011;34:93–102.

[55] Nuttelman CR, Rice MA, Rydholm AE, Salinas CN, Shah DN, Anseth KS. Macromolecular monomers for the synthesis of hydrogel niches and their application in cell encapsulation and tissue engineering. Prog Polym Sci 2008;33:167–79.

[56] Dhariwala B, Hunt E, Boland T. Rapid prototyping of tissue-engineering constructs, using photopolymerizable hydrogels and stereolithography. Tissue Eng 2004;10:1316–22.

[57] Yasuda A, Kojima K, Tinsley KW, Yoshioka H, Mori Y, Vacanti CA. *In vitro* culture of chondrocytes in a novel thermoreversible gelation polymer scaffold containing growth factors. Tissue Eng 2006;12:1237–45.

[58] Zhang S. Fabrication of novel biomaterials through molecular self-assembly. Nat Biotechnol 2003;21:1171–8.

[59] Kisiday J, Jin M, Kurz B, Hung H, Semino C, Zhang S, et al. Self-assembling peptide hydrogel fosters chondrocyte extracellular matrix production and cell division: implications for cartilage tissue repair. Proc Natl Acad Sci USA 2002;99:9996–10001.

[60] Lutolf MP, Lauer-Fields JL, Schmoekel HG, Metters AT, Weber FE, Fields GB, et al. Synthetic matrix metalloproteinase sensitive hydrogels for the conduction of tissue regeneration: engineering cell-invasion characteristics. Proc Natl Acad Sci USA 2003;100:5413–8.

[61] Kong HJ, Kaigler D, Kim K, Mooney DJ. Controlling rigidity and degradation of alginate hydrogels via molecular weight distribution. Biomacromolecules 2004;5:1720–7.

[62] Kloxin AM, Kasko AM, Salinas CN, Anseth KS. Photodegradable hydrogels for dynamic tuning of physical and chemical properties. Science 2009;324:59–63.

[63] Norotte C, Marga FS, Niklason LE, Forgacs G. Scaffold-free vascular tissue engineering using bioprinting. Biomaterials 2009;30(30):5910–7.

[64] Klein TJ, Schumacher BL, Schmidt TA, Li KW, Voegtline MS, Masuda K, et al. Tissue engineering of stratified articular cartilage from chondrocyte subpopulations. Osteoarthritis Cartilage 2003;11:595–602.

[65] Bryant SJ, Anseth KS. Hydrogel properties influence ECM production by chondrocytes photoencapsulated in poly(ethylene glycol) hydrogels. J Biomed Mater Res 2002;59:63–72.

[66] Tirella A, Vozzi F, Vozzi G, Ahluwalia A. PAM2 (piston assisted microsyringe): a new rapid prototyping technique for biofabrication of cell incorporated scaffolds. Tissue Eng Part C 2011;17:229–37.

[67] Kang K, Hockaday L, Butcher J. Quantitative optimization of solid freeform deposition of aqueous hydrogels. Biofabrication 2013;5(3):035001.

[68] Shim JH, Lee JS, Kim JY, Cho DW. Bioprinting of a mechanically enhanced three-dimensional dual cell-laden construct for osteochondral tissue engineering using a multi-head tissue/organ building system. J Micromech Microeng 2012;22(8):085014.

[69] Hockaday L, Kang K, Colangelo N, Cheung PY, Duan B, Malone E, et al. Rapid 3D printing of anatomically accurate and mechanically heterogeneous aortic valve hydrogel scaffolds. Biofabrication 2012;4(3):035005.

[70] Macario DK, Entersz I, Paul Abboud J, Nackman GB. Inhibition of apoptosis prevents shear-induced detachment of endothelial cells. J Surg Res 2008;147(2):282–9.

[71] Kretzmer G, Schügerl K. Response of mammalian cells to shear stress. Appl Microbiol Biotechnol 1991;34(5):613–6.

[72] Khalil S, Sun W. Bioprinting endothelial cells with alginate for 3D tissue constructs. J Biomech Eng 2009;131(11):111002.

[73] Dutta RC, Dutta AK. Cell-interactive 3D-scaffold: advances and applications. Biotechnol Adv 2009;27:334–9.

[74] Hutmacher DW, Cool S. Concepts of scaffold-based tissue engineering-the rationale to use solid free-form fabrication techniques. J Cell Mol Med 2007;11:654–69.

[75] Mironov V, Visconti RP, Kasyanov V, Forgacs G, Drake CJ, Markwald RR. Organ printing: tissue spheroids as building blocks. Biomaterials 2009;30(12):2164–74.

[76] Nakamura M, Iwanaga S, Henmi C, Arai K, Nishiyama Y. Biomatrices and biomaterials for future developments of bioprinting and biofabrication. Biofabrication 2010;2(1):014110.

[77] Tsuda Y, Shimizu T, Yarnato M, Kikuchi A, Sasagawa T, Sekiya S, et al. Cellular control of tissue architectures using a three-dimensional tissue fabrication technique. Biomaterials 2007;28:4939–46.

[78] Zorlutuna P, Jeong JH, Kong H, Bashir R. Tissue engineering: stereolithography-based hydrogel microenvironments to examine cellular interactions. Adv Funct Mater 2011;21(19):3597.

[79] Cao T, Ho KH, Teoh S-H. Scaffold design and *in vitro* study of osteochondral coculture in a three-dimensional porous polycaprolactone scaffold fabricated by fused deposition modeling. Tissue Eng 2003;9:s103–12.

[80] Chang R, Nam J, Sun W. Direct cell writing of 3D microorgan for *in vitro* pharmacokinetic model. Tissue Eng Part C Methods 2008;14(2):157–66.

[81] Khalil S, Sun W. Biopolymer deposition for freeform fabrication of hydrogel tissue constructs. Mater Sci Eng C 2007;27(3):469–78.

[82] Khalil S, Nam F, Sun W. Multi-nozzle deposition for construction of 3-D biopolymer tissue scaffolds. Rapid Prototyping J 2005;11:9–17.

[83] Nair K, Yan K, Sun W. A multi-level numerical model for quantifying cell deformation in encapsulated alginate structures. J Mech Mater Struct 2007;6:1121–39.

[84] Kim JY, Jin GZ, Park IS, Kim JN, Chun SY, Park EK, et al. Evaluation of solid free-form fabrication-based scaffolds seeded with osteoblasts and human umbilical vein endothelial cells for use *in vivo* osteogenesis. Tissue Eng Part A 2010;16(7):2229–36.

[85] Shim J-H, Huh J-B, Park JY, Jeon YC, Kang SS, Kim JY, et al. Fabrication of blended polycaprolactone/poly (lactic-*co*-glycolic acid)/β-tricalcium phosphate thin membrane using solid freeform fabrication technology for guided bone regeneration. Tissue Eng Part A 2013;19(3–4):317–27.

[86] Zhong S, Zhang Y, Lim CT. Fabrication of large pores in electrospun nanofibrous scaffolds for cellular infiltration: a review. Tissue Eng Part B 2012;18(2):77–87.

[87] Lee J, Jang J, Oh H, Jeong YH, Cho D-W. Fabrication of three-dimensional lattice patterned nanofibrous scaffold using a direct-write electrospinning apparatus. Mater Lett 2013;93:397–400.

[88] Brown T, Slotosch A, Thibaudeau L, Taubenberger A, Loessner D, Vaquette C, et al. Design and fabrication of tubular scaffolds via direct writing in a melt electrospinning mode. Biointerphases 2012;7(1–4):7–13.

[89] Fedorovich NE, Schuurman W, Wijnberg HM, Prins HJ, van Weeren PR, Malda J, et al. Biofabrication of osteochondral tissue equivalents by printing topologically defined cell-laden hydrogel scaffolds. Tissue Eng Part C Methods 2011;18(1):33–44.

[90] Gaetani R, Doevendans PA, Metz CHG, Alblas J, Messina E, Giacomello A, et al. Cardiac tissue engineering using tissue printing technology and human cardiac progenitor cells. Biomaterials 2012;33(6):1782–90.

[91] Kundu J, Shim J-H, Jang J, Kim S-W, Cho D-W. An additive manufacturing-based PCL–alginate–chondrocyte bioprinted scaffold for cartilage tissue engineering. J Tissue Eng Regen Med 2013; Jan. doi: 10.1002/term.1682.

[92] Gaetani R, Doevendans PA, Metz CH, Alblas J, Messina E, Giacomello A, et al. Cardiac tissue engineering using tissue printing technology and human cardiac progenitor cells. Biomaterials 2012;33(6):1782–90.

[93] Landers R, Pfister A, Hübner U, John H, Schmelzeisen R, Mülhaupt R. Fabrication of soft tissue engineering scaffolds by means of rapid prototyping techniques. J Mater Sci 2002;37(15):3107–16.

[94] Chung JH, Naficy S, Yue Z, Kapsa R, Quigley A, Moulton SE, et al. Bio-ink properties and printability for extrusion printing living cells. Biomater Sci 2013;1(7):763–73.

[95] Finger AR, Sargent CY, Dulaney KO, Bernacki SH, Loboa EG. Differential effects on messenger ribonucleic acid expression by bone marrow-derived human mesenchymal stem cells seeded in agarose constructs due to ramped and steady applications of cyclic hydrostatic pressure. Tissue Eng 2007;13:1151–8.

[96] Fedorovich NE, Kuipers E, Gawlitta D, Dhert WJ, Alblas J. Scaffold porosity and oxygenation of printed hydrogel constructs affect functionality of embedded osteogenic progenitors. Tissue Eng Part A 2011;17(19–20):2473–86.

[97] Nolan K, Millet Y, Ricordi C, Stabler CL. Tissue engineering and biomaterials in regenerative medicine. Cell Transplant 2008;17:241–3.

[98] Shi CM, Zhu Y, Ran XZ, Wang M, Su YP, Cheng TM. Therapeutic potential of chitosan and its derivatives in regenerative medicine. J Surg Res 2006;133:185–92.

[99] Heras A, Rodri'guez NM, Ramosc VM, Agullo E. *N*-Methylene phosphonic chitosan: a novel soluble derivative. Carbohydr Polym 2001;44:1–8.

[100] Wang X, Yan Y, Pan Y, Xiong Z, Liu H, Cheng J, et al. Generation of three-dimensional hepatocyte/gelatin structures with rapid prototyping system. Tissue Eng 2006;12(1):83–90.

[101] Snyder J, Hamid Q, Wang C, Chang R, Emami K, Wu H, et al. Bioprinting cell-laden Matrigel for radioprotection study of liver by pro-drug conversion in a dual-tissue microfluidic chip. Biofabrication 2011;3(3):034112.

[102] Kadler KE, Baldock C, Bella J, Boot-Handford RP. Collagens at a glance. J Cell Sci 2007;120(12):1955–8.

[103] Yan Y, Wang X, Xiong Z, Liu HX, Liu F, Lin F, et al. Direct construction of a three-dimensional structure with cells and hydrogel. J Bioact Compat Polym 2005;20(3):259–69.

[104] Chang R, Nam J, Sun W. Effects of dispensing pressure and nozzle diameter on cell survival from solid freeform fabrication-based direct cell writing. Tissue Eng Part A 2008;14(1):41–8.

[105] Schuurman W, Levett PA, Pot MW, van Weeren PR, Dhert WJ, Hutmacher DW, et al. Gelatin-methacrylamide hydrogels as potential biomaterials for fabrication of tissue-engineered cartilage constructs. Macromol Biosci 2013;13(5):551–61.

[106] Kolesky DB, Truby RL, Gladman AS, Busbee TA, Homan KA, Lewis JA. 3D bioprinting of vascularized heterogeneous cell-laden tissue constructs. Adv Mater 2014;26(19):3124–30.

[107] Zhao H, Ma L, Zhou J, Mao Z, Gao C, Shen J. Fabrication and physical and biological properties of fibrin gel derived from human plasma. Biomed Mater Eng 2008;3:15001.

[108] Gerecht S, Burdick J, Ferreira L, Townsend S, Langer R, Vunjak-Novakovic G. Hyaluronic acid hydrogel for controlled self-renewal and differentiation of human embryonic stem cells. Proc Natl Acad Sci USA 2007;104(27):11298.

[109] Das S, Pati F, Choi Y-J, et al. Bioprintable, cell-laden silk fibroin-gelatin hydrogel supporting multilineage differentiation of stem cells for fabrication of 3D tissue constructs. Acta Biomater 2015;11:233–46.

[110] Badylak SF, Freytes DO, Gilbert TW. Extracellular matrix as a biological scaffold material: structure and function. Acta Biomaterialia 2009;5:1–13.

[111] Fedorovich NE, Swennen I, Girones J, Moroni L, van Blitterswijk CA, Schacht E, et al. Evaluation of photocrosslinked lutrol hydrogel for tissue printing applications. Biomacromolecules 2009;10(7):1689–96.

[112] Hockaday L, Kang K, Colangelo N, Cheung PY, Duan B, Malone E, et al. Rapid 3D printing of anatomically accurate and mechanically heterogeneous aortic valve hydrogel scaffolds. Biofabrication 2012;4(3). 035005.

[113] Fedorovich NE, De Wijn JR, Verbout AJ, Alblas J, Dhert WJA. Three-dimensional fiber deposition of cell-laden, viable, patterned constructs for bone tissue printing. Tissue Eng Part A 2008;14:127–33.

[114] Censi R, Schuurman W, Malda J, di Dato G, Burgisser PE, Dhert W. Printable photopolymerizable thermosensitive p(HPMAlactate)-PEG hydrogel for tissue engineering. Adv Funct Mater 2011;21:1833–42.

[115] Dang F, Shinohara S, Tabata O, Yamaoka Y, Kurokawa M, Shinohara Y, et al. Replica multichannel polymer chips with a network of sacrificial channels sealed by adhesive printing method. Lab Chip 2005;5(4):472–8.

[116] Miller JS, Stevens KR, Yang MT, Baker BM, Nguyen DH, Cohen DM, et al. Rapid casting of patterned vascular networks for perfusable engineered three-dimensional tissues. Nat Mater 2012;11: 768–74.

[117] Golden AP, Tien J. Fabrication of microfluidic hydrogels using molded gelatin as a sacrificial element. Lab Chip 2007;7(6):720–5.

[118] Chang CC, Boland ED, Williams SK, Hoying JB. Direct-write bioprinting three-dimensional biohybrid systems for future regenerative therapies. J Biomed Mater Res Part B 2011;98B(1):160–70.

[119] Visser J, Peters B, Burger TJ, Boomstra J, Dhert WJ, Melchels FP, et al. Biofabrication of multi-material anatomically shaped tissue constructs. Biofabrication 2013;5:035007.

[120] Lee JS, Hong JM, Jung JW, Shim JH, Oh JH, Cho DW. 3D printing of composite tissue with complex shape applied to ear regeneration. Biofabrication 2014;6:024103.

[121] Reiffel AJ, Kafka C, Hernandez KA, Popa S, Perez JL, Zhou S, et al. High-fidelity tissue engineering of patient-specific auricles for reconstruction of pediatric microtia and other auricular deformities. PLoS ONE 2013;8(2):e56506.

[122] Skardal A, Zhang J, Prestwich GD. Bioprinting vessel-like constructs using hyaluronan hydrogels crosslinked with tetrahedral polyethylene glycol tetracrylates. Biomaterials 2010;31(24): 6173–81.

[123] Shin SR, Bae H, Cha JM, Mun JY, Chen YC, Tekin H, et al. Carbon nanotube reinforced hybrid microgels as scaffold materials for cell encapsulation. ACS Nano 2011;6(1):362–72.

[124] Erickson IE, Kestle SR, Zellars KH, Farrell MJ, Kim M, Burdick JA, et al. High mesenchymal stem cell seeding densities in hyaluronic acid hydrogels produce engineered cartilage with native tissue properties. Acta Biomaterialia 2012;8(8):3027–34.

[125] Shin H, Olsen BD, Khademhosseini A. The mechanical properties and cytotoxicity of cell-laden double-network hydrogels based on photocrosslinkable gelatin and gellan gum biomacromolecules. Biomaterials 2012;33(11):3143–52.

[126] Seliktar D. Designing cell-compatible hydrogels for biomedical applications. Science 2012;336(6085):1124–8.

[127] Suri S, Schmidt CE. Photopatterned collagen–hyaluronic acid interpenetrating polymer network hydrogels. Acta Biomaterialia 2009;5(7):2385–97.

[128] Shin MK, Kim SI, Kim SJ, Park SY, Hyun YH, Lee Y, et al. A tough nanofiber hydrogel incorporating ferritin. Appl Phys Lett 2008;93(16).

[129] Saez-Martinez V, Garcia-Gallastegui A, Vera C, Olalde B, Madarieta I, Obieta I, et al. New hybrid system: poly(ethylene glycol) hydrogel with covalently bonded pegylated nanotubes. J Appl Polym Sci 2011;120(1):124–32.

[130] Coburn JM, Gibson M, Monagle S, Patterson Z, Elisseeff JH. Bioinspired nanofibers support chondrogenesis for articular cartilage repair. Proc Natl Acad Sci 2012;109(25):10012–7.

[131] Xu T, Binder KW, Albanna MZ, Dice D, Zhao W, Yoo JJ, et al. Hybrid printing of mechanically and biologically improved constructs for cartilage tissue engineering applications. Biofabrication 2013;5:015001.

[132] Schuurman W, Khristov V, Pot MW, Weeren PRV, Dhert WJA, Malda J. Bioprinting of hybrid tissue constructs with tailorable mechanical properties. Biofabrication 2011;3:021001.

[133] Lee H, Ahn S, Bonassar LJ, Kim G. Cell(MC3T3-E1)-printed poly(ε-caprolactone)/alginate hybrid scaffolds for tissue regeneration. Macromol Rapid Commun 2013;34(2):142–9.

[134] Catros S, Guillemot F, Nandakumar A, Ziane S, Moroni L, Habibovic P, et al. Layer-by-layer tissue microfabrication supports cell proliferation in vitro and in vivo. Tissue Eng Part C Methods 2012;18(1):62–70.

[135] Pham QP, Sharma U, Mikos AG. Electrospun poly(ε-caprolactone) microfiber and multilayer nanofiber/microfiber scaffolds: characterization of scaffolds and measurement of cellular infiltration. Biomacromolecules 2006;7:2796–805.

[136] Farrugia BL, Brown TD, Upton Z, Hutmacher DW, Dalton PD, Dargaville TR. Dermal fibroblast infiltration of poly(ε-caprolactone) scaffolds fabricated by melt electrospinning in a direct writing mode. Biofabrication 2013;5:025001.

[137] Brown TD, Dalton PD, Hutmacher DW. Direct writing by way of melt electrospinning. Adv Mater 2011;23(47):5651–7.

[138] Jang J, Lee J, Seol YJ, Jeong YH, Cho DW. Improving mechanical properties of alginate hydrogel by reinforcement with ethanol treated polycaprolactone nanofibers. Composites Part B 2013;45(1):1216–21.

[139] Mironov V, Trusk T, Kasyanov V, Little S, Swaja R, Markwald R. Biofabrication: a 21st century manufacturing paradigm. Biofabrication 2009;1(2):022001.

[140] Fabien G, Vladimir M, Makoto N. Bioprinting is coming of age: report from the International Conference on Bioprinting and Biofabrication in Bordeaux (3B'09). Biofabrication 2010;2(1):010201.

[141] Park JH, Jang J, Cho DW. Three-dimensional printed 3D structure for tissue engineering. Trans Korean Soc Mech Eng B 2014;38(10):817–29.

Chapter 8

Indirect Rapid Prototyping for Tissue Engineering

Carmelo De Maria*,**, Aurora De Acutis*, and Giovanni Vozzi*,**

*Research Center E. Piaggio, University of Pisa, Pisa, Italy; **Department of Ingegneria dell'Informazione, University of Pisa, Pisa, Italy

ABSTRACT

Tissue engineering (TE) aims at producing patient-specific biological substitutes in an attempt to repair, replace and regenerate damaged tissues or organs in order to improve the current state of clinical treatments. A three-dimensional substrate, the scaffold, is a key aspect to promote cell organization to form a tissue. Recently, rapid prototyping (RP) technologies have been successfully used to fabricate complex scaffolds, thanks to the ability to create highly reproducible architecture and compositional variation across the entire structure, due to their precise controlled computer driven fabrication. The drawback of most of the fabrication principles applied in the RP processes is the requirement of particular conditions (e.g., pressure or temperature) that limit the material choice. Natural polymers, such as collagen, have underlined their superiority for TE solutions but they are challenging to be processed with RP techniques. As alternative, scaffold made of natural biomaterial can be produced by indirect fabrication techniques, casting a biomaterial into sacrificial mold realized by RP processes. So far, the indirect rapid prototyping (iRP) has emerged in a number of different approaches with promising results. The present chapter is focused on iRP multistep methods, highlighting strength and weakness and indicating possible future perspectives.

Keywords: indirect rapid prototyping; additive manufacturing; scaffold; casting

1 INTRODUCTION

Tissue engineering (TE) is an interdisciplinary field in which knowledge from physics, chemistry, biological sciences, and engineering are combined to design, build, modify, raise, and maintain living tissues [1], with the aim of overcoming the intrinsic limits of traditional approaches based on transplants (lack of donors [2] and histocompatibility problems [3]) and on artificial prosthesis (inflammatory response [4] and complexity of biologic systems [5]).

In the scaffold-based TE [6,7], the construction of a bioartificial tissue begins with an *in vitro* formation, in an environment that provides mechanical and metabolic support, of a tissue construct starting from cells seeded onto a scaffold. The scaffold usually is a porous structure realized in a suitable geometry [8] using a degradable biomaterial, often modified to be adhesive for cells or selective for a specific cell type. The scaffold provides mechanical stability to the tissue construct in the short term and guides the three-dimensional (3D) organization of cells until the tissue completion; as the tissue grows the material is degraded by the host cells with a rate similar to the biosynthesis rate of extracellular matrix (ECM). In addition to the scaffold, the realization of 3D tissue substitutes requires a biological model; that is, an appropriate source of proliferating nonimmunogenic cells, with appropriate biological functions, a protocol for cells expansion that does not

Essentials of 3D Biofabrication and Translation. http://dx.doi.org/10.1016/B978-0-12-800972-7.00008-6

alter the specific phenotype, and cell culture techniques that allow us to reach high cell densities ($>10^6$ cells/cm^3). Stem cells could be a solution, but several problems are still unsolved [9,10].

The main challenge of 3D cultures is the accomplishment of controllable and homogeneous cell behavior within the entire 3D volume. The cell behavior is characterized by both an intrinsic variability related with the properties of the cell itself and an extrinsic variability related to the microenvironmental properties. These two sources of variability can both induce heterogeneity of cell fate within the 3D domain, leading to an undesired nonhomogeneous 3D culture structure. The intrinsic variability depends only on the cell source and cannot be controlled during the culture, whereas the extrinsic variability can be minimized by strictly controlling the microenvironmental properties that depend on the scaffold features and on the experimental setup used for culturing cells. Inside a bioreactor, a "living organism" simulator, temperature, pH, O$_2$ and metabolites concentration, waste removal, and mechanical loading [11–13] can be controlled. In a second phase, the bioartificial construct can be implanted in the appropriate anatomical site to engage the *in vivo* remodeling promoting the functional architecture assembly of the desired tissue or organ. Some of the previous steps can be omitted; for example, to engineer cardiac valves or blood vessels, decellularized tissues have been used as a scaffold to attract endogenous cells [14].

Biochemical [15–17], mechanical (e.g., stiffness) [18–20], and topological [21,22] stimuli can be transmitted by the scaffold to correctly engineer a tissue and maintaining a complete functionality. Biocompatibility, biodegradability, and bioabsorbability with a controlled degradation rate balanced with cell growth *in vitro* and/or *in vivo* [23,24], sterilizability [25], and reproducibility [26] are the other fundamental properties of scaffolds.

A porous structure is necessary to guarantee appropriate mass transport (permeability and diffusion of nutrients) as well as cell in-growth and reorganization [27,28]. Pore size is critical not only because it affects the resultant mechanical properties [29] but also because it is related with the specific tissue microenvironment. In several studies it has been reported that each tissue has specific porosity features; bone tissue scaffolds have a pore size of 400–500 μm [30] while smaller pores (100–200 μm) are used, for example, to regenerate skin [31]; it should be noted that excessively small pores prevent cellular penetration [32]. However, increasing the porosity results in a decrease of scaffold rigidity and for a given porosity, different scaffold microstructures will lead to different effective stiffnesses [33].

The research on the design and realization of 3D scaffolds is directed on polymeric materials and bioceramics. Biological (natural) polymers are usually ECM components, such as collagen, elastin, and hyaluronic acid, or can be polysaccharides with vegetal origin, such as alginate and agarose [34,35]. The ECM itself was used with some success as scaffold material [36]. The main advantage of biological polymers is that they mimic ECM and so are an ideal environment for cell growth, ensuring also atoxicity and cellular adhesion. The drawback is that they are less resistant from the mechanical and chemical point of view, are less stable, and more expensive. Most of the natural polymers can be formed into hydrogels, colloids formed by long polymeric chains with high water content (up to 99%), representing an ideal class of material to realize a bioinspired scaffold for the strong analogies with ECM [24]. Synthetic (man-made) polymers are more flexible, with more predictable behavior and can be processed more into different sizes and shapes with respect to natural polymers. Their physical and chemical properties can be easily modified, and the mechanical and degradation characteristics can be altered by their chemical composition of the macromolecules; functional groups and side chains can be linked to peptides or other bioactive molecules [24]. The most extensively used synthetic polymers are poly-glycolic acid (PGA), poly-lactic acid (PLA) and their copolymers, polycaprolactone (PCL), and polyethylene glycol (PEG). Hydrogels from synthetic polymers, mainly based on PEG-based polymers, are less suitable for scaffold realization compared to the natural counterpart because of the cytotoxicity of the crosslinking agents. Bioceramic materials, mainly used in bone TE applications, include hydroxyapatite (HA), the inorganic component of bone, three calcium phosphate (TCP), bioglasses, and glass ceramics [30]. In general, ceramic biomaterials are able to form bone apatite-like material or carbonate HA on their surface, thus enhancing their osteointegration. Brittleness and slow degradation rates are disadvantages associated with their use. Examples of a combination of natural and synthetic polymers together with bioceramic materials are numerous with the aim of integrating the best features of each component [37–39].

The final properties of the scaffold are often induced by manufacturing techniques. To realize 3D scaffolds several techniques can be used, from more conventional, such as textile techniques and molding processes, to the more recent microfabrication techniques that use a rapid prototyping (RP) approach.

2 RP TECHNOLOGIES

The term RP indicates a group of technologies that allows the automatic realization of a physical model based on design data using a computer [40]. RP processes belong to the generative (or additive) production processes. Generative production process is the generic term (seldom used in practice) to indicate an additive production process. More common are the expressions "additive manufacturing," or "solid freeform manufacturing," and "solid freeform fabrication," which emphasize the ability to produce framed solids by means of freeform surfaces. In contrast to abrasive (or subtractive) processes, such as lathing, milling, drilling, grinding, eroding, and so forth in which the form is shaped by removing material, in RP processes the final object is formed by joining volume elements. This approach, when applied to TE, takes the name of bioprinting [41], defined as the use of computer-aided transfer processes for patterning and assembling living and nonliving materials with prescribed two-dimensional (2D) or 3D organization in order to produce bioengineered structures serving in regenerative medicine, pharmacokinetic, and basic cell biology studies. Bioprinting can be inserted in the broader field of computer-aided tissue engineering(CATE), the application of advanced computer-aided technologies to TE problems [42].

RP techniques follow a computer-aided design/computer-aided manufacturing (CAD/CAM) approach. Practically, the object is designed using a computer (CAD), which then sends to the machine the instructions to obtain the desired shape (CAM), fabricated layer by layer. In some cases (especially in the biomedical field), the geometry is extracted, using dedicated segmentation algorithms, from a tomographic scan of the human body. The 2D layer is shaped (contoured) in a (x–y) plane. The third dimension results from single layers being stacked up on top of each other [40]. For the implementation of the RP principle several fundamentally different physical processes are suitable:

- Solidification of liquid materials (e.g., photopolymerization process [43]);
- Generation from the solid phase:

- Incipiently or completely melted solid materials, powder, or powder mixtures (extrusion [26], ballistic [44], and sinter processes [39]);
- Conglutination of granules or powders by additional binders (3D printer approach [45]);
- Precipitation from the gaseous phase (e.g., laser chemical vapor deposition [46]).

Other classifications have also been proposed [47]. An important parameter to choose between different RP techniques is the material used, in particular its chemical–physical properties, which interact with the working principles of the fabrication process.

3   INDIRECT RAPID PROTOTYPING

Most of the fabrication techniques described earlier require particular conditions (e.g., pressure or temperature) that limit the choice of material for realizing the 3D structure to synthetic polymers; a natural polymer as collagen, which has a low degradation temperature (40°C), cannot be processed. As an alternative, scaffold realization using natural biomaterial can be obtained by indirect fabrication techniques, starting from sacrificial mold realized by RP processes. This multistep method practically involves casting of the (bio)material into a mold and then removing or sacrificing the mold to obtain the final scaffold (Fig. 8.1).

This technique, known as indirect rapid prototyping (iRP), indirect solid freeform fabrication (iSFF) [47], or rapid tooling, seems to be the most promising indirect fabrication technique, which allows in any case a good quality of

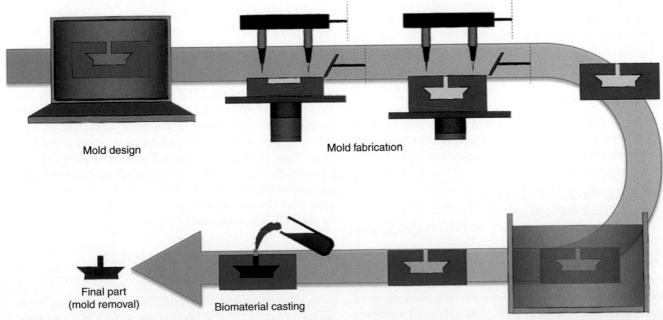

FIGURE 8.1   Indirect rapid prototyping process. Flow work of iRP process, from CAD design to realization.

Mold design

Mold fabrication

Mold postprocessing

Biomaterial casting

Final part
(mold removal)

TABLE 8.1 Examples of Indirect Rapid Prototyping

Technology	Casted Material	Dissolution Technique	References
Solidification of liquid materials			
SLA	Thermoplastic elastomer	Mechanical extraction	[48]
	HA	Pyrolysis	[49,50]
	PCL, PLLA, PLGA, chitosan, alginate, bone cement	Sodium hydroxide	[51]
Photopolymer 3DP	Paraffin wax	Sodium hydroxide	[52]
Generation from the solid phase			
Fused deposition modelling	Alumina, TCP	Pyrolysis	[53]
Ballistic	HA	Pyrolysis	[54–56]
	TCP	Pyrolysis	[57]
	HA/TCP	Pyrolysis	[29]
	PCL, CaP	Organic solvent	[58]
	Collagen	Organic solvent	[44]
	Silk	Organic solvent	[59]
	Chitosan, PLLA, HA	Organic solvent	[60]
	PCL-PLLA	Organic solvent	[61,62]
	PLA-PGA	Organic solvent	[63,64]
	PLLA	Organic solvent	[65]
3DP	PLA	Water, ethanol and calcium reagent	[66]

The table highlights the RP methods, the casted materials, and the techniques used to dissolve the mold.

the external shape and a satisfactory control on the internal scaffold architecture (Table 8.1).

So, overcoming the limits and taking the advantages of RP, the iRP processes allow the following:

- Realization of composite scaffolds, realized with a wide range of (bio)material that are incompatible with other fabrication techniques or might require conflicting processing parameters [29,58,60,62,63].
- Less waste of raw material.
- High fidelity in the scaffold realization compared to other RP techniques.

The macroporosity can be determined by the RP technology, while the microporosity in the iRP approach is usually achieved by including in the processing steps one of the following approaches:

- Critical point dried (CPD) [44,67]
- Free-drying [60,66,68]
- Leaching (e.g., wax microparticles [65], salt [63,67], sucrose [66]).

The first two processes involve just the last step of the indirect fabrication method. The freeze-dry process is considered a representative technique for the fabrication of porous foam-like scaffolds [68]. The produced scaffolds are generally in simple geometrical shape such as a disc or cylinder. Freeze-drying involves the removal of water or other solvents from a frozen product by a sublimation process. Sublimation occurs when a frozen liquid goes directly to the gaseous state without passing through the liquid phase. The pore structure is a replica of sublimated ice crystals. The mean diameter of pores is around 200 μm. The CPD can be used when an organic solvent (i.e., ethanol) is part of the production process; CPD involves the exchange of ethanol with liquid CO_2 at room temperature and a controlled pressure. Increasing the temperature 10°C above the CO_2 critical point (31°C at ~7.4 MPa) changes the liquid CO_2 to gaseous CO_2, with a respective increase in pressure. Pore dimensions in this case are around 50 μm. The leaching method, whatever porogen agent is used, is inserted in the first parts of the iRP process that is the preparation of the mixture that will be casted into the mold. The porogen is then removed by a bath into an appropriate solvent. Pore dimensions depend on the particle size (and on its distribution) of the used porogen.

4 SOLIDIFICATION OF LIQUID MATERIALS

In the photopolymerization techniques, optical energy is applied to irradiate the thin layer at the surface of liquid photopolymer resin. The irradiation areas of resin chemically

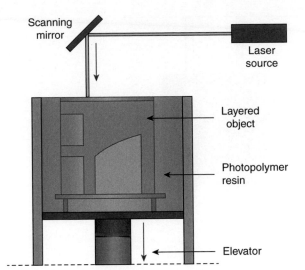

Scanning
mirror

Laser
source

Layered
object

Photopolymer
resin

Elevator

FIGURE 8.2 Stereolithography process. A layer of photopolymerizable resin is cured and solidified by a laser. Then the elevator drops to cover the solid polymer with another layer of liquid resin.

react and transform into a solid phase (Fig. 8.2). A well-known photopolymerization technique is stereolithography (SLA). In SLA, a UV laser traces out the first layer of photocurable resin, solidifying the model's cross-section while leaving the remaining areas in liquid form. An elevator then drops by a sufficient amount to cover the solid polymer with another layer of liquid resin. A sweeper recoats the solidified layer with liquid resin and the laser traces the second layer atop the first.

Chu et al. [49] produced HA-based porous implants using SLA-built epoxy molds. A thermal curable HA-acrylate suspension was cast into the mold to obtain a scaffold with interconnected channels. A thermal treatment at high temperatures (>1000°C) has a double effect of sintering the HA powder and dissolving the mold by pyrolysis. The resolution of channel width was as low as 366 μm. Despite a few fabrication inaccuracies, *in vivo* experiments demonstrated osteoconductivity and biocompatibility of these scaffolds in a minipig model [69]. Similar results in terms of fabrication outcome were also obtained using μ-SLA [50]. SLA models derived from X-ray computed tomography have been used to generate biocompatible and biodegradable heart valve scaffolds from poly-4-hydroxybutyrate and polyhydroxyoctanoate (PHOH) by thermal processing [48]. The use of thermoplastic elastomers, P4HB and PHOH, allowed molding of a complete trileaflet heart valve scaffold without the need for suturing or other postprocessing steps, and demonstrated the ability of (iRP) to reproduce complex anatomical structures. The development of an alkali soluble photopolymer [70] allowed the use of this technology with geliform material. Kang et al. [51], applying indirect SLA, produced a 3D-organ shaped scaffold using a large selection of biomaterial, including PCL, PLLA, PLGA, chitosan, alginate, and bone cement. In the process, 3D porous

sacrificial molds were made of alkali-soluble photopolymer. This technology allowed us to obtain a high resolution with a minimum pore or strut size on a scale of several tens of micrometers; in addition, it was also combined with conventional technologies, such as salt leaching and phase inversion, to fabricate dual-pore scaffolds.

5 GENERATION FROM THE SOLID PHASE

In a typical melt-dissolution deposition system, each layer is created by extrusion of a strand of material through an orifice while it moves across the plane of the layer cross-section. The material cools, solidifying itself and fixing to the previous layer. Successive layer formation, one atop another, forms a complex 3D solid object.

5.1 Fused Deposition Modeling

A representative system using melt-dissolution deposition is fused deposition modeling (FDM) (Fig. 8.3). This method spun off several other systems that operate under similar principles. In the FDM method, a filament of a suitable material is fed and melted inside a heated liquefier before being extruded through a nozzle. The system operates in a temperature-controlled environment to maintain sufficient fusion energy between each layer. Drawbacks of the FDM technique include the need as input a filament with a specific diameter and with appropriate physical/chemical properties to feed through the rollers and the nozzle. As a result, FDM has a narrow processing window. The typical resolution of FDM is 250 μm. In FDM, a limited range of materials can be used with almost complete exclusion of natural polymers, as the material used must be made into filaments and melted into a semiliquid phase before extrusion. The operating temperature of the system is too high to incorporate biomolecules into the scaffold, hence limiting the biomimetic aspect of the produced scaffold [6].

Bose et al. [53] produced, using the iRP approach with FDM technology, alumina and β-TCP ceramic scaffolds with pore sizes in the range of 300–500 μm and porosity of 25–45%. The molds were made using a standard thermoplastic polymer. Their research aimed at investigating the effect of pore size and porosity on the mechano-biological responses. Miller et al. [71], using a modified RepRap 3D printer (an open-source FDM system), printed a rigid filament network of carbohydrate glass adapted for the creation of a densely populated tissue construct with perfusable vascular channel and junctions. These filaments were encapsulated in ECM along with living cells; after crosslinking the ECM, the filaments were dissolved in cell media; to avoid a potential osmotic shock to living cells, the filaments were coated by a thin layer of poly-D-lactide-*co*-glycolide. The rigid filaments network of carbohydrate glass lattice worked as a mold, leaving the tissue construct with vascular

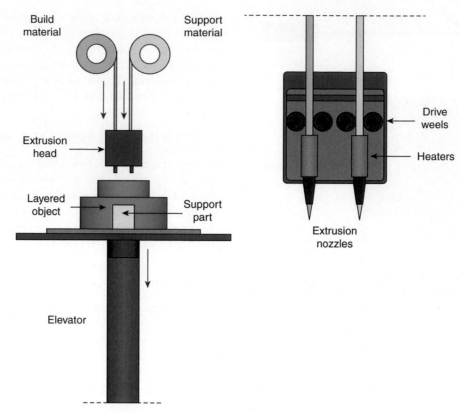

FIGURE 8.3 Fused deposition modeling process. Filaments of a suitable material are fed and melted inside a heated liquefier before being extruded through nozzles, while the extruder or the deposition plate trace trajectories in the *x–y* plane. A complete 3D object is fabricated by repeating this procedure for the other layers.

architecture, with a vessel lumen of about 200 μm, which matched the original lattice.

5.2 3D Bioplotter

Another technology, which belongs to "generation from solid phase," is the 3D bioplotter, which is based on the 3D dispensing of liquids and pastes into a liquid medium with matched density. The plotting material leaves the nozzle and solidifies in the plotting medium after bonding to the previous layer. The liquid medium compensates for gravity and hence no support structure is needed (Fig. 8.4). Hydrogel structures with well-defined internal geometry can be prepared [72].

A border-line iRP application following this approach was proposed by Norotte et al. [73], where the bioplotter system was used to fabricate agarose rods, which served as a template and for supporting a vascular tree. Agarose was then mechanically removed at the end of the maturation process of the blood vessel wall.

5.3 Ballistic Technology

A very common fabrication method used in the iRP field is the so-called ballistic technology [40]. This technique is based on inkjet printing technology. A stream of molten

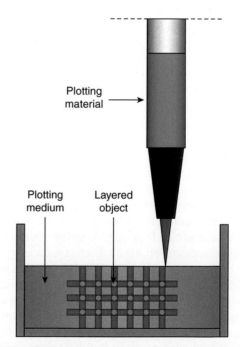

FIGURE 8.4 3D bioplotter system. The plotting material is dispensed into a liquid medium layer by layer.

thermoplastic droplets is deposited on a working surface. Thermal energy in the deposited droplet causes local

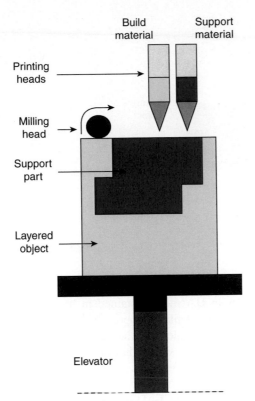

FIGURE 8.5 Ballistic technology. A stream of molten thermoplastic droplets is deposited on a working surface, a milling head is passed over the layer to ensure that a uniform thickness has been achieved, and then another layer is deposited.

melting on the previous layer and then solidifies as a single piece. A representative system is the Model-MakerII (Solidscape, Inc.). This machine uses a single jet each for a plastic building material and a wax-like support material, which are held in a melted liquid state in reservoirs. The printer head ejects droplets of the materials as they are moved in the x–y plane. After an entire layer of the object has hardened, a milling head is passed over the layer to ensure that a uniform thickness has been achieved (Fig. 8.5).

This technology is selected by researchers to fabricate calcium phosphate scaffolds because the building material has a very low thermal expansion coefficient (not provided by the manufacturer, but for analogy other wax materials can be around $\sim 10^{-4}\,K^{-1}$) [6,74] that closely matches that of ceramics (e.g., for HA is $\sim 10^{-5}\,K^{-1}$ [75]), thus minimizing the risk of fracture during sintering. The sintering process, also in this case, is used to remove the mold by pyrolysis. As an example, Limpanuphap et al. [57] fabricated TCP scaffolds with controlled internal porosity using a suspension of TCP in an acrylate binder. A similar route was used to produce a polymer-TCP scaffold. Schumacher et al. [29], following the same method, casted into a wax mold a biphasic mixture of HA and TCP at different percentages. Pore diameters of $\sim 340\ \mu m$ have been obtained; shrinkage and compressive mechanical properties have been analyzed. Wilson et al. [54]

fabricated HA scaffolds with a defined macroarchitecture. Achieved channels were ~ 350–$400\ \mu m$ wide. The authors suggested that the inherent surface texture obtained from ballistic-built mold increased the surface area and led to a higher degree of calcium and phosphate release, which is advantageous to bone formation, reporting ectopic bone formation for all scaffold and cell constructs. Similar results, with the same approach, were obtained also in other works [55,56].

The use of ballistic technology is not limited to calcium–phosphate-based scaffolds. Manufacturing of collagen-based scaffolds by using the ballistic technology to fabricate a mold has also been reported. Sachlos et al. [44] have successfully produced collagen scaffolds with predefined and reproducible internal channels. The smallest channel width achieved was reported to be around $135\ \mu m$. In this case an organic solvent (ethanol) was used to dissolve the mold; the collagen scaffold was then critical point dried with liquid CO_2. Detailed analysis has been carried out to determine if any trace of the solvents used was entrapped in the scaffolds. In a variation to Sachlos' work, Yeong et al. [6] produced chitosan-collagen scaffolds. In the work of Taboas et al. [63] biphasic scaffolds with mechanically interdigitated PLA-PGA and sintered HA regions were fabricated for bone TE. These biphasic scaffolds were obtained by casting PLA-PGA into an HA mold, previously fabricated using a wax molds. The fabricated scaffolds had interconnected pores ranging from $500\ \mu m$ to $800\ \mu m$ as specified by the prefabricated mold, while micropores varied from 10 to $300\ \mu m$ depending on the microporosity fabrication method. Several other groups worked with PLA and PGA [64,65]. Li et al. [60] fabricated a PLLA/chitosan scaffold filled with HA microspheres by casting a mixture of these components into a wax mold (with macropore size of $500\ \mu m$), which was removed at the end of the process using an organic solvent. The physicochemical characterization (eSEM, phase composition, mechanical test) and cell culture tests with MC3T3-EI osteoblasts (eSEM, alkaline phosphatase activity, histology) indicated that these scaffolds could be used for nonload-bearing applications. Another application that involves the use of HA mixed with a polymer (PCL in this case) was presented by Mondrinos et al. [58], who obtained pore sizes of ~ 200 to $\sim 600\ \mu m$. A composite scaffold for cell transplantation of genetically modified human tooth dentin–ligament–bone complex *in vivo* was presented in the work of Park et al. [61]. The newly- formed tissues showed the interfacial generation of parallel and obliquely oriented fibers that grew and traversed within the PCL–PGA construct, characterized by the presence of a well-oriented guidance channel. Also in this case, scaffolds were built by casting into a wax mold, dissolved into an organic solvent. The application *in vivo* (rat model) to repair the soft–hard tissue interfaces of periodontal defects promoted the triphasic tissue regeneration with the compartmentalization of hard and soft tissue structures [62]. Silk fibroin protein

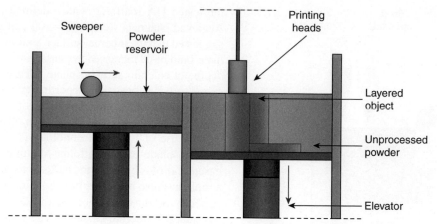

FIGURE 8.6   3D printing. Thin layers of powder are bonded one upon another by a stream of adhesive droplets expelled by inkjet print head.

is another suitable material for the application of iRP. Liu et al. [59], using the ballistic technology and mineral oil to dissolve the mold, fabricated tissue scaffolds with macro- and micromorphology, qualitatively examined by spectral-domain optical coherence tomography. Using this advanced imaging technique, they were able to differentiate the cells and scaffold material by producing high contrasting images. The macropore size is, in this application, ~700 μm.

5.4   3D Printing

In the conglutination of granules or powders by additional binders, particles are selectively bonded in a thin layer of powder material. Thin 2D layers are bonded one upon another to form a complex 3D solid object. During fabrication, the object is supported by and embedded in the unprocessed powder. Typical systems in this category include 3D printing (Fig. 8.6). In this process a stream of adhesive droplets is expelled through an inkjet print head, selectively bonding a thin layer of powder particles to form a solid shape, achieving a resolution of ~300 μm.

This technology was used by Lee et al. [66] to realize PLGA scaffolds with villi-shaped features, using a calcium sulfate hemihydrate mold. Sucrose was used as porogen agent to obtain microporosity. The mold removal process was carried out with successive rinsing in water, ethanol, and calcium reagent. The final resolution obtained is around ~800 μm. The scaffolds were tested with IEC-6 cells, an appropriate biological model of intestinal tract, showing promising results. From the engineering point of view, the feasibility of the methodology proposed was tested realizing zygoma-shaped scaffolds.

6   iRP^2

Some groups introduced in the realization process an additional step, using a secondary mold built by casting into a primary RP-made mold, thus named here as iRP^2. Saito

et al. [76] fabricated 50:50 PLGA scaffolds by casting the polymer into an HA mold, obtained in turn from a wax mold built with the ballistic approach. The HA mold was dissolved using a decalcifier. The final pore size was predicted considering the shrink due to sintering of the HA, and corresponded to a value of ~800 μm. A similar work was carried out by Mantila-Roosa et al. [77] whose final scaffold was made in PCL. Different materials were used by Uchida et al. [67] where a human carotid artery was fabricated starting from a wax mold, which was used as a template for the coating with PVA. In turn, the PVA structure was used as a mold for PCL, which constitutes the bulk material for the final scaffold. Salt leaching completed the fabrication process. A "two-step" mold was investigated also using other techniques, such as photopolymer 3D printing (3DP) [52], which combines the photopolymerization with the printing technology. In this case a wax structure was obtained by casting wax into a mold made by a photopolymer. This mold was dissolved using a sodium hydroxide solution. The wax structure was used as a mold for gelatin/PLLA and then dissolved in an organic solvent.

7   DISCUSSION AND CONCLUSIONS

Taken together, these studies demonstrated that iRP adds a greater degree of versatility and detail to scaffold design. The previous restriction on casting was the inability of molds to produce complex geometry and the internal architecture. The RP molds can meet specific TE requirements, including the need for mechanical integrity and customized shapes [66]. In addition, iRP allows the use of a wider range of biomaterials or combination of materials (composites and/or copolymers) [29,58,60,62,63] and at the same allows to obtain a good resolution (200/300 μm).

However, some drawbacks still exist with this approach, including resolutions of the RP methods, the casting procedure, and the extraction methods.

Obviously, the first requirement for a reliable molded part is the absence of interaction between the casted material and the mold; purity and composition of the final object can be evaluated using Fourier transform infrared spectroscopy, X-ray diffraction, and UV–Vis spectrometry [44–54].

The casted model inherits errors and defects from the mold, such as cracks and dimensional changes. The casted material must completely fill the mold. An optimal filling can be ensured by using injection techniques [70], but in this case a hopper must be purposely designed. The filling problem is also extremely linked to the wettability of the mold internal surfaces: an unwettable surface results in higher capillary forces that can prevent the penetration of the casted material into the narrower channels and features. Chemical treatments to change the surface wettability with respect to a given casted material can be a solution.

A safer method to remove the mold precisely while preserving the scaffold intact without compromising its properties must be developed (Table 8.1).

Mechanical extraction may result in damage due to scaffold's fragile structure. The approach of dissolving the mold in a solvent (usually organic) and washing the scaffold to eliminate traces of the solvent and of the mold brings with itself a potential toxicity, and does not allow an immediate extraction of the piece from the mold (dissolving time is around 12 h). Long procedures are needed to ensure the absence of toxic solvent entrapped in the scaffold [44,54]. The pyrolysis process is limited to ceramic-based scaffolds, requires a specific instrumentation, and also in this case several hours are needed for fabrication. A possible solution is to realize the mold with low melting point materials, dissolving the mold at the end of the process without damaging the scaffold. Moving in this route, low melting point molds were fabricated with the PAM^2 system [78,79].

The remodeling of a graft generated by scaffold-based TE can be considered analogous to guided wound healing [42]; constructs become extensively remodeled as part of the normal tissue repair process. The iRP techniques can provide a CAD supported manufacturing processes that are capable of fabricating reproducible and affordable scaffolds from a variety of biomaterials with different physical and biochemical properties, which can guide this repair process. Several iRP scaffolds were tested on animals with a good success, to engineer the bone–ligament complex [62] and bone [69,77].

The preparation of a scaffold using the iRP approach should even be performed directly in a surgery room if a most suitable extraction method can be designed. This could solve the off-the-shelf availability problem [80]; molds can be stored separately from the casting material, which eventually can be ordered when necessary. These techniques offer the right balance of capability, practical, and cost benefits that make them suitable for fabrication of constructs in sufficient quantity and quality for clinically driven TE applications.

GLOSSARY

Casting A manufacturing process to form solid shape out of a fluid (slurry, molten or liquid) material. The material poured into a cavity or a mold of the desired shape, and then allowed to solidify and it is ejected from the mould once solidified

Extrusion A continuous shaping of material by forcing it under pressure through a die

Hydrogels A three-dimensionally cross-linked polymeric network capable of imbibing considerable amounts of water or biological fluids without undergoing dissolution

Indirect rapid prototyping The use of a rapid prototyped structure as a mould for casting

Inkjet A number of differing printing technologies, all of which accomplish the delivering streams of ink drops, which are deflected onto the substrate, based on data contained in digital files

Mould A shaped cavity used to give a definite form to fluid material

Photopolymerisation The bonding of two or more monomers of a photocurable resin to form a polymer after the exposure to a curing light

Pyrolysis Chemical decomposition of compounds caused by high temperatures

Porosity The measure of void space in a material

Rapid prototyping A group of technologies that allows the automatic realization of physical model based on design data using a computer. The final object is formed by joining volume elements with a layer-by-layer approach

Scaffold A substrate derived from natural or synthetic material that is intended to mimic the extracellular matrix of a living tissue and giving support and appropriate physical and biochemical stimuli to seeded cells

Sintering The process of compacting and forming a solid mass of material by heat and/or pressure without melting

ABBREVIATIONS

2D	Two-dimensional
3D	Three-dimensional
3DP	3D printing
CAD	Computer-aided design
CAM	Computer-aided manufacturing
CATE	Computer-aided tissue engineering
eSEM	Environmental scanning electron microscope
ECM	Extracellular matrix
FDM	Fused deposition modelling
HA	Hydroxyapatite
iRP	Indirect rapid prototyping
PCL	Polycaprolactone
PEG	Polyethylene glycol
PGA	Poly-glycolic acid
PHOH	Polyhydroxyoctanoate
PLA	Poly-lactic acid
RP	Rapid prototyping

SLA	Stereolithography
TCP	Three calcium phosphate
TE	Tissue engineering

REFERENCES

[1] Lanza RP, Langer R, Vacanti J. Principles of tissue engineering. San Diego: Academic Press; 2007.

[2] Fukumitsu K, Yagi H, Soto-Gutierrez A. Bioengineering in organ transplantation: targeting the liver. Transplant 2011;43(6):2137–8.

[3] Valenzuela NM, Reed EF. The link between major histocompatibility complex antibodies and cell proliferation. Transplant Rew 2011;25(4):154–66.

[4] Catelas I, Wimmer M, Utzschneider S. Polyethylene and metal wear particles: characteristics and biological effects. Semin Immunopathol 2011;33:257–71.

[5] Pezaris JS, Eskandar EN. Getting signals into the brain: visual prosthetics through thalamic microstimulation. Neurosurg Focus 2009;27(1):E6.

[6] Yeong WY, Chua CK, Leong KF, Chandrasekaran M. Rapid prototyping in tissue engineering: challenges and potential. Trends Biotechnol 2004;22(12):1–10.

[7] Ortinau S, Schmich J, Block S, Liedmann A, Jonas L, Weiss D, et al. Effect of 3D-scaffold formation on differentiation and survival in human neural progenitor cells. Biomed Eng OnLine 2010;9(1):70.

[8] Vozzi G, Previti A, Ciaravella G, Ahluwalia A. Microfabricated fractal branching networks. J Biomed Mater Res A 2004;71(2):326–33.

[9] Re'em T, Cohen S. Microenvironment design for stem cell fate determination. Adv Biochem Eng Biotechnol 2012;126:227–62.

[10] Beeson D, Lippman A. Egg harvesting for stem cell research: medical risks and ethical problems. Reprod BioMed Online 2006;13(4):573–9.

[11] Mazzei D, Guzzardi MA, Giusti S, Ahluwalia A. A low shear stress modular bioreactor for connected cell culture under high flow rates. Biotechnol Bioeng 2010;106(1):127–37.

[12] De Maria C, Giusti S, Mazzei D, Crawford A, Ahluwalia A. SQPR: a hydrodynamic bioreactor for non-contact stimulation of cartilage constructs. Tissue Eng Part C Meth 2011;17(7):757–64.

[13] Knight MM, Toyoda T, Lee DA, Bader DL. Mechanical compression and hydrostatic pressure induce reversible changes in actin cytoskeletal organisation in chondrocytes in agarose. J Biomech 2006;39(8):1547–51.

[14] Mendelson K, Schoen FJ. Heart valve tissue engineering: concepts, approaches, progress, and challenges. Ann Biomed Eng 2006;34(12):1799–819.

[15] Itano N. Simple primary structure, complex turnover regulation and multiple roles of hyaluronan. J Biochem 2008;144(2):131–7.

[16] Singh M, Berkland C, Detamore MS. Strategies and applications for incorporating physical and chemical signal gradients in tissue engineering. Tissue Eng Part B Rev 2008;14(4):341–66.

[17] Zhang X, Meng L, Lu Q. Cell behaviors on polysaccharide-wrapped single-wall carbon nanotubes: a quantitative study of the surface properties of biomimetic nanofibrous scaffolds. ACS Nano 2009;3(10):3200–6.

[18] Gray DS, Tien J, Chen CS. Repositioning of cells by mechanotaxis on surfaces with micropatterned Young's modulus. J Biomed Mater Res A 2003;66(3):605–14.

[19] Lo CM, Wang HB, Dembo M, Wang YL. Cell movement is guided by the rigidity of the substrate. Biophys J 2000;79(1):144–52.

[20] Baker SC, Rohman G, Southgate J, Cameron NR. The relationship between the mechanical properties and cell behaviour on PLGA and PCL scaffolds for bladder tissue engineering. Biomaterials 2009;30(7):1321–8.

[21] Bettinger CJ, Langer R, Borenstein JT. Engineering substrate topography at the micro- and nanoscale to control cell function. Angew Chem Int Ed Engl 2009;48(30):5406–15.

[22] Biggs MJP, Richards RG, Gadegaard N, McMurray RJ, Affrossman S, Wilkinson CDW, et al. Interactions with nanoscale topography: adhesion quantification and signal transduction in cells of osteogenic and multipotent lineage. J Biomed Mater Res A 2009;91(1):195–208.

[23] Ding X, Takahata M, Akazawa T, Iwasaki N, Abe Y, Komatsu M, et al. Improved bioabsorbability of synthetic hydroxyapatite through partial dissolution-precipitation of its surface. J Mater Sci Mater M 2011;22(5):1247–55.

[24] Cheung HY, Lau KT, Lu TP, Hui D. A critical review on polymer-based bio-engineered materials for scaffold development. Compos Part B Eng 2007;38(3):291–300.

[25] Temenoff JS, Mikos AG. Injectable biodegradable materials for orthopedic tissue engineering. Biomaterials 2000;21(23):2405–12.

[26] Zein I, Hutmacher DW, Tan KC, Teoh SH. Fused deposition modeling of novel scaffold architectures for tissue engineering applications. Biomaterials 2002;23(4):1169–85.

[27] Zeltinger J, Sherwood JK, Graham DA, Mueller R, Griffith LG. Effect of pore size and void fraction on cellular adhesion, proliferation, and matrix deposition. Tissue Eng 2001;7:557–72.

[28] O'Brien FJ, Harley BA, Yannas IV, Gibson LJ. The effect of pore size on cell adhesion in collagen-gag scaffolds. Biomaterials 2005;26: 433–41.

[29] Schumacher M, Deisinger U, Detsch R, Ziegler G. Indirect rapid prototyping of biphasic calcium phosphate scaffolds as bone substitutes: influence of phase composition, macroporosity and pore geometry on mechanical properties. J Mater Sci Mater M 2010;21(12):3119–27.

[30] Karageorgiou V, Kaplan D. Porosity of 3D biomaterial scaffolds and osteogenesis. Biomaterials 2005;26(27):5474–91.

[31] Ma L, Gao C, Mao Z, Zhou J, Shen J, Hu X, et al. Collagen/chitosan porous scaffolds with improved biostability for skin tissue engineering. Biomaterials 2003;24(26):4833–41.

[32] Oh SH, Park IK, Kim JM, Lee JH. *In vitro* and *in vivo* characteristics of PCL scaffolds with pore size gradient fabricated by a centrifugation method. Biomaterials 2007;28(9):1664–71.

[33] Cheah CM, Chua CK, Leong KF, Chua SW. Development of a tissue engineering scaffold structure library for rapid prototyping. Part 2: Parametric library and assembly program. Int J Adv Manuf Tech 2003;21(4):302–12.

[34] Drury JL, Mooney DJ. Hydrogels for tissue engineering: scaffold design variables and applications. Biomaterials 2003;24(24):4337–51.

[35] Cabral J, Moratti SC. Hydrogels for biomedical applications. Future Med Chem 2011;3(15):1877–88.

[36] Badylak SF, Freytes DO, Gilbert TW. Extracellular matrix as a biological scaffold material: structure and function. Acta Biomater 2009;5(1):1–13.

[37] Suzawa Y, Funaki T, Watanabe J, Iwai S, Yura Y, Nakano T, et al. Regenerative behavior of biomineral/agarose composite gels as bone grafting materials in rat cranial defects. J Biomed Mater Res 2010;93(3):965–75.

[38] Peter M, Ganesh N, Selvamurugan N, Nair SV, Furuike T, Tamura H, et al. Preparation and characterization of chitosan-gelatin/

nanohydroxyapatite composite scaffolds for tissue engineering applications. Carbohyd Polym 2010;80(3):687–94.

[39] Tan KH, Chua CK, Leong KF, Cheah CM, Cheang P, Abu Bakar MS, et al. Scaffold development using selective laser sintering of polyetheretherketone-hydroxyapatite bio- composite blends. Biomaterials 2003;24(18):3115–23.

[40] Gebhardt A. Rapid prototyping. Carl Hanser Verlag GmbH & Co. KG; 2003. p. 394.

[41] Guillemot F, Mironov V, Nakamura M. Bioprinting is coming of age: report from the International Conference on Bioprinting and Biofabrication in Bordeaux (3B'09). Biofabrication 2010;2:010201.

[42] Sun W, Darling A, Starly B, Nam J. Computer-aided tissue engineering: overview, scope and challenges. Biotechnol Appl Biochem 2004;39(1):29–47.

[43] Melchels FPW, Feijen J, Grijpma DW. A review on stereolithography and its applications in biomedical engineering. Biomaterials 2010;31(24):6121–30.

[44] Sachlos E, Reis N, Ainsley C, Derby B, Czernuszka JT. Novel collagen scaffolds with predefined internal morphology made by solid freeform fabrication. Biomaterials 2003;24(8):1487–97.

[45] Bassoli E, Gatto A, Iuliano L, Violante MG. 3D printing technique applied to rapid casting. Rapid Prototyping J 2007;13(3):148–55.

[46] Endo J, Ito A, Kimura T, Goto T. High-speed deposition of dense, dendritic and porous SiO_2 films by Nd:YAG laser chemical vapor deposition. Mater Sci Eng B 2010;166(3):225–9.

[47] Hutmacher DW, Sittinger M, Risbud MV. Scaffold-based tissue engineering: rationale for computer-aided design and solid freeform fabrication systems. Trends Biotechnol 2004;22(7):354–62.

[48] Sodian R, Loebe M, Hein A, Martin D, Hoerstrup S, Potapov E, et al. Application of stereolithography for scaffold fabrication for tissue engineered heart valves. ASAIO J 2002;48(1):12–6.

[49] Chu TMG, Halloran JW, Hollister SJ, Feinberg SE. Hydroxyapatite implants with designed internal architecture. J Mater Sci Mater M 2001;12:471–8.

[50] Seol YJ, Kim JY, Park EK, Kim SY, Cho DW. Fabrication of a hydroxyapatite scaffold for bone tissue regeneration using microstereolithography and molding technology. Microelectron Eng 2009;86(4–6):1443–6.

[51] Kang HW, Cho DW. Development of an indirect stereolithography technology for scaffold fabrication with a wide range of biomaterial selectivity. Tissue Eng Part C Meth 2012;18(9):719–29.

[52] Tan JY, Chua CK, Leong KF. Indirect fabrication of gelatin scaffolds using rapid prototyping technology. Virtual Phys Prototyping 2010;5(1):45–53.

[53] Bose S, Darsell J, Kintner M, Hosick H, Bandyopadhyay A. Pore size and pore volume effects on alumina and TCP ceramic scaffolds. Mat Sci Eng C 2003;23(4):479–86.

[54] Wilson CE, de Bruijn JD, van Blitterswijk CA, Verbout AJ, Dhert WJA. Design and fabrication of standardized hydroxyapatite scaffolds with a defined macro-architecture by rapid prototyping for bone tissue engineering research. J Biomed Mater Res A 2004;68A(1):123–32.

[55] Deisinger U, Hamisch S, Schumacher M, Uhl F, Detsch R, Ziegler G. Fabrication of tailored hydroxyapatite scaffolds: comparison between a direct and an indirect rapid prototyping technique. Key Eng Mat 2008;361–363:915–8.

[56] Detsch R, Uhl F, Deisinger U, Ziegler G. 3D-cultivation of bone marrow stromal cells on hydroxyapatite scaffolds fabricated by dispense-plotting and negative mould technique. J Mater Sci Mater M 2008;19:1491–6.

[57] Limpanuphap S, Derby B. Manufacture of biomaterials by a novel printing process. J Mater Sci Mater 2002;13:1163–6.

[58] Mondrinos MJ, Dembzynski R, Lu L, Byrapogu VKC, Wootton DM, Lelkes PI, Zhou J. Porogen-based solid freeform fabrication of polycaprolactone-calcium phosphate scaffolds for tissue engineering. Biomaterials 2006;27(25):4399–408.

[59] Liu MJJ, Chou SM, Chua CK, Tay BCM, Ng BK. The development of silk fibroin scaffolds using an indirect rapid prototyping approach: morphological analysis and cell growth monitoring by spectral-domain optical coherence tomography. Med Eng Phys 2013;35(2):253–62.

[60] Li LH, Kommareddy KP, Pilz C, Zhou CR, Fratzl P, Manjubala I. *In vitro* bioactivity of bioresorbable porous polymeric scaffolds incorporating hydroxyapatite microspheres. Acta Biomater 2010;6(7):2525–31.

[61] Park CH, Rios HF, Jin Q, Bland ME, Flanagan CL, Hollister SJ, et al. Biomimetic hybrid scaffolds for engineering human tooth-ligament interfaces. Biomaterials 2010;31(23):5945–52.

[62] Park CH, Rios HF, Jin Q, Sugai JV, Padial-Molina M, Taut AD, et al. Tissue engineering bone-ligament complexes using fiber-guiding scaffolds. Biomaterials 2012;33(1):137–45.

[63] Taboas JM, Maddox RD, Krebsbach PH, Hollister SJ. Indirect solid free form fabrication of local and global porous, biomimetic and composite 3D polymer-ceramic scaffolds. Biomaterials 2003;24(1):181–94.

[64] Moore MJ, Friedman JA, Lewellyn EB, Mantila SM, Krych AJ, Ameenuddin S, et al. Multiple-channel scaffolds to promote spinal cord axon regeneration. Biomaterials 2006;27(3):419–29.

[65] Wang P, Hu J, Ma PX. The engineering of patient-specific, anatomically shaped, digits. Biomaterials 2009;30(14):2735–40.

[66] Lee M, Dunn JCY, Wu BM. Scaffold fabrication by indirect three-dimensional printing. Biomaterials 2005;26:4281–9.

[67] Uchida T, Ikeda S, Oura H, Tada M, Nakano T, Fukuda T, et al. Development of biodegradable scaffolds based on patient-specific arterial configuration. J Biotechnol 2008;133(2):213–8.

[68] Yeong WY, Chua CK, Leong KF, Chandrasekaran M, Lee MW. Comparison of drying methods in the fabrication of collagen scaffold via indirect rapid prototyping. J Biomed Mater Res Part B Appl Biomater 2007;82(1):260–6.

[69] Chu TMG, Orton DG, Hollister SJ, Feinberg SE, Halloran JW. Mechanical and *in vivo* performance of hydroxyapatite implants with controlled architectures. Biomaterials 2002;23(5):1283–93.

[70] Liska R, Schwager F, Maier C, CanoVives R, Stampfl J. Water-soluble photopolymers for rapid prototyping of cellular materials. J Appl Polym Sci 2005;97(6):2286–98.

[71] Miller JS, Stevens KR, Yang MT, Baker BM, Nguyen D-HT, Cohen DM, et al. Rapid casting of patterned vascular networks for perfusable engineered three-dimensional tissues. Nat Mater 2012;11(9):768–74.

[72] Landers R, Hubner U, Schmelzeisen R, Mulhaupt R. Rapid prototyping of scaffolds derived from thermoreversible hydrogels and tailored for applications in tissue engineering. Biomaterials 2002;23:4437–47.

[73] Norotte C, Marga FS, Niklason LE, Forgacs G. Scaffold-free vascular tissue engineering using bioprinting. Biomaterials 2009;30(30):5910–7.

[74] Bennett H. Industrial waxes. New York: Chemical Publishing Company Inc.; 1975.

[75] Miyazaki H, Ushiroda I, Itomura D, Hirashita T, Adachi N, Ota T. Thermal expansion of hydroxyapatite between $-100°C$ and $50°C$. Mat Sci Eng C 2009;29:1463–6.

[76] Saito E, Kang H, Taboas JM, Diggs A, Flanagan CL, Hollister SJ. Experimental and computational characterization of designed and fabricated 50:50 PLGA porous scaffolds for human trabecular bone applications. J Mater Sci Mater M 2010;21(8): 2371–83.

[77] Mantila-Roosa SM, Kemppainen JM, Moffitt EN, Krebsbach PH, Hollister SJ. The pore size of polycaprolactone scaffolds has limited influence on bone regeneration in an *in vivo* model. J Biomed Mater Res A 2010;92(1):359–68.

[78] Tirella A, De Maria C, Criscenti G, Vozzi G, Ahluwalia A. The PAM^2 system: a multilevel approach for fabrication of complex three-dimensional microstructures. Rapid Prototyping J 2012;18(4):299–307.

[79] De Acutis A, Maria C, Vozzi G. Indirect microfabrication using PAM^2 system. Proceedings of the GNB III national conference, Rome; June 26–29, 2012.

[80] Burg T, Cass CA, Groff R, Pepper M, Burg KJ. Building off-the-shelf tissue-engineered composites. Philos Trans A Math Phys Eng Sci 2010;368(1917):1839–62.

Chapter 9

Bioprinting Using Aqueous Two-Phase System

Brendan M. Leung*, Joseph M. Labuz*, Christopher Moraes*, and Shuichi Takayama*,**,†

*Department of Biomedical Engineering, College of Engineering, University of Michigan, Ann Arbor, MI, USA; **Macromolecular Science and Engineering Center, College of Engineering, University of Michigan, Ann Arbor, MI, USA; †Biointerfaces Institute, University of Michigan, Ann Arbor, MI, USA

Chapter Outline

ABSTRACT

A key for successful bioprinting is the design and formulation of a suitable "ink" that maintains cell viability while being patternable. Here, we describe the uses of aqueous two-phase systems (ATPS) to bioprint or micropattern cells as well as reagents. The unique advantage of the method compared to other bioprinting methods is that one can pattern while fully immersed in aqueous solutions without diffusion or dispersion of the aqueous ink. The fully aqueous environment is advantageous for cell printing where even brief drying can be lethal. Additionally, bioprinting with ATPS is typically performed in a noncontact manner allowing printing over delicate materials such as living cells, tissues, and hydrogels straightforward. While bioprinting generally implies additive fabrication, sculpting of existing biological structures can also serve to create cellular patterns. This chapter thus provides an overview of the use of ATPS bioinks to perform both additive and subtractive fabrication.

Keywords: additive printing; ATPS; microcollagen gel; biofilm; apoptosis

1 BRIEF INTRODUCTION TO ATPS

Before jumping into specific applications to bioprinting, we briefly describe the basics of aqueous two-phase systems (ATPS). It is well know that aqueous and organic solvents form two immiscible phases when mixed (e.g., mixture of water and oil and other nonpolar organic solvents), and most mixtures of aqueous solution form one continuous phase. Less known is that when two aqueous solutions containing the appropriate concentrations of certain polymers or salts (kosmotropics and chaotrophic) are mixed, two immiscible phases appear. This is known as an aqueous two-phase system, or ATPS [1]. A commonly used ATPS consists of polyethylene glycol (PEG) and dextran (DEX). ATPS has been extensively employed as a nondenaturing and gentle alternative to water-organic solvent extraction of biomolecules. Separation is possible because solutes preferentially partition to one of the two phases. Cells and other biomolecules also experience similar portioning and therefore tend to be confined in either PEG or DEX and can be patterned onto a surface. Now let us proceed with descriptions of the applications of ATPS in the field of bioprinting.

2 ADDITIVE PRINTING USING ATPS

One of the attractive features of ATPS for cell biologists is that the technique allows the deposition of materials, including cells and bioactive reagent, directly over delicate

Essentials of 3D Biofabrication and Translation. http://dx.doi.org/10.1016/B978-0-12-800972-7.00009-8

biological samples. Typically these substrates would be hydrogels, cell monolayer or intact tissue samples. The nature of ATPS bioprinting allows the ink and substrate to remain in an aqueous environment and conditions that are compatible with cell survival. Using this technique, complex biological microenvironments can be created layer-by-layer by additive method much like conventional 3D bioprinting. The following sections will discuss some of the applications of additive ATPS bioprinting.

2.1 Mammalian Cell Patterning Using ATPS

Seeding arrays of single cell or cluster of cells in defined geometric patterns is a powerful way to study cell–cell interactions and complex collective cell motions. Generally, this can be achieved on rigid substrates using microcontact printing of extracellular matrix (ECM) protein [2] and stencils [3] to direct cell seeding pattern on tissue culture substrate, which is usually glass or polystyrene. However, these methods have some shortcomings. First, the stiffness of conventional plastic cell culture vessels is several orders of magnitude higher than most physiologic substrate. For example, the elastic modulus of most soft tissue ranges from 0.5 kPa to 5 kPa, while harder tissues like bone can be as stiff as 20 kPa. In comparison, the stiffness of most cell culture vessels, typically made of plastic or glass, are in the GPa range [4]. The difference in mechanical property has been shown to affect cell phenotypes, such as cell function and differentiation [5–7]. Therefore, building a complex 3D tissue model using substrates with the appropriate mechanical properties is crucial for recreating its functions. It would also be necessary to maintain direct cell–cell contacts to faithfully replicate cellular microenvironments. To overcome this, the cell can be deposited directly using noncontact methods to prevent damages to delicate substrates. Several methods exist for seeding cells on delicate substrates, including the use of cell-loaded solid pin [8] or stacking of prefabricated cell layers [9,10]; however, resolution and fidelity of the pattern necessary for building a complex tissue niche can be difficult to achieve. An alternative method for printing cells came from repurposing existing inkjet printing technology and adapting it for direct cell printing using piezoelectric inkjet print nozzles [11–13]. This method provides precise control of deposition volume and location methods require specialized equipment and may negatively affect cell viability.

A convenient alternative to print cells over a delicate substrate is to use ATPS consisting of two aqueous polymer solutions. Tavana et al. demonstrated this technique and its applications using a polytheleneglycol (PEG) and DEX based systems [14]. The affinity of cell partitioning to the DEX phase or the PEG phase depends on the difference in interfacial tension $\Delta\gamma$ between the cell and the two phases, respectively. Since it is difficult to measure the interfacial tension between the cell membrane and the respective phases, one can calculate $\Delta\gamma$ using the interfacial tension γ_{12} and the contact angle θ between the two phases. Using Young's equation, $\Delta\gamma$ is defined as

$$\gamma_{12}\cos\theta = \Delta\gamma$$

For a PEG/DEX system, contact angles θ above 90° favors cell partitioning to the PEG phase, between 0° and 90° favors partitioning to the DEX phase, and when θ equals to 90° cells have similar affinity toward both phases and tend to partition at the interface. As cell patterning would benefit from cell partitioning to the DEX "bioink," the interfacial tension between PEG and DEX should be kept to a minimum. The example of how interfacial tension can affect cell partitioning is shown in Fig. 9.1a. Two-phase separating DEX/PEG pairs were tested, and while cells were confined in the DEX droplet in both formulation, the pattern was more uniform when using the PEG/DEX pair with lower interfacial tension.

Cells may also be introduced to the PEG phase to achieve exclusion patterning [15]. In this case, DEX droplet can be deposited on the cell culture substrate and allowed to dry, leaving behind a DEX polymer in a desired pattern. Subsequently, cell and PEG mixture can be laid over these patterns. As the DEX rehydrates, a droplet of DEX will reappear and cells will be excluded from those areas and not seeded to the substrate. After washing off the PEG/DEX solution, the cell-free area can be tracked to observe cell migration in a no-scratch healing assay (Fig. 9.1d and e). ATPS exclusion patterning can be high throughput and multiplex compatible, all at a relatively low cost, thus making it an attractive alternative to other cell migration assays such as scratch wounds model and stenciling techniques.

Direct cell-on-cell patterning can also be a useful technique to uncover spatially dependent paracrine signaling effects on cell fate that are difficult to observe with uniform coculture. For example, Tavana et al. printed mouse embryonic stem cells (mESC) over a monolayer of PA6 feeder cell and found that mESC colony size affects their ability to differentiate into neurons [14]. Larger mESC colonies resulted in an increase of neuron differentiation marker (TuJ1) expression on a per cell basis (Fig. 9.2a and b). Using the same technique but instead printing feeder cells over the mESC layer, the group also observed enhanced neuronal differentiation at the interspace between the PA6 feeder cell colonies (Fig. 9.2c) [16]. Combined with commutation models, the group showed that there exist optimal regions of cytokine concentrations for neuronal differentiation within the combined cytokine gradient fields. The ATPS patterning techniques provide control of colony size and spacing of stem cell and feeder colony, which in turn facilitates the observation of the effects of physical parameters on cell fate.

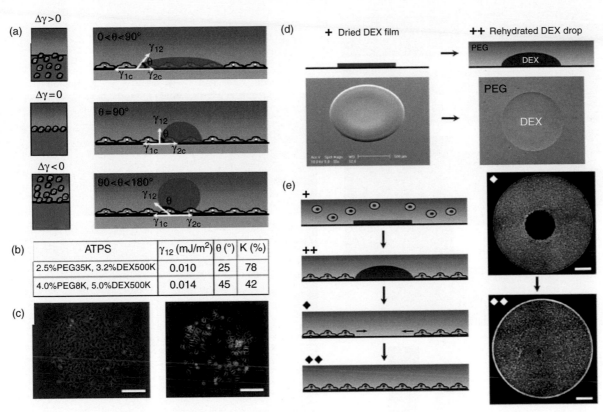

FIGURE 9.1 Effects of polymer concentrations on cell partitioning in a PEG/DEX-based ATPS system. (a) Partitioning ratio of cell in an ATPS system is determined by the difference in interfacial tension of cells between the two phases, expressed as $\Delta\gamma = \gamma 1c - \gamma 2c$, where $\gamma 1c$ and $\gamma 2c$ are the interfacial tensions between cell and PEG or DEX phase, respectively. The efficiency of ATPS printing is also determined by the contact angle of DEX phase droplet on cell monolayer, and this can be used to predict $\Delta\gamma$ using Young's equation, where $\gamma_{12}\cos\theta = \Delta\gamma$. Generally smaller γ_{12} and lower contact angle favors effective cell printing. (b, c) Two formulations were empirically tested to confirm this prediction. Spots of C2C12 cells were printed using the ATPS comprising 2.5%PEG35K and 3.2%DEX500K (left) or 4.0%PEG8K and 5.0%DEX500K (right). As predicted, cells seeded more densely and evenly with lower contact angle and lower surface tension between the two phases. Scale bar 200 μm in Fig. 9.1c. (d, e) In the absence of a cell, DEX droplet can be patterned and dried on a surface. Subsequent seeding of cells in PEG solution rehydrates DEX and prevents cells from attaching in those areas. After washing, cells in seeded areas migrate into the cell-free zone to close the gap, thus serving as a healing assay. Scale bar 1 mm in Fig. 9.1e. *(Figures reprinted with permission from Tavana et al. [14,15].)*

2.2 Patterned Modification of Cell Phenotype With Bioactive Molecules Using ATPS

Cells in multicellular organisms are constantly responding to their environment and communicating with neighboring cells. The chemical and physical niche generated by heterogeneous populations of cells is responsible for the overall function of the tissue they made up. In the laboratory, inter- and intracellular signaling pathways can be manipulated using small molecules, recombinant proteins, and genetic materials. However, delivering these reagents in a spatially controlled fashion can be challenging.

Taking advantage of the fact that ATPS can be formulated so that most nucleic acids, transfection agents, and viruses preferentially partition to one of the two phases, Tavana et al. demonstrated that nucleic acids could be delivered to a spatially defined area on a cell monolayer [17]. Using a PEG/DEX system, fluorescently labeled RNA together with

Lipofectamine™ were mixed in the DEX phase where it remained confined for at least 4 h. With the same formulation, enhanced green fluorescent protein (eGFP) and dsRed constructs were transfected into a distinct area on a single HEK293 monolayer (Fig. 9.3a). The transfected cell expressed these exogenous genes in a dose-dependent manner (Fig. 9.3b). On the other hand, protein levels can also be manipulated to reduce expression by a genetic tool, one of which is RNA interference [18]. To demonstrate this, regions of an eGFP expressing MDA-MB-231 monolayer were exposed to DEX droplets containing shRNA against eGFP. As expected, eGFP expression levels were reduced in the spotted area but not in adjacent regions (Fig. 9.3c). Using a slot pin, the sizes and locations of these spots can be tuned to yield a dense array, with each 340 μm diameter spot containing as little as 10 ng of plasmid covered by a 500 nL DEX droplet. In theory, this would allow any number of defined regions on the monolayer to be genetically modified and observed simultaneously. In addition, other molecules, such as enzymes

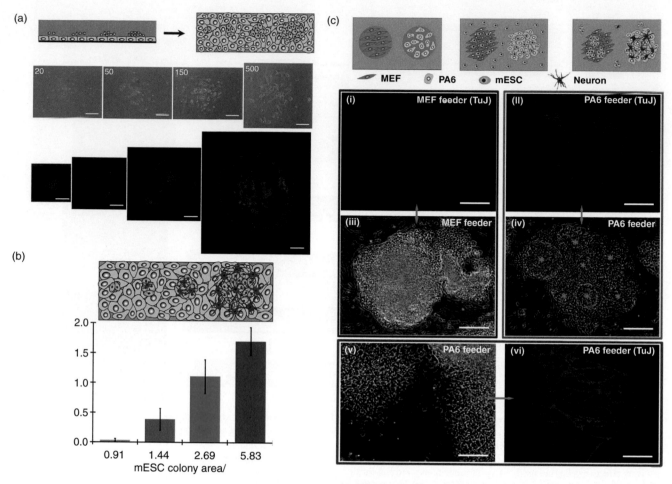

FIGURE 9.2 Contribution of physical factors to PA6 mediated neuronal differentiation of mESC colonies. (a) Schematics and micrographs (brightfield and fluorescence) mESC differentiation into neurons on PA6 feeder layer. Differentiated mESC were stained with anti-TuJ1 antibodies (red). Scale bar 250 μm in brightfield image, 500 μm in fluorescence image. (b) When normalized to cell number, larger colonies of mESC maintained higher level of TuJ1 expression, suggesting that differentiation efficiency is related to colony size. (c) Conversely, feeder cells can be printed onto a layer of mESC to produce mixed niche patterns. Mouse embryonic fibroblast (MEF) and PA6 were printed over mESC using ATPS. As expected, PA6 drove neuronal differentiation of mESC, while MEF did not. Interestingly extensive intercolony neuronal clusters were formed around printed PA6 colony, thus illustrating the potential effect of colony spacing and the utility of ATPS to perform further mechanistic studies. *(Figures reprinted with permission from Tavana et al. [14,16].)*

and growth factors, can be delivered using similar methods. Some of these applications in subtractive patterning will be discussed in the following sections.

2.3 Microcollagen Gel Patterning Using ATPS

The ECM surrounding cells is now being recognized as a critical modulator of cell function [19]. By providing cells with structural, topographic, and chemical cues across a range of length scales, the ECM can be an important feature in tissue engineering, particularly in building tissues by additive printing technologies [20,21]. Hence, the ability to pattern cells within an ECM will be of critical importance in tissue engineering and in fundamental studies of cell-matrix biology.

A classic example of cell-matrix interactions is the collagen contraction assay [22,23]. Contraction is a well-established measure of cellular ability to remodel the surrounding ECM. When cultured in a collagen matrix, certain cells compact the matrix, making it denser and mechanically stiffer. This contraction is a characteristic feature of developing biological systems, and of fibrotic disease progression, and is hence a critically important measure of cellular activity. Conventionally, this assay requires at least 100 μL of collagen containing between 10,000 and 100,000 cells [24], making it challenging to use this assay on populations of rare cells, or to incorporate it into a high-throughput screening system. The large sample volume also limits diffusion of nutrients and soluble cues, which further complicates analysis of biological response to externally applied soluble signals.

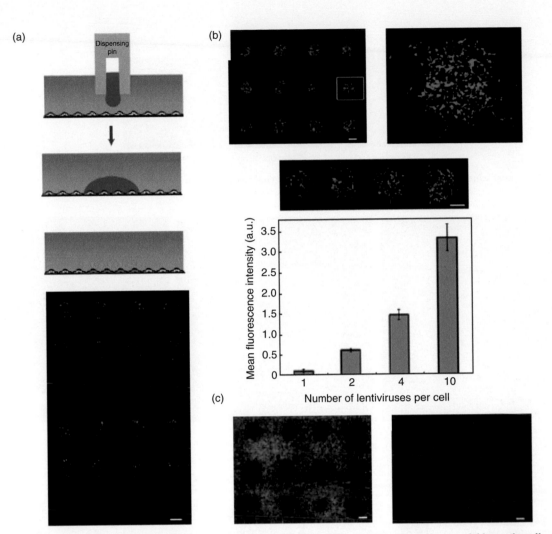

FIGURE 9.3 **Spatially controlled delivery of nucleic acid on cell monolayer with ATPS.** (a) PEG-rich medium was laid over the cell monolayer. A dispensing pin containing DEX mixed with lentivirus or Lipofectamine carrying nucleic acid was patterned over the cell monolayer. Multiple genes can be codelivered at varying combinations, in a dose-dependent manner. (b) Gene expressions can also be suppressed by transfecting cells with shRNA. (c) Cells coexpressing eGFP and mPlum were transfected with shRNA loaded DEX in a 3 × 4 array pattern. Only eGFP signal was suppressed while mPlum expression remained unchanged. This confirms that the changes in gene expression are due specifically to the presence of shRNA and not the ATPS printing procedure. *(Figures reprinted with permission from Tavana et al. [17].)*

Attempts to reduce the volume of the collagen gel are challenging [25], primarily due to restrictions involved in the thermal gelation process. Collagen is isolated from a variety of sources including rat tails and bovine skin, using an enzymatic acid-extraction process. While the solution remains acidic, collagen is soluble. In most experiments, the acidic collagen is first neutralized with a base, and then incubated at 37°C where it crosslinks to form a hydrogel [26]. This process takes time, and most protocols suggest 30–40 min before the collagen is reliably gelled. During this time, evaporation occurs, which can change the osmolarity of the solution, killing any encapsulated cells. The evaporation rates are typically low enough that the percent change in volume of a large gel is negligible. However, when making the gel smaller, this can be a significant factor, and work published by Moraes et al. [27] demonstrated that cells polymerized within small collagen gels have greatly reduced viability. Presumably, this effect may be minimized by carefully monitoring and controlling the humidity of the environment, which may decrease evaporation rates and improve cell viability. However, this can be challenging and would require specialized equipment and protocols.

In order to address these issues, Moraes et al. developed an ATPS able to generate small "pockets" of cell-laden collagen gels within an all-aqueous environment (Fig. 9.4) [27]. The aqueous environment eliminates evaporation as a concern, and enables the formation of small collagen gels with excellent cell viability. Briefly, collagen components were added to the DEX phase of a typical PEG-DEX ATPS, described earlier in this chapter. At final concentrations of 3% DEX T-500, and 2 mg/mL neutralized collagen, an aqueous two-phase system is formed in 5.2% PEG (35k).

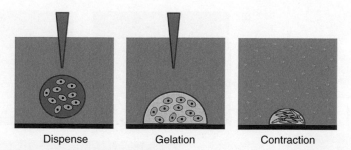

FIGURE 9.4 Conceptual schematic of aqueous-two-phase bioprinting of contractile collagen microgels.

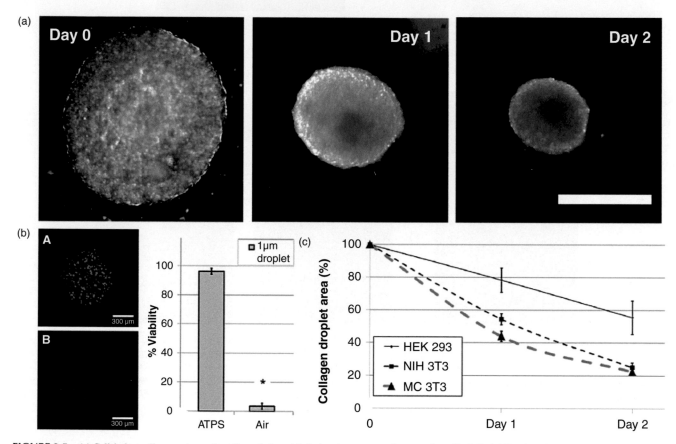

FIGURE 9.5 (a) Cell-laden collagen microgels at Days 0, 1, and 2 during the process of contraction. (b) Cell viability is maintained when microgels are formed in an aqueous two-phase system, as opposed to microgels polymerized in air using the conventional technique. (c) Contraction rates depend on a number of factors, including cell type. *(Figures reprinted with permission from Moraes et al. [27].)*

The ATPS localizes the collagen components to a small dome-shaped volume, where they undergo thermal gelation and form a collagen "microgel" (Fig. 9.5a). Analysis of the partition coefficient suggests that the collagen components are primarily localized at the interface, and partially in the DEX phase. When the droplet size is small enough, however, the collagen forms a hemispherical dome when it is allowed to cure. The PEG phase can then be washed away, while the microgel adheres temporarily to the bottom of the container. Using this technique, microgels as small as 0.5 μL have been formed. Although it is possible to generate smaller microgels using other techniques, such

as micromolding in micromachined wells, the ATPS-based technique can be used to create low-volume, high-viability microgels (Fig. 9.5b), with no special equipment, and can be performed using only a standard lab micropipette and conventional microscopic imaging techniques. For high-throughput applications, the technology is compatible with liquid handling robots and other more advanced technologies.

Once formed, the microgel can be released from the bottom of the container by pipetting a directed flow of media at it – the small forces involved are sufficient to dislodge the gel. The microgel, now free of external constraints, can be remodeled by the encapsulated cells (Fig. 9.5c), creating a miniaturized and potentially high-throughput collagen contraction assay. The use of microgels presents some significant advantages over their large-scale counterparts. In addition to the conventionally cited advantages in the microengineering literature of reduced reagent costs, and possibilities for high-throughput screening, the reduction in gel size enables rapid diffusion of large molecules throughout the volume of the gel. This means that applied chemical signals, such as transforming growth factor (TGF)-β1 (MW = 25 kDa), rapidly diffuse through the gel uniformly stimulating cells within them. In contrast, adding TGF-β1 to a conventional collagen contraction assay usually requires about 24 h for the assay to fully equilibrate. As such, cells are nonuniformly stimulated, and may not even see TGF-β1 until the end stages of the assay.

The collagen contraction assay is a well-established assay, first reported 30 years ago. It provides one of the few methods to study functional mechanical remodeling of tissues, and is hence applicable to a wide variety of biological systems, and continues to be a standard technique in labs today. This work with ATPS systems enables effective miniaturization of this assay, and leverages the benefits of scaling to increase ease and throughput of the assay, while minimizing confounding transport effects and assay costs.

2.4 Biofilm Coculture With Living Tissue

Aside from mammalian cell culture, ATPS bioprinting has also found its application in the spatially controlled deposition of bacteria colony known as biofilm. Biofilm bacteria have distinct functions compared to their planktonic states and play a crucial role in human diseases. The structure and function of the biofilm depends greatly on its constituent and surrounding microenvironment, such as nutrient availability and presence of adjacent biofilms. There is a substantial interest in developing biofilm models *in vitro* because they play an important role in human health [28–31]. Bacteria can form biofilms spontaneously *in vitro* by substrate surface-directed patterning methods, including stenciling [32], soft hydrogel stamping [33], and surface modification [34]. To study the interaction between heterogeneous biofilms, however, a bioprinting method with spatial and temporal control over bacteria deposition would be preferred.

To this end, Yaguchi et al. patterned two separate populations of bacteria directly over each other using ATPS bioprinting [35]. Using a PEG/DEX ATPS system, *Escherichia coli* suspended in DEX were patterned over PDMS substrate and formed a biofilm over 24 h (Fig. 9.6 a–d). Extracellular structural DNA can also be detected suggesting a mature biofilm can be formed using this method. On top of this mature biofilm, a second biofilm with a different bacteria strain can be seeded in such way that it completely or partially overlays the underlying biofilm (Fig. 9.6e–g).

In this configuration the interaction between the two colonies of bacteria can be examined quantitatively. When an ampicillin-sensitive strain of *E. coli* was seeded on top of an ampicillin-resistant stain, the secreted β-lactamase from the resistant strain conferred protection to the sensitive strain and the overall toxicity of the antibiotic on both strains were reduced (Fig. 9.6h). This model demonstrated the commensalistic benefit of one strain of bacteria with another that are often seen in biofilm formations found in nature [36–38]. The main advantage of ATPS biofilm coculture is that the spatial (size and location) as well as temporal interactions of biofilms can be decoupled and investigated separately thus improving the utility of providing mechanistic insight on bacterial function in an *in vitro* system. It may also serve as a physiologically relevant platform to screen for novel antibiotics against specific biofilm combinations.

To extend this technique further, the same research group has applied the ATPS bioprinting technique to pattern a biofilm over a mammalian cell monolayer [39]. This is of interest because the relationship between bacteria and humans plays a crucial role in our health. Each person carries with them an abundant and diverse flora of bacteria on their external surfaces, that is, skin, oral cavity, intestine, and mucous membranes [40]. Many of these bacteria exist in biofilm form, and the balance between beneficial and pathogenic bacterial populations can be influenced by many factors [41]. Therefore, a cocultured model of bacteria and mammalian cells can be a powerful tool to understand these interactions.

In general, bacteria cultured with mammalian cells *in vitro* do not form a stable biofilm, but instead propagate in planktonic form in cell culture medium and quickly deprive the mammalian cell of nutrients causing massive cell death. This problem can be circumvented using ATPS bioprinting technique because bacteria seed in the DEX phase droplet can be trapped within the droplet due to interfacial tension, which would allow sufficient time for them to mature to form biofilm. Dwidar et al. demonstrated that biofilm from several species of bacteria commonly found in humans can be established and maintained over a mammalian epithelial monolayer [39]. When cocultured by direct mixing in the culture medium, suspended bacteria quickly cause widespread death on the MCF10a monolayer (Fig. 9.7a–c). Using a bacterial culture optimized, PEG/DEX-based ATPS system, biofilm from nonpathogenic *E. coli* stain was maintained over human mammary epithelial cells MCF10a monolayer for 24 h with minimal harm to the underlying cells, while its isogenic strain expressing the Invasin gene isolated from *Yersinia pseudotuberculosis* caused massive cell death on the MCF10a layer directly below the biofilm. In nature, many pathogenic biofilm secretes cytotoxic substances that may cause local or distal cell death. An emerging strategy to remove these biofilms is to use probiotics, such as *Bdellovibrio bacteriovorus* (BALO), to

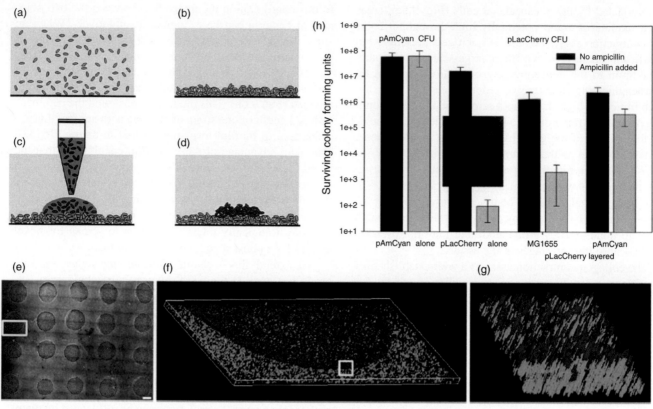

FIGURE 9.6   Patterning of biofilm over existing biofilm. (a–d) Schematics illustrating the deposition of a secondary bacteria population (red) over an existing biofilm (blue) using PEG/DEX-based ATPS. (e) *E. coli* DH5α/pHKT3 in LB medium containing DEX was patterned on top of the primary biofilm consisting of *E. coli* MG1655/pAMCyan. (f, g) Confocal microscope images of a circular array of *E. coli* DH5α/pHKT3 biofilms patterned over *E. coli* MG1655/pAMCyan biofilm layer. A magnified 3D view revealed the structure of the layered bacterial biofilms. (h) Antibiotic-resistant biofilm consisting of *E. coli* pAmCyan conferred ampicillin resistance to a sensitive strain *E. coli* str. MG1655/pLacCherry in a biofilm coculture model. The inset shows the *E. coli* str. MG1655/pLacCherry cells after ampicillin treatment, showing the abnormally elongated and filamentous morphology. *(Figures reprinted with permission from Yaguchi et al. [35].)*

predate on the biofilm. This strategy is particularly useful as an alternative to pharmacologic agents against antibiotic resistant biofilms. Indeed, the presence of BALO in PEG phase was able to remove patterned biofilm without harming the underlying MCF10a layer (Fig. 9.7d). Both Gram-negative and Gram-positive bacteria can be predated using the appropriate BALO strain [39,42]. The versatility in cell sources and microenvironment makes this a suitable model to study the interactions between human cells and bacteria communities. Overall, patterning of bacteria biofilm over living substrates using ATPS provides techniques to create a bacteria coculture system with other bacteria or mammalian cell with qualitative and quantitative consistency.

localize either chemical (e.g., enzymes) or physical (e.g., ultrasound microbubbles) reagents to control these subtractive processes. Current state-of-the-art methods, such as, mechanical scratching or picking [43,44], laser passaging [45], ultrasound [46], and enzymatic digestion [47] suffer from a variety of shortcomings including high cost, low efficiency, lack of spatial control in both the X–Y as well as Z dimensions, physical removal of tissue (which may hinder wound healing or invasion studies), and disturbance of possibly precious samples. When tuned to deliver various payloads, ATPS technology can be leveraged to address these deficiencies, realize new engineering techniques, and enable advanced biological studies.

3   SUBTRACTIVE PRINTING USING ATPS

3.1   Introduction

Beyond patterning cell deposition, ATPSs can also be used to facilitate controlled removal or disruption of cells from already growing cultures and constructs. ATPSs can

3.2   ATPS-Mediated Enzymatic Digestion

ATPS can be used to localize digestive enzymes – such as collagenase, dispase, or trypsin – for removal of cells *in situ*. Of course, the success of this (or any ATPS-based) approach depends critically on the partitioning behavior of the enzyme selected. Frampton et al. investigated this

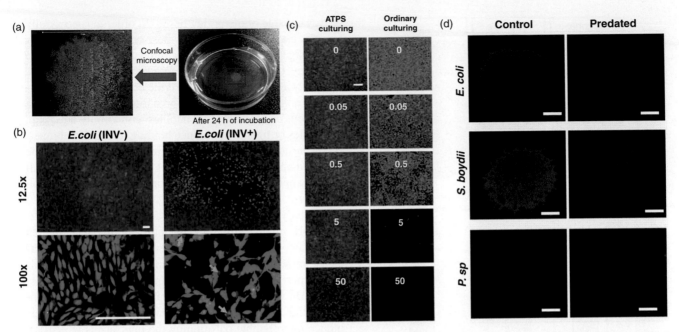

FIGURE 9.7   Coculture of biofilm on human mammary epithelial cells (MCF10a). (a) Two strains of bacteria, *E. coli* MG1655/pLacCherry (INV−) and its isogenic counterpart, *E. coli* MG1655/pINVCherry (INV+) were deposited on an MCF10a layer using ATPS. (b) Microscopic analysis of MCF10a cell stained with Calcein AM cell tracker. Areas underneath biofilm containing INV+ were devoid of MCF10a cells, compared to INV− control where the bacteria biofilm did not harm the MCF10a monolayer. (c) Coculture of INV− bacteria directly in cell culture medium led to the death of MCF10a in a dose-dependent manner, whereas the same concentration of bacteria confined in a DEX droplet in an ATPS system did not cause similar damage to the MCF10a monolayer. (d) Bacteria biofilm can be removed by adding *Bdellovibrio bacteriovorus* (BALO), a probiotic that predate only bacteria. *(Figures reprinted with permission from Dwidar et al. [39].)*

partitioning behavior in detail for systems composed of poly(ethyleneglycol) with molecular weight 35,000 kDa (PEG 35k) and either 10,000 kDa Dextran (DEX T10) or 500,000 kDa Dextran (DEX T500) [48]. Consistent with previous reports regarding proteins in general [7], collagenase, dispase, and trypsin all preferred the lower DEX-rich phase as evidenced by partition coefficients ($K_{part} = C_{top\ phase}/C_{bottom\ phase}$) less than unity (Fig. 9.8). In each case the partition coefficient was significantly ($p < 0.01$) lower for the DEX T10 system indicating that lower molecular weight DEX systems induce stronger partitioning behavior – a finding supported by theoretical treatments [8] as well as empirical observation [1].

In conjunction with a custom-built microcapillary injector capable of dispensing DEX droplets <10 pL, Frampton et al. applied these ATPSs for digesting select cellular attachments – a process they termed localized enzymatic microdissection (LEM) (Fig. 9.9a). First studying the process on a monolayer of HEK 293 cells, they showed that a DEX droplet alone did not disturb cell attachment, but DEX-trypsin treatment disrupted cell attachments within the boundary of the droplet (Fig. 9.9b).

The researchers then used this system to isolate entire colonies of induced pluripotent stem cells from a monolayer of feeder cells (Fig. 9.9c) [48]. To prevent premature differentiation, many stem cells are cultured on a layer of support cells (typically irradiated murine embryonic fibroblasts) [49]. Although feeder-free systems have been developed [50] support cells typically result in better viability, maintain SC potency for longer, and more cost effective [50]. However, purifying stem cells from the supporting culture is necessary, time-consuming, and difficult. Since the process gently lifts off the colony, feeder cell contamination is minimized, as evidenced by an absence of smooth muscle actin-positive cells postisolation, while the larger induced pluripotent stem cells colony remains intact and as viable as an enzyme-free control leading to faster postisolation growth. Upon stimulation with activinA, BMP4, and bFGF, this low feeder cell contamination contributed to a more efficient cardiomyocyte differentiation – approximately 90% compared to conventional results ranging from 2% to 20% – as measured by α-actinin, MLC2v, and cardiac troponin T expression as well as spontaneous beating [48].

3.3   Ultrasonic Induction of Apoptosis

Ultrasound can be used to destroy or remove cultured tissue *in situ* [51]. For this purpose, commercial perfluorocarbon- or gas-core microbubbles, often used as contrast agents for clinical imaging, provide a cavitation source that transforms the ultrasonic energy into mechanical and thermal

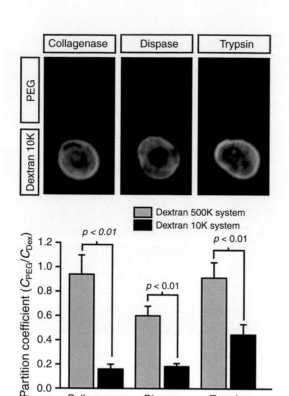

FIGURE 9.8 **Protein blots establish partitioning behavior of various digestive enzymes in ATPSs composed of different molecular weight DEX.** Generally, proteins partition more strongly to the bottom DEX phase for lower molecular weight DEX. Data expressed as mean +/− SEM. *(Figures reprinted with permission from Frampton et al. [48].)*

energy capable of disrupting tissue. In order to successfully transfer energy, these bubbles must be in close proximity to the target tissue [52] – a requirement complicated by the buoyancy of the lipid- or protein-encapsulated particles. Typically, researchers either coat the cells/tissue to be treated with a thin layer of fluid to force the bubbles close to the target cells or decorate these bubbles with affinity tags (e.g., antibodies or ligands) to induce binding to cell membranes. By using affinity tags, some degree of specificity can be achieved among cell types, but the microbubbles must still be bath applied if they are to facilitate ultrasonic destruction of tissues.

A study by Frampton et al., separate from the previous work on digestive enzyme partitioning, interrogated the behavior of these microbubbles in ATPS. The researchers examined three different commercially available, lipid-coated perfluorocarbon bubbles in three different ATPSs: 3500 kDa Dextran (DEX T3.5), DEX T10, and T500 all with PEG 35k. According to their results, most systems induce microbubble partitioning the interface (Fig. 9.10a). Definity brand microbubbles were found to prefer the PEG phase to the lower DEX phase (speculatively due to the lipid coating) while Targestar-SA brand microbubbles partitioned primarily to the DEX phase rather than the PEG phase (likely due to the streptavidin decoration present on the microbubbles tested) [53]. In both cases, surface chemistry, buoyancy, size, and other factors also affect microbubble behavior [1]. The results were further confirmed

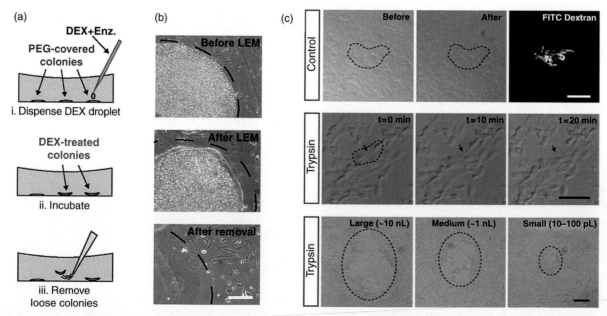

FIGURE 9.9 **LEM can be used to disrupt cell adhesions.** (a) DEX-enzyme solution is deposited over a select colony of cells using a custom microcapillary injector. The surrounding PEG phase confines the enzymes to the DEX droplet where they release the edges of the colony, which can then be gently blown off. (b) Trypsin-DEX droplets can digest individual cell attachments (small droplets) or many cell attachments (medium and large droplets). (c) Phase contrast images of the LEM process. *(Figures reprinted with permission from Frampton et al. [48].)*

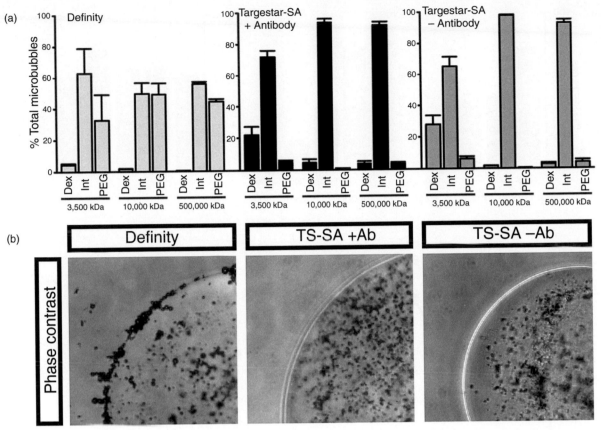

FIGURE 9.10   Ultrasound microbubbles generally reside at the ATPS interface. (a) Partitioning experiments revealed that Definity (left) microbubbles generally prefer the top phase and interface while Targestar microbubbles (middle and right) generally partition to the bottom phase and interface. (b) Video microscopy experiments where the bubbles are placed in a DEX droplet and observed over time confirm the results in (a). *(Figures reprinted with permission from Frampton et al. [53].)*

by video microscopy of microbubbles placed in the DEX phase and allowed to equilibrate over a period of time (Fig. 9.10b). Since the majority of microbubbles are confined to the interface or the DEX phase, this means that the ATPS is not only useful for spatially patterning microbubble deposition, but also restricting microbubble location in the *z*-dimension. Since ATPS droplets spread considerably (depending on the composition) over a given substrate due to low interfacial tension [14], this means that most of the microbubbles are trapped at or near the cell surface – a necessary condition for the successful application of ultrasound. Essentially, use of an ATPS allows scientists to control microbubble location in *three* dimensions, enabling more efficient and more tightly localized ultrasonic tissue destruction despite bath application of ultrasound to a culture or construct.

Conventional methods such as enzymatic, mechanical, or optical disruption and their derivatives remove tissue; microbubble-mediated sonoporation does not [51]. Since microbubble-mediated sonoporation only kills and does not necessarily remove cells or tissue, it can be a valuable tool for studying development, migration, and apoptosis *in situ*.

To that end, Frampton et al. investigated how changing the parameters of the ATPS used alter the efficacy of the ultrasonic disruption process. First, the researchers tested different DEX and microbubble concentrations to determine the optimal values for killing cells. According to their results, high levels of DEX and (perhaps unsurprisingly) cavitation-inducting microbubbles produce the largest decreases in cell viability. Furthermore, most cell lines seemed equally sensitive to the process with only HUVECs exhibiting increased death, likely due to a higher sensitivity to chemical and mechanical perturbations (Fig. 9.11a). Buffer composition was also shown to affect cell death. Both propidium iodine (PI), which indicates a loss of membrane integrity, and Annexin V, which demonstrates the presence of phosphatidylserine on the outside of the cell membrane, showed that increasing extracellular Ca^{2+} as well as increasing ultrasonic power enhanced cell death. Interestingly, at low Ca^{2+} concentrations there was a much larger proportion of PI-positive cells versus Annexin V-positive cells indicating that, under appropriate conditions, this method is capable of disrupting cell membranes without inducing apoptosis (Fig. 9.11b) [53].

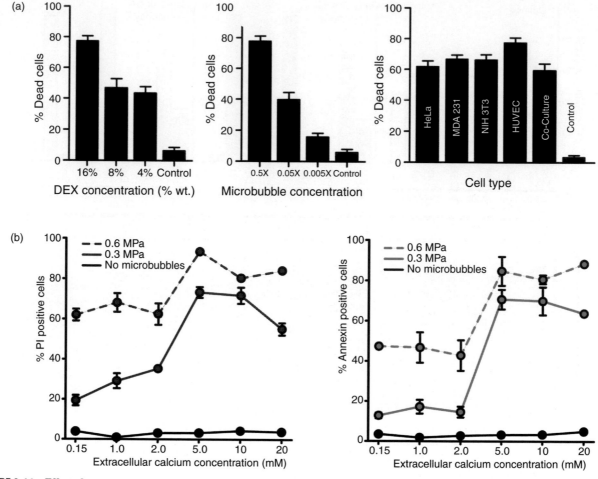

FIGURE 9.11 **Effect of treatment parameters on ultrasound-induced apoptosis.** (a) Increasing concentrations of DEX (left panel) and microbubbles (middle) induced higher levels of apoptosis in tested cells. Most cells responded to treatment in a similar fashion with the exception being HUVEC cells, which are more sensitive to chemical and mechanical perturbations (right). (b) Increasing extracellular Ca^{2+} also resulted in more cell death for both high- and low-power ultrasound treatments. Interestingly, Annexin V staining (right) indicated lower levels of apoptosis than PI dye at low Ca^{2+} concentrations. *(Figures reprinted with permission from Frampton et al. [53].)*

4 CONCLUSIONS

The successful realization of 3D biofabrication to produce organ replacements and biomimetic tissue models requires the integration of several enabling technologies, and of different elements, the choice of materials and the related delivery methods are the most crucial. The ways in which biomaterials and cells are delivered and combined has a direct effect on their qualities, and by extension their function in the final product. The aqueous and noncontact nature of ATPS patterning technique discussed in this chapter alleviates the problems of dehydration and substrate/cell damage, two commonly encountered problems associate with bioprinting. When used with a robotic stabilized delivery platform, ATPS can generate patterns with sufficient resolution and fidelity for *in vitro* tissue model applications. In addition, ATPS requires few specialized equipment and can be easily integrated with existing liquid handling and bioprinting technologies, thus making this technique valuable for a wide variety of applications.

GLOSSARY

ATPS Aqueous two phase system, consisting of two immiscible aqueous solutions made of two polymers, one polymer and one kosmotropic salt, or one chaotropic salt and one kosmotropic salt.

Kosmotropic Describes a solute that increases water–water interactions and stabilize intermolecular interactions between macromolecules in aqueous solution.

Chaotrophic Describes a solute that decrease water–water interactions and destabilize polymer and macromolecule structures in aqueous solutions.

Bioprinting Deposition of biological components, including cells and ECM, in a spatially controlled fashion.

Bioink A solution consisting of living cells and ECM that is printed onto a substrate to produce ordered structures.

Hydrogel A hydrophilic network of polymer chains containing large fraction of water (80–90%) as dispersed medium.

Microcontact printing A method to physically pattern ECM proteins onto a surface using a stamp made of an elastomer, such as polydimethylsiloxane (PDMS).

Elastic modulus A measure of an object's resistance to elastic deformation, defined as the ratio of stress applied to an object to the resulting stain

mESC Mouse embryonic stem cells are pluripotent stem cells derived from the inner cell mass of a mouse blastocyst.

Feeder cells Mitotically inactivated mouse embryonic fibroblast that serves as a substrate for mESC culture to maintain stemness and differentiation capacity of mESC.

Partition coefficient The ratio of concentration of a solute at equilibrium between two immiscible solvents.

ABBREVIATIONS

ATPS	Aqueous two phase system
BALO	*Bdellovibrio bacteriovorus*
DEX	Dextran
ECM	Extracellular matrix
eGFP	Enhanced green fluorescent protein
LEM	Localized enzymatic microdissection
mESC	Mouse embryonic stem cell
PEG	Polyethylene glycol
TGF-β1	Transforming growth factor beta 1

REFERENCES

[1] Albertsson PÅ. Partition of cell particles and macromolecules: separation and purification of biomolecules, cell organelles, membranes, and cells in aqueous polymer two-phase systems and their use in biochemical analysis and biotechnology. 3rd ed. New York: Wiley; 1986. 346.

[2] Chen CS, Mrksich M, Huang S, Whitesides GM, Ingber DE. Geometric control of cell life and death. Science 1997;276(5317):1425–8.

[3] Park J, Cho CH, Parashurama N, Li Y, Berthiaume F, Toner M, et al. Microfabrication-based modulation of embryonic stem cell differentiation. Lab Chip 2007;7(8):1018–28.

[4] Butcher DT, Alliston T, Weaver VM. A tense situation: forcing tumour progression. Nat Rev Cancer 2009;9(2):108–22.

[5] Page-McCaw A, Ewald AJ, Werb Z. Matrix metalloproteinases and the regulation of tissue remodelling. Nat Rev Mol Cell Biol 2007;8(3):221–33.

[6] Krieg M, Arboleda-Estudillo Y, Puech PH, Kafer J, Graner F, Muller DJ, et al. Tensile forces govern germ-layer organization in zebrafish. Nat Cell Biol 2008;10(4):429–36.

[7] Czirok A, Rongish BJ, Little CD. Extracellular matrix dynamics during vertebrate axis formation. Dev Biol 2004;268(1):111–22.

[8] Lee MY, Kumar RA, Sukumaran SM, Hogg MG, Clark DS, Dordick JS. Three-dimensional cellular microarray for high-throughput toxicology assays. Proc Natl Acad Sci USA 2008;105(1):59–63.

[9] Yang J, Yamato M, Sekine H, Sekiya S, Tsuda Y, Ohashi K, et al. Tissue engineering using laminar cellular assemblies. Adv Mater 2009;21(32–33):3404–9.

[10] Harimoto M, Yamato M, Hirose M, Takahashi C, Isoi Y, Kikuchi A, et al. Novel approach for achieving double-layered cell sheets co-culture: overlaying endothelial cell sheets onto monolayer hepatocytes utilizing temperature-responsive culture dishes. J Biomed Mater Res 2002;62(3):464–70.

[11] Xu T, Jin J, Gregory C, Hickman JJ, Boland T. Inkjet printing of viable mammalian cells. Biomaterials 2005;26(1):93–9.

[12] Saunders RE, Gough JE, Derby B. Delivery of human fibroblast cells by piezoelectric drop-on-demand inkjet printing. Biomaterials 2008;29(2):193–203.

[13] Nakamura M, Kobayashi A, Takagi F, Watanabe A, Hiruma Y, Ohuchi K, et al. Biocompatible inkjet printing technique for designed seeding of individual living cells. Tissue Eng 2005;11(11–12):1658–66.

[14] Tavana H, Mosadegh B, Takayama S. Polymeric aqueous biphasic systems for non-contact cell printing on cells: engineering heterocellular embryonic stem cell niches. Adv Mater 2010;22(24):2628–31.

[15] Tavana H, Kaylan K, Bersano-Begey T, Luker KE, Luker GD, Takayama S. Polymeric aqueous biphasic system rehydration facilitates high throughput cell exclusion patterning for cell migration studies. Adv Funct Mater 2011;21(15):2920–6.

[16] Tavana H, Mosadegh B, Zamankhan P, Grotberg JB, Takayama S. Microprinted feeder cells guide embryonic stem cell fate. Biotechnol Bioeng; 2011;108(10):2509–2516.

[17] Tavana H, Jovic A, Mosadegh B, Lee QY, Liu X, Luker KE, et al. Nanolitre liquid patterning in aqueous environments for spatially defined reagent delivery to mammalian cells. Nat Mater 2009;8(9):736–41.

[18] Fire A, Xu S, Montgomery MK, Kostas SA, Driver SE, Mello CC. Potent and specific genetic interference by double-stranded RNA in *Caenorhabditis elegans*. Nature 1998;391(6669):806–11.

[19] Das RK, Zouani OF. A review of the effects of the cell environment physicochemical nanoarchitecture on stem cell commitment. Biomaterials 2014;35(20):5278–93.

[20] Eng G, Lee BW, Parsa H, Chin CD, Schneider J, Linkov G, et al. Assembly of complex cell microenvironments using geometrically docked hydrogel shapes. Proc Natl Acad Sci USA 2013;110(12):4551–6.

[21] Pataky K, Braschler T, Negro A, Renaud P, Lutolf MP, Brugger J. Microdrop printing of hydrogel bioinks into 3D tissue-like geometries. Adv Mater 2012;24(3):391–6.

[22] Bell E, Ivarsson B, Merrill C. Production of a tissue-like structure by contraction of collagen lattices by human fibroblasts of different proliferative potential *in vitro*. Proc Natl Acad Sci USA 1979;76(3):1274–8.

[23] Bell E, Ehrlich HP, Buttle DJ, Nakatsuji T. Living tissue formed *in vitro* and accepted as skin-equivalent tissue of full thickness. Science 1981;211(4486):1052–4.

[24] Gullberg D, Tingstrom A, Thuresson AC, Olsson L, Terracio L, Borg TK, et al. Beta 1 integrin-mediated collagen gel contraction is stimulated by PDGF. Exp Cell Res 1990;186(2):264–72.

[25] Raghavan S, Shen CJ, Desai RA, Sniadecki NJ, Nelson CM, Chen CS. Decoupling diffusional from dimensional control of signaling in 3D culture reveals a role for myosin in tubulogenesis. J Cell Sci 2010;123(Pt 17):2877–83.

[26] McGuigan AP, Leung B, Sefton MV. Fabrication of cell-containing gel modules to assemble modular tissue-engineered constructs [corrected]. Nat Protoc 2006;1(6):2963–9.

[27] Moraes C, Simon AB, Putnam AJ, Takayama S. Aqueous two-phase printing of cell-containing contractile collagen microgels. Biomaterials 2013;34(37):9623–31.

[28] Al Safadi R, Abu-Ali GS, Sloup RE, Rudrik JT, Waters CM, Eaton KA, et al. Correlation between *in vivo* biofilm formation and virulence gene expression in *Escherichia coli* O104:H4. PloS One 2012;7(7):e41628.

[29] Hall-Stoodley L, Costerton JW, Stoodley P. Bacterial biofilms: from the natural environment to infectious diseases. Nat Rev Microbiol 2004;2(2):95–108.

[30] Peters G, Locci R, Pulverer G. Microbial colonization of prosthetic devices. II. Scanning electron microscopy of naturally infected intravenous catheters. Zentralbl Bakteriol Mikrobiol Hyg B 1981;173(5):293–9.

[31] Marrie TJ, Nelligan J, Costerton JW. A scanning and transmission electron microscopic study of an infected endocardial pacemaker lead. Circulation 1982;66(6):1339–41.

[32] Eun YJ, Weibel DB. Fabrication of microbial biofilm arrays by geometric control of cell adhesion. Langmuir 2009;25(8):4643–54.

[33] Weibel DB, Lee A, Mayer M, Brady SF, Bruzewicz D, Yang J, et al. Bacterial printing press that regenerates its ink: contact-printing bacteria using hydrogel stamps. Langmuir 2005;21(14):6436–42.

[34] Bos R, van der Mei HC, Gold J, Busscher HJ. Retention of bacteria on a substratum surface with micro-patterned hydrophobicity. FEMS Microbiol Lett 2000;189(2):311–5.

[35] Yaguchi T, Dwidar M, Byun CK, Leung B, Lee S, Cho YK, et al. Aqueous two-phase system-derived biofilms for bacterial interaction studies. Biomacromolecules 2012;13(9):2655–61.

[36] Christensen BB, Haagensen JA, Heydorn A, Molin S. Metabolic commensalism and competition in a two-species microbial consortium. Appl Environ Microbiol 2002;68(5):2495–502.

[37] Ryan RP, Dow JM. Diffusible signals and interspecies communication in bacteria. Microbiology 2008;154(Pt 7):1845–58.

[38] Periasamy S, Kolenbrander PE. Mutualistic biofilm communities develop with *Porphyromonas gingivalis* and initial, early, and late colonizers of enamel. J Bacteriol 2009;191(22):6804–11.

[39] Dwidar M, Leung BM, Yaguchi T, Takayama S, Mitchell RJ. Patterning bacterial communities on epithelial cells. PloS One 2013;8(6):e67165.

[40] Morgan XC, Huttenhower C. Chapter 12: Human microbiome analysis. PLoS Comput Biol 2012;8(12):e1002808.

[41] Round JL, Mazmanian SK. The gut microbiota shapes intestinal immune responses during health and disease. Nat Rev Immunol 2009;9(5):313–23.

[42] Monnappa AK, Dwidar M, Seo JK, Hur JH, Mitchell RJ. *Bdellovibrio bacteriovorus* inhibits *Staphylococcus aureus* biofilm formation and invasion into human epithelial cells. Sci Rep 2014;4:3811.

[43] Kent L. Culture and maintenance of human embryonic stem cells. J Vis Exp 2009;(34).

[44] Joannides A, Fiore-Heriche C, Westmore K, Caldwell M, Compston A, Allen N, et al. Automated mechanical passaging: a novel and efficient method for human embryonic stem cell expansion. Stem Cells 2006;24(2):230–5.

[45] Hohenstein Elliott KA, Peterson C, Soundararajan A, Kan N, Nelson B, Spiering S, et al. Laser-based propagation of human iPS and ES cells generates reproducible cultures with enhanced differentiation potential. Stem Cells Int 2012;2012:926463.

[46] Marmottant P, Hilgenfeldt S. Controlled vesicle deformation and lysis by single oscillating bubbles. Nature 2003;423(6936):153–6.

[47] Thomson A, Wojtacha D, Hewitt Z, Priddle H, Sottile V, Di Domenico A, et al. Human embryonic stem cells passaged using enzymatic methods retain a normal karyotype and express CD30. Cloning Stem Cells 2008;10(1):89–106.

[48] Frampton JP, Shi H, Kao A, Parent JM, Takayama S. Delivery of proteases in aqueous two-phase systems enables direct purification of stem cell colonies from feeder cell co-cultures for differentiation into functional cardiomyocytes. Adv Healthc Mater 2013;2(11):1440–4.

[49] Thomson JA, Itskovitz-Eldor J, Shapiro SS, Waknitz MA, Swiergiel JJ, Marshall VS, et al. Embryonic stem cell lines derived from human blastocysts. Science 1998;282(5391):1145–7.

[50] Villa-Diaz LG, Brown SE, Liu Y, Ross AM, Lahann J, Parent JM, et al. Derivation of mesenchymal stem cells from human induced pluripotent stem cells cultured on synthetic substrates. Stem Cells 2012;30(6):1174–81.

[51] Kudo N, Okada K, Yamamoto K. Sonoporation by single-shot pulsed ultrasound with microbubbles adjacent to cells. Biophys J 2009;96(12):4866–76.

[52] Ohl CD, Arora M, Ikink R, de Jong N, Versluis M, Delius M, et al. Sonoporation from jetting cavitation bubbles. Biophys J 2006;91(11):4285–95.

[53] Frampton JPFZ, Simon A, Chen D, Deng CX, Takayama S. Aqueous two-phase system patterning of microbubbles: localized induction of apoptosis in sonoporated cells. Adv Funct Mater 2013;23(27):3366.

Chapter 10

Bioprinting of Organs for Toxicology Testing

Hyun-Wook Kang*, Anthony Atala**, and James J. Yoo**

*Biomedical Engineering, School of Life Sciences, Ulsan National Institute for Science and Technology, Ulsa, Korea; **Wake Forest Institute for Regenerative Medicine, Wake Forest School of Medicine, Winston-Salem, NC, USA

ABSTRACT

Toxicology testing of new drugs is generally performed in animal models. While this traditional approach is instrumental in gaining valuable information from an *in vivo* model, a number of issues arise with this type of testing such as questionable results, a large financial investment, and ethical considerations. Alternative approaches to animal testing include the use of organoids, which are biological models for specific organs or tissues that are composed of living cells or biomaterials. The organoid should have similar functionality with the native organ or tissue for toxicology testing to be successful. Bioprinting allows for the production of a designed architecture consisting of multiple types of living cells in combination with a variety of biomaterials and can be used in the construction of an organoid that mimics the native organ or tissue. In this chapter, bioprinting technology and its application for the development of new organoids will be discussed.

Keywords: toxicology testing; organoid; animal model; bioprinting

1 INTRODUCTION

Considerable financial investment is required to develop a new drug [1,2]. The development procedure involves testing of metabolic functions and toxicity to determine therapeutic effect and potential risk [3,4], which are generally assessed through animal studies. While preclinical testing in animal models provides useful information, a number of issues arise [1,5,6]. One factor that affects the usefulness of animal studies includes the variability in the response of certain species to a new drug compared to humans. Financial and ethical considerations surrounding animal testing of drugs further add to the challenges faced by this type of testing.

Consequently, many researchers are focusing on alternatives to animal studies using technologies such as cell chips for *in vitro* toxicology testing [1,7]. Recent studies have shown that *in vitro* cellular tissue can be used for drug-screen applications and may significantly reduce cost and time associated with new drug discovery [1,7–9]. The cellular tissue or organoid should have similar cellular responses that resemble *in vivo* behavior for testing drug candidates. Native organs or tissues are composed of multiple types of cells and extracellular matrix (ECM); moreover, they often have a complex microarchitecture. Tissue composition and architecture are highly related to function [10–12]. Therefore, researchers are attempting to construct architectures that mimic native organs or tissues in the development of new organoids. Bioprinting technology has gained prominence as a viable option in the development of new organoids through the production of computer-designed, complex structures composed of multiple types of living cells and biomaterials [13–15] (Fig. 10.1). This capability makes it possible to produce biomimetic architectures using

Essentials of 3D Biofabrication and Translation. http://dx.doi.org/10.1016/B978-0-12-800972-7.00010-4

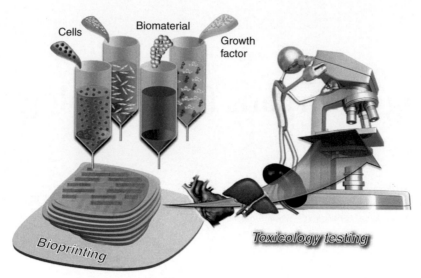

FIGURE 10.1 Construction of organoid for toxicology testing.

living cells. A number of bioprinting technologies have been introduced and applied to the construction of organoids. In this chapter, a review of these bioprinting technologies and their application will be presented.

2 BIOPRINTING TECHNOLOGIES FOR ORGANOID CONSTRUCTION

Production of two- or three-dimensional (3D) structures composed of single or multiple types of cells and biomaterials are possible through bioprinting [13–15]. This technology can be divided into two categories, direct and indirect cell patterning (Fig. 10.2). The indirect method induces cell adhesion to a desired area through the use of biological materials [15–18]. With direct patterning, an external force is applied to deposit cells into specific sites [13,14,19,20]. The

working principal and features of each of these methodologies is presented in the proceeding sections.

2.1 Indirect Cell Patterning Technology

The process for indirect cell patterning begins with the generation of a pattern designed on the surface of a plate with chemically sensitive biological materials. If the patterned material contains adherent cells, then the cells will be patterned as negative shape. Therefore, the biological property should be carefully examined when designing micropatterns.

Diverse methods for patterning of chemically bioactive materials have been applied to the indirect technology (Fig. 10.2). Among them, photolithography [21], microcontact printing [22], and stencil patterning [23] are the most

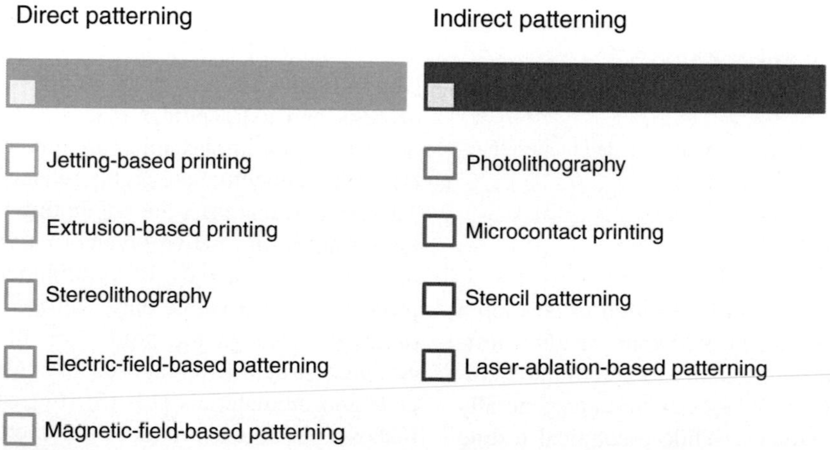

Direct patterning

☐ Jetting-based printing

☐ Extrusion-based printing

☐ Stereolithography

☐ Electric-field-based patterning

☐ Magnetic-field-based patterning

Indirect patterning

☐ Photolithography

☐ Microcontact printing

☐ Stencil patterning

☐ Laser-ablation-based patterning

FIGURE 10.2 Bioprinting technology.

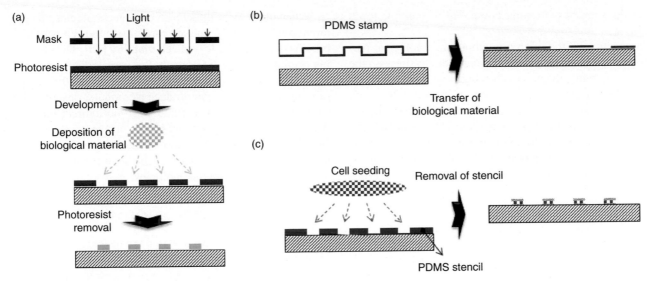

FIGURE 10.3   Indirect cell patterning methods. (a) Photolithography, (b) microcontact printing, and (c) stencil patterning.

frequently used methods (Fig. 10.3). Figure 10.3a illustrates the procedure for cell patterning based on the photolithography process. Following construction of a photoresistant pattern by the photolithography process, a bioactive material is coated over the surface. The final pattern is obtained by removing the uncured photoresistant material with a corresponding biological material. A wide range of bioactive materials has been used for cell patterning through this process. Ilic et al. [24] applied antibodies, poly-L-lysine, and aminopropyltriethoxysilane to create a self-assembled monolayer (SAM) for patterning *Escherichia coli* serotype O157:H7 bacteria cells, rat basophilic leukemia cells, and aldehyde-sulfate coated fluorescent polystyrene beads (20 nm diameter), respectively. The result was a cell patterning having several tenths of a micrometer in width. Additionally, Wang et al. [25] developed an integration method to combine a bioprinted organoid and microfluidic device. Using poly-L-lysine as the bioactive material, h-TERT-RPE1 cells were patterned to demonstrate the usefulness of the developed technology. The photolithography process has been applied to cell sheet engineering. Okano and colleagues have introduced several methods for production of a sheet having designed cell patterns by patterning PIPAAm material on the surface of culture dish [26–28].

Microcontact printing technology uses a stamp with a micropattern to transfer a pattern on to a surface (Fig. 10.3b). The stamp is usually manufactured through a molding process with poly(dimethylsiloxane) (PDMS). The patterning process using the stamp is relatively simpler and faster than other methods, and the stamp can be used repeatedly. Luk et al. [29] introduced a method for construction of a pattern with a SAM terminated in the mannitol group by using microcontact printing technology. Following the construction of the monolayer pattern, the surface is coated with protein, which can bind as a negative pattern, where the SAM pattern suppresses the absorption of protein. Using this method, 3T3 fibroblasts were plated on the pattern and an induced cell pattern having a negative shape was obtained. Interestingly, cell patterns obtained by this method can be maintained for a long time in culture, greater than 25 days. The microcontact printing method is usually not suitable for producing 3D patterns. Dusseiller et al. [30] attempted to solve this issue by combining microcontact printing with molding technology. They demonstrated that the technology is suitable to fabricate a 2.5-dimensional cellular tissue. Conversely, Rhee et al. [31] introduced an integration method using a microfluidic device and microcontact printing. Following placement of a PDMS stamp on the surface of glass substrate or Petri dish, the uncovered area was treated with plasma-based dry etching. Next, the microfluidic device was bonded on the surface and successfully demonstrated cell patterning and flow results with the constructed microfluidic device.

Another method for indirect cell patterning, demonstrated by Yen et al. [32], includes a laser writing process. A CO_2 laser was used to construct a micropattern on a silane coated substrate followed by plating cells. This study showed that cells can only attach to the ablated region; moreover, the process can produce microfluidic channels without the need for any additional device. Thus, this method can produce cell patterning along with a microfluidic device through a single process. In another study, Cheng et al. [33] introduced a combined technology of cell sheet engineering and photolithography. A microheater was prepared by using a photolithography process. A thermo responsive plasma film was formed by surface coating with poly(*N*-isopropylacrylamide). Finally, the localization of cell adhesion could be controlled by adjusting the temperature, and the results showed patterning composed of multiple types of cell.

As mentioned, the indirect method uses biological materials for cell patterning that eliminates direct stress on the cell. Consequently, the process has high cell viability. In addition, a higher patterning resolution is achieved than with the direct method. Resolution down to several micrometers was reported with the indirect method [15]. Although the process becomes more complex and is not ideal for creating 3D structures, some modified technologies have made it possible to construct composite patterns with multiple types of cells [33–35].

2.2 Direct Cell Patterning Technology

Direct cell patterning technology uses external force such as pneumatic pressure, electric force, and magnetic force. Currently, researchers apply direct patterning technology in the development of biomimetic organoids. Inkjet printing, extrusion-based printing, and stereolithography are typical examples of the direct patterning technology, which can produce 3D cellular tissues [13,14]; moreover, these technologies are described in the other chapters. This chapter will focus on additional methods of direct cell patterning.

Several researchers are studying the dielectric phenomenon to obtain specific cell patterns by designing an electric field [36,37]. Dielectric particles in nonuniform electric field are forced into specific direction by dielectrophoresis (DEP) phenomenon [38]. A precisely designed microelectrode induces the desired form of electric-field that induces cell patterns. Usually, photo-lithography technology is used to construct the electrodes. Various kinds of electrode design have been proposed to induce specific cell patterns. Hsiung et al. [39] introduced an electrode design for uniform patterning of cells. Hunt et al. [40] developed an integrated circuit that can freely program the movement of dielectric particles; furthermore, the developed circuit can be combined with a microfluidic device. Most studies use DEP to produce a 2D structure because the lithography technology is not suitable for construction of 3D structures. As an example, Albrecht et al. [41] combined electropatterning technology with photopatterning to produce 2.5-dimensional patterning.

Conversely, a magnetic field, instead of an electric field, was also applied for cell patterning [42,43]. Ino et al. [42] fabricated micropatterns on the magnetic soft iron plate by direct machining with wire-EDM. Magnetite cationic liposomes along with labeled NIH/3T3 fibroblasts were applied to demonstrate cell patterning results. Tsang et al. [44] introduced an additive photopatterning technology that used a photocurable hydrogel with the lithography process and showed that a 3D cell patterning can be achieved by stacking the 2D patterns. This technology was applied to produce 3D hepatic tissues.

When using these direct patterning technologies, the external force should be applied carefully in the cell pattering process to reduce the loss of cell viability [45]. The patterning resolution of the direct method is usually lower than the indirect method [15]; however, various kinds of direct bioprinting technologies can produce a computer designed 3D structure, which is not possible with the indirect method. Additionally, a variety of living cells and biomaterials can be applied into single structures by the direct cell patterning method. As a result, many researchers are favoring the use of direct cell patterning technologies.

3 BIOPRINTED ORGANOIDS

A number of studies have reported the development of new organoids. The majority of these studies were focused on the development of liver and muscle organoids. Table 10.1 provides a list of bioprinted organoids.

3.1 Liver Organoid

Liver is the principal organ for detoxification of the human body. Consequently, studies describing liver organoids are of considerable interest to researchers focused on toxicology testing. Several researchers applied DEP-based cell patterning technology to produce a liver organoid [46–48]. Ho et al. [46] developed biomimetic liver-like architecture through copatterning human hepatoma HepG2 cell and human umbilical vein endothelial cells (HUVEC) using the DEP technology (Fig. 10.4a). Puttaswamy et al. [48] investigated the effect of medium conductivity and voltage level on cell viability and adhesion during negative-DEP cell patterning process. The results showed that low conductivity medium and optimal patterning condition can significantly enhance cell viability and adhesion rate. Ho et al. [47] reported copatterning results of HepG2 and HUVEC, and cell viability greater than 95% was observed. Activity of CYP450-1A1 enzyme was measured to determine the functionality of the liver organoid, which was shown to be improved by 80% in comparison with nonpatterned samples.

Kidambi et al. [35] applied a microcontact printing process to produce liver organoids by copatterning primary rat hepatocytes and fibroblasts and measured functionality by urea and albumin secretion. The results showed that the copatterned organoids had enhanced activity. Khetani et al. [49] reported a liver organoid composed of human liver cells and stromal cells that was used for a drug screening test (Fig. 10.4b). The organoid was prepared by using a PDMS stencil to produce dot patterns with hepatocytes, which were placed at regular intervals. Functionality, toxicity, and phenotype test were conducted to demonstrate efficacy. The results showed that the functionality of the liver organoid was maintained well for more than several weeks and could be used in drug development.

TABLE 10.1 Bioprinted Organoids

Target Organ	Cells	Bioprinting Technology	Study Features	Reference
Liver	Human liver cell line HepG2	Extrusion-based bioprinting	Combining 3D cell printing and microfluidic device	[51]
	Primary human hepatocyte	Stencil patterning	Long term test for more than several weeks	[49]
	Human liver cell line HepG2 Human umbilical vein endothelial cell	Dielectrophoresis-based cell patterning	Lobule-mimetic liver-cell patterning	[46,47]
	Primary rat hepatocyte NIH3T3 fibroblast	Microcontact printing	Coculturing system and its functionality test	[35]
	Human liver cell line HepG2	Dielectrophoresis-based cell patterning	Enhanced cell viability and adhesion rate	[48]
	Primary rat hepatocyte	Additive photopatterning	3D patterning	[44]
Muscle	Mouse C2C12 myoblast	Inkjet printing	Combining muscle organoid and microsized cantilever	[53]
	Human vascular smooth muscle cell Neonatal rat ventricular myocyte	Microcontact printing	Thin film form to measure contractile form	[34]
	Mouse C2C12 myoblast	Stencil patterning	Combining muscle organoid and microfluidic device Drug localization	[52]
Blood microvessel	Human umbilical vein endothelial cell Normal human dermal fibroblast	Magnetic force-based cell patterning	Vascular network	[54]

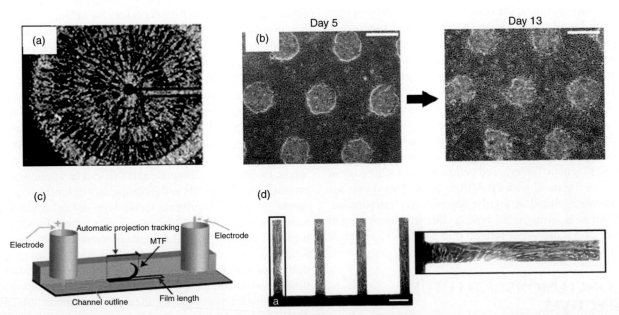

FIGURE 10.4   Bioprinted organoids. (a) Liver organoid composed of HepG2 (green) and HUVECs (red) [46]; (b) liver organoid composed of primary human hepatocytes and 3T3-J2 fibroblasts [49]; (c) photograph of a muscle chip having muscle organoids produced by muscular thin film technology [34]; and (d) bioprinted C2C12 on the microcantilever (scale bar : 200 μm) [53].

Extrusion-based printing process was also applied in the development of liver organoids. Chang et al. [50] demonstrated patterning results with HepG2 to produce a printed liver organoid that could be combined with a microfluidic system. Snyder et al. [51] used a temperature-controlled syringe to print a pattern with Matrigel that contained HepG2 cells and human mammary epithclials (M10), which showed feasibility in radiation testing. Moreover, this study showed that the system can produce a dual-tissue organoid through the use of multiple cartridges.

3.2 Muscle Organoid

Muscle tissue generates physical force required for activity of the human body. In the development of artificial muscle organoids, therefore, a measurement of physical force is necessary to estimate functionality. A number of studies have been reported for muscle organoid fabrication. Thin film form has typically been used for development of muscle organoids. Tourovskaia et al. [52] introduced a muscle organoid that was fabricated with indirect cell patterning using an elastomeric mask and oxygen plasma etching and combined with a microfluidic system. They demonstrated that the combined system could be used for the study of muscle differentiation to form multinucleated myotube. Grosberg et al. [34] developed a muscular thin film technology for the construction of a heart organoid (Fig. 10.4c). After patterning the organoid on the thin film with a microcontact printing method, the functionality was estimated by measuring the bending of film using charge-coupled device camera. Cui et al. [53] applied an inkjet printing technology to produce a muscle organoid. After loading C2C12 cells into the inkjet cartridge, the cells were directly printed onto the silicon-based microcantilever. The muscle function of the organoid was assessed by using an electric stimulator. A laser was used to measure the displacement of the cantilever to estimate the functionality of the muscle organoid.

3.3 Vascular Organoid

In addition to the reports for liver and muscle organoids, a study of vascular chip technology was introduced. Ino et al. [54] used magnetic force-based cell pattering technology to produce an organoid with HUVECs. Spot shaped cell aggregates were placed at regular intervals on microscales. The result was an organoid having 250 μm intervals and a density of 5.9 cells/spot formed into a cord-like structure after *in vitro* culture of 8.5 h.

4 CONCLUSIONS AND FUTURE PERSPECTIVES

Recently, a number of researchers have applied bioprinting technologies toward the construction of organoids for toxicology testing. Copatterned organoids with multiple cell types have been described that have improved functionality when compared to unpatterned organoids [35, 47]. While considerable advancements have been made in the bioprinting of organoids for testing, the technology is relatively new and requires further development to overcome obstacles that inhibit production of organoids that can represent native organs or tissues.

First, the resolution of 3D cell printing needs to be improved. Organs or tissues in the human body have complex 3D microarchitecture that is highly related to function. Many researchers believe biomimetic architecture will enhance the functionality of organoids. Several studies have introduced 3D cell culture environments that possess different cellular responses compared to a two-dimensional system [10–12]. High resolution of several micrometers is required to mimic native organs or tissues; however, the resolution of current 3D bioprinting technologies does not allow for the precision necessary to create the native microarchitecture. Resolution of several hundreds of micrometers using 3D printing has been reported by several research groups [15], and additional studies could lead to improvements in the patterning resolution. Novel biomaterials that have good cell compatibility and could be applied to 3D printing should be investigated; furthermore, the material used in organoid construction can influence cell viability and overall functionality. For example, a study by Skardal et al. [55] prepared a liver ECM-laden hydrogel and compared it with a number of hydrogels for 3D culture of human primary hepatocytes. Their results showed that the liver ECM enhanced the viability and functionality of the 3D construct. As shown in this result, choice of material is among the critical factors necessary to achieve enhanced functionality of an organoid. Printability of a biomaterial is another property that should be considered when using a printing mechanism. In the case of inkjet printing technology, materials used in the printer should have low viscosity to allow for smooth jetting motion. Alternatively, materials used with extrusion-based printing technology should have high viscosity that allows for layers to be stacked for 3D patterning. Development of new bioink should be focused on materials with cell compatibility and printability. Currently, the number of biomaterials that can be applied to 3D printing are limited due to the complexity of bioink requirements. To move bioprinting technology toward toxicology testing, a wide range of bioinks are required. Finally, an organoid used for toxicology testing should exhibit similar functionality with target organ or tissue as well as proper design for easy measurement. Therefore, a study of interaction of cell-to-cell or cell-to-ECM and 3D design of organoid considered functionality measurement should be conducted to develop an organoid for toxicology testing.

A variety of advanced bioprinting technologies have been introduced that have been used to produce 2D and

3D cellular tissues composed of multiple types of living cells and biomaterials. Moreover, such advances have allowed production of a designed microarchitecture that mimics native tissues and organs. Researchers in a number of disciplines including bioengineering and regenerative medicine have applied these bioprinting technologies to produce artificial organoids that have yielded promising results. Although numerous obstacles have yet to be overcome, we believe bioprinting technologies will provide a new paradigm for toxicology testing of drugs that may eliminate the need for preclinical testing in animal models, reduce the financial burden, and provide reliable and accurate assessments of new drug treatments for use in humans.

GLOSSARY

Direct cell patterning A process that uses external force such as pneumatic pressure, electric force, and magnetic force to produce 2D or 3D patterns with living cells.

Extracellular matrix (ECM) A substance containing collagen, elastin, proteoglycans, glycosaminoglycans, and fluid, produced by cells and in which the cells are embedded.

Extrusion-based bioprinting A bioprinting method that dispenses continuous filaments of a material consisting of cells mixed with hydrogel through a micronozzle to fabricate 2D or 3D structures.

Indirect cell patterning A process that begins with the generation of a pattern designed on the surface of a plate with chemically sensitive biological materials.

Microcontact printing A form of soft lithography that uses a stamp with a micropattern to transfer a pattern on to a surface.

Microfluidics A technology based on geometrically constrained minute volume transport through channels in a glass or plastic chip.

Organoid Any structure that resembles an organ in appearance or function.

Photolithography A process involving the photographic transfer of a pattern to a surface for etching (as in producing an integrated circuit).

Stencil patterning A process of applying a pattern or design to a surface, consisting of a thin sheet of cardboard, metal, or other material from which figures are cut out.

Toxicology testing Testing to determine the degree to which a substance can damage a living or nonliving organisms.

ABBREVIATIONS

3D Three-dimensional
ECM Extracellular matrix
PDMS Poly(dimethylsiloxane)
SAM Self-assembled monolayer
DEP Dielectrophoresis
HUVEC Human umbilical vein endothelial cells

ACKNOWLEDGMENTS

The authors would like to thank Dr Heather Hatcher for editorial assistance.

REFERENCES

[1] Esch M, King T, Shuler M. The role of body-on-a-chip devices in drug and toxicity studies. Annu Rev Biomed Eng 2011;13: 55–72.

[2] Rawlins MD. Cutting the cost of drug development? Nat Rev Drug Discov 2004;3(4):360–4.

[3] Giuliano KA, Haskins JR, Taylor DL. Advances in high content screening for drug discovery. Assay Drug Dev Technol 2003;1(4):565–77.

[4] Littman BH, Williams SA. The ultimate model organism: progress in experimental medicine. Nat Rev Drug Discov 2005;4(8):631–8.

[5] Lebonvallet N, Jeanmaire C, Danoux L, Sibille P, Pauly G, Misery L. The evolution and use of skin explants: potential and limitations for dermatological research. Eur J Dermatol 2010;20(6):671–84.

[6] Mitra A, Wu Y. Use of *in vitro–in vivo* correlation (IVIVC) to facilitate the development of polymer-based controlled release injectable formulations. Recent Pat Drug Deliv Formul 2010;4(2):94–104.

[7] Williamson A, Singh S, Fernekorn U, Schober A. The future of the patient-specific body-on-a-chip. Lab Chip 2013;13(18):3471–80.

[8] El-Ali J, Sorger PK, Jensen KF. Cells on chips. Nature 2006;442(7101):403–11.

[9] Starkuviene V, Pepperkok R, Erfle H. Transfected cell microarrays: an efficient tool for high-throughput functional analysis. Expert Rev Proteomics 2007;4(4):479–89.

[10] Cheema U, Brown R, Alp B, MacRobert A. Spatially defined oxygen gradients and vascular endothelial growth factor expression in an engineered 3D cell model. Cell Mol Life Sci 2008;65(1):177–86.

[11] Feder-Mengus C, Ghosh S, Reschner A, Martin I, Spagnoli GC. New dimensions in tumor immunology. what does 3D culture reveal? Trends Mol Med 2008;14(8):333–40.

[12] Yamada KM, Cukierman E. Modeling tissue morphogenesis and cancer in 3D. Cell 2007;130(4):601–10.

[13] Boland T, Xu T, Damon B, Cui X. Application of inkjet printing to tissue engineering. Biotechnol J 2006;1(9):910–7.

[14] Derby B. Printing and prototyping of tissues and scaffolds. Science 2012;338(6109):921–6.

[15] Guillotin B, Guillemot F. Cell patterning technologies for organotypic tissue fabrication. Trends Biotechnol 2011;29(4):183–90.

[16] Khademhosseini A, Langer R, Borenstein J, Vacanti JP. Microscale technologies for tissue engineering and biology. Proc Natl Acad Sci USA 2006;103(8):2480–7.

[17] Park TH, Shuler ML. Integration of cell culture and microfabrication technology. Biotechnol Prog 2003;19(2):243–53.

[18] Xu F, Wu J, Wang S, Durmus NG, Gurkan UA, Demirci U. Micro-engineering methods for cell-based microarrays and high-throughput drug-screening applications. Biofabrication 2011;3(3):034101.

[19] Hannachi IE, Yamato M, Okano T. Cell sheet technology and cell patterning for biofabrication. Biofabrication 2009;1(2):022002.

[20] Lin RZ, Ho CT, Liu CH, Chang HY. Dielectrophoresis based-cell patterning for tissue engineering. Biotechnol J 2006;1(9):949–57.

[21] Britland S, Clark P, Connolly P, Moores G. Micropatterned substratum adhesiveness: a model for morphogenetic cues controlling cell behavior. Exp Cell Res 1992;198(1):124–9.

[22] Kumar A, Whitesides GM. Features of gold having micrometer to centimeter dimensions can be formed through a combination of stamping with an elastomeric stamp and an alkanethiol "ink" followed by chemical etching. Appl Phys Lett 1993;63(14): 2002–4.

[23] Folch A, Jo BH, Hurtado O, Beebe DJ, Toner M. Microfabricated elastomeric stencils for micropatterning cell cultures. J Biomed Mater Res 2000;52(2):346–53.

[24] Ilic B, Craighead H. Topographical patterning of chemically sensitive biological materials using a polymer-based dry lift off. Biomed Microdevices 2000;2(4):317–22.

[25] Wang L, Lei L, Ni X, Shi J, Chen Y. Patterning bio-molecules for cell attachment at single cell levels in PDMS microfluidic chips. Microelectron Eng 2009;86(4):1462–4.

[26] Tsuda Y, Kikuchi A, Yamato M, Nakao A, Sakurai Y, Umezu M, et al. The use of patterned dual thermoresponsive surfaces for the collective recovery as co-cultured cell sheets. Biomaterials 2005;26(14):1885–93.

[27] Yamato M, Konno C, Utsumi M, Kikuchi A, Okano T. Thermally responsive polymer-grafted surfaces facilitate patterned cell seeding and co-culture. Biomaterials 2002;23(2):561–7.

[28] Yamato M, Kwon OH, Hirose M, Kikuchi A, Okano T. Novel patterned cell coculture utilizing thermally responsive grafted polymer surfaces. J Biomed Mater Res 2001;55(1):137–40.

[29] Luk Y-Y, Kato M, Mrksich M. Self-assembled monolayers of alkanethiolates presenting mannitol groups are inert to protein adsorption and cell attachment. Langmuir 2000;16(24):9604–8.

[30] Dusseiller MR, Schlaepfer D, Koch M, Kroschewski R, Textor M. An inverted microcontact printing method on topographically structured polystyrene chips for arrayed micro-3-D culturing of single cells. Biomaterials 2005;26(29):5917–25.

[31] Rhee SW, Taylor AM, Tu CH, Cribbs DH, Cotman CW, Jeon NL. Patterned cell culture inside microfluidic devices. Lab Chip 2005;5(1):102–7.

[32] Yen MH, Cheng JY, Wei CW, Chuang YC, Young TH. Rapid cell-patterning and microfluidic chip fabrication by crack-free CO_2 laser ablation on glass. J Micromech Microeng 2006;16(7):1143.

[33] Cheng X, Wang Y, Hanein Y, Böhringer KF, Ratner BD. Novel cell patterning using microheater-controlled thermoresponsive plasma films. J Biomed Mater Res Part A 2004;70(2):159–68.

[34] Grosberg A, Nesmith AP, Goss JA, Brigham MD, McCain ML, Parker KK. Muscle on a chip: *in vitro* contractility assays for smooth and striated muscle. J Pharmacol Toxicol Methods 2012;65(3):126–35.

[35] Kidambi S, Sheng L, Yarmush ML, Toner M, Lee I, Chan C. Patterned co-culture of primary hepatocytes and fibroblasts using polyelectrolyte multilayer templates. Macromol Biosci 2007;7(3):344–53.

[36] Gray DS, Tan JL, Voldman J, Chen CS. Dielectrophoretic registration of living cells to a microelectrode array. Biosens Bioelectron 2004;19(7):771–80.

[37] Taff BM, Voldman J. A scalable addressable positive-dielectrophoretic cell-sorting array. Anal Chem 2005;77(24):7976–83.

[38] Pohl HA, Pohl H. Dielectrophoresis: the behavior of neutral matter in nonuniform electric fields. Cambridge: Cambridge University Press; 1978.

[39] Hsiung LC, Yang CH, Chiu CL, Chen CL, Wang Y, Lee H, et al. A planar interdigitated ring electrode array via dielectrophoresis for uniform patterning of cells. Biosens Bioelectron 2008;24(4): 869–75.

[40] Hunt TP, Issadore D, Westervelt RM. Integrated circuit/microfluidic chip to programmably trap and move cells and droplets with dielectrophoresis. Lab Chip 2008;8(1):81–7.

[41] Albrecht DR, Tsang VL, Sah RL, Bhatia SN. Photo- and electropatterning of hydrogel-encapsulated living cell arrays. Lab Chip 2005;5(1):111–8.

[42] Ino K, Okochi M, Konishi N, Nakatochi M, Imai R, Shikida M, et al. Cell culture arrays using magnetic force-based cell patterning for dynamic single cell analysis. Lab Chip 2008;8(1):134–42.

[43] Shimizu K, Ito A, Honda H. Enhanced cell-seeding into 3D porous scaffolds by use of magnetite nanoparticles. J Biomed Mater Res B Appl Biomater 2006;77(2):265–72.

[44] Tsang VL, Chen AA, Cho LM, Jadin KD, Sah RL, DeLong S, et al. Fabrication of 3D hepatic tissues by additive photopatterning of cellular hydrogels. FASEB J 2007;21(3):790–801.

[45] Nair K, Gandhi M, Khalil S, Yan KC, Marcolongo M, Barbee K, et al. Characterization of cell viability during bioprinting processes. Biotechnol J 2009;4(8):1168–77.

[46] Ho CT, Lin RZ, Chang WY, Chang HY, Liu CH. Rapid heterogeneous liver-cell on-chip patterning via the enhanced field-induced dielectrophoresis trap. Lab Chip 2006;6(6):724–34.

[47] Ho CT, Lin RZ, Chen RJ, Chin CK, Gong SE, Chang HY, et al. Liver-cell patterning lab chip: mimicking the morphology of liver lobule tissue. Lab Chip 2013;13(18):3578–87.

[48] Puttaswamy SV, Sivashankar S, Chen RJ, Chin CK, Chang HY, Liu CH. Enhanced cell viability and cell adhesion using low conductivity medium for negative dielectrophoretic cell patterning. Biotechnol J 2010;5(10):1005–15.

[49] Khetani SR, Bhatia SN. Microscale culture of human liver cells for drug development. Nat Biotechnol 2007;26(1):120–6.

[50] Chang R, Emami K, Wu H, Sun W. Biofabrication of a three-dimensional liver micro-organ as an *in vitro* drug metabolism model. Biofabrication 2010;2(4):045004.

[51] Snyder J, Hamid Q, Wang C, Chang R, Emami K, Wu H, et al. Bioprinting cell-laden matrigel for radioprotection study of liver by pro-drug conversion in a dual-tissue microfluidic chip. Biofabrication 2011;3(3):034112.

[52] Tourovskaia A, Figueroa-Masot X, Folch A. Differentiation-on-a-chip: a microfluidic platform for long-term cell culture studies. Lab Chip 2005;5(1):14–9.

[53] Cui X, Gao G, Qiu Y. Accelerated myotube formation using bioprinting technology for biosensor applications. Biotechnol Lett 2013;35(3):315–21.

[54] Ino K, Okochi M, Honda H. Application of magnetic force-based cell patterning for controlling cell–cell interactions in angiogenesis. Biotechnol Bioeng 2009;102(3):882–90.

[55] Skardal A, Smith L, Bharadwaj S, Atala A, Soker S, Zhang Y. Tissue specific synthetic ECM hydrogels for 3-D *in vitro* maintenance of hepatocyte function. Biomaterials 2012;33(18):4565–75.

Chapter 11

High Throughput Screening with Biofabrication Platforms

Carlos Mota and Lorenzo Moroni

Department of Complex Tissue Regeneration (CTR), Institute for Technology-Inspired Regenerative Medicine (MERLN), Maastricht University, Maastricht, The Netherlands

Chapter Outline

ABSTRACT

High throughput screening (HTS) has been applied during the last decades for compound libraries screening. Pioneered by pharmaceutical companies for drug discovery, it has been largely adopted by several other research fields such as stem cell, biomaterials, tissue engineering, and regenerative medicine. Whether conventional HTS encompasses the assessment of experimental conditions mostly in two-dimensional (2D) cultures, the advent of biofabrication technologies allowed screening to take over the third dimension, thus resulting in culture models better representing the physiological environment. Three-dimensional (3D) screening platforms span from classical spheroids to bioprinted constructs where the spatial positioning of cells can be precisely controlled. In this book chapter, we introduce the general need to pass from 2D to 3D screening and the potential that biofabrication technologies hold for this purpose. We will focus our attention on bioprinting, as this set of biofabrication technologies is more apt to be implemented in HTS assays for different biomedical applications.

Keywords: high throughput screening; biofabrication technologies; bioprinting; *in vitro*; *in vivo*

1 OUTLINE

High throughput screening (HTS) has drawn attention in the past three decades and has been exploited in different research fields. HTS techniques in conjugation with combinatorial chemistry concepts were largely adopted by pharmaceutical companies to improve drug discovery processes [1]. These techniques have migrated from pharmaceutical companies to academia and research institutes, resulting in the emerging new public–private partnerships [2]. Despite still being largely used by pharmaceutical companies and academia for drug discovery [3–5], HTS techniques have started to be implemented in several other research fields such as diagnosis [6], cancer research [7], tissue engineering (TE) or regenerative medicine (RM) (e.g., stem cell driven TE/RM research [8–10], extracellular microenvironment [11]), biomaterials development (e.g., for drug delivery [12], stem cell fate [13], or other TE/RM applications [14,15]), proteomics [3,16], and genomics [17,18].

With the growing use of HTS and assays miniaturization, it is becoming possible to reduce reagents costs while

Essentials of 3D Biofabrication and Translation. http://dx.doi.org/10.1016/B978-0-12-800972-7.00011-6

obtaining a large amount of information per screening process [3,19]. The higher the well number, the lower the volume available per well, resulting in further challenges in the manual dispensing of components to screen. The need for a higher number of conditions/elements to screen, therefore, imposed the development of liquid handling automated equipment. These robotic liquid dispensing technologies, also commonly used to handle fluids on HTS assays, have been used to reduce processing times, generally observed in low-volume dispensing, increasing the accuracy and reproducibility and thus reducing sample variability or manual-induced dispensing errors [20]. Several other fundamental aspects, such as the selection of (1) the technique for assay manufacturing, (2) the screening platforms, (3) the screening compounds, and (4) the cells, need to be carefully considered according to the ultimate goal of the screen.

The necessity of manufacturing miniaturized assays drove the customization of available biofabrication technologies or the development of new ones, according to the different necessities and challenges observed for each field of research. Bioprinting techniques are a group of computer-controlled biofabrication technologies that allow for an accurate dispensing of cells, extracellular matrix (ECM), synthetic or natural biomaterials, biochemical factors, and natural or synthetic drugs to form previously designed patterns or arrays [10]. Commonly, bioprinting technologies used for the manufacture of HTS arrays are classified as (1) contact technologies, where the dispensing element touches the target platform to deliver the compound, and (2) noncontact technologies, where the compound is normally ejected toward a collection platform [21]. Contact technologies allow the dispensing of solutions in the attolitter to microliter range with almost no blockage of the solution on the dispensing tips. Although, when the tips come into physical contact with the target platform, mechanical stresses and subsequent damage of the dispensing tip might occur. In case of noncontact technologies, the volumes dispensed can vary from the picoliter to microliter range. Normally, these technologies are limited to low-viscosity solutions and the main drawback that can be observed while dispensing is the obstruction or blockage of the dispensing cartridge.

With new research fields adopting HTS, the development of novel HTS platforms or the customization of the commercially available ones was necessary to explore new *in vitro* screening approaches. New implantable HTS screening platforms were also developed and applied *in vitro* and *in vivo* [22,23]. Among the several experimental conditions that have become nowadays of interest to screen, a large number of natural [24] and synthetic [1,25] chemical compounds are available and aided in the possibility of discovering new drug applications. Several compound libraries have become commercially available, allowing the possibility of making drug discovery faster and the discovery of new applications for already known drugs possible. A careful selection of the optimal library according to the aims of the HTS studies dictate the successful achievement of readouts [26]. Several specialty libraries are available and are generally classified as [26] (1) target class libraries, (2) known pharmacologically active drugs collections, (3) fragments-based screening collections, and (4) diversity-oriented synthesis collections. Online databases containing information over hundreds of thousands of small molecules approved by the Food and Drug Administration (FDA) regulatory agency are currently available (Table 11.1) [2].

The use of primary cells or immortalized cell lines is also a key aspect to be considered in HTS assays, especially in the drug discovery field [27]. Commonly immortalized cell lines are used on primary HTS studies. Yet, the results obtained in some cases fail on subsequent phases of the drug discovery process [27], likely due to modifications that have occurred in the cell lines during their continuous *in vitro* manipulation. It is believed that the utilization of primary cells will improve the reliability of the results

TABLE 11.1 Small Molecules Databases Containing FDA Approved Drugs

Database Name	Content and Details	URLs
PubChem	≥30 M molecules includes FDA approved drugs	http://pubchem.ncbi.nlm.nih.gov/
NPC Browser	~10,000 compounds includes FDA approved drugs	http://tripod.nih.gov/npc/
ToxCast	≥1000 compounds includes some drugs and drug like molecules	http://epa.gov/ncct/toxcast/
DailyMed	≥31,942 labels – many labels for the same drug	http://dailymed.nlm.nih.gov/dailymed/about.cfm
ChemIDplus	>295,000 structures including many FDA small molecule approved drugs	http://chem.sis.nlm.nih.gov/chemidplus/
DrugBank	6707 drug entries including 1436 FDA-approved small molecule drugs (this may be underestimated)	http://www.drugbank.ca/

Reprinted with permission from Ekins et al. [2].

attained with HTS, making the drug discovery process more accurate and consequently faster than with cell lines [27]. Recently, induced pluripotent stem cells (iPSCs) have also been proposed as an optimal alternative cell source for HTS [4,27].

In this chapter, we highlight the main bioprinting techniques used for the preparation of HTS assays, paying attention to key aspects such as assays relevance, data acquisition, and treatment. The differences between technologies and approaches developed for HTS are presented and discussed, focusing on the development of assays for drug discovery, stem cell fate, and biomedical polymer screening. Platforms with the potential for *in vitro* and *in vivo* screening are also discussed.

2 TWO-DIMENSIONAL AND THREE-DIMENSIONAL HTS

Conventional HTS comprises the evaluation of experimental conditions mostly in two dimensions [28]. Initially, screening studies were performed on two-dimensional (2D) cell culture monolayers and became the gold standard for drug discovery and cell biology research. Cell-based assays are normally used in pharmacology studies to identify new targets, evaluate drug efficacy, and study absorption, distribution, metabolism, and excretion/toxicity (ADME/Tox) *in vitro* [29]. Small-molecule HTS may highlight several hits of drug candidates for the investigated hypothesis. Positive hits normally identified may subsequently be tested in animal models but this raises scientific and ethical challenges [30]. Furthermore, animal models are expensive, time consuming and the results obtained with this preclinical screening in rodents is often discordant with the results observed in human clinical trials [4,31]. Several reported cases found in literature highlight significant differences between the results obtained with 2D screening when compared with three-dimensional (3D) *in vitro* models [32–34]. Normally, high efficacy drug hits identified with assays performed with cell monolayers often fail in terms of toxicology when screened in 3D. Due to this factor, it is believed that 3D models will become a more reliable alternative avoiding misleading data. The translation from 2D screening assays to 3D is gradually being introduced in diverse research fields. New *in vitro* screening approaches have been developed over the last years in order to mimic the 3D microenvironment, thus resulting in culture models that better represent the *in vivo* physiology [29,32,35]. The development of these 3D approaches was mainly driven by the drug discovery field aiming for the acquisition of more reliable data in order to prevent drawbacks at later stages (e.g., failure at *in vivo* animal studies and clinical trials phases). Methods involving the application of spheroids manufactured by force-floating, hanging drop, or agitation-based methods are promising models for 3D screening and have been extensively investigated [28]. Despite these models having already shown how cell activity can dramatically change in 3D, they are still confined to mono- or cocultures in spherical shapes, which does not entirely recapitulate the spatial distribution of native tissues and organs. Therefore, 3D *in vitro* cell culture models exploiting membranes, sponges/gels, and microcarriers started to be introduced in an attempt to replicate the results normally obtained *in vivo* [35].

3 HIGH-CONTENT SCREENING

When millions of molecules are screened with HTS assays, generally hundreds of positive hits might be unveiled [36]. Due to this fact, high content screening (HCS) approaches are normally used to perform analysis at the single-cell level and have been increasingly used in the drug discovery field, allowing different assay types while increasing the throughput [37]. HCS assays are exploited as a bridge between the positive hits that better perform on HTS assays and the *in vivo* animal studies normally performed prior to clinical trials [36]. HCS technologies have become main stream in several steps of drug discovery, namely primary compound screening, postprimary screening, ADME/Tox evaluation, and multivariate drug profiling [38]. Multicolor fluorescence microscopy is one of the technologies largely used for HCS due to the possibility of automation, the large amount of staining solutions (e.g., organic dyes, immunoreagents, genetically encoded fluorescent proteins, or quantum dots), and the possibility of extracting quantitative data from the acquired images by means of image analysis software [38]. HCS has been gradually implemented in more complex analysis for the different fields. An example is 3D-engineered tissue models that can mimic physiological functions used for compound screening where positive and negative effects on tissue function are analyzed on the changes observed in multiple intracellular pathways [36].

4 BIOPRINTING TECHNOLOGIES

With the advent of HTS, several bioprinting technologies have been developed to ensure the proper manufacturing of the screening platforms hosting a number of conditions varying from hundreds to thousands. HTS bioprinting systems can be either custom-developed or adapted from other industries, and are normally classified as contact and noncontact technologies [21]. These HTS technologies should meet some fundamental requirements such as linearity, precision, accuracy, throughput, reliability, and flexible dynamic range [19]. Despite the large number of technologies developed over the last decades, bioprinting technologies are still generally limited to a narrow window of materials that can be processed (Table 11.2).

TABLE 11.2 Resume Table of Contact and Noncontact Bioprinting Techniques Used for High Throughput Screening Research Studies Performed in Different Research Fields

Technology Used	Research Field	Equipment	Platform	Other Relevant Details	References
Contact Techniques					
Pin-based system	Biomaterials development, Cell-biomaterial interaction	Pixsys 5500 (Cartesian) with CMP9B and CMP6B (Telechem international)	Epoxy-coated glass slides (dip-coated with poly hydroxyethyl methacrylate)	**Ct*:** 576 different conditions screened (25 monomers and one photoinitiator); **Cells:** hESCs and mouse muscle myoblasts; **C*:** chemical characterization of the different conditions and immunohistochemistry; **Pp*:** nL volume range, spot diameter ~300 μm, 1728 spots/ slide	[40,43]
Pin-based system	Biomaterials development (substrates for clonal growth)	Pixsys 5500 (Cartesian) with CMP9B and CMP6B (Telechem international)	Epoxy-coated glass slides (dip-coated with poly hydroxyethyl methacrylate)	**Ct:** 22 monomers (combined in a combinatorial fashion); **Cells:** hESCs (BG01 and WIBR3) and iPSCs; **C:** physio-chemical characterization of the different conditions and immunostainings; **Pp:** spot diameter 300 μm, height ~15 μm, center-to-center 740 μm	[41]
Pin-based system	Biomaterials development (TE applications)	Qarray mini with 16 aQu solid pins (Genetix)	Solutions prepared in a 384 well microtiter plate and printed onto microscope slides coated with aminoalkylsilane.	**Ct:** 7 polymers (135 binary polymer blends) **Cells:** fetal skeletal cells, human skeletal stem cell (STRO-1 +), early osteoblast-like (MG-63) cell line and mature osteoblast-like (SaOs) cell line **C:** thermal and morphological characterization of polymer spots, cell adhesion **Pp:** spot diameters 300–400 μm, center-to-center 750 μm (x-direction) and 900 μm (y-direction), 960 spots/slide	[42]
Pin-based system	Drug discovery	GeneMachine OmniGrid (Genomic Instrumentation Services) or the Molecular Dynamics GenIII (Amersham Biosciences).	Poly-L-lysine coated slides or plain glass Slides	**Ct:** library of over 150,000 compounds **Cells:** n/a **C:** fluorescence **Pp:** 1 nL or 1.6 nL volume, spot diameter 180 μm, center-to-center 500 μm, 6600 conditions/slide	[3]
Pin-based system	Drug discovery	OmniGrid contact microarrayers (Gene Machines)	Glass slides	**Ct:** a 352 compound library **Cells:** n/a **C:** fluorescence microcopy **Pp:** 1.6 nL spot volume, spot diameter 200 μm, center-to-center 500 μm, 400 spots/cm^2	[45]

TABLE 11.2 Resume Table of Contact and Noncontact Bioprinting Techniques Used for High Throughput Screening Research Studies Performed in Different Research Fields *(cont.)*

Technology Used	Research Field	Equipment	Platform	Other Relevant Details	References
Pin-based system	Cell micro-environment microarrays	SpotArray 24 (PerkinElmer) equipped with TeleChem Stealth SMP3 pins (ArrayIt)	Microscope slides coated with polyacrylamide gel	**Ct:** ECM proteins (human collagen type I, III, IV, V, fribronectin, laminin, matrigel, vitronectin), growth factors (human recombinant FGF2, BMP-4, TGFβ, mouse recombinant Wnt3a, Wnt5a), glycans or proteoglycans (human recombinant biglycan and bovine heparin sulphate) and small molecules (retinoic acid or SB203580). **Cells:** postseeded hPSCs (HUES9) **C:** endpoint imaging with micro array scanner or live imaging with confocal microscope **Pp:** spot diameter 150 μm, center-to-center 450 μm, 1600 spots (320 conditions)/slide	[44]
Pin-based system	Stem cell research	SpotBot 3 contact microarrayer with four pinheads (Arrayit, CA).	glass slides functionalized with 3-(trimethoxysilyl) propylmethacrylate (TMSPMA)	**Ct:** combinatorial screen of gelatin, fibronectin, laminin, and osteocalcin. Printed hydrogels UV cross-linked ((2-hydroxy-1-(4-(hydroxyethoxy) phenyl)-2-methyl-1-propanone, Irgacure 2959 photoinitiator). Cells were cultured with different medium. **Cells:** printed hMSCs **C:** fluorescence and optical microscopy **Pp:** spot height 75 μm, 400 spots/slide	[9]
Pin-based system	Reverse transfection siRNA arrays	ChipWriter "Compact" and "Pro" (Bio-Rad Laboratories equipped with solid pins (PTS 400 or PTS 600)	Chambered coverglass tissue culture dishes (LabTeks)	**Ct:** two genome-wide siRNA screens (constitutive protein secretion or mitosis and cell-cycle progression) **Cells:** postseeded HeLa cells or human primary fibroblasts cells **C:** immunofluorescence or time-lapse microscopy **Pp:** spot diameter 270 or 400 μm; center-to-center of 900 μm, 1125 μm, 1500 μm or 2250 μm; 600, 384, 216 or 96 samples/LabTek slide	[18,46]
Dip-pen nanolithography	Proteins screening	NanoeNabler™	Octyltrichlorosilane (OTS) treated glass slides	**Ct:** cellular attachment and spreading to different proteins (laminin and collagen) **Cells:** mouse muscle myoblasts **C:** fluorescence **Pp:** spot diameters 6–9 μm, center-to-center of 18–42 μm in 6-μm increments, 16,800 spots/slide	[50]

(Continued)

TABLE 11.2 Resume Table of Contact and Noncontact Bioprinting Techniques Used for High Throughput Screening Research Studies Performed in Different Research Fields *(cont.)*

Technology Used	Research Field	Equipment	Platform	Other Relevant Details	References
Contact Techniques *(cont.)*					
Dip-pen nanolithography	ECM microenvironment study	NanoInk, Inc.	Epoxy functionalized glass surfaces	**Ct:** fibronectin **Cells:** fibroblasts (NIH 3T3) and mouse muscle myoblasts **C:** fluorescence **Pp:** spot diameter 6–8 μm, center-to-center 10–15 μm	[51]
Dip-pen nanolithography	Drug discovery – drug delivery	NanoInk, Inc.	Poly-L-lysine coated glass sides	**Ct:** valinomycin and taxotere® (Docetaxel). **Cells:** (NIH 3T3) **C:** bright-field and fluorescence **Pp:** four-by-four spot patterns, center-to-center 15 μm, pattern spacing 35 μm	[12]
Noncontact Techniques					
Piezoelectric-assisted inkjet	Biomaterials development	Microdrop inkjet printer (Microdrop Technologies) and sciFLEXARRAYER S5 (Scienion AG)	Glass slide with sucrose mask treated with tride-cafluoro-1,1,2,2-tetrahydrooctyl)-1-dimethylchlorosilane and coated with 3-(trimethyoxysilyl) propylmethacrylate	**Ct:** two monomers (acrylate/acrylamide) and one photoinitiator (1-hydroxycyclohexyl phenyl Ketone) **Cells:** human erythroleukaemic (K562) and human cervical cancer cells (HeLa) **C:** cell adhesion **Pp:** gradient lines with 5 × 0.5 mm, 84 gradients/slide	[14]
Piezoelectric-assisted inkjet	Biomaterials development (Hydrogel array for cell culture)	Microdrop MDE-401 printer with AD-K-501 autodrop pipette	Glass slides	**Ct:** 18 monomers and a redox initiator (ammonium persulfate) **Cells:** human cervical cancer (HeLa), mouse fibroblast (L929), mouse melanoma (B16F10) and human embryonic kidney (HEK-293T) cells **C:** cellular adhesion, proliferation and thermal release. **Pp:** 20 nL volume, spot diameter ~300 μm, center-to-center ~800 μm, 2436 conditions/slide	[58]
Piezoelectric-assisted inkjet	Drug discovery (miniaturized toxicity analysis)	Dimatix inkjet printer (DMP-2800, Fujifilm) and microarray spotter (piezoarray™, Perkin–Elmer)	Glass slides coated with PEG-silane	**Ct:** collagen and poly-L-lysine **Cells:** embryo mouse fibroblasts (NIH-3T3 ATCC) and primary rat hepatocytes **C:** cell viability, cytochrome P450 activity measurement **Pp:** 10 pL and 330 pL volumes, spot diameter 100–180 μm	[59]
Piezoelectric-assisted inkjet	Drug discovery	Dimatix inkjet printer (DMP-2800, Fujifilm)	Antireflecting silicon oxide substrates	**Ct:** ketoconazole, erythromycin, TS51 and TS28 at different concentrations **Cells:** n/a **C:** luminescence (luciferase) **Pp:** 480 pL final volume, spot diameter 155 ± 31 μm, center-to-center ~500 μm, 72 conditions/slide	[60]

TABLE 11.2 Resume Table of Contact and Noncontact Bioprinting Techniques Used for High Throughput Screening Research Studies Performed in Different Research Fields *(cont.)*

Technology Used	Research Field	Equipment	Platform	Other Relevant Details	References
Piezoelectric-assisted inkjet	Diagnosis	Dimatix inkjet printer (DMP-2800, Fujifilm)	Polyimide film.	**Ct:** gold nanoparticle, poly(amic acid) **Cells:** n/a **C:** electrochemical detection **Pp:** 56 units of eight-electrode array	[6]
Piezoelectric-assisted inkjet	Drug discovery	Piezorrayer (PerkinElmer)	Custom-made slide-size microwell array and PDMS pillar array	**Ct:** library of 320 natural compound and verapamil **Cells:** breast cancer cells (MCF-7) **C:** cell viability (fluorescence) **Pp:** 2 nL, 2100 conditions/slide	[5]
Piezoelectric-assisted inkjet	Growth factor screening	(MicroFab, Inc., Plano, TX) mounted to *xyz* motion control stages (Aerotech, Inc., Pittsburgh, PA).	Fibrin-coated glass slides	**Ct:** FGF-2 [61], bone morphogenetic protein-2 (BMP-2) [62] **Cells:** MG-63 [61], mouse primary muscle derived stem cells and mouse muscle myoblasts [62] **C:** Cell proliferation [61], immunocytochemistry and alkaline phosphatase activity [62] **Pp:** 14 ± 4 pL volumes, spot diameter ~75 μm, 4 squares (2,12, 22, and 32 spots)/slide	[61,62]
Piezoelectric-assisted inkjet	Drug discovery	microarray spotter (Samsung Electro-Mechanics Company, Ltd., South Korea).	Custom made glass-slide size microwell and micropillar chips	**Ct:** 24 anticancer drugs **Cells:** U251 brain cancer cell line and three primary brain cancer cells from patients **C:** cell staining (fluorescence) **Pp:** 30 nL, 532 conditions/chip	[93]
Thermal-assisted inkjet	Drug discovery (titration studies and IC$_{50}$ determination)	Hewlett-Packard (HP) D300 Digital Dispenser (Tecan, Durham, NC)	96 and 384 microtiter plates	**Ct:** nine chemical compounds **Cells:** HeLa Cell Line (CellSensor MMTV-bla) **C:** enzyme assay, reporter gene assay and cytotoxicity assay **Pp:** 13 pL to 0.26 μL volumes	[65]
Thermal-assisted inkjet	Tissue engineering (organ printing)	Hewlett Packard (HP 550C) printer	Glass coverslips coated with soy agar gel or collagen gel.	**Ct:** printing of different cell line **Cells:** Chinese Hamster Ovary (CHO) and embryonic motoneuron cells [67], and primary embryonic hippocampal and cortical neurons [68] **C:** cell viability and immunostaining **Pp:** n/a	[67,68]
Thermal-assisted inkjet	Drug discovery	Hewlett Packard (HP 5360) compact disc printer	Glass slide with bioprinted soy agar spots	**Ct:** three layers printed: (1) soy agar and cells, (2) alginate and (3) antibiotics and CaCl$_2$ **Cells:** printed *Escherichia coli* (*E. coli*) DH5α cells **C:** optical and fluorescence microscopy **Pp:** 180 ± 26 pL volume, spot diameter 150–240 μm, center-to-center 420 μm, 123 dots/slide	[66]

(Continued)

TABLE 11.2 Resume Table of Contact and Noncontact Bioprinting Techniques Used for High Throughput Screening Research Studies Performed in Different Research Fields *(cont.)*

Technology Used	Research Field	Equipment	Platform	Other Relevant Details	References
Noncontact Techniques *(cont.)*					
Accoustic-assisted bioprinting	Stem cell research	Custom made equipment	Glass substrate	**Ct:** cells encapsulated in sucrose–dextrose solutions **Cells:** mESCs, fibroblasts, hepatocytes (AML-12), human Raji cells, and cardiomyocytes (HL-1) **C:** optical and fluorescence microscopy **Pp:** spot size 37 μm, 1 to 10,000 droplets/second	[70]
Laser-assisted bioprinting	Tissue engineering (organ printing)	Custom made equipment	Quartz disk	**Ct:** sodium alginate, nanosize hydroxyapatite (HA) and cells **Cells:** human endothelial cells (EA.hy926) **C:** optical and fluorescence microscopy **Pp:** droplets of 50–80 μm	[75]
Valve-assisted bioprinting, liquid handling system	Stem cell fate research.	Microsolenoid valve (MicroSys™ 5100-4SQ)	Borosilicate glass slides coated with poly(styrene-*co*-maleic anhydride) or methyltrimethoxysilane	**Ct:** two small molecules: tretinoin and FGF-4; **Cells:** mouse embryonic stem (46C) encapsulated inside alginate; **C:** results detection: immunofluorescence assay; **Pp:** 60 nL volume, spot diameter <800 μm, 560 spots/slide.	[13]
Valve-assisted bioprinting, liquid handling system	Cell-based drug screening.	Microsolenoid valve (MicroSys™ 5100-4SQ)	Borosilicate glass slides coated with poly(styrene-*co*-maleic anhydride) or methyltrimethoxysilane	**Ct:** two small molecules (hypoxia dysregulators): carbobenzoxy-L-leucyl-L-leucyl-L-leucinal (MG-132, hypoxia inducer) and 2-methoxyestradiol (2ME2, hypoxia inhibitor); **Cells:** human pancreatic tumor cells (MIA PaCa-2) encapsulated inside alginate; **C:** results detection: immunofluorescence assay; **Pp:** 60 nL volume, spot diameter ~800 μm, center-to-center ~1200 μm, 560 spots/slide.	[77]
Valve-assisted bioprinting, liquid handling system	Drug discovery (cancer cell migration assay)	liquid handler (SRT Bravo, Agilent)	96 well plate	**Ct:** two natural compounds (fisetin and quercetin), polyethylene glycol (PEG) and DEX **Cells:** two metastatic breast cancer cells lines (MDA-MB-231 and MDA-MB-157) **C:** fluorescence **Pp:** 96 conditions/microtiter plate	[78]
Valve-assisted bioprinting, liquid handling system	Drug discovery	Microsolenoid valve (MicroSys 5100-4SQ)	glass slides	**Ct:** nine compounds and their metabolites **Cells:** human breast cancer cells (MCF-7) **C:** fluorescence **Pp:** 60 nL volume, 1080 conditions/slide	[94]

TABLE 11.2 Resume Table of Contact and Noncontact Bioprinting Techniques Used for High Throughput Screening Research Studies Performed in Different Research Fields *(cont.)*

Technology Used	Research Field	Equipment	Platform	Other Relevant Details	References
Valve-assisted bioprinting, liquid handling system	Cancer research	microsolenoid valve system (TechElan, G100-150300NJ) controlled with a pulse generator (Hewlett Packard, 8112A)	growth factor-reduced Matrigel™ substrate	**Ct:** cell-encapsulated droplets **Cells:** human epithelial ovarian cancer cells (OVCAR-5) and human fibroblast cells (MRC-5) **C:** optical and fluorescence microscopy **Pp:** spot diameter 510 ± 26 μm	[7]
Valve-assisted bioprinting, liquid handling system	Spheroid aggregates formation	Solenoid valve (VHS 25+ Nanolitre Dispense Valve, Lee Products Ltd)	60 well Terasaki plates (653102, Greiner bio-one)	**Ct:** medium or cells suspended in medium **Cells:** human embryonic kidney cell line (HEK293) or hESCs **C:** optical and fluorescent microscopy **Pp:** droplet volume 0–1.5 μL, 110 droplet/well	[79]
Valve-assisted bioprinting, liquid handling system	Drug discovery	microarray spotter (Omnigrid Micro, Digilab Inc., Holliston, MA) equipped with ceramic tip (Digilab)	Borosilicate glass slides coated with 3-aminopropyl-triethoxysilane and poly(styrene-comaleic anhydride)	**Ct:** two known antifungal agents (fluconazole and amphotericin B) [80], combinatorial screening with 28 antifungal compounds in the presence of immunosuppressant tacrolimus [82] **Cells:** *Candida albicans* yeast cells encapsulated in alginate or collagen **C:** optical and fluorescent microscopy **Pp:** 30–50 nL spot volume, spot diameter 400–700 μm, center-to-center 1.2 mm, spot height 140–150 μm, 750–1200 conditions/slide	[80-82]
Hybrid microfluidic pin driven printer	Biomaterial dispensing	Microfluidic impact printer equipped with dot-matrix printer head (Panasonic KX-P1150)	planar PDMS substrates	**Ct:** DMSO and aqueous solutions, fluorescent proteins (BSA–FITC, BSA–TR, and streptavidin–MB), agarose and cells **Cells:** human leukemic monocyte lymphoma cell line (U937) **C:** optical microscopy for droplet characterization **Pp:** spot diameter ~80 μm	[83]

*Conditions tested, Ct; printed properties, Pp; characterization, C; not applicable, n/a.

4.1 Contact Technologies

Contact technologies rely on the transfer of the materials from small tips to a destination platform upon physical contact (Fig. 11.1). Frequently, two different techniques, namely, pin-based systems and dip-pen nanolithography (DPN), are used. These techniques are mostly used to produce microarrays due to their high precision and low droplet volumes. The advantages and disadvantages for the pin-based and DPN systems are reported in (Table 11.3).

4.1.1 Pin-Based Systems

Technologies based on the transfer of liquids from a source plate to a destination substrate generally employ the application of pin-based technology (Fig. 11.1a–c). The liquid transfer involves the dipping of a pin or pin array into the solution of a donor plate (e.g., 96 or 384 microtiter plates). Transfer to the destination surface, normally a glass slide, occurs thereafter upon physical contact between the droplet formed at the pin's tip and the surface. The droplet formation

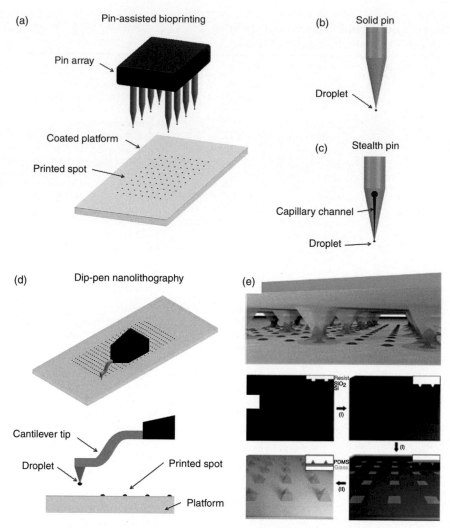

FIGURE 11.1 **Schematics of the different contact bioprinting technologies.** (a) Pin-based systems, which can be equipped with (b) solid or (c) stealth pins; (d) DPN; and (e) PPL. Pin-based systems are equipped with a pin or pin array that might be assembled according to the array spacing to be printed. Solid and stealth pins are commercially available allowing the transport of different volumes. Normally, with solid pins a limited amount of solution can be transferred from a donor plate to the microarray surface. Alternatively, a stealth pin allows the transfer of a larger volume of solution due to the presence of a capillary channel making it possible to print large-sized single droplets or multiple small droplets. DPN is composed of a cantilever tip that allows the transfer down to attolitter scale volume. PPL is composed of an array of pins custom-designed and manufactured with hard PDMS. PPL allows the printing in parallel fashion combinatorial microarrays. *(Fig. 11.1e reprinted with permission from Eichelsdoerfer et al. [52].)*

observed at the tip of the pin after being withdrawn from the donor plate depends on the capillary and fluid surface tension forces [39]. Several pin types (solid or stealth) are commercially available, allowing the manipulation of different volumes with the possibility to generate multiple droplets per loading with the aid of a small capillary filled channel (Fig. 11.1a–c). Stealth and solid pin arrays also allow simultaneous droplets dispensing in order to improve microarray manufacturing time (Fig. 11.1a).

Pin-based technologies have been used mainly to produce microarrays for screening new biomaterials, chemical compounds, to investigate cellular microenvironment, or to perform reverse transfection. Biomaterial microarrays, manufactured by pin bioprinting technology, have been

developed aiming at a faster screening of new biomaterials for TE or cell therapy applications and to better understand polymer–cell interactions [40–42]. Anderson et al. [40] manufactured a biomaterial microarray composed of 25 different monomers combined with one photoinitiator (576 conditions) to evaluate human embryonic stem cells (hESCs) and mouse muscle myoblast cells attachment, proliferation, and differentiation. The surface chemistry of these microarrays composed of different biomaterials were characterized by means of static X-ray photoelectron spectroscopy, time-of-flight secondary ion mass spectrometry, and water contact angle [43]. In a similar study, 22 acrylate monomers printed in a combinatorial fashion were used to prepare biomaterial microarrays and investigate the clonal

TABLE 11.3 Comparison Between Bioprinting Technologies

Technologies	Advantages	Disadvantages	Resolutions/Density
Contact Techniques			
Pin-based bioprinting	High density and high precision; easy transfer; fewer blockages.	Need of many pins or pin array to meet high throughput; high standard deviation between printed spots can occur; mechanical stress on the pin upon physical contact.	150–400 μm; 16 spots/mm^2
Dip-pen nanolithography	Attoliter scale volume control; nano-resolution and high density.	Restricted detection limits; lack of massive multiplexing capabilities.	25–200 nm; 1M spots/mm^2
Noncontact Techniques			
Piezoelectric-assisted systems	Fair precision; picoliter to microliter volumes; volume control; additive operation.	Obstruction and blockage; sensitive to bubbles and precipitation; limited to low-viscosity materials.	15–50 μm; 80 spots/mm^2
Thermal-assisted systems	Low investment possible;	Obstruction and blockage; thermal degradation of the material in direct contact with the thermal film resistor; limited to low-viscosity materials.	10–240 μm; 50 spots/mm^2
Acoustic-assisted systems	Thermal and mechanical stresses are normally absent; optimal for sensitive biomaterials and cells.	Thermal effects must be considered when high power pulses are used.	240 nm to 5 μm; 500 spots/mm^2
Laser based	Printing of high viscosity materials; ultra HT; high resolution and density; additive operation.	Laser source might have adverse effect on cellular genetic material.	1–3 μm; 100 spots/mm^2
Valve-assisted systems	Nanoliter to microliter volumes;	Fluids/compounds are exposed to high pressures.	500 nm to 800 μm; 250 spots/mm^2

Reprinted with permission from Rodríguez-Dévora et al. [21].

growth of hESCs and iPSCs [41]. Results showed that within the large range of monomer combinations tested, the optimal substrate for hESCs and iPSCs colony formation had a moderate wettability and high acrylate content. Polymers frequently studied for the manufacturing of TE scaffolds, namely, chitosan, agarose, poly(ethylenimine), poly(ε-caprolactone) (PCL), poly(L-lactic acid) (PLLA), poly(2-hydroxyethylmethacrylate), poly(vinyl acetate), and poly(ethylene oxide), were used to manufacture microarrays with 135 binary mixtures [42]. Fetal skeletal cells, human skeletal stem cells (STRO-1 +), human preosteoblastic osteosarcoma cells (MG-63) an early osteoblast-like cells, and mature osteoblast-like (SaOs) cell lines were cultured on top of the polymer microarrays. A hit for the PLLA/PCL (20/80) polymer combination was selected due to the porous morphology, mechanical stability, and cell growth behavior of STRO-1+ cells. Based on this selection, 3D scaffolds were manufactured and evaluated *in vitro* and *in*

vivo. In vitro studies performed with PLLA/PCL scaffolds cultured with STRO-1+ cells showed a good cell attachment, viability, and osteogenic differentiation. Furthermore, *in vivo* precultured scaffolds presented significant new bone formation when compared to controls (cell-free scaffolds and empty defect) in a critical size femur defect in mice. Other microarrays have been developed to screen different cell microenvironments and are termed arrayed cellular microenvironments (ACMEs) [44]. These pin-printed microarrays were prepared by spotting ECM proteins (human collagen type I, III, IV, V, fribronectin, laminin, Matrigel, vitronectin), growth factors (human recombinant fibroblast growth factor-2 (FGF-2), bone morphogenetic protein-4 (BMP-4), transforming growth factor beta (TGFβ), mouse recombinant proteins (Wnt3a and Wnt5a), glycans or proteoglycans (human recombinant biglycan and bovine heparin sulfate), and small molecules (retinoic acid or SB203580) on a polyacrylamide gel-coated glass slide.

A deoxyribonucleic acid (DNA) microarray scanner and an automated confocal microscope were used to image human pluripotent stem cells (hPSCs) cultured ACMEs arrays, and cell adhesion and growth were quantified for the different printed conditions. With ACMEs technology it was already possible to identify candidate growth conditions that withstand maintenance and growth of hPSCs. Dolatshahi-Pirouz et al. [9] manufactured a 3D microarray with a combinatorial cell-laden gel approach to study the osteoinduction capability of the different printed gel microenvironments encapsulating human mesenchymal stem cells (hMSCs). The optimal 3D microenvironment gel condition or hit identified with the microarray capable of inducing the formation of mineralized tissue was reproduced on large macroscale gels and a good correlation with previously screen miniaturized conditions was observed.

Chemical compound microarrays composed of nanoliter-sized reaction droplets have been produced with pin-based bioprinters [3,45]. In this case, small-size chemical molecules generally dissolved in dimethyl sulfoxide (DMSO) are printed on a glass substrate. Chemical microarrays with a nanoliter reaction volume per condition have been developed by Ma et al. [3] to screen up to 6600 conditions per glass slide. Plain or slides coated with poly-L-lysine (PLL) and printed with the chemical compounds and/or peptides were subsequently coated with aerosolized biological substances (e.g., samples containing enzymes), substrates or detection materials and finally analyzed by means of fluorescence signal. With these nanoliter-size reaction microarrays, it is possible to determine accurately the half maximal inhibitory concentration (IC_{50}) values for the different compounds screened, thus reducing the targets and reaction chemistries by 40-fold and the consumption of compounds over 10,000-fold when compared to a 384 well plate assay [3]. Gosalia and Diamond [45] investigated a 352-compound combinatorial library to identify caspases inhibitors using nanoliter reaction glass slide arrays with 400 spots/cm^2. With these nanoliter reactions being 10^3–10^4 smaller than normal HTS assays, it was possible to identify a compound that inhibited caspase 2, 4, and 6, thrombin, and chymotrypsin with a lower compound consumption ($<$1 nanomole).

Microarrays for reverse transfection were also developed and pin-based bioprinting used to dispense a previously prepared mixture of transfection cocktails and small interfering ribonucleic acids (siRNAs) [46]. Microarrays with up to 600 different conditions were spotted onto a chambered coverglass tissue culture dishes (LabTeks) with a siRNAs library prepared on a 348-well microtiter plate. HeLa cells and four other cell lines seeded on the microarrays were successfully transfected and immunofluorescence or time-lapse microscopy were used to analyze protein secretion, or cell cycle progression and mitosis, respectively. Similar microarrays were used to study gene function in HeLa cultured cells and the suppression of 49 endogenous genes was evaluated by early and late phenotyping characterized by means of time-lapse imaging, allowing the automatic identification of primary (e.g., prometaphase arrest) and secondary (e.g., apoptotic) phenotypes with an image processing software [18].

4.1.2 Dip-Pen Nanolithography

DPN is a technique that allows the direct printing of materials with a resolution generally below 100 nm [47,48]. With a similar principle to the previously mentioned pin-based systems, DPN is based on the physical transport and transfer of materials previously loaded on a cantilever scanning probe tip to a surface by contact (Fig. 11.1d) [47]. The transfer of the materials is controlled by physical phenomena, such as molecular diffusion and fluid flow, and chemical interaction phenomena between the processed materials, the tip, and the surface [47]. DPN has become a widespread technique because any atomic force microscope equipment can be used to print materials with this approach. Several materials, such as organic molecules, polymers, DNA, proteins, peptides, colloidal nanoparticles, metal ions, and other solutions, have been printed with DPN [48].

New commercially available DPN equipment, such as NanoEnabler™ [49,50] (BioForce Nanoscience) and NanoInk [12,51] (Nanoink, USA), have been used to print HTS microarrays. The NanoEnabler™ is equipped with special tips of individual pin-like or multiple pin-like heads to print materials. These tips are composed of a small material reservoir that is connected to the tip of the cantilever with a microchannel. Mei et al. [50] used the NanoEnabler™ equipment to produce multiple-component protein arrays with spot size between 1 μm and 10 μm. Microarrays were produced with laminin, collagen or the combination of both, and mouse muscle myoblast attachment and spreading was evaluated on the printed arrays with different spacing between the printed spots. Cell morphology was significantly affected by the spot-to-spot distance and spread cells were observed for smaller interspot distances, while less spreading was observed in areas with larger interspot distances.

The Nanoink system is equipped with diverse types of proprietary ink wells and dispensing tips. As an example, the M-Type ink wells are composed of 12 well reservoirs with a capacity of 1 μL solution that is driven by capillarity to the tip loading area [51]. An array of 12 cantilever tips (M-type) is then dipped into the ink wells loading area. The liquid transferred to the tips is subsequently printed on the destination glass substrate. Collins et al. [51] used the Nanoink system to print fibronectin and mixtures with polymers to study drug release substrates. Arrays with different dimensions were produced and the response of mouse fibroblasts (NIH 3T3) and mouse muscle myoblast cells to the printed microenvironment was evaluated. Calcein AM, Calcein Red AM, and quantum dots, used as drug models, were selectively delivered to a single or a limited amount of cells when these were in direct contact with the printed spots. This approach allows the screening of different conditions by exposing single cells to the different printed compounds. Kusi-Appiah et al. [12] used the Nanoink system to prepare subcellular-sized pattern microarrays composed of phospholipid multilayers encapsulating small molecules to investigate drug delivery and screening. With this microarray system, it was possible to selectively deliver the compounds to the cells that were in direct contact with the printed drug-loaded phospholipid spots. Furthermore, it was also possible to tune the dosage delivered to the cells, depending on the spot size [12].

In order to increase the manufacturing throughput, alternative dispensing tips or tip arrays with similar concept of the ones used for DPN have been developed. One example is a cantilever-free imprinting approach that allows the simultaneous dispensing of multiple droplets, which was developed by Eichelsdoerfer et al. [52]. This technology, termed polymer pen lithography (PPL), is composed of an array of pins manufactured with hard polydimethylsiloxane (PDMS). The PDMS is casted on an etched master array of pyramidal pits, prepared by conventional lithography microfabrication technique (Fig. 11.1e). PPL arrays can be used to print in a parallel fashion combinatorial microarrays of a wide range of materials, such as alkanethiols, polymers, lipids, and proteins, which have already been printed onto both hydrophilic and hydrophobic surfaces [52].

Other techniques with a similar concept to pin-based or DPN bioprinting (i.e., liquid transfer upon contact), normally termed microcontact printing techniques, generally involve a PDMS stamp with small features that resemble small pin arrays. These stamps are used to transfer compounds between surfaces [53,54]. Despite being termed as printing techniques they will not be included in this book chapter, as they are less automatable than the printing solutions just mentioned. More details about microcontact printing principles, stamps, inks, and applications can be found elsewhere [55–57].

4.2 Noncontact Technologies

Noncontact technologies, also termed inkjet technologies, are a second group of bioprinting technologies, used for the manufacture of HTS platforms, where no contact between the dispensing system and the target platform occurs (Fig. 11.2). Examples of these techniques are inkjet (thermal, piezoelectric, acoustic) bioprinting (Fig. 11.2a–e), laser-assisted bioprinting (LaBP) (Fig. 11.2f), and valve-assisted bioprinting (Fig. 11.2g), which are hereafter described. These noncontact techniques allow the dispensing of the different conditions on commercial or custom developed platforms, such as microtiter plates and other microwell platforms.

The advantages and disadvantages are reported in Table 11.2.

4.2.1 Piezoelectric-Assisted Bioprinting

Piezoelectric-assisted printers contain normally a cartridge that is composed of a small piezo ceramic element connected to a vibration plate. Upon electrical stimulation, the plate deforms and applies mechanical stress to the fluid present in a small chamber where the ceramic element is located, thus inducing the ejection of the formed droplet from the nozzle (Fig. 11.2b). The piezo-assisted technology is commonly used in cartridges that equip industrial printers. The application of this technology has been investigated over several years and is currently applied for the manufacturing of microarrays for new biomaterial development, for drug discovery, diagnosis, and to investigate cell fate.

Polymer gradient arrays with 84 different combinations of two monomers (one acrylate and 22 acrylamide) were manufactured with a piezoelectric-assisted microarrayer (sciFLEXARRAYER S5, Scienion AG) and cell adhesion was evaluated [14]. Hydrogel arrays containing 2436 conditions were also manufactured by piezoelectric bioprinting, combining 18 different monomers with a redox initiator [58]. Different cell lines were seeded on the prepared hydrogel arrays and cell adhesion, proliferation, and thermal-induced detachment of the cells from the hydrogels were evaluated in order to obtain an optimal hydrogel formulation for cell culture. Protein microarrays were prepared with a simple and reliable process by printing collagen and PLL on top of poly(ethylene glycol) (PEG) functionalized glass slides [59]. These microarrays, cultured with primary rat hepatocytes, showed to be a simple method for drug metabolism screening by cytochrome P450 (CYP450) characterization. Luminometric arrays, capable of target phase I enzymatic drug metabolism (CYP3A4 enzyme) combined with ketoconazole and erythromycin inhibitors, were used to validate the array [60]. Furthermore, the inhibitory potential of two CYP450 known inhibitors (TS51 and TS28) was tested at different concentrations on CYP3A4 arrays and the IC_{50} obtained by the enzymatic activity expressed by luminescence.

Microarrays to investigate the effect of growth factors on cells were also prepared by piezoelectric-based printing [61,62]. FGF-2 was printed on fibrin-coated glass slides. Arrays with different spot numbers were seeded with MG-63 cells and their proliferation evaluated by time-lapse microscopy images [61]. Bone morphogenetic protein-2 (BMP-2) arrays were printed and cultured with mouse primary muscle-derived stem cells and mouse muscle myoblast cells. Alkaline phosphatase activity and immunocytochemistry showed the effect of BMP-2 onto cell differentiation toward the osteogenic lineage [62].

Piezoelectric-assisted bioprinting was used for the manufacturing of HTS microarrays for reverse transfection by

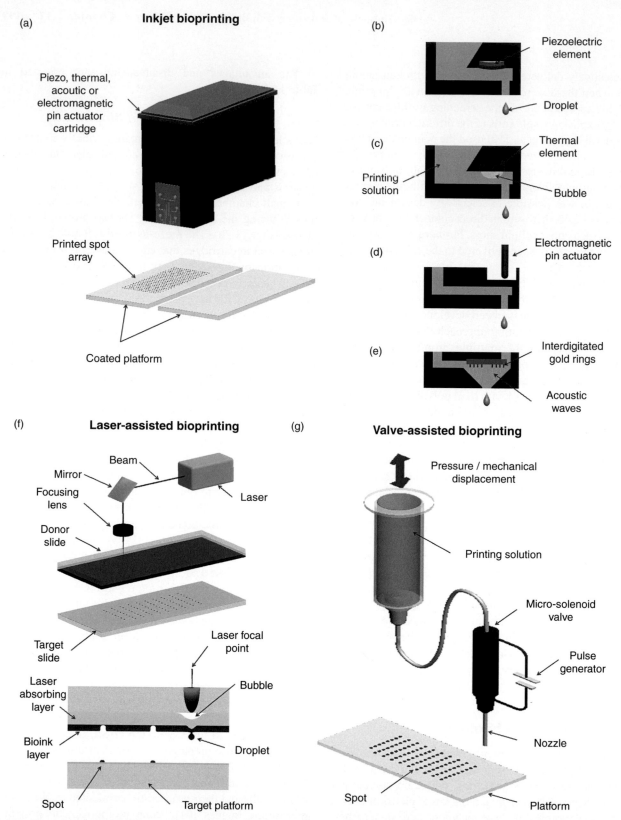

FIGURE 11.2 **Schematics of different noncontact bioprinting technologies.** (a) Inkjet bioprinting; (b) piezo-assisted cartridge; (c) thermal-assisted cartridge; (d) electromagnetic pin actuator cartridge; (e) acoustic-assisted cartridge; (f) laser-assisted system; and (g) valve-assisted system. An inkjet printing system is normally equipped with a cartridge that can include different dispensing principles (piezoelectric element, thermal film resistor, electromagnetic pin actuator, or acoustic ejector). A piezo-assisted cartridge contains a piezoelectric element that when electrically stimulated, deforms and induces mechanical stress to the nearby fluid inducing the ejection of the formed droplet at the nozzle tip. A thermal-assisted cartridge contains a thin film resistor that induces high temperature to form a gas bubble at the interface with the fluid inducing the formation and ejection of a droplet at the nozzle's tip. The electromagnetic pin actuator cartridge induced the ejection of the material, by action of pin displacement, without direct contact with the processed fluid, thus avoiding contamination. Acoustic-assisted cartridge is composed of a piezoelectric substrate that contains interdigitated gold rings; when stimulated with a sinusoidal electrical signal at a resonance frequency the rings induce acoustic waves that promote the ejection of small picoliter-sized droplets. LaBP is composed of a laser that when focused onto a metallic absorption layer induces evaporation at the focal points, resulting in the formation of a bubble on a subjacent layer that induces the ejection of the material toward a target platform. Valve-assisted or automated liquid dispensing systems generally employ a microsolenoid valve controlled at high frequency by a pulse generator, allowing an accurate dispensing of pressurized fluids.

printing two plasmids (encoding Venus or mCherry fluorescent protein), ECM proteins (fibronectin or type I collagen), and other materials necessary for the transfection [63]. A high density array with a 50 μm spot diameter and 150 μm spot-to-spot distance was printed on a glass substrate, previously grafted with PEG. HeLa cells cultured on the microarrays adhered to the spot regions and were transfected when in contact with the surface of each printed spot. Proteome screening microarrays were optimized with piezoelectric printing by dispensing fluorescently labelled proteins over previously printed antibody microarray spots [64]. With the developed approach, fewer amounts of protein extracts are necessary when compared to the conventional antibody microarrays screening approach that is normally performed by flooding the antibody microarray surface. Electrochemical array sensors for the detection of interleukin-6 as a cancer biomarker have been produced by piezo-assisted bioprinting [6]. Gold nanoparticle arrays composed of eight electrodes were printed on a polyimide substrate and used to detect interleukin-6 in calf serum. With the manufactured array it was possible to measure with a clinically relevant detection limit (20 pg/mL).

4.2.2 Thermal-Assisted Bioprinting

Thermal-assisted bioprinting is composed of a thin film resistor that induces high temperature to form a gas bubble at the interface with the fluid that is dispensed (Fig. 11.2c). The confined liquid to vapor phase conversion of the dispensed fluid induces the formation and ejection of the droplet at the nozzle's tip.

The Hewlett-Packard (HP) inkjet thermal printing technology (HP D300, Tecan, Switzerland) was adapted to accurately dispense small molecules dissolved in DMSO [65]. This equipment is composed of a disposable dispense head cassette with eight channels that allow an accurate dispensing of volumes down to 13 pL. Furthermore, it is fully optimized for titration screening of small molecules in 96 or 384 microtiter plates for drug discovery [65]. Rodríguez-Dévora et al. [66] also used a desktop HP thermal-assisted printer (HP D5360 CD printer) to produce HTS arrays for drug discovery. Arrays were manufactured layer-by-layer on glass slides by printing a first layer of soy agar containing E. coli cells, followed by a second layer of alginate and a final layer of calcium chloride ($CaCl_2$) and antibiotics (penicillin/streptomycin, antimycotic, and kanamycin sulphate).

The utilization of thermal inkjet printers to process biomaterials or cells in HTS studies is still limited in the literature. This is probably related to the concerns about thermal degradation of the biomaterials in the region near the thermal resistor. Furthermore, cell viability is to some extent affected after printing, as demonstrated by Xu et al. [67–69]. A significant amount of cells printed with a modified thermal-assisted HP desktop printer (HP 550C) were damaged during the printing process reaching 8% of cell death for the conventionally robust Chinese hamster ovary cells [67], 11% for insulin-producing beta cells (beta-TC6), and 15% for neural cells [68].

4.2.3 Acoustic-Assisted Bioprinting

Acoustic-assisted bioprinting is equipped with an ejector, composed of a piezoelectric substrate that contains interdigitated gold rings. When stimulated with sinusoidal electrical signal at a resonance frequency, the gold rings induce the ejection of the fluid [70]. The induced acoustic waves follow the ejection path and a small picoliter-sized droplet is formed at the focal point between the air and the fluid at the nozzle's exit (Fig. 11.2e). The advantage of this technology is that the fluid is ejected without inducing thermal and mechanical stresses generally observed in other inkjet techniques, being optimal for sensitive biomaterials and cells. Acoustic-assisted bioprinting already allows the manufacturing of microarrays with a single cell per spot and in the future may also allow to perform research studies at a single cell level, possibly improving the understanding of aspects, such as stem cell differentiation, by performing single-cell DNA and ribonucleic acid (RNA) isolation [70].

New microfluidic devices using surface acoustic waves were developed by Collins et al. [71]. In this technology, the waves are generated at the interface of a patterned interdigital transducers manufactured on a piezoelectric substrate, allowing the dispensing of picoliter-scale droplets on-demand. The application of such a surface acoustic waves device might allow the dispensing at high frequency of encapsulated cells or other biomolecules in HTS studies without using surfactants, emulsifiers, or other fluid treatments [71].

4.2.4 Laser-Assisted Bioprinting

LaBP was developed as an alternative to the cartridge-based inkjet systems and is mainly used for the manufacturing of tissue-engineered scaffolds [72]. LaBP is based on a laser-induced forward transfer technique in which a laser, when focused onto a metallic absorption layer that evaporates at the focal points, induce the formation of a bubble on a subjacent layer that is normally composed of a cell-containing hydrogel precursor (Fig. 11.2f). The bubble formation triggers the ejection of a droplet that is propelled toward a lower collector (hydrogel coated slide) [73]. Several other similar techniques, termed absorbing film-assisted laser-induced forward transfer, biological laser processing, matrix-assisted pulsed laser evaporation direct writing, and laser-guided direct writing, were developed with similar principles and were reviewed by Schiele et al. [74]. With LaBP technology it is possible to dispense high cell densities and very small volumes down to a few hundred femtoliters [72]. New-generation LaBP equipment have been developed to produce arrays in a high throughput manner of sodium alginate, nanosized hydroxyapatite, and human endothelial cells with droplet sizes of 70 μm [75].

Since this technology allows an accurate deposition with high resolution of high viscosity materials, it might be used in the future for the manufacturing of high throughput cell microarrays, polymer microarrays and/or drug microarrays to screen a more widespread range of biomaterials, stem cell fate, or for more complex drug discovery applications.

4.2.5 Valve-Assisted Bioprinting Techniques

Other noncontact systems are based on automated liquid dispensing equipment generally using a microsolenoid valve (Fig. 11.2g). These normally closed valves are controlled with a pulse generator and can open at high frequency allowing an accurate dispensing of pressurized fluids with the volume dispensed varying from the nanoliter to the microliter range [39]. These valve systems are highly sensitive to variations of rheological parameters (e.g., fluid viscosity) and environmental parameters (e.g., temperature) [76]. Fernandes et al. [13,77] used a noncontact microarrayer equipped with a high-speed microsolenoid valve that allowed an accurate control of the fluid dispensing with a dynamic range of 50 nL to 8 μL (MicroSys 5100-4SQ). Microarrays were developed to evaluate stem cell fate. Two glass slides were used: (1) one previously coated with poly(styrene-co-maleic anhydride) (PSMA) was used to prepare cell microarrays; (2) the other one coated with methyltrimethoxysilane was used to manufacture small-molecule microarrays. Mouse embryonic stem cells (mESCs) mixed with alginate were printed on top of previously dispensed PLL/BaCl$_2$ spots (560 spots/slide). Alginate spots with a diameter below 800 μm allowed the mESCs cells to be maintained in their undifferentiated state. To test the mESCs pluripotency, a microarray with 60 nL spots of all-trans retinoic acid (RA) and fibroblast growth factor-4 (FGF-4), dispensed at different concentrations, was placed in contact with the cell microarray. Overnight stamping with a mixture of RA and FGF-4 resulted in an irreversible cell differentiation and a consequent reduction of the expression of pluripotency markers [13].

A similar study, performed by the same authors [77], aimed at the development of an HTS assay to evaluate cellular protein levels by means of immunofluorescence. In this case, human pancreatic tumor cells were encapsulated in alginate and chemically stimulated with a microarray spotted with 2-methoxyestradiol and carbobenzoxy-L-leucyl-L-leucyl-L-leucinal, which are known dysregulators of the hypoxia-inducible factor. Lemmo et al. [78] developed a cancer cell migration assay on a 96-microtiter plate by exploiting aqueous biphasic principles. Aqueous solutions of PEG and dextran (DEX) formed two distinct phases and cell migration from the PEG phase to the DEX region was evaluated in order to understand the metastatic progression

of cancer. Two natural compounds (fisetin and quercetin) were screened with this assay and the cell migration evaluated. Cells were stained after 48 h in culture and the migration was calculated and compared to controls (without the compounds). Results showed that the migration of the cells was inhibited by the two compounds at the tested concentration highlighting a high robustness of the assay. Xu et al. [7] used a microsolenoid valve system (TechElan, G100-150300NJ) assisted with nitrogen gas pressure and controlled with a pulse generator (Hewlett Packard, 8112A) to print cell-encapsulated droplets on top of a growth factor-reduced Matrigel™ substrate. Micronodules similar to those found in ovarian cancer were replicated in vitro by printing human epithelial ovarian cancer cells (OVCAR-5) and human fibroblast cells (MRC-5). The printed 3D cell aggregates were compared to aggregates prepared by manual pipetting and growth kinetics remained unchanged. A custom assembled bioprinter with a valve-assisted system was used to print hESCs to form spheroid aggregates [79]. The printer, composed of two independent channels controlled by two solenoid valves, was used to produce in a first step a gradient of droplets of medium containing cells followed by the printing of the inverse gradient of medium in order to produce droplets with equal volume. After printing, the plates containing the printed droplets with cells were inverted and the formation of the spheroids induced by gravity was observed in the lower region of the hanging droplets. This simple approach allowed the production of cell spheroids with different dimensions depending on the initial number of cells printed. Immunofluorescence staining revealed the presence of a pluripotency marker on the printed hESCs indicating that the process did not affect the pluripotency of the cells.

Biofilm microarrays from antifungal drug discovery were developed and prepared by means of microsolenoid valve-assisted bioprinting (Omnigrid Micro, Digilab Inc., Holliston, MA) [80–82]. These microarrays were prepared by printing C. albicans yeast cells encapsulated in collagen [80,81] or alginate [82] on borosilicate glass slides previously coated with 3-aminopropyltriethoxysilane and PSMA. For cell-laden alginate studies, a previous printing step of PLL was required to ensure a stable attachment of alginate gels while bound to PSMA-coated glass slides [82]. These microarrays were validated with two known antifungal agents (fluconazole and amphotericin B) [80] and further used to perform a combinatorial screening of 28 antifungal compounds in the presence of immunosuppressant tacrolimus (FK506) [82].

To overcome most of the limitations commonly observed in the previously described noncontact bioprinting techniques, an alternative technique, termed microfluidic impact printer, was developed [83] (Fig. 11.2d). This printer is composed of a low-cost PDMS disposable and interchangeable microfluidic cartridge attached to a dot-matrix printer head (Panasonic KX-P1150). By electromagnetic pin actuator, the cartridge channel is deformed and induces the ejection of the material to be dispensed without direct contact with the printer head, thus avoiding contamination. Furthermore, this bioprinting platform allows the dispensing, with high accuracy, of aqueous and nonaqueous fluids.

5 HTS PLATFORMS

Diverse platforms have been used to perform HTS in diverse fields of research (Table 11.4). The most commonly used platforms in research laboratories in academia or industries are still the microtiter plates. Yet, with the advent of new technologies capable of increasing the throughput, microtiter platforms are gradually being replaced. New-generation platforms, such as microarrays, generally involve lower amounts of compounds, lower fabrication costs, ultrathroughput and higher speed of analysis [3,19]. Despite these advantages, the equipment generally used for the preparation and analysis of these platforms are still expensive. Although most of the commonly used platforms for HTS were developed for *in vitro* screening, a new gen-

eration of implantable platforms have also been developed allowing the increase of the throughput of *in vivo* screening. The *in vitro* and *in vivo* HTS platforms are hereafter described and some examples of recently developed solutions are presented.

5.1 *In Vitro* HTS Platforms

Different microtiter plates have been developed and have become a gold standard applied in pharmaceutical and biotech companies, as well as in research laboratories around the world [25]. These plates are available in several well configurations (e.g., flat and round bottom) and number (e.g., 6, 24, 48, 96, 384, 1536, and 3456). Despite the large number of wells of commercially available microtiter plates, a solution with 9600 wells has also been developed but not commercialized [84]. Microtiter plates are generally associated with a 2D screening system, since it generally involves a monolayer sheet of cells cultured on each well. New-generation plates allow the manufacture of 3D cell aggregates by means of hanging drops technology, which was developed as an alternative to 2D (Fig. 11.3) [85–87].

TABLE 11.4 Comparison Between the Different Platforms Used for HTS

Platform	Advantages	Disadvantages	Number of Conditions
In vitro **Screening**			
Microtiter plates	Available in different well number and well types; optimized for robotic liquid handling devices; optimized for laboratory readout equipment.	Generally only monolayer sheet of cells (2D) are screened.	6, 12, 48, 96, 384, 1536, and 3456
Hanging drop plates	Improved mimicking of the *in vivo* microenvironment; 3D cell spheroids formation with high reproducibility; optimized for robotic liquid handling devices.	Manual pipetting possible but cumbersome; complete change of the medium not possible.	96 or 384 [85]
Microarrays	Fully standardize for proteomics and genomics screening.	Generally only 2D screening are performed; new 3D screening approaches are being developed.	100–17,000 [50]
In vivo **Screening**			
Vers 3D™	Available with different well number for different animal models; manufactured with biocompatible polymers; adapted for manual and robotic dispensing of the different conditions; appropriate for *in vitro* and *in vivo* screenings.	Possible cross-contamination of the screen conditions.	36 (mouse model) up to 4,096 (goat model) [22,23]
Array for biomaterials' *in vivo* response	Flat and flexible implantable films.	Possible cross-contamination of the screen conditions; biomaterials detachment from the array upon implantation.	36 (Wistar rats model) [97]

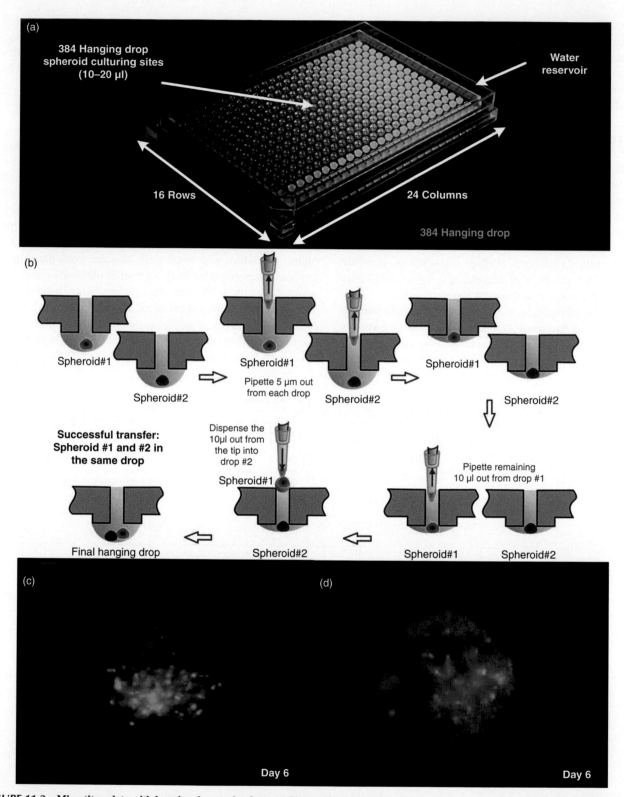

FIGURE 11.3 **Microtiter plate with hanging drop technology.** (a) Picture of the 384 spheroid microtiter-size plate; (b) cell culture technique showing spheroid formation and the process of spheroid transfer between two plates to form coculture spheroids; (c) example of side-by-side spheroid fusion obtained by combining CellTracker™ green and red labeled monkey kidney fibroblast (COS7) spheroids after 6 days of culture; and (d) example of concentric layers spheroid obtained after 6 days of culture prepared by coating green labelled spheroid with red labelled COS7 cells. *(Figure reprinted with permission from Hsiao et al. [86].)*

With these plates it is possible to easily prepare cell aggregates by dispensing 10–20 μL of a cell suspension in each hole that the plate contains. Within hours cells start to aggregate and normally only one aggregate is obtained per hole. With these plates, it is also possible to perform coculture spheroids by joining aggregates previously prepared (Fig. 11.3b–d) or by sequential preparation of layers to obtain fully coated spheroids with concentric layers (Fig. 11.3d) [86]. One of the limitations of the hanging drops technology is the impossibility to perform a complete change of the culture medium without removing the spheroids. Furthermore, despite it being possible to manually handle the content of each well, this approach might be cumbersome.

The 3D microspheroids normally obtained with this technology can better represent the native environment when compared to 2D cell culture. Yet, alternatives are being developed aiming at a better reproduction of the native environment possibly improving the results obtained. Custom microtiter plates have been developed by applying a novel technique termed cells-in-gels-in-paper (CiGiP) that comprise the stacking of 200 μm thick sheets of paper to allow the preparation of 3D cell cultures in a microtiter-like 96-well plate (Fig. 11.4) [88].

The sheets of paper were printed with wax using a commercially available printer (Phaser 8560DN, Xerox) and hydrophobic barriers were created outside the well areas.

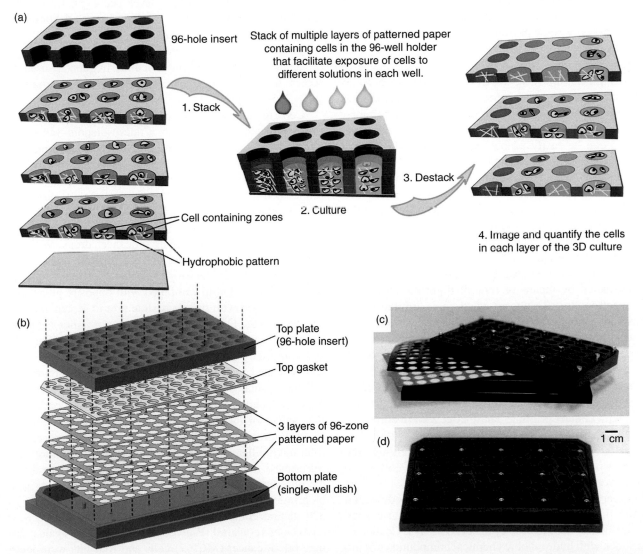

FIGURE 11.4 Microtiter plate applying the CiGiP technique. (a) Schematic representation of a custom 96-well plate developed for compound screening with 3D cell cultures composed of sheets of paper seeded with mammalian cells; (b) assembly of the CiGiP bottom plate with the three layers of paper; (c and d) photographs of a 96-well holder in Delrin before and after assembly. A 96-well holder contains three sheets of paper, separated by printed hydrophobic barriers, where mammalian cells are cultured. The sheets are stacked to form 3D cultures, which are exposed to different media containing the compounds to be screened. After culture, the 2D paper layers cultured with cells are detached and analyzed by microscope, fluorescent scanner, or a plate reader. *(Figure reprinted with permission from Deiss et al. [88].)*

After printing, the paper sheets were heated up to 120°C in order to impregnate them throughout their thickness with the printed wax. Cells were cultured with a hydrogel on the remaining hydrophilic zones of the printed paper and three layers were stacked between a single-well dish and a 96-hole insert to form the 3D cell culture constructs. After cell culture with a different medium, paper layers were detached and stained to evaluate cell viability in each zone of the paper-based layers. The CiGiP technique allows to perform screening in 3D with the convenience of being compatible with commercial laboratory instruments (e.g., automated liquid dispensing systems and plate readers). Furthermore, the simple process of layer separation allows an easier characterization of the different layers by means of optical or fluorescence analysis. Despite the advantages of this system, in some cases the diffusion of small molecules between wells can occur through the pores present in the wax areas.

Microarrays have been developed as an alternative to microtiter plates, applying different strategies in order to reduce the amount of chemical compounds, cells, and other biomolecules normally used in HTS. Furthermore, the small size of microarrays (normally with a glass-slide footprint) can be easily adapted to HCS equipment, such as microscopes with automated moving stage, allowing a fast acquisition of higher amounts of data. Polymer microarrays, produced by means of a microcontact printer and a piezoelectric inkjet printer, have also been developed in order to perform HTS and allow the development of new materials [89]. Printed microarrays produced with different acrylate-based monomers (95 monomers) were polymerized *in situ* and the chemical species present at the surface were evaluated by means of time-of-flight secondary ion mass spectrometry. Results revealed some limitations on the printability of some monomers, due to the wide range of viscosities and surface tensions of the monomers used. The presence of air inside the piezoelectric nozzle of the inkjet printing impaired spot dispensing for some of the monomers tested. Noncircular spots and spot spreading phenomena were observed for some of the monomers printed with contact printing. These effects were correlated with the chemistry of the processed monomers [90]. Despite the drawbacks observed for the two different printing techniques used, spots with a size varying from 250 μm to 400 μm were achieved.

Microarray substrates functionalization or modification generally involves several cumbersome steps that can directly influence the outcome of the HTS assays. Coating of glass slides with hydrophobic chemical compounds that aid the droplet stability after printing is normally performed. The preliminary preparation of the microarrays substrates can be a long process although alternative approaches have been investigated. Lee et al. [91] applied a new microwave-assisted surface chemistry procedure to develop microarrays for ligand screening, improving the manufacturing speed, reproducibility, and scalability. The microwave-assisted microarrays, capable of screening up to 5000 conditions prepared by pin-assisted bioprinting, could be prepared 42-fold faster when compared to conventional techniques [91]. Another drawback identified in some microarrays is the nonspecific binding inducing incorrect fluorescence measurements and false positive hits. Several surface functionalization methods were investigated for the development of microarrays for proteomics screening allowing the reduction of nonspecific interactions of proteins of interest to enhance signal-to-noise ratio and to return stronger fluorescence signals on the surface of small-molecule microarrays [16]. To avoid complex functionalization procedures, Vallès-Miret and Bradley [92] developed a generic strategy allowing the immobilization of any compound library on printed fluorous-functionalized glass slides.

The combination of microarrays, termed "sandwich" or stamping technique, has become frequently used to screen small molecules with bioprinted cell microarrays [5,13,77,93–95] (Fig. 11.5).

Cell microarrays (datachip) and drug microarrays (metachip) were printed with a microsolenoid valve system and combined by stamping to screen nine drug candidates and their cytochrome P450-generated metabolites (Fig. 11.5a) [94]. Datachip containing 1080 individually printed cell spots were stamped with a metachip containing nine compounds and their metabolites (CYP2D6, CYP1A2, CYP3A4 and a mixture of the three P450s) emulating the human liver in order to screen anticancer drugs. With the combination of the datachip and metachip it was possible to determine the IC_{50} values for the tested compounds. Miniaturized microwell chip combined with opposite pillar chip were developed with the glass-slide size to screen anticancer drugs (Fig. 11.5b) [93]. Human brain tumor cells encapsulated in alginate were printed on the micropillar chip and stamped in a microwell chip containing growth medium and incubated during 1 day. The micropillar chip containing cells encapsulated in alginate was then stamped with a new microwell chip containing the different drug conditions. IC_{50} values of the drugs tested were calculated according to the fluorescence measured for the live cells present on each micropillar. A similar stamping concept combining a cell-seeded microwell with PDMS micropillar containing the different drug conditions printed with a piezo-assisted bioprinter (Fig. 11.5c) was investigated [5]. A library of 320 natural compounds was screened in the presence or absence of verapamil (P-glycoprotein inhibitor) against human breast cancer (MCF-7) cells and four hits were identified by means of cell viability analysis. A sponge-like microscaffold array chip was developed to test drugs in a more biomimetic 3D environment [33]. Polymethylmethacrylate microchips were manufactured by CO_2 laser ablation of the wells, which served as containers for casting a gelatin

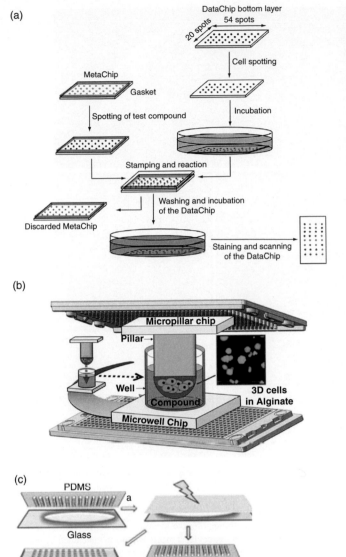

FIGURE 11.5 Schematics of different microarray platforms used for high-throughput compound screening by means of "sandwich" or stamping technique. (a) Datachip printed by spotting hydrogel encapsulating cells is stamped with a metachip array containing printed compound spots [94]. (b) Miniaturized micropillar printed with hydrogel encapsulating cells is stamped with microwell chip platforms containing the different compounds to screen [93]. (c) A micromolded polyethylene glycol diacrylate microwell seeded with cells is stamped with a PDMS array previously printed with a chemical library on to the tips of the posts [5]. *(Fig. 11.5a reprinted with permission from Lee et al. [94]; Fig. 11.5b reprinted with permission from Lee et al. [93]; Fig. 11.5c reprinted with permission from Evenou et al. [54].)*

gel that after cross-linking and lyophilization formed porous scaffolds. Another polymethylmethacrylate layer was produced with the same array spacing, glued to the scaffold microchip, and a cell suspension was seeded onto the scaffold chip side. Culture medium was added to the array chip and a previously prepared drug-loaded chip was stamped on the medium. After drug exposure, live/dead staining was performed with calcein-AM and propidium iodide and cell viability was measured by means of a microplate reader. The results obtained with this platform highlighted the differences between the 3D screening on the prepared scaffolds and 2D controls performed on 96-well plates.

5.2 *In Vivo* HTS Platforms

The *in vivo* microenvironment is complex. Despite all the progresses in attempting to replicate *in vivo* conditions *in vitro* with the previously described platforms, it is still nowadays impossible to fully predict the response to biomaterials, chemical or biochemical molecules with accuracy. As an example, the complexity of the drug discovery process generally involves animal studies prior to clinical human trials. A common example of tests that require animal experimentation is the validation of the ADME/Tox properties of new drugs [96]. Despite being largely used in drug discovery and other fields of research, animal model experimentation raises several ethical questions, but remains essential to understanding key aspects of diagnosis, prevention, and treatment of diseases [30]. The replacement, refinement, and reduction of these *in vivo* tests is nowadays essential [30].

We have recently developed a new concept proposed for the reduction of animals while screening different conditions [23]. For this purpose, novel multiwell array devices were manufactured with different biocompatible polymers. These can be customized to screen a variable number of conditions, according to the animal model used (e.g., 36 conditions in a mouse, 4096 in a goat model) (Fig. 11.6).

The devices, which resemble miniaturized cell culture plates, are adapted for manual pipetting, automated liquid dispensing or noncontact bioprinting techniques, allowing the loading of different screening conditions into the wells. A first *in vivo* study with this implantable HTS multiwell array exploited the cocultures of hMSCs and bovine primary chondrocytes (bPCs) with different ratios in a mouse model. The optimal coculture condition was evaluated according to the newly formed tissue and the 80:20 (hMSCs:bPCs) cell culture ratio revealed to be the best for new cartilage tissue formation [23]. Recently, we used the multiwell array devices to perform *in vitro* screening of small chemotherapeutic drugs (e.g., 5-Fluorouracil) (unpublished data). Two cell lines (MG-63 and human colon adenocarcinoma cell line) were seeded at different densities in each microwell and aggregate formation was analyzed. Cell viability was

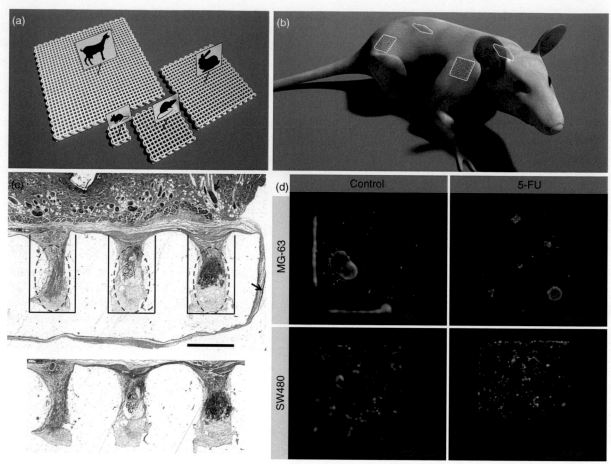

FIGURE 11.6   Multiwell array platform for *in vivo* and *in vitro* HTS. (a) Different arrays with a variable number of conditions in accordance to the animal study, varying from 36 conditions in a mouse up to 4096 in a goat model. (b) *In vivo* implantation strategy in a mouse model with randomization of the conditions. (c) Example of histological sections after subcutaneous implantation on a mice showing a three-well platform where host tissue can be observed by hematoxylin counterstained with eosin showing cell cytoplasm (red) and nuclei (dark blue), respectively. Red arrow points to hair growth on the skin of nude mice, black arrow points to the area of the screening device stained with India ink to pinpoint the location of conditions (rows and columns) after histological processing. Wells are delineated with black lines and the material of the screening device appears white and enveloped by the host tissue. The dotted oval displays the area of interest for screening purposes, where implanted conditions interact with the host tissue. The three wells are shown after Masson's tri-chrome staining: keratin (red), collagen (blue), and nuclei (dark brown) to evaluate the ECM in the wells. (d) Example of *in vitro* screening of small molecules for chemotherapeutic applications. Human osteosarcoma cell line (MG-63) and human colon adenocarcinoma cell line (SW480) were seeded onto the microwell arrays and cultured with 5-Fluorouracil supplemented medium. Live/dead images were acquired after 5 days in culture (unpublished data). *(Fig. 11.6 a-c reprinted with permission from Higuera et al. [23].)*

evaluated by live/dead staining (Fig. 11.6d) and aggregate formation was evaluated by HCS imaging techniques. We envision that with this platform, it will be possible to screen a large number of conditions, both *in vitro* and *in vivo*, allowing a better understanding of the factors influencing the low reproducibility of the results normally obtained on comparative studies (2D versus 3D and *in vitro* versus *in vivo*).

A similar implantable array device was developed to screen biomaterials for TE applications [97]. These arrays were prepared with 36 combinations of biomaterials implanted subcutaneously in rats and the inflammatory response evaluated in the different biomaterial regions. These first examples show that the HTS platform can be also developed for direct screening *in vivo*, which holds the promise to directly test the targeted experimental conditions

in preclinical models. This might be useful to complement *in vitro* screening, where a larger number of "positive hit" candidates can be taken along in the *in vivo* platforms, thus potentially solving the issues of false positives.

6   HTS AND ORGAN BIOPRINTING

The endeavor to produce replacement organs in the laboratory has drawn particular attention after some breakthrough in bioprinting technologies. Still at its infancy, organ bioprinting presents some fundamental questions that need to be addressed before the complete manufacture of an organ *in vitro* becomes a reality. One approach that has been investigated to produce organs *in vitro* rely on the production of tissue spheroids that are used as building blocks. After bioprinting, these building blocks will aid in the new organ tissue formation [98]. There are still some aspects that need to be mastered before achieving tissue maturation after printing [99]. For this purpose, Hajdu et al. [99] performed an HTS study with spheroid building blocks to understand the kinetics of tissue spheroid fusion and tissue rearrangement upon fusion in order to understand the maturation of the bioprinted tissues.

Another combination of HTS and organ bioprinting envisions the development of organ sections or organs on a 96-microtiter-plate platform for drug discovery in an attempt to make the screening conditions as closely mimicking as possible what occurs in the body [100]. Nowadays, it is already possible to print diverse tissue models, such as liver, kidney, and even cancer tissue models. Despite the advances in organ bioprinting, producing large-sized tissues or even organs is still impossible due to the absence of vascular, neural, and in some cases lymphatic networks [100].

7 DATA ACQUISITION TREATMENT AND ANALYSIS

Detailed methods and protocols have become available to perform small-molecule [101], cell-based [102], and protein [103] microarrays [44,46], among others. The data from an HTS assay are normally acquired by imaging techniques with radioactive and nonradioactive labelling techniques [39,104]. Nowadays, mainly nonradioactive labelling solutions based on fluorescence [105] or luminescence [60] are used. Image acquisition is generally performed with microscopy, plate readers, and other HCS devices [39]. Several commercial and open-source informatics tools have been used for image processing, generation of database platforms, visualization, and workflow preparation of HTS and HCS assays [38].

Data reported in scientific literature does not normally follow a standard and no official standards have been proposed. Envisioning a correct evaluation and interpretation of the data retrieved from small-molecule HTS assays, Inglese et al. [106] proposed a set of guidelines. These guidelines include a detailed description of the assay (nature, strategy, protocol, reagents, and sources), the library screened (nature, size, source, concentration tested, quality control, and details), the HTS process (format, plate controls, plate number and duration, dispensing system, output, detection, software, correction factors, normalization, and performance), the post-HTS analysis (selection of hits, retesting, structure, and compound purification/resynthesis), and screen result (list of screening hits, list of validated compound, and comments on the hit compound selection) [106]. Another fundamental factor that needs to be taken in consideration is the statistical analysis of the data retrieved from HTS assays [107]. Several factors should be considered when performing statistical analysis, namely, the positional effect, the choice for the hit thresholds, the correct minimization of false-positives and false-negatives, and the necessity of replicating measurements [107]. All these factors, if not carefully addressed, will most likely compromise the robustness of the statistical analysis and subsequently the reliability of the determined hits.

8 CONCLUSIONS AND FUTURE PERSPECTIVES

HTS started as a simple, fast, and cheap solution to perform the screening of large libraries with thousands of small molecules, making the drug discovery process easier. With the maturation of the technology, new limitations have been found and new solutions have been developed in order to make these screening techniques more reliable. Several bioprinting techniques have been used to dispense a wide range of materials with high accuracy, aiding significantly the results obtained in HTS assays. Due to the advantages of HTS methods, these have been adopted in several research fields such as stem cell research and biomaterials development. In order to screen a large number of conditions simultaneously *in vivo*, new implantable HTS platforms have also been developed.

Despite the numerous different approaches developed over the last years, it is still not possible to fully simulate the *in vivo* microenvironment *in vitro*, making it difficult to ensure the success of the hits normally identified with *in vitro* screenings. 3D models are gaining robustness and might aid in the better understanding of the *in vivo* microenvironment. Despite the complexity that can be achieved with 3D *in vitro* models, it is still impossible to fully replicate animal models and ultimately the human physiology. Integration of different technological platforms could be a solution in the near future to add complexity where needed and achieve an improved mimicry of human physiology. In this respect, integration of HTS techniques with microfluidic, sensing, and actuating platforms is expected to generate devices where it would be possible to react to certain sensed conditions and study dynamically the efficacy of new possible treatments. Translating these approaches to clinically relevant 3D models is of key importance and will require significant advances in imaging and data analysis. Future developments in HTS should address these limitations, while solutions that allow the screening of *in vitro* and *in vivo* on the same platform might help in a further understanding of these factors.

GLOSSARY

Biofabrication technologies Technologies developed for the manufacturing of complex living or nonliving bioengineered microarrays, scaffolds, tissues or organs.

Bioprinting A group of computer-controlled biofabrication technologies that allows an accurate dispensing of synthetic or natural biomaterials, ECM, cells, and drugs on previously designed patterns or arrays.

Combinatorial chemistry Rapid chemical synthesis of a large number of compounds that can be used in HTS assays to enable the identification of new drugs.

Target class libraries Libraries composed of chemotypes that target kinases, nuclear hormone receptors, ion channels, G protein-coupled receptors, and central nervous system targets.

Known pharmacologically active drugs Collections approved drug library, normally used for drug repurposing (new drug applications) and for the validation of new targets and pathways.

Fragments-based screening collections Library with thousands of agents with low molecular weight and with a millimolar concentration.

Diversity-oriented synthesis collections Library composed of compounds with complex architectures that are designed to explore new areas of biological space.

Hit Selected compound/condition that presented positive results over the performed screening.

ABBREVIATIONS

2D	Two-dimensional
3D	Three-dimensional
ADME/Tox	Absorption, distribution, metabolism and excretion/toxicity
RA	All-trans retinoic acid
ACMEs	Arrayed cellular microenvironments
BMP-2	Bone morphogenetic protein-2
BMP-4	Bone morphogenetic protein-4
bPCs	Bovine primary chondrocytes
CiGiP	Cells-in-gels-in-paper
DNA	Deoxyribonucleic acid
DEX	Dextran
DMSO	Dimethyl sulfoxide
DPN	Dip-pen nanolithography
ECM	Extracellular matrix
FGF-2	Fibroblast growth factor-2
FGF-4	Fibroblast growth factor-4
FDA	Food and Drug Administration
IC50	Half maximal inhibitory concentration
HP	Hewlett-Packard
HCS	High content screening
HTS	High throughput screening
hESCs	Human embryonic stem cells
hMSCs	Human mesenchymal stem cells
hPSCs	Human pluripotent stem cells
MG-63	Human preosteoblastic osteosarcoma cells
iPSCs	Induced pluripotent stem cells
LaBP	Laser-assisted bioprinting
mESCs	Mouse embryonic stem cells
PCL	Poly(3-caprolactone)
PLL	Poly-L-lysine
PEG	Poly(ethylene glycol)
PLLA	Poly(L-lactic acid)
PSMA	Poly(styrene-co-maleic anhydride)
PDMS	Polydimethylsiloxane
PPL	Polymer pen lithography
RM	Regenerative medicine
RNA	Ribonucleic acid
siRNAs	Small interfering ribonucleic acids
TE	Tissue engineering
TGFβ	Transforming growth factor beta

ACKNOWLEDGMENTS

The authors would like to acknowledge funding from the MERLN Institute for Technology-Inspired Regenerative Medicine at Maastricht University and financial support of the Dutch Province of Limburg.

REFERENCES

[1] Gribbon P, Andreas S. High-throughput drug discovery: what can we expect from HTS? Drug Discov Today 2005;10(1):17–22.

[2] Ekins S, Waller CL, Bradley MP, Clark AM, Williams AJ. Four disruptive strategies for removing drug discovery bottlenecks. Drug Discov Today 2013;18(5-6):265–71.

[3] Ma H, Horiuchi KY, Wang Y, Kucharewicz SA, Diamond SL. Nanoliter homogenous ultra-high throughput screening microarray for lead discoveries and IC$_{50}$ profiling. Assay Drug Dev Technol 2005;3(2):177–87.

[4] Astashkina A, Mann B, Grainger DW. A critical evaluation of *in vitro* cell culture models for high-throughput drug screening and toxicity. Pharmacol Ther 2012;134(1):82–106.

[5] Wu J, Wheeldon I, Guo Y, Lu T, Du Y, Wang B, et al. A sandwiched microarray platform for benchtop cell-based high throughput screening. Biomaterials 2011;32(3):841–8.

[6] Jensen GC, Krause CE, Sotzing GA, Rusling JF. Inkjet-printed gold nanoparticle electrochemical arrays on plastic. Application to immunodetection of a cancer biomarker protein. Phys Chem Chem Phys 2011;13(11):4888–94.

[7] Xu F, Celli J, Rizvi I, Moon S, Hasan T, Demirci U. A three-dimensional *in vitro* ovarian cancer coculture model using a high-throughput cell patterning platform. Biotechnol J 2011;6(2):204–12.

[8] Mei Y, Goldberg M, Anderson D. The development of high-throughput screening approaches for stem cell engineering. Curr Opin Chem Biol 2007;11(4):388–93.

[9] Dolatshahi-Pirouz A, Nikkhah M, Gaharwar AK, Hashmi B, Guermani E, Aliabadi H, et al. A combinatorial cell-laden gel microarray for inducing osteogenic differentiation of human mesenchymal stem cells. Sci Rep 2014;4:3896.

[10] Tasoglu S, Demirci U. Bioprinting for stem cell research. Trends Biotechnol 2013;31(1):10–9.

[11] Ankam S, Teo BK, Kukumberg M, Yim EK. High throughput screening to investigate the interaction of stem cells with their extracellular microenvironment. Organogenesis 2013;9(3):128–42.

[12] Kusi-Appiah AE, Vafai N, Cranfill PJ, Davidson MW, Lenhert S. Lipid multilayer microarrays for *in vitro* liposomal drug delivery and screening. Biomaterials 2012;33(16):4187–94.

[13] Fernandes TG, Kwon SJ, Bale SS, Lee MY, Diogo MM, Clark DS, et al. Three-dimensional cell culture microarray for high-throughput studies of stem cell fate. Biotechnol Bioeng 2010;106(1):106–18.

[14] Hansen A, Zhang R, Bradley M. Fabrication of arrays of polymer gradients using inkjet printing. Macromol Rapid Commun 2012;33(13):1114–8.

[15] Yliperttula M, Chung BG, Navaladi A, Manbachi A, Urtti A. High-throughput screening of cell responses to biomaterials. Eur J Pharm Sci 2008;35(3):151–60.

[16] Lee HY, Park SB. Surface modification for small-molecule microarrays and its application to the discovery of a tyrosinase inhibitor. Mol Biosyst 2011;7(2):304–10.

[17] Moon S, Kim YG, Dong L, Lombardi M, Haeggstrom E, Jensen RV, et al. Drop-on-demand single cell isolation and total RNA analysis. PLoS ONE 2011;6(3):e17455.

[18] Neumann B, Held M, Liebel U, Erfle H, Rogers P, Pepperkok R, et al. High-throughput RNAi screening by time-lapse imaging of live human cells. Nat Methods 2006;3(5):385–90.

[19] Rose D. Microdispensing technologies in drug discovery. Drug Discov Today 1999;4(9):411–9.

[20] Gaisford W. Robotic liquid handling and automation in epigenetics. J Lab Autom 2012;17(5):327–9.

[21] Rodríguez-Dévora JI, Shi ZD, Xu T. Direct assembling methodologies for high-throughput bioscreening. Biotechnol J 2011;6(12):1454–65.

[22] Higuera GA, Moroni L, inventors. High throughput multiwell system for culturing 3D tissue constructs *in vitro* or *in vivo*, method for producing said multiwell sytem and methods for preparing 3D tissue constructs from cells using said multiwell system; 2012.

[23] Higuera GA, Hendriks JA, van Dalum J, Wu L, Schotel R, Moreira-Teixeira L, et al. *In vivo* screening of extracellular matrix components produced under multiple experimental conditions implanted in one animal. Integr Biol 2013;5(6):889–98.

[24] Gu J, Gui Y, Chen L, Yuan G, Lu HZ, Xu X. Use of natural products as chemical library for drug discovery and network pharmacology. PLoS ONE 2013;8(4):e62839.

[25] Mayr LM, Bojanic D. Novel trends in high-throughput screening. Curr Opin Pharmacol 2009;9(5):580–8.

[26] Dandapani S, Rosse G, Southall N, Salvino JM, Thomas CJ. Selecting, acquiring, and using small molecule libraries for high-throughput screening. current protocols in chemical biology. John Wiley & Sons, Inc.; 2009.

[27] Eglen R, Reisine T. Primary cells and stem cells in drug discovery: emerging tools for high-throughput screening. Assay Drug Dev Technol 2011;9(2):108–24.

[28] Breslin S, O'Driscoll L. Three-dimensional cell culture: the missing link in drug discovery. Drug Discov Today 2013;18(5-6):240–9.

[29] Gidrol X, Fouque B, Ghenim L, Haguet V, Picollet-D'hahan N, Schaack B. 2D and 3D cell microarrays in pharmacology. Curr Opin Pharmacol 2009;9(5):664–8.

[30] Workman P, Aboagye EO, Balkwill F, Balmain A, Bruder G, Chaplin DJ, et al. Guidelines for the welfare and use of animals in cancer research. Br J Cancer 2010;102(11):1555–77.

[31] Pampaloni F, Stelzer EH, Masotti A. Three-dimensional tissue models for drug discovery and toxicology. Recent Pat Biotechnol 2009;3(2):103–17.

[32] Kimlin L, Kassis J, Virador V. 3D *in vitro* tissue models and their potential for drug screening. Expert Opin Drug Discov 2013;8(12):1455–66.

[33] Li X, Zhang X, Zhao S, Wang J, Liu G, Du Y. Micro-scaffold array chip for upgrading cell-based high-throughput drug testing to 3D using benchtop equipment. Lab Chip 2014;14(3):471–81.

[34] Horning JL, Sahoo SK, Vijayaraghavalu S, Dimitrijevic S, Vasir JK, Jain TK, et al. 3-D tumor model for *in vitro* evaluation of anticancer drugs. Mol Pharmaceutics 2008;5(5):849–62.

[35] Justice BA, Badr NA, Felder RA. 3D cell culture opens new dimensions in cell-based assays. Drug Discov Today 2009;14(1–2):102–7.

[36] Vandenburgh H. High-content drug screening with engineered musculoskeletal tissues. Tissue Eng, Part B 2010;16(1):55–64.

[37] Gasparri F. An overview of cell phenotypes in HCS: limitations and advantages. Expert Opin Drug Discov 2009;4(6):643–57.

[38] Zanella F, Lorens JB, Link W. High content screening: seeing is believing. Trends Biotechnol 2010;28(5):237–45.

[39] Wölcke J, Ullmann D. Miniaturized HTS technologies – uHTS. Drug Discov Today 2001;6(12):637–46.

[40] Anderson DG, Levenberg S, Langer R. Nanoliter-scale synthesis of arrayed biomaterials and application to human embryonic stem cells. Nat Biotech 2004;22(7):863–6.

[41] Mei Y, Saha K, Bogatyrev SR, Yang J, Hook AL, Kalcioglu ZI, et al. Combinatorial development of biomaterials for clonal growth of human pluripotent stem cells. Nat Mater 2010;9(9):768–78.

[42] Khan F, Tare RS, Kanczler JM, Oreffo RO, Bradley M. Strategies for cell manipulation and skeletal tissue engineering using high-throughput polymer blend formulation and microarray techniques. Biomaterials 2010;31(8):2216–28.

[43] Urquhart AJ, Anderson DG, Taylor M, Alexander MR, Langer R, Davies MC. High throughput surface characterisation of a combinatorial material library. Adv Mater 2007;19(18):2486–91.

[44] Brafman DA, Chien S, Willert K. Arrayed cellular microenvironments for identifying culture and differentiation conditions for stem, primary and rare cell populations. Nat Protoc 2012;7(4):703–17.

[45] Gosalia DN, Diamond SL. Printing chemical libraries on microarrays for fluid phase nanoliter reactions. Proc Natl Acad Sci USA 2003;100(15):8721–6.

[46] Erfle H, Neumann B, Liebel U, Rogers P, Held M, Walter T, et al. Reverse transfection on cell arrays for high content screening microscopy. Nat Protoc 2007;2(2):392–9.

[47] Brown KA, Eichelsdoerfer DJ, Liao X, He S, Mirkin CA. Material transport in dip-pen nanolithography. Front Phys 2013;9(3):385–97.

[48] Salaita K, Wang Y, Mirkin CA. Applications of dip-pen nanolithography. Nat Nano 2007;2(3):145–55.

[49] Birch HM, Clayton J. Cell biology: close-up on cell biology. Nature 2007;446(7138):937–40.

[50] Mei Y, Cannizzaro C, Park H, Xu Q, Bogatyrev SR, Yi K, et al. Cell-compatible, multicomponent protein arrays with subcellular feature resolution. Small 2008;4(10):1600–4.

[51] Collins JM, Lam RTS, Yang Z, Semsarieh B, Smetana AB, Nettikadan S. Targeted delivery to single cells in precisely controlled microenvironments. Lab Chip 2012;12(15):2643–8.

[52] Eichelsdoerfer DJ, Liao X, Cabezas MD, Morris W, Radha B, Brown KA, et al. Large-area molecular patterning with polymer pen lithography. Nat Protoc 2013;8(12):2548–60.

[53] Wang YM, Cui Y, Cheng ZQ, Song LS, Wang ZY, Han BH, et al. Poly(acrylic acid) brushes pattern as a 3D functional biosensor surface for microchips. Appl Surf Sci 2013;266:313–8.

[54] Evenou F, Di Meglio JM, Ladoux B, Hersen P. Micro-patterned porous substrates for cell-based assays. Lab Chip 2012;12(9):1717–22.

[55] Quist AP, Pavlovic E, Oscarsson S. Recent advances in microcontact printing. Anal Bioanal Chem 2005;381(3):591–600.

[56] Kaufmann T, Ravoo BJ. Stamps, inks and substrates: polymers in microcontact printing. Polym Chem 2010;1(4):371.

[57] Alom Ruiz S, Chen CS. Microcontact printing: a tool to pattern. Soft Matter 2007;3(2):168.

[58] Zhang R, Liberski A, Sanchez-Martin R, Bradley M. Microarrays of over 2000 hydrogels – Identification of substrates for cellular trapping and thermally triggered release. Biomaterials 2009;30(31):6193–201.

[59] Zarowna-Dabrowska A, McKenna EO, Schutte ME, Glidle A, Chen L, Cuestas-Ayllon C, et al. Generation of primary hepatocyte

microarrays by piezoelectric printing. Colloids Surf B Biointerfaces 2012;89:126–32.

[60] Arrabito G, Galati C, Castellano S, Pignataro B. Luminometric sub-nanoliter droplet-to-droplet array (LUMDA) and its application to drug screening by phase I metabolism enzymes. Lab Chip 2013;13(1):68–72.

[61] Miller ED, Fisher GW, Weiss LE, Walker LM, Campbell PG. Dose-dependent cell growth in response to concentration modulated patterns of FGF-2 printed on fibrin. Biomaterials 2006;27(10):2213–21.

[62] Phillippi JA, Miller E, Weiss L, Huard J, Waggoner A, Campbell P. Microenvironments engineered by inkjet bioprinting spatially direct adult stem cells toward muscle- and bone-like subpopulations. Stem Cells 2008;26(1):127–34.

[63] Fujita S, Onuki-Nagasaki R, Fukuda J, Enomoto J, Yamaguchi S, Miyake M. Development of super-dense transfected cell microarrays generated by piezoelectric inkjet printing. Lab Chip 2013;13(1):77–80.

[64] Nagaraj VJ, Eaton S, Wiktor P. NanoProbeArrays for the analysis of ultra-low-volume protein samples using piezoelectric liquid dispensing technology. J Lab Autom 2011;16(2):126–33.

[65] Jones RE, Zheng W, McKew JC, Chen CZ. An alternative direct compound dispensing method using the HP D300 digital dispenser. J Lab Autom 2013;18:367–74.

[66] Rodríguez-Dévora JI, Zhang B, Reyna D, Shi ZD, Xu T. High throughput miniature drug-screening platform using bioprinting technology. Biofabrication 2012;4(3).

[67] Xu T, Jin J, Gregory C, Hickman JJ, Boland T. Inkjet printing of viable mammalian cells. Biomaterials 2005;26(1):93–9.

[68] Xu T, Gregory CA, Molnar P, Cui X, Jalota S, Bhaduri SB, et al. Viability and electrophysiology of neural cell structures generated by the inkjet printing method. Biomaterials 2006;27(19):3580–8.

[69] Xu T, Kincaid H, Atala A, Yoo JJ. High-throughput production of single-cell microparticles using an inkjet printing technology. J Manuf Sci Eng 2008;130(2):021017.

[70] Demirci U, Montesano G. Single cell epitaxy by acoustic picolitre droplets. Lab Chip 2007;7(9):1139–45.

[71] Collins DJ, Alan T, Helmerson K, Neild A. Surface acoustic waves for on-demand production of picoliter droplets and particle encapsulation. Lab Chip 2013;13(16):3225–31.

[72] Mota C, Puppi D, Chiellini F, Chiellini E. Additive manufacturing techniques for the production of tissue engineering constructs. J Tissue Eng Regen Med 2012;9(3):174–90.

[73] Guillotin B, Catros S, Guillemot F. Laser assisted bio-printing (LAB) of cells and bio-materials based on laser induced forward transfer (LIFT). In: Schmidt V, Belegratis MR, editors. Laser technology in biomimetics. Berlin Heidelberg: Springer; 2013. p. 193–209.

[74] Schiele NR, Corr DT, Huang Y, Raof NA, Xie Y, Chrisey DB. Laser-based direct-write techniques for cell printing. Biofabrication 2010;2(3):032001.

[75] Guillemot F, Souquet A, Catros S, Guillotin B, Lopez J, Faucon M, et al. High-throughput laser printing of cells and biomaterials for tissue engineering. Acta Biomater 2010;6(7):2494–500.

[76] Bammesberger S, Ernst A, Losleben N, Tanguy L, Zengerle R, Koltay P. Quantitative characterization of non-contact microdispensing technologies for the sub-microliter range. Drug Discov Today 2013;18(9-10):435–46.

[77] Fernandes TG, Kwon SJ, Lee MY, Clark DS, Cabral JM, Dordick JS. On-chip, cell-based microarray immunofluorescence assay for high-throughput analysis of target proteins. Anal Chem 2008;80(17):6633–9.

[78] Lemmo S, Nasrollahi S, Tavana H. Aqueous biphasic cancer cell migration assay enables robust, high-throughput screening of anti-cancer compounds. Biotechnol J 2013;9(3):426–34.

[79] Faulkner-Jones A, Greenhough S, King JA, Gardner J, Courtney A, Shu W. Development of a valve-based cell printer for the formation of human embryonic stem cell spheroid aggregates. Biofabrication 2013;5(1):015013.

[80] Srinivasan A, Uppuluri P, Lopez-Ribot J, Ramasubramanian AK. Development of a high-throughput Candida albicans biofilm chip. PLoS ONE 2011;6(4):e19036.

[81] Srinivasan A, Lopez-Ribot JL, Ramasubramanian AK. Candida albicans biofilm chip (CaBChip) for high-throughput antifungal drug screening. J Vis Exp 2012;(65):e3845.

[82] Srinivasan A, Leung KP, Lopez-Ribot JL, Ramasubramanian AK. High-throughput nano-biofilm microarray for antifungal drug discovery. mBio 2013;4(4).

[83] Ding Y, Huang E, Lam KS, Pan T. Microfluidic impact printer with interchangeable cartridges for versatile non-contact multiplexed micropatterning. Lab Chip 2013;13(10):1902–10.

[84] Oldenburg KR, Zhang J-H, Chen T, Maffia A, Blom KF, Combs AP, et al. Assay miniaturization for ultra-high throughput screening of combinatorial and discrete compound libraries: a 9600-well (0.2 microliter) assay system. J Biomol Screening 1998;3(1):55–62.

[85] Tung YC, Hsiao AY, Allen SG, Torisawa YS, Ho M, Takayama S. High-throughput 3D spheroid culture and drug testing using a 384 hanging drop array. Analyst 2011;136(3):473–8.

[86] Hsiao AY, Tung YC, Qu X, Patel LR, Pienta KJ, Takayama S. 384 hanging drop arrays give excellent Z-factors and allow versatile formation of co-culture spheroids. Biotechnol Bioeng 2012;109(5):1293–304.

[87] Hsiao A, Tung YC, Kuo CH, Mosadegh B, Bedenis R, Pienta K, et al. Micro-ring structures stabilize microdroplets to enable long term spheroid culture in 384 hanging drop array plates. Biomed Microdevices 2012;14(2):313–23.

[88] Deiss F, Mazzeo A, Hong E, Ingber DE, Derda R, Whitesides GM. Platform for high-throughput testing of the effect of soluble compounds on 3D cell cultures. Anal Chem 2013;85(17):8085–94.

[89] Celiz AD, Hook AL, Scurr DJ, Anderson DG, Langer R, Davies MC, et al. ToF-SIMS imaging of a polymer microarray prepared using ink-jet printing of acrylate monomers. Surf Interface Anal 2013;45(1):202–5.

[90] Hook AL, Scurr DJ, Burley JC, Langer R, Anderson DG, Davies MC, et al. Analysis and prediction of defects in UV photo-initiated polymer microarrays. J Mater Chem B 2013;1(7):1035.

[91] Lee JH, Hyun H, Cross CJ, Henary M, Nasr KA, Oketokoun R, et al. Rapid and facile microwave-assisted surface chemistry for functionalized microarray slides. Adv Funct Mater 2012;22(4):872–8.

[92] Vallès-Miret M, Bradley M. A generic small-molecule microarray immobilization strategy. Tetrahedron Lett 2011;52(50):6819–22.

[93] Lee DW, Choi YS, Seo YJ, Lee MY, Jeon SY, Ku B, et al. High-throughput screening (HTS) of anticancer drug efficacy on a micropillar/microwell chip platform. Anal Chem 2014;86(1):535–42.

[94] Lee MY, Kumar RA, Sukumaran SM, Hogg MG, Clark DS, Dordick JS. Three-dimensional cellular microarray for high-throughput toxicology assays. Proc Natl Acad Sci USA 2008;105(1):59–63.

[95] Fernandes TG, Diogo MM, Clark DS, Dordick JS, Cabral JM. High-throughput cellular microarray platforms: applications in drug discovery, toxicology and stem cell research. Trends Biotechnol 2009;27(6):342–9.

[96] Pellegatti M. Preclinical *in vivo* ADME studies in drug development: a critical review. Expert Opin Drug Metab Toxicol 2012;8(2):161–72.

[97] Oliveira MB, Ribeiro MP, Miguel SP, Neto AI, Coutinho P, Correia IJ, et al. *In vivo* high-content evaluation of three-dimensional scaffolds biocompatibility. Tissue Eng, Part C 2014;20(11):851–64.

[98] Mehesz AN, Brown J, Hajdu Z, Beaver W, Da Silva JVL, Visconti RP, et al. Scalable robotic biofabrication of tissue spheroids. Biofabrication 2011;3(2):025002.

[99] Hajdu Z, Mironov V, Mehesz AN, Norris RA, Markwald RR, Visconti RP. Tissue spheroid fusion-based *in vitro* screening assays for analysis of tissue maturation. J Tissue Eng Regener Med 2010;4(8):659–64.

[100] Dutton G. 3D printing may revolutionize drug R&D. Genet Eng Biotechnol News 2013;33(20):10, 2.

[101] Uttamchandani M, Yao S. Small molecule microarrays methods and protocols. New York, N.Y.: Humana Press, 2010.

[102] Palmer E. Cell-based microarrays: methods and protocols. New York, N.Y.: Humana Press; 2011.

[103] McWilliam I, Kwan MC, Hall D. Inkjet printing for the production of protein microarrays. In: Korf U, editor; 2011. p. 345–361.

[104] Sundberg SA. High-throughput and ultra-high-throughput screening: solution- and cell-based approaches. Curr Opin Biotechnol 2000;11(1):47–53.

[105] Ge X, Eleftheriou NM, Dahoumane SA, Brennan JD. Sol-gel-derived materials for production of pin-printed reporter gene living-cell microarrays. Anal Chem 2013;85(24):12108–17.

[106] Inglese J, Shamu CE, Guy RK. Reporting data from high-throughput screening of small-molecule libraries. Nat Chem Biol 2007;3(8):438–41.

[107] Malo N, Hanley JA, Cerquozzi S, Pelletier J, Nadon R. Statistical practice in high-throughput screening data analysis. Nat Biotechnol 2006;24(2):167–75.

Chapter 12

Biosensor and Bioprinting

Jeung Soo Huh*, Hyung-Gi Byun**, Hui Chong Lau†, and Grace J. Lim†

*Department of Materials Science and Metallurgy, Kyungpook National University, Korea; **Division of Electronics, Information and Communication Engineering, Kwangwon National University, Korea; †Department of Biomedical Science, Joint Institute for Regenerative Medicine, Kyungpook National University, Korea

Chapter Outline

ABSTRACT

Advanced biosensors for diagnostic and clinical applications will require capabilities such as real-time detection or monitoring of physiological changes *in vivo* in addition to being light weight, small size, highly sensitive, selective, have ease in signal transmission, and stable. The fabrication of a high-performance biosensor can be realized by the development of bioprinting technology. Bioprinting enables high throughput, digital control, and accurate delivery of target material to the specific locations. Also, the sophisticated sensors fabricated by advanced science and technology, including nanosensors, wearable sensors, implantable sensors, and wireless sensors, have opened huge possibilities in understanding phenomena that occur in the body, and can be widely used for translational research. In this chapter, we will review various cutting-edge sensor technologies and bioprinting from the viewpoint of their potential application to biomedical engineering and therapeutics.

Keywords: sensor; biosensor; bioprinting; clinical translation; electronic nose; disease patterning; breath sensor; noninvasive monitoring

1 INTRODUCTION

From the Wikipedia encyclopedia, a sensor is referred to as a device that responds to an input quantity by generating a functionally related output usually in the form of an electrical or optical signal. In other words, a sensor measures a physical or chemical quantity of a substance and converts it into a signal that can be presented to the readers or analysts. Sensors are used in everyday objects to know today's temperature, blood pressure, flow, viscosity and density, acceleration, position, humidity, vibration and shock, oxygen, carbon monoxide, odor, body structure, and so on. Also, a variety of sensor technologies and their applications are available including manufacturing and machinery, automotive, medicine, and robotics. There are innumerable applications for sensors of which most people are never aware.

Among the sensing systems, a biosensor is a device for the detection of substances, which combines a biological component with a physicochemical detector component. In general, the sensor part is sensitive to biological elements and made of a biological material such as tissue, microorganisms, organelles, cells, receptors, enzymes, antibodies, nucleic acids, and sometimes, biologically derived or biomimetic materials. Biosensors have sought to mimic such natural processes by fixing and connecting isolated cells or organs to a transducing and detecting system. Biosensors involve confronting new aspects of detection and data processing. The quantity detected is always a measure of active molarity of the analyte, whose calibration is correlated with quantities such as the activity molarity of the interfering species, the pH and temperature of the sample, and ionic strength relevant for charged and uncharged analytes.

For biotechnological and medical applications, the analyte activity delivers only the biologically relevant information when measured in the specimen directly. Future

Essentials of 3D Biofabrication and Translation. http://dx.doi.org/10.1016/B978-0-12-800972-7.00012-8

biosensors for diagnostic and clinical application will require capabilities of real-time detection or monitoring of physiological changes *in vivo*. Such functionality will necessitate biosensors with increased sensitivity, specificity, and throughput, as well as the ability to simultaneously detect multiple analytes. The fabrication of such high-performance biosensor can be realized by the development of bioprinting technology. Bioprinting enables high throughput, digital control, and accurate delivery of target material to the specific locations. Also, the sophisticated sensors fabricated by advanced science and technology, including nanosensors, wearable sensors, implantable sensors, and wireless sensors, have opened huge possibilities in understanding phenomena that occur inside the body and can be widely used for translational research. In this chapter, we will review various cutting-edge sensor technologies and bioprinting from the viewpoint of their potential application to biomedical engineering and therapeutics.

2 SENSORS

A sensor is a device that measures a physical or chemical quantity of a substance and converts it into a signal that can be demonstrated to the readers or analysts. It is composed of sensing materials, a transducer, and associated electronics. The sensor material is the part that directly interacts with substances. Conventional sensors use metals or inorganic materials as a sensor material while biological molecules are increasingly used these days. The transducer transforms the signal obtained from the sensor part into another form of signal that can be more easily measured and quantified such as optical, piezoelectric, electrochemical, and mechanical. An associated electronics is made of electronic circuits and signal processing algorithms that are primarily responsible for acquisition and visualization of data to the end users.

Sensors can be classified by transducer type according to the type of energy transfer that they detect [1]. The thermal sensors, such as thermometer, thermocouple, temperature sensitive resistor, and thermostat, detect the change of temperature using different modalities. Mechanical sensors include pressure, flow, viscosity and density, acceleration, position, humidity, vibration, and shock sensors.

The chemical sensors include oxygen sensor, ion-selective electrode, pH electrode, redox electrode, carbon monoxide detector, and odor sensor [1,2]. There are different types of chemical sensors available, which are shown in Fig. 12.1.

A biosensor is a device for the detection of substances that combines a biological component with a physicochemical detector component. In general, the sensor part is sensitive to biological elements and is made of biological materials such as tissue, microorganisms, organelles, cells, receptors, enzymes, antibodies, nucleic acids, and sometimes, biologically derived or biomimetic materials [1]. One of the well-known commercial biosensors is the blood glucose test kit, which uses the glucose oxidase as the sensing part. During the enzymatic reaction, current is produced and indicates the concentration of glucose, as shown in Fig. 12.2. Application of diverse biological molecules into sensing parts could produce various types of biosensors having broad detection spectrum suitable for medical use.

The biosensor industry is rapidly growing and the major market need is from the healthcare industry. For example, the number of diabetic patients is increasing all over the world and most of them use blood glucose kits for its rapid and simple characteristics. Food quality testing and environment monitoring are the other important fields that require biosensors. Since the research and development in the biosensor area are broad, multidisciplinary collaboration is necessary including biochemistry, bioreactor science, physics, physical chemistry, electrochemistry, electronics, information processing, and software programming. With the progress of sensor technology, new types of sensing modalities will be presented in the near future. The conceptual diagram for a biosensor is shown in Fig. 12.3 [3,4].

In the past, the size of sensors was huge and only suitable for the laboratory level. But current development in technology allows sensors to be manufactured on a microscopic scale. Most of the microsensors possess higher speed and sensitivity compared to macroscopic sensors. With the development in materials science and IT, we expect the development of nanosensors as shown in Fig. 12.4 [5]. This implies that we can measure the microenvironment using very tiny sensors in the near future.

Quartz microbalance   Metal oxide   Conducting polymer

FIGURE 12.1 Chemical sensors.

- Glucose oxidase reaction:
  Glucose + O_2 + H_2O → Gluconic acid + H_2O_2
- Anode: H_2O_2 → $2H^+$ + O_2 + $2e^-$

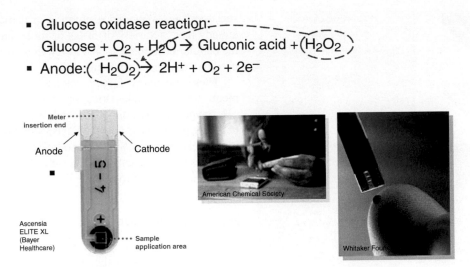

Meter insertion end

Anode

Cathode

Ascensia
ELITE XL
(Bayer
Healthcare)

Sample
application area

American Chemical Society

Whitaker Found

FIGURE 12.2   **Oxidase immobilized glucose biosensor.**

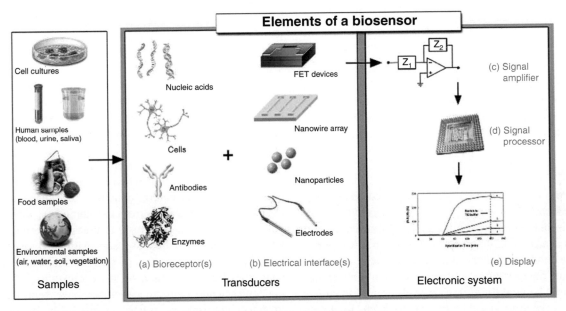

Elements of a biosensor

Cell cultures

Human samples
(blood, urine, saliva)

Food samples

Environmental samples
(air, water, soil, vegetation)

Samples

Nucleic acids

Cells

Antibodies

Enzymes

(a) Bioreceptor(s)

FET devices

Nanowire array

Nanoparticles

Electrodes

(b) Electrical interface(s)

Transducers

(c) Signal
amplifier

(d) Signal
processor

(e) Display

Electronic system

FIGURE 12.3   **Conceptual diagram of a biosensor: sample source and elements of a biosensor.**

3 SENSOR TECHNOLOGY FOR TRANSLATIONAL RESEARCH ON REGENERATIVE THERAPY

Researches on regenerative therapeutics have achieved a remarkable progress in clinical medicine during the last few decades. The development of a tissue-engineered urinary bladder and production of inducible pluripotent stem cells are considered as the landmarks in the clinical translation of regenerative medicine. Even though big steps were made,

we only have the slightest idea on the fundamental issues on how cells proliferate, differentiate, and migrate to form new tissue or remodel during the regeneration process *in vivo* and many scientific questions are yet to be exploited. Based on the longstanding clinical experiences, we assume that the environment surrounding the cells or tissues influences their remodeling process and final fate. But there is little confirmative data supporting such hypothesis since current assessment tools to monitor changes at the cellular level *in vivo* are not fully developed yet. We expect that current

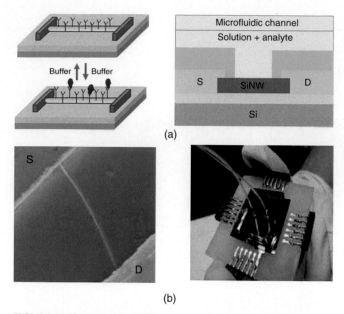

FIGURE 12.4 Nanowire FET sensor. (a) Schematic of a Si nanowire-based FET device configured as a sensor with antibody, where S, D, and SiNW correspond to source, drain, and silicon nanowire, respectively. (b) Cross-sectional diagram and scanning electron microscopy image of a single Si nanowire sensor device, and a photograph of a prototype nanowire sensor biochip with integrated microfluidic sample delivery.

advances in sensor engineering and nanotechnology will ultimately enable the achievement of reliable data through sensing microenvironment change accurately in regenerating cells and tissues *in vivo*.

3.1 Microenvironment

Microenvironment has been a commonly used term in the cancer research field recently [6,7]. The microenvironment of a tumor is an integral part of its physiology, structure, and function. It is an essential part of the tumor, since it supplies a nurturing condition for the malignant process. Monitoring the tumor microenvironment via molecular and cellular profiles would be vital for identifying cell or protein targets for cancer prevention and therapy [8]. As such, for tissue regeneration *in vivo* the microenvironment is crucially important. In addition to monitoring the microenvironment of the tissue itself, monitoring of exogenous substances in the tissue or cellular level would be essential. One recent anticancer drug study demonstrates that nanomolar concentrations of arginylglycylaspartic acid-mimetic $\alpha v \beta 3$ and $\alpha v \beta 5$ inhibitors can paradoxically stimulate tumor growth and tumor angiogenesis [9]. In the regeneration process of damaged tissue after injury, the recovery or regeneration is influenced significantly by adjacent environment changes in a very small and fast scale. If we have specific and sensitive tools to monitor the environment in a microscale and real-time mode, we can diagnose or predict and control the microenvironment, then we can improve the environment for

successful recovery and restoration of damaged tissue. We already have a variety of drugs or tools that can modulate the cellular or tissue responses even though they are usually used in a macroenvironment.

There are many factors regulating the microenvironment, which includes ions or electrolytes, temperature, pH, oxygen, carbon dioxide, metabolites, growth factors or peptides, hormones, heavy metals, stretching or contracting forces, vibration, and many other factors. Thus, we need various sensing systems to precisely detect these phenomena in the microenvironment.

3.2 Nanosensor

Nanosensors are any sensor used to convey information about nanoparticles to the macroscopic world. Their use mainly includes clinical medicine or as tools for other nanoproducts, such as semiconductor chip or nanoscale machines. Even though human beings developed the synthetic nanosensors just recently, nanosensors have been employed by living creatures since the beginning of life on earth. The most common mass-produced functioning nanosensors exist in the world as receptors. The sense of smell in animals is acquired by hundreds of different olfactory receptors that sense nanosized molecules.

Some plants also use nanosensors to detect sunlight. Fishes use nanosensors to detect vibrations in the water and many insects detect sex pheromones using pheromone receptors. One of the first examples of synthetic nanosensors was built by researchers at the Georgia Institute of Technology [10]. This sensor detects the change of vibrational frequency of the nanotube when a single particle is attached onto the end of the carbon nanotube. The discrepancy between the two frequencies from the nanotubes, with or without a particle attached, is used to measure the mass of the attached particle. Chemical sensors using nanotubes is used to detect various properties of gaseous molecules. Carbon nanotubes have been used to sense ionization of gaseous molecules while titanium nanotubes have been employed to detect atmospheric concentrations of hydrogen at the molecular level. Many of these chemical nanosensors are designed to have a small pocket or binding site for specific small substances. Therefore, even the sensor is surrounded by a mixture of different substances; only specific molecules fit into the binding site of the nanosensor.

As we mentioned in the beginning, nature has been using diverse nanosensors in many species. With the progress of biology and biological engineering, many of the hidden nanosensors are being discovered and waiting to be applied in other fields. These bionanosensors are not developed yet, but they will be the major part of the sensor system in the medical or biological field in the near future. For example, propyl hydroxylase domain (PHD) proteins serve as oxygen sensors and may regulate oxygen delivery. Mazzone et al.

[11] reported the role of endothelial PHD2 in vessel shaping by implanting tumors in PHD2 (+/−) mice. Haplodeficiency of PHD2 did not affect tumor vessel density or lumen size, but normalized the endothelial lining and vessel maturation. This resulted in improved tumor perfusion and oxygenation. Haplodeficiency of PHD2 redirected the specification of endothelial tip cells to a more quiescent cell type, lacking filopodia and arrayed in a phalanx formation. This transition relied on hypoxia-inducible factor-driven up-regulation of (soluble) vascular endothelial growth factor receptor-1 and vascular endothelial-cadherin. Thus, decreased activity of an oxygen sensor in hypoxic conditions prompts endothelial cells to readjust their shape and phenotype to restore oxygen supply. Based on these data, we expect that PHD protein can be used for sensing and therapeutic purposes. By applying PHD protein as the biological part of the sensor, it may be possible to monitor the concentration of oxygen in the regenerating tissue. This also can be used in cancer research to monitor the hypoxic atmosphere after chemotherapy.

Biosensors have major advantages over chemical or physical sensors with regard to specificity, sensitivity, and portability. Recently, many types of whole-cell bacterial biosensors have been developed using recombinant DNA technology. The bacteria are genetically engineered to respond to the presence of chemicals or physiological stresses by synthesizing a reporter protein, such as luciferase, β-galactosidase, or green fluorescent protein [12].

In addition to the biological nanosensors, novel nanosensors enabling the demonstration of important intracellular or extracellular factors in a real-time mode are also being developed. Hydrogen concentration, or pH, in the cell or tissue is the essential factor determining the activity of many biological reactions. Uchiyama and Makino [13] demonstrated digital-type fluorescent pH sensors based on the incorporation of a water-sensitive fluorophore into a pH-responsive polymer. It is noteworthy that many digital signal transductions occur in living organisms in which the digital processors are biological versions of macromolecules.

Iron is the most abundant transition metal in cellular systems holding an outstanding biological importance because of its presence in the structures of numerous enzymes and proteins [14,15]. Iron-sensitive fluorescent chemosensors in combination with digital fluorescence spectroscopy were developed to identify a distinct subcellular compartment of intracellular redox-active labile iron [16].

Another essential ion in the cell system is calcium. This ion is one of the key factors of the intracellular signal systems and also contributes to mineralized tissue homeostasis. Calmodulin (CaM) is a ubiquitous protein involved in Ca^{2+}-mediated signal transduction. On Ca^{2+} influx, CaM acquires a strong affinity to various cellular proteins with one or more CaM recognition sequences, resulting in the onset or termination of Ca^{2+}-regulated cascades. Truong et al. [17] created protein-based Ca^{2+} sensors using CaM complexes and green fluorescent proteins, previously named "chameleon." The major advantage of chameleons is that they can be expressed in single cells and targeted to the specific organelles or tissues to measure localized Ca^{2+} changes.

The recent developments in the field of nanoelectronics, with transducers progressively shrinking down to smaller sizes through nanotechnology and carbon nanotubes, are expected to result in innovative biomedical instrumentation possibilities, with new therapies and efficient diagnostic tools. The use of integrated systems, smart biosensors, and programmable nanodevices are advancing nanoelectronics, enabling the progressive research and development of molecular machines. It should provide high-precision pervasive biomedical monitoring with real-time data transmission [18].

3.3 Cell-Based Biosensor

The current biosensors used for monitoring are primarily based on biological recognition elements, such as enzymes, binding proteins, and antibodies, which are coupled to electrical or optical transducers [19–21]. With the progress of cell engineering, a new type of biosensor is being developed. Cell-based biosensors use whole cells as the biorecognition elements. Contrary to biosensors using pure proteins, cell-based sensors can detect agents functionally [22,23]. This means cell-based biosensors could offer physiological monitoring advantages over protein-based biosensors. The cells used for this purpose are varied including tumor cells, neurons, and many other cell types.

Cells are already equipped with receptors that can transduce chemical signals into electrical or physical ones. If efficiently coupled to an electronic device, cells can work as excellent biosensors with versatile functions for many different purposes [24]. Cell-based biosensors are able to respond to many biological substances and detect functionally and biologically complicated information. When the detection system is improved, cells can be brought into being a sensitive unit of biosensors for applications ranging from new drug screening to environmental monitoring.

3.4 Cell-Based Biosensor for Disease Patterning

The cell-based biosensor has long been reported in biosensing technology. This type of sensor uses living cells as sensing substrate to detect the functional biological analytes. Several types of cells have been used as the sensing element in biosensors. This includes the use of human cells, insect cells, yeast, bacteria cells, and others [25–27]. These cells are often genetically modified in order to possess specific properties of the targets such as olfactory and gustatory receptors. These engineered cells can then be used in such biosensors with the capability to detect odors, toxins, and sugars [28–31]. Hence, the cell-based biosensor has demonstrated its specificity and uniqueness in biosensing applications.

Several studies have reported especially on the use of olfactory cell-based biosensors that showed a high sensitivity, specificity, and rapid response. These olfactory cell-based biosensors have been developed on the basis of the cells expressing the chemosensory receptor of interest. The use of cells carrying the sensory receptors as the sensing substance on the sensor device could generate signals that correspond directly to the biological analytes. The generated signals from the interactions between the receptor and analytes are measured using several types of biosensors. Currently, there are several transducers used for the measurement of biosensors. These transducers measure the change in chemical composition, light, weight, sound, and heat. Such designs include surface plasmon resonance, field-effect transistors (FETs), surface acoustic wave, and others [25]. The choice of transducer depends on the application and purpose of the study.

The discovery and characterization of the multigenes family encoding the sensory receptors in different types of vertebrate or invertebrate models accelerated the development of biosensors. Sensory receptors vary among organisms in terms of structure, signaling pathways, and applications. Thus, the choice of organism for sensory receptors depends on the application and design of the biosensor device. For instance, olfactory receptors from dogs are used in sensing narcotics, people, or explosive products. In contrast, in clinical diagnostics, sensory receptors from *Drosophila* are considered to be a good biosensing candidate due to their size and reduced cell numbers. *Drosophila* cells are relatively easy to obtain and grow at ambient temperature. Moreover, the basic genetics, molecular biology, and neurology of *Drosophila* olfactory and gustatory systems have been well investigated and reported. This could provide better insight into the biosensing applications of the *Drosophila* sensory systems.

In order to maintain the viability of cells, *Drosophila* cells are easily maintained at room temperature without a supply of carbon dioxide. The response of cells expressing gustatory receptor (Gr5a) to trehalose was found to be specific and sensitive. Although the use of cells could mimic the genuine response in nature, maintaining the same density of cells for use over the long term is tedious and time consuming [31]. Hence, in the long run, a robust and easy-to-use biological molecule should be targeted.

A research group in Korea has demonstrated the use of *Drosophila* cells and an ion-sensitive field-effect transistor (ISFET) for the detection of sugar, in particular trehalose (Fig. 12.5) [29].

The cells were genetically modified to allow the abundant expression of Gr5a. The gustatory receptors expressed in

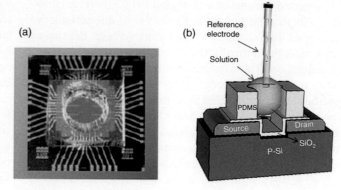

FIGURE 12.5 (a) Top view image of ISFETs device, and (b) schematic diagram of an ISFETs with a SiO$_2$ membrane.

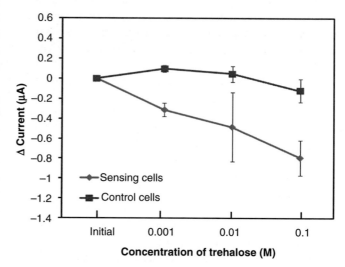

FIGURE 12.6 Response voltage of *Drosophila* cells expressing Gr5a and S2, respectively to trehalose at various concentrations.

Drosophila cells were used to detect sugar, in particular trehalose. By using the cell-oriented ISFET biosensor that our group developed, the response of cells to trehalose was found to be specific and sensitive up to 0.001 M (Fig. 12.6) [29,32].

In light of the demands for an improved diagnostic method for Alzheimer's disease, we developed a diagnostic tool to screen for potential Alzheimer's disease patients using their saliva. The method developed was based on the use of a cell-oriented ISFET biosensor. We also use the developed ISFET sensor to detect the trehalose found in human saliva of persons with Alzheimer's disease and the normal person. Human saliva has been studied for the detection of glucose and beta amyloid 42 [33–35]. Thus, it is possible that some molecules present in saliva could be the potential novel biomarkers for Alzheimer's disease.

The cell-oriented ISFET was designed to investigate the variation in the level of sugar detected in the saliva of persons with Alzheimer's disease and a normal person. The cell-based biosensor exhibited varying response voltages

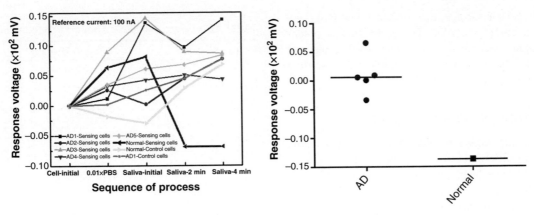

FIGURE 12.7 Response voltages from sensing and control cells to saliva of patients and normal person (left) and median of response voltages from sensing cells after being normalized with control cells (right).

when reacted with saliva from patients and a normal person. Significantly, higher response voltages were obtained from all the saliva samples of Alzheimer's disease patients compared to the normal person. These results indicate that sugar, such as trehalose, can be a potential biomarker for Alzheimer's disease (Fig. 12.7) [29,32].

The results of the varying sugar level found in the saliva could be used as an alternative screening method for Alzheimer's disease diagnosis. In addition, it could also be utilized as an indicator of the risk of Alzheimer's disease development. The present investigation provides a new approach toward the development of a cell-based biosensor by utilizing functions of a biological substance for screening or diagnosis of Alzheimer's disease. The findings of this proof-of-concept study suggest that noninvasive screening using saliva and cell-based biosensors can be a promising patient-friendly screening method for Alzheimer's disease.

Moreover, some researchers have monitored bidirectional, noninvasive communication between external electronics and cells cultured on a chip [36]. Even though this cell-based biosensor system seems fascinating, there are many limitations up to now. The most important obstacle is producing quality cells for the sensor system. Since cell-based biosensor technology has been developed from the point of engineering at the beginning, support from cell biologists has not been enough and this hindered the development of high-standard sensor cells.

The use of cells as a biosensing substrate is not always practical. Cells require consistent monitoring in order to maintain the desired cell densities for optimum sensing purposes. Besides, most cells, such as mammalian cells, must be maintained at 37°C in the presence of 5% carbon dioxide. Upon exposure to a change in environment, the viability of the cells will be altered, and consequently, the sensitivity and specificity of detection could be affected.

In order to maintain the viability of cells, insect cells, such as *Drosophila* cells, can be used. *Drosophila* cells are easily maintained at room temperature without a supply of

carbon dioxide. The response of cells expressing Gr5a to trehalose was found to be specific and sensitive [29]. Although the use of cells could mimic the genuine response in nature, maintaining the same density of cells for use over the long term is tedious and time consuming. Hence, in the long run, a robust and easy-to-use biological molecule should be targeted.

In contrast to the patch-clamp method, most of the cell-based sensor systems use extracellular measurement. At present, there are many potential extracellular detection methods, such as field-effect transistor array cell-based biosensor or microelectrode array cell-based biosensor. Microelectrode arrays and field-effect transistors typically consist of multiple recording sites, presenting a tremendous lead for data acquisition from networks of electrically living cells. Such recordings offer a noninvasive and long-term method to measure multiple parameters of cells and tissue, and sometimes offer a measure of cell coupling that is virtually inaccessible using standard intracellular recording techniques.

Even though the patch–clamp technique is an excellent experimental tool, it still needs well-trained scientists. Since microelectrode arrays or field-effect transistors can be applied to measure the extracellular action potential, the transmission path of ionic channels, and the transmission velocity of biological signals along the layer of neurons, microelectrode arrays or field-effect transistors will be an easy tool for the electrophysiologist [37,38]. However, microelectrodes and FETs have many weak points. They are confined to measuring the potential only at a limited number of active measuring sites, such as the tip of individual microelectrodes and the gate-electrode of individual field-effect transistors, which make it difficult to culture cells on where the suitable sites are designed.

The light-addressable potentiometric sensor is a commonly used semiconductor device that overcomes the limitation of microelectrodes or field-effect transistors [39]. By scanning the light-pointer along the light-addressable potentiometric sensor surface, the surface potential at any desired position can be recorded. Therefore, surface potential measurements may no longer be restricted to discrete active sites. But light-addressable potentiometric sensors also need more progress.

3.5 Electronic Nose for Diagnosis

For the last decade, arrays of electronic sensors, capable of detecting and differentiating complex mixtures of volatile compounds, have been utilized to differentiate gases and aromas. These sensor arrays have been dubbed "electronic nose." Electronic nose technology has been around for many years and used by various research groups in Europe, but medical applications, especially in diagnostics, are relatively recent. Electronic nose technology is still in its development phase, both in respect to hardware and software development. Electronic nose has three components: (1) sample handling system, (2) detection system, and (3) data processing system. The instruments contain an array of multiple sensors, using a variety of different sensor technologies such as organic polymers, metal oxides, and microbalances [40].

Currently available commercial electronic noses do not use biological elements and most of the electronic noses used for medical diagnosis are based on the principle of the disruption of an electrical current as it passes through a chip. The electronic nose from Cyrano Sciences uses composite materials housed in a conducting polymer matrix. The KAMINA electronic nose designed at the Karlsruhe Research Center uses a metal oxide-based sensor array. Another category of electronic noses now undergoing medical application testing is based on the quartz microbalance. This system measures the beat frequency between two quartz-controlled oscillators, one of which is coated with metalloporphyrins. Regardless of the mechanism, electronic nose technologies are expected to make great strides in the field of medical diagnostics [41].

The major application of electronic nose in medicine has been the real-time identification of bacterial species in infected patients [42]. Table 12.1 gives examples of the medical areas using electronic nose and research groups who developed the systems.

Many diseases are accompanied by characteristic odors, and their recognition can provide diagnostic clues, guide the laboratory evaluation, and affect the choice of immediate therapy. In the last few years the noninvasive analysis of exhaled breath has emerged as an innovative tool for diagnostics using electronic nose technology. This technology is a source of valuable contribution to the field of exhaled biomarkers.

A research group in Korea contributed significant experimental results for disease prescreening techniques using gas sensor technology. They developed a micro gas

TABLE 12.1 Electronic Nose Applied to Medical Uses

Research Group	Target Disease
Illinois Institute of Technology	Tuberculosis
Cranfield Center for Analytical Science	Urinary tract infection
Karisruhe Research Center	Bad breath, body odor
Technobiochip	Kidney disease, lung cancer
Osmetech	Bacterial vaginosis, urinary tract infection, pneumonia, pharyngitis
Cyrano Sciences	*H. pylori* infection, upper respiratory infection
Manchester University, School of chemical engineering and analytical science	Bacterial and/or fungal infections from burns and chronic skin ulcers
Kyungpook National University	Breath analysis for stomach disease, diabetes, and COPD

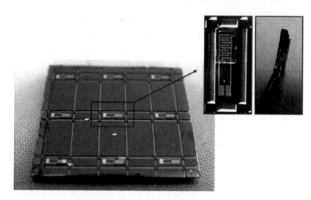

FIGURE 12.8 Photograph of fabricated microgas sensor used for stomach disease.

sensor for medical application using the silicon process in order to detect the ammonia odor inside the stomach, which would be used as an indicator for stomach diseases [43]. The fabricated sensor can detect the variation of ammonia concentration that is generated by *Helicobacter pylori* living in the stomach. It is possible to use this system for clinical applications since it demonstrates high sensitivity at room temperature and it is proved to be safe for humans. The picture of the fabricated micro gas sensor is shown in Fig. 12.8.

Also, they implemented a portable analyzer system using a conducting polymer sensor array to analyze a diabetic patient's breath. A statistically significant difference was found between the diabetic patients and a normal person [44]. The sensing characteristics of acetone using a conducting polymer sensor array were investigated and the results indicated the possibility of diagnosis of diabetic patients by exhaled breath. Figures 12.9 and 12.10 illustrate typical results for

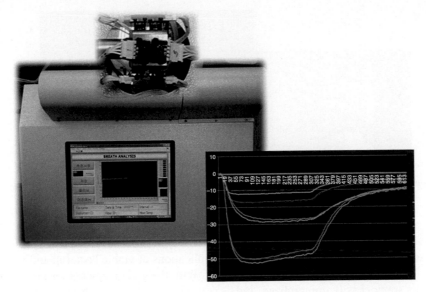

FIGURE 12.9 Exhaled breath sensing system.

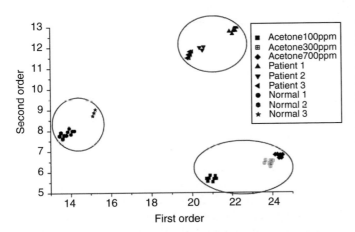

FIGURE 12.10 Analyzed data from diabetic patients' breaths with conducting polymer sensor array.

data analysis obtained from diabetic patients. In addition, the group carried out collaborative research with Manchester University to find a biomarker for chronic obstructive pulmonary diseases (COPD) and to develop a health care monitoring system.

In most cases, electronic noses that are being developed for medical diagnostics measure chemicals released by bacteria, cancer cells, or diseased tissue, through either natural metabolites or compounds generated by the organisms or cells in response to the chemistry of their environment. Even though the electronic nose developed so far is not suitable for microenvironment monitoring because of its large size and some other technical limitations, we expect that the hurdle can be overcome by the development of a nanosized electronic nose system (Fig. 12.11).

4 BIOPRINTING TECHNOLOGY FOR ADVANCES OF BIOSENSOR

Advanced biosensing devices can be fabricated by applying bioprinting technologies. Bioprinting as an enabling technology possesses the advantages of high throughput, digital controlled patterning, rapid deposition, and highly accurate transfer of various biological factors to the desired locations for numerous applications. There are many bioprinting technologies that can be adapted for use in fabricating such as deposition of metal nanoparticles [45] or nanowires [46] to a substrate via an electric field. Printing thin films of metals can be utilized to create circuits that may be an integral part of a biosensor, as well as for some immunoassays or microarrays [47,48]. Electrodeposition has even been applied for printing thin films of biological material such as proteins, enzymes, nucleic acids, polysaccharides, and bacterial cells [49–51]. Furthermore, with layer-by-layer printing technology, various cell types inside the bulk material can be spatially arranged to produce a signal cascade, where one cell type acts as a transducer and gives feedback to a secondary cell type to perform a desired action. When faster analyte detection is desired, channels can be constructed to facilitate movement of the analyte to the transducer. This capability can be useful in a combined biosensor and therapeutic application. For example, a cell-based sensor could be fabricated for diabetics, in which sugar levels could be detected rapidly by the insulin-producing beta-cells embedded in an implanted construct.

Moreover, spatial precision can enable multiplexing and high-throughput analysis to rapidly screen and detect multiple signals. For instance, rapidly patterning multiple proteins at different concentrations can enable detection of

FIGURE 12.11 Electronic nose and device: conceptual model and a realistics status for electronic nose system.

threshold levels to elicit a cellular response, or to promote cellular adhesion for parallel experiments.

In order for a biosensor to function properly, it may require the immobilization of a biological element on a transducer. However, living cells are not easily immobilized, and may migrate away from the transducer on a homogenous substrate. Similar issues arise when trying to localize cells into a particular area. Cell encapsulation is one possible solution to this problem. Microbeads and microcapsules are popular microencapsulation technologies, at a scale appropriate for use in a biosensor. Traditional microbead and microcapsule fabrication technologies are unable to precisely place the fabricated structures in specific locations, which would be necessary to transduce multiple signals. However, a new method allowing for the one-step fabrication and patterning of cell-containing microbeads could solve this problem.

5 MICROPRINTING AND PATTERNING OF LIVING BIOSENSOR APPLICATION

The combination of microscale printing and patterning of living cells has been recently studied for biomedical application. This type of 3D printing technique has been applied in several areas such as tissue engineering, cell signaling assays, and fabrication of cell-based biosensors. 3D printing of live-cell-based biosensors has raised the interest among researchers because of its advantages of patterning cells that allow quick and direct observation and tracking of individual cell response. With this technique, variables, such as cell position, density and signal concentration, and gradient formation, can be well controlled. In addition, real-time monitoring of cell physiology, spatial control of cell, or cell signaling factor distribution to determine the correlation between signal and cell response can also be verified.

Several kinds of bioprinting technologies have been developed and used to pattern the viable cells. This includes the use of thermal, piezoelectric inkjet printers, ablative laser printers, and extrusion-based printers. In addition, several types of patterning for microarrays-based printing systems exist such as SpotBot (Arrait Corp., Sunnyvale, CA, USA) and bioforce's Nano eNabler™ (Bioforce

Nanosciences, Inc., Ames, IA, USA). Especially for bioforce's Nano eNabler, studies have been used for printing applications of viable bacterial and mammalian cells.

With the great challenge of increasing the anchoring of cells onto solid surfaces for printing, the hydrogel-based anchoring matrix can be used to improve the existing problem in cell printing. Consequently, the commercially available HyStem®-C, which is comprised of thiolated hyaluronic acid and porcine gelatin cross-linked using polyethylene glycol diacrylate, has been developed. This type of hydrogel bead matrix was previously tested with the function of providing a life-like stem cell niche and also 2D and 3D proliferation of different types of cells such as stem, primary, and cancer cells.

With the existing findings, Cady and coworkers successfully developed a novel method of microprinting using quill pen lithography. They reported the use of Nano eNabler spotting and the HyStem-C hydrogel for developing cell-based biosensors to quantify the toxicant within the cellular microenvironment [52]. The existing Bioforce Nano eNabler was modified to enable larger printing channels of the surface patterning tools with 10 μm diameter for mammalian cells. In addition, HyStem-C was improved using thiol-en chemistry to provide a hydrogel matrix that is suitable for bioprinting application.

With these modifications of the printer and components composition of HyStem-C, a functioning live-cell-based reactive oxygen species (ROS) sensor was successfully developed. The developed cell printer with robust fibroblast line printed on the HyStem-C/PEG norbornene hydrogel was able to show progressive growth of cells up to 100 h (Fig. 12.12) Cells survived and proliferated during long-term culture. In addition, this cell-based bioprinter also supports 3D cellular organization of mESC cells. mESC cells were found with high survivability during the printing process and is suitable for microcell bioprinting.

By using such construct, a functional and living ROS cell-based biosensor system was tested using genetically modified MHS-roGFP-R12 cells, which express green fluorescent protein. This type of cell responds to oxidizing and reducing agents with emission of fluorescence. The oxidative and reductive reactions were quantified and fluorescence emission

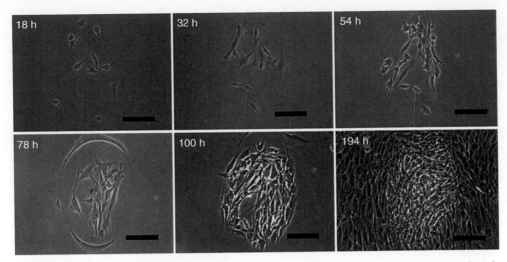

FIGURE 12.12   Growth of NIH/3T3 cells within the printed spots up to 8 days. Images were taken at 18, 32, and 54 h to show the consistent progression of the same spots.

of cells was also observed using a confocal scanning laser microscope. The printed cells were observed with direct and effective response to hydrogen peroxide and dithiothreitol at 5 min and 13 min, respectively (Fig. 12.13). Such live-cell-based ROS biosensor allows easy observation and quantification of the intensity, which can represent the changes of environmental redox. With this quill pen lithography method, more sophisticated devices with wider function and usage in the microenvironment of cells can be developed in the near future.

6   FUTURE PERSPECTIVES

Even though many modern sensors are fabricated, we still have many scientific and technological issues to overcome before achieving a clinically applicable sensor system. First of all, most of the sensor technologies developed so far are designed for and used *in vitro*. But for microenvironment monitoring in regenerative medicine, a sensor part should be implanted inside the human body. To address important issues related to the implantation of sensors, we should carefully consider biocompatibility, controllability, and size of the sensors. Second, we have to compile a list of clinically meaningful sensing targets and specific sensors for targets. Without any preclinical and clinical evidences, the sensor cannot be used even though it is cutting-edge technology. To overcome this barrier, there should be more information from the biological and medical research field on tissue regeneration. Third, the size of the sensor should be reduced to micro- or nanosized systems. Considering the current IT development, this can be achieved without much difficulty.

As the demands and application of biosensing advance, the incorporation of biofabrication technologies into biosensor elements will be paramount. In order to multiplex a variety of signals and evaluate cellular responses, sophisticated transducers must be able to separate and quantify analytes of interest. There have been numerous recent advances in

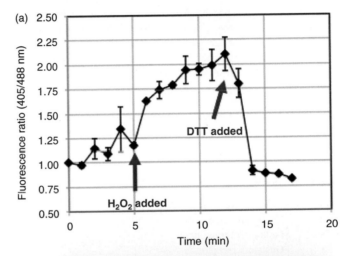

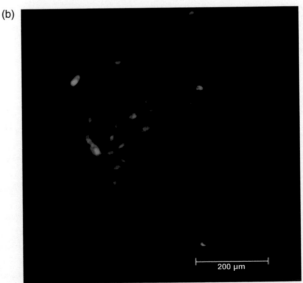

FIGURE 12.13   (a) Living biosensor function is examined by plotting the ration of 405 nm fluorescence over the reduced intensities (488 nm). (b) Fluorescence micrograph of a representative printed spot of MHS-roGFP-R12 cells.

transducer technology. By combining these advances with the latest enabling biofabrication approaches, even further advances in biosensing technology can be achieved. This combination of sensing and fabrication advances can lead to the next generation of biosensors, with a greater degree of sensitivity, throughput, and dynamic range within a single sensor for future sensing, research, diagnostic, and therapeutic applications.

GLOSSARY

Electronic nose An instrument which comprises an array of electronic chemical sensors with partial sensitivity and an appropriate pattern recognition system capable of recognizing simple and complex odors.

Disease patterning Represents actual heath state by supporting modules such as bio- and chemical sensors system for monitoring biomarkers and/or chemical marks regarding characteristics of diseases.

Breath sensor A device which can be fabricated various sensors technologies, for breath testing to detect exhaled breath for monitoring physiological state of human.

Noninvasive monitoring A technology to monitor and/or diagnose human diseases not involving the making of a relatively large incision in the body or the insertion of instruments into the patients.

Quartz microbalance A sensor for measuring the change in frequency or quartz crystal resonator in gas phase. It is valuable to use for bio- or chemical sensors and also called QCM (quartz crystal microbalance).

ABBREVIATIONS

COPD	Chronic obstructive pulmonary diseases
ESCs	Embryonic stem cells
FETs	Field-effect transistors
Gr5a	Gustatory receptor 5a
ISFET	Ion-sensitive field-effect transistor
KAMINA	Karlsruhe Micronose
PEG	Polyethylene glycol
PHD	Propyl hydroxylase domain
ROS	Reactive oxygen species

REFERENCES

[1] Pohanka M, Skládal P. Electrochemical biosensors – principles and applications. J Appl Biomed 2008;6:57–64.

[2] Eggins BR. Biosensors: an introduction. Chichester, UK: Wiley; 1996.

[3] Hierlemann A, Baltes H. CMOS-based chemical microsensors. Analyst 2003;128:15–28.

[4] Hierlemann A, Brand O, Hagleitner C, Baltes H. Microfabrication techniques for chemical/biosensors. Proc IEEE 2003;91:839–63.

[5] Patolsky F, Lieber CM. Nanowire nanosensors. Mater Today 2005;8:20–8.

[6] Liotta LA, Kohn EC. The microenvironment of the tumour–host interface. Nature 2001;411:375–9.

[7] Hanahan D, Weinberg RA. The hallmarks of cancer. Cell 2000;100:57–70.

[8] Mbeunkui F, Johann DJ Jr. Cancer and the tumor microenvironment: a review of an essential relationship. Cancer Chemother Pharmacol 2009;63:571–82.

[9] Reynolds AR, Hart IR, Watson AR, Welti JC, Silva RG, Robinson SD, et al. Stimulation of tumor growth and angiogenesis by low concentrations of RGD-mimetic integrin inhibitors. Nat Med 2009;15:392–400.

[10] Poncharal P, Wang Z, Ugarte D, De Heer WA. Electrostatic deflections and electromechanical resonances of carbon nanotubes. Science 1999;283:1513–6.

[11] Mazzone M, Dettori D, Leite de Oliveira R, Loges S, Schmidt T, Jonckx B, et al. Heterozygous deficiency of *PHD2* restores tumor oxygenation and inhibits metastasis via endothelial normalization. Cell 2009;136:839–51.

[12] Yagi K. Applications of whole-cell bacterial sensors in biotechnology and environmental science. Appl Microbiol Biotechnol 2007;73:1251–8.

[13] Uchiyama S, Makino Y. Digital fluorescent pH sensors. Chem Commun 2009;(19):2646–8.

[14] Kaim W, Schwederski B. Bioinorganic chemistry, inorganic element. The chemistry of life. Chichester, UK: John Wiley and Sons; 1994.

[15] Lippard SJ, Berg JM. Principles of bioinorganic chemistry. Mill Valley, CA: University Science Books; 1994.

[16] Fakih S, Podinovskaia M, Kong X, Collins HL, Schaible UE, Hider RC. Targeting the lysosome: fluorescent iron (III) chelators to selectively monitor endosomal/lysosomal labile iron pools. J Med Chem 2008;51:4539–52.

[17] Truong K, Sawano A, Miyawaki A, Ikura M. Calcium indicators based on calmodulin-fluorescent protein fusions. Protein engineering protocols. Totowa, NJ: Humana Press Inc., Springer; 2007.

[18] Cavalcanti A, Shirinzadeh B, Zhang M, Kretly LC. Nanorobot hardware architecture for medical defense. Sensors 2008;8:2932–58.

[19] Blake DA, Jones RM, Blake RC II, Pavlov AR, Darwish IA, Yu H. Antibody-based sensors for heavy metal ions. Biosens Bioelectron 2001;16:799–809.

[20] Bontidean I, Berggren C, Johansson G, Csöregi E, Mattiasson B, Lloyd JR, et al. Detection of heavy metal ions at femtomolar levels using protein-based biosensors. Anal Chem 1998;70:4162–9.

[21] Chouteau C, Dzyadevych S, Durrieu C, Chovelon J-M. A bi-enzymatic whole cell conductometric biosensor for heavy metal ions and pesticides detection in water samples. Biosens Bioelectron 2005;21:273–81.

[22] Bousse L. Whole cell biosensors. Sens Actuators, B 1996;34:270–5.

[23] Stenger DA, Gross GW, Keefer EW, Shaffer KM, Andreadis JD, Ma W, et al. Detection of physiologically active compounds using cell-based biosensors. Trends Biotechnol 2001;19:304–9.

[24] Neher E. Molecular biology meets microelectronics. Nat Biotechnol 2001;19:114.

[25] Glatz R, Bailey-Hill K. Mimicking nature's noses: from receptor deorphaning to olfactory biosensing. Prog Neurobiol 2011;93:270–96.

[26] Liu Q, Ye W, Yu H, Hu N, Du L, Wang P, et al. Olfactory mucosa tissue-based biosensor: a bioelectronic nose with receptor cells in intact olfactory epithelium. Sens Actuators, B 2010;146:527–33.

[27] Misawa N, Mitsuno H, Kanzaki R, Takeuchi S. Highly sensitive and selective odorant sensor using living cells expressing insect olfactory receptors. Proc Natl Acad Sci USA 2010;107:15340–4.

[28] Liu Q, Wu C, Cai H, Hu N, Zhou J, Wang P. Cell-based biosensors and their application in biomedicine. Chem Rev 2014;114(12):6423–61.

[29] Lau HC, Bae TE, Jang HJ, Kwon JY, Cho WJ, Lim JO. Biomimetic trehalose biosensor using gustatory receptor (Gr5a) expressed in *Drosophila* cells and ion-sensitive field-effect transistor. Jpn J Appl Phys 2013;52. 04CL2.

[30] Fang Y. Label-free biosensors for cell biology. Int J Electrochem 2011;2011:1–16.

[31] Du L, Wu C, Liu Q, Huang L, Wang P. Recent advances in olfactory receptor-based biosensors. Biosens Bioelectron 2013;42:570–80.

[32] Lim J, Yu J, Kwon J, Byun H, Huh J, Cho W. Development of sugar sensitive *Drosophila* cell based ISFET sensor for Alzheimer's disease diagnosis. J Korean Sensor Soc 2013;22:281–5.

[33] Abikshyeet P, Ramesh V, Oza N. Glucose estimation in the salivary secretion of diabetes mellitus patients. Diabetes Metab Syndr Obes 2012;5:149.

[34] Bermejo-Pareja F, Antequera D, Vargas T, Molina JA, Carro E. Saliva levels of Abeta1-42 as potential biomarker of Alzheimer's disease: a pilot study. BMC Neurol 2010;10:108.

[35] Naik VV, Satpathy Y, Pilli GS, Mithilesh NM. Comparison and correlation of glucose levels in serum and saliva of patients with diabetes mellitus. Indian J Public Health Res Dev 2011;2:103–5.

[36] Maher M, Pine J, Wright J, Tai YC. The neurochip: a new multielectrode device for stimulating and recording from cultured neurons. J Neurosci Methods 1999;87:45–56.

[37] Yeung CK, Ingebrandt S, Krause M, Offenhäusser A, Knoll W. Validation of the use of field effect transistors for extracellular signal recording in pharmacological bioassays. J Pharmacol Toxicol Methods 2001;45:207–14.

[38] Denyer M, Riehle M, Britland S, Offenhauser A. Preliminary study on the suitability of a pharmacological bio-assay based on cardiac myocytes cultured over microfabricated microelectrode arrays. Med Biol Eng Comput 1998;36:638–44.

[39] Hafeman DG, Parce JW, McConnell HM. Light addressable potentiometric sensor for biochemical systems. Science 1988;240:1182–5.

[40] Harper WJ. The strengths and weaknesses of the electronic nose. Headspace analysis of foods and flavors. New York, NY: Kluwer Academic/Plenum Publishers, Springer; 2001.

[41] Willis R, Lesney M. Diagnosis: medicine. Mod Drug Discov 2001;4:49–58.

[42] Bruins M, Bos A, Petit P, Eadie K, Rog A, Bos R, et al. Device-independent, real-time identification of bacterial pathogens with a metal oxide-based olfactory sensor. Eur J Clin Microbiol Infect Disease 2009;28:775–80.

[43] Lee YS, Song KD, Huh JS, Chung WY, Lee DD. Fabrication of clinical gas sensor using MEMS process. Sens Actuators B 2005;108:292–7.

[44] Yu J-B, Byun H-G, So M-S, Huh J-S. Analysis of diabetic patient's breath with conducting polymer sensor array. Sens Actuators B 2005;108:305–8.

[45] Dias AD, Kingsley DM, Corr DT. Recent advances in bioprinting and applications for biosensing. Biosensors 2014;4:111–36.

[46] Cui X, Gao G, Qiu Y. Accelerated myotube formation using bioprinting technology for biosensor applications. Biotechnol Lett 2013;35:315–21.

[47] Gruene M, Pflaum M, Deiwick A, Koch L, Schlie S, Unger C, et al. Adipogenic differentiation of laser-printed 3D tissue grafts consisting of human adipose-derived stem cells. Biofabrication 2011;3. 015005.

[48] Serra P, Fernández-Pradas JM, Colina M, Duocastella M, Dominguez J, Morenza JL. Laser-induced forward transfer: a direct-writing technique for biosensors preparation. J Laser Micro/Nanoen 2006;1:236–42.

[49] Lee TK, Sokoloski TD, Royer GP. Serum albumin beads: an injectable, biodegradable system for the sustained release of drugs. Science 1981;213:233–5.

[50] Zaman MH, Trapani LM, Sieminski AL, Mackellar D, Gong H, Kamm RD, et al. Migration of tumor cells in 3D matrices is governed by matrix stiffness along with cell-matrix adhesion and proteolysis. Proc Natl Acad Sci USA 2006;103:10889–94.

[51] Cukierman E, Pankov R, Stevens DR, Yamada KM. Taking cell-matrix adhesions to the third dimension. Science 2001;294: 1708–12.

[52] Hynes WF, Doty NJ, Zarembinski TI, Schwartz MP, Toepke MW, Murphy WL, et al. Micropatterning of 3D microenvironments for living biosensor applications. Biosensors 2014;4:28–44.

Chapter 13

Polymers for Bioprinting

James K. Carrow*, Punyavee Kerativitayanan*, Manish K. Jaiswal*, Giriraj Lokhande*, and Akhilesh K. Gaharwar*,**
*Department of Biomedical Engineering, Texas A&M University, College Station, TX, USA; **Department of Materials Science and Engineering, Texas A&M University, College Station, TX, USA

Chapter Outline

ABSTRACT

Bioprinting is a process of precisely designed scaffolds using three-dimensional printing technologies for functional tissue engineering utilizing cell-laden biomaterials as bioink. A range of polymers can be used as bioink to stimulate favorable cellular interactions, leading to enhanced cell motility, proliferation, and subsequent differentiation. Both natural and synthetic polymers have been considered for various bioprinting applications, each with a corresponding set of advantages and limitations. Natural polymers more aptly mimic the native extracellular matrix, leading to more favorable cellular responses, while synthetic polymers can be more easily tailored for more efficient printing. Because many of these bioink materials are rooted in traditional tissue engineering scaffold design, bioprinting optimization remains a challenge; however, emerging trends in bioink development have begun to circumvent these issues, providing bioprinting research with a very promising future in regenerative medicine. Further investigation into the interplay of polymer type and fabrication technique will help to formulate new polymer bioinks that can expedite the process from printing to implantation.

Keywords: bioprinting; natural polymers; synthetic polymers; polymer hybrids; polymer blends; tissue engineering; 3D printing

1 INTRODUCTION

Bioprinting is a process of precisely designed scaffolds using three-dimensional (3D) printing technologies for functional organ engineering. Due to the pressing need for functional organ engineering, precisely designed scaffolds for tissue repair and organ replacement are needed. The emergence of nano- and microscale printing technologies resulted in the development of 3D-printed scaffolds consisting of spatially controlled cell patterns that may be loaded

Essentials of 3D Biofabrication and Translation. http://dx.doi.org/10.1016/B978-0-12-800972-7.00013-X

with appropriate biological moieties to control or direct cell fate [1–4]. The rationale for such significant control over spatially driven design is to better coordinate cellular arrangements into tissues and organs of interest and therefore lead to the successful production of functional and implantable constructs. Bioprinting remains in its early stages of development but continues to gain popularity amongst regenerative medicine researchers due to its immense potential in the field (Fig. 13.1). A steady increase in the number of publications and citations in the area of bioprinting indicate its huge potential in biomedical applications including tissue engineering, drug development, and organ-on-chip platforms. While challenges exist to maintain the intended shape and cell distribution of the construct over time, researchers have employed a variety of novel methods and technologies to improve upon the bioprinting process [5–8].

A vital aspect and bottleneck to the design and implementation of a bioprinting system is the consideration of a bioink. The bioink mainly comprises of a polymer matrix loaded with cells and bioactive signals [9]. By controlling the physical and chemical properties of the extracellular matrix (ECM), cell behavior can be regulated to accelerate tissue integration and functional recovery. Thus, it is important to carefully select polymers for bioprinting to support and enhance tissue regeneration in a temporospatial manner [10]. A range of polymers can be used as a bioink to provide tunability to stimulate favorable cellular interactions, leading to enhanced cell motility, proliferation, and subsequent differentiation (Fig. 13.2). In order to mimic native tissue microenvironment and to drive regeneration of tissue, many researchers are focusing on natural polymers to fabricate 3D printed scaffolds [11]. This is mainly due to the chemical and structural similarities of natural polymers to native ECM. Some of the natural polymers that are currently explored for bioprinting include collagen/

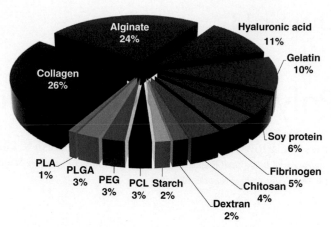

FIGURE 13.2 **Chart diagramming natural (red) and synthetic (blue) polymer distributions for use as bioinks compiled from relevant literature.** Hybrid systems split into polymer constituents for consideration.

gelatin, alginate, fibrin, hyaluronan, and dextran. Synthetic polymers, however, can be more easily tailored for a given application by optimizing mechanical properties, degradation rates, or functionalized with a variety of bioactive factors [12]. Some of the synthetic polymers that are currently used in bioprinting include poly(ethylene glycol) (PEG), poly(lactide-*co*-glycolide) (PLGA), poly(ε-caprolactone) (PCL), and poly(L-lactic acid) (PLLA) (Fig. 13.2). Blends of natural and synthetic polymers can result in a combination of benefits in an attempt to enhance and tailor cellular responses within the 3D fabricated scaffolds. For example, synthetic polymers can be copolymerized with enzymatic degradation sites found in natural polymers to enhance cell migration and aggregation within the scaffold [13]. These polymer combinations used in conjunction with the spatial resolution of many bioprinting technologies enable a more efficacious tissue and organ regeneration process.

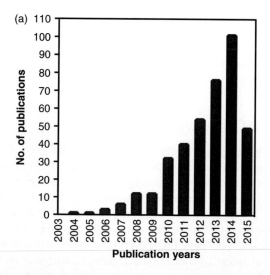

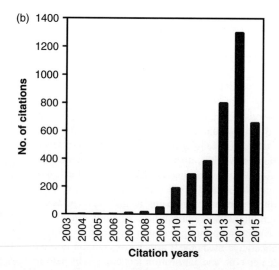

FIGURE 13.1 **Publication trends over the past decade demonstrating a significant rise in the field of bioprinting.** (a) Trends in number of publications, and (b) number of citations for articles pertaining to bioprinting (Data obtained on June 2015 from ISI Web of Science).

In this chapter, we focus on various natural and synthetic polymer systems for bioprinting. We will discuss some of the important physical and chemical properties of polymers that play a crucial role in the bioprinting process. Different types of polymers, their physical, chemical, and biological properties that require optimization for specific bioprinting methods will also be discussed. Specifically, polymer properties as they apply to synthetic, natural, and hybrid systems, will be examined. Emerging trends, such as polymer-nanocomposite bioinks, growth factor incorporation, and multinozzle polymer systems, along with future directions in bioink research, are highlighted. Additionally, some of the most critical challenges associated with the selection, fabrication, and evaluation of polymeric systems for bioprinting are emphasized.

2 POLYMER PROPERTIES FOR BIOPRINTING

A true material science approach toward bioink design is essential for bioink success. This material design approach is required to understand how polymer characteristics influence printing efficacy and cytocompatibility. Because these bioinks will be subjected to the printing process and then expected to maintain structural integrity, there is an inherent complexity of bioink production. Properties like viscoelasticity of polymer bioinks, which affects printed structure outcome, or polymer behavior in solution, which impacts printing ability, are common design criteria. Similarly, polymers demonstrating creep and stress relaxation can alter material shape and lead to impeded functionality. Other considerations like shelf-life and cost of polymer bioinks are also important determinants of material choice [14]. Many of the important properties relevant for bioprinting are illustrated in Fig. 13.3. Additionally, while the availability of natural polymers is more limited than synthetics, the majority of research on bioinks has focused on naturally derived materials. The challenge to simultaneously optimize polymer–polymer interactions as well as polymer–cell interactions while maintaining printability persists in current bioprinting research. A need exists to develop bioink-specific materials as opposed to investigators transposing traditional tissue engineering biomaterials into forced bioprinting roles.

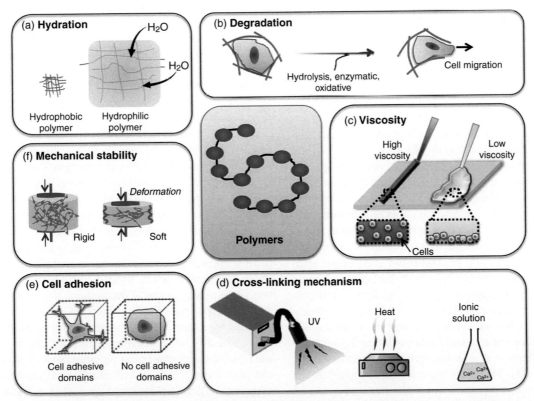

FIGURE 13.3 Polymer properties considered for bioprinting applications. (a) Hydration of a polymer system enables nutrient and waste transport to encapsulated cells deep within a printed construct. (b) Degradation mechanism can be modified to influence cellular migration and tissue regeneration. (c) Viscous solutions can better suspend and shield cells from shear forces within the nozzle, yet are more likely to obstruct flow, while low viscosity solutions can avoid clogging however can suffer from cell settling. (d) Cross-linking mechanism is dependent on polymer type, cell viability considerations, and desired construct properties as cross-links can exist permanently or reversibly. (e) Cell adhesion influences cell fate and can be controlled through polymer type or processing. (f) Stability of a printed material is vital to the success as it should mimic native tissue mechanically as well as maintain its shape to organize cell growth.

2.1 Cross-Linking Mechanism

The type of polymer selected for bioprinting mainly depend on the type of cross-linking techniques employed to generate 3D networks. Various cross-linking mechanisms to obtain structurally stable polymeric network include photo-, ion-, electrostatic-, pH-, and temperature- based cross-linking mechanisms. Care must be taken when choosing the cross-linking method, as the viability of the encapsulated cells will significantly depend on the environmental conditions such as heat or pH. Similarly, the cross-linking technique influences construct outcomes with considerations of gelation via reversible or irreversible processes. Specifically, thermal gelation mechanisms result in constructs with a temporary shape that can vary with temperature, while chemical gelation can provide a permanent shape with a lower chance of deformation over time. The trade-off in many instances, however, is that the time required for chemical-dependent solidification exceeds that of temperature-dependent gelation. For example, a variety of polymers can be modified chemically via introduction of acrylate or methacrylate groups, which can subsequently be photocrosslinked in the presence of a photoinitiator (PI) to form a covalently cross-linked network [15,16]. For example, reacting poly(ethylene glycol) (PEG) with methacryloyl chloride in the presence of triethylamine and modifying gelatin with unsaturated methacrylamide side groups yield photocrosslinkable poly(ethylene glycol) dimethacrylate (PEGDMA) and gelatin methacrylamide, respectively [17,18]. Physical cross-links can be similarly induced ionically via an additional cation solution (e.g., Ca^{2+}) to produce natural polymer-based networks [4,5,16]. For example, sodium alginate solution gels when in contact with Ca^{2+}, a divalent cation, tethers neighboring G-blocks of alginate chains, forming a reversible electrostatic bridge. The degree of gelation can be tuned by adjusting the concentration of $CaCl_2$ solution [19–22]. Thermoresponsive gelation of gelatin can also be employed in bioprinting since it aids in retaining the shape of printed constructs. However, native gelatin has not been used alone for bioprinting because its reversible sol–gel transition (i.e., upper critical solution temperature (UCST) material behavior) poses difficulties in optimizing printing temperature and viscosity. These cross-links can then influence the mechanical strength of the fabricated tissue as well as influence cellular responses of encapsulated stem cells through a more efficient distribution of mechanical stresses or reducing migration [23–25]. The resulting networked materials can display thixotropic properties where the stress of extrusion induces a quasi-liquid state until exiting the nozzle, where it displays a solid-like behavior once more [5]. Depending on the cross-linking method (e.g., ionic, photonic, or electrostatic) and the polymer concentration, printing efficiency can be significantly impacted. Lastly, many extrudable polymers maintain their structural integrity by using a high curing temperature or solvents toxic to cells for polymerization [26]. These conditions (e.g., extreme temperature) could be too harsh for cell survival. As a result, not all polymers used for 3D printing, where cells are seeded after scaffold fabrication, can be used for bioprinting. For example, heat treatment at 100°C is necessary to dry and maintain structural integrity of printed starch constructs. This extreme temperature prevents the use of starch blend for bioprinting [27].

2.2 Viscosity

Determination of the polymer structure and its rheological properties are also important for smooth extrusion of ink in contact of high shear rate ($\sim$20–200 s^{-1}) and shape retention [28]. The most critical location during extrusion is the dispensing tip where clogging and fracture can occur if the viscosity is too high while an inadequate viscosity will impart shear forces on the cells, resulting in death [29]. The printed polymer should demonstrate multiple physical phases: first like a fluid within the printer and then a solid once dispensed. This will maintain the extruded filament-like shape. Polymers with shear thinning behavior promote this solidification process since viscosity will increase significantly as shearing wanes after extrusion [26,28]. Many bioprinted polymers have been shown to exhibit shear thinning behavior. For example, silk fibroin molecules align and the friction resistance between adjacent fibers decreases as shear rate increases [30]. These shear-thinning capabilities reduce nozzle-clogging effects and provide researchers with an alternative over the use of low-viscosity polymer solutions. Higher viscosity polymer bioinks from increased cross-linking density also reduce the likelihood of nozzle obstruction from cell aggregates as cells are better suspended in the bioink; however, printing time and risks of cell death or gel fracture increases with viscosity among other challenges to this approach [5,10]. Regarding the rheological properties, one challenge posed by natural polymers is variability of viscosity from batch to batch, affecting optimal printing parameters including nozzle size, shear rate, and pressure needed [28]. Particularly important is the ability to shield the biological components from the forces associated with printing. Cell suspensions surrounded by a more gel-like material will be more likely to retain viability as opposed to more fluid-like bioinks. The use of semi-interpenetrating polymer networks (semi-IPN) can produce a hydrogel with mechanical stability as well as being sufficiently hydrated to enable nutrient transfer and construct shape due to the presence of a secondary uncross-linked polymer web. In some cases, charged domains on the cross-linked network can provide electrostatic repulsion for increased hydration and strength, although future research is needed for the practicality of such a bioink.

2.3 Hydration Properties

One property that can greatly alter the viscosity of the bioink is hydration of the material. Hydration and porosity provide control over mechanical strength, tunable viscoelasticity and in the design of optimized pressure through the micronozzle during the fabrication process. Apart from that, oxygen and nutrient transport to the laden cells within the hydrogel network are prerequisites for any successful engineered tissue. Many natural polymers and hydrophilic synthetic polymers, such as PLGA and PEG, are finding expanded applicability in designing complex tissue structures like blood vessels, where on one hand the fluidity of the material allows for easy fabrication while hydration enables the material to successfully mimic natural tissue. There is a trade-off, however, between porosity or hydration and mechanical strength. With increased pore size, the modulus of the hydrogel decreases because of increased water-to-hydrogel contact, and water acts as a plasticizer within the polymer chains. Thermo-responsive hydrogels in particular are known to have more aqueous uptake and therefore slower deswelling kinetics, which could be important for applications in which the hydrogel is heated or cooled upon extrusion.

2.4 Biological Interaction Properties

Physical properties of polymers, such as hydrophilicity and surface energy, influence cellular behavior and are important parameters in determining the applicability of polymers for generating specific tissues. For example, polymers with cell adhesive site are required to enhance survival and proliferation of adhesive types of cells such as osteoblasts. However, chondrocytes would prefer to be encapsulated in hydrogels with minimum cell adhesion sites. Polymer matrices that interact on similar length scales to mechanoreceptors present on cell membranes have also been documented to improve differentiation of seeded stem cells. For example, collagen and gelatin hold Arg–Gly–Asp (RGD) sequences, which promote cell adhesion via integrin receptors [31,32]. Polymers without cell binding moieties, such as synthetic polymers and some natural polymers like gelatin, may prevent cell adhesion and proliferation. Apoptosis could be induced as a result of the lack of cell adhesion, which is termed "anoikis." This drawback could be overcome by adding cell adhesive molecules, for example, by blending synthetic polymers with natural polymers and blending alginate with gelatin [33–35]. It should be noted that many synthetic polymers that are widely used for tissue engineering, such as PCL and PLLA, have been shown to exhibit good cytocompatibility and promote cell adhesion and proliferation even in the absence of cell-binding moieties [34,36,37]. Nevertheless, their biological properties could be further enhanced by an introduction of cell adhesive molecules.

2.5 Mechanical Properties

The mechanical functions of natural tissues are vital in the replication process by any proposed biomaterial considered for bioprinting. Invariably these natural tissues are composed of a stiff acellular environment surrounded by softer cellular matrices. Several cell-laden thermoplastic polymers have been explored to construct the well-organized acellular content with a range of rigidities and desired mechanical properties. Along these same lines, the mechanical properties of a chosen polymer are heavily reliant on chemical composition. Consequently, there is an intimate relationship between biomaterial robustness and molecular arrangement of the included polymer. Therefore, an understanding of the factors that influence polymer mechanical behavior is required to effectively design a bioprinting system. By varying cross-link density and therefore the tensile modulus of the hydrogel, stem cell interactions with the environment, through actin-mediated pathways for example, can be controlled. This leads to control over differentiation and subsequent formation of tissue.

2.6 Polymer Chain Considerations

Aside from those polymer properties previously described, there are multiple interactions specific to the polymer backbone that alter response to the environment. One such interplay stems from varying temperature. Because of the thermal sensitivity of polymers through chain mobility, temperature control during the bioprinting process could be required, especially for those with low glass transition temperatures (T_g). Solvent interactions also affect bulk material behavior; for example, polymer hydrophilicity and concentration can significantly impact hydrogel swelling. The presence of water not only affects mechanical properties of the material, but also transport of nutrients and metabolic waste products within the polymer matrix, which influences cell proliferation. Another crucial consideration of printed polymers is the mechanism and rate of degradation. Hydrolyzable domains, like in polyanhydrides, polyorthoesters, polyamides, and polyesters, enable variable degradation rates of synthetic polymer constructs. Enzymatically cleavable sites, like those found in naturally based polymers, are important to consider particularly for cell migration through a bioprinted hydrogel as biological agents in the surrounding environment cause localized mass loss of the material through the release of specific enzymes. Once these linkage sites have been cleaved, the polymer can incur mass loss, either at the surface of the bulk material or internally, depending on the relative rates of water penetration and chain scission.

2.7 Fabrication Techniques

A multitude of microscale technologies have emerged that employ preprocessing of polymers to generate functional strands capable of cross-linking [16,38]. Some of the fabrication methods include drop-on-demand bioprinting with inkjet nozzles utilizing thermal or piezoelectric technology to generate displacement forces, stereolithography, or laser-assisted printing [16,39,40]. Due to the variation between printing methods and physiological environments associated with various applications, the development of tailored bioinks is necessary to ensure tissue and organ functionality. Individual building blocks comprised of a cell-laden biomaterial are one such microscale technology to materialize. Hydrogel droplets, or "tissue spheroids," containing large amounts of cells can fuse together over time, resulting in a highly controllable tissue construct that does not require a scaffold [8]. Multiple technologies have also emerged to fabricate constructs with a combination of polymer inks as well as cell types. These methodologies are particularly useful for therapies like osteochondral regeneration where the tissue adjoins multiple cell types comprising regions of different mechanical strength and stiffness [41]. As the complexity of new printing technologies rises, so too will the intricacy of bioink interactions. Interpolymer interactions between separate bioinks could result in better construct stability or enhanced cellular behavior.

3 NATURAL POLYMERS FOR BIOPRINTING

The selection of polymers is critically important for the bioprinting process. Natural polymers can avoid coarse fabrication conditions like temperature and organic solvents, thereby allowing the printing of cells and bioactive components [28]. However, there are three key challenges associated with printing natural polymers. First, innate characteristics, such as viscosity, can fluctuate over a group of natural polymers, making printing reproducible scaffolds with a precise structure a challenge. Also, the presence of moisture can be inconsistent in polymer groups, varying pressure needed to print the polymer inks. The other challenges include strut solidification and the interplay of charged components between polymer chains and fluid if the polymer ink is plotted into a solution. Solidification in air requires the polymer to dry at an optimum rate to support and merge to subsequent printed layers. If printed into liquid media, density of the solution must be kept constant to preserve the strut's shape and prevent dissolution of printed struts [25].

Owing to the aforementioned criteria and challenges, the number of natural polymers that have been successfully bioprinted is limited. These include collagen, gelatin, alginate, hyaluronic acid, dextran, and fibrin. Some of them have difficulties when printed alone and need to combine with other polymers for more suitable rheological properties and enhanced mechanical integrity. Chitosan, silk fibroin, starch-based polysaccharides, and soy protein have been used for 3D printing but not bioprinting due to non-cell-friendly processing conditions. These polymers will be discussed later in this section.

3.1 Natural Polymers for Bioprinting

Of the bioinks currently established for research, many stem from naturally derived polymers. The following polymers alone provide significant cytocompatibility through simulating a natural environment for tissue remodeling. A multitude of cross-linking methods are capable of generating a solidified construct with varying degrees of mechanical integrity.

3.1.1 Collagen/Gelatin

Collagen contains a large quantity of glycine, proline, and hydroxyproline residues, and contains a small amount of aromatic and sulfur-containing amino acids. Pyrrolidine (i.e., proline and hydroxyproline) is responsible for the stabilization of the tertiary super-helix structure through steric hindrance. Collagen significantly constitutes the ECM and participates in numerous physiological interactions. Its contacts with cells modulate cell adhesion, migration, proliferation, and differentiation [32]. As a result, collagen is widely used to fabricate tissue-engineering scaffolds. However, the inferior mechanical integrity limits its application in bioprinting [42].

Losing the secondary structure as well as tertiary and primary structures found in collagen results in the denaturalized form of gelatin. As a result, the helical arrangement of collagen is lost to the randomness present in gelatin. Gelatin is coiled at temperatures above 40°C in aqueous environments; and it reversibly forms an alpha helix when the solution is cooled to below 30°C. At diluted concentrations, chain mobility can produce intramolecular bonds. When the concentration increases to above 1%, chain association and three-dimensional networks are induced. The manner and extension of this reversible fold of the triple-helix structure appears to depend on solvent, concentration, and temperature [43]. Gelatin is widely used for tissue engineering applications since it has almost identical composition to collagen, the main component of natural ECM. Specifically, it holds RGD sequences, which promotes cell adhesion via integrin receptors. In addition, since gelatin is a denatured polymer, concerns of immunogenicity and pathogen transmission associated with collagen can be circumvented.

Thermoresponsive transition of gelatin could be employed in bioprinting since it aids in structure maintenance upon printing. However, native gelatin has not been used alone for bioprinting because this temperature-dependent, but reversible sol–gel transition (i.e., UCST material behavior), poses difficulties in optimizing printing temperature and viscosity. This shortcoming can be overcome by chemical modification and/or combining gelatin with other polymers. For example, gelatin can be modified with unsaturated methacrylamide side groups to yield a photosensitive gelatin derivative, gelatin methacrylamide [17]. A stable cross-linked scaffold can be obtained by exposing to UV radiation in the presence of a PI. After photocrosslinking of gelatin chains through methacrylamide double bonds, thermal gelation of physical crosslinks are no longer temperature responsive (i.e., become irreversible) [44].

In a similar approach, Billiet et al. reported bioprinting of cell-laden gelatin methacrylamide using a pneumatic-based bioplotter [17]. They observed that the types of PI have profound effects on cell viability. In this study, water soluble halogen-free azo initiator (2,2′-azobis[2-methyl-*N*-(2-hydroxyethyl)propionamide], also known as VA-086 (Wako Chemicals, USA)) gave rise to enhanced hepatocarcinoma cell viability (<97%) compared to the conventional alpha-hydroxy ketone-based PI (2959 (2-hydroxy-1-[4-(2-hydroxyethoxy) phenyl]-2-methyl-1-propanone, also known as Irgacure). The prepolymer solution containing gelatin methacrylamide showed shear thinning behavior that is preferable for bioprinting. The scaffolds with a 100% interconnected pore network could be produced using 10–20% w/v% gelatin methacrylamide. Another research group reported impaired construct formation with gelatin methacrylamide less than 20 w/v% due to severe viscosity issues caused by difficult temperature control. They solved this problem by blending gelatin methacrylamide with hyaluronic acid to increase the solution viscosity [44]. On the other hand, Billiet et al. were able to produce gelatin-only scaffolds by varying gelatin acrylamide concentrations [17]. This was achieved by optimizing multiple printing parameters including concentration, temperature, pressure, nozzle type and diameter, and plotting speed. Also, device adaptations were necessary to ensure homogenous plotting temperature and enable the cooling of the platform to a temperature far below the gelling point.

Similar to gelatin methacrylamide, gelatin methacrylate (GelMA) is electrostatically charged and has intrinsic adhesive properties owing to its uncured acrylate groups. Cell-laden gelatin methacrylate was bioprinted followed by photocrosslinking under UV light [45]. Printability declined with reducing UV exposure time, GelMA concentration, and increased cell density. Printed HepG2 cells could be preserved at viability levels higher than 80%. In addition, cell proliferation in printed constructs was found to be higher than control hydrogel blocks. This could be attributed to easier access to nutrients in precisely designed bioprinted scaffolds.

Double chemical functionalization could further enhance control over bioprinting. For example, methacrylation and acetylation of gelatin allowed control over solution viscosity, gelling behaviors, and photochemical cross-linking simultaneously [31]. Particularly, the degree of methacrylation determined the mechanical properties of cross-linked constructs while additional acetylation influ-

enced rheological properties of the gelatin solution. The degree of methacrylation could be adjusted by molar excess of methacrylic anhydride during the methacrylation process. Highly methacrylated gelatin made from 10-fold molar excess had low viscosities within inkjet-printable range (3.3 ± 0.5 mPa•s, 37°C) and resulted in cross-linked constructs with high storage moduli G' (15.2 ± 6.4 kPa). Twofold molar excess was suitable for the preparation of soft hydrogels, but its solution was highly viscous and could result in nozzle clogging. The viscosity could be reduced by additional acetylation of GelMA. In this manner, a two-fold functionalized gelation solution could be inkjet printed and could result in a soft gel with a low degree of cross-linking. It should be noted that unmodified gelatin and GelMA with low degrees of methacrylation have high viscosity even above their melting temperature and gel at room temperature. Consequently, they are prone to clogging nozzles. The printed constructs were found to be cytocompatible. Porcine chondrocytes printed with two-fold-modified gelatin had high viability and normal functionality.

3.1.2 Alginate

Alginate is an anionic polysaccharide obtained from brown seaweed. The raw material extracted from seaweed is known as sodium alginate. The terms alginate and sodium alginate are often used interchangeably. Alginate is a linear block copolymer composed of β-D-mannuronic acid monomers (M-blocks) in sequence with α-L-guluronic acid blocks (G blocks), and intermixed M and G domains. Alginate solutions gel with divalent cations due to ionic bridge formation between G-blocks [19,21,22,46]. Structural similarities to natural ECM, an excellent biocompatibility, viscosity, and the ease of gelation that takes place at room temperature make it attractive for bioprinting [19,21,22,46].

Nakamura et al. printed 3D alginate cell-laden constructs using an inkjet printer [33]. Alginate solution was ejected onto a calcium chloride ($CaCl_2$) solution and each droplet formed a homogenous gel bead. The beads then fused together forming gel fibers and finally 3D gel constructs. The printed constructs were found to have good mechanical integrity and biocompatibility. However, since alginate does not have cell-binding moieties, such as RGD sequences, it may prevent cell adhesion and proliferation. Apoptosis could be induced due to the lack of cell adhesion, which is termed "anoikis." This drawback could be overcome by adding cell adhesive molecules, for example, by blending alginate with gelatin or other polymers with cell adhesive sites.

In addition, alginate biomaterials encapsulating endothelial cells were printed using a multinozzle deposition system [47]. Suitable fabrication parameters were found to be 1.5% w/v alginate and 0.5% w/v $CaCl_2$. The elastic modulus of printed scaffolds increased from day 0 to day 1, and then gradually decreased over time. This could be attributed to continued alginate cross-linking by ions present in the cell culture media. The ions diffused and interchanged within scaffolds, allowing alginate chains to detach from the main cross-linked constructs and diffuse out. Eventually, the structures degraded and lost their mechanical integrity over time. The viability of encapsulated cells ranged between 76% and 83% for printing shear stresses of 100–1150 kPa, respectively [47]. Similarly, a solution with 1% w/v alginate and 1% $CaCl_2$ was successfully printed using a multinozzle system [46]. However, dragging between printed layers caused by viscous fraction made the pattern slant, hindering accurate shaping of structures. The possible solutions to this problem are optimizing gelling duration and extent, and adopting a feedback control system of printing speed using a vision system that can monitor the printed construct in real time [46].

There are two main challenges associated with bioprinting of alginate. First, it is difficult to print 3D cell-laden alginate scaffolds with completely interconnected pores due to the difficulty in controlling the gelation process. Second, the thickness of constructs that can be printed is limited because alginate has favorable interactions with water and minimal viscosity hindering thick structures printing. Specifically, since alginate is highly soluble in aqueous solution, dispensing alginate directly in $CaCl_2$ solution can weaken the constructs. To circumvent these challenges, Ahn et al. proposed a new printing system consisting of a dispensing method and an aerosol-spraying method [19]. A cell-laden alginate solution was printed using a dispensing system followed by an aerosol spray of $CaCl_2$. After printing, the construct was immersed in $CaCl_2$ solution for a second curing. Before an aerosol spray, viability of encapsulated preosteoblast cells (MC3T3-E1) was as high as 97%. With increasing weight fractions of $CaCl_2$, cell viability decreased. To further incorporate bioactive clues, researchers have incorporated proteins/drug-loaded microparticles within the printed alginate scaffold [48]. For example, bone morphogenic protein 2 (BMP-2)-loaded gelatin microparticles were embedded in cell-laden alginate and showed osteogenicity *in vivo* [48]. The encapsulation of BMP-2 within the gelatin microparticles results in sustained release of the protein. It is expected that by incorporating such bioactive clues cellular process can be controlled.

3.1.3 Fibrin

Fibrin is formed by the interaction between fibrinogen and thrombin, the mechanism known for blood coagulation. It is also a component of natural ECM. Fibrinogen is a glycoprotein consisting of multiple pairs of polypeptide chains: Aα, Bβ, and γ. It contains a cell-signaling domain including protease degradation and cell adhesion motifs. To form fibrin gel, thrombin cleaves Aα and Bβ chains to fibrinopeptide Λ and B [33,49]. Then, these fibrin monomers spontaneously polymerize to form protofibrils, which associate laterally to form fibrin fibers. Finally, fibrin fibers associate to form fibrin gel [33,49]. Due to the presence of cell adhesion motifs and the ease of gelation, fibrin has a potential for bioprinting. Cells were found to adhere and proliferate well in cell-laden printed fibrin scaffolds [33]. In comparison to alginate-only gel-laden constructs, fibrin has an advantage in terms of cytocompatibility. This is mainly due to the presence of cell adhesion moieties within the fibrin structure. However, fibrin gel, formed by an inkjet printer, was shown to be soft and fragile, and has difficulty in maintaining its 3D structure [33]. Some of these drawbacks can be overcome by combining fibrin with different natural and synthetic polymers. Although fibrin-based materials are promising, very limited work has been reported using fibrin-based bioink.

3.2 Cell-Laden Polymer Blends for Bioprinting

As opposed to a single naturally derived polymer for use as a bioink, multiple polymers may be blended together to improve printing efficacy or construct performance. We will discuss these polymer systems in particular in the following section.

3.2.1 Fibrin/Collagen

Fibrin and collagen have been used for bioprinting. Both of them are ECM components and have excellent biocompatibility. Fibrin/collagen blend solutions containing amniotic-fluid-derived stem cells and bone-marrow-derived mesenchymal stem cells were bioprinted onto full-thickness skin wounds [50]. Gelation was achieved by alternate printing of fibrin/collagen layer and thrombin. Migration and integration of cells into regenerated tissues were not observed, suggesting that the mobility of printed cells was limited due to fibrin/collagen struts. Fibrin and collagen contain cell adhesive motifs that might prevent the cell from being mobile. This could potentially limit therapeutic success. In addition, collagen naturally contracts when cross-linked, which could be detrimental to the healing process, resulting in fibrosis and scarring. In another study, bioprinting of collagen and fibrin gel loaded with VEGF have been used for culturing neural stem cells. In this study, collagen type I was chosen as a main scaffold material, while fibrin gel was used for VEGF delivery and to promote migration and proliferation of embedded cells. With combined biological effects of collagen and VEGF release, neural stem cells showed signs of differentiation after 2 days.

3.2.2 Gelatin/Alginate

Gelatin/alginate blend combines thermoresponsive qualities of gelatin with the chemical cross-linking ability of alginate [29,51]. The main role of gelatin is to alter the flow characteristics of the solution for advantageous departure from the printing nozzle and to improve the initial stability of printed constructs before the chemical cross-linking of alginate [29,51]. The polymer blend instantaneously gel once it is cooled below 10°C due to the thermoresponsive behavior of gelatin. Chemical cross-linking of alginate via $CaCl_2$ takes longer than the temperature-driven gelation mechanism of gelatin, requiring several minutes. This cross-linking should occur while the printed material is solidified in order to maintain structural integrity; otherwise, movement of the interface would result in unstable and inaccurate geometry [29].

Gelatin/alginate solutions containing sinus smooth muscle cells and aortic valve leaflet interstitial cells were bioprinted to form aortic valve conduits [51]. Increased alginate concentration caused poor mechanical integrity of the printed struts, while higher gelatin concentration resulted in high viscosity of bioink, which impaired deposition process. Both printed SMCs and aortic valve leaflet interstitial cells had cell viability more than 80% over 7 days. Cell spreading was found to increase over time that could be attributed to time-dependent dissociation of alginate by exchanges between Na^+ and Ca^{2+}. Tensile stress and modulus also decreased with culture time as a consequence of ion exchanges and the early release of gelatin [51]. Mechanical properties of printed constructs can be modified via changing the gelatin:alginate ratio, but the temperature has to be accurately controlled to avoid premature gelation of gelatin [29,51] Also, there should be a balance between the amounts of gelatin and alginate accountable for immediate geometry stability and enhancing long-term stability.

3.2.3 Gelatin/Hyaluronan

Skardal et al. reported on the bioprinting of a polymer blend consisting of photocrosslinkable methacrylated hyaluronan (HA-MA) and gelatin ethanolamine methacrylate (GE-MA) [26]. HA-MA is a promising material for bioprinting since its cross-linking degree can be easily controlled during the photopolymerization process. However, most cells cannot attach to HA-MA alone due to the lack of cell adhesive sites. Blending of HA-MA with a gelatin, which contains cell adhesion motifs, can help to enhance the cell viability. In this study, photocrosslinkable GE-MA was blended with HA-MA for bioprinting of cells [26]. Higher gelatin concentration gave rise to enhanced cell attachment compared to HA-MA only hydrogels. However, decrease in the modulus of the constructs was observed due to addition of GE-MA to HA-MA. The optimal composition was found to be 80% HA-MA 20% GE-MA, which provided adequate cell adhesion sites while maintaining the structural integrity of cell-laden bioprinted construct [26].

3.2.4 Hyaluronic Acid and Dextran

Hyaluronic acid is a linear polysaccharide component of ECM composed of β-1,4-linked D-glucuronic acid (β-1,3) and N-acetyl-D-glucosamine disaccharide units. Due to its

viscoelasticity, excellent biocompatibility, and biodegradability, hyaluronic acid is one of the promising candidates for bioprinting. However, one drawback of bioprinting unmodified hyaluronic acid is the low stability of the construct due to its high water solubility [52]. Strategies aiming to reduce hydrophilicity of hyaluronic acid have been reported in literature, including derivatizing polysaccharide chains with hydrophobic moieties and/or cross-linkable chemical groups. Nevertheless, cell-laden constructs composed of only hyaluronic acid have not been successfully bioprinted [52].

Pescosolido et al. circumvented the instability of printed hyaluronic acid constructs by blending viscoelastic bioactive hyaluronic acid with photocrosslinkable dextran derivative, hydroxyethyl methacrylate derivatized dextran (dex-HEMA) [52]. Dextran is a bacteria-derived polysaccharide consisting of α-1,6 linked D-glucopyranose units with some α-1,2-, α-1,3-, and α-1,4,-linked side chains. Synthesis of dextran hydrogels can be achieved by radical polymerization of dextran derivatives with a reactive group [53,54]. Dex-HEMA can polymerize to form a gel in which hydrolytically sensitive esters are present in cross-links [53,54]. Since dex-HEMA is photocrosslinkable, a stable hydrogel can be formed after UV irradiation [52]. Mechanical properties of printed hyaluronan/dex-HEMA can be controlled by varying degrees of dextran derivative substitution as well as the concentration of dex-HEMA in the solution [52]. The solution of hyaluronan/dex-HEMA blend had a high viscosity at a low shear rate that is favorable for bioprinting. The rheological property of the polymer blend was mainly dominated by hyaluronan due to its high molecular weight (MW) and stiff polymer chains. Entangled hyaluronan chains efficiently dissipated deformation energy and consequently retarded network collapse. The printed constructs showed high cell viability of chondrocytes [52].

3.3 Natural Polymers for 3D Printing

Naturally derived bioinks are advantageous due to their intrinsic cytocompatible properties; however, limitations exist to tailor these polymers for biomedical applications. Some natural polymers, such as chitosan, silk fibroin, starch, and soy protein, have not been used for bioprinting. This is mainly attributed to printing and postprinting conditions, which are harmful to cells. Current and future research will attempt to circumvent these issues in order to better regenerate functional tissue. Their ability to form cell-encapsulated hydrogels with shear-thinning capabilities make them an attractive material for bioprinting bioinks. Improvements on mechanical integrity will be crucial for naturally derived bioink success, especially in larger constructs for organ replacement, although research seems to be on the right track. Some of these natural polymers that are used for 3D printing include chitosan, silk fibroin,

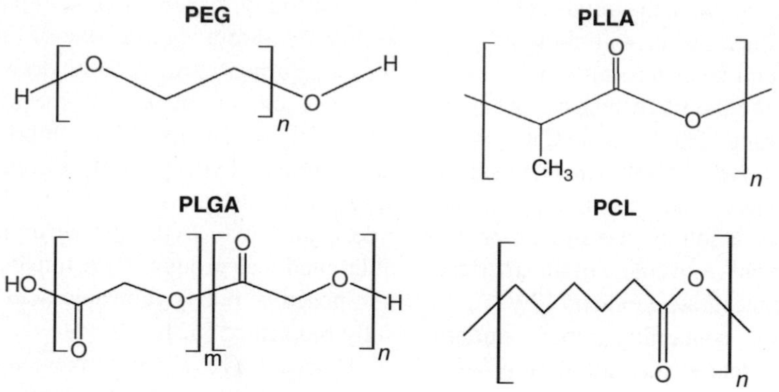

FIGURE 13.4 Polymer structures of naturally derived polymers for use as bioinks that more aptly mimic the components of the ECM, providing an attractive route for bioink design. Polymer blends are not shown.

starch, and soy protein that will be highlighted in this section (Fig. 13.4).

3.3.1 Chitosan

Chitosan is a linear amino-polysaccharide composed of β(1–4) linked D-glucosamine residues and randomly located N-acetyl-glucosamine groups. It is a semicrystalline polymer that cannot dissolve in aqueous environments above neutral pH. In diluted acid, protonation of free amino acid groups generates a fully soluble molecule below pH 5. This solubility dependency allows chitosan to be used for bioprinting [55,56]. Viscous chitosan solution can be extruded and gelled by neutralization of acetic acid by sodium hydroxide (NaOH). To leach out residual NaOH, printed scaffolds are soaked in ethanol before being kept in deionized water. To remove excess water, the scaffolds are heated in the oven and then freeze-dried. The optimal range of NaOH concentrations was found to be between 0.75% and 1.5% v/v. Higher NaOH concentrations led to rapid gelation and consequently, little or no attachment between printed layers. On the other hand, low NaOH concentrations resulted in undesirable spreading of gel into the path of parallel struts, causing dragging of the overall constructs. Cells seeded on the scaffolds spread well and showed high viability. However, due to the harsh processing conditions, cells cannot be printed with the gel [55,56].

3.3.2 Silk Fibroin

Silk fibroin is a fibrous protein derived from the *Bombyx mori* silk worm. It is an amphiphilic block copolymer with a heavy chain composed of 12 repetitive domains predominated by the sequence G-X-G-X-G-X (G = glycine; X = alanine or serine). Eleven amorphous regions, consisting of more hydrophilic peptides, separate the dominating hydrophobic repetitive clusters [30,57]. Silk fibroin has excellent biocompatibility and robust mechanical properties. The molecular organization can transition between random coils to β-sheet through the addition of a poor solvent, such as methanol, which induces aggregation. This property allows silk fibroin to be used for 3D printing, that is, extruded struts crystallize when deposited into a methanol reservoir [30]. The system is optimized using 86% methanol to generate fibers that were elastic enough to maintain shape yet soft enough to stick to the layers underneath. Fibroin displayed shear-thinning behavior, that is, molecules aligned and the friction resistance between adjacent fibers decreased as shear rate increased. The optimal working concentration was found to be 29 wt%. At higher concentrations, chain mobility was suppressed and consequently hindered the transition into an aggregated structure [30].

However, printing silk-only solutions frequently led to clogged nozzles from shear-induced β-sheet crystallization. The unconstrained molecular mobility could be preserved by blending fibroin with gelatin [28]. These two polymers have opposite charges at physiological conditions (pH 7.2–7.4), yielding binding, which facilitated the smooth flow through nozzles [28]. Fibroin fractions in methanol transition from random coils to β-sheets, while gelatin fractions could undergo helix-random coil transition in the range of atmospheric and body temperatures. Consequently, structural integrity was maintained even at elevated temperatures. In addition, since fibroin does not have cell adhesion motifs, blending with gelation introduces integrin-recognizing RGD sequence that could enhance cell attachment and proliferation. Chondrocytes seeded on silk-only printed scaffolds were dedifferentiated whereas those seeded on fibroin/gelatin scaffolds exhibited round shapes indicating redifferentiation and maintenance of chondrocytic phenotype [28].

3.3.3 Starch

Starch is a polysaccharide produced by plants as energy storage. It is composed of two types of D-glucose: amylose and amylopectin. The ratio between the two glucoses varies according to the origin of the starch [27,58]. 3D printing of a blend of cornstarch, dextran, and gelatin powders in water was reported [27]. Using water as a binder is advantageous as problems of solvent residues and toxic fabrication environments can be eliminated. Starch granules do not dissolve in water but merely form a suspension. But, these starch granules swell and gelatinize to form a paste as the suspension is heated. Then, the amylose fraction separates from amylopectin and forms a continuous phase surrounding swollen granules. As the starch cools down, the amylose phase separates leading to gel formation and rapid retrogradation. This results in a semicrystalline structure that is highly resistant to hydrolysis. Heat treatment at 100°C is necessary to dry and maintain structural integrity of printed constructs. This extreme temperature prevents the use of starch blend for bioprinting [27].

3.3.4 Soy Protein

Soy protein is plant-based with controllable properties through variable processing treatments. Its thermoplastic nature and biocompatibility allows the use of soy protein for 3D printing. Chien et al. reported on the 3D printing of this polymer for tissue engineering [59]. The printed constructs were treated with ethanol and freeze-dried to dehydrate proteins and increase structural stability. During this process, electrostatic interactions between proteins, PBS, and media ions promoted aggregates of protein. This is imperative for scaffold shape preservation as noncrosslinked constructs dissolve in water. Although soy protein can be printed at room temperature lacking organic solvents, which enables the incorporation of cells and growth factors, postprinting treatments create harsh conditions for cells, which hinder its use in bioprinting.

4 SYNTHETIC POLYMERS

While natural polymers provide a positive cell environment through mimicking native components of the ECM, synthetic polymers facilitate chemical manipulation of the structure to improve mechanical, biocompatible, and degradation properties. This chemical processing also enables researchers to generate cross-linkable structures for tissue engineering. In order to approach the biological recognition found in natural materials, these synthetic polymers can be equipped with molecular agents to enhance bioactivity as well as can be spatially deposited to stimulate cellular responses via mechanotransduction pathways [60]. These capabilities will be addressed in the following sections, along with relevant physical characteristics of the synthetic bioinks that have been developed thus far. Translatability of these materials used in 3D printing into bioprinting will also be considered as the utilization of toxic solvents or significant heat render them inhospitable to biological agents; however, concurrent printing of hydrogel bioinks is achievable. Some of these synthetic polymers that are used for bioprinting include PEG, PLGA, poly(ε-caprolactone), and PLLA that will be highlighted in this section (Fig. 13.5).

4.1 Poly(ethylene glycol)

Poly(ethylene glycol) is a hydrophilic, biocompatible, and FDA approved polymer that is extensively used in biomedical engineering through tissue engineering, drug delivery, and biosensors. PEG coating on polymeric nano- or microparticles results in significantly increased blood circulation time due to the nonfouling nature of PEG. In tissue engineering, the surface modifications of biomaterials surface

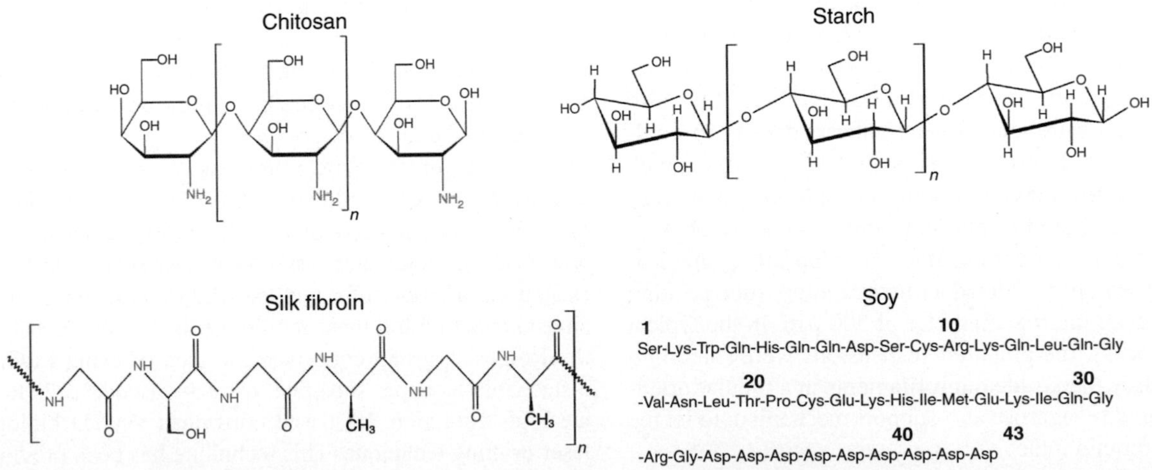

FIGURE 13.5 **Polymer structures of synthetic polymers that have emerged as bioinks and additionally as 3D supportive structures capable of simultaneous printing with cell-laden hydrogels in hybrid systems.** Fabrication strategies employing these polymers may be inhospitable to cells and may limit use as a pure bioink. Shown are three polyesters (PLLA, PLGA, and PCL) with one polyether (PEG).

via PEGylation facilitates easy control over cell-adhesion properties. PEG coating also results in biologically inactive surfaces that limit protein adhesion. Because of several astounding tunable properties, it has witnessed an unprecedented growth in tissue engineering applications in recent decades. At a structural level, the polyether is charge neutral; it has beneficial optical qualities, and it can be easily conjoined with multiple materials. Alterations to the MW and extent of cross-linking make it versatile for investigations of substrate elasticity and its effect on cell behavior, like in stem cell differentiation, which enables possible use in tissue engineering. Some of the resounding applications of PEG and its derivatives in bioprinting of scaffolds in recent times are discussed here.

PEG's solubility in water makes it an attractive material for cell encapsulation, yet the polymer alone cannot form physical or chemical networks to result in a hydrogel. Thus, the polymer must be chemically modified prior to use as a bioink. Acrylation of this polymer has proven to be a key element toward achieving this goal of gel formation and, in this networked state, can stimulate cell differentiation. The general approach to accomplish this task is to cross-link the polymer chains via photoinitiator (PI)-induced polymerization under UV exposure. The use of UV light can be limited to preserve cell viability while still enabling sufficient free-radical generation for proper cross-linking. Prestwich et al. demonstrated the utility of acrylated PEG in the bioprinting of vascular grafts. In a typical experiment, they synthesized PEG-based multiarmed acrylated hydrogels of different chain lengths; TetraPEG-8 and TetraPEG-13 of 2 and 3.4 kDa MW, respectively, with acrylate group activated on both chain ends. The synthesized TetraPEGs were cross-linked with thiolated-hyaluronic acid and gelatin derivatives to form extrudable hydrogels (TetraPAc) for bioprinting applications. The extraordinary stiffness of these cross-linked hydrogel structures renders them suitable for printing tissue constructs especially vascular grafts as they retain physical integrity as well as fluidity during the high-pressure microcapillary pathway. The cellular viability of these hydrogels were examined using murine fibroblast (NIH3T3) cell lines before the cells were loaded at an optimal density in a calculated weight of TetraPAc13-cross-linked hydrogel, which could provide the maximum cell concentration with easy gelation and diffusion of nutrients and cellular metabolites in the constructs. To demonstrate the bioprinting applications, the researchers utilized a microcapillary tube printing process with an internal diameter of 500 μm. In the typical printing process, the prepared suspensions were filled into microcapillary to extrude microfilaments in a tubular orientation covered by agarose as a support mechanism to let the constructs remain intact.

There is an immense need for research in osteochondral repair and current implant therapies have not sufficiently addressed the exponential growth in the number of patients worldwide. D'Lima and coworkers exploited acrylated PEG for the quick and durable solution in this direction realizing the importance of UV-assisted cross-linked PEGDMA [18]. A typical cell-laden bioink based on PEGDMA was prepared using human articular chondrocytes stained with green and orange fluorescent dyes via UV exposure induced by photoinitiator Irgacure 2959. Here the simple methodology of PEG to generate terminal double bonds allowed rapid cross-linking in a solution conducive to cell viability. While these hydrogels would not be as stiff as the TetraPEG counterparts, they maintain the positive characteristics of the polymer, while also demonstrating favorable printing qualities. It is important to note that due to the lack of strong secondary forces present in the molecule, the T_g of PEG is fairly low, even when cross-linked in a network. This results in greater chain mobility and thus a more rubbery material at body temperature, providing cells with a soft tissue mimicking material. The extent of cross-linking, therefore, will raise the T_g and result in a stiffer bulk material [18]. Furthermore, a novel 3D thermal inkjet-based layer-by-layer deposition method was tested using prepared PEG bioink to repair osteochondral plugs serving as a 3D-biopaper. The obtained compressive modulus of bioink in the range of 400 kPa was found to be quite close to that measured for human articular cartilage. Further survival studies confirmed that cell-loaded hydrogels remained attached to both the adjacent cartilage and subchondral bone even after 6 weeks in culture and was supporting the larger proteoglycan production at their interface.

4.2 Poly(lactide-*co*-glycolide)

Despite the advent of strong cross-linking chemistry, most of the biopolymers applicable for bioprinting suffer from uncontrolled degradation and poor mechanical properties. PLGA being the copolymer of lactide and glycolide, obtained via ring-opening polymerization, has fetched wide attention as an alternative to overcome the listed drawbacks existing with other polymers. The popular condensation polymerization of D- and L-configurations can yield D,L-lactide, which is frequently used due to its improved toughness and easy manipulation of degradation rates. Some of the complex biostructures require immediate vasculature networks and involve cells like human umbilical vein endothelial cells (HUVEC), which need precise fluid flow control, invariable viscoelasticity, and fast solvent evaporation to be fabricated via a bioprinting method. PLGA emerges as a possible choice to fulfill these requirements. To accomplish this, the Ringeisen group conceived the idea of using PLGA as a stackable biopaper substrate to stack vascular cells to create high-resolution 3D tissue constructs via 2D biological laser printing technique. This technique has been previously used to print a variety of cell types including osteosarcoma cells [61], olfactory ensheathing cells [62], carcinoma cells [63], and bovine aortic endothelial cells [64], among several others, and has been established as a hallmark in bioprinting techniques. The researchers in a typical experiment dissolved PLGA (MW 40–75 kDa) in chloroform and poured into PDMS molds with salt to fix it. After salt washing and solvent drying, the obtained scaffolds were filled with collagen and/or Matrigel, which provide the construct with biological sources. The HUVEC cells were then deposited in a stacked manner, which demonstrated the role of PLGA as a biopaper supporting the printing and transfer of 2D cell patterns and allowing the stacking of vascular cells to form 3D tissue structures. Their further experiments suggested that stacked HUVEC cells were more volatile to migrate or remained on the surface forming a network depending on whether the PLGA scaffold is either loaded with collagen or Matrigel, respectively. The employment of this material as a biopaper distinguishes this polymer from those previously presented as it is not strictly speaking a bioink due to its inability to incorporate cells directly into the solution. However, the rapid solvent evaporation and structural enhancements allow PLGA to enter the bioprinting realm as an additive biomaterial. Its ability to hydrolytically degrade ensures only temporary involvement in the tissue regeneration process and will allow cell mobility and remodeling to transpire.

4.3 Poly(ε-caprolactone)

PCL is a well-known biodegradable synthetic polyester, often used as plasticizer with several other polymers to design daily use household materials. It is a high MW semicrystalline polymer and imparts the additional elongation to the material by overcoming the brittle nature of conjugating copolymers like poly(lactic acid) and polyurethanes. PCL has been utilized in tissue engineering scaffolds for multiple applications due to its thermoplastic behavior, respectable mechanical strength, hydrolysis-induced biodegradation profile, and low melting point (approx. 60°C), which allows its easy processing. As a polyester, it exhibits nonenzymatic degradation and hydrolyzes to undergo sequential fragmentation in the primary and intermediate stages leading to bulk erosion eventually. Researchers have begun to introduce this polymer into 3D printing strategies to better spatially control PCL interconnectivity and porosity. However, the drawback with previously existing printing methods, like fused deposition modeling and precision extrusion deposition, are their dependency on extreme extrusion pressure on nozzle diameter to achieve maximum resolution. In that scenario, working with a viscous thermoplastic polymer like PCL would require very high pressure that sometimes goes beyond practical limits. Wei and Dong proposed the eletrohydrodynamic jet (EHD-jet) technique where the polymer melt is subjected to an electrostatic field to form a conical structure releasing the fine jet with significantly better resolution than obtained with previous existing methods [65]. PCL being quite stable thermally, with better rheological properties, qualifies for fabrication in its melting phase using this technique. In a typical experiment PCL plates (MW 45 kDa) were used to form the conical jet under electrostatic field where the generated temperature gradient quickly solidifies the jetted PCL and form mechanically stable 3D constructs with resolution as good as 10 μm [65].

Again, similar to PLGA, the fabrication process limits the use of this polymer as a cell-laden bioink. The melting temperature of PCL (approx. 60°C) is too high to sustain cell viability; therefore, printing PCL using this method requires cell seeding after scaffold fabrication or via a separate bioink. Hence, PCL is not strictly a bioink that can encapsulate cells, but rather an additive network to provide hydrogel bioinks with a supportive structure. Because the PCL is utilized purely to reinforce the printed material rather than to mimic the ECM like in the traditional tissue-engineering paradigm, the construct is classified as a "scaffold-free" system. Scaffold-free systems aim to take advantage a cell's inherent ability to organize themselves into complex structures via cell–cell interactions with minimal scaffold presence [66]. Researchers developed scaffold-free cell printing technology wherein the layer-by-layer deposition of cells can enable the construction of 3D organs. A novel modified printing technique with six dispensing heads is suggested using thermoplastic PCL and PLGA biomaterials to print the desired organ [41]. To demonstrate the system's efficiency, heterogeneous cell lines, chondrocytes, and osteoblasts, were encapsulated within alginate solutions and infused within the PCL framework to construct 3D osteochondral plugs. The lamellae within the semicrystalline structure of this polyester provide the solidified form with mechanical properties relevant for soft tissue engineering. Amorphous regions also enable water penetration for hydrolytic cleavage of the polymer chain to result in suitable biodegradation as new tissue begins to inherit physiological stresses. It is noteworthy that the separately dispensed chondrocytes and osteoblasts remained viable for the next 7 days without significant fusion as confirmed by fluorescence microscopy, indicating a possible role of PCL frameworks in the regeneration of heterogeneous tissue constructs.

4.4 Poly(L-Lactic Acid)

PLLA is a well-known established aliphatic polymer. Its noteworthy mechanical properties (elastic modulus 1.5–2.7 GPa) and glass transition around 60°C, offers easy blending with many plasticizers to achieve desired rigidity. Furthermore, it exhibits process-related viscosity where low viscosity during extrusion allows adequate flow through an inkjet nozzle, while after printing, due to fast evaporation of the solvent, it becomes stiff and the printed structure stabilizes [67]. It is semicrystalline, biodegradable, biocompatible, and has found use in several medical applications like orthopedic implants, drug delivery systems, and biofabrication.

Therriault and coworkers selected PLLA for the fabrication of different microchannels using SC-DW (solvent-cast direct-write) method because of its resistance to thermal degradation at high temperatures and easy shape retention [67,68]. In the current work the researchers demonstrated the use of PLLA in fabricating square spiral, circular spiral, and microcup-shaped architectures following layer-by-layer extrusion technique where the extruded filaments transition from fluid state to viscous-solid state after extrusion, thus retaining the shape of the microarchitecture. It is imperative to mention that PLLA fluids exhibited shear-thinning behavior, which allows it to flow smoothly through the nozzle, and at the same time with faster jet speed, the solvent evaporation of extrudates could be enhanced. The researchers further proposed to extend the advanced 2D and 3D printing applications with similar thermoplastic polymers.

On a comparative note to PCL and PLGA, fabrication environments necessary for polymer printing once again inhibit simultaneous cellular interaction. PLLA has improved mechanical properties over PCL due to a shorter vinyl backbone and additional methyl pendant groups, decreasing the rotational mobility of the backbone. The T_g of ~60°C is evidence of the increased backbone stiffness and overall chain motion restriction relative to PCL. Again, the semicrystalline morphology offers mechanical integrity while access for hydrolysis of the polyester, although hydrophobicity of the material may somewhat limit the presence of water, extend the lifetime of the material. Solidification kinetics and rheological properties of these inks alone have been investigated, although incorporation into bioprinted constructs has yet to be researched [67].

5 SUMMARY

Due to biocompatibility concerns around fabrication methods involving synthetic polymers, cell-laden bioinks based purely from these biomaterials are uncommon. Synthetic polymers can include cell binding domains to enhance proliferation or be chemically modified for mechanical and degradation control. In order to take advantage of the mechanical strength found in many synthetics, further research is needed to enable cell manipulation with simultaneous printing through introducing natural materials or improving printer designs. 3D printing may provide some supplementary methods that bioprinting researchers could expand upon to realize the possibilities of synthetic bioinks. Furthermore, collaborative efforts are needed between printer design engineers and material scientists to generate synthetic polymer systems specifically for bioink applications.

6 POLYMER HYBRIDS

Similar to traditional tissue engineering, some bioprinting researchers have attempted to merge synthetic and natural polymer systems to more aptly control material properties. These hybrid polymer designs strive to incorporate the benefits of both types of polymers, for example, the tunability of synthetic materials with the biomimetic characteristics of natural polymers. Many of the synthetic polymers previously described, while useful for 3D printing, have limited use in bioprinting as the only biomaterial component; therefore, they can be an adjunct material when printed with another cell-laden ink. Due to the inherent complexity of native tissues, the usage of both types may be warranted for successful regeneration [25]. As previously mentioned, one group explored the usage of natural-based hyaluronan hydrogels cross-linked with tetrahedral PEG for the bioprinting of blood vessels [35]. Hyaluronic acid (HA) hydrogels alone demonstrate good biocompatibility and have been utilized for vessel repair [69]. PEG was introduced as a cross-linker due to its bioinert characteristic and its ability to covalently cross-link with natural polymers; in this case, thiolated HA once it has been processed to have terminal acrylate groups. Synthesized as a four-arm star polymer, this tetraPEG provided greater mechanical stiffness over HA cross-linked with PEG diacrylate (PEGDA) linkers. Interestingly, the chemical cross-linking between acrylate and thiol groups seemed to be hindered above a threshold value of cell density.

Copolymerization of synthetic and natural polymers has been utilized by tissue engineers to avoid the shortcomings of single polymer type systems. While the previously described synthetic polymers demonstrate acceptable mechanical strength and biocompatibility, they lack cell-recognizable binding sites to improve adhesion; however, hybridization of these polymers with natural polymers can better mimic the ECM, leading to superior cellular outcomes [70]. A fair amount of research has targeted natural/synthetic copolymers for traditional tissue engineering scaffolds; however, development of an extrusion system for these polymers has been limited [71,72]. In some cases, both polymer types are present in the deposition solution, but are not chemically conjugated to one another. For example, the presence of the natural polymer could increase bioink viscosity for better extrusion control during printing [73].

Rather than a single bioink containing both synthetic and natural polymers, groups have developed hybrid systems that apply synthetic polymers as scaffolding for structure and shape, with the cell-laden naturally based bioink as a filler. This method has multiple advantages. First, synthetic thermoplastic polymers provide mechanical strength that typical hydrogel materials cannot provide [25,74]. Second, it does not limit the use of a single hydrogel, allowing for multiple hydrogels loaded with multiple cell types to be printed in a single construct [25]. PCL is popular as a supportive scaffolding due to its stiffness and degradation capabilities [6,25]. The thermoplastic polymer can be printed using an XYZ-controlled nozzle similar to applied

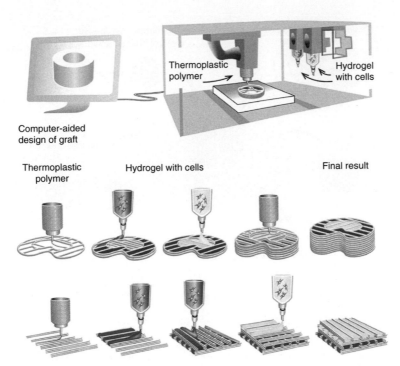

FIGURE 13.6 Schematic of a hybrid layer-by-layer printing system utilizing a thermoplastic supportive scaffold with multiple cell-infused polymer bioinks to better replicate the structural and cellular complexity of native tissue. *(Figure reprinted with permission from Schuurman et al. [25].)*

hydrogel to enable additional control over the final structure or through electrospinning; layers of randomly aligned PCL fibers can separate sections of hydrogels, which could allow variation of printed cell type and hydrogel material at different layers, as seen in Fig. 13.6. Additionally, lower viscosity hydrogels can be printed due to the mechanical strength provided by the PCL, enabling researchers to print with a greater amount of bioink materials.

7 EMERGING TRENDS AND FUTURE DIRECTIONS

While bioprinting itself has only just emerged as an exciting field of regenerative medicine research, recent trends have surfaced that attempt to propel these technologies into areas of even greater clinical relevance (Fig. 13.7). To overcome limitations associated with purely polymeric systems (e.g., insufficient mechanical strength and inefficient cellular stimulation), nanocomposites have been introduced as possible alternatives to improve upon these lacking characteristics [75–78]. Nanomaterials used in conjunction with polymer systems enable additional sites for cross-linking for mechanical stability or provide the cells with an alternate stimulus to motivate differentiation [79]. Multinozzle printing systems can also enhance mechanical integrity as well as inductivity through the inclusion of multiple materials acting in conjunction with one another. Both of these aspects are crucial for bioprinting design to more suitably mimic the native ECM.

An emerging trend in bioprinting is the development of multinozzle systems to print several bioinks within a single construct. Multiple bioinks allow investigators to integrate multiple cell types as well as polymer hydrogels to capture the complexity of the intended regenerated tissue [8,41,46,80,81]. Polymer bioinks could also include different growth factors that are spatially controlled in order to motivate stem cells into different lineages depending on their location within the printed material. While complex designs are a major advantage of these systems, they have the drawback of greater associated costs due to more advanced apparatuses [39]. As these multinozzle systems become more feasible for basic bioprinting research, polymer bioinks will be developed for specific printing applications as opposed to current bioinks, some of which were created with no considerations of bioprinting.

In order to better replicate this microenvironment, one group chemically functionalized 3D printed PLLA with multiwalled carbon nanotubes (MWCNTs), which have been shown to mimic collagen through similar size and shape, while also inducing stem cell differentiation into osteogenic and chondrogenic lineages. Polymer–nanocomposite interactions boosted mechanical strength of the modified scaffold, with a Young's modulus similar to that of subchondral bone (30–50 MPa) [82]. By exposing the MWCNTs with poly-L-lysine after a H_2 treatment, the MWCNTs became more hydrophilic and thus more biocompatible. Stem cells seeded directly onto the scaffold demonstrated increased proliferation due to the

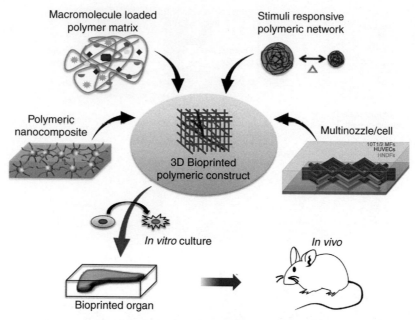

FIGURE 13.7 Multiple methods have emerged to improve upon initial bioink designs for the transplantation *in vivo* of *in vitro* printed tissues including polymer materials loaded with macromolecules or nanocomposites to improve bioactivity, stimuli responsive networks for enhanced printing and cellular outcomes, and the simultaneous printing of multiple cell/polymer bioinks using several nozzles in a single system. *(Multinozzle schematic reprinted with permission from Kolesky et al. [78].)*

bioinspired design. If instead of directly seeding cells onto the scaffold, they were encapsulated in a hydrogel bioink and printed layer-by-layer simultaneously with the functionalized PLLA-MWCNT scaffold, tissue formation could be improved further. Similarly, nanotitania has been dispersed in a printed PLGA scaffold to introduce surface roughness comparable to native bone [79]. Not only was osteoblast adhesion greater on well-dispersed scaffolds, tensile modulus also increased, which is vital for long-term construct success. Last, these nanoparticles shield the scaffold from the acidic degradation products of PLGA, reducing autocatalysis effects. Again, printing a polymer hydrogel bioink concurrently with the nanocomposite 3D scaffold could provide the correct microenvironment for tissue formation *in vivo*. These demonstrate the possible avenues of translation between the similar technologies of 3D printing and bioprinting.

In addition to polymer interactions with other nanomaterials, polymer bioinks can be integrated with growth factors to result in a bioactive scaffold. Due to the sensitivity to temperature, solvent, and conformation of these biological agents, they may be limited by fabrication method. Growth factors could be included directly in the polymer solution and then encapsulated during cross-linking [83]. Another mechanism to incorporate growth factors into a bioprinted material is through the use of degradable microspheres [84]. Charged domains on amino acids in growth factors and gelatin enable the formation of polyion complexes within gelatin microspheres. Growth factors like BMP-2, transforming growth factor beta 1, or basic fibroblast growth factor can

be contained within the microspheres, leading to a versatile carrier system for multiple regenerative applications. The microspheres are then dispersed within an alginate hydrogel containing a suspension of stem cells that can be printed with a defined architecture. Microspheres can be modified to vary release rates of the growth factors; however, size considerations are necessary to avoid clogging the nozzle during printing. The effect of extended BMP-2 exposure in scaffolds motivates osteogenic differentiation and therefore bone formation compared to immediate factor release. While the addition of the growth factor resulted in the desired cellular response, scaffold mechanical properties dissipated too quickly, which can be problematic if new stable tissue has yet to form [84].

To improve bioprinting materials and bring us closer toward functional tissue and organ replacement, several directions have been imagined for polymer bioinks. One of these proposed designs utilizes RGD peptides to augment cell adhesion within synthetic printed material, which could lead to better cellular fusion and enhanced function [14]. Bioinks composed of amphiphilic polymers that are capable of functionalization with bioactive agents could significantly improve upon cellular outcomes. These polymers would have the versatility to be printed as a gel or as microsphere cell encapsulation vehicles. "Active materials" are a novel type of polymer-based printing material that has recently been explored for device design, although they have not been employed for bioprinting usage [85]. As environmentally responsive materials, they demonstrate shape memory characteristics, which provide an extra dimension of printing properties. While the true practicality of these "four-dimensional" materials for bioprinting would need to be demonstrated, one could envision a support system for a bioink, poly(*N*-isopropylacrylamide)for example, that transitions to a specific shape once implanted within the body or bioreactor to mold tissue formation. A more general outlook on future endeavors aims to determine optimal polymer "recipes" that provide a specific microenvironment to optimally stimulate encapsulated cells for a specific application. These formulations would be unique for each targeted physiological environment as well as a bioprinting method. Semi-IPN hydrogels have also been developed as bioinks to improve mechanical integrity to effectively mimic native tissue. The presence of one complete polymer network interlaced with non-cross-linked polymer strands distributes mechanical stress more effectively across the construct, while remaining highly swollen [86]. Last, one of the ultimate goals of bioprinting is to repair the body using *in situ* printing. A bioink capable of maintaining its complex structure and withstanding biological and mechanical conditions within the body will be imperative. Additionally, nutrient transfer will be vital for continuous cell growth and function, particularly for larger organs that require greater depths of penetrating vasculature for nutrient and waste transfer.

8 CONCLUSIONS

The properties of bioink are very important in the bioprinting process to precisely design scaffolds for functional tissue engineering. Polymer bioinks should satisfy the following criteria: first, its rheological properties should allow smooth and uniform extrusion through fine nozzles without any choking or fractures. The printed polymer should shift from a fluid-like phase to solid-like phase soon once it exits the nozzle, maintaining the extruded form. In order to achieve this, the elastic modulus of the polymer bioink should be lower than the viscous modulus prior to printing. Second, once deposited the polymer should maintain its structural integrity. This means its elastic modulus should be greater than the viscous modulus. Polymer ink with shear thinning behavior also promotes this solidification process since the viscosity will increase significantly as shear disappears after extrusion. Third, it must provide cytocompatible environments for cells before, during, and after printing. Many bioinks fall short in regards to these guidelines. Some polymers require cytotoxic chemical cross-linkers or post-printing processing to enhance their mechanical integrity or remove coagulation solvents. These conditions (e.g., extreme temperature) could be too harsh for cell survival. As a result, not all polymers that are used for 3D printing, where cells are seeded after scaffold fabrication, can be used for bioprinting, where cells are printed together with the polymer bioink.

Polymer bioinks have great potential to revolutionize the field of tissue engineering. As a carrier of biological material that can be applied in complex structures at high resolution, bioinks provide bioprinting technologies with advantages over traditional scaffold fabrication. Specific regeneration applications require variable construct parameters and thus the consideration of bioink materials is vital to design success. Polymers can be functionalized to optimize design characteristics leading to enhanced regeneration outcomes. While many polymer bioinks have not yet reached their full potential, researchers have begun to pinpoint key properties necessary for cell viability and tissue formation. Similarly, new methods to print multiple polymer bioinks within a single design have improved outcomes. Further investigation into the interplay of polymer type and fabrication technique will help to formulate new polymer bioinks that can expedite the process from printing to implantation.

GLOSSARY

Biodegradation Chemical scission of covalent bonds via hydrolytic, enzymatic, or oxidative cleavage in a biological environment.

Bioink Printing material comprised of a noncytotoxic biomaterial utilized as a vehicle for cell deposition in a spatially controlled structure.

Biomimetic The engineering of biological structures based off of the structure or function of natural materials.

Extracellular matrix (ECM) A combination of structural components of tissues external to the cell.

Glass transition temperature (T_g) Temperature range that results in increased segmental mobility of the polymer backbone.

Hydrogel A lightly chemically or physically cross-linked water swollen polymer network that retains its three-dimensional shape.

Polymer A material, both naturally and synthetically derived, from repeating unit structures, resulting in the formation of a macromolecule.

Scaffold-free tissue engineering The arrangement of stem cells without structural components to guide growth or differentiation for tissue regeneration.

Thermoplastic Polymer type that softens into a more pliable phase upon heating, while becoming more rigid during cooling.

Viscoelasticity Type of deformation and recovery exhibiting the mechanical characteristics of viscous flow and elastic deformation.

ABBREVIATIONS

3D	Three-dimensional
BMP-2	Bone morphogenic protein-2
ECM	Extracellular matrix
GE	Gelatin ethanolamine
HA	Hyaluronan/hyaluronic acid
HEMA	Hydroxyethyl methacrylate
MW	Molecular weight
MWCNTs	Multiwalled carbon nanotubes
PCL	Poly(ε-caprolactone)
PEGDMA	Poly(ethylene glycol) dimethacrylate
PI	Photoinitiator

PLGA	Poly(lactic-*co*-glycolic acid)
PLLA	Poly(L-lactic acid)
RGD	Arg–Gly–Asp
Semi-IPN	Semi-interpenetrating polymer network
TetraPEG	Four-arm PEG derivative
T_g	Glass transition temperature
UCST	Upper critical solution temperature
VEGF	Vascular endothelial growth factor

REFERENCES

[1] Mironov V, Reis N, Derby B. Review: bioprinting: a beginning. Tissue Eng 2006;12(4):631–4.

[2] Hutmacher DW, Sittinger M, Risbud MV. Scaffold-based tissue engineering: rationale for computer-aided design and solid free-form fabrication systems. Trends Biotechnol 2004;22(7):354–62.

[3] Nakamura M, Iwanaga S, Henmi C, Arai K, Nishiyama Y. Biomatrices and biomaterials for future developments of bioprinting and biofabrication. Biofabrication 2010;2(1):014110.

[4] Mironov V, Visconti RP, Kasyanov V, Forgacs G, Drake CJ, Markwald RR. Organ printing: tissue spheroids as building blocks. Biomaterials 2009;30(12):2164–74.

[5] Ferris CJ, Gilmore KJ, Beirne S, McCallum D, Wallace GG, Panhuis MIH. Bio-ink for on-demand printing of living cells. Biomater Sci 2013;1(2):224–30.

[6] Xu T, Binder KW, Albanna MZ, Dice D, Zhao W, Yoo JJ, et al. Hybrid printing of mechanically and biologically improved constructs for cartilage tissue engineering applications. Biofabrication 2013;5(1):015001.

[7] Wust S, Godla ME, Muller R, Hofmann S. Tunable hydrogel composite with two-step processing in combination with innovative hardware upgrade for cell-based three-dimensional bioprinting. Acta Biomaterialia 2014;10(2):630–40.

[8] Ozbolat IT, Yu Y. Bioprinting toward organ fabrication: challenges and future trends. IEEE Trans Biomed Eng 2013;60(3):691–9.

[9] Boland T, Xu T, Damon B, Cui X. Application of inkjet printing to tissue engineering. Biotechnol J 2006;1(9):910–7.

[10] Chung JHY, Naficy S, Yue ZL, Kapsa R, Quigley A, Moulton SE, et al. Bio-ink properties and printability for extrusion printing living cells. Biomater Sci 2013;1(7):763–73.

[11] Mano JF, Silva GA, Azevedo HS, Malafaya PB, Sousa RA, Silva SS, et al. Natural origin biodegradable systems in tissue engineering and regenerative medicine: present status and some moving trends. J R Soc Interface 2007;4(17):999–1030.

[12] Gunatillake PA, Adhikari R. Biodegradable synthetic polymers for tissue engineering. Eur Cell Mater 2003;5:1–16.

[13] West JL, Hubbell JA. Polymeric biomaterials with degradation sites for proteases involved in cell migration. Macromolecules 1999;32(1):241–4.

[14] Murphy SV, Skardal A, Atala A. Evaluation of hydrogels for bioprinting applications. J Biomed Mater Res A 2013;101(1):272–84.

[15] Nguyen KT, West JL. Photopolymerizable hydrogels for tissue engineering applications. Biomaterials 2002;23(22):4307–14.

[16] Selimovic S, Oh J, Bae H, Dokmeci M, Khademhosseini A. Microscale strategies for generating cell-encapsulating hydrogels. Polymers (Basel) 2012;4(3):1554.

[17] Billiet T, Gevaert E, Schryver TD, Cornelissen M, Dubruel P. The 3D printing of gelatin methacrylamide cell-laden tissue-engineered constructs with high cell viability. Biomaterials 2014;35:49–62.

[18] Cui XF, Breitenkamp K, Finn MG, Lotz M, D'Lima DD. Direct human cartilage repair using three-dimensional bioprinting technology. Tissue Eng Part A 2012;18(11–12):1304–12.

[19] Ahn S, Lee H, Bonassar LJ, Kim G. Cells (MC3T3-E1)-laden alginate scaffolds fabricated by a modified solid-freeform fabrication process supplemented with an aerosol spraying. Biomacromolecules 2012;13:2997–3003.

[20] Song SJ, Choi J, Park YD, Hong S, Lee JJ, Ahn CB, et al. Sodium alginate hydrogel-based bioprinting using a novel multinozzle bioprinting system. Artif Organs 2011;35(11):1132–6.

[21] Stevens MM, Qanadilo HF, Langer R, Shastri VP. A rapid-curing alginate gel system: utility in periosteum-derived cartilage tissue engineering. Biomaterials 2004;25:887–94.

[22] LeRoux MA, Guilak F, Setton LA. Compressive and shear properties of alginate gel: effects of sodium ions and alginate concentration. John Wiley & Sons, Inc.; 1999.

[23] Bryant SJ, Anseth KS, Lee DA, Bader DL. Crosslinking density influences the morphology of chondrocytes photoencapsulated in PEG hydrogels during the application of compressive strain. J Orthop Res 2004;22(5):1143–9.

[24] Ehrbar M, Sala A, Lienemann P, Ranga A, Mosiewicz K, Bittermann A, et al. Elucidating the role of matrix stiffness in 3D cell migration and remodeling. Biophys J 2011;100(2):284–93.

[25] Schuurman W, Khristov V, Pot MW, van Weeren PR, Dhert WJA, Malda J. Bioprinting of hybrid tissue constructs with tailorable mechanical properties. Biofabrication 2011;3(2):021001.

[26] Skardal A, Zhang J, McCoard L, Xu X, Oottamasathien S, Prestwich GD. Photocrosslinkable hyaluronan-gelatin hydrogels for two-step bioprinting. Tissue Eng Part A 2010;16(8):2675–85.

[27] Lam CXF, Moa XM, Teoh SH, Hutmacher DW. Scaffold development using 3D printing with a starch-based polymer. Mater Sci Eng 2002;20:49–56.

[28] Das S, Pati F, Chameettachal S, Pahwa S, Ray AR, Dhara S, et al. Enhanced redifferentiation of chondrocytes on microperiodic silk/gelatin scaffolds: toward tailor-made tissue engineering. Biomacromolecules 2013;14:311–21.

[29] Wüst S, Godla ME, Müller R, Hofmann S. Tunable hydrogel composite with two-step processing in combination with innovative hardware upgrade for cell-based three-dimensional bioprinting. Acta Biomaterialia 2014;10:630–40.

[30] Ghosh S, Parker ST, Wang X, Kaplan DL, Lewis JA. Direct-write assembly of microperiodic silk fibroin scaffolds for tissue engineering applications. Adv Funct Mater 2008;18:1883–9.

[31] Hoch E, Hirth T, Tovarab GuEM, Borchersa K. Chemical tailoring of gelatin to adjust its chemical and physical properties for functional bioprinting. J Mater Chem B 2013;1:5675–85.

[32] Liu CZ, Xia ZD, Han ZW, Hulley PA, Triffitt JT, Czernuszka JT. Novel 3D collagen scaffolds fabricated by indirect printing technique for tissue engineering. J Biomed Mater Res B Appl Biomater 2008;85(2):519–28.

[33] Nakamura M, Iwanaga S, Henmi C, Arai K, Nishiyama Y. Biomatrices and biomaterials for future developments of bioprinting and biofabrication. Biofabrication 2010;2(014110).

[34] Schuurman W, Khristov V, Pot MW, Weeren PRv, Dhert WA, Malda J. Bioprinting of hybrid tissue constructs with tailorable mechanical properties. Biofabrication 2011;3(021001).

[35] Skardal A, Zhang JX, Prestwich GD. Bioprinting vessel-like constructs using hyaluronan hydrogels crosslinked with tetrahedral polyethylene glycol tetracrylates. Biomaterials 2010;31(24):6173–81.

[36] Seyednejad H, Gawlitta D, Kuiper RV, de Bruin A, van Nostrum CF, Vermonden T, et al. *In vivo* biocompatibility and biodegradation of 3D-printed porous scaffolds based on a hydroxyl-functionalized poly(ε-caprolactone). Biomaterials 2012;33:4309–18.

[37] Guo SZ, Gosselin F, Guerin N, Lanouette AM, Heuzey MC, Therriault D. Solvent-cast three-dimensional printing of multifunctional microsystems. Small 2013;9(24):4118–22.

[38] Pataky K, Braschler T, Negro A, Renaud P, Lutolf MP, Brugger J. Microdrop printing of hydrogel bioinks into 3D tissue-like geometries. Adv Mater 2012;24(3):391–6.

[39] Burg T, Cass CAP, Groff R, Pepper M, Burg KJL. Building off-the-shelf tissue-engineered composites. Philos Tran R Soc A 2010;368(1917):1839–62.

[40] Atala A, Allen AJ, Yoo JJ, Binder KW. Drop-on-demand inkjet bioprinting: a primer. Gene Ther Regul 2011;06(01):33–49.

[41] Shim JH, Lee JS, Kim JY, Cho DW. Bioprinting of a mechanically enhanced three-dimensional dual cell-laden construct for osteochondral tissue engineering using a multihead tissue/organ building system. J Micromech Microeng 2012;22(8).

[42] Lee YB, Polio S, Lee W, Dai G, Menon L, Carroll RS, et al. Bioprinting of collagen and VEGF-releasing fibrin gel scaffolds for neural stem cell culture. Exp Neurol 2010;223:645–52.

[43] Nijenhuis KT. Thermoreversible networks - viscoelastic properties and structure of gels - introduction. Adv Polym Sci 1997;130:1–12.

[44] Schuurman W, Levett PA, Pot MW, van Weeren PR, Dhert WJ, Hutmacher DW, et al. Gelatin-methacrylamide hydrogels as potential biomaterials for fabrication of tissue-engineered cartilage constructs. Macromol Biosci 2013;13:551–61.

[45] Bertassoni LE, Cardoso JC, Manoharan V, Cristino AL, Bhise NS, Araujo WA, et al. Direct-write bioprinting of cell-laden methacrylated gelatin hydrogels. Biofabrication 2014;6(024105):11.

[46] Song SJ, Choi J, Park YD, Hong S, Lee JJ, Ahn CB, et al. Sodium alginate hydrogel-based bioprinting using a novel multinozzle bioprinting system. Artif Organs 2011;35(11):1132–6.

[47] Khali S, Sun W. Bioprinting endothelial cells with alginate for 3D tissue constructs. J Biomech Eng 2009;131:111002.

[48] Poldervaart MT, Wang H, van der Stok J, Weinans H, Leeuwenburgh SC, Öner FC, et al. Sustained release of BMP-2 in bioprinted alginate for osteogenicity in mice and rats. PLoS ONE 2013;8(8):0072610.

[49] Blombäck B, Hessel B, Hogg D, Therkildsen L. A two-step fibrinogen–fibrin transition in blood coagulation. Nature 1987;275:501–5.

[50] Skardal A, Mack D, Kapetanovic E, Atala A, Jackson JD, Yoo J, et al. Bioprinted amniotic fluid-derived stem cells accelerate healing of large skin wounds. Stem Cells Transl Med 2012;1:792–802.

[51] Duan B, Hockaday LA, Kang KH, Butcher JT. 3D Bioprinting of heterogeneous aortic valve conduits with alginate/gelatin hydrogels. J Biomed Mater Res A 2013;101A(5):1255–64.

[52] Pescosolido L, Schuurman W, Malda J, Matricardi P, Alhaique F, Coviello T, et al. Hyaluronic acid and dextran-based semi-IPN hydrogels as biomaterials for bioprinting. Biomacromolecules 2011;12:1831–8.

[53] Lévesque SpG, Lim RM, Shoichet MS. Macroporous interconnected dextran scaffolds of controlled porosity for tissue-engineering applications. Biomaterials 2005;26:7436–46.

[54] Dijk-Wolthuis WNEv, Tsang SKY, Bosch JJK-vd, Hennink WE. A new class of polymerizable dextrans with hydrolyzable groups: hydroxyethyl methacrylated dextran with and without oligolactate spacer. Polymer 1997;38(25):6235–42.

[55] Geng L, Feng W, Hutmacher DW, Wong YS, Loh HT, Fuh JYH. Direct writing of chitosan scaffolds using a robotic system. Rapid Prototyping J 2005;11(2):90–7.

[56] Ang TH, Sultana FSA, Hutmacher DW, Wong YS, Fuh JYH, Mo XM. Fabrication of 3D chitosan–hydroxyapatite scaffolds using a robotic dispensing system. Mater Sci Eng 2002;20:35–42.

[57] Marsh RE, Corey RB, Pauling L. An investigation of the structure of silk fibroin. Biochimica et Biophysica Acta 1955;16(1):1–34.

[58] Tester RF, Karkalas J, Qi X. Starch – composition, fine structure and architecture. J Cereal Sci 2004;39:151–65.

[59] Chien KB, Makridakis E, Shah RN. Three-dimensional printing of soy protein scaffolds for tissue regeneration. Tissue Eng Part C 2013;19(6):417–26.

[60] Lutolf MP, Hubbell JA. Synthetic biomaterials as instructive extracellular microenvironments for morphogenesis in tissue engineering. Nat Biotechnol 2005;23(1):47–55.

[61] Barron JA, Wu P, Ladouceur HD, Ringeisen BR. Biological laser printing: a novel technique for creating heterogeneous 3-dimensional cell patterns. Biomed Microdevices 2004;6(2):139–47.

[62] Othon CM, Wu XJ, Anders JJ, Ringeisen BR. Single-cell printing to form three-dimensional lines of olfactory ensheathing cells. Biomed Mater 2008;3(3):034101.

[63] Ringeisen BR, Kim H, Barron JA, Krizman DB, Chrisey DB, Jackman S, et al. Laser printing of pluripotent embryonal carcinoma cells. Tissue Eng 2004;10(3–4):483–91.

[64] Chen CY, Barron JA, Ringeisen BR. Cell patterning without chemical surface modification: cell–cell interactions between printed bovine aortic endothelial cells (BAEC) on a homogeneous cell-adherent hydrogel. Appl Surf Sci 2006;252(24):8641–5.

[65] Wei C, Dong JY. Direct fabrication of high-resolution three-dimensional polymeric scaffolds using electrohydrodynamic hot jet plotting. J Micromech Microeng 2013;23(2):025017.

[66] Billiet T, Vandenhaute M, Schelfhout J, Van Vlierberghe S, Dubruel P. A review of trends and limitations in hydrogel-rapid prototyping for tissue engineering. Biomaterials 2012;33(26):6020–41.

[67] Guo SZ, Heuzey MC, Therriault D. Properties of polylactide inks for solvent-cast printing of three-dimensional freeform microstructures. Langmuir 2014;30(4):1142–50.

[68] Guo SZ, Gosselin F, Guerin N, Lanouette AM, Heuzey MC, Therriault D. Solvent-cast three-dimensional printing of multifunctional microsystems. Small 2013;9(24):4118–22.

[69] Mironov V, Kasyanov V, Markwald RR, Prestwich GD. Bioreactor-free tissue engineering: directed tissue assembly by centrifugal casting. Expert Opin Biol Ther 2008;8(2):143–52.

[70] Chen GP, Sato T, Ushida T, Ochiai N, Tateishi T. Tissue engineering of cartilage using a hybrid scaffold of synthetic polymer and collagen. Tissue Eng 2004;10(3–4):323–30.

[71] Hoffman AS. Hydrogels for biomedical applications. Adv Drug Deliv Rev 2002;54(1):3–12.

[72] Fedorovich NE, Alblas J, de Wijn JR, Hennink WE, Verbout AJ, Dhert WJ. Hydrogels as extracellular matrices for skeletal tissue engineering: state-of-the-art and novel application in organ printing. Tissue Eng 2007;13(8):1905–25.

[73] Lixandrao AL, Noritomi PY, da Silva JVL, Colangelo N, Kang H, Lipson H, et al. Construction and adaptation of an open source rapid prototyping machine for biomedical research purposes – a multinational collaborative development. Innovative Developments in Design and Manufacturing 2010;469–73.

[74] Kundu J, Shim JH, Jang J, Kim SW, Cho DW. An additive manufacturing-based PCL-alginate-chondrocyte bioprinted scaffold for cartilage tissue engineering. J Tissue Eng Regen Med 2013.

[75] Kim K, Dean D, Lu A, Mikos AG, Fisher JP. Early osteogenic signal expression of rat bone marrow stromal cells is influenced by both hydroxyapatite nanoparticle content and initial cell seeding density in biodegradable nanocomposite scaffolds. Acta Biomater 2011;7(3):1249–64.

[76] Gaharwar AK, Peppas NA, Khademhosseini A. Nanocomposite hydrogels for biomedical applications. Biotechnol Bioeng 2014; doi: 10.1002/bit.25160.

[77] Carrow JK, Gaharwar AK. Bioinspired polymeric nanocomposites for regenerative medicine. Macromol Chem Phys 2015;216(3):248–64.

[78] Kerativitayanan P, Carrow JK, Gaharwar AK. Nanomaterials for engineering stem cell responses. Adv Healthcare Mater 2015; doi: 10.1002/adhm.201500272.

[79] Liu HN, Webster TJ. Enhanced biological and mechanical properties of well-dispersed nanophase ceramics in polymer composites: from 2D to 3D printed structures. Mater Sci Eng C 2011;31(2):77–89.

[80] Kolesky DB, Truby RL, Gladman A, Busbee TA, Homan KA, Lewis JA. 3D Bioprinting of vascularized, heterogeneous cell-laden tissue constructs. Adv Mater 2014;26(19):3124–30.

[81] Yan YN, Xiong Z, Hu YY, Wang SG, Zhang RJ, Zhang C. Layered manufacturing of tissue engineering scaffolds via multi-nozzle deposition. Mater Lett 2003;57(18):2623–8.

[82] Holmes B, Zhang L. Enhanced human bone marrow mesenchymal stem cell functions in 3D bioprinted biologically inspired osteochondral construct. San Diego, CA: ASME IMECE; 2013. p. 7.

[83] Lee YB, Polio S, Lee W, Dai G, Menon L, Carroll RS, et al. Bioprinting of collagen and VEGF-releasing fibrin gel scaffolds for neural stem cell culture. Exp Neurol 2010;223(2):645–52.

[84] Poldervaart MT, Wang H, van der Stok J, Weinans H, Leeuwenburgh SC, Öner FC, et al. Sustained release of BMP-2 in bioprinted alginate for osteogenicity in mice and rats. PLoS ONE 2013; 8(8):e72610.

[85] Ge Q, Qi HJ, Dunn ML. Active materials by four-dimension printing. Appl Phys Lett 2013;103(13).

[86] Pescosolido L, Schuurman W, Malda J, Matricardi P, Alhaique F, Coviello T, et al. Hyaluronic acid and dextran-based semi-IPN hydrogels as biomaterials for bioprinting. Biomacromolecules 2011;12(5):1831–8.

Chapter 14

Hydrogels for 3D Bioprinting Applications

Tyler K. Merceron*,** and Sean V. Murphy*

*Wake Forest Institute for Regenerative Medicine, Wake Forest School of Medicine, Winston-Salem, NC, USA; **Vanderbilt University School of Medicine, Vanderbilt University, Nashville, TN, USA

ABSTRACT

Hydrogels are highly hydrated polymeric networks used in tissue engineering to homogenously encapsulate cells and other biological molecules. This class of biomaterials is of particular interest because of their structural similarity to a cell's natural extracellular matrix. Hydrogels can be derived from various sources, including natural and synthesized derivatives. Hydrogels can be induced to quickly solidify using a number of methods to introduce cross-links and covalent bonding between polymer strands. Due to their high biocompatibility and processability, hydrogels have become the choice medium to pattern cells in a volumetric space using three-dimensional (3D) bioprinting, an additive manufacturing process that deposits biomaterials in a layer-by-layer fashion to fabricate a 3D tissue construct. In this chapter, we will review important general principles that make a hydrogel useful for bioprinting, followed by a discussion of the specific hydrogels used for bioprinting applications.

Keywords: hydrogel; 3D bioprinting; cell encapsulation; biomaterial; extracellular matrix; cross-linking; polymer

1 INTRODUCTION

Three-dimensional (3D) printing was first described by Charles Hull in 1986, when he patented a method of printing ultraviolet (UV)-curable materials on top of each other in a layer-by-layer fashion called stereolithography. This technique has served as the foundation upon which other 3D printing (also called rapid prototyping or additive manufacturing) systems have been built. Broadly speaking, 3D printing involves the sequential deposition of any material capable of solidification in a cross-sectional pattern for the building up of a three-dimensional object. These systems involve a printing head that can extrude, transfer, or convert starting materials in the x- and y-planes, with a stage that moves in the z-plane upon completion of each layer. Once one layer is printed, it is given time to solidify before the stage descends for printing of the next layer. Thus, 3D printing proceeds in a "layer-by-layer" fashion such that each printed layer serves as the foundation upon which the next printed layer is deposited.

Over the past 30 years, 3D printing has been applied in various industries, including manufacturing, engineering,

Essentials of 3D Biofabrication and Translation. http://dx.doi.org/10.1016/B978-0-12-800972-7.00014-1

art, and medicine. Recently, there has been a convergence of 3D printing technology with the biological and materials sciences, leading to a novel tissue engineering technique called 3D bioprinting. This process involves the layer-by-layer deposition of biomaterials, cells and other bioactive molecules to fabricate precise, complex 3D tissue-mimicking constructs. Using this method, cells and bioactive molecules are printed using "bioinks," which serve the dual roles of printing substrate and tissue engineering scaffold.

The principal role of scaffolds in tissue engineering and regenerative medicine is to provide a temporary framework for the delivery of cells in a three-dimensional space [1]. In the body, cells are arranged into complex three-dimensional arrangements by a supporting matrix of proteins, glycosaminoglycans, and proteoglycans known as the extracellular matrix (ECM). The ECM has myriad functions, including roles as an adhesive substrate, providing macroscopic shape and microscope architecture, sequestering and presenting growth factors in a spatially and temporally regulated fashion, and sensing and transducing mechanical signals to cells [2]. Thus, the ECM is a highly dynamic contributor to tissue development and morphogenesis. As tissue engineering scaffolds function as temporary ECM to cells, these materials have functions beyond serving as inert supporting materials. Instead, scaffolds should be designed as transient, biologically active, cell-instructive microenvironments that aid in the development of the specific tissue being engineered. Over time, it is expected that the scaffold will degrade at a similar rate to which cells produce their own ECM, resulting in a functional regenerated tissue equivalent.

Hydrogels have emerged as an attractive medium for cell delivery because of their hydrophilicity and ability to encapsulate cells and bioactive molecules, thus mimicking many of the characteristics of natural ECM [3]. As a material, hydrogels are shape-retentive polymeric networks swollen with a high percentage of water [4]. The polymers that make up the backbone of a hydrogel can either be naturally derived proteins or glycosaminoglycans (e.g., collagen, gelatin, fibrin, and hyaluronic acid) or synthetic polymers (e.g., poly(ethylene glycol) (PEG) and Pluronic®). These molecules can be mixed with cells and other bioactive factors in aqueous solution and then be manipulated to form an insoluble, cross-linked meshwork, resulting in a cell-laden hydrogel. Manipulation from the monomeric/un-cross-linked form to the polymeric/cross-linked form is accomplished by inducing physical or chemical bonding through environmental changes (such as pH, temperature, and ionic concentration), enzymatic initiation, or photopolymerization.

In general, hydrogels are highly biocompatible systems that can homogeneously incorporate cells and bioactive factors. They have good porosity for diffusion of oxygen, nutrients, and metabolites; can be processed under mild cell-friendly conditions; and produce little to no irritation, inflammation, or products of degradation [5]. Thus, hydrogels are highly adaptable biomaterials for 3D printing technology and tissue engineering due to their physical characteristics and biocompatibility, respectively.

2 GENERAL PRINCIPLES

There are several criteria that are important in making a hydrogel suitable for tissue development. First, it should be biocompatible, nontoxic to cells, and produce little to no immune response upon implantation. Second, it should have favorable biodegradation kinetics that matches and supports the cells' intrinsic ability to produce a tissue-specific ECM. Third, it should exhibit some degree of biomimicry to the intended tissue type, such that the implanted construct can serve some interim functionality as the cells within develop into fully functional tissue. Fourth, the hydrogel should have sufficient structural and mechanical properties in order to retain its 3D structure and volume over a period of time relevant to tissue regeneration. It should be noted that these four properties are often interdependent. For example, the structural and mechanical properties are integral to many tissues' functionality, and these properties change as the biomaterial degrades.

While the biological considerations are essential for the broad application of hydrogels in tissue engineering, there are additional processing requirements for the use of hydrogels in bioprinting. Broadly, these include rheological properties, such as viscosity and shear-thinning, and the mechanism by which the hydrogel is cross-linked to form a stable matrix. These characteristics contribute to the ultimate "printability" of a hydrogel by determining a hydrogel's ability to be handled, deposited, stacked, and retain volume over time. Because bioprinting utilizes a layer-by-layer additive manufacturing approach, hydrogels should be relatively high-viscosity and have rapid cross-linking for the fabrication of volumetric constructs.

Finally, one must consider other practical features in terms of feasibility for clinical application such as ease of use, cost, shelf life, regulatory status, and translatability. In this section, we will cover each of these properties individually to develop a framework for the discussion of hydrogels currently in use and the development of novel hydrogels for future bioprinting applications. Thus, for hydrogels to be successful for bioprinting applications, they must meet certain biological, printing, and practical criteria (Fig. 14.1).

2.1 Biological Properties

2.1.1 Biocompatibility

Prior to the use of any material for biomedical purposes, one must consider its biocompatibility and overall safety for use in an organism. A universal definition of *biocompatibility* does not currently exist; however, for the purposes

Base hydrogel	Source	Bioprinting methods used	Gelation/cross-linking mechanisms used for bioprinting	Biodegradation mechanism	Notes
Collagen I	Natural peptide (mammals)	• Inkjet[44–47] • Extrusion[13, 48]	• pH neutralization + thermal (37°C)[13, 44–47] • With fibrin[48]	• Enzymatic (MMP)	• Excellent biocompatibility • Poor mechanical properties • Poor gelation/cross-linking kinetics • Easily degraded
Gelatin	Natural peptide (mammals)	• Extrusion[56–66]	• Thermal (room temperature)[63–66] • Enzymatic (gluteraldehyde)[56,57] • Photopolymerization (UV)[58,62] • With alginate[57,64] • With alginate/fibrinogen[63] • With fibrinogen[85]	• Dissolution at 37°C • Enzymatic (MMP)	• Excellent biocompatibility • Requires modification to be useful in terms of gelation/cross-linking and mechanical properties • Quick dissolution in aqueous media at physiologic temperature
Fibrin	Natural peptide (mammals)	• Inkjet[84–85] • Extrusion[48, 63, 86]	• Enzymatic (thrombin)	• Enzymatic (plasmin)	• Excellent biocompatibility • Quick gelation/cross-linking • Quick degradation • Poor mechanical properties
Hyaluronic acid	Natural carbohydrate (mammals)	• Extrusion[62, 103, 104]	• Photopolymerization (UV)[62] • PEG[103] • Gold nanoparticles[104]	• Enzymatic (hyaluronidase)	• Good biocompatibility • Requires modification to be useful in terms of gelation/cross-linking and mechanical properties
Alginate	Natural carbohydrate (algae)	• Inkjet[108–111] • Laser[112–116] • Extrusion[29, 63, 64, 117–123]	• Ionic (Ca^{2+})	• Ionic displacement (Na$^+$ for Ca^{2+})	• Good biocompatibility • Quick gelation/cross-linking • Easy to use and high tailorability
Agarose	Natural carbohydrate (seaweed)	• Inkjet[10] • Extrusion[122, 125, 127]	• Thermal (32°C)[10,122, 126] • Submerged bioprinting	• Nonbiodegradable	• Moderate biocompatibility • Difficult to print

FIGURE 14.1   Different properties of specific hydrogels used for bioprinting.

of this chapter, biocompatibility refers to the ability of a biomaterial to perform its desired function with respect to a medical therapy, without eliciting any undesirable local or systemic effects in the recipient or beneficiary of that therapy, but generating the most appropriate beneficial cellular or tissue response in that specific situation, and optimizing the clinically relevant performance of that therapy [6]. As tissue engineering scaffolds, a bioprinting hydrogel must thus be cytocompatible, weakly immunogenic and have nontoxic byproducts of degradation that can be metabolized and secreted without eliciting detrimental local or systemic effects. Under this paradigm, we can cover the biocompatibility of a bioprinting hydrogel from the time of fabrication and *in vitro* maturation (cytocompatibility) to implantation (immunogenicity) and long-term effects during and after scaffold resorption (byproducts of degradation).

First, let us consider the biocompatibility of a hydrogel through the fabrication stage. Cell viability during the printing process depends largely on the printing modality used, with laser bioprinting methods achieving cell survival rates from 90% to 100% [7–9], inkjet methods between 75% and 90% [10,11], and extrusion methods varying between 40%

and 98% [12,13]. Depending on the method used, certain hydrogel properties may have an effect on the cells during the printing process. For example, the thermal conductivity of a hydrogel may affect cells in thermal inkjet printing and laser-based printing; while viscosity may play a major role in the development of shear stresses cells are exposed to in extrusion-based printing. After cells have been deposited, hydrogel composition may play a more important role in supporting cell viability and proliferation. The ability of a cell to adhere to a substrate is a major determinant in its ability to maintain its phenotype and proliferate. Thus, naturally derived polymers, which have cell-adhesive peptide sequences (e.g., RGD-, IKVAV-, GFOFER-domains), may provide a microenvironment more conducive to cell maintenance relative to synthetic polymers [14]. Efforts to modify synthetic PEG hydrogels with cell-adhesive sequences have been reported [15–17] resulting in greater cell viability, growth, and differentiation [18].

After fabrication *in vitro*, bioprinted constructs are expected to fare in a much different environment *in vivo*. In addition to cells as a potential antigenic source, the immune reaction to hydrogels is an important consideration. In very broad terms, the immune response to a foreign body can be separated into two major divisions: (1) innate immunity principally mediated through macrophages, neutrophils, and natural killer cells; and (2) acquired immunity principally mediated through B and T lymphocytes. The innate immune system may cause a nonspecific foreign body reaction resulting in the infiltration of fibroblasts, endothelial cells and macrophages that form granulation tissue and an ensuing fibrotic capsule to isolate the foreign material from the body, while the acquired immune system may generate a more targeted, antigen-specific reaction. Thus, naturally derived biomaterials are susceptible to acquired immunity (due to the presence of antigens), while synthetic biomaterials are usually subject to innate immunity and foreign body reactions. One must consider the immunogenicity of a biomaterial since a more intense immune response will lead to quicker scaffold degradation times, potential attack on the cells embedded within, and an increased likelihood of fibrosis rather than tissue regeneration.

Finally, for a hydrogel to be biocompatible, it must break down into monomers that are water-soluble and nontoxic such that they can be further metabolized by the liver and/or excreted through the kidney. Furthermore, the mechanisms of degradation and byproducts formed should not elicit potentially detrimental changes outside of normal physiologic pH, temperature, and electrolyte balance, as to not cause damage to the regenerating tissue and/or surrounding tissues.

2.1.2 Biodegradation Kinetics

Like the body's own exctracellular matrix, hydrogels are dynamic materials that can be remodeled and degraded over time. Degradation of these polymeric scaffolds occurs primarily via enzymes (most natural polymers and functionally modified synthetic polymers), hydrolytic reactions (synthetic polymers), or ion exchange (e.g., alginate). The kinetic profile for each of these mechanisms of degradation varies. Hydrolytic reactions and ion exchange occur in a bulk manner, resulting in a constant rate of degradation *in vivo* and *in vitro*. On the other hand, enzymatic reactions are more specific and controllable, as the rate of enzymatic degradation depends on the number of cleavage sites on the polymer and concentration of enzymes in the surrounding environment [19]. In the body, there are whole classes of enzymes specialized for matrix degradation and reorganization, including matrix metalloproteinases (MMPs), a disintegrin and metalloproteinase with thrombospondin motifs (ADAMTS1), serine proteases, hyaluronidase, and heparinase. Therefore, one must consider the environment into which the hydrogel is placed. If a hydrogel is to be placed in a fresh wound bed, for example, there will be a much higher concentration of matrix remodeling enzymes present, which will result in a faster rate of degradation than if the hydrogel were *in vitro* or placed subcutaneously in a facilitating

environment. In addition to the specific polymer used and how the chemical bonds of that particular polymer are broken down, degradation kinetics also depends on the concentration of polymer, degree of cross-linking, the cell type and concentration of cells embedded within the hydrogel, and final location of the hydrogel construct.

In order for a hydrogel to be successfully applied in a regenerative medicine setting, the degradation kinetics of the polymer must match the rate to which the cells embedded within produce their own supporting ground substance and ECM. If degradation is too slow, the polymer meshwork may disrupt normal tissue development and limit diffusion of oxygen and nutrients to cells. Conversely, if the hydrogel degrades too quickly, the construct will become mechanically weak, lose its shape, and cells will lose their anchoring substrate, precluding tissue development. Thus, there is a delicate equilibrium that must be achieved between the degradation of the hydrogel and the synthesis of ECM by the cells encapsulated (Fig. 14.2). Study of a cell type's ability to secrete ECM, specifically the rate of synthesis, may be an important investigation to help in the fine-tuning of a hydrogel's degradation profile.

There are many ways one can alter the degradation kinetics of a hydrogel polymer. Because cells are the source of matrix remodeling proteases *in vitro*, perhaps the simplest method to controlling matrix degradation is by optimizing the cell/polymer ratio. However, while relatively lower cell densities and higher polymer concentrations may slow degradation times, this may also result in poor tissue development that is impractical for engineering purposes. Another way to modulate the degradation rate of hydrogels is by controlling the degree of cross-linking between polymer stands. Increasing the amount of polymer in solution, concentration of cross-linking agent, and exposure time of the cross-linking agent are all ways to achieve higher cross-linking between polymer chains. One may also consider controlling the degradation rate by inhibiting the mechanism of degradation itself. For example, small molecule inhibitors can be supplemented within the hydrogel itself to delay enzymatically mediated degradation (e.g., aprotinin and/or galardin for fibrin hydrogels [20] or tissue inhibitors of metalloproteinases-releasing hydrogels [21]), and ions can be supplemented in the media for polymers that dissolve secondary to ion exchange (e.g., daily Ca^{2+} replenishment for alginate hydrogels). Finally, strategies can be employed to encourage ECM synthesis. Examples include the addition of ascorbic acid, which may cause up to an eightfold increase in collagen synthesis [22], and mechanical stimulation resulting in ECM remodeling and production [23–25].

2.1.3 Biomimicry

There are four basic tissue types in the human body: (1) epithelial, (2) muscular, (3) nervous, and (4) connective.

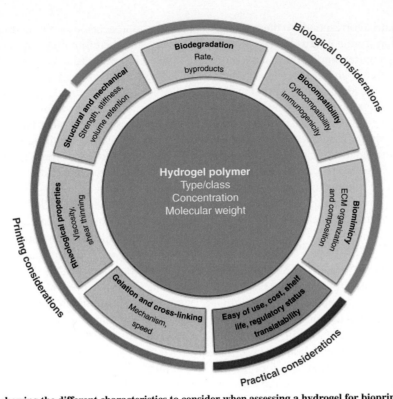

FIGURE 14.2 Concept map showing the different characteristics to consider when assessing a hydrogel for bioprinting applications.

Each of these tissue types possesses a balance between the cellular components and the extracellular environment that is in accordance with that tissue's primary function. In a broad sense, epithelial tissues are specialized for secretion and absorption, muscle generates force through contraction, nerve conducts the transmission of electrical signals to and from the peripheral and central nervous systems, and connective tissues provide mechanical support and shape. Because of their diverse functions, each of these tissue types has distinct histological arrangements that correspond to the overall role of the tissue. Epithelial, muscular and nervous tissues have relatively high cell/ECM ratios; however, it is the organization of this relatively scant ECM that provides a framework upon which function is completely dependent. For example, epithelial tissues are often polarized, having an apical membrane for the interaction with a lumen or external environment and a basal membrane that anchors the cell into the underlying connective tissue for integration with the rest of the body; skeletal muscles, on the other hand, are linearly arranged by a collagenous hierarchy (endo-, peri-, and epimysium) for the efficient transduction of force; and nerve axons have ECM arranged for the insulation of electric conduction. In contrast to the aforementioned tissue classes, connective tissues (including tendon, ligament, cartilage, and bone) are abundant in ECM and have relatively low cell/ECM ratios. The cells in these tissues are specialized for ECM production, secreting large amounts of protein for the maintenance of structure and mechanical integrity to the body.

While the native structure and function of a tissue as defined by the arrangement of its ECM is an important consideration for the engineering of biomimetic tissues, one must also consider the ECM as more than simply an inert scaffold for cells. The ECM is an extremely dynamic environment, composed of hundreds of proteins and carbohydrates, that is constantly remodeling itself, expressing cell-instructive factors [26], and presenting factors for growth, migration, and differentiation [2]. All of these factors serve to create an active cellular microenvironment, which has a large influence on cell attachment, shape, and ability to proliferate and differentiate into the appropriate phenotype. Due to their organic source, naturally derived hydrogels already have many of these complexes "pre-engineered" into their structure, thus promoting a favorable cellular response. While synthetic hydrogels lack these naturally occurring motifs and structures, they can be modified to contain components such as cell adhesion molecules [15,16] and enzyme-specific sequences [27], thereby enhancing their biomimicry.

As a relatively simple example of how hydrogels might be modified to take on a more biomimetic composition, consider the transition from soft to hard tissue at the tendon/ligament–bone interface. This interface, known as the enthesis, can be divided into four distinct regions by ECM composition: (1) tendon/ligament (demineralized collagen I), (2) uncalcified fibrocartilage (demineralized collagen II and glycosaminoglycans (GAGs)), (3) calcified fibrocartilage (mineralized collagen II and GAGs), and (4) bone (mineralized collagen I) [28]. The change from a demineralized matrix to a mineralized matrix corresponds to the mechanical and functional transition from a relatively elastic material more attuned to stretching (tendon/ligament) to a more stiff material that resists stretch and compression (bone). Bioprinting hydrogels have already begun to be designed in this fashion to create a biomimetic environment. Fedorovich et al. created a bioprinted construct consisting of chondrocytes in an alginate hydrogel and mesenchymal stem cells in an alginate-hydroxyapatite-tricalcium phosphate hydrogel to create a heterogeneous osteochondral graft [29]. These changes in ECM composition are a good example of how hydrogels can be customized to mimic the natural environment they attempt to regenerate.

2.1.4 Structural and Mechanical Properties

The structural and mechanical properties of a hydrogel must be considered in the context of the target tissue and also in the context of stability as a substrate for printing. As already noted, tissues have a wide range of mechanical properties, from soft tissues such as fat, skin, and muscle, to harder tissues such as cartilage and bone. These mechanical properties are essential to the function of the tissue, and are important to consider when selecting a hydrogel. All hydrogels are by nature viscoelastic materials, and thus more applicable for soft tissue engineering; however, this limitation can be addressed by using synthetic hydrogels such as PEG, which have more tailorable mechanical properties and by combining hydrogels with thermoplastic materials [30,31].

In addition to matching the mechanical properties of the target tissue, a hydrogel must be a suitable material for fabrication. Because bioprinting involves a layer-by-layer deposition process for the construction of a volumetric tissue construct, each layer must provide the structural integrity for the subsequently printed layer. This property has much to do with the polymer chosen, molecular weight and concentration of polymer used, and degree of gelation/cross-linking attained.

Last, one must consider that the structure and mechanical properties of a hydrogel are not fixed. Indeed, degradation of the hydrogel structure leads to a progressively weaker scaffold. Furthermore, depending on hydrogel composition, cell type, and cell density, there can be a significant amount of swelling or contraction. According to one study, collagen gels can contract up to 50% and PEGDA gels can swell up to 200% in vitro [32]. The amount of deformation is an important parameter when considering the final structure of a bioprinted construct.

2.2 Printability

Before discussing the printability of hydrogels for bioprinting, we must first briefly discuss how the three main types of bioprinting utilize these biomaterials. Each of these printing approaches uses the hydrogel in a slightly different way, and therefore may require different properties for a hydrogel to be considered printable. For example, inkjet-based and extrusion-based systems make use of a nozzle for deposition of the hydrogel material, making rheological properties, such as viscosity and shear forces, important considerations. On the other hand, laser-based printers are nozzle-free systems, requiring different properties for the transfer of a hydrogel material from the laser-absorbing donor slide to the recipient slide.

2.2.1 Rheological Properties

Rheology is the study of material flow in response to an applied force. While this is a complex field involving a deep understanding of fluid mechanics and non-Newtonian physics, two basic concepts that should be considered when attempting to use a hydrogel for bioprinting are the material's viscous and shear thinning properties. These two properties are especially important to the nozzle-based inkjet and extrusion printing methods.

Viscosity is a material's resistance to flow when a force is applied. For printing technologies, this force is pressure. In inkjet printing, small amounts of air pressure are created within the printing syringe via vaporization (thermal inkjet printing) or acoustic pulsation (piezoelectric inkjet printing) of the cell-laden suspension within. These high-frequency thermal or acoustic pulses create enough air pressure to propel liquid through the printing nozzle in a "drop-on-demand" fashion. Inkjet printers deposit very small volumes (<100 pL) and require hydrogels in their precursor (uncross-linked), low-viscosity state (ideally below 0.1 Pa·s) [3]. Extrusion-based systems work by applying pressure in a pneumatic, piston-driven or screw manner to deposit continuous filaments (rather than droplets) of hydrogel material. These systems extrude through nozzles on the order of tens to hundreds of microns, and often require hydrogels that are significantly more viscous (30 to 6×10^6 mPa·s) [33] than inkjet systems to prevent material leakage. Although hydrogels are non-Newtonian fluids, one may consider the Poiseuille equation as a guideline when optimizing hydrogels for nozzle-based dispensing:

$$\Delta P = 8\mu LQ/\pi r^4,$$

where ΔP is the difference between the pressure applied to the fluid and the ambient pressure, μ is the material's viscosity, L is the length of the nozzle tip, Q is the flow rate (or "scan speed"), and r is the radius of the nozzle tip.

Laser-based systems do not require ejection through a nozzle and thus can employ the use of hydrogels with a wide range of viscosities (1–300 mPa·s) [34].

Shear thinning is a non-Newtonian behavior of some fluids that refers to the inverse relationship between shear rate and viscosity. Polymers are shear-thinning materials because when they experience shear stress, their entanglements stretch and become more uniformly aligned parallel to the applied stress, thus decreasing the viscosity [3]. Due to this phenomenon, hydrogels may have similar (high) viscosities within a printing syringe and after deposition, which provide good mechanical strength while also possessing a transitional (low) viscosity state that allows for the extrusion of the material through the nozzle. Because shear thinning is a property of high-viscosity polymer networks through an orifice, this is of particular importance in extrusion-based printing systems.

Other rheological properties that may contribute to the printability of a hydrogel include yield stress, surface tension, and thermal conductivity. As bioprinting technology advances, rheology will become an even more important parameter for the optimization of hydrogels. With a more thorough understanding of these properties, we may be able to design hydrogels with higher printing fidelity, decrease total fabrication time, and make more informed modifications to hydrogel composition.

2.2.2 Gelation and Cross-linking

Hydrogels form their semisolid state by complex interactions between adjacent and overlapping polymer strands. These cross-links result in an entangled polymeric meshwork that can ultimately be used to encapsulate cells and other biological compounds. Thus, the mechanisms of gelation and cross-linking may be the most important feature contributing to a hydrogel's overall function. In order for a hydrogel to be applied to bioprinting, it must have a relatively quick gelation time such that subsequent layers can be immediately deposited on top; the mechanism of cross-linking should be noncytotoxic; and the cross-links formed should be stable in an aqueous environment at physiologic temperature and pH.

The methods of cross-linking can be broadly divided into physical cross-linking (which includes ionic interactions and hydrogen bonding) and chemical cross-linking (which forms covalent bonds secondary to photoinitiation or enzyme catalysis). There are thus four major classes of hydrogels in terms of cross-linking: (1) thermosensitive hydrogels, (2) ionically cross-linking hydrogels, (3) enzymatically cross-linking hydrogels, and (4) photopolymerizable hydrogels. Thermosensitive hydrogels are formed secondary to conformational changes in their constituent monomers at a certain temperature. This temperature is called the critical solution temperature. Depending on the hydrogel, it can have a lower or upper critical solution temperature, above which or below which an insoluble semisolid gel forms. For example, gelatin has an upper critical solution temperature around 30°C. This means that gelatin solutions are soluble liquids above 30°C and insoluble gels below 30°C. The solution-to-gel transition is usually rather gradual depending on distance from the critical solution temperature. Other examples of thermosensitive hydrogels are collagen I, agarose, and Pluronic®.

Ionically cross-linking hydrogels are formed by physical interactions between ions and polymer side chains. The only ionically cross-linking hydrogel we will discuss in this chapter is alginate. Alginate forms cross-links via hydrogen bonding. When divalent cations (e.g., calcium, barium and strontium) are introduced into the environment, they compete with monovalent cations (e.g., sodium) to allow for the cross-linking of carboxylic acid groups on adjacent glucuronic acid blocks. Cross-linking using this method is instantaneous.

Enzymatically cross-linking hydrogels are formed by enzyme-mediated covalent bonding, often between lysine and glutamine residues on adjacent polymer strands. A number of molecules can catalyze covalent bonding including transglutaminase, glutaraldehyde, lysyl oxidase, and genipen. Transglutaminase and glutaraldehyde have been used to induce cross-linking for collagen I and gelatin hydrogels; however, this may be associated with some toxicity. The main enzymatically cross-linking hydrogel is fibrin, which is cross-linked with the serine protease thrombin. Enzymatic cross-linking occurs rather quickly, usually less than a minute.

Last, photopolymerizable hydrogels are formed when free radicals propagate through reactive functional groups (often C=C bonds) to form stable cross-links. In order to catalyze this reaction, photoiniators (i.e., Igracure 2959) are often employed to generate free radicals when exposed to UV-wavelength light. Reactive functional groups, such as thiols and acrylates, can be substituted onto polymers for participation in covalent bonding. This method, which induces polymerization on the order of seconds to minutes, has been employed for polymers that would not readily form cross-links via other means, such as hyaluronic acid, gelatin, and PEG.

2.3 Practical Considerations

2.3.1 Ease of Use

For any product to become widely utilized, especially if scale-up is desired, it must be relatively easy to use. The ease of use of a hydrogel can be dictated by the availability of the precursor gel, the complexity of the gel, the preparation time and effort, and the degree to which the gel can be modified. An ideal gel would be easily acquired and/or commercially available, have minimal mixing components, require little to no preparation time, and be highly customizable to the user's applications. The ability to customize a hydrogel is particularly important because its utility is dependent on both its printability (which may differ between printers) and its ability to be used for widespread applications (i.e., the fabrication of tissue types with different structural, mechanical, and biological properties). At the present moment, no hydrogel meets all of these criteria, resulting in the need to make compromises when it comes to biomaterial choice.

2.3.2 Cost and Shelf Life

The cost and shelf life of a hydrogel material should also be evaluated as practical points. A material should not be cost-prohibitive and should be relatively stable such that its use is reliable and consistent from one use to the next. In general, naturally derived biomaterials are more expensive and have shorter shelf lives than synthetic biomaterials due to an involved extraction, isolation, and purification process and susceptibility to denaturation *ex vivo*. Conversely, synthetic materials can be readily manufactured (making them cheaper) and are often more physically and chemically stable products.

2.3.3 Regulatory Status and Translatability

Finally, we must consider the regulatory status of polymers as potentially translatable products for use in humans. This point is largely related to the biocompatibility of the biomaterial. Most of the hydrogels covered here for bioprinting applications have FDA-approved commercially available products, have been used in clinical trials or are sufficiently close to FDA-approved products such that they can be regarded as potentially translatable. One exception is the Matrigel™ hydrogel, which is derived from a mouse tumor cell line, thus rendering it unsuitable for use in humans.

It should also be noted here that there is a high translatability for many of the hydrogels currently being used for bioprinting applications. Most of the hydrogels being used, including those that are collagen-, fibrin-, and hyaluronic acid-based, are derived from mammals. While the sources of these hydrogels are principally xenogenic for use in animal studies, it is feasible to derive these components from cadaveric, allogenic, or autologous sources for use in the clinic. This creates an exciting potential for species-specific, patient-specific, and even tissue-specific sourcing for hydrogel development.

3 COMMONLY USED HYDROGELS

The polymeric biomaterials that comprise the backbone of hydrogels belong to one of two major classes: naturally derived or synthetic. In tissue engineering, particularly in the field of biofabrication, the main function of these hydrogels is to position cells in a defined 3D space during and after fabrication. Because of their role as provisional extracellular matrices, the majority of hydrogels on the market are derived from natural extracellular matrix sources, including collagen (type I), gelatin, fibrin, hyaluronic acid, alginate, agarose, and the basement membrane (Matrigel™). As naturally derived materials, these polymers are well suited as biomaterials because they have intrinsic abilities to support cell attachment and proliferation and can be easily degraded and metabolized by the body. Despite these biological advantages, naturally derived biomaterials have many limitations, including high potential for batch-to-batch variability, potential for rejection and/or immune-related sequelae, relatively quick degradation times, and poor mechanical properties. Synthetic hydrogels, such as PEG and Pluronic®, can be modified to address many of these disadvantages; however, these hydrogels are themselves biologically inert and nonbiodegradable. This section discusses, in detail, the different types of hydrogels currently used in bioprinting applications (Fig. 14.3). For the purposes of this review, we will only cover hydrogels that can be directly used with cells for bioprinting applications and that do not require preprocessing before they can be seeded with cells.

3.1 Naturally Derived Hydrogels

3.1.1 Collagen, Type I

Collagen is the single most abundant protein in the human body, accounting for up to 25% of an individual's dry protein mass [35]. While there are a multitude of collagens in

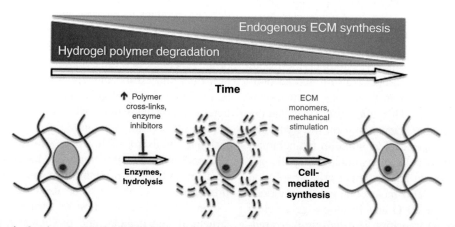

FIGURE 14.3 **Schematic showing the relationship between hydrogel degradation and cell ECM synthesis from time of fabrication to time of complete tissue regeneration.** At the time of fabrication, cells are encapsulated in a cross-linked hydrogel polymer matrix. As time passes, polymers degrade via hydrolysis and cell-secreted enzymes. At the same time, cells are producing ECM that will eventually take the place of the temporary hydrogel polymer.

the collagen superfamily, type I is predominant, accounting for approximately 90% of total body collagen [35]. Structurally, collagen I is a trimeric fibrillar protein comprised of three α helical chains, each of which contains a repeating amino acid sequence $(G-X-Y)_n$ where G is glycine, and X and Y are most commonly proline and hydroxyproline, respectively [36]. Collagen molecules are modular in nature in that they create a supramolecular quaternary structure via end-to-end and lateral organization. This structural hierarchy extends from the basic collagen molecule (1.5 nm diameter) to collagen fibrils (30–300 nm diameter) in a process called fibrillogenesis. The degree to which collagen organizes is a major determinant of the mechanical properties of a tissue [37] (i.e., fibrils can organize into fibers, which can organize into fascicles, which can ultimately form connective tissues such as tendon/ligament/bone). Collagen serves as a major component of the extracellular matrix, especially in fibrous connective tissues where it can be the main tissue component by weight. Because of its essential role in native tissue organization, collagen I is an attractive biomaterial for tissue engineering purposes.

Collagen I hydrogels are an ideal hydrogel in terms of biocompatibility. Once prepared, they are able to support cell growth, proliferation, and differentiation; they have relatively low immunogenicity compared to other naturally derived biomaterials, likely due to the fact that the collagen molecule has high evolutionary stability among mammals (although there is still a potential for antigenicity in the nonconserved terminal regions of the molecule) [38]; and they are readily degraded and remodeled by MMPs (particularly MMP-1/8) [39]. Due to the numerous cell adhesion and enzymatically active sites present on collagen molecules, the cells within collagen hydrogels are able to quickly remodel the matrix. Shape fidelity of collagen hydrogels is a significant limitation for the engineering of anatomically correct structures. According to one study, contraction can be up to 1/28th of the original area of the hydrogel after 24 h, and the amount of deformation is dependent on collagen concentration, cell number and concentration of the microtubule inhibitor Colcemid [40]. In addition to having low volume retention, collagen hydrogels are often made using 0.1–0.5% w/v collagen solutions which is, conservatively, one order of magnitude less than many tissues depending on tissue type [41]. Thus, the collagen present in collagen hydrogels does not translate to the mechanical strength found *in vivo* because of the relatively low concentrations (often <3 mg/mL) and low supramolecular organization (i.e., collagen cross-linking in a collagen hydrogel is not equal to the fibrillogenesis that occurs in tissue). This results in collagen hydrogels having an elastic modulus less than 5 kPa [42,43], which is much lower than many tissues in the body.

Collagen hydrogels are thermosensitive and pH-sensitive. Commonly prepared from stock solutions derived from bovine skin or rat-tail tendon, these preparations are often kept chilled at 4°C and dissolved in an acidic solution to prevent precipitation. Upon neutralizing the pH to a physiologic range (pH = 7.4) and incubating the solution at 37°C for 15–30 min, a viscoelastic hydrogel forms as functional groups become available for hydrogen and covalent bonding. As with many hydrogels, the uncross-linked collagen hydrogel precursor solution can be utilized for inkjet-based printing because of its low viscosity, while a partially or fully cross-linked hydrogel is more viscous for use in extrusion methods.

The first use of collagen hydrogels was in an inkjet printing approach by Boland et al. in 2003 when bovine aortic endothelial cell aggregates were printed within a prepared collagen I hydrogel [44]. This was the first use of a hydrogel for bioprinting applications and thus paved the road for three-dimensional bioprinting. Similarly, the first use of an extrusion-type printer used bovine aortic endothelial cells mixed with a collagen I hydrogel [13]. The cells were mixed with a 3 mg/mL collagen I solution and kept at 10°C in the syringe tip to prevent polymerization (and thus needle clogging), while a warming lamp was directed at the construct after deposition to increase polymerization of the printed construct. Using this technique, Smith et al. were able to print a five-layer construct in the form of the left anterior descending artery that exhibited up to 86% viability, maintained cell proliferation and shape fidelity after 35 days in culture, and developed cordlike structures resembling blood vessels.

In 2009, Lee et al. reported on the printing of neural constructs consisting of rat embryonic neurons and astrocytes using an inkjet printing approach and collagen I hydrogel [45]. Similar to Boland et al. in 2003, the collagen I hydrogel was printed first with subsequent printing of cells; however, this group performed a novel cross-linking step by introducing a nebulized $NaHCO_3$ buffer to induce polymerization. By nebulizing the cross-linking agent, this group was able to make one of the first multilayer bioprinted constructs in the following fashion, performed in tandem: (1) printing of collagen I hydrogel precursor (1.12 mg/mL, pH 4.5), (2) nebulization of $NaHCO_3$ (cross-linking), and (3) printing of cells. This group and others have expanded this methodology to the printing of 3D-printed skin constructs [46,47].

Because of the slow gelation time of collagen (i.e., it takes pH neutralization and warming the suspension at 37°C for approximately 15–30 min to achieve stable gelation), researchers have mixed the collagen hydrogel with other hydrogels that have quicker cross-linking kinetics. Skardal et al. used a fibrin-collagen (50 mg/mL–2.2 mg/mL, pH 7.0) hydrogel to achieve quick gelation by first printing a layer of thrombin (20 IU/mL) (the cross-linking agent for fibrin), followed by a printed cell-laden fibrin-collagen hydrogel, followed by another layer of thrombin [48]. By gelling each layer in this fashion, they were able to deposit multiple layers of cells (amniotic fluid stem cells and mesenchymal stem cells) in an *in situ* extrusion-based printing approach.

Collagen hydrogels are attractive for tissue engineering purposes because they comprise such a large percentage of the natural extracellular matrix. Thus, they exhibit excellent cytocompatibility, can be readily modified by encapsulated or infiltrating cells, and have relatively low immunogenicity. Furthermore, these hydrogels are relatively easy to use (for cross-linking, one need only modulate the pH and warm to physiologic conditions), have good shelf lives (as long as they are kept at 4°C), and are readily available at a low cost. Furthermore, many collagen products are already on the market for human use, including Evolence® Collagen Filler, AlloDerm® Tissue Matrix, and Integra™ Meshed Bilayer Wound Matrix. While collagen hydrogels are advantageous for the aforementioned reasons, they do have significant limitations. The principal disadvantage for using collagen hydrogels is their poor mechanical properties due to a relatively low protein concentration in comparison to natural tissues. This is likely contributed to the fact that collagen hydrogels undergo rapid modification by cells *in vitro*, causing them to contract and lose their original shape (due to cells attempting to rearrange the molecules into a more condensed, tissue-like form). Collagen hydrogels are also relatively slow when it comes to cross-linking time, taking between 15 min and 30 min to achieve a solidified state. In addition to the cross-linking methods just described for bioprinting applications, additional cross-linking steps may help with the creation of a final construct with superior mechanical properties and quicker cross-linking times, including the utilization of transglutaminase [49], lysyl oxidase [50], and PEGylation [51]. Ultimately, while the collagen hydrogel is ideal for its biological properties, translatability, and ease-of-use, its relatively weak mechanical properties and lengthy cross-linking time are significant limitations to its use as a scalable hydrogel for printing applications.

3.1.2 Gelatin

Gelatin is the denatured, partially-hydrolyzed form of collagen I. Because of its derivation from natural collagen, these hydrogels are highly translatable and are identical in terms of biocompatibility to collagen hydrogels (see Section 3.1.1). Gelatin belongs to a class of polymers known as "sol-gels," which form semisolid states in aqueous solution by phase transition rather than by chemical reactions or other external stimuli [52]. In many cases, the hydrogel forms simply in response to temperature, usually with lower temperatures stabilizing the molecule's tertiary structure (a triple helix in the case of gelatin) and allowing for the creation of physical junctions, which result in gelation [52]. For gelatin, gelation occurs at room temperature, making this hydrogel particularly useful for bioprinting. While these thermal gelation properties make it a good hydrogel in terms of manu-

facturing, gelatin has a melting point below natural body temperature (around 27–33°C depending on concentration [53]), which renders it unsuitable for *in vivo* applications. Thus, in order to be suitable for a tissue-engineering scaffold, gelatin must be chemically modified or be mixed with another cross-linking polymer (where the gelatin component provides the viscous and bulk properties for optimal printing and the cross-linking component provides the final cross-linked polymer network). The most widely used ways to induce cross-linking in gelatin hydrogels are by using glutaraldehyde [54] or via the addition of methacrylamide groups followed by UV irradiation [55]. Alternatively, the thermosensitive properties of gelatin can be exploited for use as a "sacrificial material." In additive manufacturing, a sacrificial material is a material that can be deposited to help build up a particular shape or structure, but is then washed away after fabrication, leaving holes or pores where that material was located. Gelatin can be employed in this way because when it is placed in aqueous media at 37°C it will solubilize, leaving only the other cross-linked matrix or hydrogel behind. Thus, for a bioprinting hydrogel, gelatin can be used in one of three ways: (1) as the polymer itself after direct cross-linking by glutaraldehyde [56,57] or UV-cross-linking after methacrylation [58–62], (2) as a component of the hydrogel to tailor the overall dispensing viscosity [63–65], or (3) as a sacrificial material [66].

Due to the highly viscous nature of gelatin at room temperature, bioprinting using gelatin hydrogels has occurred almost exclusively using extrusion-based printers. While gelatin can be cross-linked using glutaraldehyde for biofabrication purposes [56,57], this method has fallen out of favor due to the potential for glutaraldehyde-induced cytotoxicity [67,68]. In these early bioprinting studies, hepatocyte-laden hydrogels were exposed to low concentrations (0.25–2.5%) of glutaraldehyde for a maximum of 5 s to ensure good cell viability and shape retention after 8 weeks. More recent attempts using gelatin have focused on using a methacrylated form of gelatin (Gel-MA) as a noncytotoxic means of achieving thermally stable bioprinted constructs [58–62,69]. Bertassoni et al. attempted to optimize the fabrication conditions using Gel-MA, finding that 10% w/v Gel-MA irradiated for 15–60 s using a 6.9 mW/cm^2 UV ($\lambda = 360$–480 nm) light source could produce scalable 3D layered structures using a cell concentration of 1.5×10^6/mL [58]. Furthermore, this group showed that a cell could be printed with good cell viability through 8 days and that the mechanical properties of the gel correlated with gelatin concentration and UV exposure time ($E = 2.6$ kPa after 10 s to 60.3 kPa after 60 s exposure time for 15% w/v gels) [58]. These Gel-MA hydrogels have been employed for the fabrication or the engineering of cardiac valves [60], cartilage [61], and vessel-like structures [62].

Outside of using gelatin as the polymer, its viscous and thermosensitivity make it useful in other ways too. For example, several groups have used gelatin as a component of a more complex hydrogel system such as alginate/gelatin [57,64] and alginate/fibrinogen/gelatin [63]. In these complex hydrogels, the alginate and/or fibrinogen serve as cross-linking agents that can hold the structure of the construct *in vitro*, while the gelatin serves to provide the correct viscosity for stability during fabrication. Because the gelatin in these systems is not cross-linked it will ultimately dissolve when placed in an incubator or *in vivo*, leaving behind only the cross-linked hydrogel component. The thermosensitivity of gelatin can be further exploited by using it as a sacrificial material. In this way, Lee et al. coprinted a cell-laden collagen hydrogel adjacent to a gelatin hydrogel to create a multilayer block structure [66]. In so doing, they demonstrated that they were able to create internal porous interconnectivity by creating microfluidic channels throughout the construct through the dissolution of gelatin, and that the cell viability throughout the scaffold (particularly in the center) was significantly greater than a nonporous scaffold [66].

In conclusion, gelatin is an extremely versatile molecule for use as a hydrogel in bioprinting. Its various properties allow it to play many different roles, including being a scaffold itself, being used as an additive for processing, or being a temporary material for the scaling up and induction of porosity within a scaffold. Because gelatin is naturally derived from collagen, gelatin-based hydrogels share many of the advantages and limitations of collagen hydrogels (i.e., they have favorable biologic properties, but lack structural and mechanical stability). The mechanical limitation of gelatin is largely secondary to its molecular instability at physiologically relevant temperatures. These disadvantages are beginning to be addressed by chemically modifying individual gelatin monomers to be cross-linkable and by making gelatin one component of a composite hydrogel.

3.1.3 Fibrin

The formation of a fibrin clot is the body's first step in natural wound repair. Following any vascular insult, a coagulation cascade occurs resulting in a cross-linked fibrin matrix, which entraps platelets and other blood components to create a stable hemostatic plug. This meshwork serves as the substrate upon which endogenous cells transmigrate and intrinsically regenerate the lesion. Surgeons have exploited the fibrin matrix in the form of fibrin glues (which can be sourced from allogenic plasma under the commercial names Tissucol/Tisseel®, Beriplast®, and Quixil™; or even be autologous-derived using Cryoseal®-FS or Vivostat® systems) because of their ability to promote tissue apposition, hemostasis, wound healing, and protection from bacterial invasion [70].

A tissue-engineered fibrin hydrogel is produced in the same fashion that it is formed *in vivo*. That is, a solution of fibrinogen monomers is activated to form a polymeric fibrin matrix upon cleavage of fibrinopeptides of their N-terminal ends by the serine protease thrombin [71]. In the body, thrombin also activates Factor XIII, which catalyzes lysine–glutamine covalent bonding to further cross-link the fibrin matrix such that it is more resistant to protease degradation. *In vitro*, genipin has been used in the same way to create more robust fibrin hydrogels [72].

Fibrinolysis is the breakdown of a fibrin clot, primarily by the enzyme plasmin (although other MMPs can be involved). *In vivo* clot resolution can begin as soon as a few hours after insult; however, *in vitro*, cells must produce these enzymes themselves, with hydrogels usually being completely dissolved after 1 week. Thus, fibrin hydrogels alone lack the structural and mechanical stability for tissue engineering applications due to their rapid dissolution. Efforts to enhance the mechanical properties include modifying thrombin concentration to produce larger fibrin bundles [73] and adding protease inhibitors, such as ε-aminocaporic acid [74], tranexamic acid [75], aprotinin [20,76], and galardin [20], to preserve the matrix for several weeks *in vitro*.

While fibrin hydrogels have been exploited for the engineering of various tissue types, including cartilage [72,77,78], muscle [79,80], and skin [81–83], it has been less explored as a bioprinting medium. Cui and Boland utilized an inkjet printing approach by first depositing an un-cross-linked fibrinogen monomer sheet. This hydrogel precursor solution was then printed with human microvascular endothelial cells embedded in a thrombin-calcium solution for rapid cross-linking and encapsulation of the cells [84]. Using this approach, optimal parameters for printing were denoted as 60 mg/mL fibrinogen, 50 IU/mL thrombin, and 80 mM CaCl$_2$ with a polymerization time of 6 min for layer formation. Similarly, Xu et al. used a hybrid electrospinning-inkjet printing approach for the fabrication of enhanced cartilage constructs by alternating electrospun polycaprolactone/Pluronic® layers with a printed a layer of chondrocyte-laden fibrinogen-collagen hydrogel followed by a printed thrombin layer for cross-linking [85]. The construct was allowed to polymerize for 15 min prior to the next electrospun PCL/Pluronic® layer being deposited. As described earlier, Skardal et al. used a codeposition strategy with one syringe containing a cell-laden collagen-fibrin hydrogel and another cross-linking syringe containing thrombin for the *in situ* printing of amniotic-fluid stem cells and mesenchymal stem cells on open wounds [48]. In order to avoid the potentially cytotoxic effects of gelatin cross-linkers, Xu et al., created a gelatin-fibrinogen hydrogel for use in an extrusion-based printing system. In this way, gelatin provided the increased viscosity needed for controlled extrusion and fibrin provided a cross-linked matrix upon placement in a thrombin solution immediately following printing [86]. Separately, in search of a more biomimetic tissue environment to model metabolic syndrome, Xu et al. devised a gelatin–alginate–fibrinogen custom hydrogel. In this composite hydrogel system, each component had a specific role: gelatin was used to facilitate extrusion by creating an optimal viscosity, and alginate and fibrinogen were sequentially cross-linked (alginate to provide primary structural integrity and fibrin as an adjuvant) [63].

Like the gelatin hydrogel, fibrin has many advantages and disadvantages that make it more useful as an ingredient rather than the principal hydrogel component. Fibrin hydrogels are attractive because they are a natural part of wound repair. They have been shown to promote angiogenesis, neurite extension, and cell recruitment [70]. Furthermore, fibrin is a highly translatable substance in that fibrinogen monomers can be procured in a patient-specific manner from the peripheral blood. One of the most attractive features of the fibrin hydrogel is its quick gelation time, as low as 15 s [32]. High biocompatibility and quick cross-linking time are important factors for bioprinting applications; however, the primary setback for widespread adoption of the fibrin hydrogel is its relatively poor mechanical properties and quick degradation time. Methods to create stiffer matrices and slow down degradation have been described *in vitro*, but it is uncertain how these efforts will translate to the *in vivo* environment. Ultimately, fibrin hydrogels may be effective for *in vitro* experiments up to 1 week or as a rapidly cross-linked component in a more complex hydrogel.

3.1.4 Hyaluronic Acid

Hyaluronic acid (HA) is an anionic, nonsulfated glycosaminoglycan composed of repeating units of the disaccharide: β-1,4-D-glucuronic acid- β-1,3-N-acetyl-D-glucosamine. In the body, HA is ubiquitous as a component of the extracellular matrix, vitreous humor, and synovial fluid of articulating joints. Thus, HA is highly biocompatible with roles in cell signaling, wound repair, regulation of cell morphogenesis, and matrix organization, making HA-based biomaterials attractive clinical products [87]. Furthermore, HA has been shown to have significant immunomodulatory and anti-inflammatory effects, sparking investigation as an anti-adhesion agent during surgeries involving the abdominal cavity and connective tissues [88–91].

Owing to its highly negative charge, HA attracts cations – and water through osmosis – to create a gel-like substance [92]. While HA does form a water-swollen matrix, cross-linking between polymer strands occurs rather infrequently, making HA highly soluble at room temperature and limiting its structural integrity as a scaffolding material. Because the degree of cross-linking is a determinant of matrix degradation, the relatively simple matrix organization HA forms increases its susceptibility to hyaluronidase, resulting in a matrix half-life of 0.5–3 days (2–5 min upon exposure to blood) [93]. Thus, a number of methods have been employed to modify HA that introduce functional side groups

capable of forming covalent bonds including the use of carboiimide [94], hydrazides [95], tyramine substitution [96], thiolation [97–99], and methacrylation [100].

The thiolated and methacrylated derivatives of HA have been the most widely adopted HA-based hydrogels for tissue engineering because of their ease of use and versatility. Thiolated HA (thiol-HA) forms stable cross-links slowly in air through oxidation and formation of disulfide bonds [97–99]. This process can be accelerated with the use of PEG as a thiol-reactive cross-linking agent [101]. Methacrylated HA (HA-MA), on the other hand, forms stable cross-links when exposed to UV light in the presence of a free-radical-generating photoinitiator. The structural and mechanical properties for thiol-HA and HA-MA hydrogels can be easily modified by changing the concentration of functional side groups available for cross-linking and/or concentrations of PEG [102] or exposure time to UV light [62], respectively. For better biomimetic properties, HA hydrogel networks have been mixed with gelatin hydrogels in an attempt to recreate the collagen-glycosaminoglycan composition of natural ECM, and these hydrogels have been shown to have greater cell retention and bioactivity than HA-only hydrogels [62,99]. HA-gelatin hydrogels, which are commercially available under the trade names Extracel™ and HyStem™, have found use in bioprinting because of their ability to be solidified at a rate compatible with the printing of layers in quick succession. Because HA-gelatin hydrogels contain gelatin, they are highly viscous, making them particularly useful for extrusion-based printers. HA-gelatin hydrogels consisting of thiolated HA and gelatin were cross-linked with four-armed PEG tetra-acrylates (tetraPAcs) for extrusion of cell-laden filaments into the form of a tubular construct [103]. The use of tetraPAcs as a cross-linker outperformed PEG-diacrylate (PEGDA) in terms of cross-linking time (10 vs 30 min), rheological properties ($G' = 800$ vs 100 Pa), and swelling ratio (PEGDA > tetraPAcs), and these constructs were able to retain good structural fidelity and cell viability after 4 weeks *in vitro* [103]. In a different method, gold nanoparticles were used as cross-linking agents for thiolated HA and gelatin to create a "dynamically cross-linking hydrogel" that had enough intragel cross-linking for cell encapsulation and extrusion from the printing nozzle and subsequently formed intergel cross-links for enhanced structural integrity [104]. This "dynamic cross-linking" was due to the relatively slow kinetics of multivalent gold nanoparticles-thiol reactions, allowing for a material that has the less stiff rheological properties conducive for printing at early time points and subsequently gains strength at later time points. This hydrogel system was able to support good cell viability, proliferation, and ECM formation after 4 weeks *in vitro* [104]. Finally, photocrosslinkable HA-gelatin hydrogels were created by synthesizing HA-MA and gelatin, which were subsequently cross-linked upon exposure to UV light in the presence of the photoinitiator Irgacure 2959 [62]. Because the degree of cross-linking in methacrylated hydrogels is a function of exposure time to UV light, hydrogels were partially cross-linked (exposure time of 120 s) to obtain a printable gel-like consistency and then irradiated for an additional 60 s after layer deposition to create the stiffness necessary for printing of the following layer [62]. Similar to the aforementioned HA-gelatin hydrogels, this hydrogel was printable and supported cell proliferation and ECM formation after 4 weeks *in vitro* [62].

In conclusion, modified HA-based hydrogels (particularly when they are mixed with modified gelatin) are useful systems for bioprinting applications because they are relatively easy to work with, highly customizable in terms of structural and mechanical properties, replicate the native ECM well, and have a favorable biocompatibility profile. While HA by itself is weak and quickly degraded, cross-linkable HA hydrogels have good strength and are more resistant to enzymatic degradation. Thus, the chemical modification of naturally derived hydrogels may be an important step toward a "best of both worlds" hydrogel that has all the biological advantages of a natural hydrogel combined with the mechanical tailorability of a synthetic hydrogel.

3.1.5   Alginate

Alginate is a natural polysaccharide found in the cell walls of brown algae. Structurally, alginate is very similar to HA. Specifically, it is a linear homopolymer composed of (1,4)-linked β-D-mannuronate (M) and α-L-glucuronic acid (G). Its high carboxylic acid content causes massive water absorption, and the addition of divalent cations (e.g., calcium, barium, strontium) results in rapid cross-linking between the G blocks of adjacent polymer strands. Thus, the mechanical strength and dissolution time of alginate hydrogels is directly related to the G/M ratio within the gel: a higher G/M content will create a stiffer, slower dissolving hydrogel [101]. Alginate hydrogels do not undergo enzymatic degradation by mammalian cells (because mammals do not produce alginase), but rather degrade via replacement of divalent cations with monovalent cations (e.g., sodium) present in the surrounding environment. This presents an important consideration in terms of biocompatibility because alginate polymer strands exceed the filtration size for renal clearance [105] and large displacements of calcium may lead to transient local hypercalcemia. It should be noted that this has not caused any observed adverse events in animal studies using alginate hydrogels, and that these hydrogels have been shown to have low to no immunogenicity [106]. Furthermore, alginate hydrogels can be modified to degrade in aqueous media when they are partially oxidized [107], should this be deemed necessary.

Alginate hydrogels have found much use in bioprinting applications because they are so easily cross-linked. The first use of alginate hydrogels to fabricate 3D bioprinted structures was by Boland et al. in 2006. Using their modified inkjet printing system, the printing hydrogel substrate was made up of a 2% (w/v) solution of alginic acid that was subsequently printed with a 0.25 M $CaCl_2$ solution for cross-linking. After one layer was printed, the next layer of hydrogel was prepared by lowering the stage into un-cross-linked alginic acid solution followed by printing another layer of $CaCl_2$ and the process was repeated to build up a three-dimensional structure [108]. This group also tested different concentrations of alginate and $CaCl_2$, and found that an 8% alginic acid and 0.4 M $CaCl_2$ solution produced the best printed channels [108]. Many bioprinting investigations have followed this methodology using inkjet [109–111] and laser-based [112–116] bioprinting systems.

More recently, alginate hydrogels have been applied in extrusion-based printing regimes [117–119]. These applications have been mainly targeted at musculoskeletal tissues, such as cartilage [120], the osteochondral interface [29], and bone [121,122]. The extrusion approach provides interesting applications by allowing for *in situ* printing as shown by Cohen et al.'s direct printing onto chondral and osteochondral defects [123] and dual-phase biomimetic hydrogel systems utilizing gelatin for its thermosensitive properties for processing, alginate for its mechanical properties after ionic cross-linking, and hydroxyapatite as an osseoinductive agent for bone tissue engineering [64].

Alginate hydrogels have been used extensively in tissue engineering applications and to a large degree in the more specialized fields of biofabrication and 3D bioprinting. They are low cost, abundantly available, and have shown good biocompatibility with negligible inflammatory effect after implantation *in vivo*. The ability to easily modify viscosity by concentration of precursor solution and degree of cross-linking (by altering calcium concentration and G/M ratio) before and after deposition makes this hydrogel highly versatile in terms of processing and mechanical characteristics for application in bioprinting. Finally, alginate hydrogels are also good systems because of their shape fidelity and volume retention *in vitro* [120] and *in vivo* [124]. Although alginate hydrogels have significant advantages (particularly in terms of processing), an important point of contention is the fact that they are derived from a xenogenic source that is significantly removed from humans on the phylogenetic tree. Although it is an organic, naturally derived biomaterial, alginate cannot be considered biomimetic for mammalian cells because mammals do not utilize this polymer in their ECM. Another important consideration is the high calcium content in alginate hydrogels. If these hydrogels are to be used for large tissue replacements, there is the potential for a large release of calcium into the blood stream (leading to transient hypercalcemia and possible renal damage) or into the surrounding tissue (possibly leading to heterotopic ossification).

3.1.6 Agarose

Agarose is a natural polysaccharide found in seaweed. Structurally, it is a linear polymer of agarbiose, a disaccharide consisting of D-galactose and 3,6-anhydro-L-galactopyranose. Like alginate, agarose is not biodegradable in mammals because we lack the enzyme, thus limiting its use in *in vivo* applications. Agarose hydrogels are thermosensitive hydrogels that have a melting temperature at around 40°C and a gelling temperature at approximately 32°C, making this hydrogel suitable for *in vitro* and *in vivo* applications. Because of these thermal gelation properties, agarose has the added advantage of requiring no additional cross-linking step, an ideal manufacturing characteristic.

Agarose hydrogels were first used by Xu et al. who printed Chinese hamster ovary cells and embryonic motor neuron cells onto the hydrogel via an inkjet method [10]. More recently, agarose hydrogels have been used as the substrate for printing via extrusion-type bioprinters. Fedorovich et al. tested agarose using their organ printing system but found a 1–5% agarose solution to solidify too slowly such that neighboring fibers fused, which caused the printed construct to lose its shape [122]. Other groups, using slightly different methods, have used agarose successfully. Norotte et al. utilized agarose rods as support structures for constructing cylindrical vascular structures [125]. In this scheme, the agarose was not used as a cell-encapsulation method, but as a sacrificial mold for multicellular spheroids that fused after approximately 3 days postprinting. Conversely, Duarte Campos and coworkers have devised a system of "submerged 3D bioprinting" that utilizes agarose as the cell-supporting hydrogel being printed within a high-density fluorocarbon (in this case, perfluorotributylamine ($C_{12}F_{27}N$)) chamber [126,127]. Using this methodology, this group has been able to fabricate precise structures by using the fluorocarbon as a support structure for the printed cell-laden agarose hydrogel. They have shown that this system results in a retained shape, as well as good cell viability, proliferation, and ECM production after 3 weeks in culture.

Agarose has not been used extensively for cell printing applications, possibly because of the difficulty of making the hydrogel printable. Furthermore, because this hydrogel is derived from a plant source, it is not biomimetic for mammalian cell types. This makes agarose a difficult polymer for cells to attach to, proliferate on, and degrade over time. While this hydrogel has been challenging to print, its gelation properties still make this an attractive potential hydrogel for bioprinting applications. Innovative methods, such as those presented by Duarte Campos et al., provide interesting new ways that this hydrogel can be processed to create constructs with complex geometries.

3.1.7 Matrigel™

Matrigel™ is an extract of proteins and small molecules (primarily collagen IV, laminin, perlacan, and various growth factors) derived from the Engelbreth–Holm–Swarm mouse tumor, a poorly differentiated chondrosarcoma that more closely resembles the basement membrane than cartilage [128]. This extract is often solubilized in serum-free media and kept frozen until use. At the time of use, it can be kept chilled at 4°C for use as a liquid and achieves a gel-like consistency at room temperature. At 37°C, a solidified matrix forms after approximately 30 min in a similar fashion to thermosensitive collagen I-based hydrogels. Matrigel™ has been used extensively for 3D cell culture because of its record of strongly promoting cell proliferation and differentiation.

The thermal gelation qualities of Matrigel™ make it a good potential hydrogel for bioprinting applications. Like the collagen I hydrogel, a 3D sheet of Matrigel™ can be prepared as a supporting matrix onto which cells are dropped using inkjet- or laser-based approaches; or the hydrogel can be kept chilled to retain its more soluble form for extrusion-based printing. The thermal gelation kinetics of Matrigel™, however, makes this a difficult medium for extrusion-based printing because it requires a system that can provide a chilled syringe for dispensing and a warmed ambient environment for gelation, and there is a required waiting of approximately 30 min between successive layers to allow for enough solidification. There is sparse literature using Matrigel™ as a bioprinting hydrogel. Xu et al. prepared a Matrigel™ hydrogel onto which ovarian cancer cells and fibroblast droplets were printed to develop a more biomimetic 3D coculture system for the *in vitro* study of ovarian cancer [129]. In a similar fashion, Matrigel™ has been used as the receiving substrate onto which fibroblasts and keratinocytes were laser bioprinted for skin tissue engineering *in vitro* [130,131] and *in vivo* [132]. Finally, a custom printing system was devised for the extrusion printing of Matrigel™. In this system, a cooling chamber was fashioned around the syringe containing hepatic carcinoma and epithelial cells in Matrigel™ for low-pressure (12.3 Pa) deposition onto a PDMS substrate to create a microfluidic system for *in vitro* cancer studies [133]. While a Matrigel™ cell suspension was extruded in this system, it should be noted that only one layer was printed and then the chip was incubated for solidification.

In conclusion, Matrigel™ has significant advantages in terms of its cytocompatibility, biomimicry to native ECM, and ability to promote cell adhesion and differentiation across a number of cell types. In the future, Matrigel™'s superior biological properties may serve as a model system for the creation of new tissue-specific, biomimetic hydrogels. Although it provides a great ECM substitute for *in vitro* studies, the malignant source and inconsistent molecular composition of Matrigel™ bars any potential clinical translatability. Additionally, while Matrigel™ has important biological characteristics, it has not been widely applied for bioprinting because its gelation kinetics requires a relatively long incubation time for the printing of multilayered structures. Thus, bioprinting applications using Matrigel™ have been limited to single-layered structures for *in vitro* models or very thin skin substitutes. Future tissue-specific ECM-derived hydrogels may suffer from similar technical processing challenges for bioprinting; however, novel systems to control thermosensitive hydrogels building on the cooling system described earlier [133] can be envisaged.

3.2 Synthetic Hydrogels

3.2.1 Poly(ethylene glycol) (PEG)

PEG is a highly versatile synthetic compound that has been exploited in biomedical research because of its excellent biocompatibility and chemical labiality for modification with various functional groups [134]. In its simplest unmodified form, PEG is a polymer of ethylene oxide monomers. Based on the level of polymerization, this molecule can take different names including PEG (M_w < 20 kDa), poly(ethylene oxide) (PEO) (M_w > 20 kDa) or poly(oxyethylene) (any M_w). These distinctions are important because the rheological properties and mechanical characteristics can be modified significantly by simply adjusting the molecular weight of the molecule. Additionally, the morphology of the PEG molecule can be modified to take on various configurations from linear to multiarmed "star" PEGs. The latter star-PEG molecules have multiple side chains emanating from their central core, creating increased functional groups available for modification. This last point is important because PEG itself is biologically inert. However, with relatively simple chemistry, PEG can incorporate bioactive molecules including cell-adhesion peptides, enzyme-sensitive peptide sequences, and growth factor-binding domains to create a more biomimetic environment for cells [134].

PEG can polymerize to form hydrogels by a number of mechanisms, including step growth, chain growth, and mixed-mode polymerization depending on the monomers present [135]. Polymerization is induced by free-radical generation, which can occur secondary to thermal energy, redox reactions or photocleavage of photoinitiator molecules [135]. Photopolymerization is the most popular form of PEG polymerization because of spatiotemporal control over polymerization, rapid curing rates (seconds to minutes) at room or physiologic temperatures, and minimal heat production that has negligible toxic effects on cells and tissues [136]. Upon exposure to UV light, photoinitiators produce free radicals that can in turn propagate through the highly reactive $C = C$ bonds of functionalized PEG molecules (often PEG-methacrylate (PEGMA), PEG-dimethacrylate (PEGDMA) or PEG-diacrylate (PEGDA)) to yield covalently bonded macromolecules. The hydrophilic nature of the PEG molecule, in turn, results in a highly hydrated cross-linked polymeric network that can be used for cell encapsulation.

Due to the linear or multiarmed nature of the PEG monomer, PEGMA and PEGDA can serve as a cross-linker to join adjacent polymer strands of other methacryl- or thiol-functionalized polymers during bioprinting, for example in gelatin-based [58,59,61] and HA:gelatin-based [60,62,103] hydrogels. While PEG participated as a cross-linker in these hydrogels, it was not the backbone molecule. Recently, however, PEG-based hydrogels have been bioprinted with PEG as the primary hydrogel constituent (instead of a cross-linking additive). Using an inkjet-based system, Cui et al. bioprinted 5×10^6 chondrocytes/mL of PEGDMA (M_w = 3400) directly onto osteochondral defects, resulting in good cell viability and proliferation, a compressive modulus close to that seen in native articular cartilage, and enhanced ECM production in comparison to 2D implants [137]. In order to efficiently use photopolymerizable PEGDMA as a printing medium, this group utilized the photoinitiator Irgacure 2959, positioned the UV light 25 cm above the printed material with an intensity of 4.5 mW/cm², and shielded the printing syringe using aluminum foil to avoid premature polymerization and nozzle clogging. Hockaday et al. used an extrusion-based printing system to create a mechanically heterogeneous aortic valve structure using porcine aortic valve interstitial cells in a PEGDA-alginate hydrogel [138]. In this system, alginate was used to tailor viscosity for consistent extrusion and photopolymerizable PEGDA was the main polymerizing agent. In addition to two hydrogel materials, this group tested mixtures of different molecular weight PEGDA molecules to create a mechanically heterogeneous heart valve model. Ultimately, they used a stiffer gel (E = 74.6 kPa) consisting of 20%:0% PEG-DA(M_w = 700):PEG-DA(M_w = 8000) for the aortic root and a more elastic gel (E = 5.3 kPa) consisting of 0%:10% PEG-DA(M_w = 700):PEG-DA(M_w = 8000) for the valve leaflets. Furthermore, this group found that, using their system, a 1–2% w/w Irgacure 2959 concentration was optimal to induce rapid gelation for bioprinting. This study illustrates the mechanical diversity that can be engineered into PEG hydrogels and that this system is applicable in extrusion-based systems with good cell viability and shape fidelity after 21 days.

PEG is a low-cost, readily available, synthesizable hydrogel system that already has regulatory approval for use in humans for various clinical applications. This hydrogel is highly customizable in terms biocompatibility, bioactivity, and mechanical properties. Furthermore, by using photoinitiation, cross-links can form in a time frame that is very useful for the printing of multilayered structures. While the bioprinting studies using PEG to date have not employed modifications with bioactive molecules such as RGD-peptide to enhance cell adhesion or enzyme-sensitive sequences to enhance biodegradation, these methods have been explored with encouraging results [134]. Indeed, the paucity of literature using PEG-based hydrogels for bioprinting may be secondary to its biological inertness resulting in poor cell adhesion leading to poor tissue development. Nonetheless, the versatility of PEG as a hydrogel system makes it a very interesting candidate in the search for an optimal bioprinting hydrogel.

3.2.2 Poloxamers

Poloxamers, more commonly known by the trade names Pluronic® and Lutrol®, are a class of amphiphilic triblock copolymers with the base molecular structure polyoxyethylene–polyoxypropylene–polyoxyethylene (PEO–PPO–PEO). Due to the amphiphilic nature of these polymers (where PPO is hydrophobic and PEO is hydrophilic), this class of polymers has been widely exploited for drug delivery as they can form soluble micelles, which can carry nanoscale therapeutics [139]. Pluronics® are thermosensitive polymers in that the PPO side chains become less soluble above a threshold temperature (known as the critical micelle temperature, which is usually somewhere between 22°C and 37°C, depending on polymer concentration), above which self-association occurs resulting in a gel-like substance [140]. Due to their synthetic nature, this class of polymers shares many of the biological disadvantages that PEG hydrogels have, including low cell adhesion and inability to be enzymatically degraded. In fact, due to the relatively weak mode of polymerization (i.e., micelle formation), Pluronic® remains quite soluble in aqueous solution and readily dissolves after 1 week *in vitro* [141]. In addition to poor structural properties over the long term, Pluronic®'s cytocompatibility profile is questioned due to potential disruption of the cell membrane [142], which may be due to its polar–nonpolar structure closely resembling that of phospholipids.

Despite these disadvantages, Pluronic® has been used for bioprinting as a cell carrier and as a sacrificial molding agent. Due to its high viscosity, Pluronic®'s principal advantage is creating accurate structures with good shape fidelity immediately after printing. While Pluronic® has been principally used in extrusion-based bioprinters, one study added 0.05% Pluronic® to a 1.5 mg/mL collagen hydrogel for subsequent inkjet printing with HepG2 cells in order to create droplets with better shape and gelation [143]. Although this produced better-printed constructs, cell viability was significantly decreased after 1 and 2 weeks in comparison to collagen-only hydrogels. All other bioprinting studies using Pluronic® have utilized extrusion-based systems. While some early work using extrusion-based systems used Pluronic® as a cell encapsulation and delivery medium, it was quickly found that this hydrogel produced suboptimal results in terms of cell viability, noted between 4% and 60% after only a few days [13,122]. Such low cell viability may be due to a number of effects including potential membrane disruption, inability of cells to adhere to the polymer matrix, and rapid dissolution of the hydrogel. Indeed, these studies

did note excellent printability of the cell-laden Pluronic®, but ultimately low mechanical strength of the resulting construct. One group attempted using a photocrosslinkable Pluronic® for cell printing, and found significantly enhanced structural integrity after 3 weeks; however, cell viability remained low at 50% [144]. Due to the rapid dissolution of Pluronic® in aqueous solution and its potential toxic effect on cells, it can be used as a sacrificial molding material. Chang et al. have utilized such a "mold-and-fill" strategy where high-viscosity 40% Pluronic® F-127 (w/v) is first printed as a bio-inert sacrificial mold followed by the printing of a low-viscosity cells-laden hydrogel (in this case 3 mg/mL collagen I) [33]. In this way, the Pluronic® only serves as a temporary structure that ultimately washes away, leaving a completely biological construct behind. Sacrificial channels can be printed throughout a construct to leave way for a microfluidic system that can be potentially vascularized *in vivo* [145].

In conclusion, while Pluronic® lacks many of the biological and mechanical qualities to be a cell printing material, it may serve potential use as a "sacrificial material." The highly viscous nature of Pluronic® makes it a very easily printed hydrogel that can aid in the fabrication of complex multilayered 3D structures with complicated shapes and/or internal architecture. As such, the biological qualities of Pluronic® may be irrelevant as it is only a mildly toxic material that is only transiently in contact with cells.

4 CONCLUSIONS AND FUTURE OUTLOOK

Current 3D bioprinting technology allows researchers to precisely deposit cells and other biomaterials for the fabrication of volumetric tissue constructs. For this purpose, hydrogels have become important "bioinks" because: (1) they are highly biocompatible; (2) they can form crosslinked polymeric networks that are capable of cell encapsulation; and (3) they can be induced to solidify such that each layer can serve as the foundation for subsequently printed layers. An ideal bioprinting hydrogel would be one that has high biocompatibility and biomimicry, degrades at a rate equal to that of ECM synthesis by the neo-tissue, provides appropriate structural and mechanical support during and after printing, and forms stable cross-links on the order of seconds to minutes. Thus, hydrogels for bioprinting applications are must meet biological and processing requisites that are often in conflict with one another. While no currently available hydrogel meets all of these criteria, the hydrogels in use today have important qualities that make them amenable to the printing process and tissue regeneration. In an attempt to remedy the shortcomings of certain hydrogels, many groups created combination hydrogel systems [48,57,60,62–65,85,86,103] or make chemical modifications to base polymers [58–62,

103,137,144] to provide specific properties for their specific applications.

Most of the hydrogels currently used for bioprinting are derived from natural ECM components. These naturally derived hydrogels possess excellent biological advantages, such as high biocompatibility and the presence of cell adhesion sites, enzyme-cleavage sites, and growth factors; however, their poor mechanical properties and quick degradation rates limit the use of these polymers for long-term applications. While naturally derived polymers can be chemically modified or mixed with each other to achieve more desirable properties, this creates more complicated systems that are difficult to reproduce. Synthetic hydrogels are thus appealing alternatives because they can be readily modified to express bioactive domains and because the user has greater control over gelation kinetics, degradation rates, and mechanical properties. Although synthetic hydrogels are highly adaptable to specific applications, our capacity to manipulate the synthetic microenvironment is limited by our ability to engineer and recreate the highly complex functions of ECM generated during billions of years of evolution using relatively simple molecules and chemical reactions.

At the present time, hydrogels used in bioprinting are relatively simple in composition, have low functionality outside of cell encapsulation, and are plagued by poor mechanical properties. While the development of printable hydrogels that are able to support cell viability and proliferation is a feat in itself, there is a significant need to engineer systems that go beyond these basics to create more sophisticated tissue constructs. Some examples of these "smart bioink" systems include printing thermoplastic polymers alongside hydrogels to generate higher-order architecture and more mechanically robust structures [30,31] and printing growth factors [146–149] or plasmids for gene transfection [150] into the hydrogel to enhance tissue-specific development. These methods are an important first step in the design of bioprinted constructs with enhanced functionality and applicability. Other novel "smart hydrogels" that may some day be used for bioprinting include electroconductive hydrogels [151], which could be potentially used for the printing of nerve or muscle constructs and tissue-specific ECM-derived hydrogels [152–155], which may provide the precise milieu of ECM components that is optimal for the development of the particular tissue type being printed.

In the end, 3D bioprinting is a field that is ripe for innovation. The use of hydrogels as "bioink" provides a versatile system that is highly processable, biocompatible, and easily manipulated for the purpose of printing a tissue construct. As the fields of engineering, materials science, biology, and medicine converge, it will not be long before we are able to develop precise and highly sophisticated tissue-specific constructs using a 3D bioprinter.

GLOSSARY

3D printing The sequential deposition of any material capable of solidification in a cross-sectional pattern for the building up of a three-dimensional object

Biocompatibility The ability of a biomaterial to perform its desired function with respect to a medical therapy, without eliciting any undesirable local or systemic effects in the recipient or beneficiary of that therapy, but generating the most appropriate beneficial cellular or tissue response in that specific situation, and optimizing the clinically relevant performance of that therapy

Bioink Materials that serve the dual roles of printing substrate and tissue engineering scaffold

Biomimicry The imitation of the models, systems, and elements of nature for the purpose of solving complex human problems

Bioprinting The process of generating spatially controlled cell patterns using 3D printing technologies, where cell function and viability are preserved within the printed construct

Cross-link Inducing physical or chemical bonding through environmental changes (such as pH, temperature, and ionic concentration), enzymatic initiation, or photopolymerization

Hydrogel A group of polymeric materials, the hydrophilic structure of which renders them capable of holding large amounts of water in their three-dimensional networks

Polymer A large molecule, or macromolecule, composed of many repeated subunits

Polymerization Process of reacting monomer molecules together in a chemical reaction to form polymer chains or three-dimensional networks

Thermoplastic Become pliable or moldable above a specific temperature and solidify upon cooling

Viscosity A material's resistance to flow when a force is applied

ABBREVIATIONS

3D	Three-dimensional
ECM	Extracellular matrix
Gel-MA	Methacrylated gelatin
HA-MA	Methacrylated HA
MMP	Matrix metalloproteinase
PEG	Poly(ethylene glycol)
PEGDA	PEG-diacrylate
PEGDMA	PEG-dimethacrylate
PEGMA	PEG-methacrylate
PEO	Poly(ethylene oxide)
TetraPAcs	Four-armed PEG tetra-acrylates
UV	Ultraviolet

REFERENCES

[1] Langer R, Vacanti JP. Tissue engineering. Science 1993;260(5110): 920–6.

[2] Rozario T, DeSimone DW. The extracellular matrix in development and morphogenesis: a dynamic view. Dev Biol 2010;341(1): 126–40.

[3] Malda J, Visser J, Melchels FP, Jüngst T, Hennink WE, Dhert WJ, et al. 25th Anniversary article: engineering hydrogels for biofabrication. Adv Mater 2013;25(36):5011–28.

[4] Coury A, Miller R, et al. Bioresorbable hydrogels for medical thera-pies. Transactions of the Knowledge Foundation Workshop 2002.

[5] Fedorovich NE, Alblas J, Wijn JR, Hennink WE, Verbout AJ, Dhert WJ. Hydrogels as extracellular matrices for skeletal tissue engineer-ing: state-of-the-art and novel application in organ printing. Tissue Eng 2007;13:1905–25.

[6] Williams DF. On the mechanisms of biocompatibility. Biomaterials 2008;29(20):2941–53.

[7] Ringeisen BR, Kim H, Barron JA, Krizman DB, Chrisey DB, Jack-man S, et al. Laser printing of pluripotent embryonal carcinoma cells. Tissue Eng 2004;10(3-4):483–91.

[8] Barron JA, Krizman DB, Ringeisen BR. Laser printing of single cells: statistical analysis, cell viability, and stress. Ann Biomed Eng 2005;33(2):121–30.

[9] Hopp B, Smausz T, Kresz N, Barna N, Bor Z, Kolozsvári L, et al. Survival and proliferative ability of various living cell types after laser-induced forward transfer. Tissue Eng 2005;11(11–12):1817–23.

[10] Xu T, Jin J, Gregory C, Hickman JJ, Boland T. Inkjet printing of vi-able mammalian cells. Biomaterials 2005;26(1):93–9.

[11] Wilson WC, Boland T. Cell and organ printing 1: protein and cell printers. Anat Rec A Discov Mol Cell Evol Biol 2003;272(2):491–6.

[12] Chang R, Nam J, Sun W. Effects of dispensing pressure and nozzle diameter on cell survival from solid freeform fabrication-based direct cell writing. Tissue Eng Part A 2008;14(1):41–8.

[13] Smith CM, Stone AL, Parkhill RL, Stewart RL, Simpkins MW, Ka-churin AM, et al. Three-dimensional bioassembly tool for generat-ing viable tissue-engineered constructs. Tissue Eng 2004;10(9–10): 1566–76.

[14] Melchels FPW, Domingos MAN, Klein TJ, Malda J, Bartolo PJ, Hutmacher DW. Additive manufacturing of tissues and organs. Prog Polym Sci 2012;37(31):1079–104.

[15] Burdick JA, Anseth KS. Photoencapsulation of osteoblasts in inject-able RGD-modified PEG hydrogels for bone tissue engineering. Bio-materials 2002;23(22):4315–23.

[16] Hern DL, Hubbell JA. Incorporation of adhesion peptides into non-adhesive hydrogels useful for tissue resurfacing. J Biomed Mater Res 1998;39(2):266–76.

[17] Lee HJ, Lee JS, Chansakul T, Yu C, Elisseeff JH, Yu SM. Collagen mimetic peptide-conjugated photopolymerizable PEG hydrogel. Biomaterials 2006;27(30):5268–76.

[18] Peyton SR, Raub CB, Keschrumrus VP, Putnam AJ. The use of poly(ethylene glycol) hydrogels to investigate the impact of ECM chemistry and mechanics on smooth muscle cells. Biomaterials 2006;27:4881-L4893.

[19] Drury JL, Mooney DJ. Hydrogels for tissue engineering: scaf-fold design variables and applications. Biomaterials 2003;24(24): 4337–51.

[20] Ahmed TA, Griffith M, Hincke M. Characterization and inhibition of fibrin hydrogel-degrading enzymes during development of tissue engineering scaffolds. Tissue Eng 2007;13(7):1469–77.

[21] Purcell BP, Lobb D, Charati MB, Dorsey SM, Wade RJ, Zellars KN, et al. Injectable and bioresponsive hydrogels for on-demand matrix metalloproteinase inhibition. Nat Mater 2014;13(6):653–61.

[22] Murad S, Grove D, Lindberg KA, Reynolds G, Sivarajah A, Pinnell SR. Regulation of collagen synthesis by ascorbic acid. Proc Natl Acad Sci U S A 1981;78(5):2879–82.

[23] Gupta V, Grande-Allen KJ. Effects of static and cyclic loading in regulating extracellular matrix synthesis by cardiovascular cells. Car-diovasc Res 2006;72(3):375–83.

[24] Kjaer M. Role of extracellular matrix in adaptation of tendon and skeletal muscle to mechanical loading. Physiol Rev 2004;84(2): 649–98.

[25] Skutek M, van Griensven M, Zeichen J, Brauer N, Bosch U. Cyclic mechanical stretching modulates secretion pattern of growth factors in human tendon fibroblasts. Eur J Appl Physiol 2001;86(1):48–52.

[26] Hynes RO, Naba A. Overview of the matrisome – an inventory of extracellular matrix constituents and functions. Cold Spring Harb Perspect Biol 2012;4(1). a004903.

[27] Park Y, Lutolf MP, Hubbell JA, Hunziker EB, Wong M. Bovine primary chondrocyte culture in synthetic matrix metalloproteinase-sensitive poly(ethyelen glycol)-based hydrogels as a scaffold for car-tilage repair. Tissue Eng 2004;10:515.

[28] Benjamin M, Toumi H, Ralphs JR, Bydder G, Best TM, Milz S. Where tendons and ligaments meet bone: attachment sites ('en-theses') in relation to exercise and/or mechanical load. J Anat 2006;208(4):471–90.

[29] Fedorovich NE, Schuurman W, Wijnberg HM, Prins HJ, van Weeren PR, Malda J, et al. Biofabrication of osteochondral tissue equivalents by printing topologically defined, cell-laden hydrogel scaffolds. Tis-sue Eng Part C Methods 2012;18(1):33–44.

[30] Cornock R, Beirne S, Thompson B, Wallace GG. Coaxial additive manufacture of biomaterial composite scaffolds for tissue engineer-ing. Biofabrication 2014;6. 025002.

[31] Schuurman W, Khristov V, Pot MW, Weeren PR, Dhert WJ, Malda J. Bioprinting of hybrid tissue constructs with tailorable mechanical properties. Biofabrication 2011;3. 021001.

[32] Murphy SV, Skardal A, Atala A. Evaluation of hydrogels for bio-printing applications. J Biomed Mater Res Part A 2013;101(1): 272–84.

[33] Chang CC, Boland ED, Williams SK, Hoying JB. Direct-write bioprinting three-dimensional biohybrid systems for future regen-erative therapies. J Biomed Mater Res B Appl Biomater 2011;98(1): 160–70.

[34] Guillotin B, Guillemot F. Cell patterning technologies for organo-typic tissue fabrication. Trends Biotechnol 2011;29(4):183–90.

[35] Alberts B, Johnson A, Lewis J, Raff M, Roberts K, Walter P. Col-lagens are the major proteins of the extracellular matrix. 5th edn Mo-lecular biology of the cell. Garland Sci; 2008. 1184-6.

[36] van der Rest M, Garrone R. Collagen family of proteins. FASEB J 1991;5(13):2814–23.

[37] Banos CC, Thomas AH, Kuo CK. Collagen fibrillogenesis in tendon development: current models and regulation of fibril assembly. Birth Defects Res C Embryo Today 2008;84(3):228–44.

[38] Lynn AK, Yannas IV, Bonfield W. Antigenicity and immunogenic-ity of collagen. J Biomed Mater Res B Appl Biomater 2004;71(2): 343–54.

[39] Lauer-Fields JL, Juska D, Fields GB. Matrix metalloproteinases and collagen catabolism. Biopolymers 2002;66(1):19–32.

[40] Bell E, Ivarsson B, Merrill C. Production of a tissue-like structure by contraction of collagen lattices by human fibroblasts of different proliferative potential *in vitro*. Proc Natl Acad Sci USA 1979;76(3): 1274–8.

[41] Brown RA. In the beginning there were soft collagen-cell gels: towards better 3D connective tissue models? Exp Cell Res 2013;319:2460–9.

[42] Chan G, Mooney DJ. New materials for tissue engineering: towards greater control over the biological response. Trends Biotechnol 2008;26(7):382–92.

[43] Lau YK, Gobin AM, West JL. Overexpression of lysyl oxidase to increase matrix crosslinking and improve tissue strength in dermal wound healing. Ann Biomedical Eng 2006;34(8):1239–46.

[44] Boland T, Mironov V, Gutowska A, Roth EA, Markwald RR. Cell and organ printing 2: fusion of cell aggregates in three-dimensional gels. Anat Rec A Discov Mol Cell Evol Biol 2003; 272(2):497–502.

[45] Lee W, Pinckney J, Lee V, Lee JH, Fischer K, Polio S, et al. Three-dimensional bioprinting of rat embryonic neural cells. Neuroreport 2009;20(8):798–803.

[46] Lee W, Debasitis JC, Lee VK, Lee JH, Fischer K, Edminster K, et al. Multi-layered culture of human skin fibroblasts and keratino-cytes through three-dimensional freeform fabrication. Biomaterials 2009;30(8):1587–95.

[47] Lee V, Singh G, Trasatti JP, Bjornsson C, Xu X, Tran TN, et al. Design and fabrication of human skin by three-dimensional bioprinting. Tissue Eng Part C Methods 2014;20(6):473–84.

[48] Skardal A, Mack D, Kapetanovic E, Atala A, Jackson JD, Yoo J, et al. Bioprinted amniotic fluid-derived stem cells accelerate healing of large skin wounds. Stem Cells Transl Med 2012;1(11): 792–802.

[49] Orban JM, Wilson LB, Kofroth JA, El-Kurdi MS, Maul TM, Vorp DA. Crosslinking of collagen gels by transglutaminase. J Biomed Mater Res Part A 2004;68(4):756–62.

[50] Elbjeirami WM, Yonter EO, Starcher BC, West JL. Enhancing mechanical properties of tissue-engineered constructs via lysyl oxidase crosslinking activity. J Biomed Mater Res Part A 2003;66(3): 513–21.

[51] Liang Y, Jeong J, DeVolder RJ, Cha C, Wang F, Tong YW, et al. A cell-instructive hydrogel to regulate malignancy of 3D tumor spheroids with matrix rigidity. Biomaterials 2011;32(35):9308–15.

[52] Jeong B, Kim SW, Bae YH. Thermosensitive sol-gel reversible hydrogels. Adv Drug Delivery Rev 2002;54(1):37–51.

[53] Michon C, Cuvelier G, Launay B. Concentration dependence of the critical viscoelastic properties of gelatin at the gel point. Rheol Acta 1993;32:94–103.

[54] Olde Damink LH, Dijkstra PJ, Van Luyn MJ, van Wachem PB, Nieuwenhuis P, Feijen J. Glutaraldehyde as a cross-linking agent for collagen-based biomaterials. J Mater Sci Mater Med 1995;6: 460–72.

[55] Van Den Bulcke AI, Bogdanov B, De Rooze N, Schacht EH, Cornelissen M, Berghmans H. Structural and rheological properties of methacrylamide modified gelatin hydrogels. Biomacromolecules 2000;1(1):31–8.

[56] Wang X, Yan Y, Pan Y, Xiong Z, Liu H, Cheng J, et al. Generation of three-dimensional hepatocyte/gelatin structures with rapid prototyping system. Tissue Eng 2006;12(1):83–90.

[57] Yan Y, Wang X, Pan Y, Liu H, Cheng J, Xiong Z, et al. Fabrication of viable tissue-engineered constructs with 3D cell-assembly technique. Biomaterials 2005;26(29):5864–71.

[58] Bertassoni LE, Cardoso JC, Manoharan V, Cristino AL, Bhise NS, Araujo WA, et al. Direct-write bioprinting of cell-laden methacrylated gelatin hydrogels. Biofabrication 2014;6(2). 024105.

[59] Billiet T, Gevaert E, De Schryver T, Cornelissen M, Dubruel P. The 3D printing of gelatin methacrylamide cell-laden tissue-engineered constructs with high cell viability. Biomaterials 2014;35(1): 49–62.

[60] Duan B, Kapetanovic E, Hockaday LA, Butcher JT. Three-dimensional printed trileaflet valve conduits using biological hydrogels and

[61] Schuurman W, Levett PA, Pot MW, van Weeren PR, Dhert WJ, Hutmacher DW, et al. Gelatin-methacrylamide hydrogels as potential biomaterials for fabrication of tissue-engineered cartilage constructs. Macromol Biosci 2013;13(5):551–61.

[62] Skardal A, Zhang J, McCoard L, Xu X, Oottamasathien S, Prestwich GD. Photocrosslinkable hyaluronan-gelatin hydrogels for two-step bioprinting. Tissue Eng Part A 2010;16(8): 2675–85.

[63] Xu M, Wang X, Yan Y, Yao R, Ge Y. An cell-assembly derived physiological 3D model of the metabolic syndrome, based on adipose-derived stromal cells and a gelatin/alginate/fibrinogen matrix. Biomaterials 2010;31(14):3868–77.

[64] Wust S, Godla ME, Muller R, Hofmann S. Tunable hydrogel composite with two-step processing in combination with innovative hardware upgrade for cell-based three-dimensional bioprinting. Acta Biomaterialia 2014;10(2):630–40.

[65] Duan B, Hockaday LA, Kang KH, Butcher JT. 3D bioprinting of heterogeneous aortic valve conduits with alginate/gelatin hydrogels. J Biomed Mater Res Part A 2013;101(5):1255–64.

[66] Lee W, Lee V, Polio S, et al. On-demand three-dimensional freeform fabrication of multi-layered hydrogel scaffold with fluidic channels. Biotechnol Bioeng 2010;105(6):1178–86.

[67] Gough JE, Scotchford CA, Downes S. Cytotoxicity of glutaraldehyde crosslinked collagen/poly(vinyl alcohol) films is by the mechanism of apoptosis. J Biomed Mater Res 2002;61(1): 121–30.

[68] Sun HW, Feigal RJ, Messer HH. Cytotoxicity of glutaraldehyde and formaldehyde in relation to time of exposure and concentration. Pediatr Dent 1990;12(5):303–7.

[69] Schuurman W, Levett PA, Pot MW, van Weeren PR, Dhert WJ, Hutmacher DW, et al. Gelatin-methacrylamide hydrogels as potential biomaterials for fabrication of tissue-engineered cartilage constructs. Macromol Biosci 2013;13(5):551–61.

[70] Ahmed TA, Dare EV, Hincke M. Fibrin: a versatile scaffold for tissue engineering applications. Tissue Eng Part B Rev 2008;14(2): 199–215.

[71] Mosesson MW, Siebenlist KR, Meh DA. The structure and biological features of fibrinogen and fibrin. Ann New York Acad Sci 2001;936:11–30.

[72] Dare EV, Griffith M, Poitras P, Kaupp JA, Waldman SD, Carlsson DJ, et al. Genipin cross-linked fibrin hydrogels for *in vitro* human articular cartilage tissue-engineered regeneration. Cells Tissues Organs 2009;190(6):313–25.

[73] Rowe SL, Lee S, Stegemann JP. Influence of thrombin concentration on the mechanical and morphological properties of cell-seeded fibrin hydrogels. Acta Biomaterialia 2007;3(1):59–67.

[74] Kupcsik L, Alini M, Stoddart MJ. Epsilon-aminocaproic acid is a useful fibrin degradation inhibitor for cartilage tissue engineering. Tissue Eng A 2009;15(8):2309–13.

[75] Cholewinski E, Dietrich M, Flanagan TC, Schmitz-Rode T, Jockenhoevel S. Tranexamic acid–an alternative to aprotinin in fibrin-based cardiovascular tissue engineering. Tissue Eng Part A 2009; 15(11):3645–53.

[76] Smith JD, Chen A, Ernst LA, Waggoner AS, Campbell PG. Immobilization of aprotinin to fibrinogen as a novel method for controlling degradation of fibrin gels. Bioconjug Chem 2007;18(3): 695–701.

[77] Eyrich D, Brandl F, Appel B, Wiese H, Maier G, Wenzel M, et al. Long-term stable fibrin gels for cartilage engineering. Biomaterials 2007;28(1):55–65.

[78] Dare EV, Vascotto SG, Carlsson D, Hincke MT, Griffith M. Differentiation of a fibrin gel encapsulated chondrogenic cell line. Int J Artif Organs 2007;30(7):619–27.

[79] Huang YC, Dennis RG, Larkin L, Baar K. Rapid formation of functional muscle *in vitro* using fibrin gels. J Appl Physiol 2005;98: 706–13.

[80] Nieponice A, Maul TM, Cumer JM, Soletti L, Vorp DA. Mechanical stimulation induces morphological and phenotypic changes in bone marrow-derived progenitor cell within a three-dimensional fibrin matrix. J Biomed Mater Res 2007;81:523–30.

[81] Geer DJ, Swartz DD, Andreadis ST. Fibrin promotes migration in a three-dimensional *in vitro* model of wound regeneration. Tissue Eng 2002;8(5):787–98.

[82] Hojo M, Inokuchi S, Kidokoro M, Fukuyama N, Tanaka E, Tsuji C, et al. Induction of vascular endothelial growth factor by a fibrin as dermal substrate for cultured skin substitutes. Plast Reconstr Surg 2003;111(5):1638–45.

[83] Balestrini JL, Billiar KL. Equibiaxial cyclic stretch stimulates fibroblasts to rapidly remodel fibrin. J Biomech 2006;39(16):2983–90.

[84] Cui X, Boland T. Human microvasculature fabrication using thermal inkjet printing technology. Biomaterials 2009;30(31):6221–7.

[85] Xu T, Binder KW, Albanna MZ, Dice D, Zhao W, Yoo JJ, et al. Hybrid printing of mechanically and biologically improved constructs for cartilage tissue engineering applications. Biofabrication 2013;5(1). 015001.

[86] Xu W, Wang X, Yan Y, Zhen W, Xiong Z, Lin F, et al. Rapid prototyping three-dimensional cell/gelatin/fibrinogen constructs for medical regeneration. J Bioact Compat Polym 2007;22:363–77.

[87] Prestwich GD. Hyaluronic acid-based clinical biomaterials derived for cell and molecule delivery in regenerative medicine. J Controlled Release 2011;155(2):193–9.

[88] Pederzini LA, Milandri L, Tosi M, Prandini M, Nicoletta F. Preliminary clinical experience with hyaluronan anti-adhesion gel in arthroscopic arthrolysis for posttraumatic elbow stiffness. J Orthop Traumatol 2013;14(2):109–14.

[89] Smit X, van Neck JW, Afoke A, Hovius SE. Reduction of neural adhesions by biodegradable autocrosslinked hyaluronic acid gel after injury of peripheral nerves: an experimental study. J Neurosurg 2004;101(4):648–52.

[90] Riccio M, Battiston B, Pajardi G, Corradi M, Passaretti U, Atzei A, et al. Efficiency of Hyaloglide in the prevention of the recurrence of adhesions after tenolysis of flexor tendons in zone II: a randomized, controlled, multicentre clinical trial. J Hand Surg Eur Vol 2010;35(2):130–8.

[91] McGonagle L, Jones MD, Dowson D, Theobald PS. The bio-tribological properties of anti-adhesive agents commonly used during tendon repair. J Orthop Res 2012;30:775–80.

[92] Allison DD, Grande-Allen KJ. Review. Hyaluronan: a powerful tissue engineering tool. Tissue Eng 2006;12(8):2131–40.

[93] Fraser JR, Laurent TC, Laurent UB. Hyaluronan: its nature, distribution, functions and turnover. J Intern Med 1997;242(1):27–33.

[94] Tomihata K, Ikada Y. Crosslinking of hyaluronic acid with water-soluble carbodiimide. J Biomed Mater Res 1997;37(2):243–51.

[95] Vercruysse KP, Marecak DM, Marecek JF, Prestwich GD. Synthesis and *in vitro* degradation of new polyvalent hydrazide cross-linked hydrogels of hyaluronic acid. Bioconjug Chem 1997;8(5):686–94.

[96] Darr A, Calabro A. Synthesis and characterization of tyramine-based hyaluronan hydrogels. J Mater Sci Mater Med 2009;20(1): 33–44.

[97] Liu Y, Zheng Shu X, Prestwich GD. Biocompatibility and stability of disulfide-crosslinked hyaluronan films. Biomaterials 2005;26(23):4737–46.

[98] Shu XZ, Liu Y, Luo Y, Roberts MC, Prestwich GD. Disulfide cross-linked hyaluronan hydrogels. Biomacromol 2002;3:1304–11.

[99] Shu XZ, Liu Y, Palumbo F, Prestwich GD. Disulfide-crosslinked hyaluronan-gelatin hydrogel films: a covalent mimic of the extracellular matrix for *in vitro* cell growth. Biomaterials 2003;24(21): 3825–34.

[100] Baier Leach J, Bivens KA, Patrick CW Jr, Schmidt CE. Photo-crosslinked hyaluronic acid hydrogels: natural, biodegradable tissue engineering scaffolds. Biotechnol Bioeng 2003;82(5): 578–89.

[101] Vanderhooft JL, Mann BK, Prestwich GD. Synthesis and characterization of novel thiol-reactive poly(ethylene glycol) cross-linkers for extracellular-matrix-mimetic biomaterials. Biomacromolecules 2007;8(9):2883–9.

[102] Ouasti S, Donno R, Cellesi F, Sherratt MJ, Terenghi G, Tirelli N. Network connectivity, mechanical properties and cell adhesion for hyaluronic acid/PEG hydrogels. Biomaterials 2011;32(27): 6456–70.

[103] Skardal A, Zhang J, Prestwich GD. Bioprinting vessel-like constructs using hyaluronan hydrogels crosslinked with tetrahedral polyethylene glycol tetracrylates. Biomaterials 2010;31(24): 6173–81.

[104] Skardal A, Zhang J, McCoard L, Oottamasathien S, Prestwich GD. Dynamically crosslinked gold nanoparticle – hyaluronan hydrogels. Adv Mater 2010;22(42):4736–40.

[105] Al-Shamkhani A, Duncan R. Radioiodination of alginate via covalently-bound tyrosinamide allows monitoring of its fate *in vivo*. J Bioact Compat Polym 1995;10:4–13.

[106] Orive G, Ponce S, Hernandez RM, Gascon AR, Igartua M, Pendaz JL. Biocompatibility of microcapsules for cell immobilization elaborated with different type of alginates. Biomaterials 2002;23: 3825–31.

[107] Orive G, Ponce S, Hernandez RM, Gascon AR, Igartua M, Pedraz JL. Biocompatibility of microcapsules for cell immobilization elaborated with different type of alginates. Biomaterials 2002;23(18):3825–31.

[108] Boland T, Xu T, Damon B, Cui X. Application of inkjet printing to tissue engineering. Biotechnol J 2006;1(9):910–7.

[109] Arai K, Iwanaga S, Toda H, Genci C, Nishiyama Y, Nakamura M. Three-dimensional inkjet biofabrication based on designed images. Biofabrication 2011;3(3). 034113.

[110] Nishiyama Y, Nakamura M. Development of a three-dimensional bioprinter: construction of cell supporting structures using hydrogel and state-of-the-art inkjet technology. J Biomed Eng 2009;131. 035001-1-6.

[111] Xu T, Zhao W, Zhu JM, Albanna MZ, Yoo JJ, Atala A. Complex heterogeneous tissue constructs containing multiple cell types prepared by inkjet printing technology. Biomaterials 2013;34(1): 130–9.

[112] Catros S, Guillemot F, Nandakumar A, Ziane S, Moroni L, Habibovic P, et al. Layer-by-layer tissue microfabrication supports cell proliferation *in vitro* and *in vivo*. Tissue Eng Part C Methods 2012;18(1):62–70.

[113] Guillemot F, Souquet A, Catros S, Guillotin B, Lopez J, Faucon M, et al. High-throughput laser printing of cells and biomaterials for tissue engineering. Acta biomaterialia 2010;6(7):2494–500.

[114] Guillotin B, Souquet A, Catros S, Duocastella M, Pippenger B, Bellance S, et al. Laser assisted bioprinting of engineered tissue with high cell density and microscale organization. Biomaterials 2010;31(28):7250–6.

[115] Phamduy TB, Raof NA, Schiele NR, Yan Z, Corr DT, Huang Y, et al. Laser direct-write of single microbeads into spatially-ordered patterns. Biofabrication 2012;4. 025006.

[116] Yan J, Huang Y, Chrisey DB. Laser-assisted printing of alginate long tubes and annular constructs. Biofabrication 2013;5(1). 015002.

[117] Gaetani R, Doevendans PA, Corina HG, Metz CH, Alblas J, Messina E, et al. Cardiac tissue engineering using tissue printing technology and human cardiac progenitor cells. Biomaterials 2012;33: 1782–90.

[118] Song SJ, Choi JS, Park YD, Hong S, Lee JJ, Ahn CB, et al. Sodium alginate hydrogel-based bioprinting using a novel multinozzle bioprinting system. Artif Organs 2011;35(11):1132–6.

[119] Williams SK, Touroo JS, Church KH, Hoying JB. Encapsulation of adipose stromal vascular fraction cells in alginate hydrogel spheroids using a direct-write three-dimensional printing system. Biores Open Access 2013;2(6):448–54.

[120] Cohen DL, Malone E, Lipson H, Bonassar LJ. Direct freeform fabrication of seeded hydrogels in arbitrary geometries. Tissue Eng 2006;12(5):1325–35.

[121] Diogo GS, Gaspar VM, Serra IR, Fradique R, Correia IJ. Manufacture of beta-TCP/alginate scaffolds through a Fab@home model for application in bone tissue engineering. Biofabrication 2014;6(2). 025001.

[122] Fedorovich NE, De Wijn JR, Verbout AJ, Alblas J, Dhert WJ. Three-dimensional fiber deposition of cell-laden, viable, patterned constructs for bone tissue printing. Tissue Eng Part A 2008;14(1): 127–33.

[123] Cohen DL, Lipton JI, Bonassar LJ, Lipson H. Additive manufacturing for *in situ* repair of osteochondral defects. Biofabrication 2010;2(3). 035004.

[124] Hwang CM, Ay B, Kaplan DL, Rubin JP, Marra KG, Atala A, et al. Biomed Mater. Assessments of injectable alginate particle-embedded fibrin hydrogels for soft tissue reconstruction 2013;8(1). 014105.

[125] Norotte C, Marga FS, Niklason LE, Forgacs G. Scaffold-free vascular tissue engineering using bioprinting. Biomaterials 2009;30(30):5910–7.

[126] Blaeser A, Duarte Campos DF, Weber M, Neuss S, Theek B, Fischer H, et al. Biofabrication under fluorocarbon: a novel freeform fabrication technique to generate high aspect ratio tissue-engineered constructs. Biores Open Access 2013;2(5):374–84.

[127] Duarte Campos DF, Blaeser A, Weber M, Jäkel J, Neuss S, Jahnen-Dechent W, et al. Three-dimensional printing of stem cell-laden hydrogels submerged in a hydrophobic high-density fluid. Biofabrication 2013;5(1). 015003.

[128] Kleinman HK, Martin GR. Matrigel: basement membrane matrix with biological activity. Semin Cancer Biol 2005;15(5): 378–86.

[129] Xu F, Celli J, Rizvi I, Moon S, Hasan T, Demirci U. A three-dimensional *in vitro* ovarian cancer coculture model using a high-throughput cell patterning platform. Biotechnol J 2011;6:204–12.

[130] Koch L, Deiwick A, Schlie S, Michael S, Gruene M, Coger V, et al. Skin tissue generation by laser cell printing. Biotechnol Bioeng 2012;109(7):1855–63.

[131] Koch L, Kuhn S, Sorg H, Gruene M, Schlie S, Gaebel R, et al. Laser printing of skin cells and human stem cells. Tissue Eng Part C Methods 2010;16(5):847–54.

[132] Michael S, Sorg H, Peck CT, Koch L, Deiwick A, Chichkov B, et al. Tissue engineered skin substitutes created by laser-assisted bioprinting form skin-like structures in the dorsal skin fold chamber in mice. PloS ONE 2013;8(3):e57741.

[133] Snyder JE, Hamid Q, Wang C, Chang R, Emami K, Wu H, et al. Bioprinting cell-laden Matrigel for radioprotection study of liver by pro-drug conversion in a dual-tissue microfluidic chip. Biofabrication 2011;3(3). 034112.

[134] Zhu J. Bioactive modification of poly(ethylene glycol) hydrogels for tissue engineering. Biomaterials 2010;31(17):4639–56.

[135] Lin CC, Anseth KS. PEG hydrogels for the controlled release of biomolecules in regenerative medicine. Pharm Res 2009;26(3): 631–43.

[136] Nguyen KT, West JL. Photopolymerizable hydrogels for tissue engineering applications. Biomaterials 2002;23(22):4307–14.

[137] Cui X, Breitenkamp K, Finn MG, Lotz M, D'Lima DD. Direct human cartilage repair using three-dimensional bioprinting technology. Tissue Eng A 2012;18(11-12):1304–12.

[138] Hockaday LA, Kang KH, Colangelo NW, Cheung PY, Duan B, Malone E, et al. Rapid 3D printing of anatomically accurate and mechanically heterogeneous aortic valve hydrogel scaffolds. Biofabrication 2012;4(3). 035005.

[139] Batrakova EV, Kabanov AV. Pluronic block copolymers: evolution of drug delivery concept from inert nanocarriers to biological response modifiers. J Controlled Release 2008;130(2):98–106.

[140] Klouda L, Mikos AG. Thermoresponsive hydrogels in biomedical applications. Eur J Pharm Biopharm 2008;68(1):34–45.

[141] Weinand C, Pomerantseva I, Neville CM, Gupta R, Weinberg E, Madisch I, et al. Hydrogel-beta-TCP scaffolds and stem cells for tissue engineering bone. Bone 2006;38(4):555–63.

[142] Khattak SF, Bhatia SR, Roberts SC. Pluronic F127 as a cell encapsulation material: utilization of membrane-stabilizing agents. Tissue Eng 2005;11:974–83.

[143] Parsa S, Gupta M, Loizeau F, Cheung KC. Effects of surfactant and gentle agitation on inkjet dispensing of living cells. Biofabrication 2010;2(2). 025003.

[144] Fedorovich NE, Swennen I, Girones J, Moroni L, van Blitterswijk CA, Schacht E, et al. Evaluation of photocrosslinked Lutrol hydrogel for tissue printing applications. Biomacromolecules 2009;10(7):1689–96.

[145] Wu W, DeConinck A, Lewis JA. Omnidirectional printing of 3D microvascular networks. Adv Mater 2011;23(24):H178–83.

[146] Campbell PG, Miller ED, Fisher GW, Walker LM, Weiss LE. Engineered spatial patterns of FGF-2 immobilized on fibrin direct cell organization. Biomaterials 2005;26(33):6762–70.

[147] Ker ED, Chu B, Phillippi JA, Gharaibeh B, Huard J, Weiss LE, et al. Engineering spatial control of multiple differentiation fates within a stem cell population. Biomaterials 2011;32(13):3413–22.

[148] Lee YB, Polio S, Lee W, Dai G, Menon L, Carroll RS, et al. Bioprinting of collagen and VEGF-releasing fibrin gel scaffolds for neural stem cell culture. Exp Neurol 2010;223(2):645–52.

[149] Miller ED, Fisher GW, Weiss LE, Walker LM, Campbell PG. Dose-dependent cell growth in response to concentration modulated

patterns of FGF-2 printed on fibrin. Biomaterials 2006;27(10): 2213–21.

[150] Xu T, Rohozinksi J, Zhao W, Moorefield EC, Atala A, Yoo JJ. Inkjet-mediated gene transfection into living cells combined with targeted delivery. Tissue Eng Part A 2009;15(1):95–101.

[151] Guiseppi-Elie A. Electroconductive hydrogels: synthesis, characterization and biomedical applications. Biomaterials 2010;31(10): 2701–16.

[152] Medberry CJ, Crapo PM, Siu BF, Carruthers CA, Wolf MT, Nagarkar SP, et al. Hydrogels derived from central nervous system extracellular matrix. Biomaterials 2013;34(4):1033–40.

[153] Sawkins MJ, Bowen W, Dhadda P, Markides H, Sidney LE, Taylor AJ, et al. Hydrogels derived from demineralized and decellularized bone extracellular matrix. Acta Biomaterialia 2013;9(8): 7865–73.

[154] Skardal A, Smith L, Bharadwaj S, Atala A, Soker S, Zhang Y. Tissue specific synthetic ECM hydrogels for 3-D *in vitro* maintenance of hepatocyte function. Biomaterials 2012;33(18): 4565–75.

[155] Wolf MT, Daly KA, Brennan-Pierce EP, Johnson SA, Carruthers CA, D'Amore A, et al. A hydrogel derived from decellularized dermal extracellular matrix. Biomaterials 2012;33(29):7028–38.

Chapter 15

Bioprinting of Organoids

Carlos Kengla*,**, Anthony Atala*,**, and Sang Jin Lee*,**

*Wake Forest Institute for Regenerative Medicine, Wake Forest School of Medicine, Winston-Salem, NC, USA; **School of Biomedical Engineering and Sciences, Wake Forest University Virginia Tech, Winston-Salem, NC, USA

Chapter Outline

ABSTRACT

Three-dimensional (3D) bioprinting technology was developed to allow the construction of biological tissue constructs that mimic the structure and function of native tissues or organs. This technology makes it possible to precisely place various cell types with hydrogel-based bioinks in a 3D architecture. The hypothesis driving the development of 3D bioprinted tissue constructs is that by precisely placing cells in relation to each other, an environment that encourages physiologically relevant cues can be created, resulting in a tissue construct with appropriate functionality. Consequently, the bioprinted organoids could provide more anatomical and functional similarity of human tissues or organs for pharmacokinetics and pharmacodynamics models. The bioprinted tissue organoid structures will play an increasing role in drug discovery and therapeutics over the next decade. This chapter will discuss these bioprinting technologies and their applications in organoid printing.

Keywords: bioprinting; organoid; bioinks; hydrogel; drug screening; tissue engineering

1 INTRODUCTION

The use of three-dimensional (3D) cell culture has grown in utility and has developed into a field of investigation to deepen our understanding of cell differentiation, tissue maturation and organization, and organogenesis [1,2]. Organoids are the products of 3D culture techniques that result in tissue-like masses that serve as representatives of native tissue in one or more aspects. Tissue engineering and tissue models are two obvious and potentially useful applications for organoids. Tissue engineering seeks to recapitulate tissues and organs in functional ways for the purpose of replacing or repairing injured tissues and organs [3,4]. Tissue models have various uses in toxicology and pharmacokinetic studies for the purpose of representing native tissues and organs while not involving human subjects and minimizing reliance on nonhuman mammals [5,6]. A limitation is that of cell homogeneity or unguided heterogeneity. Guiding the 3D architecture and cell placement within the organoids has the potential to enable the production of enhanced complexity more closely mimicking native tissues, which is the expressed aim of bioprinting. In this chapter we present the strategy of bioprinting for producing tissue organoids with various biomaterials, mostly hydrogels, as bioinks. Availability of biomaterials that can serve as cell delivery bioinks but which also provide mechanical support, cell-specific cues, and negligible cytotoxicity is limited. Advances in the field of suitable cell-compatible bioink materials for bioprinting of organoids are necessary for the long-term success of this technology.

Essentials of 3D Biofabrication and Translation. http://dx.doi.org/10.1016/B978-0-12-800972-7.00015-3

2   STRATEGY OF BIOPRINTING TECHNOLOGY FOR PRODUCING ORGANOID STRUCTURES

Many efforts to develop regular and simplified 3D tissue organoids, which are suitable for biological and pathological investigation, have been performed; however, it may not yet be feasible. Bioprinting technology is a recent advanced tissue engineering approach that has the potential to fabricate clinically applicable tissue or organ constructs appropriate in shape and size. The feasibility of producing complex tissue constructs that are anatomically and functionally applicable has been demonstrated [7,8]. Through spatial combinations of tissue-specific cell types and biomaterials in complex, meaningful, 3D architecture, we can better harness the regenerative capacity innate to cells and thereby generate needed tissues or organs. In the bioprinting of organoids, the dimensions of the nozzle allow microscale control over the volume and the location of the dispensed droplets containing live cells. Moreover, the geometric and compositional structure of the organoids can be controlled (Fig. 15.1). Therefore, the bioprinted organoids could provide more anatomical and functional similarity to human tissues or organs for pharmacokinetic and pharmacodynamics models. The bioprinted tissue organoid structures will play an increasing role in drug discovery and therapeutics over the next decade.

3   BIOPRINTING MODALITIES

Hydrogel-based bioinks have been used as a carrier for tissue-specific cells for organoid printing. The choice of bioinks is dependent on the printing methods; inkjet printing, extrusion-based printing, and laser-induced forward transfer (LIFT). Each method requires very specific characteristics of the bioinks for building a 3D construct. The suitability of hydrogel-based bioinks is mainly subject to their physicochemical properties under the 3D bioprinting process [9]. The major physiochemical properties of bioinks can be determined by their rheological properties and cross-linking mechanism, which reflect "printability."

3.1   Inkjet Printing

Various printing and rapid prototyping technologies have been adapted for use with biologics. The use of inkjet printers and graphics plotters to "cytoscribe" fibronectin patterns on substrates for cell adhesion marks the beginnings of what is now referred to as "Bioprinting [10]." The most common bioprinting modalities for printing cells and encouraging organoid formation are inkjet, plotting, and forward transfer. Inkjet printers utilize an ink reservoir that feeds a chamber with some extrusion mechanism that forces the ink through an orifice generating small droplets. Various designs have found success for printing chromatic inks as well as bioinks, liquids or gels containing biological compounds with or

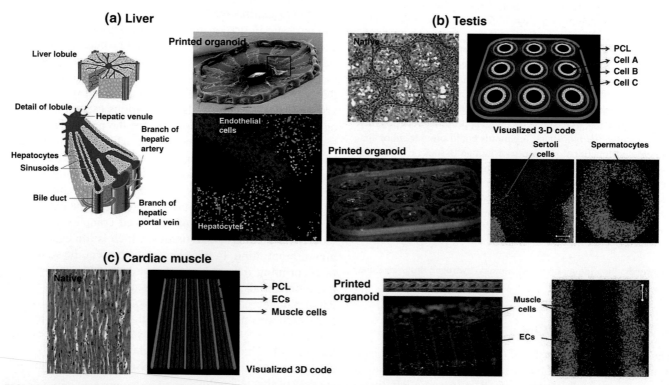

FIGURE 15.1   Bioprinting of organoid structures. (a) Liver, (b) testis, and (c) cardiac muscle organoids *(unpublished data)*.

without living cells. Typical mechanisms for ejection are pneumatic, piezoelectric, and thermal. Pneumatic actuation involves air pressure applied to the ink in concert with a valve regulating the orifice. Piezoelectric ejection chambers have a surface made of a piezoelectric material that deforms with electric current resulting in a rapid volumetric change forcing ink out of the chamber through the orifice. Thermal inkjet systems have an ink chamber with a heating element providing rapid, localized heating to induce bubble formation, which propels the bioink through the orifice.

The inkjet printing method has been mostly applied to cell deposition in a precise manner. The resolution of the printed patterns using the inkjet printing method is about 20–100 μm [11]. In order to achieve solidification in a desired 3D architecture, the biopaper (layer substrate material) induces solidification of the droplet of bioink material [12] or the bioink initiates solidification of the biopaper material [13]. The former approach can result in a solid spherical structure that becomes a building block within the 3D construct. The latter approach can form spherical porous structures. For instance, a 3D architecture can be fabricated by printing the cell-laden alginate solution (as a bioink) on $CaCl_2$ solution (as a biopaper) [12]. On the other hand, $CaCl_2$ solution (as a bioink) can be printed on the cell-mixed alginate solution (as a biopaper) [13]. Even though the inkjet printing method has many advantages, such as 3D freeform fabrication, high resolution, multiple cartridge capability, and low cost, it also has several limitations. A major limitation is that only bioink with low viscosity can be reasonably used for the inkjet printing method.

3.2 Extrusion-Based Printing

Extrusion-based printing technology uses a micronozzle (few microns to several hundred microns) and precise pressure controller or syringe pump for building a 3D microscale architecture in a layer-by-layer fashion. The cell-laden hydrogel biomaterials, bioink, in the cartridge can be precisely dispensed by controlling the actuating pressure or the piston of the syringe pump. This method can also construct a hybrid composite structure using a multiple-cartridge system capable of dispensing different cell types and biomaterials. When this method is compared with the other printing technologies, it offers a wider selection of bioink materials and the producible construct size is scalable. Moreover, biologically active and structural protein molecules can also be incorporated into the bioink; however, their impact on the bioink properties must be considered. In contrast, extrusion-based printing has comparatively low resolution (200–400 μm).

Scaffold-free dispensing is another extrusion-based method that also utilizes the idea of biopaper and bioink. Generally, the bioink carries the cells or cell aggregates being used [14]. The biopaper forms the cell environment

and incorporates structural components to form a hydrogel that mimics aspects of an extracellular matrix (ECM). The bioink or biopaper can also incorporate cytokines or other biomolecules. The resulting construct is formed by guided self-assembly driven by the innate tendencies of the biological components.

3.3 Laser-Induced Forward Transfer

Forward transfer refers to the method of bioink delivery used in thermal transfer printing or other printing technologies where a ribbon loaded with ink is selectively stimulated to deposit ink onto the substrate. In the bioprinting community, LIFT is used by several groups as a method for depositing bioink onto a substrate and has several other names (BioLP, LAB, MAPLE-DW). The ribbon is a silicate slide with a metallic absorptive coating. The bioink is evenly loaded on top of the metallic surface and oriented face-down above the intended substrate. A high-speed laser is focused through the backside of the slide targeting the absorptive layer. Excitation of this absorptive layer produces localized, rapid heating causing bubble formation under the bioink film resulting in a small droplet being ejected from the surface of the ribbon toward the substrate [15]. This method is capable of precisely printing cells and bioink materials in relatively small 3D constructs while maintaining cell viability and function [15]. It is a nozzle-free approach and therefore is not affected by clogging issues seen in other methods. It has also been used with a wide range of viscosities of bioink materials. However, the process requires rapid gelation of bioink materials to achieve high resolution of the printed patterns, resulting in low flow rate.

4 BIOINKS FOR ORGANOID PRINTING

Proper gelation (e.g., mechanism and time) of a printed bioink structure is required to build a 3D organoid structure. Especially, gelation mechanisms of bioink materials are critical based on the printing process and target tissue organoid structure (Fig. 15.2). 3D printing allows for the deposition of 3D patterns and dots that are already in a gel state. Various techniques have been employed for increasing the stability of these 3D hydrogel-based bioinks. One approach is the use of thermo-sensitive hydrogels, like gelatin and Pluronic F127, to maintain the structure until a cross-linkable bioink can be cured for longer periods of gel stability at physiological conditions (temperature, pH, etc.). This section reviews various hydrogel-based bioinks and their applications in organoid printing (Table 15.1).

4.1 Alginate-Based Bioinks

Alginate is a naturally derived anionic polysaccharide exhibiting gelation in the presence of bivalent ions such as

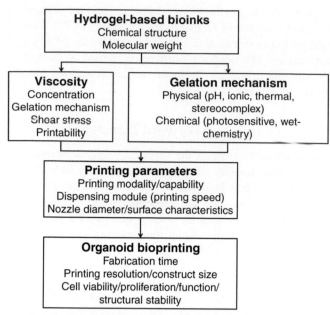

FIGURE 15.2 Schematic diagram of variables critical to 3D organoid bioprinting strategy. The hydrogel-based bioinks determine the viscosity, gelation mechanism, and printing parameters, eventually, bioprinted organoid structures.

Ca^{2+} [16]. This robust hydrogel has served as a cell delivery material for many tissue engineering and cell encapsulation approaches due to ease of preparation and good cell compatibility. The primary drawback for most tissue engineering applications is the lack of mammalian enzymatic degradation, which limits tissue remodeling. In addition, there is limited cell attachment to alginate chains without chemically altering the polymer [17]. Alginate holds an important position as a relatively simple biocompatible hydrogel. As such, it serves as a good starting point to test cell printing conditions and cell vitality. In the early stage, an inkjet printing set up was customized for 3D printing by printing patterns of a Ca^{2+} solution into a reservoir of feline cardiomyocytes mixed with alginate solution [13]. The printing of the $CaCl_2$ solution induced gelation to form a hollow shell architecture in the desired pattern with each shell having an average outer diameter of 25 μm. An elevator system moves the gelled construct down to expose fresh alginate solution to the printed solution to allow the formation of a 3D construct in the shape of a two-chambered heart-like structure. The final printed product demonstrated both cardiomyocyte beating as well as concerted construct beating, which is interesting considering the limited cell attachment typical to alginate.

TABLE 15.1 Hydrogel-Based Bioinks for Organoid Printing

Organoids	Cell Types	Printing Methods	Outcomes	References
Alginate-based Bioinks				
Blood vessel	NIH/3T3	Inkjet	Tubular structure of fibroblasts with designed bend.	[40]
Blood vessel	Ovine ECs, SMCs	LIFT + stereolithography	Concentric positioning of SMCs outer ring and ECs inner ring mimicking vessel.	
Blood vessel	HUVECs	LIFT	Cells printed in lines that show coalescence.	
Bone, cartilage	Porcine MSCs	LIFT	ALP activity, calcium ion concentration, osteocalcin expression in bone-differentiated cells. Collagen II and aggrecan expression, positive GAG staining in cartilage differentiated cells.	[20]
Bone	MSCs	Extrusion	Over 80% viability, ALP staining	[41]
Bone	Human bone progenitor cells	LIFT	Positive staining of ALP and osteocalcin.	[19]
Osteochondral interface	MSCs, chondrocytes	Extrusion	Bone region expression of osteocalcin, cartilage region expression of collagen II and VI.	[42]
Heart	Cardiomyocytes	Inkjet	Simulated beating	[13]
Skin	NIH/3T3, HaCaT, hMSCs	LIFT	Viability averaging 90%, maintained stem cell surface markers.	[21]
Collagen-based Bioinks				
Bone	MG63	Extrusion	Gene expression: Runx-2, ALP for bone and collagen II, aggrecan for cartilage. Alizarin Red staining of bone and Alcian Blue staining of cartilage.	[43]
Neural	Neural stem cells	Inkjet	Viability of 93%, differentiation and proliferation capacity.	[44]
Neural	Neuron, astrocytes	Inkjet	Histological markers.	[45]

TABLE 15.1 Hydrogel-Based Bioinks for Organoid Printing *(cont.)*

Organoids	Cell Types	Printing Methods	Outcomes	References
Skin	Keratinocytes, fibroblasts	Inkjet	>90% viability, *N*-cadherin stained keratinocytes, stratified morphology.	[26]
Skin	Human dermal fibroblasts, human keratinocytes	Inkjet	Viability 95% for fibroblasts, 85% for keratinocytes. Stratified construct. Keratin production in epidermal layer.	
Smooth muscle	SMCs	Inkjet	> 93% viability, connexin-43 staining of confluent printed patch for tissue engineering.	[46]
Smooth muscle	SMCs	Inkjet + seeding	Cell alignment to printed patterns and microstructures.	[24]
Fibrin-based Bioinks				
Microvasculature	HUVECs	Inkjet	Lumen formation and maintenance shown by dextran exclusion by day 21.	[29]
Vasculature	ECFCs, ASCs	LIFT	Network formation seen after ASC migration toward clusters of ECFCs.	[30]
Gelatin-based Bioinks				
Vasculature	HUVECs, HNDFs, 10T1/2	Extrusion	Lumen creation with HUVEC lining. Demonstration of advanced pattern generation with multiple cell types and vasculature.	[8]
Liver	HepG2 cells	Extrusion	GelMa – preserved cell viability for at least 8 days.	[47]
HA-based Bioinks				
Blood vessel	NIH/3T3	Extrusion	Lumen maintenance of simple structure.	[31]
Cartilage	Chondrocytes	Extrusion	Chondrocytes in HA region express aggrecan and collagen II, positive staining of GAGs.	[43]
ECM-based Bioinks				
Bone	MSCs, EPCs	Extrusion	Bone-like staining with Goldner's trichrome in region with BCP, safranin-O, collagen II stain in region without BCP. Increased vWF staining in ECP region.	[48]
Neural	Olfactory ensheathing cells	LIFT	3D patterns with axon-like extensions and positive for p75^{NTR} surface marker.	[49]
Vasculature	HUVECs, HUVSMCs	LIFT	Network formation, lumen formation, SMC support.	[37]
Vasculature	HUVECs	LIFT	Network formation along printed pattern. Thick constructs were not viable.	[38]

Alginate hydrogel can be combined with other biomaterials for printing bone structures. Catros et al. moved toward coprinting a cell-laden bioink along with other biomaterials, specifically nanohydroxyapatite (nHA). Hydroxyapatite is known to support bone tissue formation when used with osteoblasts and osteoprogenitor cell types as it mimics the native bone mineral calcium phosphate phase [18]. A solution containing nHA (10% w/v) was used as a bioink for depositing the mineral layer. The mineral layer was tested first by seeding human osteosarcoma cell line (MG63) cells in 2% alginate on printed nHA and was shown to promote typical morphology and cell viability. In a similar study, the printed cells tested positive for alkaline phosphatase (ALP) activity indicating that the cells began early differentiation to the bone-forming lineage; however, experiments were not carried out to the point of seeing mineral deposition

[19]. Another printing approach used a single mesenchymal stem cell (MSC) source and showed evidence for bone and cartilage formation thereby demonstrating the ability to print a cell niche capable of supporting progenitor cells [20]. Toward this end, porcine MSCs were printed to assess further differentiation potential (Fig. 15.3a–d). Once printed, cells encapsulated in 2% alginate hydrogel were cultured in chondrogenic or osteogenic differentiation media. Printed, differentiated cells were validated by comparison to unprinted, differentiated cells as well as to articular chondrocytes. The cell density in the tissue formation experiments was calculated to be 46,000 ± 23,000 cells/μL, which the authors hypothesize enabled the *in vitro* tissue formation and differentiation.

It has been demonstrated that the LIFT printing of multiple cell types relevant to skin using alginate to hold the cells

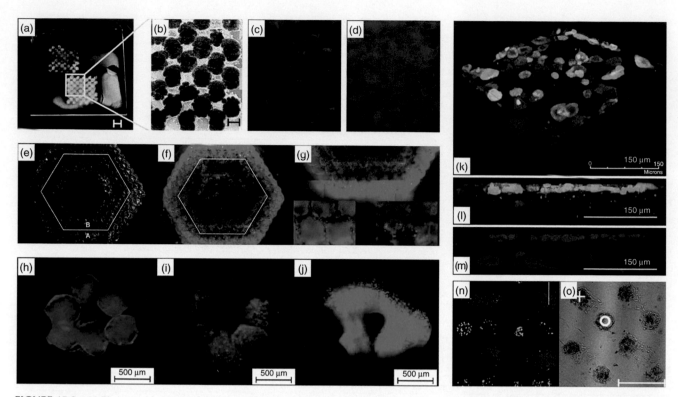

FIGURE 15.3 (a) Photographic image of the printed MSCs, predifferentiated in osteogenic lineage for 7 days, printed in two chessboard patterns and cultivated for 17 days under osteogenic conditions (scale bar: 2 mm), (b) transmission image (scale bar: 800 μm), (c) fluorescence images of nuclei stained with Hoechst 33342 (blue staining), (d) membranes stained with DIL (red staining) of the chessboard section [20], (e–g) scaffold seeded with cells by means of LIFT: dark field image (e). The white hexagon indicates the border between the two scaffold areas seeded with SMCs (a) and ECs (b), respectively; fluorescence image indicating the location of different cell types after the LIFT procedure (f); detailed image of the border area (g). The insets demonstrate that a sharp transition from SMCs to EC-seeded regions is present along the entire thickness of the scaffold [23]. Cross-sectional views of the bioprinted construct taken immediately after printing (h) with encapsulated fluorescent HA-BODIPY tracer for increased visualization, at 14 days (i), and at 28 days (j) of culture using LIVE/DEAD staining to highlight viable and dead cells. Green fluorescence indicates calcein AM-stained live cells and red fluorescence indicates ethidium homodimer-1-stained dead cells [31]. (k–m) Cell images after multilayered printing of fibroblasts and keratinocytes on the tissue culture dish. Volume rendered immunofluorescent images (k) of multilayered printing of keratinocytes and fibroblasts and its projection of keratin-containing keratinocytes layer (l) and β-tubulin-containing keratinocytes and fibroblasts (m) [26]. (n,o) 3D Reconstruction of the cell array. ECFCs were stained with calcein (green) and ASCs were stained with TAMRA-5 (red) (n), light microscopic images of ECFCs (○) and ASCs (+) at 3 days in culture (o) [30].

in place for 2D patterning on Matrigel-coated substrate can be printed successfully [21]. Keratinocytes (HaCaT), fibroblasts (NIH/3T3), and human adipose-derived stem cells (hASCs) were printed and assessed for viability, proliferation, apoptosis, DNA damage, and hASC phenotype. Results indicate over 98% cell viability, proliferation of each cell type, no significant increase in apoptosis or DNA damage, and no difference in selected hASC surface markers. This study validates the printing method and the hydrogel carrier for safe patterning of multiple cell types. Another group demonstrated the use of alginate gel to print high cell density patterns with the LIFT approach, depositing 10 and 100 million human umbilical vein endothelial cells (HUVEC) per milliliter of bioink [22]. Several concentrations of alginate were evaluated (0.1, 0.5, or 1.0%) to achieve appropriate resolution. Low concentrations of alginate resulted in slashing, spreading, and loss of resolution; however, splashing was eliminated and resolution maintained at 1.0% alginate.

The resulting printed patterns demonstrated cell viability and the printing method's ability to print the high viscosity gel along with high cell density while maintaining a desired resolution for 2D patterns. Moreover, modulating the laser and stage parameters allowed for the deposition of droplets with a single cell and then increasing the deposition of coalescing lines with high cell density from the same bioink source. The substrate was coated in a fibrinogen solution allowing for printed patterns to be grown in 3D culture conditions of fibrin and alginate gels after treatment with $CaCl_2$ solution and thrombin to cross-link the substrates. Ovsianikov et al. showed that the LIFT printing system can be combined with other 3D fabrication techniques by printing alginate-based bioink into a PEGDA scaffold fabricated by stereolithography [23]. The highly porous scaffold was designed to accept the printed bioink and had a doughnut shape for the purpose of acting as a potential vascular graft. Two bioinks were used: (1) ovine endothelial cells

(ECs) in alginate and (2) ovine vascular smooth muscle cells (SMCs) in alginate. Laser energy parameter deposited droplets containing 20–30 cells. Bioprinting was utilized in order to position each cell-type into the appropriate areas of the scaffold in order to form an endothelium ensheathed by vascular smooth muscle. Results of this study reiterated the ability of the LIFT technique to safely deposit cells of multiple types into predetermined regions of an arbitrary substrate (Fig. 15.3e–g).

4.2 Collagen-Based Bioinks

Collagen is found in most mammalian tissues in various configurations, type 1 being the most abundant. The collagen molecules self-assemble into interlocking fibers and fibrils or web-like networks. Collagen type 1 is the focus of this section as it is the type used for printing and cell encapsulation. Under the appropriate temperature and pH, a collagen solution will undergo gelation to form a gel with properties dependent on its concentration. Cells can attach to collagen through integrin binding and can enzymatically degrade fibers allowing for cell migration and ECM remodeling. Many reactive moieties allow for chemical alteration and cross-linking to biological and mechanical properties. Primary sources are porcine, bovine, and equine while medical-grade collagen is from cadaveric sources [17]. Collagen bioinks can be difficult to work with when printing as the solution must be handled with care to prevent premature setting, usually kept below 4–10°C. The work of Roth et al. is an example of Klebe's "cytoscribing" approach [24]. A 1% solution of collagen is printed with a modified inkjet printer into the desired patterns on agarose-coated glass coverslips. The collagen solution for this approach was kept slightly acidic to prevent clogging. After printing, the gel must be aseptically dried and reconstituted before cells could be cultured on the patterns. SMC cell line (CRL-1476) isolated from rat aorta were cultured and shown to adhere to and self-align on the collagen patterns. In the same year, Smith et al. reported the printing of aortic ECs of bovine origin via a pneumatically actuated bioplotting system [25]. The cells were mixed with cold 3 mg/mL collagen solution titrated to 7–7.4 pH, which was subsequently maintained at 10°C. ECs were mixed with the collagen solution at concentrations ranging from 5×10^6 cell/mL to 20×10^6 cell/mL, printed, and cultured as printed patterns as well as collected for viability testing. The results of the viability study indicate that cells printed with the small diameter 33-gauge tip had lower viability (46%) than those printed with a 25-gauge tip (86%). Cell proliferation after printing was confirmed after 24 h while printed patterns were maintained for 35 days in culture. The formation of "cord-like" structures was interpreted as evidence of phenotype development, that is, possible vessel-like structure. It is also posited that this approach to patterning cords of ECs could one day serve as the means

of repairing damaged coronary vessels since the authors demonstrated the pattern of vessels seen in an angiogram printed with ECs in collagen bioink.

More recently, an approach for printing collagen and cells together from separate nozzles via inkjet style microvalve dispensing was performed [26]. The collagen solution was prepared at 2 mg/mL, but remained acidic and chilled for printing. To mimic skin tissue, layers of collagen bioink were printed then treated with aerosolized $NaHCO_3$ (sodium bicarbonate) to buffer the pH toward neutral and induce gelation. Once gelled, another layer of collagen or cells could be printed on top. Primary human dermal fibroblasts and epidermal keratinocytes were cultured and loaded into the printer suspended in media at 1×10^6 cells/mL. Interdroplet distance was set at a distance of 300 μm in order to see the effect of resolution due to rapid confluency achieved after 10 days in culture. A layer of fibroblasts was sandwiched between layers of collagen, followed by six more layers of collagen, then a sandwiched layer of keratinocytes to fabricate a skin-like structure (Fig. 15.3k–m). In addition to layering, the approach to wound treatment was modeled in a t-shaped PDMS mold. Viability tests showed no statistically significant difference between printed and control samples at 1 day postprinting for both keratinocytes and fibroblasts. Immunohistochemical staining with pan-keratin and β-tubulin antibodies showed separation of cellular layers with β-tubulin staining throughout, but keratin staining was limited to the top layer of keratinocytes. This research demonstrates the survival of cells and spatial control of the printing approach, which is needed to produce functional skin replacement for wounded patients. Another approach advanced a method for using a collagen solution and cell suspensions to print layered skin structures [27]. The micronozzle system was set to dispense droplets of chilled, acidic collagen solution that can coalesce to form a sheet, then aerosolized $NaHCO_3$ is sprayed on the surface to induce gelation. Subsequent layers were printed in a similar fashion such that three layers of fibroblasts (HFF-1) were separated by two collagen layers and the structure was capped by two layers of HaCaT. A range of cell densities and droplet spacing distances were tested in an attempt to maximize cell viability; however, many conditions resulted in high cell viability. The results of this optimization allowed choosing parameters that would reflect average cell distribution found in the epidermis and dermis of normal skin (2×10^6 fibroblasts/mL and 5×10^6 keratinocytes/mL with droplets spaced 500 μm). Another enhancement to this study was the inclusion of the air/liquid interface culture method and a control based on "conventional 3D skin constructs." The printed structures were cultured in submerged conditions immediately after printing, followed by culture with the top surface exposed to air, which promoted terminal differentiation and stratification of the keratinocytes in the epidermis. The results showed evidence of keratinocyte stratification and cornification after 14 days of air/liquid interface culture. Distinct dermal- and epidermal-like layers were maintained throughout the experiment and also maintained the shape of the printed structures as compared to the distortion of conventionally fabricated structures. The epidermal region of printed constructs stained positive for *N*-cadherin confirming the formation of tight junctions.

4.3 Gelatin-Based Bioinks

Gelatin, denatured collagen, forms a thermoreversible hydrogel with strength dependent on solution concentration. Many sites for cell attachment and degradation are shared with collagen, allowing for good degradation and remodeling. Raof et al. utilized a LIFT approach to print arrays of droplets containing embryonic stem cells (ESCs) [28]. The ribbon was coated with 20% gelatin solution, allowed to set, then an ESC suspension of 2–5 × 10^6 cells was placed on the gelatin coating and allowed to adhere. Excess fluid was removed such that the cells were partially embedded and faced the receiving substrate during printing. The receiving substrate was coated with 10% gelatin solution to mitigate forces from transfer and allow localization of the printed droplet. The array maintained a distinct pattern for 24 h and demonstrated proliferation and embryoid body formation after 7 days in culture indicating the printing approach was able to maintain the phenotype and vitality of the ESCs as confirmed by markers octamer-binding transcription factor 4 immediately after printing, and nestin, Myf-5, and PDX-1 after 7 days culture without leukemia inhibitory factor.

Also like collagen, additional modification can be made to enhance cross-linking and bioactivity. Kolesky et al. reported the use of methacrylated gelatin (GelMA) for bioprinting complex architectures incorporating cells and vasculature [8]. To achieve this aim, a bulk volume of pure GelMA was prepared and set in block form. This block would be the volumetric substrate for printing. Human dermal fibroblasts and 10T1/2 fibroblasts were mixed with GelMA bioink composed of 15% w/v GelMA powder, 1:1 DMEM:EGM-2 media, and 0.3 wt% Irgacure 2959 photoinitiator. Aqueous 40% Pluronic F127 gel was used as a sacrificial ink for the purpose of generating printed paths with open lumen for vascular engineering. The concept allows for printing to be done in a temperature window between 4°C and 22°C where both GelMA and Pluronic F127 are set hydrogels. During printing, cell-laden GelMA and Pluronic F127 are dispensed and embedded within the GelMA block in predetermined 3D patterns. After printing, the entire block is exposed to a brief UV illumination to induce photoinitiator-dependent cross-linking of the GelMA. The temperature is reduced below 4°C to produce the phase transition of Pluronic F127 to a liquid, which is then removed from the system leaving open channels within the GelMA block. After flushing with EGM-2 media, a

suspension of 1 × 10^7 HUVEC per milliliter was injected into the open channels. The construct was transferred to an incubator for 30 min at 37°C flipped, incubated for an additional 30 min and flipped again. After 5 h of incubation the construct was moved to a rocker oscillating at 1 Hz for further culture. Results show that this method of fabrication allows for the viable deposition of cells in 3D with the incorporation of microvessel-like channels that can be coated with ECs maintaining a lumen for provision of nutrients to surrounding tissue. The study demonstrates the power of bioprinting and clever use of thermoreversible hydrogels.

4.4 Fibrin-Based Bioinks

Fibrin is a naturally occurring protein network similar to collagen. Fibrinogen molecules are cleaved enzymatically by thrombin and the fibrin network self-assembles from the straight chain products [17]. Similar to collagen, fibrin has many motifs allowing for cell attachment and vulnerability to proteases for remodeling. Fibrinogen or thrombin can be incorporated into the cell suspension being printed. Fibrin will be formed when treated with the counterpart agent during or after printing. Cui and Boland reported the use of a printed, cell-laden thrombin solution onto a substrate coated with fibrinogen solution resulting in a fibrin pattern encapsulating the cells [29]. This particular study found that a 60 mg/mL fibrinogen solution, 50 U/mL thrombin, and 80 mM CaCl$_2$ solution and a 6 min cross-linking time resulted in the highest resolution and continuous fibrin patterns. Human microvascular ECs were suspended in PBS and mixed with the same volume of 2× thrombin/CaCl$_2$ solution for a resulting bioink with 1–8 × 10^6 cells/mL. Results indicated that after 21 days of culture, the printed pattern was cellularized with confluent ECs that showed evidence of lumen formation and excluded labeled dextran molecules from the 3D structure.

Fibrin-based bioink has also been used to fabricate a 3D multicellular array using laser printing process [30]. Hyaluronic acid (HA) was added to fibrinogen to stabilize the viscosity for the printing of cell arrays. Endothelial colony-forming cells (ECFCs) were printed alongside ASCs in 3D configurations such that a 9 × 9 array of ASC droplets were printed followed by an inset 8 × 8 array of ECFCs in which printed droplets had an average of 54 cells. Droplet arrays were printed onto a layer of fibrinogen-HA, which was spray-treated with a solution of thrombin and CaCl$_2$ to initiate fibrin formation. The cell-laden droplets would convert to fibrin-HA as they encountered the treated substrate with residual thrombin solution. The arrays were produced with ASCs and ECFCs printed coplanar or separated by a plane (Fig. 15.3n and o). Structures were designed to show the potential of 3D printing of cells in complex configurations as a tool for probing biological phenomena in a way 2D culture and conventional 3D culture struggle to accomplish. The results of this study show that ASCs initially migrate toward ECFCs without evidence of ECFC sprouting or migrating at all. Once ACSs contact the ECFC mass, an explosion of ECFC network sprouts begin to extend from the initial droplet position and remain as stable networks for several weeks.

4.5 Hyaluronic Acid-Based Bioinks

HA is a glycosaminoglycan (GAG) found in most tissues in the human body, especially the skin, vitreous humor, and synovial fluid. The high molecular weight and large amount of branching allows for intermolecular hydrogen bonding and viscous solution formation. HA gels are not stable enough for 3D structures so a chemical modification is necessary to increase the amount of cross-linking. Similar to other polysaccharides, HA supports cell viability but has poor cell attachment characteristics. Skardal et al. utilized the naturally derived HA as a robust hydrogel material by the addition of poly(ethylene glycol) (PEG)-based arms for cross-linking by photoinitiated acrylate polymerization [31]. Four-armed PEG linkers of 8 or 13 repeat units were used for modification and formed TetraPAc cross-linker molecules. The cross-linkers were reacted with thiolated HA, CMHA-S, and thiolated gelatin, Gtn-DTPH, to create a cross-linkable printable hydrogel allowing cell attachment. Evaluation was conducted on NIH/3T3, HepG2, and Int 407 cell lines representing fibroblast, hepatocyte, and intestinal epithelial cells. Cells were grown in PEG-DA, TetraPAc8, and TetraPAc13 cross-linked hydrogels, which demonstrated biocompatibility. It has been demonstrated that cross-linker efficiency was obstructed in the TetraPAc8 hydrogels, so bioprinting experiments were conducted using 4:1 CMHA-S:Gtn-DTPH and 4:1 hydrogel:TetraPAc13 to form a 2% w/v hydrogel mixture. NIH/3T3 were printed at a cell density of 25×10^6 cells/mL by mixing cell pellet with hydrogel and loading into a microcapillary for printing after cross-linking. The cross-linked hydrogel was dispensed as cylindrical filaments, stacked to form a tubular shape, and covered with agarose to maintain structure and orientation of filaments. Results indicated the maintenance of cell viability, position, and structural orientation with a lumen out to 4 weeks (Fig. 15.3h–j).

4.6 Extracellular Matrix-Based Bioinks

ECM is the network of proteins, GAGs, and other biomolecules produced by cells that support the function of cells within a given tissue. Matrigel™ is the secreted ECM from murine Engelbreth–Holm–Swarm tumors producing a basement membrane-like material rich in laminin, collagen type IV, and heparan sulfated proteoglycan [32]. Once set, Matrigel provides a rich matrix with growth factor content that supports many cell types including the undifferentiated

growth of stem cells [33]. Each tissue has a specific ECM composition suited to the functional needs of the tissue and metabolic needs of the cells. The use of decellularized ECM in combination with cells, both used in a tissue-specific fashion, has been shown to be beneficial for recapitulating desired tissue features [34]. Matrigel bioink by LIFT approach has been used to print arrays of MG63 osteoblast cells and GFP expressing ECs [35]. Matrigel bioink shows good viability, control over droplet size and therefore cell density printed, and 3D layering configurations. Another approach shows similar experiments using a variation on the LIFT method for transferring the biological material [36]. A Matrigel or glycerol solution was used for printing MG63 cells or CRL-1764 rat cardiac cells where the receiving substrate was coated with Matrigel. Results showed good cell compatibility of the system with multiple mammalian cell types.

Several attempts showed the use of Matrigel-based bioinks for printing microvasculature structures. Wu and Ringeisen printed stem and branch patterns of human vascular ECs and SMCs onto Matrigel [37]. HUVECs connected with each other in the printed pattern and also connected in patterns similar to the veins of a leaf. The SMCs did not show the same propensity for interconnection. Printed patterns of ECs were covered with SMCs, which seem to migrate to the EC pattern and proliferate. Pirlo et al. conducted a similar experiment printing HUVECs from a bioink composed of 0.125% methylcellulose in medium containing an estimated 8×10^7 cells/mL then collected on a Matrigel-coated or -uncoated poly(glycolide-co-lactide) biopaper [38]. Results demonstrated that the 2D patterns of HUVECs survived well and maintained the printed pattern on Matrigel-coating while conforming more to the topography of the thin or uncoated biopaper. Biopaper with printed HUVECs were then also stacked to attempt a vascular network in a thick construct.

ECM-based bioinks can also be derived from decellularized tissues. Pati et al. showed that ECM from decellularized tissues can be powdered and made into a gel that can serve as a bioink [39]. Bioinks were generated from decellularized heart, adipose, and cartilage ECM forming heart-ECM, adipose-ECM, and cartilage-ECM, respectively. Rat myoblasts were printed with heart-ECM to test the ECM ability to support heart-like tissue formation. To test adipose-ECM, ASCs were printed in a construct along with supporting poly(ε-caprolactone) structure having a filament diameter of 100 μm for minimal stiffness, followed by adipogenic medium culture. MSCs isolated from inferior turbinate were used to print structures in cartilage-ECM and 200 μm poly(ε-caprolactone) filaments for enhanced mechanical strength, followed by chondrogenic culture to promote cartilage formation. Tissue-specific gene expression demonstrated evidence for tissue-specific differentiation and tissue formation based on ECM source. These tissue-specific ECM-based bioinks are capable of providing crucial cues for cells engraftment, survival, and long-term function.

5 SUMMARY AND FUTURE DIRECTIONS

Bioprinting is a new and exciting field being advanced along several avenues of technology. Experimental biology, tissue engineering, and pharmacology/toxicology are a few fields that are expected to greatly benefit from the development of the field. Presented here was representative work performed in several labs around the world in which cells were printed and shown to remain viable and maintain cell function. The resulting printed-cell-containing bioinks can be considered microtissue organoids with potential for further culture and use as a screening tool, experimental configuration for sophisticated biological systems, or a tissue construct to aid in regenerative medicine.

The printing resolution (layer thickness) is a key element of 3D bioprinting technology. Therefore, there are many attempts to improve the printing resolution for organoid printing. An inkjet method for cell-contained bioink patterning has been shown to achieve relatively high resolutions on the order of approximately 20–100 μm, but this method has been limited by structural stability of the construct. The extrusion-based method has been used to make microfilaments down to approximately 200–400 μm in layer thickness, but this is met with considerable trade-offs. At higher resolutions, the flow resistance and shear stress within the bioink through the nozzle can increase dramatically due to the smaller diameter of the nozzle, resulting in severe cell damage. Related to the printing resolution limitations, patterning intricacies become challenging, as complex tissue organoids with many coordinating cell types need to be arranged in close proximity. In order to attain the juxtaposition of many cell types, high resolutions may be required. More importantly, a new bioink system needs to be developed to improve printability with high-resolution capability. Availability of currently available biomaterials that can suffice as cell delivery bioinks but which also provide mechanical support, cellular interaction cues, and negligible cytotoxicity is limited. Advances in the field of suitable cell-compatible hydrogel-based bioinks are necessary for the long-term success of organoid printing.

GLOSSARY

Bioink A substance composed of proteins and/or glycoproteins and/or cells, which is dispensed in a desired shape by printing mechanism.

Biopaper A substance that can receive bioink in such a way as to preserve intended printed pattern.

Cell aggregate A cluster of cells typically formed by allowing cell suspension to settle into a well or droplet of media.

Confluence The percent coverage of a cell culture surface by cells as they attach and spread across the substrate.

Cornification The process by which certain cell types, that is, keratinocytes, produce cytoplasmic keratin to the point that the cell filled with keratin and dies.

Cross-linking The chemical process of molecules in a material forming covalent bonds with neighboring molecules to form a network of covalently bonded molecules throughout the material.

Cytotoxicity The property of a substance which proves toxic to cells.

Decellularization The process by which cells are removed from a tissue but structural ECM is left intact.

Degradation The process of breaking down an implanted material which often occurs by hydrolysis or enzymatic scission.

Engraftment Particularly speaking of cells used for tissue engineering and their ability to functionally integrate with the host tissue as opposed to migrating away or being destroyed.

Fibrinogen A protein network produced by the clotting cascade and used in its purified form as a degradable hydrogel.

Fibronectin A protein commonly found in basement membranes to which many cells bind.

Filament A continuous strand of material extruded by certain bioprinting system.

Gene expression The process by which genes are activated, interpreted, and utilized to produce the associated protein.

Immunohistochemistry The staining procedure using antibody specificity to highly specific proteins in a histological section.

Methacrylation Verb connoting the addition of a methacrylate group to molecules of a substance giving it the ability to crosslink when it was previously unable.

Nozzle An attachment for some bioprinting systems allowing for interchangeable orifice size and variable conduit geometries leading to the orifice.

Pharmacokinetic/pharmacodynamics The properties related to the uptake, affect, degradation, and elimination of a drug in a biological system.

Photoinitiation The process of using photon energy to initiate a chain reaction, often for crosslinking.

Physicochemical The properties of a substance related to its material and chemical properties.

Physiological Those conditions experienced by a tissue *in situ* which can reflect states of injury, healing, or homeostasis.

Pneumatic Employing pressurized fluid as motive force.

Powderize The process of converting solid material to a powder by a mechanical action.

Remodeling The process of tissues repairing and reordering to reflect new homeostatic conditions.

Rheology The study of material properties related to flow.

Stereolithography Fabrication technique using laser energy to crosslink material in localized positions allowing two-dimensional patterns to be built up into three-dimensional structures.

Viscosity Rheological property of a material quantifying the resistance to flow based on internal friction.

ABBREVIATIONS

LIFT	Laser-induced forward transfer
ECM	Extracellular matrix
GAGs	Glycosaminoglycans

ALP	Alkaline phosphatase
nHA	Nanohydroxyapatite
HA	Hyaluronic acid
GelMA	Methacrylated gelatin
PEG	Poly(ethylene glycol)
MG63	Human osteosarcoma cell line
NIH/3T3	Mouse embryo fibroblast cell line
HaCaT	Human keratinocyte cell line
CRL-1476	Rat aorta myoblast cell line
CRL-1764	Rat embryo fibroblast cell line
HFF-1	Human fibroblast cell line
MSC	Mesenchymal stem cells
hASCs	Human adipose-derived stem cells
HUVEC	Human umbilical vein endothelial cells
EC	Endothelial cells
ECFC	Endothelial colony-forming cells
SMC	Smooth muscle cells
ESCs	Embryonic stem cells
Myf-5	Myogenic factor 5
PDX-1	Pancreatic and duodenal homeobox 1
DMEM	Dulbecco's modified Eagle's medium
EGM-2	Endothelial cell growth medium-2

ACKNOWLEDGMENTS

We would like to thank Dr John D. Jackson for editorial assistance. This chapter was supported by the Armed Forces Institute of Regenerative Medicine (W81XWH-13-2-0052).

REFERENCES

[1] Zegers MM. 3D in vitro cell culture models of tube formation. Semin Cell Dev Biol 2014;31C:132–40.

[2] Page H, Flood P, Reynaud EG. Three-dimensional tissue cultures: current trends and beyond. Cell Tissue Res 2012;352:123–31.

[3] Atala A, Kasper FK, Mikos AG. Engineering complex tissues. Sci Transl Med 2012;4:160rv12.

[4] Withers GS. New ways to print living cells promise breakthroughs for engineering complex tissues in vitro. Biochem J 2006;394: e1–2.

[5] Li Y, Xu C, Ma T. In vitro organogenesis from pluripotent stem cells. Organogenesis 2014;10(2):159–63.

[6] Prestwich GD. Evaluating drug efficacy and toxicology in three dimensions: using synthetic extracellular matrices in drug discovery. Acc Chem Res 2008;41:139–48.

[7] Michael S, Sorg H, Peck CT, Koch L, Deiwick A, Chichkov B, et al. Tissue engineered skin substitutes created by laser-assisted bioprinting form skin-like structures in the dorsal skin fold chamber in mice. PLoS One 2013;8:e57741.

[8] Kolesky DB, Truby RL, Gladman AS, Busbee TA, Homan KA, Lewis JA. 3D bioprinting of vascularized, heterogeneous cell-laden tissue constructs. Adv Mater 2014;26:3124–30.

[9] Malda J, Visser J, Melchels FP, Jungst T, Hennink WE, Dhert WJ, et al. 25th anniversary article: engineering hydrogels for biofabrication. Adv Mater 2013;25:5011–28.

[10] Klebe RJ. Cytoscribing: a method for micropositioning cells and the construction of two- and three-dimensional synthetic tissues. Exp Cell Res 1988;179:362–73.

[11] Melchels FPW, Domingos MAN, Klein TJ, Malda J, Bartolo PJ, Hutmacher DW. Additive manufacturing of tissues and organs. Prog Polym Sci 2012;37:1079–104.

[12] Nakamura M, Iwanaga S, Henmi C, Arai K, Nishiyama Y. Biomatrices and biomaterials for future developments of bioprinting and biofabrication. Biofabrication 2010;2:014110.

[13] Xu T, Baicu C, Aho M, Zile M, Boland T. Fabrication and characterization of bio-engineered cardiac pseudo tissues. Biofabrication 2009;1:035001.

[14] Norotte C, Marga FS, Niklason LE, Forgacs G. Scaffold-free vascular tissue engineering using bioprinting. Biomaterials 2009;30: 5910–7.

[15] Guillotin B, Guillemot F. Cell patterning technologies for organotypic tissue fabrication. Trends Biotechnol 2011;29:183–90.

[16] Jin R, Dijkstra PJ. Hydrogels for tissue engineering applications. In: Ottenbrite RM, Park K, Okano T, editors. Biomedical applications of hydrogels handbook. New York: Springer; 2010. p. 203–25.

[17] Nair LS, Laurencin CT. Biodegradable polymers as biomaterials. Prog Polym Sci 2007;32:762–98.

[18] Samavedi S, Whittington AR, Goldstein AS. Calcium phosphate ceramics in bone tissue engineering: a review of properties and their influence on cell behavior. Acta Biomater 2013;9:8037–45.

[19] Catros S, Fricain JC, Guillotin B, Pippenger B, Bareille R, Remy M, et al. Laser-assisted bioprinting for creating on-demand patterns of human osteoprogenitor cells and nano-hydroxyapatite. Biofabrication 2011;3:025001.

[20] Gruene M, Deiwick A, Koch L, Schlie S, Unger C, Hofmann N, et al. Laser printing of stem cells for biofabrication of scaffold-free autologous grafts. Tissue Eng Part C Methods 2011;17(1):79–87.

[21] Koch L, Kuhn S, Sorg H, Gruene M, Schlie S, Gaebel R, et al. Laser printing of skin cells and human stem cells. Tissue Eng Part C Methods 2009;16:847–54.

[22] Guillotin B, Souquet A, Catros S, Duocastella M, Pippenger B, Bellance S, et al. Laser assisted bioprinting of engineered tissue with high cell density and microscale organization. Biomaterials 2010;31:7250–6.

[23] Ovsianikov A, Gruene M, Pflaum M, Koch L, Maiorana F, Wilhelmi M, et al. Laser printing of cells into 3D scaffolds. Biofabrication 2010;2:014104.

[24] Roth EA, Xu T, Das M, Gregory C, Hickman JJ, Boland T. Inkjet printing for high-throughput cell patterning. Biomaterials 2004;25: 3707–15.

[25] Smith CM, Stone AL, Parkhill RL, Stewart RL, Simpkins MW, Kachurin AM, et al. Three-dimensional bioassembly tool for generating viable tissue-engineered constructs. Tissue Eng 2004;10:1566–76.

[26] Lee W, Debasitis JC, Lee VK, Lee JH, Fischer K, Edminster K, et al. Multi-layered culture of human skin fibroblasts and keratinocytes through three-dimensional freeform fabrication. Biomaterials 2009;30:1587–95.

[27] Lee V, Singh G, Trasatti JP, Bjornsson C, Xu X, Tran TN, et al. Design and fabrication of human skin by three-dimensional bioprinting. Tissue Eng Part C Methods 2013;20:473–84.

[28] Raof NA, Schiele NR, Xie Y, Chrisey DB, Corr DT. The maintenance of pluripotency following laser direct-write of mouse embryonic stem cells. Biomaterials 2010;32:1802–8.

[29] Cui X, Boland T. Human microvasculature fabrication using thermal inkjet printing technology. Biomaterials 2009;30:6221–7.

[30] Gruene M, Pflaum M, Hess C, Diamantouros S, Schlie S, Deiwick A, et al. Laser printing of three-dimensional multicellular arrays for

studies of cell–cell and cell–environment interactions. Tissue Eng Part C Methods 2011;17:973–82.

[31] Skardal A, Zhang J, Prestwich GD. Bioprinting vessel-like constructs using hyaluronan hydrogels crosslinked with tetrahedral polyethylene glycol tetracrylates. Biomaterials 2010;31:6173–81.

[32] Kleinman HK, McGarvey ML, Liotta LA, Robey PG, Tryggvason K, Martin GR. Isolation and characterization of type IV procollagen, laminin, and heparan sulfate proteoglycan from the EHS sarcoma. Biochemistry 1982;21:6188–93.

[33] Hughes CS, Postovit LM, Lajoie GA. Matrigel: a complex protein mixture required for optimal growth of cell culture. Proteomics 2010;10:1886–90.

[34] Badylak SF. The extracellular matrix as a scaffold for tissue reconstruction. Semin Cell Dev Biol 2002;13:377–83.

[35] Barron JA, Wu P, Ladouceur HD, Ringeisen BR. Biological laser printing: a novel technique for creating heterogeneous 3-dimensional cell patterns. Biomed Microdevices 2004;6:139–47.

[36] Barron JA, Ringeisen BR, Kim HS, Spargo BJ, Chrisey DB. Application of laser printing to mammalian cells. Thin Solid Films 2004;453:383–7.

[37] Wu PK, Ringeisen BR. Development of human umbilical vein endothelial cell (HUVEC) and human umbilical vein smooth muscle cell (HUVSMC) branch/stem structures on hydrogel layers via biological laser printing (BioLP). Biofabrication 2010;2:014111.

[38] Pirlo RK, Wu P, Liu J, Ringeisen B. PLGA/hydrogel biopapers as a stackable substrate for printing HUVEC networks via BioLP. Biotechnol Bioeng 2011;109:262–73.

[39] Pati F, Jang J, Ha DH, Won Kim S, Rhie JW, Shim JH, et al. Printing three-dimensional tissue analogues with decellularized extracellular matrix bioink. Nat Commun 2014;5:3935.

[40] Xu C, Chai W, Huang Y, Markwald RR. Scaffold-free inkjet printing of three-dimensional zigzag cellular tubes. Biotechnol Bioeng 2012;109:3152–60.

[41] Fedorovich NE, De Wijn JR, Verbout AJ, Alblas J, Dhert WJ. Three-dimensional fiber deposition of cell-laden, viable, patterned constructs for bone tissue printing. Tissue Eng Part A 2008;14:127–33.

[42] Fedorovich NE, Schuurman W, Wijnberg HM, Prins HJ, van Weeren PR, Malda J, et al. Biofabrication of osteochondral tissue equivalents by printing topologically defined, cell-laden hydrogel scaffolds. Tissue Eng Part C Methods 2011;18:33–44.

[43] Park JY, Choi JC, Shim JH, Lee JS, Park II, Kim SW, et al. A comparative study on collagen type I and hyaluronic acid dependent cell behavior for osteochondral tissue bioprinting. Biofabrication 2014;6:035004.

[44] Lee YB, Polio S, Lee W, Dai G, Menon L, Carroll RS, et al. Bioprinting of collagen and VEGF-releasing fibrin gel scaffolds for neural stem cell culture. Exp Neurol 2010;223:645–52.

[45] Lee W, Pinckney J, Lee V, Lee JH, Fischer K, Polio S, et al. Three-dimensional bioprinting of rat embryonic neural cells. Neuroreport 2009;20:798–803.

[46] Moon S, Hasan SK, Song YS, Xu F, Keles HO, Manzur F, et al. Layer by layer three-dimensional tissue epitaxy by cell-laden hydrogel droplets. Tissue Eng Part C Methods 2009;16:157–66.

[47] Bertassoni LE, Cardoso JC, Manoharan V, Cristino AL, Bhise NS, Araujo WA, et al. Direct-write bioprinting of cell-laden methacrylated gelatin hydrogels. Biofabrication 2014;6:024105.

[48] Fedorovich NE, Wijnberg HM, Dhert WJ, Alblas J. Distinct tissue formation by heterogeneous printing of osteo- and endothelial progenitor cells. Tissue Eng Part A 2011;17:2113–21.

[49] Othon CM, Wu X, Anders JJ, Ringeisen BR. Single-cell printing to form three-dimensional lines of olfactory ensheathing cells. Biomed Mater 2008;3:034101.

Chapter 16

Bioprinting of Three-Dimensional Tissues and Organ Constructs

Young-Joon Seol, James J. Yoo, and Anthony Atala

Wake Forest Institute for Regenerative Medicine, Wake Forest School of Medicine, Winston-Salem, NC, USA

Chapter Outline

ABSTRACT

Three-dimensional (3D) bioprinting technology has been utilized as a method to engineer complex tissues and organs. This rapidly growing technology allows for precise placement of multiple types of cells, biomaterials, and biomolecules in spatially predefined locations within 3D structures. Many researchers are focusing on the further development of bioprinting technology and its applications. In this chapter, we introduce the general principles and limitations of widely used bioprinting systems and applications for tissue and organ regeneration. In addition, the current challenges facing the clinical applications of bioprinting technology are addressed.

Keywords: bioprinting; jetting-based bioprinting; extrusion-based bioprinting; tissue engineering; regenerative medicine

1 INTRODUCTION

Three-dimensional (3D) porous scaffolds fabricated from biomaterials have been an essential component used for tissue and organ regeneration [1–5]. They provide a 3D environment for cells to form tissues or organs and designed to degrade over time as regenerated tissues gradually replace the scaffold. To produce 3D scaffolds for tissue engineering applications, many fabrication techniques have been developed and applied in tissue engineering. These include solvent casting, particulate leaching, gas foaming, phase separation, and freeze-drying. Considerations for selecting biomaterials for tissue regeneration include biocompatibility, biodegradability, and porosity [6,7]. However, production of 3D scaffolds with uniform shape, microarchitecture, and interconnected pores has limited the utility of these fabrication methods. Therefore, additive manufacturing was introduced as a method to produce complex 3D constructs by selectively adding biomaterials during fabrication of 3D scaffolds. This is categorized into stereolithography, fused deposition manufacturing, and selective laser sintering, and can be used to control the size, shape, and interconnectivity of pores of 3D scaffolds. Based on the rapid prototyping (RP) technology, biomimetic-shaped 3D structures can be fabricated from medical images generated by computerized tomography (CT) and magnetic resonance imaging using computer-aided design and manufacturing (CAD/CAM) technologies [8–10].

Tissue-engineered scaffolds have been used successfully for a number of clinical applications; however, several limitations exist in the regeneration of complex and composite tissues and organs. The inability to adequately control cell distribution in the 3D scaffolds, delivery of multiple cell types at specific locations, and availability of biomaterials for tissue construction [11–13] are some of the limitations in tissue engineering. Manufacturing that allows for greater precision and well-defined 3D cell constructs is critical for future progress in tissue engineering. The introduction of 3D bioprinting technology is an advanced

Essentials of 3D Biofabrication and Translation. http://dx.doi.org/10.1016/B978-0-12-800972-7.00016-5

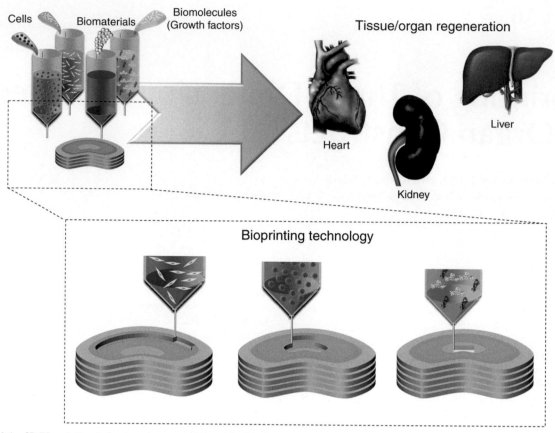

FIGURE 16.1 **3D Bioprinting technology is a powerful tool for fabrication of tissue and organ constructs using cells, biomaterials, and biomolecules.** It allows for precise placement of multiple types of cells, biomaterials, and biomolecules in a spatially predefined location within 3D tissue and organ constructs. *(Figures reprinted with permission from Seol et al. [14].)*

tissue engineering approach that has the potential to create complex tissues and organs. This technology allows for precise placement of multiple types of cells, biomaterials, and biomolecules in spatially predefined locations within 3D structures (Fig. 16.1). Furthermore, tissue and organ constructs can be formed by self-assembling of printed cells [12–14]. Although bioprinting technology is a relatively new approach in tissue engineering, it has rapidly gained favor among researchers due to the feasibility of 3D spatial positioning of living cells and constructions.

Bioprinting technology that is commonly used in tissue engineering can be categorized according to two working principles: jetting- and extrusion-based methods (Figs 16.2 and 16.3). The jetting-based method uses a commercially available inkjet printer head to deposit a droplet of cell-laden hydrogel onto a substrate. Alternatively, the extrusion-based method uses a syringe pump or pneumatic pressure to flow continuous filaments of cell-laden hydrogel through a micron-size nozzle. These methods of bioprinting have different characteristics with respect to fabrication resolution, hydrogel materials, and printing speed. Bioprinting can be combined with CAD/CAM technology to fabricate a structure that can be tailored to the anatomical shape of an individual patient since both methods are based on rapid

prototyping technology. Therefore, bioprinting technology has emerged as a new engineering tool for the fabrication of 3D tissue and organ constructs. In this chapter, the general principles of bioprinting technology and current research in tissue and organ regeneration will be discussed.

2 THREE-DIMENSIONAL BIOPRINTING TECHNOLOGY

2.1 Jetting-Based Bioprinting

The noncontact technique of the jetting-based bioprinting system represents the initial version of bioprinting technology [15–18]. Early research relied upon a modified commercially available desktop inkjet printer to generate a small-volume droplet. The cell-laden hydrogel was placed in a printer cartridge, and a stage system was designed to fabricate 3D structures. This approach could easily generate droplets onto a substrate.

Further developments resulted in the use of the jetting-based bioprinting system in research for tissue regeneration especially 3D bioprinting. A small droplet is generated by various jetting mechanisms that include the thermal method, piezoelectric actuator, laser-induced forward transfer,

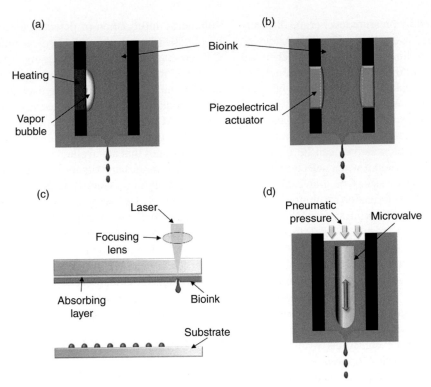

FIGURE 16.2 Jetting-based bioprinting system generates picoliter droplets including cells using (a) thermal, (b) piezoelectrical actuator, (c) focused laser energy, or (d) pneumatic pressure and valve.

and pneumatic pressure [18–22]. In the thermal method (Fig. 16.2a), localized heating generates a bubble in the bioink chamber that yields a small droplet deposited through a micron-sized nozzle. The piezoelectric actuator (Fig. 16.2b) uses a pressure pulse generated by the voltage-mediated actuation of piezocrystal film to produce a droplet. Both of these jetting mechanisms are widely used for inkjet bioprinting systems. The laser-assisted bioprinting system uses a laser-induced forward transfer mechanism (Fig. 16.2c). A focused laser beam induces vaporization of the metal film and generates a droplet. This printing system produces relatively high resolution patterns; however, cell viability in the printed hydrogel is decreased when compared to other jetting mechanisms. In the microvalve printing system using pneumatic pressure (Fig. 16.2d), the droplets are generated by opening and closing of a small valve under constant pneumatic pressure.

Using this jetting-based bioprinting system, droplets of picoliter volume can be generated and high fabrication resolution of 10–100 μm can be achieved [23,24]. The bioink material with high viscosity, however, cannot be used to generate droplets. Therefore, low-viscous bioinks, such as thrombin, fibrinogen, and collagen, have been used for jetting-based printing systems. Due to the low mechanical properties of low-viscous bioinks that yield weak printed structures, fabrication of durable 3D structures that can maintain their shape and withstand external stress after implantation are difficult to produce.

2.2 Extrusion-Based Bioprinting

The extrusion-based bioprinting system dispenses continuous filaments of a material that is a mixture of cells and hydrogel through a micron-sized nozzle to fabricate two-dimensional (2D) patterns. To dispense the hydrogel/cells mixture, the extrusion-based bioprinting system uses pneumatic pressure or a syringe pump (Fig. 16.3), and the amount of dispensed

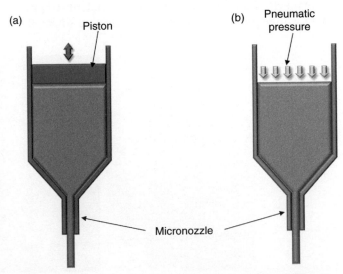

FIGURE 16.3 Extrusion-based bioprinting system dispenses continuous filaments of cells incorporating hydrogel using (a) piston or (b) pneumatic pressure.

materials can be controlled by pressure level or the displacement of the piston pump, respectively [25–29]. After dispensing the hydrogel including cells, physical or chemical treatments are applied to solidify the cell-laden hydrogel; furthermore, 2D patterns are stacked layer by layer to fabricate 3D structures. This system can be used to fabricate hybrid structures consisting of various cell types and materials using a multiple-cartridge system to dispense different biomaterials. Moreover, high-viscous biomaterials can be dispensed through the micron-nozzle to yield a wider selection of biomaterials and scalable construct size. One drawback to this system is a comparatively low fabrication resolution.

While the extrusion-based bioprinting system can fabricate 3D structures using cell-laden hydrogel, these structures lack sufficient mechanical strength for implantation due to low mechanical properties and structural stability. To overcome these limitations, an integrated bioprinting system has been developed [30]. Based on the extrusion-based bioprinting system, the integrated bioprinting system can concurrently print a synthetic biopolymer and cell-laden hydrogel to fabricate 3D tissue constructs with high mechanical properties. The synthetic biopolymer has the mechanical properties necessary to support 3D constructs and cell-laden hydrogel to promote tissue regeneration.

3 ANATOMICALLY SHAPED 3D CONSTRUCTS

Bioprinting systems can be combined with CAD/CAM technology to place cells, biomaterials, and biomolecules within a 3D construct with precision [8,9,31–33]. Using CAD/CAM technology, fabrication procedures can be generated using 3D volumetric information obtained from defected tissues and organs, which allows for fabrication of biomimetic 3D tissues and organs. This automated procedure will be important for clinical application where 3D

volumetric information of defected tissues and organs can be obtained from a patient's medical imaging data such as CT and magnetic resonance imaging. The scanned 2D data are stored in a digital imaging and communications in medicine format. Subsequently, these data are converted to a 3D CAD model by a reverse-engineering process. In this step, the 3D CAD model of the defected tissues and organs can be created by extraction of the localized volumetric data. Software packages that are available to create a 3D CAD model of defected tissues and organs include Mimics (Marerialise, Leuven, Belgium), Geomagic Studio® (Geomagic, Morrisville, NC, USA), Simpleware (Simpleware Ltd., Exeter, UK), and Analyze (AnalyzeDirect Inc., Overland Park, KS, USA). A fabrication code directs the printing system to follow a designed path generated by CAM technology. For efficient regeneration of complex tissues and organs, a well-organized fabrication code that can generate tool paths and construct inner architecture for multiple cell types is required for automated printing.

4 APPLICATIONS OF BIOPRINTING

4.1 Musculoskeletal Tissue

Musculoskeletal tissue (bone, muscle, cartilage, tendon, and ligament) provides support, stability, and movement to the human body. To restore defected and damaged musculoskeletal tissue, cell-seeded 3D scaffolds have been used, and small grafts have been successfully applied in animal models. To improve the functionality of 3D tissue constructs with clinically relevant-sized grafts, fabrication of 3D constructs with anatomical geometry, spatial organization, and microenvironment of the cells is needed [34]. Therefore, 3D bioprinting technology is one of the methods that has been developed to meet these demands, and many researchers have demonstrated the capability of 3D bioprinting technology to regenerate tissue (Fig. 16.4).

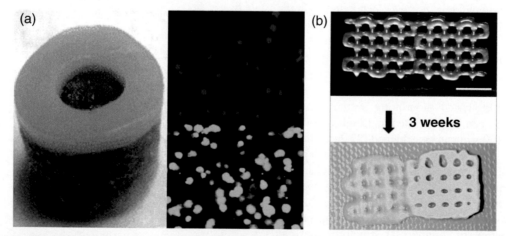

FIGURE 16.4   Musculoskeletal tissue regeneration using bioprinting system. (a) Macroscopic image of printed 3D osteochondral plug for cartilage tissue regeneration (left) and fluorescence image of printed human chondrocytes in poly(ethylene glycol) dimethacrylate hydrogel. (b) Macroscopic images of printed osteochondral constructs with an MSC-laden compartment on the left and a chondrocyte-laden compartment on the right. (top) Construct after printing and (bottom) construct after 3 weeks of culture. *(Figures reprinted with permission from Fedorovich et al. [37] and Cui et al. [40].)*

Phillippi et al. used an inkjet printing system to pattern bone morphogenic protein-2 (BMP-2) on a fibrin substrate; moreover, they demonstrated that printed constructs sustained BMP-2 expression for 6 days. In addition, osteogenic differentiation of muscle-derived stem cells was observed on the BMP-2 printed substrate [35]. This study verified that spatially controlled multilineage differentiation of stem cells is possible using this printing system. Fedorovich et al. noted the capability of the extrusion-based bioprinting system to fabricate a spatially organized cell-laden hydrogel using various hydrogels [36,37]. Bone marrow stromal cells were mixed with agarose, alginate, methylcellulose, and Lutrol F127 hydrogels and printed using Bioplotter, which is a commercialized system. The viability of printed cells was similar to that of unprinted cells, and printing conditions did not significantly affect cell survival. Moreover, the printed cells could differentiate to osteogenic lineage. Elmer and coworkers used an inkjet printing system to pattern growth factors on polystyrene fibers for regeneration of a muscle–tendon–bone structure [38]. Prior to seeding C2C12 myoblasts and C3H10T1/2 mesenchymal fibroblasts, fibroblast growth factor-2, and BMP-2 were patterned on the structure to promote tendon and bone tissue, respectively. Thus, it was demonstrated that printed fibroblast growth factor-2 and BMP-2 promoted tenocyte and osteoblast fates, respectively. Schuurman et al. demonstrated the fabrication of hybrid constructs consisting of thermoplastic polymer polycaprolactone (PCL) and cell-laden hydrogel (alginate) for musculoskeletal tissues [39]. An extrusion-based bioprinting system was used for hybrid construct fabrication. PCL structure provided the mechanical properties of hybrid constructs, which could be tailored by changing the inner-architecture of the PCL structure. Moreover, viability of printed cells was similar to that of nonprinted cells. Cui et al. fabricated 3D constructs for osteochondral tissue regeneration using a jetting-based bioprinting system. To repair articular cartilage tissue, human chondrocytes were suspended in poly(ethylene glycol) dimethacrylate and printed. Printed chondrocytes maintained their initial deposited positions and regenerated cartilage tissue that attached firmly to surrounding native tissue; furthermore, greater proteoglycan deposition was observed [40]. In another study, Keriquel et al. applied jetting-based bioprinting in the perspective of computer-assisted medical interventions [41]. They used a bioprinting system for bone tissue regeneration; however, they did not use a cell-laden hydrogel. Hydroxyapatite was directly printed in the defect of mice calvaria, and bone healing was observed for 3 months. This study demonstrated that *in vivo* printing is possible.

4.2 Nerve Tissue

Bioprinting technology was used to fabricate biological nerve grafts for regeneration of neural tissue. Clinically, the best method for repair of neural tissue involves replacing the injured segment with an autologous graft. Limitations of this method include the shortage of donor tissue, the sacrifice of functioning nerves, and morbidity at the donor site. To overcome these limitations, tissue engineering approaches have been used in neural tissue regeneration with many researchers using printed neural cells to fabricate biological nerve grafts (Fig. 16.5).

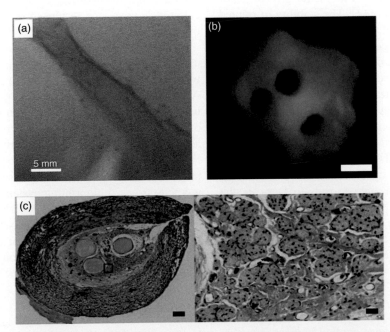

FIGURE 16.5 Neural tissue regeneration using bioprinting system. (a) Photography of a printed neural 3D sheet fabricated by printing fibrin gels and NT2 cells in the culture medium. (b) Cross-section of the nerve graft with fluorescently labeled Schwann cells. (c) Bielschowsky's staining of its histological sections. *(Figures reprinted with permission from Xu et al. [42] and Owens et al. [44].)*

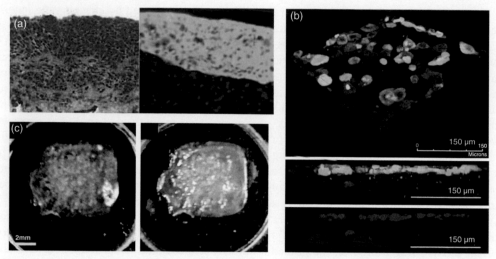

FIGURE 16.6 **Skin tissue regeneration using bioprinting system.** (a) Skin mimicking bilayered construct consisting of fibroblasts and keratinocytes: microscopic images of (left) H and E staining and (right) immunohistochemical staining of pan-reticular fibroblast (red) and keratin-14 (green). (b) Confocal microscope images of printed skin construct: (top) volume rendered image, (middle) keratin-containing keratinocyte layer labeled in green, (bottom) tubulin-containing keratinocyte and fibroblast labeled in red. (c) Skin construct (left) after printing and (right) after 7 days culture. *(Figures reprinted with permission from Lee et al. [21,45] and Koch et al. [47].)*

Xu et al. applied a jetting-based bioprinting system in neural tissue regeneration [42]. Primary embryonic hippocampal and cortical neurons, which yielded fabricated controlled patterns and structures using a modified inkjet printing system. The maintenance of neuronal phenotypes and the basic electrophysiological functions of the printed cells were demonstrated. Additionally, fabrication of 3D neural sheet constructs was achieved using alternate printing of hydrogel and cells. Lee et al. used the jetting-based bioprinting system to print murine neural stem cells (C17.2), collagen hydrogel, and vascular endothelial growth factor for neural tissue regeneration [43]. The stability of printed constructs and the effects of growth factor on the cell behavior was examined. Printed cells showed high viability; moreover, growth factor incorporated in hydrogel showed sustained release and induced cell migration and proliferation. These results demonstrated that bioprinting technology could be used in the development of neural tissue regeneration applications. Owens et al. fabricated a biological nerve conduit using the extrusion-based bioprinting system [44]. Bone marrow stem cells and Schwann cells were used as bioink, and multicellular cylindrical units with composing cells were printed with agarose rods to fabricate the nerve conduit. The printed structure was supported by an array of agarose rods, which held the conduit in place, and a printed multicellular bioink cylinder became the nerve graft by self-assembly. Recovery of motor and sensory function using printed nerve conduit was demonstrated.

4.3 Skin Tissue

Bioprinting technology has been applied to regenerate skin tissue damaged by burns, lacerations, and diabetic wounds

(Fig. 16.6). Because this tissue consists of multiple cell types and layers, partially controlled placement of cells and biomaterials are required for 3D reconstruction of the multiple skin layers. Lee et al. fabricated a biomimetic multilayered skin construct consisting of human skin fibroblasts and keratinocytes using the jetting-based bioprinting system [21,45]. Prior to fabrication of 3D skin constructs, printing parameters were optimized for maximum cell viability and concentration in the dermis and epidermis layers. A skin construct was fabricated in a layer-by-layer assembly process using cells and collagen hydrogel. Printed 3D skin constructs were cultured *in vitro*, and results showed the formation of distinctive dual-layered tissues resembling human skin tissue.

Binder et al. developed an *in situ* skin printer system using a jetting-based bioprinter [46]. Cells and biomaterials were deposited on the defected skin tissue for initial repair. A skin printer was developed and could successfully deposit keratinocytes and fibroblasts onto full-thickness wounds of pigs. The results showed rapid re-epithelialization and accelerated wound healing of the printed wound.

Koch et al. arranged the keratinocytes and fibroblasts in predefined patterns using laser-assisted bioprinting, which is a jetting-based bioprinting system [47]. Cells were embedded in collagen hydrogel to fabricate skin constructs having a layered configuration. Histological and immunohistochemical staining showed that cells developed intercellular adhesion and communication through adherens and gap junctions, which indicated tissue formation.

4.4 Vascular Tissue

In the field of tissue engineering, the transfer of oxygen and nutrients to large volume tissue constructs for cell

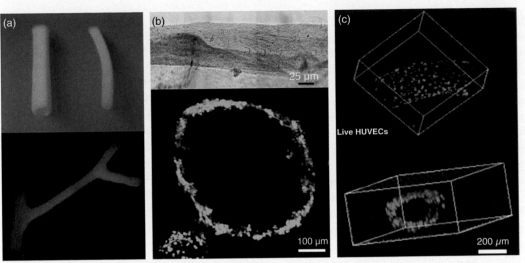

FIGURE 16.7 **Vascular tissue regeneration using a bioprinting system.** (a) (top) Printed smooth muscle cell tubes of distinct diameter after 3 days culture and (bottom) printed vascular branched construct after 6 days culture. (b) (top) Well-aligned microvasculature structure consisting of fibrin gel and endothelial cells, and (bottom) fluorescence image of microvascular structure. (c) Confocal microscope images of printed live HUVECs lining the microchannel walls. *(Figures reprinted with permission from Norotte et al. [28], Cui and Boland [49], and Kolesky et al. [50].)*

proliferation and differentiation is difficult. Cell survival is dependent upon a diffusion penetration depth of these components of 100 μm within tissue constructs; furthermore, fabrication of the vascular structure within 3D constructs for prevascularization presents challenges to regeneration of 3D tissues and organs effectively (Fig. 16.7). Without vascular structure in the 3D construct and vascular connection, the printed cells cannot survive after implantation.

To fabricate the microvasculature, Nakamura et al. used a jetting-based bioprinting system with an electrically driven inkjet system [48] in which a droplet could be made without generating heat that can cause cell damage. The bioink consisted of a suspension of bovine vascular endothelial cells that were patterned onto culture disks using the printing system. Cui et al. fabricated microvasculature whereby a mixture of human microvascular endothelial cells and fibrin hydrogel was printed [49]. A thermal inkjet bioprinting system was used to fabricate this structure on a 10 μm scale. After *in vitro* culture, printed cells aligned themselves inside of the channel structure and a confluent lining of cells could be observed. The 3D tubular structure was found in the printed structure. Jakab et al. printed multicellular spheroids consisting of embryonic cardiac cells and human umbilical vein endothelial cells (HUVEC) on a collagen substrate using an extrusion-based bioprinting system [27]. In this study, the self-organizing capacity of cells was exploited to regenerate functional living structures. The results showed that the printed endothelial cells organized into vessel-like conduits. Norrote et al. demonstrated the successful fabrication of vascular structures using an extrusion-based bioprinting system [28]. The printed cell aggregation consisted of human umbilical vein smooth muscle cells and human skin fibroblasts, and the fusion of the printed

cell aggregation resulted in a vascular tube. Moreover, a vascular tube with multiple layers and branching geometry could be fabricated. Kolesky et al. printed HUVEC and human neonatal dermal fibroblasts using an extrusion-based bioprinting system [50]. This study demonstrated that HUVECs lined the lumens within the embedded 3D microvascular network and allowed fabrication of predefined shaped 3D vascular structures.

5 CURRENT LIMITATIONS AND FUTURE PERSPECTIVES

Bioprinting technology has been widely adopted by researchers for tissue and organ regeneration, and many studies have been performed to fabricate and regenerate 3D tissue and organ constructs. Construction of more complex and composite tissue and organ structures remains a challenge. Also, printing fully functioning tissues and organs remains immature at the present time. Bioprinting technologies demonstrate various abilities and have great promise to become powerful tools for tissue and organ regeneration in the future.

Biological structures within the human body have uniquely complex architectures that are difficult to replicate. Many researchers have attempted to improve the resolution of 3D bioprinting systems. Jetting-based and extrusion-based bioprinting systems have been shown to achieve high resolution of between 10 μm and 100 μm, respectively. With respect to jetting-based bioprinting systems, the ability to fabricate clinically applicable tissue and organ-sized structures with proper mechanical properties is lacking. The jetting-based printing system cannot use high-viscous bioink materials to generate droplets and printed structures made from them are weak. Thus, it is difficult to fabricate durable 3D structures that can maintain their shape and withstand external stress after implantation. Moreover, fabrication of 3D tissue and organ-sized structures is lengthy using a tiny droplet size of 10 μm. With the extrusion-based bioprinting system, filaments can be made with a maximum resolution of 100 μm. This system uses a smaller diameter nozzle to achieve higher resolution; however, the smaller-sized nozzle affects cell printing by flow resistance and shear stress within the bioink and results in decreased cell viability [51, 52]. Also, printing process time is increased by high resolution. As mentioned earlier, resolution of the 3D bioprinting system is a crucial factor to creating complex tissue and organ structures. A lengthy printing process can lead to drying of the bioink, which results in cell death, and large tissue and organ constructs may not survive in an open environment. Thus, a printing process with high resolution and suitable bioink materials needs to be improved and developed, respectively, to maintain cell viability and assure long-term success of the bioprinting system.

Another challenge facing bioprinting systems is cell aggregation and sedimentation in the print cartridge reservoir and syringe. The bioprinting system makes cell-encapsulated droplets and filaments through a small nozzle. As a result, cell aggregation and sedimentation can occur in the nozzle during cell printing, which is an inherent problem facing bioprinting systems that require cells to be suspended in liquid bioink [53]. To address this issue, Parsa et al. mixed low-viscous surfactants and humectants in bioink; however, this method may damage cells and create additional challenges in printing tissue constructs for implantation [54]. Therefore, future development should be focused on new bioink materials that could prevent cell sedimentation and nozzle clogging during the printing process.

The pre-eminent challenge to creating tissues and organs in the laboratory remains the early vascularization to support the newly generated tissue and organ. To regenerate large-volume tissue and organ structures in the future, transportation of oxygen and nutrients inside 3D structures will be necessary to fabricate prevascularized 3D tissue and organ structures. Without a vascular structure in 3D tissue and organ constructs and vascular connection, printed cells cannot survive within the 3D construct to regenerate tissues and organs after implantation. Several researchers have fabricated vascular structures using 3D bioprinting systems and are reported in the literature [27,28,48,49]. Several studies have demonstrated the abilities of bioprinting systems to fabricate vascular structures that are larger than several millimeters; however, incorporating vascular structures into tissue and organ constructs has not yet been achieved. To develop and apply bioprinting technology for clinical use, many of challenges will need to be addressed.

6 SUMMARY

Using bioprinting technology, multiple cells, biomaterials, and biomolecules can be placed simultaneously in defined spatial locations. Because of its advantages, bioprinting technology has great potential to regenerate 3D tissue and organ constructs, and has gained favorable attention as a fabrication technology to produce 3D tissue and organ constructs. Many challenges remain with respect to fabrication of complex tissues consisting of multiple cell types that can be used to generate clinically applicable tissue and organ-sized structures with proper mechanical properties.

GLOSSARY

Additive manufacturing A manufacturing process through which three-dimensional solid objects are created using a series of additive or layered framework.

Stereolithography A 3D printing process that makes a solid object from a computer image by using a computer-controlled laser to draw the shape of the object onto the surface of liquid plastic.

Fused deposition manufacturing A type of additive manufacturing that enables the construction of 3D objects, prototypes, and products through a computer-aided or -driven manufacturing process.

Selective laser sintering An additive manufacturing technique that uses a laser to sinter powdered materials to create a solid structure.

Rapid prototyping Techniques used to rapidly fabricate a full-scale model using 3D CAD data.

CAD/CAM (computer-aided design/computer-aided manufacturing) Software used to design products and program manufacturing process for machines that make them in computer.

Jetting-based bioprinting A noncontact technique in which 2D and 3D structures are generated using picoliter bioink droplets layered onto a substrate.

Extrusion-based bioprinting A bioprinting method that dispenses continuous filaments of a material consisting of cells mixed with hydrogel through a micronozzle to fabricate 2D or 3D structures.

ABBREVIATIONS

2D	Two-dimensional
3D	Three-dimensional
BMP-2	Bone morphogenic protein-2
CAD	Computer-aided design
CAM	Computer-aided manufacturing
CT	Computerized tomography
HUVEC	Human umbilical vein endothelial cell
PCL	Polycaprolactone

REFERENCES

[1] Langer R, Vacanti JP. Tissue engineering. Science 1993;260(5110): 920–6.

[2] Hutmacher DW. Scaffold design and fabrication technologies for engineering tissues – state of the art and future perspectives. J Biomat Sci Polym E 2001;12(1):107–24.

[3] Leong KF, Cheah CM, Chua CK. Solid freeform fabrication of three-dimensional scaffolds for engineering replacement tissues and organs. Biomaterials 2003;24(13):2363–78.

[4] Hollister SJ. Porous scaffold design for tissue engineering. Nat Mater 2005;4(7):518–24.

[5] Yang SF, Leong KF, Du ZH, Chua CK. The design of scaffolds for use in tissue engineering. Part 1. Traditional factors. Tissue Eng 2001;7(6):679–89.

[6] Lo H, Ponticiello MS, Leong KW. Fabrication of controlled release biodegradable foams by phase separation. Tissue Eng 1995;1(1):15–28.

[7] Mooney DJ, Baldwin DF, Suh NP, Vacanti LP, Langer R. Novel approach to fabricate porous sponges of poly(D,L-lactic-co-glycolic acid) without the use of organic solvents. Biomaterials 1996;17(14):1417–22.

[8] Seol YJ, Kang TY, Cho DW. Solid freeform fabrication technology applied to tissue engineering with various biomaterials. Soft Matter 2012;8(6):1730–5.

[9] Melchels FPW, Domingos MAN, Klein TJ, Malda J, Bartolo PJ, Hutmacher DW. Additive manufacturing of tissues and organs. Prog Polym Sci 2012;37(8):1079–104.

[10] Peltola SM, Melchels FPW, Grijpma DW, Kellomaki M. A review of rapid prototyping techniques for tissue engineering purposes. Ann Med 2008;40(4):268–80.

[11] Marga F, Jakab K, Khatiwala C, Shepherd B, Dorfman S, Hubbard B, et al. Toward engineering functional organ modules by additive manufacturing. Biofabrication 2012;4(2):022001.

[12] Jakab K, Norotte C, Marga F, Murphy K, Vunjak-Novakovic G, Forgacs G. Tissue engineering by self-assembly and bio-printing of living cells. Biofabrication 2010;2(2):022001.

[13] Ferris CJ, Gilmore KG, Wallace GG, Panhuis MIH. Biofabrication: an overview of the approaches used for printing of living cells. Appl Microbiol Biot 2013;97(10):4243–58.

[14] Seol YJ, Kang HW, Lee SJ, Atala A, Yoo JJ. Bioprinting technology and its applications. Eur J Cardiothorac Surg 2014;46(3):342–8.

[15] Mironov V, Boland T, Trusk T, Forgacs G, Markwald RR. Organ printing: computer-aided jet-based 3D tissue engineering. Trends Biotechnol 2003;21(4):157–61.

[16] Xu T, Baicu C, Aho M, Zile M, Boland T. Fabrication and characterization of bio-engineered cardiac pseudo tissues. Biofabrication 2009;1(3):035001.

[17] Mohebi MM, Evans JRG. A drop-on-demand ink-jet printer for combinatorial libraries and functionally graded ceramics. J Comb Chem 2002;4(4):267–74.

[18] Boland T, Xu T, Damon B, Cui X. Application of inkjet printing to tissue engineering. Biotechnol J 2006;1(9):910–7.

[19] Le HP. Progress and trends in ink-jet printing technology. J Imaging Sci Technol 1998;42(1):49–62.

[20] Odde DJ, Renn MJ. Laser-guided direct writing for applications in biotechnology. Trends Biotechnol 1999;17(10):385–9.

[21] Lee W, Debasitis JC, Lee VK, Lee JH, Fischer K, Edminster K, et al. Multi-layered culture of human skin fibroblasts and keratinocytes through three-dimensional freeform fabrication. Biomaterials 2009;30(8):1587–95.

[22] Nahmias Y, Schwartz RE, Verfaillie CM, Odde DJ. Laser-guided direct writing for three-dimensional tissue engineering. Biotechnol Bioeng 2005;92(2):129–36.

[23] Chang CC, Boland ED, Williams SK, Hoying JB. Direct-write bioprinting three-dimensional biohybrid systems for future regenerative therapies. J Biomed Mater Res B 2011;98B(1):160–70.

[24] de Gans BJ, Schubert US. Inkjet printing of well-defined polymer dots and arrays. Langmuir 2004;20(18):7789–93.

[25] Khalil S, Sun W. Bioprinting endothelial cells with alginate for 3D tissue constructs. J Biomech Eng 2009;131(11):111002.

[26] Landers R, Hubner U, Schmelzeisen R, Mulhaupt R. Rapid prototyping of scaffolds derived from thermoreversible hydrogels and tailored for applications in tissue engineering. Biomaterials 2002;23(23):4437–47.

[27] Jakab K, Norotte C, Damon B, Marga F, Neagu A, Besch-Williford CL, et al. Tissue engineering by self-assembly of cells printed into topologically defined structures. Tissue Eng Part A 2008;14(3):413–21.

[28] Norotte C, Marga FS, Niklason LE, Forgacs G. Scaffold-free vascular tissue engineering using bioprinting. Biomaterials 2009;30(30):5910–7.

[29] Smith CM, Stone AL, Parkhill RL, Stewart RL, Simpkins MW, Kachurin AM, et al. Three-dimensional bioassembly tool for generating viable tissue-engineered constructs. Tissue Eng 2004;10(9–10):1566–76.

[30] Kang HW, Lee SJ, Atala A, Yoo JJ., inventors. Integrated organ and tissue printing methods, system, and apparatus. USA; 2012.

[31] Lee SJ, Kang HW, Park JK, Rhie JW, Hahn SK, Cho DW. Application of microstereolithography in the development of three-dimensional cartilage regeneration scaffolds. Biomed Microdevices 2008;10(2):233–41.

[32] Melchels F, Wiggenhauser PS, Warne D, Barry M, Ong FR, Chong WS, et al. CAD/CAM-assisted breast reconstruction. Biofabrication 2011;3(3):034114.

[33] Sun W, Starly B, Nam J, Darling A. Bio-CAD modeling and its applications in computer-aided tissue engineering. Comput Aided Des 2005;37(11):1097–114.

[34] Fedorovich NE, Alblas J, Hennink WE, Oner FC, Dhert WJA. Organ printing: the future of bone regeneration? Trends Biotechnol 2011;29(12):601–6.

[35] Phillippi JA, Miller E, Weiss L, Huard J, Waggoner A, Campbell P. Microenvironments engineered by inkjet bioprinting spatially direct adult stem cells toward muscle- and bone-like subpopulations. Stem Cells 2008;26(1):127–34.

[36] Fedorovich NE, Dewijn JR, Verbout AJ, Alblas J, Dhert WJA. Three-dimensional fiber deposition of cell-laden, viable, patterned constructs for bone tissue printing. Tissue Eng Part A 2008;14(1):127–33.

[37] Fedorovich NE, Schuurman W, Wijnberg HM, Prins HJ, van Weeren PR, Malda J, et al. Biofabrication of osteochondral tissue equivalents by printing topologically defined, cell-laden hydrogel scaffolds. Tissue Eng Part C Methods 2012;18(1):33–44.

[38] Ker EDP, Nain AS, Weiss LE, Wang J, Suhan J, Amon CH, et al. Bioprinting of growth factors onto aligned sub-micron fibrous scaffolds for simultaneous control of cell differentiation and alignment. Biomaterials 2011;32(32):8097–107.

[39] Schuurman W, Khristov V, Pot MW, van Weeren PR, Dhert WJA, Malda J. Bioprinting of hybrid tissue constructs with tailorable mechanical properties. Biofabrication 2011;3(2):021001.

[40] Cui XF, Breitenkamp K, Finn MG, Lotz M, D'Lima DD. Direct human cartilage repair using three-dimensional bioprinting technology. Tissue Eng Part A 2012;18(11–12):1304–12.

[41] Keriquel V, Guillemot F, Arnault I, Guillotin B, Miraux S, Amédée J, et al. *In vivo* bioprinting for computer- and robotic-assisted medical intervention: preliminary study in mice. Biofabrication 2010;2(1):014101.

[42] Xu T, Gregory CA, Molnar P, Cui X, Jalota S, Bhaduri SB, et al. Viability and electrophysiology of neural cell structures generated by the inkjet printing method. Biomaterials 2006;27(19):3580–8.

[43] Lee YB, Polio S, Lee W, Dai G, Menon L, Carroll RS, et al. Bioprinting of collagen and VEGF-releasing fibrin gel scaffolds for neural stem cell culture. Exp Neurol 2010;223(2):645–52.

[44] Owens CM, Marga F, Forgacs G, Heesch CM. Biofabrication and testing of a fully cellular nerve graft. Biofabrication 2013;5(4):045007.

[45] Lee V, Singh G, Trasatti JP, Bjornsson C, Xu X, Tran TN, et al. Design and fabrication of human skin by three-dimensional bioprinting. Tissue Eng Part C Methods 2014;20(6):473–84.

[46] Binder KW, Allen AJ, Yoo JJ, Atala A. Drop-on-demand inkjet bioprinting: a primer. Gene Ther Regul 2011;06(01):33–49.

[47] Koch L, Deiwick A, Schlie S, Michael S, Gruene M, Coger V, et al. Skin tissue generation by laser cell printing. Biotechnol Bioeng 2012;109(7):1855–63.

[48] Nakamura M, Kobayashi A, Takagi F, Watanabe A, Hiruma Y, Ohuchi K, et al. Biocompatible inkjet printing technique for designed seeding of individual living cells. Tissue Eng 2005;11(11–12): 1658–66.

[49] Cui XF, Boland T. Human microvasculature fabrication using thermal inkjet printing technology. Biomaterials 2009;30(31):6221–7.

[50] Kolesky DB, Truby RL, Gladman AS, Busbee TA, Homan KA, Lewis JA. 3D Bioprinting of vascularized, heterogeneous cell-laden tissue constructs. Adv Mater 2014;26(19):3124–30.

[51] Chang R, Sun W. Effects of dispensing pressure and nozzle diameter on cell survival from solid freeform fabrication-based direct cell writing. Tissue Eng Part A 2008;14(1):41–8.

[52] Nair K, Gandhi M, Khalil S, Yan KC, Marcolongo M, Barbee K, et al. Characterization of cell viability during bioprinting processes. Biotechnol J 2009;4(8):1168–77.

[53] Khatiwala C, Law R, Shepherd B, Dorfman S, Csete M. 3D cell bioprinting for regenerative medicine research and therapies. Gene Ther Regul 2012;7(1):1230004.

[54] Parsa S, Gupta M, Loizeau F, Cheung KC. Effects of surfactant and gentle agitation on inkjet dispensing of living cells. Biofabrication 2010;2(2):025003.

Chapter 17

Bioprinting of Bone

Michael Larsen, Ruchi Mishra, Michael Miller, and David Dean

Department of Plastic Surgery, The Ohio State University, Columbus, OH, USA

Chapter Outline

Keywords: additive manufacturing; 3D printing; bioprinting; nano-technology; tissue engineering; bone reconstruction; regenerative medicine

ABSTRACT

Bone is a vital tissue of the body and a key component of the musculoskeletal system, which routinely overcomes insult and degeneration through its highly regenerative capabilities. However, in certain circumstances bone fails to properly heal and treatment is required. The current standard of care therapies have significant shortcomings. Advances in bioprinting and tissue engineering hold much promise for improving both the quality and efficiency of bone reconstructive therapies. Bioprinting, technically, does not yet occur because custom resorbable scaffolds are not currently constructed with growth factors or autologous cells; however, recent research has been rapidly expanding the 3D printing techniques as well as 3D printable materials with properties finely tuned to reconstruct patient-specific defects. This research seems to be leading toward a future of resorbable scaffolds that incorporate appropriate growth factors and osteoprogenitor cells. Time will reveal whether these therapies will become widely available and standard-of-care.

1 INTRODUCTION

Bone is a vital tissue of the body, performing important functions such as hematopoiesis in the bone marrow region as well as homeostasis and storage of minerals such as calcium and phosphorus [1]. Bone is also a key component of the musculoskeletal system, where it acts as a structural support, aiding in movement and protecting internal organs. In these roles, bone is subject to insult and degeneration, which it routinely overcomes through its high regenerative capability. However, in certain circumstances bone fails to properly heal and medical or surgical treatment is required. Bone and joint regenerative medicine consists mainly of

Essentials of 3D Biofabrication and Translation. http://dx.doi.org/10.1016/B978-0-12-800972-7.00017-7

traditional reconstructive surgical procedures with an increasing penetration of alloplastic hardware, biomaterials, and bone substitutes. The increased use of these therapies highlights the shift of regenerative research and innovation toward defect-specific therapies in recent years and away from whole organ engineering. Tissue engineering research is active in the elucidation of techniques in order to fully synergize bone replacement scaffolds, growth factors, and osteoprogenitor cells in the restoration of bone defects. Bioprinting consists of constructing custom scaffolds with these biological components. This currently does not occur; however, products that combine growth factors with scaffolds, such as Infuse® (Medtronic, Minneapolis, MN), are frequently used in the clinic. Furthermore, the development of biomaterials with various properties that can be fabricated using 3D printers is an intense area of research. This research seems to be leading toward a future of resorbable scaffolds that incorporate appropriate growth factors and osteoprogenitor cells.

Section 2 will assess the need for bone grafts and substitutes, Section 3 is a review of the relevant bone biology, Section 4 is a look at wound healing and the range of bone regenerative therapies, and finally we state our conclusions from this review of bone bioprinting.

2 CLINICAL NEED FOR BONE GRAFT AND BONE SUBSTITUTE

The human skeletal system is susceptible to injury, disease, and degeneration. In fact, musculoskeletal problems make up three of the four most commonly reported medical conditions in the United States, including low back pain (62 million people), chronic joint pain (61.6 million), and arthritis (51.2 million) [2]. Musculoskeletal problems represent the most common cause of reduced quality of life and impairment and lead to the most amount of lost work days and medical bed days. The total direct (treatment and care) and indirect (lost wages) costs of musculoskeletal conditions are estimated to be $287 billion [3]. Improved bone regenerative treatment strategies could have dramatic economic benefits.

2.1 Causes of Bone Deficits

Bone deficits arise from various causes, including trauma, cancer, congenital and developmental defects, arthritis and other rheumatic disease, and infection and decay. Each of these will be presently addressed.

2.1.1 Trauma

Approximately 16.2 million fractures were treated by healthcare professionals in 2006–2007. The majority of these occurred in the upper extremity, with the forearm

bones (radius or ulna) being the most commonly fractured and accounting for about 15% of all fractures [3]. Approximately 1.5 million fractures are secondary to osteoporosis, with a lifetime risk of an osteoporotic fracture in women of 40–50% and in men 13–22% [4,5]. Instrumentation is often needed to stabilize the fracture and promote proper healing. Failure of fracture healing (nonunion) occurs in about 2.5% of fractures, necessitating further treatment, often with bone grafting [6]. Nonunion is usually secondary to vascular compromise, instability (mobile fracture pieces), infection, or intervening soft-tissue. Additionally, many high-impact traumatic injuries result in a critical size [7] defect that will not heal on its own and require grafting or a bone substitute.

2.1.2 Cancer

Bone and joint cancers are fairly rare, with an estimated 3020 new cases in the United States in 2014, representing just 0.2% of all new cancers. These cancers have a large economic and emotional impact as they are frequently diagnosed in younger persons, with 26.9% of new cases occurring in patients younger than 20 years old. Also, soft tissue sarcomas (>12,000 per year) and oral and pharyngeal cancers (>40,000 per year) can be in close proximity to bone, requiring bone resection for complete ablation and then reconstruction for restoration of structure [8]. Furthermore, an additional 280,000 cases of bone metastases were estimated to have occurred in the United States in 2008 [9]. These patients can suffer from pathologic fractures that may require surgical fixation or fusion.

2.1.3 Congenital and Developmental Deformities

It is estimated that 38,000 infants in the United States undergo surgery yearly to treat congenital defects. Craniofacial and skeletal abnormalities make up a large number of these defects, including skeletal dysplasias (1.2–2.2 of every 10,000 infants in the United States), limb deficiencies (3.9–5.7), clubfoot (8.9–11.5), spina bifida (2–3.3), developmental dysplasia of the hip (4.9–6.9), cleft lip and palate (~10), craniosynostosis or early fusion or nonformation of cranial sutures (~5–10), and scoliosis or abnormal curvature of the spine (~10) [10]. Many other children go on to develop deformities, such as idiopathic scoliosis during their infancy, childhood, or adolescence (approximately 2–3% of 10–16 year olds). Still others can form pediatric spondylosis, a deterioration of a vertebra (approximately 4% of children), or adult spinal deformities or degenerative scoliosis. Not all patients with such deformities require intervention, but approximately 18,000 undergo a surgical procedure for scoliosis each year, which usually consists of placing titanium rod fixation to correct and/or stabilize the patient's posture [3].

2.1.4 Arthritis and Other Rheumatic Diseases

Arthritis, also known as degenerative joint disease, affects the cartilage and other joint tissues and is a heterogeneous disease composed of osteoarthritis, rheumatoid arthritis, and greater than 100 other forms of arthritis. Arthritis and other rheumatic diseases affect about 22.7% (52.5 million) of adults in the United States, and it is projected that this number will rise to 25% by 2030 [11]. It is the leading cause of disability in adults, and resulted in $128 billion in medical care costs and lost wages in 2003 [12,13]. Over 1 million joint replacement surgeries are performed yearly, with knee and hip replacements accounting for 96% of the cases [3].

2.1.5 Infection and Decay

Osteomyelitis, an infection of the bone, occasionally occurs in patients with vascular compromise secondary to diabetes or athcrosclerosis, the bedridden that are susceptible to pressure ulcers, patients with sepsis or burns, patients with traumatic bone injuries, or in patients with bone fixation hardware or implants (e.g., knee replacement). Treatment includes antibiotic therapy and often surgical debridement and reconstruction.

Additionally, dental caries are ubiquitous, affecting all adults. In the United States dental decay and periodontal disease lead to tooth loss in 15–20% of 35–44 year olds and to full edentulism in 30% of those 65–74 years old. Furthermore, alveolar process bone mass is maintained by anatomic masticatory forces; thus, tooth loss subsequently causes alveolar resorption. Bone regeneration and/or augmentation may be needed for alveolar reconstruction to have sufficient mass to accept dental implants.

2.2 Bone Biologic Industry Size

The global bone biologic market was estimated to be $4.3 billion in 2009, with the United States representing the largest share at $2.3 billion. This market is currently dominated by Medtronic (19% market share), Genzyme (11%), and Depuy (10%). Because of the increasing incidence of arthritis, fueled by higher rates of obesity and sedentary lifestyle, this market is forecasted to grow 12% yearly, reaching an estimated value of $9.6 billion globally and $6 billion in the United States by 2016 (Orthobiologics Market Report).

2.2.1 Health Policy and Industry

Healthcare spending is greater in the United States than in any other country, accounting for 17.9% of the US gross domestic product (GDP) in 2012. The countries with the next largest GDP spend considerably less: China 5.4%, Japan 10.1%, Germany 11.3%, France 11.7%, and the UK 9.4% [14]. In 2011, the United States spent $8508 per capita on healthcare, approximately two-and-a-half times more than the $3339 per capita average of other developed countries

[15]. Despite this elevated spending, healthcare quality is poorer than expected, with the United States ranking 37th in overall health system performance, 72nd in level of the population's health, and 35th in life expectancy [16–19]. This disconnect between spending and quality has led to the political will for health policy change and the passing of the Patient Protection and Affordable Care Act (ACA) of 2010. It is uncertain how these reforms will affect future healthcare markets, including the regenerative therapy and medical device markets. Included in the ACA reforms is a change from fee-for service payments, where volume is incentivized, to pay-for-performance and bundled payments, where quality and fewer interventions are incentivized. In effect, this turns repeated surgical interventions into costs rather than a profit for providers, driving the medical device industry to innovate more biocompatible and longer lasting devices and implants. An additional market force imposed by the ACA is a 2.3% medical device tax. Medical device companies argue that this will stifle innovation and lead to the loss of American biomedical jobs. Lawmakers and analysts counter that the increased business secondary to the expansion of healthcare coverage to more Americans will offset the tax. This will likely have varying benefits on different devices, increasing volume for elective devices, like knee replacements, more than for devices like cardiac stents and pacemakers that have been traditionally placed in an emergency setting, irrespective of whether a patient has insurance (Wells Fargo Analysis) [20]. Time will reveal the full impact of the policy changes; however, because of the vast need for bone regenerative therapies, a market will continue to exist, regardless of the shuffling that takes place.

3 BONE BIOLOGY

Bone is composed of minerals – mainly calcium phosphate (85%) and calcium carbonate (10%), a complex matrix of intricately intertwined collagen fibers, periosteum, blood vessels, and bone cells [21]. In this section we discuss the growth, development, and healing patterns and processes of bone tissue.

3.1 Embryology

The process of bone formation is orchestrated by multiple cascades of events involving the activity of mesenchymal cells, cytokines, and growth factors [22,23].

3.1.1 Ossification and Bone Anatomy

During embryonic development, the first step toward bone formation is condensation of the mesenchyme. Thereafter, mesenchymal cells either directly differentiate into osteoblasts via intramembranous ossification to form the flat bones of skull, clavicle, scapula, and so on. Alternatively, mesenchymal cells differentiate into osteoblasts through an intermediate chondrocytic stage to form long bones via endochondral ossification. The newly formed collagenous bone is known as *woven bone*, which further mineralizes and becomes organized into *lamellar bone* [24,22,25]. The bone formed via these processes can be broadly categorized into three main components: (1) bone cells, (2) bone periosteum and extracellular matrix, and (3) bone marrow [26,27]. The cells present in bone are osteoblasts (immature, produce extracellular matrix), osteocytes (mature, trapped in mineralized extracellular matrix), osteoclasts (bone resorbing cells), bone lining cells (these cells maintain the blood–bone barrier), and osteoprogenitors (stem cells committed toward osteoblastic lineage) [28]. In compact bone, osteocytes and the vascular tissue forming Haversian and Volkmann's canals make up the osteons of lamellar bone (Fig. 17.1).

These components are arranged in two different structural organizations leading to cortical (dense) bone, which is present at the peripheral region and trabecular (porous) bone, which is present internal to cortical bone [28]. Furthermore, the thickening of these bones differs according to the requirements at different locations of bone in the body. Thick areas of less strong trabecular bone are found internal to the lamellar bone at the ends of the long bones. Deep in both of those spaces is the marrow space, which includes hemopoietic tissue (i.e., "red" marrow) (Fig. 17.2a). Large cavities of red marrow are found in the long bones of children. The deep space of adult long bones is likely to be filled with more fat than red marrow. Pools of red marrow are found in the adult pelvis, cranium, vertebrae, and sternum (Fig. 17.2b).

3.1.2 Bone Periosteum and Extracellular Matrix

The external surface of bone is covered by a connective tissue referred to as the periosteum. Periosteal membranes are known to house bone progenitor cells such as mesenchymal stem cells (MSCs). The extracellular matrix of bone is unique. Unlike other tissues it contains a significant amount (50–70%) of inorganic content, consisting largely of calcium-phosphate-based hydroxyapatite [$Ca_{10}(PO_4)_6(OH)_2$] crystals along with some magnesium, carbonate, and acid phosphate. The organic component is comprised of Type I collagen and noncollagenous proteins such as bone sialoprotein, osteopontin, osteocalcin, and osteonectin [29].

3.1.3 Bone Marrow

Marrow fills much of the medullary cavities present at the innermost region of bones and performs the function of hematopoiesis, that is, formation of blood cells. It is a vascular environment containing stromal cells (i.e., connective tissue including pericytes and fibroblasts) and stromal stem cells (i.e., MSCs) (Fig. 17.3) [30].

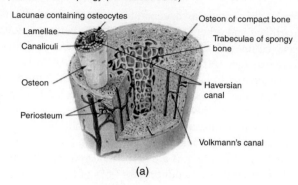

Compact bone and spongy (cancellous bone)

(a)

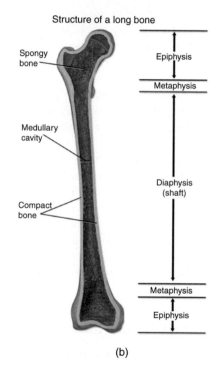

Structure of a long bone

(b)

FIGURE 17.1 Bone architecture. (a) Thick areas of trabecular bone are found internal to the cortical bone at the ends of long bones (*from Wikipedia: http://en.wikipedia.org/wiki/File:Illu_compact_spongy_bone.jpg [accessed 05.11.14]*). (b) Osteons are arrayed in layers or lamellae in the cortex (i.e., outer shell), the strongest part, of bones in the adult skeleton (*from Wikipedia: http://en.wikipedia.org/wiki/File:Structure_of_a_Long_Bone.png [accessed 05.11.14].*)

4 WOUND HEALING

Bone is a key constituent of the musculoskeletal system along with other tissues such as cartilage, muscle, tendon, and ligaments. This system transmits permissible amounts of strain from one part of the body to another and then to the environment during motion and rest. In instances where the load (stress) exceeds the permissible limit (i.e., the bone is not sufficiently stiff), strain can no longer be successfully transmitted and the bone yields by fracturing [31,32].

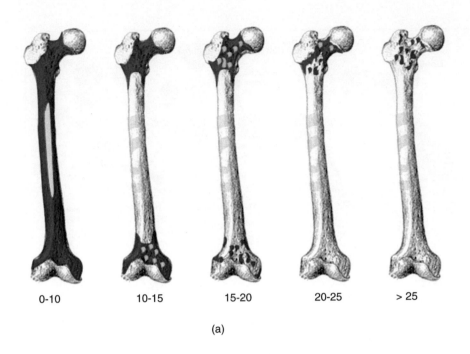

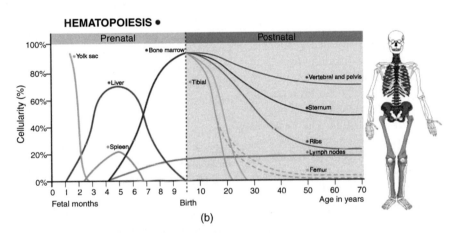

FIGURE 17.2 **Hemopoietic tissue (red marrow) is found in the deep marrow space of bones.** Red marrow is initially concentrated in the long bones, especially the femur and tibia in children. The bulk of the red marrow is found in the pelvis, vertebrae, sternum, and ribs in adults *((a) reproduced with permission from: http://posterng.netkey.at/esr/viewing/index.php?module=viewing_poster&task=viewsection&ti=348191 [accessed 05.11.14]; (b) reproduced with permission from: http://en.wikipedia.org/wiki/File:Hematopoesis_EN.svg [accessed 05.11.14].)*

Generally, if a fracture site is immobilized, it heals on its own within 6–8 weeks following an injury [33].

Fracture healing is a very complex and chronologically organized physiological process. It involves three phases: (1) an inflammatory phase, (2) a reparative phase, and (3) a remodeling phase [33]. The inflammatory phase starts between 12 h and 14 h following a fracture, peaks at 48 h, and subsides at one week. It is accompanied by the formation of a hematoma, the activation of clotting factors and the migration and infiltration of various cells like endothelial cells, inflammatory cells, and fibroblasts from the perivascular spaces. The macrophages at the fracture site release various cytokines and growth factors, which initiate the healing process [34]. The second phase, the reparative phase, commences within 7–9 days and lasts for weeks. Initially, an avascular soft callus is formed via cartilaginous tissue. The callus is later replaced by vascularized woven bone tissue known as a bony or hard callus. Finally, in the third phase, woven bone is remodeled to form lamellar bone. This remodeling phase lasts for months to years [26,35].

These phases of bone healing are analogous to the wound healing of other bodily tissues. However, bone has the unique ability to truly regenerate, healing without a scar. This ability naturally occurs in fractures and defects with minimal space between the bone ends. However, if the gap is too large, also known as a critical size defect, or if there is mobility between the fracture ends, the bone will not regenerate but will either form a fibrous scar or fail to unite.

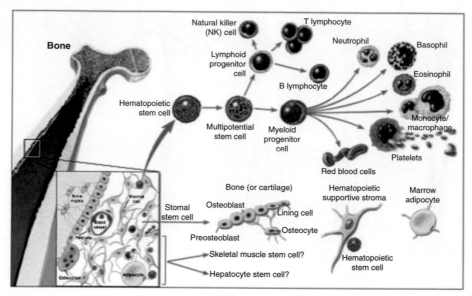

FIGURE 17.3 **Hematopoietic stem cells and stromal stem cells.** *(From: http://beyondthedish.wordpress.com/2013/04/14/the-nooks-and-crannies-in-bone-marrow-that-nurture-stem-cells/ [accessed 05.11.14].)*

4.1 Procedures

The ultimate goal of osseous reconstructive interventions is to augment the mentioned regenerative capabilities of bone in order to restore the body to preinsult status or to make whole a congenital deformity. Current treatment modalities often fall short of this, but aim to at least aid the patient in regaining as much function and quality of life as possible. The treatment of bone deficits depends upon location, vascular supply, presence of infection, and the size of the defect (i.e., fracture vs bone gap vs amputation). Treatment options include autogenous graft (autograft), allogeneic graft (allograft), prostheses, and/or alloplast. These are individually discussed next.

4.1.1 Autograft

Autografts have traditionally been considered the gold standard treatment for critical size defects because they have the properties of an ideal biomaterial: biocompatible, nontoxic, nonallergenic, noninflammatory, mechanically reliable, resistant to microorganism growth, provide permanency, and consistently reproducible results [36]. Autogenous bone also provides three components essential for bone regeneration: (1) a scaffold to organize and support bone growth (osteoconduction), (2) growth factors to recruit and induce osteoprogenitor cells (osteoinduction), and (3) stem and osteoprogenitor cells to lay down bone (osteogenesis). Nonvascularized bone grafts can be used in small defects with ideal wound bed conditions and can be taken from the iliac crest, tibial plateau, outer table of the calvarium, or the olecranon. Nonvascularized bone grafts are frequently used for joint fusions (e.g., spinal fusion), also called arthrodesis.

Vascularized bone grafts are used to fill defects >6 cm or defects with a compromised wound bed, such as when infection is present or when vascularity has been damaged by trauma or radiation from cancer treatment. Common donor sites are the fibula, rib, iliac crest, and radius [37–42]. Distraction osteogenesis or bone transport, while not exactly autografting, is an ingenious technique that uses bone's regenerative capacity to fill critical-size defects, especially in the extremities, or to reconstruct hypoplastic bone (e.g., treating retrognathia, an "overbite," by distracting/elongating the mandible). An additional osteotomy is made and these segments are distracted 1 mm/day, slow enough to allow bone to be formed in the distraction zone. This is continued until the original defect is closed [43,44].

Although an autograft has ideal properties, it is not without its disadvantages, which include a limited supply, a significant increase in operating time, variability and inaccuracy of graft shaping and inset, requirement for larger operative teams, and a donor site with its complications (e.g., infection, pain, delayed wound healing) [45]. Therefore, improved regenerative options can vastly improve upon the inefficiencies and difficulties of autografting.

4.1.2 Allograft

Allografting is an alternative regenerative option that challenges the long-held paradigm of autografting as the gold standard. Bone allografts are derived from cadaveric tissue and come in three main variants: (1) fresh bone, (2) freeze-dried bone allograft, and (3) demineralized freeze-dried bone allograft. With regards to the three components of regenerative therapies, allografts act as a good scaffold, retain some osteoinductive capabilities (especially demineralized bone), but lack osteogenic capabilities as the cells are removed during tissue processing. In order to theoretically augment the osteoinduction of allografts, autologous products with growth factors or osteoprogenitor cells, such as platelet-rich plasma or bone marrow aspirate, have been added. These additives may be beneficial, but results have varied, demanding higher-powered, well-designed clinical trials [46,47].

An analogous procedure to allografting is allotransplantation, where the donated tissue retains live cells and, therefore, is capable of osteogenesis. Upper extremity, lower extremity, and face transplants have been successfully performed. A consequence of the live cells in the transplant is that the patient must be on lifelong immunotherapy to suppress rejection of the transplanted tissue. For large craniofacial defects or amputations, obturators and prostheses are viable noninvasive options. Exciting advances in the biointegration of prostheses are being made that allow prostheses to respond to the patient's nerve impulses.

4.1.3 Alloplast: History of Bone Substitute Materials

With the disadvantages of autografts, allografts, and prostheses, advances in alloplasts hold promise for improving patient care and containing healthcare costs. This section surveys the biological and material properties of the metals, ceramics, and polymers currently used to fabricate bone implants (Table 17.1). Section 4.2 introduces the tissue engineering framework and the role these materials play in it. Section 4.3 discusses the traditional fabrication processes used with these materials. Section 4.4 discusses the additive manufacturing (AM) (3D printing) fabrication processes that are currently in use or in development for bone implants.

4.1.3.1 Metals

Pure metals are rarely used in implants because they usually degrade more quickly and high concentrations of metal ions cause adverse reactions in the body. Rather, alloys are usually used as implant metals. Except for the skull, where polymers are often used, most metal musculoskeletal implants and many fixation devices are currently fabricated from either surgical grade 5 titanium (Ti-6Al-4V) or nitinol (NiTi). Titanium of grades 1–4 are commercially pure and used much less often in biomedical implants. Implants continue to be made from stainless steel, cobalt–chromium, and less commonly tantalum. These metals are used because their stiffness is reasonably close to bone [49] and they are corrosion resistant [50] in the body. Surgical grade 5 titanium is almost half as stiff as stainless steel, cobalt–chromium, and tantalum. NiTi, used in stents and orthodontic wire, is almost half as stiff as Ti-6Al-4V. NiTi also exhibits shape memory properties. Shape memory allows NiTi implants to be deformed after refrigeration and then return to their original shape at body temperature, a useful property for the deployment of stents [51]. Recently, there has been research into the possibility of controlled corrosion and degradation of primarily magnesium alloy metallic implants [52]. In the future these implants might allow bone immobilization hardware or stents

TABLE 17.1 Three Major Classes of Biocompatible Implant Materials are Polymers, Ceramics, and Metals

Implant Materials
Metals
Stainless steel
Vitallium (cobalt-chromium)
Titanium
Gold
Calcium ceramics
Hydroxyapatite
Tricalcium phosphate
Hydroxyapatite cement
Bioactive glass
Polymers
Silicone
Polymethylmethacrylate
Hard tissue replacement (HTR) polymer
Polyesters (Dacron, Mersilene)
Biodegradable polyesters (polyglycolic acid, Poly-L-lactic acid)
Polyamides (Supramid, Nylamid)
Polyethylene (Medpor)
Polypropylene (Prolene, Marlex)
Cyanoacrylates
Polytetrafluoroethylene (Teflon, Gore-Text)
Biologic materials
Collagen
AlloDerm
Discontinued materials
Polyurethane
Proplast

Some of the most commonly used biomaterials, followed by materials of biological origin and materials that are no longer in use.
Reproduced with permission from Ref. [48].

to resorb after performing the desired function, thereby saving a surgical intervention to remove the device. The major musculoskeletal applications for metals are as fixation plates and screws for surgical reconstruction or trauma repair.

4.1.3.2 Ceramics

The high ductility and strength of ceramics have made them useful as joint surfaces in some implants. Ceramics have three phases: (1) crystalline, (2) glass, and (3) porous [53]. Introduction of porosity can reduce strength and ductility, but it may also provide opportunities for the inclusion of absorbable materials that are subsequently released over time [54]. Porosity can also provide texture, which can facilitate tissue infusion or attachment. Ceramics used as bone implants, such as those used for dental alveoli exposed during tooth extractions, are often resorbable. It has been found that tissue infilling can provide a positive feedback, increasing the resorption rate [55].

4.1.3.3 Polymers

Polymers are the most extensively used alloplastic materials [48]. Polymers may be of natural or synthetic origin. Polymeric resorbable sutures are extensively used in medicine. Polymer chains are created from repeating units, monomers, in a process termed polymerization. The resulting polymer chains or strands can be further assembled by forming cross-links between them. Once cross-linked, previously liquid polymer strands may take the form of a gel or a solid. Polymers can be water soluble and may degrade in the body by hydrolysis, enzymatically, or by a stimulus-response mechanism [56].

The artificially occurring polymers that have been used most frequently are polyesters such as polylactides, polyglycolides, polycaprolactone, and poly(propylene fumarate). They are hydrophobic with properties that vary widely depending on molecular weight and crystallinity. One of the most-studied polyesters in bone tissue engineering research is poly(propylene fumarate) [57]. Another group of well-studied polymers utilizes poly(ethylene glycol) (PEG). PEGs resist adsorption of proteins and are nonimmunogenic. It is also relatively easy to add functional molecules to PEG. Poly(acrylates) and poly(methacrylate)s have traditionally been used to prepare insoluble cements and sealants, but more recently have been used in tissue engineering [58]. Naturally occurring polymers, such as collagen, chitosan, and polypeptides including extracellular matrix components, have also been studied as source materials for tissue engineering scaffolds [59].

4.2 Bone Tissue Engineering Strategies

Tissue engineering, as previously stated, is a three-pronged approach: scaffolds, cells, and growth factors. It is possible to use any of these to recruit tissue from the host into a defect site. In small, subcritical-size wounds, the host will naturally recruit any of these components that are missing. However, the larger the defect, the more necessary scaffolding and prompt vascularization become. While incoming tissue and vascularity are critical, large bone wounds often require protection of the space from invasion of soft, especially fibrous, tissue to fully regenerate. Thus, immobilization and strategic barrier membranes may be useful. It is possible to supply both of these needs by growing extracellular matrix on a scaffold that incorporates sufficient growth factors and attracts host cells. That scaffold can also house an artificial or natural graft that has been prevascularized *in vitro* in an incubator-sited bioreactor or in an *in vivo* (implanted) bioreactor (Fig. 17.4).

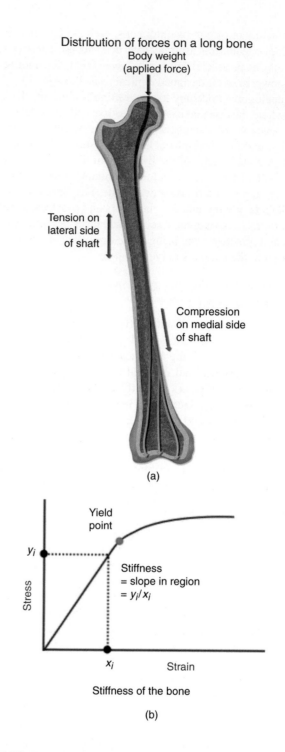

(a)

(b)

FIGURE 17.4 Bone biomechanics. (a) Areas of cortical bone respond to the normal stress-strain trajectories encountered by the body in motion and at rest *(from: http://upload.wikimedia.org/wikipedia/commons/0/09/Blausen_0401_Femur_DistributionofForces.png [accessed 06.02.15].)* (b) The stiffness of a bone is a function of its ability to accept stress and transmit strain. As loads increase a point is reached where it will yield (fracture) *(from: http://www.pt.ntu.edu.tw/hmchai/Biomechanics/BMmaterial/BMbone.htm [accessed 06.02.15].)*

4.2.1 Scaffolds

After the tissues' initial differentiation (Fig. 17.5a), bone tissue is unlike many other tissues in the body in that it needs significant remodeling to become and remain strong [60]. The signaling pathways important to conversion of MSCs to osteoblasts and then to mature (strong) bone are shown in Fig. 17.5a [61]. The remodeling process responds to local forces, as described by Wolff [62]. Unfortunately,

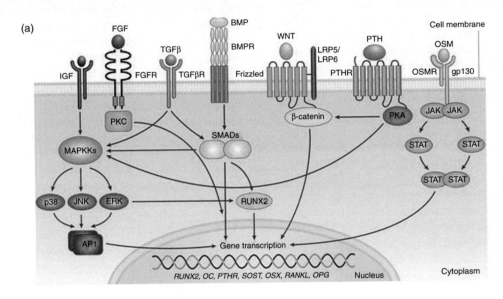

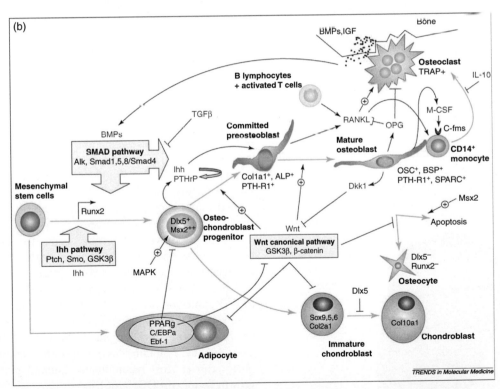

FIGURE 17.5 Bone formation and bone remodeling. (a) Bone formation requires activation of osteoblasts, which requires bone morphogenetic proteins (BMP) and their receptors via SMAD proteins. SMAD proteins activate runt-related transcription factor 2. The wingless-related MMTV frizzled pathway uses β-catenin and parathyroid hormone (PTH) and its receptor signaling that is mediated by protein kinase-A *(from Ref. [60])*. (b) Following a bone fracture or injury mesenchymal stem cells actually home to the site. IHH induces the expression of Runx2. BMP-2 is necessary to cause maturation of the MSCs into osteoblasts through Runx2 and Dlx5 induction. Mitogen-activated protein kinase can phosphorylate Runx2 and Dlx5. Osteochondroblast progenitors express IHH that induces secretion of PTHrP that causes pre-osteoblasts (positive for collagen 1a1, alkaline phosphatase, and PTH-R1) to mature further. Msx2 is found in proliferative progenitors whereas Dlx5 leads to maturation; both Msx2 and Dlx5 compete for DNA binding. Thus their ratio drives bone maturation. Osteoclasts are necessary for bone remodeling of cortical bone to produce strong bone *(from Ref. [63])*.

scaffolds that attempt to provide rigid support on implantation may physically block osteoclast-driven remodeling (Fig. 17.5), that needs to occur for bone to become strong, or shield regions of the construct from strain [63]. The lack of strain, referred to as stress-shielding, will also prevent the remodeling needed to cause formation of osteons within new lamellae of the cortex. As bone tissue engineering begins to contribute to regenerative medicine, scaffolds will play a role in immobilization and/or permanent structural hardware and alloplastic joints.

All of the material types reviewed in Section 4.1.6, metals, ceramics, and polymers, have been studied as resorbable bone tissue engineering scaffolds. Many scaffolding materials have been studied *in vitro* in terms of cell attachment, and a large percentage of those materials have been studied *in vivo* in terms of reconstructing particular defects. Summarizing that work would be akin to summarizing all bone tissue-engineering research, a topic that is far beyond the scope of this chapter. However, there is very little work published on how scaffolds would be fabricated for patients, the subject of this chapter. It is possible that this work is occurring in a corporate setting, perhaps protected as a trade secret. However, to date, there has not been an effusion of bone tissue engineering products.

Indeed, very few bone tissue-engineering products have been introduced to the clinic. The most prominent product is Medtronic's (Minneapolis, MN) Infuse®. This product places a biologically derived polymer scaffold, a bovine collagen carrier of recombinant human bone morphogenetic protein 2 (rhBMP-2), into a titanium intervertebral cage (Fig. 17.6). This product has seen the most widespread *in vivo* use of a recombinant human growth factor in a bone tissue-engineering product.

FIGURE 17.6 Medtronic Infuse® titanium cage, bovine collagen sponge, and rhBMP-2 used for intervertebral fusion. *(From: Medtronic spine surgery patient brochure, https://www.infusebonegraft.com/spine_patient_brochure.pdf [accessed 05.12.14].)*

4.2.2 Growth Factors

The term "growth factors" is synonymous with signaling molecules. The signaling pathways important to the conversion of MSCs to osteoblasts and then to mature (strong) bone are shown in Fig. 17.5a. Bone formation requires activation of osteoblasts, which require bone morphogenetic proteins (BMP) and their receptors via SMAD proteins. The term SMAD refers to proteins that are similar to either *Drosophila* "mothers against decapentaplegic" protein [MAD] or *Caenorhabditis elegans* protein SMA [61]. Following a bone fracture or injury, MSCs actually home to the site. Indian hedgehog (IHH) induces the expression of Runx2. BMP-2 is necessary to cause maturation of the MSCs into osteoblasts through Runx2 and distal-less homeobox 5 (Dlx5) induction. Mitogen-activated protein kinase can phosphorylate Runx2 and Dlx5. Osteochondroblast progenitors express IHH, which induces secretion of parathyroid hormone-related protein (PTHrP) that causes preosteoblasts (positive for collagen 1a1, alkaline phosphatase, and parathyroid hormone 1 receptor (PTH-R1)) to mature further. Msh-like 2, homeobox (Msx2) is found in proliferative progenitors whereas Dlx5 leads to maturation; Msx2 and Dlx5 compete for DNA binding. Thus, their ratio drives bone maturation. Osteoclasts are necessary for bone remodeling of cortical bone to produce strong bone [60,63]. Many of the same signaling pathways are involved in the bone remodeling (Fig. 17.5b) that follows the initial cellular response seen when bone is produced.

The Infuse product and other Food and Drug Administration (FDA) approved products allow the administration of rhBMP-2. Similarly, Medtronics' rhBMP-2 has received indications for use elsewhere in the body for the purpose of generating bone. There have been concerns about the safety and expense of this product [64]. There are other bone growth factors under investigation within and outside the United States; however, none have been approved by the FDA for use in the United States.

There are two interesting bone tissue-engineering products that have been studied in-depth. The first uses collagen I as a carrier for BMP-7 and has been registered as OP-1®. OP-1 was originally developed by Stryker Biotech (Kalamazoo, MI) and has been acquired by Olympus (Hopkinton, MA) [65]. A second bone substitute product that has been investigated, Augment® Bone Graft (BioMimetic Therapeutics, Franklin, TN), uses a Beta-tricalcium phosphate carrier for recombinant human platelet-derived growth factor (rhPDGF) [66].

4.2.3 Cell-Based Therapies

According to the FDA, there are no approved cell-based bone-tissue-engineering therapies. The FDA does not approve of collecting and transplanting bone progenitor cells to a defect site if it is considered more than minimal manipulation. The FDA considers cells taken from the patient "that are minimally manipulated, labelled or advertised for homologous use only, and not combined with a drug or device" as outside their jurisdiction. However, this is unlikely to cover MSCs. The FDA has tried to regulate Regenerative Sciences, Inc. (Broomfield, CO) from providing Regenexx™, a product using autologous bone-marrow-derived stem cells for nonunion fractures. Similarly, another company is seeking to make Osteocel Plus® (NuVasive, Inc., San Diego, CA) available for clinical use. This product combines autologous bone-marrow-derived osteoprogenitor cells with allograft material composed of demineralized bone matrix and cancellous bone [67].

4.2.4 Bioreactors

Bioreactors are chambers in which a tissue-engineered construct (i.e., cells, growth factors, and scaffolding material) is cultured prior to placement in the defect site. Bioreactors give seeded cells the opportunity to attach and proliferate (i.e., coat) on the scaffold. They can also allow those cells to mature and secrete extracellular matrix, an important event in bone formation. Bioreactors can be set up outside the body in incubators or inside the body (i.e., *in vitro* or *in vivo*). Incubator-based bioreactors allow the control of gas, nutrient, and signaling molecule levels [68]. While there have been attempts to coculture bone progenitor cells (e.g., MSCs) and blood vessel progenitor cells (e.g., endothelial cells), prevascularizing an implant remains a challenge [69]. Tissue grown in a compartment inside the body can be readily associated with a blood vessel that can become a vascular pedicle for the neobone and thus facilitate transplantation to a defect site [70].

4.3 Subtractive and Other Non-AM Techniques

Traditional fabrication strategies allow the creation of internal porous geometries at the resolution needed to control cell attachment and/or host tissue invasion as well as a scaffold's resorption profile. Traditional technologies include the use of standard molds, often combining polymers and ceramics with salt crystals, referred to as porogens, which could be leached out producing irregular porous geometries.

Internal and external scaffold geometry can be designed on a computer and then rendered on an AM device also referred to as a 3D printer. If the device cannot directly print the material needed for the scaffold and control of only the external surface is needed, then the rendered part can be recast in an implantable material. AM devices are often compared with subtractive technologies that also control only the external morphologies of the rendered part. The most commonly used subtractive technology is computer numerical control. These devices are commonly used to render dental crowns for dental implants from blocks of ceramic.

Another non-AM technology that has been explored to produce sheets of polymer material is electrospinning technology. Electrospinning collects a strand of polymer on a spinning mandrill into sheets. The fiber orientation can be controlled. The orientation can be used to influence cell migration and attachment on the sheet [71].

4.4 AM of Bone Tissue-Engineered Implants

AM allows the use of computer-aided design to control the architecture of the internal porous scaffold space to guide invading host tissue and vasculature. This is true for resorbable, solid-cured polymer, [72], ceramic [73], metal [74], or hydrogel [75] scaffolds. However, the technologies for each are different (Fig. 17.7).

4.4.1 3D Metal Bone Scaffolds

The diversity of metal AM has recently increased, although all are based on metal powders with very small, uniformly sized particles. That is because the powder has to continuously flow through the machine as the build progresses. The oldest method is to combine these powders with a polymer binder and then to use focused light polymerization (e.g., single point light in stereolithography or layer projection in continuous digital light processing). The bound powder part is then sintered, which is to raise the temperature of the part to the point where the powder particles are fused. When these parts are sintered in normal atmosphere and variable humidity it affects the uniformity and material properties of the resulting part. The more accurate devices now operate in a vacuum or under argon gas. The major modalities are selective laser sintering [76], electron beam deposition [77], and selective laser melting

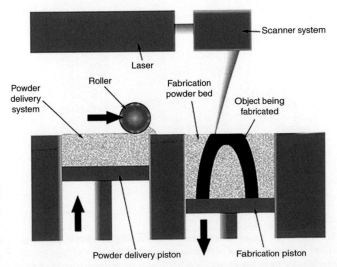

FIGURE 17.7 Powder bed metal or plastic selective laser sintering. This is the general format for most powder bed systems. (*From: http://www. makeuseof.com/tag/what-is-3d-printing-and-how-exactly-does-it-work/.*)

[78]. All of these methods have a single source of sintering energy that is aimed at a powder bed. New powder is rolled in after each layer has been sintered. Another process, laser-engineered net shaping [79] (Fig. 17.8) is able to sinter two separate powder flows simultaneously. Argon gas is used to purge other gases to control the purity of the resulting parts.

4.4.2 Ceramic Bone Scaffolds

Most 3D-printed bone scaffolds that have been studied have been prepared from tricalcium phosphate or hydroxyapatite via an inkjet method using a polymer binder that is later sintered [80,81]. Crystalline, porous, and/or glass ceramics can be combined [82]. However, it is also possible to

(a)

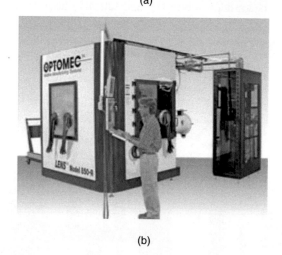

(b)

FIGURE 17.8 Laser Engineered Net Shaping 3D metal printing. (a) Two or more streams of powder are sintered simultaneously *(from: http://www.tms.org/pubs/journals/JOM/9907/Hofmeister/Hofmeister-9907.fig. 17.1.lg.gif).* (b) Atmospheric gases are purged from the system by Argon gas *(from: http://www.optomec.com/Additive-Manufacturing-Systems/Laser-Additive-Manufacturing-Systems [accessed 05.15.14].)*

produce ceramic scaffolds using a polymer binder with higher accuracy by utilizing a stereolithographic device or a continuous digital light processing device.

4.4.3 Solid-Cured Polymer Bone Scaffolds

Selective laser sintering is now frequently used to prepare polycaprolactone scaffolds for bone tissue engineering and has been used in the clinic [83]. While nanometer-level photocrosslinking can produce solid structures [84], it would be difficult and time consuming to use these techniques to produce scaffolds with the requisite 1–100 μm level accuracy needed for bone tissue engineering. Therefore, much of the study of solid cured polymer scaffolds has been achieved by stereolithography-based [85,86] and DLP-based [87] methods (Fig. 17.9) [88].

Bioreactor culturing of cell-seeded porous solid-cured scaffolds does not present great challenges. However, trying to get bone to grow within critical size and larger scaffolds, and to become sufficiently vascularized, does present challenges [89,90]. Control of scaffold resorption materials and triggering vascular response is a current area of research. The challenge of vascularization is one of the greatest motivations for *in vivo* bioreactors with vascular pedicles [70,91,92].

4.4.4 Hydrogel-Based Cell Printing and Spraying for Bone Scaffolds

Cell printing and spraying is the most recent form of tissue engineering scaffold. Most narrowly defined, "cell printing" most commonly utilizes inkjet-style printing (Fig. 17.10) [93–96], laser printing [97], or cell-spraying [98,99] devices for depositing hydrogel [100] that includes cells and possibly growth factors [101]. Implanting large hydrogel-suspension of cells is a difficult proposition for bone tissue engineering as the wound sites are often under compression. However, cell-printed and/or hydrogel-suspended cells can be placed in a cured polymer or another housing within the wound that will maintain the space during bone development [102].

Following cell printing, keeping cells alive and sufficiently perfused in large hydrogel constructs is less of a challenge than the vascularization of new bone forming in a large, implanted hydrogel construct. Microfluidic devices offer much promise for this area [103].

5 CONCLUSIONS

Bone regenerative research has made great advances in utilizing the elements of tissue engineering, namely, growth factors, scaffolds, and autologous cells. 3D-printed resorbable scaffolds and growth factors on biological scaffolds have been employed in the clinic, yet customized scaffolds are not currently printed with growth factors or autologous cells; therefore, bioprinting does not yet exist in a clinically relevant manner. Recent research has been

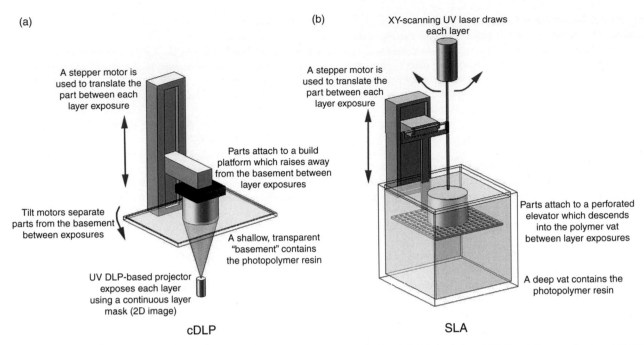

(a)

A stepper motor is used to translate the part between each layer exposure

Parts attach to a build platform which raises away from the basement between layer exposures

Tilt motors separate parts from the basement between exposures

A shallow, transparent "basement" contains the photopolymer resin

UV DLP-based projector exposes each layer using a continuous layer mask (2D image)

cDLP

(b)

XY-scanning UV laser draws each layer

A stepper motor is used to translate the part between each layer exposure

Parts attach to a perforated elevator which descends into the polymer vat between layer exposures

A deep vat contains the photopolymer resin

SLA

FIGURE 17.9 High accuracy 3D printing of polymer scaffolds. (a) Continuous digital light processing (cDLP) prints an entire layer at once. (b) Stereolithography uses a single point laser to draw the layer spot by spot. *(Reproduced with permission from Ref. [96].)*

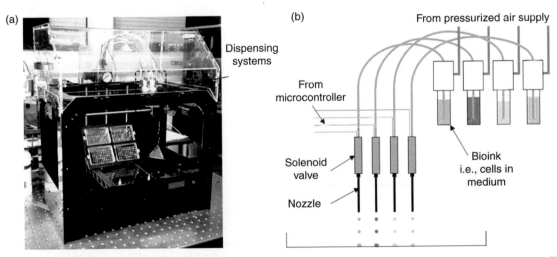

(a)

Dispensing systems

(b)

From pressurized air supply

From microcontroller

Solenoid valve

Nozzle

Bioink i.e., cells in medium

FIGURE 17.10 Cell printing. (a) A cell printer often deposits cells inside of an incubator or biosafety cabinet in order to preserve sterile conditions as well as keep the cells alive during the process. (b) Several "bioinks" or hydrogel-cell suspensions can be deposited simultaneously making it easy to deposit multiple growth factors and cells within one more layer of a construct *(http://en.paperblog.com/3d4d-live-cell-printing-732828/.)*

rapidly expanding the range of 3D printing techniques as well as 3D printable materials with properties finely tuned to reconstruct patient-specific defects and to integrate with and compliment surrounding host tissues. Fabricating these materials with viable biological components is a challenge that remains to be solved. Continued advances in bone bioprinting hold much promise for improving the quality and efficiency of bone reconstructive therapies. However, the regenerative medicine and tissue engineering paradigm is relatively new, and the recent healthcare reform in the US

is applying new market forces and provider incentives and regulations; hence, it remains to be seen whether these modalities will become widely available and supersede current therapies as standard-of-care.

GLOSSARY

AM 3D printing in a layer-by-layer fashion (ASTM F42 committee).
Allogeneic Nonautologous graft (i.e., transplanted) material.
Allograft Allogeneic graft (i.e., transplanted) material.

Alloplast Artificial, implantable materials.

Bioprinting For the purpose of this paper it is: constructing custom scaffolds with scaffolds, growth factors, and osteoprogenitor cells in the restoration of bone defects; however, in other contexts this term refers to the use of 3D printable hydrogels that perform like extracellular matrix to position organ-specific cells and possibly vascular cells for coculture and/or implantation.

Bioreactor A chamber in which a tissue engineered construct is cultured prior to placement in a defect site.

Custom Patient-specific.

Critical-size defect The size of defect that cannot heal unaided.

Distraction osteogenesis Often a metallic device is used to continuously open a surgically created bone wound in order to form new bone.

Gold standard treatment The treatment to which all other treatment methods are compared.

Immunotherapy In this context it is immune suppression to prevent a graft vs host response to an allogeneic implant.

Scaffold Resorbable construct that guides cells and tissues into place within a bioreactor, defect, and/or wound site.

Standard-of-care Accepted clinical procedures.

Subtractive manufacturing Traditional manufacturing methods that remove material from a raw material source in order to form a final product.

Stress shielding Preventing an area of bone from receiving a load that it had previously received.

Vascularization Development of a blood supply.

ABBREVIATIONS

3D	Three-dimensional
AM	Additive manufacturing
ACA	Affordable care act
BMP	Bone morphogenetic protein
COL1A1	Collagen, type 1, alpha 1
Dlx5	Distal-less homeobox 5
IHH	Indian hedgehog
MSC	Mesenchymal stem cell
Msx2	msh-like 2, homeobox
NiTi	Nitinol
OP-1	Osteogenic protein
PTH	Parathyroid hormone
PTH-R1	parathyroid hormone 1 receptor
PEG	Poly(ethylene glycol)
PTHrP	Parathyroid hormone-related protein
rhBMP-2	Recombinant human bone morphogenetic protein 2
rhPDGF	Recombinant human platelet-derived growth factor
SMAD	Homologs of both the SMA and MAD proteins
Ti-6Al-4V or	
Ti-6%Al-4%V	Surgical grade 5 titanium

REFERENCES

[1] Cohen MM JR. The new bone biology: pathologic, molecular, and clinical correlates. Am J Med Genet A 2006;140:2646–706.

[2] Centers for disease control and prevention (CDC). Prevalence of doctor-diagnosed arthritis and arthritis-attributable activity limitation – United States, 2010–2012. MMWR. Morb Mortal Wkly Rep 2013;62(44):869. Available from: <http://www.cdc.gov/mmwr/pre-view/mmwrhtml/mm6244a1.htm or http://www.ncbi.nlm.nih.gov/pubmed/24196662)>

[3] Initiative USBAJ. The burden of musculoskeletal diseases in the United States. 2nd ed. Rosemont, IL: American Academy of Orthopaedic Surgeons; 2011.

[4] Johnell O, Kanis J. Epidemiology of osteoporotic fractures. Osteoporos Int 2005;16:S3–7.

[5] U.D.O.H.A.H. Services. Bone health and osteoporosis: a report of the surgeon general; 2004.

[6] Phieffer LS, Goulet JA. Delayed unions of the tibia. J Bone Joint Surg Am 2006;88A:206–16.

[7] Schmitz JP, Hollinger JO. The critical size defect as an experimental model for craniomandibulofacial nonunions. Clin Orthop Relat Res 1986;299–308.

[8] Institute, N.C. 2014. SEER Cancer Stat Fact Sheets [Online], http://seer.cancer.gov/statfacts/ [accessed 30.05.2014].

[9] Li S, Peng Y, Weinhandl ED, Blaes AH, Cetin K, Chia VM, et al. Estimated number of prevalent cases of metastatic bone disease in the US adult population. Clin Epidemiol 2012;4:87–93.

[10] Correa A, Cragan JD, Kucik JE, Alverson CJ, Gilboa SM, Balakrishnan R, et al. Reporting birth defects surveillance data 1968–2003. Birth Defects Res A Clin Mol Teratol 2007;79:65–186.

[11] Hootman J, Bolen J, Helmick C, Langmaid G. Prevalence of doctor-diagnosed arthritis and arthritis-attributable activity limitation – United States, 2003–2005. MMWR Morb Mortal Wkly Rep 2006;55: 1089–92.

[12] Mcneil JM, Binette J, CDC. Prevalence of disabilities and associated health conditions among adults – United States, 1999. JAMA 2001;285:1571–2.

[13] Yelin E, Cisternas M, Foreman A, Pasta D, Murphy L, Helmick CG. National and state medical expenditures and lost earnings attributable to arthritis and other rheumatic conditions – United States, 2003. MMWR Morb Mortal Wkly Rep 2007;56:4–7.

[14] Worldbank, T. Health expenditure, total (% of GDP) | Data | Table [Online], http://data.worldbank.org/indicator/SH.XPD.TOTL.ZS; 2014 [accessed 30.05.2014].

[15] OECD. OECD Health Data 2013; how does the United States Compare; 2013.

[16] Mcglynn EA, Asch SM, Adams J, Keesey J, Hicks J, Decristofaro A, et al. The quality of health care delivered to adults in the United States. New Engl J Med 2003;348:2635–45.

[17] WHO. The World Health Report 2000 Health Systems – Improving performance. Geneva: WHO; 2000.

[18] Schuster MA, Mcglynn EA, Brook RH. How good is the quality of health care in the United States? Milbank Quart 1998;76: 517–63.

[19] Who, T. Life expectancy: life expectancy – data by country [Online], http://apps.who.int/gho/data/node.main.688?lang=en; 2014. [accessed 31.05.2014].

[20] Biegelsen L, Huang L, Strange K, Bijou C. Healthcare Coverage Expansion a Shot in the Arm for Med Tech [Online] 2013. Available from: <http://www.elsevierbi.com/~/media/Supporting%20Documents/The%20Gray%20Sheet/39/15/WellsFargo_benefits_Of_NewInsured_vs_DeviceTax_Impact_Report.>pdf [accessed 30.05.14]

[21] Dalen N, Olsson KE. Bone-mineral content and physical-activity. Acta Orthop Scand 1974;45:170–4.

[22] Olsen BR. Bone embryology. Washington, DC: American Society for Bone and Mineral Research; 2006.

[23] Tsiridis E, Upadhyay N, Giannoudis P. Molecular aspects of fracture healing: which are the important molecules? Injury 2007;38(Suppl. 1):S11–25.

[24] Allen MR, Burr DB. Bone modeling and remodeling. In: Matthew R, Allen DBB, editors. Basic and applied bone biology. Academic Press; 2014.

[25] Shapiro F. Bone development and its relation to fracture repair. The role of mesenchymal osteoblasts and surface osteoblasts. Eur Cell Mater 2008;15:53–76.

[26] Kalfas IH. Principles of bone healing. Neurosurg Focus 2001; 10:E1.

[27] Olszta MJ, Cheng X, Jee SS, Kumar R, Kim Y-Y, Kaufman MJ, et al. Bone structure and formation: a new perspective. Mater Sci Eng R Reports 2007;58:77–116.

[28] Doblare M, Garcia JM. On the modelling bone tissue fracture and healing of the bone tissue. Acta Cient Venez 2003;54:58–75.

[29] Clarke B. Normal bone anatomy and physiology. Clin J Am Soc Nephrol 2008;3(Suppl 3):S131–9.

[30] Nagasawa T. Microenvironmental niches in the bone marrow required for B-cell development. Nat Rev Immunol 2006;6:107–16.

[31] Martin AD, Mcculloch RG. Bone dynamics: stress, strain and fracture. J Sports Sci 1987;5:155–63.

[32] Turner CH. Bone strength: current concepts. Ann NY Acad Sci 2006;1068:429–46.

[33] Sfeir C, Ho L, Doll BA, Azari K, Hollinger JO. Fracture repair. In: Lieberman JR, Friedlaender GE, editors. Bone regeneration and repair. Totowa, NJ: Humana Press; 2005.

[34] Oryan A, Alidadi S, Moshiri A. Current concerns regarding healing of bone defects. Hard Tissue 2013;2:13.

[35] Mirhadi S, Ashwood N, Karagkevrekis B. Factors influencing fracture healing. Trauma 2013;15:140–55.

[36] Dickinson BP, Roy I, Lesavoy MA. Temporalis fascia for lip augmentation. Ann Plast Surg 2011;66:114–7.

[37] Franklin JD, Shack RB, Stone JD, Madden JJ, Lynch JB. Single-stage reconstruction of mandibular and soft-tissue defects using a free osteocutaneous groin flap. Am J Surg 1980;140:492–8.

[38] Song RY, Gao YZ, Song YG, Yu YS, Song YL. The forearm flap. Clin Plast Surg 1982;9:21–6.

[39] Taylor GI. The current status of free vascularized bone-grafts. Clin Plast Surg 1983;10:185–209.

[40] Taylor GI, Miller GDH, Ham FJ. Free vascularized bone graft – clinical extension of microvascular techniques. Plast Reconstr Surg 1975;55:533–44.

[41] Taylor GI, Townsend P, Corlett R. Superiority of the deep circumflex iliac vessels as the supply for free groin flaps – clinical-work. Plast Reconstr Surg 1979;64:745–59.

[42] Yang GF, Chen PJ, Gao YZ, Liu XY, Li J, Jiang SX, He SP, Boo-Chai K. Forearm free skin flap transplantation: a report of 56 cases. Br J Plast Surg 1997;50(3):162–165.

[43] Ilizarov GA. The principles of the Ilizarov method. Bull Hosp Jt Dis 1988;48:1–11.

[44] Paley D, Maar DC. Ilizarov bone transport treatment for tibial defects. J Orthopaedic Trauma 2000;14:76–85.

[45] Ling XF, Peng X. What is the price to pay for a free fibula flap? A systematic review of donor-site morbidity following free fibula flap surgery. Plast Reconstr Surg 2012;129:657–74.

[46] Foster TE, Puskas BL, Mandelbaum BR, Gerhardt MB, Rodeo SA. Platelet-rich plasma from basic science to clinical applications. Am J Sports Med 2009;37:2259–72.

[47] Yamada T, Yoshii T, Sotome S, Yuasa M, Kato T, Arai Y, et al. Hybrid grafting using bone marrow aspirate combined with porous beta-tricalcium phosphate and trephine bone for lumbar posterolateral spinal fusion a prospective, comparative study versus local bone grafting. Spine 2012;37:E174–9.

[48] Breitbart AS, Ablaza VJ. Implant materials. In: THORNE CH, editor. Grabb and Smith's plastic surgery. 6th ed. Philadelphia: Lippincott Williams & Wilkins; 1997.

[49] Hermawan H, Dadan R, Djuansjah JR. Metals for biomedical applications. In: Fazel-Rezai R, editor. Biomedical engineering – from theory to applications. In Tech; 2011.

[50] Manivasagam G, Dhinasekaran D, Rajamanickam A. Biomedical implants: corrosion and its prevention – a review. Recent Patents Corros Sci 2010;2:40–54.

[51] Stoeckel D, Pelton A, Duerig T. Self-expanding nitinol stents: material and design considerations. Eur Radiol 2004;14:292–301.

[52] Hermawan H. Biodegradable metals: state of the art. In: Hermawan H, editor. Biodegradable metals: from concept to applications. Berlin, Heidelberg: Springer; 2012.

[53] Kingery DW, Bowen HK, Uhlmann DR. Introduction to ceramics. 2nd edn. New York: John Wiley & Sons; 1976.

[54] Vallet-Regí M, Balas F, Colilla M, Manzano M. Bioceramics and pharmaceuticals: a remarkable synergy. Solid State Sci 2007;9:768–76.

[55] Clarke SA, Brooks RA, Lee PT, Rushton N. The effect of osteogenic growth factors on bone growth into a ceramic filled defect around an implant. J Orthop Res 2004;22:1016–24.

[56] Lucas N, Bienaime C, Belloy C, Queneudec M, Silvestre F, Nava-Saucedo JE. Polymer biodegradation: mechanisms and estimation techniques. Chemosphere 2008;73:429–42.

[57] Ratner BD, Bryant SJ. Biomaterials: where we have been and where we are going. Annu Rev Biomed Eng 2004;6:41 75.

[58] Katz JS, Burdick JA. Synthetic biomaterials. In: Fisher JP, Mikos AG, Bronzino JD, Peterson DR, editors. Tissue engineering: principles and practices. Boca Raton, FL: CRC Press (Taylor and Francis); 2013.

[59] Shi X, Henslee A, Yoon D, Kasper FK, Mikos AG. Poly(propylene fumarate). In: Hollinger JO, editor. An introduction to biomaterials. 2nd ed. CRC Press; 2011.

[60] Redlich K, Smolen JS. Inflammatory bone loss: pathogenesis and therapeutic intervention. Nat Rev Drug Discov 2012;11:234–50.

[61] Song B, Estrada KD, Lyons KM. Smad signaling in skeletal development and regeneration. Cytokine Growth Factor Rev 2009;20: 379–88.

[62] Wolff J. The law of bone remodelling. New York: Springer; 1986.

[63] Deschaseaux F, Sensebe L, Heymann D. Mechanisms of bone repair and regeneration. Trends Mol Med 2009;15:417–29.

[64] Carragee EJ, Baker RM, Benzel EC, Bigos SJ, Cheng I, Corbin TP, et al. A biologic without guidelines: the YODA project and the future of bone morphogenetic protein-2 research. Spine J 2012;12: 877–80.

[65] Vaccaro AR, Whang PG, Patel T, Phillips FM, Anderson DG, Albert TJ, et al. The safety and efficacy of OP-1 (rhBMP-7) as a replacement for iliac crest autograft for posterolateral lumbar arthrodesis: minimum 4-year follow-up of a pilot study. Spine J 2008;8: 457–65.

[66] Solchaga LA, Daniels T, Roach S, Beasley W, Snel LB. Effect of implantation of Augment((R)) bone graft on serum concentrations of platelet-derived growth factors: a pharmacokinetic study. Clin Drug Investig 2013;33:143–9.

[67] Jordaan D. Regulatory crackdown on stem cell therapy: what would the position be in South Africa? S Afr Med J 2012;102:219–20.

[68] Wallace J, Wang MO, Thompson P, Busso M, Belle V, Mammoser N, et al. Validating continuous digital light processing (cDLP) additive manufacturing accuracy and tissue engineering utility of a dye-initiator package. Biofabrication 2014;6:015003.

[69] Jiang B, Akar B, Waller TM, Larson JC, Appel AA, Brey EM. Design of a composite biomaterial system for tissue engineering applications. Acta Biomater 2014;10:1177–86.

[70] Thomson RC, Mikos AG, Beahm E, Lemon JC, Satterfield WC, Aufdemorte TB, et al. Guided tissue fabrication from periosteum using preformed biodegradable polymer scaffolds. Biomaterials 1999;20:2007–18.

[71] Lyu S, Huang C, Yang H, Zhang X. Electrospun fibers as a scaffolding platform for bone tissue repair. J Orthop Res 2013;31:1382–9.

[72] Kim K, Dean D, Wallace J, Breithaupt R, Mikos AG, Fisher JP. The influence of stereolithographic scaffold architecture and composition on osteogenic signal expression with rat bone marrow stromal cells. Biomaterials 2011;32:3750–63.

[73] Butscher A, Bohner M, Doebelin N, Hofmann S, Muller R. New depowdering-friendly designs for three-dimensional printing of calcium phosphate bone substitutes. Acta Biomater 2013;9:9149–58.

[74] Seyedraoufi ZS, Mirdamadi S. Synthesis, microstructure and mechanical properties of porous Mg–Zn scaffolds. J Mech Behav Biomed Mater 2013;21:1–8.

[75] Poldervaart MT, Wang H, Van Der Stok J, Weinans H, Leeuwenburgh SC, Oner FC, et al. Sustained release of BMP-2 in bioprinted alginate for osteogenicity in mice and rats. PLoS One 2013;8:e72610.

[76] Traini T, Mangano C, Sammons RL, Mangano F, Macchi A, Piattelli A. Direct laser metal sintering as a new approach to fabrication of an isoelastic functionally graded material for manufacture of porous titanium dental implants. Dent Mater 2008;24:1525–33.

[77] Choi JM, Kong YM, Kim S, Kim HE, Hwang CS. Formation and characterization of hydroxyapatite coating layer on Ti-based metal implant by electron-beam deposition. J Mater Res 1999;14:2980–5.

[78] Vandenbroucke B, Kruth J-P. Selective laser melting of biocompatible metals for rapid manufacturing of medical parts. Rapid Prototyping J 2007;13:196–203.

[79] Liu Q, Leu MC, Schmitt SM. Rapid prototyping in dentistry: technology and application. Int J Adv Manuf Technol 2006;29:317–35.

[80] Bose S, Vahabzadeh S, Bandyopadhyay, Amit. Bone tissue engineering using 3D printing. Mater Today 2013;16:496–504.

[81] Diogo GS, Gaspar VM, Serra IR, Fradique R, Correia IJ. Manufacture of beta-TCP/alginate scaffolds through a Fab@home model for application in bone tissue engineering. Biofabrication 2014;6:025001.

[82] Cai S, Xu GH, Yu XZ, Zhang WJ, Xiao ZY, Yao KD. Fabrication and biological characteristics of beta-tricalcium phosphate porous ceramic scaffolds reinforced with calcium phosphate glass. J Mater Sci Mater Med 2009;20:351–8.

[83] Probst FA, Hutmacher DW, Muller DF, Machens HG, Schantz JT. Calvarial reconstruction by customized bioactive implant. Handchir Mikrochir Plast Chir 2010;42:369–73.

[84] Hsieh TM, Ng CW, Narayanan K, Wan AC, Ying JY. Three-dimensional microstructured tissue scaffolds fabricated by two-photon laser scanning photolithography. Biomaterials 2010;31:7648–52.

[85] Jacobs PF. Rapid prototyping and manufacturing: fundamentals of stereolithography. Dearborn, MI: Society of Manufacturing Engineers; 1992.

[86] Melchels FP, Feijen J, Grijpma DW. A review on stereolithography and its applications in biomedical engineering. Biomaterials 2010;31:6121–30.

[87] Dean D, Mott E, Luo X, Busso M, Wang MO, Vorwald C, et al. Multiple initiators and dyes for continuous Digital Light Processing (cDLP) additive manufacture of resorbable bone tissue engineering scaffolds. Virtual Physical Prototyping 2014;9(1):3–9.

[88] Derby B. Printing and prototyping of tissues and scaffolds. Science 2012;338:921–6.

[89] Frohlich M, Grayson WL, Wan LQ, Marolt D, Drobnic M, Vunjak-Novakovic G. Tissue engineered bone grafts: biological requirements, tissue culture and clinical relevance. Curr Stem Cell Res Ther 2008;3:254–64.

[90] Tsigkou O, Pomerantseva I, Spencer JA, Redondo PA, Hart AR, O'Doherty E, et al. Engineered vascularized bone grafts. Proc Natl Acad Sci USA 2010;107:3311–6.

[91] Johnson EO, Troupis T, Soucacos PN. Tissue-engineered vascularized bone grafts: basic science and clinical relevance to trauma and reconstructive microsurgery. Microsurgery 2011;31:176–82.

[92] Liu Y, Teoh SH, Chong MS, Yeow CH, Kamm RD, Choolani M, et al. Contrasting effects of vasculogenic induction upon biaxial bioreactor stimulation of mesenchymal stem cells and endothelial progenitor cells cocultures in three-dimensional scaffolds under *in vitro* and *in vivo* paradigms for vascularized bone tissue engineering. Tissue Eng Part A 2013;19:893–904.

[93] Cui X, Boland T, D'lima DD, Lotz MK. Thermal inkjet printing in tissue engineering and regenerative medicine. Recent Pat Drug Deliv Formul 2012;6:149–55.

[94] Inzana JA, Olvera D, Fuller SM, Kelly JP, Graeve OA, Schwarz EM, et al. 3D printing of composite calcium phosphate and collagen scaffolds for bone regeneration. Biomaterials 2014;35:4026–34.

[95] Kang KH, Hockaday LA, Butcher JT. Quantitative optimization of solid freeform deposition of aqueous hydrogels. Biofabrication 2013;5:035001.

[96] Moon S, Hasan SK, Song YS, Xu F, Keles HO, Manzur F, et al. Layer by layer three-dimensional tissue epitaxy by cell-laden hydrogel droplets. Tissue Eng Part C Methods 2010;16:157–66.

[97] Koch L, Deiwick A, Schlie S, Michael S, Gruene M, Coger V, et al. Skin tissue generation by laser cell printing. Biotechnol Bioeng 2012;109:1855–63.

[98] Gerlach JC, Johnen C, Ottoman C, Brautigam K, Plettig J, Belfekroun C, et al. Method for autologous single skin cell isolation for regenerative cell spray transplantation with non-cultured cells. Int J Artif Organs 2011;34:271–9.

[99] Lee H, Ahn S, Chun W, Kim G. Enhancement of cell viability by fabrication of macroscopic 3D hydrogel scaffolds using an innovative cell-dispensing technique supplemented by preosteoblast-laden micro-beads. Carbohydr Polym 2014;104:191–8.

[100] Khademhosseini A, Langer R. Microengineered hydrogels for tissue engineering. Biomaterials 2007;28:5087–92.

[101] Kachouie NN, Du Y, Bae H, Khabiry M, Ahari AF, Zamanian B, et al. Directed assembly of cell-laden hydrogels for engineering functional tissues. Organogenesis 2010;6:234–44.

[102] Ahn S, Lee H, Kim G. Functional cell-laden alginate scaffolds consisting of core/shell struts for tissue regeneration. Carbohydr Polym 2013;98:936–42.

[103] Chung BG, Lee KH, Khademhosseini A, Lee SH. Microfluidic fabrication of microengineered hydrogels and their application in tissue engineering. Lab Chip 2012;12:45–59.

Chapter 18

Bioprinting of Cartilage

Recent Progress on Bioprinting of Cartilage

Kuilin Lai**,† and **Tao Xu***,**

*Department of Mechanical Engineering, Bio-manufacturing Center, Tsinghua University, Beijing, China; **Medprin Regenerative Technologies Co. Ltd, Guangzhou, Guangdong Province, China; †School of Materials Science and Engineering, South China University of Technology, Guangzhou, Guangdong Province, China

ABSTRACT

The number of osteoarthritis cases is booming. Due to the depth-dependent structures, complex contents, and specific properties of articular cartilage, it would be a challenge using current technologies to meet all these requirements. Instead, bioprinting shows the potential to reconstruct the fine structure of cartilage because it can selectively print cells and biomaterials to organize the spatial distribution of the extracellular matrix, signals, cells, and so on. This chapter will give a review on the recent progress of cartilage bioprinting, including some basic concepts, blueprint, cell sources, natural and synthetic materials used, key parameters such as stimulating factors and bioreactors, and preclinical animal models for bioprinted cartilage tissues.

Keywords: bioprinting; bionic; articular cartilage; depth-dependent structures; tissue reconstruction

1 INTRODUCTION

With an aging population and the growing problem of obesity, the number of osteoarthritis cases is estimated to boom in the coming years [1]. Nowadays, there are over 2.7 million knee and hip replacements performed in the United States every year for end-stage disease joint failure, and many other patients suffer from less severe cartilage damages [2,3]. Also in the United Kingdom, 8.5 million people are affected by joint pain, which may be related to osteoarthritis [4]. Moreover, with a more active adult population, cartilage damage resulting from sports injuries can often lead to premature cartilage degeneration.

The current treatments for articular cartilage repair include microfracture, mosaicplasty, autologous chondrocyte transplantation, and osteochondral allograft transplantation. Although these techniques have successfully relieved pain and improved joint function, each is plagued with their disadvantages that can deter their long-term clinical applications [5,6]. For example, cartilage produced from these techniques is often composed of type I collagen, the main composition part and characteristic of fibrocartilage. However, type I collagen is biochemically and biomechanically inferior to hyaline cartilage, leading to weaker mechanical support or less valid time. In addition, the repaired tissues often lack the native structure of normal cartilage. There are still some other drawbacks including donor site morbidity, complicated surgical procedures, risks of infection, and graft rejection. Moreover, current approaches for the engineering of osteochondral grafts suffer from poor tissue formation and compromised integration at the interface between the cartilage and bone layers [7] and between osteochondral graft and host tissue [8], advocating the need for further improvement of these techniques.

Bioprinting is the process of generating spatially controlled cell patterns in 3D, where cell function and viability are preserved within the printed construct. The precision of this technique would make it possible to control precisely not only their mechanical properties, but also their pore network architecture and their custom shape. For cartilage, bioprinting can selectively print cells and gels to control the spatial distribution of the extracellular matrix (ECM) materials, signals, cells, and so on.

2 BLUEPRINT FOR BIOPRINTING OF CARTILAGE

Articular cartilage was not a simple tissue to engineer as we thought before. It exhibits exquisite organization on several length scales. The spatial structure and composition (e.g., collagen, fluid, polygalacturonases) of the articular cartilage can be characterized by different imaging

techniques, such as microscopic methods, high-resolution magnetic resonance imaging (MRI), polarized light microscopy, and Fourier transform infrared spectroscopy (FT-IR) (Fig. 18.1) [9–13]. Over the range from nanometers to micrometers, the matrix of the cartilage extracellular can be considered as a network of collagen fibers, proteins, proteoglycans, and so on, that allow for cell adhesion, chemical and mechanical signals transduction, nutrition and metabolism product exchange, and mechanical support. Over the range of micrometers to millimeters, there are topographical differences in cartilage thickness and matrix content across the surface of the joint, which are associated with the level of load [14,15]. Furthermore, the properties of articular cartilage change with depth from the articular surface, resulting in a zonal structure that is typically divided into three zones: superficial (surface to 10–20% of thickness), middle (20–70%), and deep (70–100%). The various levels of organization *in vivo* all have importance in the mechanical, metabolic, and transport properties of the tissue [15–17]. As a result, there is a highly coordinated cell distribution within

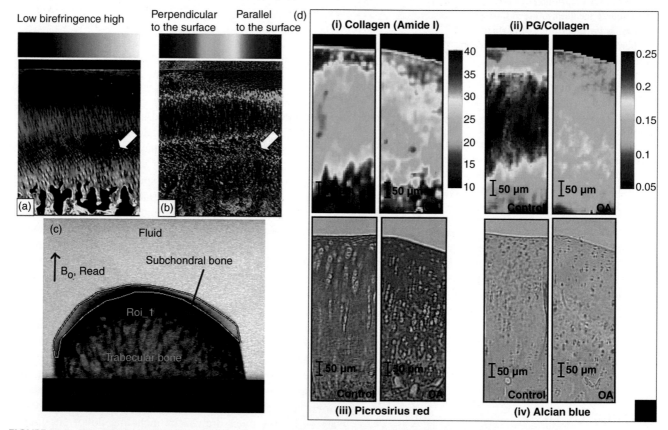

FIGURE 18.1 **The polarized light microscopic images of articular cartilage showed an extra lamina in the deep cartilage (arrow).** (a) The lamina shows a decreased birefringence signal. (b) The orientation image shows that in this region the collagen fibrils change their alignment. (c) T2-weighted MRI image of an osteochondral plug cut from the medial femoral condyle of a nonsurgical control rabbit age-matched with the 4 week postsurgical time point. Radial zone cartilage is indicated by the defined region of interest (ROI). (d) FT-IR-derived images of (i) collagen content (Amide I), and (ii) proteoglycan content (polygalacturonase/collagen), from the articular surface down to the tidemark. The histological images were acquired by staining the same tissue with picrosirius red for (iii) collagen content and Alcian blue for (iv) proteoglycan content [10,11]. *(Figure reprinted with permission from Springer and Wiley.)*

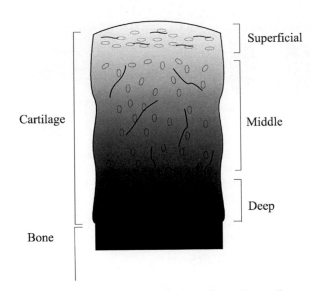

FIGURE 18.2 Zonal organization in normal articular cartilage.

these zones (Fig. 18.2) [18], which can be distinguished on the basis of morphological criteria, such as cell shape, size, and arrangement, as well as collagen, proteoglycans, and hyaluronan expression [19].

To reconstruct the articular cartilage, its structure and compositions is remodeled by the computer [20]. From the different imaging techniques listed earlier, shape models of articular cartilage is also established and shown. Additionally, finite element computer analyses have been used to study the role of local tissue mechanics on endochondral ossification patterns, skeletal morphology, and articular cartilage thickness distributions [21].

3 CELL SOURCE FOR BIOPRINTING OF CARTILAGE

The optimal cell source for cartilage tissue engineering is still being identified. Chondrocytes, fibroblasts, stem cells, and genetically modified cells have been explored for their potential as a viable cell source for cartilage repair [22]. Chondrocytes are the most obvious choice since they are found in native cartilage and have been extensively studied to assess their role in producing, maintaining, and remodeling the cartilage ECM [23]. Recent work has focused on stem cells that have multilineage potentials and can be isolated from a plethora of tissues. All of these cells can be modified genetically to induce or enhance chondrogenesis [24].

Stem cell used for cartilage repair include adult mesenchymal stem cells (MSCs), isolated from bone marrow, and allogeneic stem cells. However, we know only a little about the MSCs, and some human MSCs isolated by current methods are not homogeneous and consist of mixtures of progenitors and other cells. As a result, more exhaustive characterizations are required before and during the tissue repair process, for example, whether implantation of MSCs

would result in a matured cartilage tissue, or if MSCs could be used with other cells together to achieve native structure and better results.

4 MATERIALS FOR BIOPRINTING OF CARTILAGE

Both natural and synthetic materials are used to fabricate scaffolds for the cartilage repair. The natural materials for cartilage repair include chitosan, hyaluronic acid [25,26], collagen [27], and fibrin hydrogels [28,29]. The most widely used synthetic polymeric scaffolds in cartilage tissue engineering are the biodegradable poly-α-hydroxy esters, especially polylactic acid, polyglycolic acid, and their derivates (PLLA, PLGA, PDLA) [30]. Also, other synthetic polymers are being used in the field such as poly(ethylene glycol) [31,32], poly(N-isopropylacrylamide) [33,34], polyurethane [35], and poly(vinyl alcohol) [36]. These synthetic polymers offer relative ease of processing as well as mechanical properties suitable for this type of application [37,38] and have already been used in a variety of implantable medical devices for a decade.

Moroni et al. [39] fabricated the anatomical meniscal scaffolds using a fiber-deposited method with a bioplotter device having an *XYZ* robotic arm apparatus. The raw material to fabricate the scaffold is poly(ethylene oxide-terephthalate)-*co*-poly(butylene terephthalate), which is pressed out in molten phase by nitrogen gas.

Solid and hollow fibers were used to fabricate different scaffolds, and the fiber spacing varied from 1 mm to 1.2 mm. The scaffold was 200 μm thick with a porosity of 70~80% (Fig. 18.3). With increasing porosity of scaffolds, the extrinsic stiffness decreased. Solid fiber scaffolds varied from 495.07 ± 76.26 N/mm to 333.22 ± 26.16 N/mm when scaffold porosity increased from 70% to 80%, while hollow fiber scaffolds measured as 43.94 ± 14.71 N/mm with porosity of 70%. The equilibrium modulus of scaffolds varied from 1.4 MPa to 2.3 MPa with different fiber architectures, higher than that of the porcine menisci (0.42 ± 0.25 MPa lateral or 1.08 ± 0.56 MPa medial). Finite element analysis showed that the replacement of natural menisci with these constructs would result in a significant improvement of the strain and stress distribution on the overlying and underlying articular cartilage compared to meniscectomy.

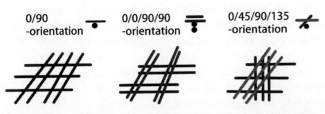

FIGURE 18.3 Fabricated anatomical meniscal scaffold with different patterns [40].

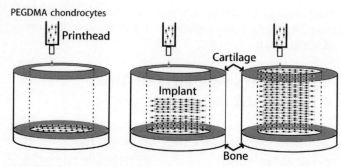

FIGURE 18.4 Schematic of bioprinting cartilage with simultaneous photopolymerization process. PEGDMA, poly(ethylene glycol) dimethacrylate; hv, UV light energy. In the bovine cartilage defect, printed cells maintained deposited positions with simultaneous photopolymerization in layer-by-layer assembly [41].

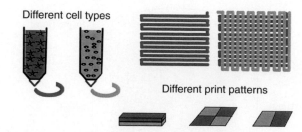

FIGURE 18.5 Tissue printing using various cell types. The design of a construct is translated to a robot arm that drives a cartridge loaded with a cell–hydrogel mixture and extrudes fibers in a layered fashion [43].

Cui et al. [40] used a thermal inkjet printer to precisely deposit human articular chondrocytes and poly(ethylene glycol) dimethacrylate (PEGDMA) layer by layer into a bovine cartilage defect (Fig. 18.4). The ink was composed of PEGDMA and chondrocytes in phosphate-buffered saline. After printing, the hydrogel was polymerized by UV exposure for less than 2 min. Cell experiments showed high cell viability (~89.2%). They also measured the gene expression of the printed cells by quantitative polymerase chain reaction after 2, 4, and 6 weeks culture *in vitro*. Interestingly, the printed cartilage implants had significantly higher collagen type II and aggrecan expression compared with the control. There were more extensive ECM production observed at the interface, and the printed cartilage implants showed excellent integration with the surrounding native cartilage and bone tissue.

Cohen et al. [41] used the solid free technology to make the cartilage. The ink contained alginate and the chondrocytes in phosphate-buffered saline. Because the mixing hydrogel had a time-dependent gelation process, the printing time was precisely controlled in ~15 min. Due to the natural gelation process without other physical/chemical methods, the cell viability was well preserved (~94%). However, the compressive elastic modulus of a typical printed disk was low (~1.8 kPa measured in ~1 h).

Fedorovich et al. [42] used human chondrocytes and osteogenic progenitors in alginate hydrogel to prepare the artificial cartilage. Interestingly, they mixed progenitor cells and chondrocytes separately with alginate solution to make the different ink. After that, the scaffold was fabricated with couples of layers having different cells directly adjacent to each other. Finally, the scaffolds were cross-linked in CaCl$_2$ solution (Fig. 18.5). The cell viabilities reached ~90% after 24 h culturing. *In vitro* studies revealed that the cells remained in the same location as the original deposition position when printed. An *in vivo* study showed that specific structural arrangements may induce heterogeneous tissue formation and could lead to a stable cellular architecture, proved by the general histology of osteocalcin and collagen.

5 OTHER PARAMETERS INCLUDE THE STIMULATING FACTORS AND BIOREACTORS

Biochemical factors have a great influence on zonal marker expression, and thus could be important to integrate into specific parts of the construct. As a result, control of biochemical signals or physical stimulation may be a key factor to produce a preimplanted tissue with appropriate structure and function. Because the ink component and print process could be precisely controlled, it is convenient to add different biochemical factors into the different parts of the scaffolds [43], and results showed the effects on cartilage matrix production and remodeling [44]. At the same time, physical stimulation is essential for the migration and differentiation of the cells. In this point, mimicking cartilage tissues should be placed under loading forces before implantation [45,46]. Normally, physical stimulation can be applied using bioreactors to impart shear, compression, tension, pressure, or a combination of these loads on the growing tissues [47]. Specifically to the cartilage tissue, dynamic compression of biphasic (bone/cartilage) constructs results in depth-varying stress distribution, and thus may provide appropriate signals for zonal matrix production and remodeling [48]. Also, it is very effective for bioreactors to supply continuous nutrients to the cells and take away the metabolism waste because there is no vessel system in the artificial cartilage.

6 PRECLINICAL ANIMAL MODELS FOR SAFETY AND EFFICACY EVALUATION OF BIOPRINTED CARTILAGE

Cartilage defects are "quality-of-life" lesions, as opposed to life-threatening conditions. Doctors should be very careful when choosing the proper therapies for different patients, and minimize the possibilities of adverse effects that might lead to worse quality of life than before surgery. *In vivo* animal studies are essential to closing the gap between *in vitro* experiments and human clinical studies. There are various animal models, such as murine [49], lapine [50], canine [51], caprine [52], and porcine [53]. Among them, the equine model is the best one mimicking the human situation [54].

To evaluate the effects of cartilage repaired after implantation, a variety of *in vitro* and *in vivo* methods have been used. For *in vitro* use, the FT-IR method is effective in the determination of depthwise proteoglycan content in articular cartilage, which was demonstrated previously. To investigate the *in vivo* situation, diffusion-weighted imaging [55,56] and MRI [57] allow for noninvasive measurement based on the soft and viscoelastic feature of the cartilage tissue with strong imaging and anisotropic mechanical properties.

By the way, some animal implantation experiments showed that 85% of implanted cells were lost over the first 4 weeks of implantation [58,59]. It seems that further *in vitro* maturation of the construct is likely to improve cell retention and zonal structure, but too mature constructs may not integrate with surrounding tissues [60].

7 DEVELOPING AREAS AND FUTURE DIRECTIONS

Considering the prevalence and importance of zonal variations in normal articular cartilage, recent studies have aimed at engineering cartilage with zonal structure, function, or both. Approaches to mimic the zonal structure and function include cell-based, scaffold-based, a combination of cells and scaffold (hybrid), and methods based on the application of depth-dependent strain fields.

For the bioprinting of cartilage, cell-laden hydrogels or thermoplastic polymer fibers nowadays face some weaknesses. The hydrogel has good biocompatibilities, but it is often weak in mechanical strength and degrades relatively fast resulting in the collapse of the scaffolds; the synthetic polymer scaffolds is easy to handle and durable, but not good enough in biocompatibility and degrades much slower. It is very important to choose the appropriate biomaterials or combinations for the various requirements of scaffolds and cell differentiation after printing.

Despite the effort, limited results have been obtained to date, as the newly formed tissue (mainly fibrocartilage) lacks the structural organization of cartilage and has inferior mechanical properties compared to native hyaline cartilage, therefore being prone to failure [61].

In conclusion, the challenges that need to be solved in the future include the generation of implants with complex shape, combinations of different cells and materials, and final delivery of the implant to the defect site with successful surgery.

GLOSSARY

Osteoarthritis A type of joint disease that results from breakdown of joint cartilage and underlying bone, of which the most common symptoms are joint pain and stiffness.

Articular cartilage A white, smooth tissue that covers the ends of bones in joints. It enables bones of a joint to easily glide over one another with very little friction.

Chondrocytes The only cells found in healthy cartilage. They produce and maintain the cartilaginous matrix, which consists mainly of collagen and proteoglycans.

Chondrogenesis The process by which cartilage is developed. It happens early even during the metal development. The adult cartilage is progressively mineralized at the junction between cartilage and bone, so has very limited repair capabilities, if damaged.

Bioreactors Engineered device or system that supports a biologically active environment, meant to grow cells or tissues in the context of cell culture for use in tissue engineering or biochemical engineering.

ABBREVIATIONS

ECM	Extracellular matrix
FT-IR	Fourier transform infrared spectroscopy
MRI	Magnetic resonance imaging
MSCs	Mesenchymal stem cells
PDLA	Poly-D-lactide
PEGDMA	Poly(ethylene glycol) dimethacrylate
PLGA	Poly(lactic-*co*-glycolic acid)
PLLA	Poly-L-lactide

REFERENCES

[1] Engel A. Osteoarthritis and body measurements. Vital Health Stat 1968;11:1–37.

[2] Kim S. Changes in surgical loads and economic burden of hip and knee replacements in the US: 1997–2004. Arthritis Rheum 2008;59:481.

[3] Lawrence RC, Felson DT, Helmick CG, Arnold LM, Choi H, Deyo RA, et al. Estimates of the prevalence of arthritis and other rheumatic conditions in the United States. Part II. Arthritis Rheum 2008;58:26.

[4] National Collaborating Centre for Chronic Conditions. Osteoarthritis: national clinical guideline for care and management in adults. London: Royal College of Physicians; 2008. pp. 4–5.

[5] Kreuz PC, Steinwachs MR, Erggelet C, Krause SJ, Konrad G, Uhl M, et al. Results after microfracture of full-thickness chondral defects in different compartments in the knee. Osteoarthr Cartil 2006;14:1119–25.

[6] Redman SN, Oldfield SF, Archer CW. Current strategies for articular cartilage repair. Eur Cell Mater 2005;9:23–32.

[7] Schaefer D, Martin I, Jundt G, Seidel J, Heberer M, Grodzinsky A, et al. Tissue-engineered composites for the repair of large osteochondral defects. Arthritis Rheum 2002;46:2524.

[8] Theodoropoulos JS, De Croos JN, Park SS, Pilliar R, Kandel RA. Integration of tissue-engineered cartilage with host cartilage: an *in vitro* model. Clin Orthop Relat Res 2011;469:2785.

[9] Arokoski JP, Hyttinen MM, Lapveteläinen T, Takacs P, Kosztaczky B, Modis L, et al. Decreased birefringence of the superficial zone collagen network in the canine knee (stifle) articular cartilage after long distance running training, detected by quantitative polarised light microscopy. Ann Rheum Dis 1996;55(4):253–64.

[10] Bi X, Yang X, Bostrom MP, Bartusik D, Ramaswamy S, Fishbein KW, et al. Fourier transform infrared imaging and MR microscopy studies detect compositional and structural changes in cartilage in a rabbit model of osteoarthritis. Anal Bioanal Chem 2007;387(5):1601–12.

[11] Rieppo J, Hallikainen J, Jurvelin JS, Kiviranta I, Helminen HJ, Hyttinen MM. Practical considerations in the use of polarized light microscopy in the analysis of the collagen network in articular cartilage. Microsc Res Tech 2008;71(4):279–87.

[12] Xia Y, Ramakrishnan N, Bidthanapally A. The depth-dependent anisotropy of articular cartilage by Fourier transform infrared imaging. Osteoarthritis Cartilage 2007;15(7):780–8.

[13] He B, Wu JP, Kirk TB, Carrino JA, Xiang C, Xu J. High-resolution measurements of the multilayer ultra-structure of articular cartilage and their translational potential. Arthritis Res Ther 2014;16:205.

[14] Brama PA, Tekoppele JM, Bank RA, Karssenberg D, Barneveld A, van Weeren PR. Topographical mapping of biochemical properties of articular cartilage in the equine fetlock joint. Equine Vet J 2000;32:19.

[15] Rogers BA, Murphy CL, Cannon SR, Briggs TW. Topographical variation in glycosaminoglycan content in human articular cartilage. J Bone Joint Surg Br 2006;88:1670.

[16] LeRoux MA, Arokoski J, Vail TP, Guilak F, Hyttinen MM, Kirivanta K, et al. Simultaneous changes in the mechanical properties, quantitative collagen organization, and proteoglycan concentration of articular cartilage following canine meniscectomy. J Orthop Res 2000;18:383.

[17] Poole AR, Kojima T, Yasuda T, Mwale F, Kobayashi M, Laverty S. Composition and structure of articular cartilage: a template for tissue repair. Clin Orthop 2001;391(Suppl):S26.

[18] Klein TJ, Malda J, Sah RL, Hutmacher DW. Tissue engineering of articular cartilage with biomimetic zones. Tissue Eng Part B 2009;15(2.):143–57.

[19] Vornehm SI, Dudhia J, Von der Mark K, Aigner T. Expression of collagen types IX and XI and other major cartilage matrix components by human fetal chondrocytes in vivo. Matrix Biol 1996;15:91–8.

[20] Brett AD, Taylor CJ. Construction of 3D shape models of femoral articular cartilage using harmonic maps. Lect Notes Comput Sci 2000;1935:1205–14.

[21] Carter Dennis R, Wong Marcy. Modelling cartilage mechanobiology. Philos Trans R Soc London B 2003;358:1461–71.

[22] Chung C, Burdick JA. Engineering cartilage tissue. Adv Drug Deliv Rev 2008;60:243–62.

[23] Genzyme. Carticel (autologous cultured chondrocytes) – package insert. Cambridge, MA; 2007.

[24] Boeuf S, Richter W. Chondrogenesis of mesenchymal stem cells: role of tissue source and inducing factors. Stem Cell Res Ther 2010;1:31.

[25] Tan H, Chu CR, Payne K, Marra KG. Injectable in situ forming biodegradable chitosan-hyaluronic acid based hydrogels for cartilage tissue engineering. Biomaterials 2009;30:2499–506.

[26] Hoemann CD, Sun J, Legare A, McKee MD, Buschmann MD. Tissue engineering of cartilage using injectable and adhesive chitosan-based cell-delivery vehicle. Osteoarthritis Cartilage 2005;13:318–29.

[27] Jancari J, Slovikova A, Amler E, Krupa P, Kecova H, Planka L, et al. Mechanical response of porous scaffolds for cartilage. Eng Physiol 2007;56:17–25.

[28] Silverman RP, Passareti D, Huang W, Randolph MA, Yaremchuk MJ. Injectable tissue-engineered cartilage using a fibrin glue polymer. Plast Reconstr Surg 1999;103:1809–18.

[29] Fussengger M, Meinhart J, Hobling W, Kullich W, Funk S, Bernatzky G. Stabilized autologous fibrin-chondrocyte constructs for cartilage repair in vivo. Ann Plast Surg 2003;51:493–8.

[30] Capito RM, Spector M. Scaffold-based articular cartilage repair. IEEE Eng Med Biol Mag 2003;22:42–50.

[31] Rakovsky A, Marbach D, Lotan N, Lanir Y. Poly(ethylene glycol)-based hydrogels as cartilage substitutes: synthesis and mechanical characteristics. J App Polym Sci 2009;112:390–401.

[32] Scholz B, Kinzelmann C, Benz K, Mollenhauer J, Wurst H, Schlosshauer B. Suppression of adverse angiogenesis in an albumin-based hydrogel for articular cartilage and intervertebral disc regeneration. Eur Cells Mat 2010;20:24–37.

[33] Ibusuki S, Iwamoto Y, Matsuda T. System-engineered cartilage using poly(N-isopropyl-acrylamide)-grafted gelatin as in situ-formable scaffold: in vivo performance. Tissue Eng 2003;9:1133–42.

[34] Park KH, Lee DH, Na K. Transplantation of poly (N-isopropylacrylamide-co-vinylimidazole) hydrogel constructs composed of rabbit chondrocytes and growth factorloaded nanoparticles for neocartilage formation. Biotechnol Lett 2009;31:334–7.

[35] Grad S, Kupcsik L, Gorna K, Gogolewski S, Alini M. The use of polyurethane scaffolds for cartilage tissue engineering: potential and limitations. Biomaterials 2003;24:5163–71.

[36] Mohan N, Nair PD, Tabata Y. Growth factor-mediated effects on chondrogenic differentiation of mesenchymal stem cells in 3D semi-IPN poly(vinyl alcohol)-poly(caprolactone) scaffolds. J Biomed Mater Res A 2010;94:146–59.

[37] Fedorovich NE, Alblas J, Wijn JR, Hennink WE, Verbout AJ, Dhert WJ. Hydrogels as extracellular matrices for skeletal tissue engineering: state of the art and novel application in organ printing. Tissue Eng 2007;13:1905–25.

[38] Munirah S, Kim SH, Ruszymah BH, Khang G. The use of fibrin and poly(lactic-co-glycolic acid) hybrid scaffold for articular cartilage tissue engineering: an in vivo analysis. Eur Cell Mater 2008;15:41–52.

[39] Moroni L, Lambers FM, Wilson W, van Donkelaar CC, de Wijn JR, Huiskes R, et al. Finite element analysis of meniscal anatomical 3D scaffolds: implications for tissue engineering. Open Biomed Eng J 2007;1:23–34.

[40] Cui X, Breitenkamp K, Finn MG, Lotz M, D'Lima DD. Direct human cartilage repair using three-dimensional bioprinting technology. Tissue Eng Part A 2012;18:1304–12.

[41] Cohen DL, Malone E, Lipson H, Bonassar LJ. Direct freeform fabrication of seeded hydrogels in arbitrary geometries. Tissue Eng 2006;12:1325–35.

[42] Fedorovich NE, Schuurman W, Wijnberg HM, Prins H-J, van Weeren RP, Malda J, et al. Biofabrication of osteochondral tissue equivalents by printing topologically defined cell-laden hydrogel scaffolds. Tissue Eng Part C 2012;18:33–44.

[43] Suciati T, Howard D, Barry J, Everitt NM, Shakesheff KM, Rose FR. Zonal release of proteins within tissue engineering scaffolds. J Mater Sci Mater Med 2006;17:1049.

[44] Li KW, Klein TJ, Chawla K, Nugent GE, Bae WC, Sah RL. In vitro physical stimulation of tissue-engineered and native cartilage. Methods Mol Med 2004;100:325.

[45] van der Kraan PM, de Lange J, Vitters EL, van Beuningen HM, van Osch GJ, van Lent PL, et al. Analysis of changes in proteoglycan content in murine articular cartilage using image analysis. Osteoarthritis Cartilage 1994;2(3):207–14.

[46] Timothy M, Tanya Farooque WICK. Bioreactor development for cartilage tissue engineering: computational modelling and experimental results. Seventh International Conference on CFD in the Minerals and Process Industries CSIRO; Melbourne, Australia, December 9–11, 2009.

[47] Li KW, Klein TJ, Chawla K, Nugent GE, Bae WC, Sah RL. In vitro physical stimulation of tissue-engineered and native cartilage. Methods Mol Med 2004;100:325.

[48] Lima EG, Mauck RL, Han SH, Park S, Ng KW, Ateshian GA, et al. Functional tissue engineering of chondral and osteochondral constructs. Biorheology 2004;41:577.

[49] Ferretti M, Marra KG, Kobayashi K, Defail AJ, Chu CR. Controlled *in vivo* degradation of genipin cross-linked polyethylene glycol hydrogels within osteochondral defects. Tissue Eng 2006;12:2657.

[50] Kawamura S, Wakitani S, Kimura T, Maeda A, Caplan AI, Shino K, et al. Articular cartilage repair. Rabbit experiments with a collagen gel-biomatrix and chondrocytes cultured in it. Acta Orthop Scand 1998;69:56.

[51] Cook SD, Patron LP, Salkeld SL, Rueger DC. Repair of articular cartilage defects with osteogenic protein-1 (BMP-7) in dogs. J Bone Joint Surg Am 2003;85(Suppl. 3):116.

[52] Kangarlu A, Gahunia HK. Magnetic resonance imaging characterization of steochondral defect repair in a goat model at 8 T. Osteoarthritis Cartilage 2006;14:52.

[53] Jiang CC, Chiang H, Liao CJ, Lin YJ, Kuo TF, Shieh CS, et al. Repair of porcine articular cartilage defect with a biphasic osteochondral composite. J Orthop Res 2007;25:1277.

[54] McIlwraith CW, Fortier LA, Frisbie D, Nixon AJ. Equine models of articular cartilage repair. Cartilage 2011;2:317–26.

[55] de Visser SK, Crawford RW, Pope JM. Structural adaptations in compressed articular cartilage measured by diffusion tensor imaging. Osteoarthritis Cartilage 2007;16(1):83–9.

[56] Deng X, Farley M, Nieminen MT, Gray M, Burstein D. Diffusion tensor imaging of native and degenerated human articular cartilage. Magn Reson Imag 2007;25(2):168–71.

[57] Potter HG, Koff MF. MR imaging tools to assess cartilage and joint structures. HSSJ 2012;8:29–32.

[58] Ostrander RV, Goomer RS, Tontz WL, Khatod M, Harwood FL, Maris TM, et al. Donor cell fate in tissue engineering for articular cartilage repair. Clin Orthop Relat Res 2001;389:228.

[59] Mierisch CM, Wilson HA, Turner MA, Milbrandt TA, Berthoux L, Hammarskjold ML, et al. Chondrocyte transplantation into articular cartilage defects with use of calcium alginate: the fate of the cells. J Bone Joint Surg Am 2003;85-A:1757.

[60] Obradovic B, Martin I, Padera RF, Treppo S, Freed LE, Vunjak-Novakovic G. Integration of engineered cartilage. J Orthop Res 2001;19:1089.

[61] Kock L, van Donkelaar CC, Ito K. Tissue engineering of functional articular cartilage: the current status. Cell Res 2012;347:613–27.

Chapter 19

Biofabrication of Vascular Networks

James B. Hoying and Stuart K. Williams

Department of Physiology, Cardiovascular Innovation Institute, University of Louisville, Louisville, KY, USA

ABSTRACT

The primary function of the mammalian vasculature is to circulate blood throughout the tissues of the body. Necessarily, the varied individual vessels comprising the vasculature are organized into a network, the topology of which establishes an effective perfusion circuit. This is most relevant in the distal portions of the vasculature where numerous, smaller-caliber flow paths must exist to properly distribute blood throughout a tissue. Whether to recreate one in the laboratory or regenerate a vasculature therapeutically, the small vessel calibers and the critical dependence on network topology of these distal microvasculatures create a number of fabrication challenges. A more complete understanding of the determinants of native vascular network form and function coupled with innovative approaches for manipulating and patterning vascular elements are now enabling the development of promising strategies for the fabrication of effective vascular networks.

Keywords: microvasculature; 3D bioprinting; vascular cells; vascular topology; endothelial cells; mural cells; microenvironment; self-assembly; vascular remodeling; perfusion; vascular networks; 3D imaging; microfluidics; soft lithography

1 INTRODUCTION

Whether intended for use as a cultured, three-dimensional (3D) multicellular model or for the repair of dysfunctional tissues, fabricated complex cell and tissue systems require a vasculature to be most effective [1]. In compact tissues, the effective diffusion distance for oxygen ranges from 20 μm to 100 μm and the distance is shorter for larger molecules with smaller diffusion coefficients. Thus, vessel densities must be such that no two capillaries in a specific vascular bed are farther apart than twice the diffusion distance to overcome this fundamental challenge to tissue health. A vasculature integrated into the tissue construct, with the means to perfuse it, can overcome this diffusion limit. While this may be less critical in small or thin cellular systems or those constructed from relatively porous extracellular matrices (e.g., collagen I) for which oxygen diffusion gradients collapse relatively quickly [2], the presence of a vasculature in these smaller systems would still be beneficial given the functional synergy between the parenchymal cells and the vascular cells [3,4]. Furthermore, given that all native vascularized tissues are integrated systems (as opposed to one with distinct components, for example, vessels + matrix + parenchymal cells, organized together) the manner in which a vasculature is incorporated into a constructed system impacts these dynamic interactions.

The complexities intrinsic to vasculatures and vascularized, multicellular systems are important to their function and must be addressed in any successful fabrication effort. The cell composition of individual vessel segments, the assembly of these cells into vessel structures, and the organization of individual vessels into a network with the proper topology are all essential to normalized, stable vasculatures and important aspects to consider [1]. Furthermore, how a vasculature is integrated into a fabricated cell system, including addressing construct architecture and incorporating accessory/stromal or parenchymal cell types, can have a significant impact on the stability and function of these systems. Finally, the biomechanics of the constructed system, usually a function of the matrix composition and physical constraints placed on the system (i.e., boundary conditions), must be such that the vasculature, accessory cells, and parenchyma are coordinately influenced. This can be challenging, particularly since single cells respond differently to matrix forces than structured cells, such as in a vessel. Ultimately, the goal is to create a tissue microenvironment that enables the constructed system to function as a unit.

2 THE VASCULATURE

In the simplest sense, the vasculature is comprised of two general contiguous vessel compartments: the conduit vessels (or macrovasculature), which enable the bulk transport of blood, and the microvasculature, which provides for blood–tissue exchange. In the normal circulatory loop, conduit arteries supply progressively narrower feed arteries, which supply still narrower arterioles. Downstream of the arterioles are the many capillaries, which drain blood into intermediate-caliber venules and subsequently veins. The network of arterioles, capillaries, and venules is collectively referred to as the microvasculature and is the aspect of the vascular tree that is the most intimately associated with tissues and the target of nearly all vascularized tissue fabrication strategies. (Box 19.1)

3 VASCULAR FORM AND FUNCTION

As mentioned, the arteries and veins function to transport large volumes of blood to the microvasculatures in the tissues. As such, their large calibers reflect this primary function (flow is proportional to the cross-sectional area). However, wall thickness varies significantly between arteries and veins due to differing hemodynamic conditions [5]. For a given intravascular pressure, a larger vessel diameter

> **BOX 19.1 The Vasculature**
> - Closed-loop circulatory system
> - Conduit vessels and microvessels
> - Each vessel comprised of endothelial cells and varying mural cell layers
> - Conduit vessels are large caliber, thick-walled, and function as single "pipes"
> - Microvessels are small caliber, thin-walled, and are only effective when part of a distribution network

experiences higher circumferential wall stress. Conversely, lower intravascular pressures reduce circumferential wall stress. Thus, on the high-pressure side of the circulation, the conduit arteries have thicker walls to counter this increased wall stress while conduit veins have much thinner walls as intravascular pressures are considerably lower. In the microvasculature, vessels are progressively smaller in diameter and thinner in wall thickness. This reflects the functional requirement of enabling blood–tissue exchange. While intravascular pressures progressively drop the more distal in the vascular bed, the smaller diameters contribute to reduced wall stress in these thin-walled vessels. The reduced diameters are important as this leads to reduced flow volumes through each microvessel. Furthermore, vessels within the different segments of the microvasculature display specialized structures reflecting unique functional roles. Capillaries, the distal-most and smallest vessels, serve as the interface for blood and tissue exchange. Consequently, capillary walls are one- or two-cell thick so as to be a minimal diffusion barrier. This primary exchange role is also reflected in the huge numbers of capillaries within a vascular bed resulting in a large surface area for exchange. Preceding the capillaries and comprising the distal end of the in-flow segments of the vasculature are the arterioles. Key to the structure of arterioles is a relatively high proportion of muscle in the wall relative to wall thickness and vessel diameter. Small changes in diameter due to relaxation or contraction of the muscle layers in the relatively small caliber arteriole have significant consequences on blood flow to the more distal capillaries (resistance to flow is inversely proportional to the 4th power of the radius) [6]. The intermediate diameters and thin walls of the venules reflect the very low pressures and increasing flow volumes (as the blood volume of many capillaries drain into fewer venules). Furthermore, the thin wall of the venules facilitates leukocyte trafficking in and out of the tissue spaces.

3.1 Vasa Vasorum

A unique feature of larger conduit vessels, which is more and more being addressed in single conduit vessel construction [7], is the presence of a microvascular blood supply feeding the vessel wall called the vasa vasorum. Given the numerous cell layers within these vessels, such as a muscular artery, diffusion distances are too large for molecules to move from the large vessel lumen to the vessel wall interior (wall thicknesses can be millimeters in width). The vaso vasorum, located within the adventitial layer [8], provides the perfusion support for the internal vessel wall cells. Little is known concerning how the vaso vasorum is established in native vessels. However, it is clear that this specialized microcirculation is under considerable regulation [9].

4 VASCULAR NETWORKS

While individual vessels function to carry blood, the vascular network distributes blood throughout the tissue. Therefore, since tissue perfusion is the goal, any approach intended to vascularize a tissue (native or fabricated) must include establishing an effective vascular network. In nearly all mammalian vascular beds, the microvasculature is arranged as a diameter-based, hierarchical network of vessels. Indeed, this ordered arrangement of vessels with a heterogeneous distribution of diameters is characteristic of functional vascular networks [10]. Loss of this diameter-based hierarchy (often coupled with vascular dysfunction) leads to perfusion deficits [11]. Important in the organization of the vascular network is a short rapid step-down in pressure preceding the distal-most vessels (terminal arterioles and capillaries; < 75 μm in diameter). This drop in pressure is important for proper capillary hemodynamic flow compatible with blood–tissue exchange. In the native circulation, this step-down involves branching into progressively smaller vessels from the feeding arteries often referred to generally as arterioles. Called resistance vessels because they impart the greatest resistance to flow across the entire vascular tree, these vessels rapidly distribute blood flow into numerous, more narrow parallel flow paths effectively dampening pressures while maintaining the same total volume of flow across this vascular compartment. The resistance vessels are highly vasoactive, acting to facilitate or impede blood flow as needed by the tissues. When dysfunctional (i.e., unable to properly vasodilate or vasoconstrict), blood flows to the tissue bed are perturbed leading to disease and end-organ failure [12].

4.1 Topology

Organized vascular branching is a central feature to an effective topology, entailing vessel bifurcations (one vessel splitting into two vessels) or the branching off of a single vessel from another (the "parent" vessel continues at the same diameter) (Fig. 19.1). Because network structure can vary from one tissue type to the next, even while maintaining an overall hierarchical topology, it is difficult to derive a specific set of "rules" that govern branching structures,

something that would prove useful in fabricating vasculatures. However, there are general descriptors that can be useful [13]. Presumably, these generic network parameters have evolved to best accommodate effective perfusion in the native tissues. A traditional means of describing network structure has involved a centripetal ordering of network features (called Horton–Strahler nomenclature) in which the most distal features (i.e., capillaries) are assigned an order of 0 and successive upstream segments are assigned order 1, 2, 3, and so on. First applied to mapping microvascular topology of a network downstream of a single feed arteriole [14], this approach has been used to describe network topologies [15,16]. However, in the mesenteric microvascular network, generation ordering (starting at an upstream feed vessel as "order 1") was considered more effective at describing the topology than the Strahler ordering [17]. Regardless, many have used Strahler ordering to assist in generating ratios of branching, segment diameters, and segment lengths, which could be used as "standard" descriptors in network fabrication. While variable, there are on average ~3 bifurcations per order, with segment diameter and length ratios of ~ 1.3–1.6 and ~1.8–2.1 (higher order/lower order), respectively, in terminal arteriole networks of skeletal muscle and mesentery [15–17]. These ratios can be larger for venular networks [17]. Interestingly, network size impacted the variability of these parameters within a network more than order position [17].

This hierarchical topology is often represented as a symmetric distribution of progressively branching vessels with each branch leading to progressively smaller caliber vessels. However, most native vascular networks are asymmetrical in organization and can include direct connections between feed arteries, as in skeletal muscle (arcades), or between in-flow (artery) and out-flow (vein) segments (i.e., shunts) (Fig. 19.2). Also, in many vascular beds, an artery supplying a tissue is often immediately adjacent to the vein draining the tissue, creating opportunities for countercurrent exchange between vascular compartments. This typical topology raises interesting questions concerning the governing factors of network structures and why flow paths that appear connecting proximal artery regions to the adjacent vein regions do not dominate the flow distribution over those more distal flow paths (which are longer and, therefore, impart more resistance to flow than the shorter paths) [10].

4.2 Network Topology Determinants

Considerable research, including computational modeling efforts, describes a variety of mechanisms governing vascular network architectures [18]. Universally, and related to the role of the vasculature, these findings indicate that hemodynamic forces play an essential role. However, other factors also dictate vessel diameter and network topology,

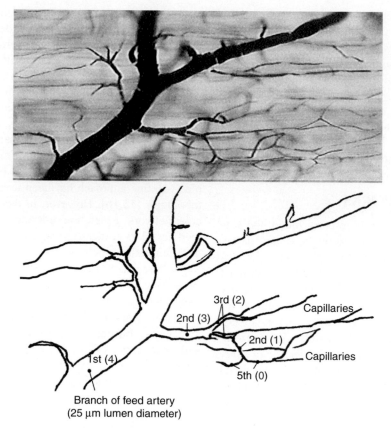

FIGURE 19.1 **Vascular ink cast of a normal, hierarchical microvasculature in skeletal muscle showing a distal branch of the feed artery and downstream vessels of the tree.** Generational ordering and Strahler ordering (red numbers) are shown for each branch order of this distal network. *Modified with permission from Ref. [1].*

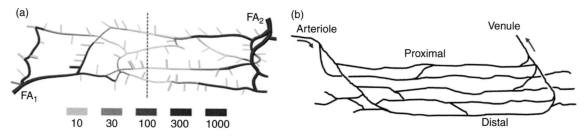

FIGURE 19.2 Examples of (a) an implanted, fabricated microvasculature visualized by India ink casting and showing the evolution of two "microvascular units" from a prevascularized implant in which the vessels were initially uniformly dispersed and (b) different symmetrical or asymmetrical networks of channels formed by soft-lithography in polydimethylsiloxane. *Parts of this figure were modified with permission from Ref. [1].*

including tissue metabolic demands, intravascular communication occurring upstream and downstream throughout the microvascular tree, and stromal environmental cures [19]. Of course, there are intrinsic, genetic programs that act to determine network architectures, particularly in establishing arterial and venous identity during development [20]. Importantly, the ability of the microvasculature to structurally adapt in response to these varied and incompletely defined influencing factors is critical to establishing a network topology optimized for a specific perfusion demand. From a fabrication perspective, it may not be necessary to generate

an exact, predetermined, final vascular topology because of this adaptability. Instead, a precursor topology could be established followed by the successive adaptation of this generalized network into a specific network topology reflective of the final, desired outcome.

4.3 Vascular Adaptation and Remodeling

In response to chronic changes in flow and or pressure, an individual vessel will structurally adapt resulting in a larger or smaller diameter or a thicker or thinner wall. It is thought that these adaptation changes reflect, in part, an effort by the vessels to normalize shear stress as a function of the intravascular pressure [21]. As an intrinsic vascular response, structural adaptation occurs in large, conduit vessels as well as microvessels. For example, a reduction of flow with constant pressure in the left carotid artery in mice results in a reduction in diameter [22]. Similarly, increases in flow and pressure within an arteriole, due to shunting of blood flow secondary to vascular occlusion, leads to the enlargement and muscularization of the artery (so called arteriogenesis) [23]. As mentioned, this general process of structural adaptation, in coordination with new vessel growth (i.e., angiogenesis), is a primary means by which microcirculation changes to meet new metabolic demands of a tissue during development and in the adult. Importantly, as individual vessels within a network adapt their structure, other vessels in the network coordinately adapt. As a consequence, overall network topology changes as individual vessels take on new roles in the network and redistribute blood [24]. Because all vascular cell types within the vessel wall appear to be involved in this adaptive process, it is important that these same cell types are present in the initial fabrication of a vascular bed.

4.4   Angiogenesis/Postangiogenesis

Native vascular networks evolve from the neovessels formed via angiogenesis (new vessels from existing vessels) and vasculogenesis (de novo assembly of vessels) during development. In the adult, microvascular beds are expanded primarily via angiogenesis coupled with those subsequent vascular processes leading to interconnected, mature vessels (postangiogenesis) [25]. For purposes of fabricating vascular networks, angiogenesis is an effective means to generate the vessel segments of the new tree. Such strategies have been explored in the context of building simple, *in vitro* microvasculatures and usually involve native "parent" vessels or vessel segments from which angiogenic neovessels arise. For example, the neovascular sprouts that spontaneously arose from the opposing ends of an artery and vein segment in a biochip system created a simple array of parallel neovessels (established by permanently constraining angiogenic sprouting) perfused via lumens contiguous with the parent artery and vein segments [26]. Similarly, channels formed by soft lithography served to pattern and direct angiogenesis from isolated, parent microvessels creating a network of defined topology [27].

Importantly, the angiogenic growth of neovessels must occur in a three-dimensional space. Thus, any strategy depending on angiogenesis to establish the vessel segments of the new, defined network involves patterning the matrix within which neovessels can grow, and not the vascular cells themselves. An added benefit is that by patterning matrix environments, the angiogenic neovessels that grow

throughout that space are able to also adapt and remodel in response to hemodynamics as needed as there are no specific constraints of the vessel walls of individual vessels (addressed in detail later). While angiogenesis is useful in generating a prenetwork, the neovasculature that develops subsequent to angiogenesis does not necessarily maintain any prepatterned architecture. Likely due to the vascular adaptation activities intrinsic to vessels, neovessels that are prepatterned during angiogenesis abandon this pattern as the neovasculature evolves and matures, unless the patterning cues are maintained during this postangiogenesis phase [28]. As discussed later, angiogenesis is one means by which vasculatures are preformed *in vitro*. Perhaps reflected in the clinical findings, other techniques are required to complete the generation of mature vascular beds from these angiogenic precursors (also discussed later).

4.5   Computational Models of Vascular Networks

In an effort to describe and predict microvasculature architectures, many have developed computational models and mathematical descriptors that may prove useful in designing and fabricating networks. Multiscaled, discretized and/or continuous models integrate a variety of putative controlling signals to predict the extent, direction, and progression of neovessel growth as well as topology outcomes of structural adaptation [29]. Some computational models incorporate the influence of extravascular matrix dynamics on vessel branching and connections [30,31]. One prediction from these varied models is that new vessel network formation depends on both deterministic and stochastic processes [32]. Still others are using physical laws, such as minimum work/energy rules or fractal dimensionality, to model microvascular branching [33,34]. Importantly, theoretical investigations of microvascular structural adaptation indicate that hemodynamic forces in combination with surrounding tissue metabolic demands and retrograde intravascular communication are required to produce realistic network architectures [35] (Box 19.2).

BOX 19.2 Vascular Networks

- Hierarchical arrangement
- Heterogeneous vessel diameters
- Tissue-specific topology (usually asymmetric)
- Downstream vessels are more numerous than upstream vessels
- Vascular adaptation
- Determined by hemodynamic forces and stromal environment
  - Intravascular fluid flow
  - Intravascular fluid pressure
  - Stroma deformation

5 VASCULAR FABRICATION

Macrovessels and microvessels display significant differences in vessel wall structure. These structural differences also reflect wall compositional differences. While all vessels are lined by a single monolayer of endothelial cells (ECs), the numbers and types of cells comprising the additional layers of the vessel wall vary considerably. For example, conduit arteries contain multiple circumferential layers of smooth muscle and stroma cells, elastin sheets, and extracellular matrix. Microvessels contain considerably fewer cells and less matrix mass, and many of the mural cell types are specialized. Furthermore, the mural cell layers represent a continuum in phenotype from the more muscular, circumferentially oriented smooth muscle cells in arterioles to the sparsely covered, muscle-like pericytes in the capillaries. Regardless of the method, fabrication of a vessel must address the type of the target vessel, which will determine the size of the vessel as well as the composition and structure of the vessel wall. In addition, when fabricating microvessels, as is the goal in most vascularized tissue constructs, it is necessary to address network design, as single microvessels alone are usually not useful. Finally, any fabrication strategy for assembling vessels and, particularly, microvascular networks, should permit vessels to adapt and networks to reorganize as the vascular system matures and stabilizes. Reflecting these design considerations, a number of general fabrication strategies using a variety of vascular cell sources have been explored. Nearly all of these strategies rely on the ability of vascular cells to self-assemble into vascular structures, either as mono- or multilayers. So far, none of these approaches result in a ready-made, fully mature vessel or vasculature at the completion of fabrication. However, many of them do establish perfusable, *in vitro* networks that are also capable of integrating into a host circulation when implanted *in vivo* (which often leads to final maturation).

5.1 Native Vascular Fabrication

The earliest approaches to vascular fabrication involved the use of biological factors to induce new vessel growth (so called angiogenic factors) within native tissues [36]. Clinically motivated as a possible means to treat tissue ischemia, angiogenic factors and/or cells generating angiogenic factors were/are delivered (usually injected directly) to the target tissue [37,38]. This induces new vessel growth thereby expanding the relevant vascular beds with the goal of increasing perfusion of the tissue. While the use of angiogenic factors is extensively used to grow vascular cells, promote vascular assembly *in vitro*, and promote new vessel growth *in vivo*, clinical outcomes have been disappointing [37]. While there are possible technical reasons why angiogenic therapies are not as effective as anticipated, one aspect reflects potential problems in postangiogenesis

maturation and adaptation of the newly formed vessels [1]. With the same clinical goal in mind, injection of vascular regenerative cells into the affected tissue is being explored as a means to build new vascular segments. Numerous different cell types are under investigation including bone marrow-derived mesenchymal cells, EC progenitors, adipose-derived cells, and pluripotent stem cells [39–41]. The consensus opinion is that these regenerative cells generate vasculatures by promoting angiogenesis [42]. However, some of the cell types, including freshly isolated adipose stromal vascular fraction cells, are able to regenerate new vessel segments *in vivo* directly via vascular assembly [4]. Whether the regenerative cell approach proves more efficacious in the clinic awaits the outcomes of numerous ongoing clinical trials.

5.1.1 Devices

A variety of device implant approaches incorporate a strategy whereby the site of implantation is prevascularized (or hyper-vascularized) using angiogenic factors prior to implantation. Along these same lines, device designs include the incorporation of angiogenic factors into the device, often as part of an external, porous material coating, which subsequently promotes angiogenesis once implanted [43,44]. Many of the vasculatures generated via this strategy have high vessel densities and often-chaotic network architectures, likely due to the relatively high concentrations of angiogenic factors employed. However, despite this, device function and survival (if living cells are also incorporated) is improved.

5.2 Vessel Assembly

The cellular building blocks commonly used in fabricating vasculatures include a variety of ECs or EC precursors (although human umbilical endothelial cells are frequently used) and mural/perivascular cells such as smooth muscle cells, pericytes, and mesenchymal smooth muscle precursors. With respect to the EC, it does not appear that the origin of the EC is critical; conduit ECs are capable of assembling into microvessels and microvessel ECs are able to establish a luminal lining in conduit vessels. ECs (or endothelial precursor cells) are necessary and sufficient for establishing the initial vessel and network structure. While not essential in the initial vessel formation, inclusion of mural cells facilitates vascular assembly, in part, by stabilizing the immature vasculature [45]. Many of the fabrication strategies, particularly for microvasculatures, rely on the intrinsic ability of these vascular cells to self-assemble into vessels, which occurs more readily in a 3D environment. This self-assembly ability likely reflects a more generalized behavior whereby cells actively aggregate with like-cells via homotypic adhesion molecules specific to ECs [46]. It also reduces considerably the challenges in vascular fabrication.

Since it is not necessary to place individual cells into specific positions within the vessel wall, the task is reduced to simply patterning the cells with the expectation that vessel assembly and network formation will happen spontaneously within the predetermined pattern.

5.3 Patterning Vascular Networks

Controlled patterning of vascular networks generally involves one or both of two general approaches: vascular cell seeding/filling of preformed channels or directed deposition of vascular cell suspensions [47]. As mentioned, nearly all of these approaches rely on the ability of the vascular cells to self-assemble into the proper structures or, in the case of EC seeding, a monolayer. In general, the intent is to establish a high-order architecture within which the vascular cells can establish appropriate vessel structures.

5.3.1 Self-Determination

It is well known that ECs (and additional support cells) randomly suspended within a 3D matrix spontaneously assemble into vascular structures and, when implanted, progress to form perfused, hierarchical vascular networks. In the absence of additional instructive cues, the individual vessels within these spontaneously assembled vasculatures appear to be randomly arranged beyond the hierarchical branching organization. Differences invariably arise in the density of segments per network, depending on the application. However, these differences may not solely reflect the numbers of vascular cells/precursors used to generate the networks as final vessel densities are influenced, in part, by the tissue environment in which the vascular construct was placed. For example, a preassembled vasculature placed in the epicardial environment formed a more vessel-dense microvasculature than one placed subcutaneously [48,49]. Interestingly, the final vessel densities within the respective networks, which were both different than the starting densities, matched closely the native vessel densities of the tissue at the implant site. In both cases, though, the network architectures

within the implants were not appreciably different (although a detailed evaluation of topologies was not performed). As implied in these experiments, these implant environment cues are sufficiently strong enough to over-ride initial patterning cues. In a separate study, a preformed vasculature in which the vessel segments were all aligned parallel to each other (an anisotropic topology) lose this organization when implanted (generated an isotropic topology) [28]. However, when constraints were placed on the implant such that the direction of implant deformation was defined, the resulting fabricated vasculature maintained anisotropy. Interestingly, the arteriole–capillary–venule composition of the isotropic and anisotropic networks were different. While difficult to assess because overall network organization was not described, implantation of multiple EC-containing spheroids, which also effectively patterns ECs in dispersed clusters, appears to generate a network that does not reflect this initial patterning when implanted (networks did not consist of interconnected clusters of microvessels) [50,51]. Thus, considering the findings from these studies and many others, the final topology of vascular networks derived from implanted vascular constructs, regardless of the initial patterning, will likely exhibit a less oriented topology reflecting the native vascular architecture of the implant site (Fig. 19.3). This may or may not be desired depending on the application (it may be ideal that the new vasculature mirrors the surrounding native vasculature). However, if a specific *in vivo* network topology is required, it appears necessary to instruct the vascular system during implantation, probably during the vascular remodeling and adaptation phases. The breadth of cues that could be employed to provide specific topology cues *in vivo* is not known. In the study mentioned earlier, the implant was mechanically constrained to influence implant deformation. In another approach, ECs were organized *in vitro* using a magnetic particle-based strategy [53]. Whether magnetic forces can be similarly used *in vivo* remains to be determined. Certainly, the presence of tissue architectural cues, such as might be provided by matrix structure or parenchymal cell organization, would influence

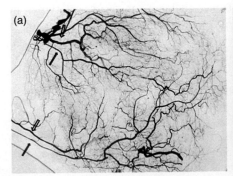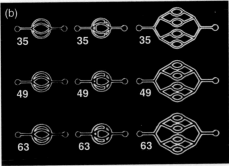

FIGURE 19.3 Examples of (a) computed pressures (color coded in units of mm Hg) for different segments of an arcade network fed by two arteries (FA) in skeletal muscle, and (b) a section of a stereotypical skeletal muscle microcirculation highlighting the shorter and longer flow paths in the proximal and distal regions of the network. *Parts of this figure were modified with permission from Ref. [52].*

overall vascular topology. For example, vascular precursors in an engineered bone matrix formed vessels within the implant only in areas not occupied by bone matrix [54]. However, the impact of the final network topology was not described.

5.3.2 Microfluidic Strategies

An often employed strategy for patterning vascular cells involves the seeding or sodding of vascular cells onto or into preformed channels with the dimensions matching individual microvessels (usually 6–200 μm in diameter). A common approach to form channels utilizes standard methods of creating complicated patterns in nonliving systems (e.g., microelectromechanical systems (MEMS) chips and circuit boards) modified to involve semisolid materials (usually polydimethylsiloxane). One such approach, soft lithography, involves creating stamps from lithography masks reflecting a desired 3D pattern and using these stamps to mold vascular-compatible matrices into 2D or 3D channels [26,27,55]. These soft-lithography approaches usually go hand-in-hand with microfluidic platforms, in which similar lithography techniques are also used to cast fluid-flow pathways to perfuse casted vessel channels (Fig. 19.3). This general strategy has been used to form predominately parallel, endothelialized channels as part of a more complex, microfluidic-based system as a "biology on a chip" or an *in vitro* engineered tissue solution in which collections of parenchymal cells are interlaced with EC-lined, perfusable channels [26,55,56]. In those situations where the matrix environment in which the channels are formed is rigid or not able to be remodeled by the cells, the "vessels" formed are merely lined walls of fixed vascular dimensions. Recognizing the need for vessel adaptability in maturing microvascular networks, more native matrices (e.g., collagen), are being employed with the idea that the vascular cells can remodel channel dimensions and/or undergo angiogenesis thereby forming new vascular connections between preformed channels [57,58]. Related approaches formed channels using fine threads or needles as mandrels for microvessel-sized channels in combination with microfluidic platforms [59,60]. In another alternate approach, a microfluidic-based strategy incorporated a native microvasculature with an attached artery and vein leash (derived from a section of the rat femoral vascular bed) to support sheets of cocultured cardiomyocytes and ECs as a means to vascularize the cardiac tissue mimics [61]. The artery and vein leash was used to connect the femoral microcirculation to an *ex vivo* perfusion system upon which the cardiac sheets were placed. Importantly, new, perfused vessels formed within the engineered cardiac sheets, but only when ECs were incorporated into the sheets during culturing. The same soft lithography approach can be used to pattern matrix compartments as opposed to creating channels within which the cells line (Fig. 19.2). In these approaches, the patterned channels are filled with a suspension of vascular cells in the matrix (often collagen or fibrin) with the expectation that vessel elements will spontaneously assemble within the 3D matrix space and adapt as needed within the patterned matrix spaces [27,62]. While there may be a benefit to providing the cells to freely assemble and adapt, there is an associated challenge with establishing intravascular perfusion *in vitro* (which is very doable in those microfluidic systems where the patterned channels serving as the vessel walls are also contiguous with in-flow and out-flow perfusion paths). Thus, the patterned matrix approach has seen real utility in implanted systems in which the preformed vascular networks are able to spontaneously inosculate with the host circulation. Recently, channels were preformed in a 3D matrix by directly dispensing carbohydrate solutions that harden into a glass to form a cast of the desired microvascular network [63]. The glass was then dissolved and seeded with vascular cells forming a perfusable, vascular cell-lined, channel network. This approach reflects the use of sacrificial materials to form channels. Once cast in the correct topology and surrounded by matrix, these materials are then flushed out of the system leaving behind open conduits, which are subsequently sodded with vascular cells. This same sacrificial approach was used to form implantable microvascular networks using patterned Pluronic hydrogel placed in a collagen matrix as the channel-forming material [58]. In this most recent approach, the prebuilt microvasculature was incorporated into the host circulation via anastomotic attachments to a feed artery and vein, thereby providing immediate perfusion of the fabricated vascular system.

5.3.3 Bioprinting Strategies

The carbohydrate glass work is an example of an emerging strategy to prepattern microvascular networks involving the directed dispensing, generally called bioprinting, of vascular cell suspensions [47,64,65]. With bioprinting, three-dimensional vascular structures and networks can be formed via the additive deposition of vascular cell suspensions in a suitable matrix (often termed bioinks) in a manner analogous to rapid prototyping [47,66]. Two general bioprinting strategies are used to fabricate vascular systems: a rasterizing (point-by-point) approach whereby discrete "packets" of vascular cells (droplets or spheroids) are deposited adjacent to each other in the desired shape or topology [67,68] and continuous extrusion-based, dispensing of vascular cell suspensions (direct-writing) [64,69,70] (Fig. 19.4). Important to the rasterizing approach is that the discrete spheroids must eventually join with each to form the final, solid structure [71]. Fortunately, such printed vascular cell systems spontaneously fuse over a short time period, perhaps reflecting the self-assembly character of vascular cells. Some of the earliest bioprinters were inkjet printers modified to dispense droplets of cell suspensions [65,67]

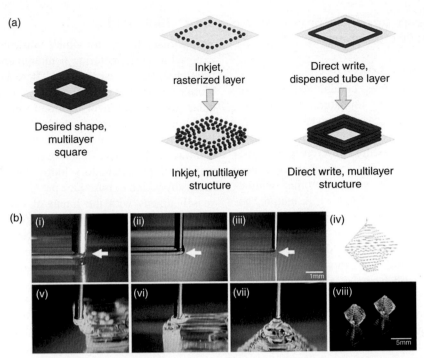

FIGURE 19.4 (a) Illustration highlighting inkjet and direct-write bioprinting of a 3D structure. (b) Examples of direct-write printing with a biomaterial. (i–iii) Still images from videos showing line morphology differences in direct writing of Pluronic gel using a (i) 20 g, (ii) 25 g, or (iii) 30 g print tip. Arrows point to the leading edge of each printed line. (iv–viii) Progressive steps in the bioprinting fabrication of a model 3D shape starting with the computer generated print line path (b-iv) and ending with the final structure (b-viii).

and have been used to successfully pattern vascular cells [72]. Similarly, small-caliber conduit vessels have been constructed one spheroid at a time employing a dispensing robot [70]. Interestingly, and immediately relevant to fabricating vasculatures, fusion of hollow spheroids containing vascular cells printed in contact with each other leads to the formation of a contiguous lumen between the spheroids via a fluid forces mechanism [73]. In continuous extrusion printing, or direct-write printing [64,74], vascular bioinks are additively extruded onto a surface using controlled x, y, and z movements of either the print head or stage of a printing robot in a manner analogous to rapid prototyping in mechanical applications [64,75]. It was this type of printing that was used to construct the sacrificial glass channels mentioned previously [63] and has been used to directly pattern ECs in a mixed-cell bone construct [76] and pre-angiogenic microvessels to form microvascular networks of defined topologies [28]. While not related to fabricating living, functional vasculatures, direct-write prototyping is also being used to create thermoplastic, anatomical models of tissues and organs, including vascular trees, derived from patient-specific image data sets [77,78]. Such models promise to guide surgical repair strategies, as the clinical team has a physical representation of the target tissue/organ "in hand" [79]. In bioprinting, the material base used in the vascular cell bioinks is the most critical aspect [47]. For effective outcomes, the vascular/matrix suspension being printed must flow well enough to be extruded but be viscous enough to hold shape once dispensed. If the bioink is not sufficiently viscous (stiff), then the printed structure will collapse making subsequent additive printing passes problematic. Importantly, appropriate material bases must also permit the vascular cells to self-assemble and microvessels to adapt. Because of this, native polymers, such as collagen or alginate, are often employed [74,80–84]. These native matrices can be modified to improve printability and functionality [85]. Many of the hydrogels and other materials used as scaffolds for cell and tissue assembly are also amenable for use in bioprinting.

5.4 Maturing Fabricated Vascular Systems

Once constructed, vascular systems, as with most fabricated tissues, require additional processing to condition and mature the construct. Whether it is to provide sufficient time for cells to proliferate and self-assemble, derive specific cell phenotypes, and/or remodel and improve construct biomechanics, postprocessing promotes a more functional vascular system. Oftentimes, the postprocessing occurs in culture involving specialized bioreactors. But, the *in vivo* environment (i.e., implantation) can also serve as a "natural bioreactor" providing those cues necessary for maturation. For the vasculature, these important cues arise primarily from intravascular perfusion: blood flow and pressure. However, recent evidence indicates that stromal deformation can also influence the phenotype of the vasculature.

5.4.1 Hemodynamic Forces

Invariably, fabricated vasculatures, conduit vessels or microvascular networks, are initially immature. As discussed, a stable mature vessel contains a monolayer of ECs intimately surrounded by one or more layers of mural cells. Both vascular cell types can exhibit different phenotypes that reflect different functional states. Furthermore, these different cell types communicate with each other facilitating vessel assembly and stability [86–88]. A mature, homeostatic EC is nonthrombogenic, nonimmunogenic, and nonleaky to macromolecules [89]. Shear stress due to fluid flow across the endothelium is a primary determinant of EC phenotype [90]. Low shear rates, such those as seen in laminar flow, are thought to establish a quiescent, stable EC phenotype. However, static conditions and high shear rates typical of turbulent flows are associated with EC dysfunction and disease. Thus, the absence of fluid flows in the static systems commonly employed in assembling vascular systems negatively impact EC activities. Even in those strategies whereby open channels are seeded with ECs by perfusing a cell suspension through the channels typically require a period of no flow to allow the cells time to adequately adhere to the channel walls. Thus, essentially all strategies intended to mature the fabricated vasculature entail exposing the ECs to a defined shear stress. The magnitude of the shear stress established in these systems is often determined by the vessel type(s) being fabricated. The non-Newtonian nature of blood, the variations in vessel wall geometry, and the challenges associated with measuring velocity profiles in vessels, make it difficult to define the shear stresses present in all segments of the native vascular tree (particularly in the microvasculature where diameters are small and rheological phenomena can drastically change blood viscosity). Generally speaking though, shear stresses acting on artery and venous ECs are calculated to range from 10 dynes/cm^2 to 70 dynes/cm^2 and 1 dynes/cm^2 to 6 dynes/cm^2, respectively [91]. In the microvasculature, shear stresses can range considerably being as high as 100 dynes/cm^2 and as low as 7 dynes/cm^2 depending on the vessel diameter and pressure [21]. Thus, due to pressure differences, ECs in arterioles tend to experience greater shear stresses than in venules and capillaries for similar diameters, while capillaries and venules are very similar in the diameter-dependent shear stress profile [21]. Coordinately, circumferential and radial stresses on the vessel wall contribute to changes in the wall structure and musculature. Cyclic stretching of smooth muscle constructs result in enhanced smooth muscle contractility and mechanical properties [92]. This is perhaps most relevant in conditioning fabricated arteries where such stresses can be high and mechanical integrity has been difficult to match with elasticity alone.

5.4.2 Bioreactors

The primary means by which fabricated vasculatures are exposed to these maturing hemodynamic forces *in vitro* is through the use of bioreactors. There are numerous *in vitro* bioreactor designs employed to condition fabricated vascular systems that involve some form of a chamber within which the vasculature is established and means to provide fluid flow into and out of the chamber. In many cases, there is intravascular perfusion as well as extravascular perfusion using multiple circulatory loops. Nearly all involve some means by which to introduce pressurized fluid flow through vessel lumens with the intent of exposing the immature vascular cells to normalizing hemodynamic forces. Given the larger diameters, this is considerably easier to do for conduit vessel constructs than microvessel systems as connections to and from circulating pumps, and so on, are more readily size-matched. For small-caliber vasculatures, the microfluidics platforms described earlier can be effective. Working from the basic design of establishing in-flow and out-flow paths contiguous with the lumens of the fabricated vasculature, new strategies are emerging in bioreactor design intended to optimize conditions and actively adapt coordinately with maturing vasculatures. Generally involving multiple sensors incorporated into the bioreactor coupled with a computer-controlled system that dynamically adjusts reactor conditions based on readouts from the sensors, these "smart" systems monitor oxygen tensions, flow rates, and so on, to continuously provide desired environmental conditions as maturation states change [93,94]. Additionally, bioreactor perfusion strategies are being developed to identify and establish specific nutrient gradients to impact cell and vessel behavior [95]. In one example, the extent of angiogenesis in the bioreactor depended on whether or not there was (1) intraconstruct perfusion or (2) intraconstruct perfusion coupled with extraconstruct perfusion [2]. Given that oxygen gradients were calculated to be uniform within the construct after a few hours in this bioreactor, other mechanisms, such as autocrine growth factor washout, were considered to contribute to the differential neovascular growth observed under these different perfusion conditions. Additional design considerations that are being explored include implementing different pressure waveforms (e.g., pulsatile vs. continuous) or establishing microgravity using rotating wall vessel chambers [96,97].

5.4.3 Implantation

As discussed, the objective of many vascular fabrication strategies is to establish a new, functional vasculature *in vivo*. While the intended outcome is perfusion of the implanted vasculature as a means to support tissue health and function, the maturation and conditioning of the vasculature naturally occurs. In this way, the recipient can be thought of as a bioreactor. Because the manner in which the vascular implant integrates with the host circulation is rarely predetermined, it is often difficult to establish a final vascular architecture that meets specific design parameters (see the earlier discussion on vascular remodeling and adaptation). However, assuming the vasculature can adapt, the stable vascular network that forms matches the hemodynamics and environmental conditions specific to the implant site. Implicit in the implantation of fabricated vasculatures is the need for integration with the host circulation. With conduit vessels, this is very straightforward as the constructed vessel is anastomosed directly to the host circulation. However, for most preformed microvasculatures, this is difficult due to the small caliber of vessels involved. One strategy employs incorporating larger caliber inflow and outflow segments (artificial and native) into the network design that can be microsurgically anastomosed into the host circulation [2,58]. In the absence of these larger segments, integration with the host circulation relies on the spontaneous inosculation between microvessel segments of the implant vasculature and surrounding host vasculature. This necessarily requires sections of the host microcirculation to undergo angiogenesis and the growing neovessels of both microvasculatures locate and interact with each other such that the respective lumens are contiguous. While this occurs spontaneously and fairly often, the underlying mechanisms are not well defined thus making it difficult to direct inosculation in a specific fashion, if desired. There is some evidence that tissue macrophages can mediate or chaperone the preconnection gap [98,99]. Whether these macrophages also guide the neovessels toward each other or serve some other role is not known. In cases where the fabricated vasculature is derived from single vascular cells, it appears that the vascular cells, in particular the ECs, themselves appear to add to the growing ends of the host neovasculature as single cells prior to assembly into an implant vessel segment [100]. This likely reflects the inherent ability of vascular cells (and vascular precursor cells) to incorporate into the neovessels of an angiogenic bed. In this regard, it may prove useful to incorporate stromal cells into a prevascularized system to promote inosculation and integration with the host circulation [4]. While hemodynamics most certainly are critical in establishing mature, stable vessel and network phenotypes, there is possibly some genetic predetermination operating in concert with or independent of hemodynamic cues, particularly during development [101]. However, there is considerable arterio-venous plasticity in ECs and microvessels [102,103], suggesting that the level of genetic predetermination might be limited in the adult.

5.5 Tissue Stromal Mechanics

Regardless of whether the vasculature is fabricated in the laboratory or in the native tissue, vessels are interacting with the surrounding microenvironment or stroma. While in many fabricated systems, this stroma may be a simple 3D matrix environment, tissue stroma is considerably more complex. While it is clear that the cells residing within the stromal space (and the parenchyma) regulate vascular form and function, there is a growing appreciation for the role stromal mechanical forces play in influencing the vasculature. This is perhaps even more relevant for *in vitro* systems in which the stromal environment is often contrived and definable. Physically constraining a collagen I-based stroma, either during angiogenesis or the postangiogenesis remodeling phase, profoundly influences vascular topology [28,104]. The extent by which stromal mechanics influences the vasculature appears to reflect the degree and orientation of stromal matrix deformation [105,106]. Forces applied to collagen-based matrices are rapidly dissipated and do not persist due to the highly viscoelastic properties of collagen and other native gels [107]. However, while they last, these forces (derived from the vascular cells themselves) deform the interstitial matrix, the extent of which is determined by the effective compliance (or stiffness) and/or physical constraints (i.e., boundary conditions) of the stroma [105,108]. It is this deformation (the direction and extent) and not directly acting stresses that influences the orientation of vessel elements such that they align perpendicular to the primary direction of compressive strain. Thus, in fabricating microvasculatures, the composition, shape, and physical constraints placed on the cell/tissue system can be manipulated to direct final vascular architectures [109].

5.6 Assessment

Any vascular fabrication effort includes an assessment of outcomes. Of course, the primary function of any vasculature is to support perfusion of the tissue bed. Often, this is indirectly assessed by measuring vascular density, reported as number of vessels/area, length density (total vascular length/area), or vascular volume density (total volume of vasculature/area). Typically, formed vessel elements are identified via specific labeling of ECs and/or mural cells using a variety of markers in histological sections or whole tissues. Platelet endothelial cell adhesion molecule (PECAM, CD31) is probably the most widely used marker. Another often used label involves tagged plant lectins that will bind to the carbohydrate moieties commonly present on ECs. *Griffonia simplicifolia* lectin (GS-1), isolectin B4 (a sub-type of the GS-1 mixed lectin preparation), and *Ulex europaeus* lectin mark rodent ECs (GS-1 and isolectin B4) and human ECs (*Ulex europaeus* lectin). Other markers that reflect more the functional state of mature ECs, such as von Willebrand Factor, have been effectively used in the past. However, as is the case with von Willebrand Factor, they do not uniformly mark all endothelium and their expression can vary depending on the phenotypic state of the ECs. In contrast to these exogenous labeling approaches, the use of vascular

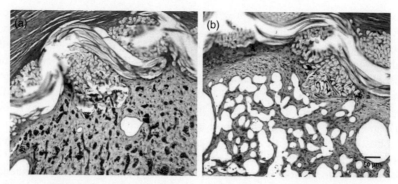

FIGURE 19.5 An example of a vasculature in which vessel elements in different regions of the same tissue do or do not contain blood cells (magenta). The tissue is an area of new vascularized tissue formation associated with a polymeric, endovascular graft. (a and b) From histology sections taken from near-adjacent regions of the same tissue block and stained with trichrome.

cells from transgenic reporter animals, such as the tie2-GFP transgenic mouse [110], in a vascular fabrication approach provides for an endogenous tag that can be used to visualize vasculatures in real-time. This strategy is also useful for tracking the location and behavior of the cells during vascular assembly and maturation. It is important to note that ECs and subsets of hematopoietic-derived cells (particularly macrophages) share many of the markers used to identify vascular structures. However, combining a positive signal from one of these tags with discrete vessel morphology indicators (e.g., lumenized structure, elongated vessel-like structure) facilitates discrimination between vascular and nonvascular entities. From these types of measurements, it may be assumed that the presence of vessel elements infers perfusion. However, this assumption is not always valid as there are a number of instances, in both normal and pathophysiological conditions, in which vessel density and perfusion density are uncoupled [1]. It is not uncommon, particularly in vascular engineering applications, for vessel segments to be not perfused, even if adjacent segments in the network/tissue are (Fig. 19.5). Even the presence of blood cells within the vessel segments, say at the time of explant, does not necessarily mean that those segments are actively perfused. While they may at one time have carried blood, perhaps as immature vessels, active vessel revision intrinsic to vascular network development can generate dead-ended vessel segments resulting in flow stasis or trapped blood components.

5.6.1 Perfusion

Common approaches to directly assess perfusion in the experimental and clinical settings generally include some aspect of imaging such as Doppler-based imaging (optical or sonic), fluorescence-based imaging, or direct visualization via microscopy [111–114]. Laser Doppler perfusion imaging (LDPI) is commonly used to generate an index of perfusion or blood flux (the product of the velocity and number of moving erythrocytes (RBCs)) in microvascular beds. Based on the Doppler effect, LDPI systems scan the tissue bed with infrared (IR) or near IR illumination (for better tissue penetration) and detect the Doppler shift in reflected light caused by the moving RBCs. Most systems can generate a color-coded map of perfusion values superimposed on the imaged tissue. LDPI is sensitive enough to assess the perfusion of small tissue volumes, including small implanted fabricated microvasculatures [49]. However, despite the use of the IR lasers common to LDPI, imaging depths are shallow, usually on the order of millimeters, depending on the tissue type. Similarly, the Doppler shifts in ultrasonic waves are used to assess flow velocities. Usually used to measure blood flow in conduit vessels, combinations of blood contrast agents and high-resolution ultrasound make it possible to generate an index of microvascular perfusion [112]. The advantage with ultrasound is that it is possible to capture tissue anatomical information simultaneous with flow data. However, resolutions are such that small tissue volumes with low perfusion are difficult to visualize and measure. Historically used to visualize the eye circulation, IR fluorescence imaging is now being used more often to evaluate tissue perfusion, particularly in surgical applications such as flap reconstruction and tissue repair [114]. Typically involving the intravascular delivery of an IR-sensitive dye, such as indocyanine green (ICG), these systems visualize the presence of the dye within the vascular space. Imaging in real-time before and after intravascular delivery of the dye can identify the distribution paths and/or a perfusion "bloom" as the dye moves into the downstream vascular network. Imaging depths are slightly deeper than with LDPI, but current systems do not include quantitative assessments. The traditional methods for assessing the microcirculation have involved intravital microscopy in which the vascular network of interest is directly visualized [113]. Coupled with the use of blood tracers, flow velocities in individual microvessels and across microvascular networks are measured with high spatial and temporal resolution. Unlike these other methods, intravital microscopy usually entails exposing the target microvascular bed and is thus more invasive. Magnetic resonance imaging (MRI) techniques, such as arterial spin labeling (ASL), provides a noninvasive measure

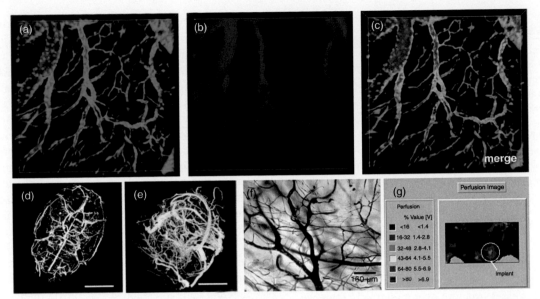

FIGURE 19.6 **Example images of vasculatures acquired via different imaging modalities.** (a–c) Confocal microscopy: a fabricated microvascular implant imaged using dual-fluorescence tagging (green = EC, red = blood tracer) and confocal microscopy and volume rendered. μCT: (d) brain and (e) tumor microvasculatures cast with a contrast agent. Scale bars = 0.5 cm. (f and g) Ink casting and LDPI: a mature, fabricated microvascular network visualized via intravascular ink casting and LDPI to assess implant perfusion. *Parts of this figure were modified with permission from Refs [49] and [118].*

of perfusion through a tissue bed by magnetically inverting a defined volume of in-flowing arterial blood to a tissue bed and assessing the rate at which this packet moves through a tissue bed. Because of the relatively low resolution of MRI, the ASL method is often used clinically or in large animal models. However, new refinements in ASL techniques are making it more useful for smaller tissue volumes, such as in the mouse [115]. Whether such a method, even with magnetic contrast agents, would be useful for the often-smaller microvascular volumes of implanted, fabricated networks remains to be seen. Related, magnetic tagging of cells for tracking by MRI is being actively investigated as a means to locate injected or implanted cells [116]. While not imaging per se, microsphere-based approaches can provide measurements of perfusion into vascular beds, even for small vascular volumes [117]. To make these types of measurements, radioactive or fluorescent beads (1–15 μm in diameter) are injected intra-arterially upstream of the target perfusion bed followed by arterial blood sampling and harvest of the target tissue. Normalized counts of microspheres trapped within the microcirculation (either via microscope, fluorescence reader, or radioactive counter) reflect the volume of blood flowing through that particular tissue and therefore can serve as a relative perfusion index. Different isotope- or fluorochrome-tagged beads can be used to evaluate vascular function in the perfusion circuit such as flow reserve or functional hyperemia [117,118].

5.6.2 Network Architecture

While vessel morphometric and perfusion measurements are important in assessing the character and function of fabricated vasculatures, they do not yet provide information about the network architecture that may have evolved. In most tissues (with the exception of sheet-like tissues such as mesentery), this invariably involves some means to capture the three-dimensionality of the fabricated vasculature across multiple scales: the individual microvessel level (microns) and the overall vascular network (≥ millimeters). In this regard, 3D imaging and reconstruction (e.g., volume rendering) methods, such as with confocal microscopy and microcomputed tomography (μCT), are proving useful. Combined with measurement algorithms that enable one to derive morphometric parameters from 3D structures, it is possible to derive vessel diameter distributions, segment length, branching architecture, and vascular volumes for a network (parameters that facilitate an analysis of network structure). However, due to the low contrast of vessel structures (to both light and X-rays), it is necessary to use vascular-targeting contrast agents. For optical assessments, fluorescent tracers are most commonly used and often involve combinations of transgenic reporters, en bloc labelling, and intravascular delivery of a tracer capable of binding ECs (e.g., a tagged lectin). It is important to recognize that in applications involving intravascular delivery of the tracer prior to evaluation, only those vessels that were perfused at the time of tracer injection are identified. Thus, combining different labeling strategies permits an evaluation of both perfused and nonperfused vessels within a network. For example, all vessels in a tissue are labeled with one fluorochrome while only perfused vessels are labeled via an intravascular tag or tracer (Fig. 19.6). The high resolution of microscopy coupled with methods for assembling

---

BOX 19.3 Vascular Fabrication and Patterning

- Variety of cellular building blocks
- Vessel self-assembly
- Self-determining topology
- Photolithography/MEMS
  - Cellularizing preformed vascular channels
- Bioprinting
  - Spheroid printing (rasterizing)
  - Direct-write printing
- Assessment
  - Vascular markers
  - Vessel morphometry
  - Network morphometry
    - 3D imaging

---

or "stitching" multiple confocal image stacks enables visualization of fine vessel structure over a relatively large vascular field. For visualizing the microvasculature via μCT, it is necessary to cast the vasculature with a contrast agent (commonly Microfil doped with lead chromate) in order to provide sufficient X-ray contrast (Fig. 19.6). Typical μCT spatial resolution is on the order of 15–50 μm over a field of view of 15–50 mm, which permits capturing terminal arterioles and venules, but not capillaries, in the image set [119]. A less technologically dependent casting method to visualize the 3D character of the network, but effective nonetheless, involves filling the vascular network with carbon ink (Fig. 19.6). Often, this is coupled with a tissue-clarifying agent, such as methyl salicylate or glycerol, in order to reduce light scattering of the surrounding tissue and improve visualization of the casted vasculature. Under a stereoscopic microscope, it is possible to visualize the ink-casted network in the context of the whole tissue (Box 19.3).

6 CONCLUSIONS

Effective tissue fabrication depends on the incorporation of an integrated vascular system into the tissue construct. The vasculature itself is a complex, multicellular system with unique but different biological requirements. At the single vessel level, the general structure entails a tube of which the walls are comprised of different cellular layers, each of which impart structural and functional characteristics to the vessel. However, a single vessel will contribute effectively to tissue perfusion only when incorporated into a vascular network. Any vascular network entails inflow and outflow vessels (arteries and veins) delivering and draining blood to and from the downstream distribution and tissue-interface vessels (the microvasculature). It is this network aspect of vasculatures that is creating new challenges in vascular fabrication and fostering innovative solutions.

Decades of work have defined the cellular components of vessels, the structures of native microvascular networks,

the dynamics of vascular self-assembly, and the rules governing blood perfusion through the microcirculation. Continuing efforts are uncovering the determinants of vessel phenotype and vascular network architecture, which are guiding fabrication design and solutions for the directed generation of vascular networks. The profound ability of vascular forming cells to self-assemble has enabled a new array of strategies for creating vascular networks of defined topologies. Of these, combinations of microfluidics and bioprinting approaches are showing good promise with respect to microvascular networks. Current challenges yet to be met are related to guiding (or enabling) vascular adaptation *in vitro* and *in vivo*, understanding how fabricated vasculatures integrate with the host circulation *in vivo* (and therefore developing methods to guide this process), incorporating unique vascular architectures (such as arcades) into design plans, and integrating vascular systems with parenchymal cells to establish synergistic interactions and native-like functionality. As new solutions are developed, subsequent efforts will likely entail scale-up strategies for properly integrating vascular networks into larger tissue systems leading to the fabrication of vascularized, functional organs.

GLOSSARY

Alginate A natural polysaccharide polymer extracted from brown seaweed used in a variety of tissue engineering applications.

Anastomosis The manual (surgical) connection of two vessels.

Angiogenesis Formation of new daughter vessels from existing parent vessels.

Anisotropic Exhibiting properties with different values when measured in different directions. Indicates directionality in orientation.

Arteriovenous plasticity The ability of blood vessels to change their arterial and venous character.

Biochip The integration of assays, reaction wells, etc. onto one, usually miniaturized, platform.

Biomechanics The collection of forces, strains, and other material properties of a biological material.

Bioprinting The guided deposition of biological components into defined structures.

Bioreactor A collection of chambers, pumps, and control systems for supporting multicellular systems.

Blood tracer A detectable molecule that is used to identify the blood compartment.

Blood–tissue exchange The exchange of molecules between the blood and tissue across the vessel wall.

Caliber Diameter of a vessel.

Cardiomyocyte Muscle cell of the heart

Cell marker A cell feature (usually a molecule) that specifically identifies a particular cell or cell type.

Cellular systems An integrated combination of cells and extracellular matrices.

Circumferential wall stress The tangential force applied to the vessel wall; hoop stress.

Collagen Interstitial matrix molecule often used in cell systems

Computational model A mathematical model of complex behavior using computational resources.

Computed tomography The use of X-ray image slices to generate a 3D image of a structure.

Conduit vessels Larger diameter vessels; arteries and veins

Continuous model A mathematical model using continuous variables.

Contrast agent A chemical entity used to enhance contrast between two or more structures for imaging.

Deterministic Not involving random chance or probability.

Direct-writing The continuous deposition of material.

Discretized model A mathematical model using discrete variables.

Doppler The change in frequency of a wave of reflected energy (light or sound) associated with a moving object.

EC Primary cell comprising small blood vessels and lining large blood vessels.

Epicardial On the surface of the heart.

Erythrocytes Red blood cells

Extracellular matrix The collection of biomolecules that surrounds and underlies cells in a tissue or tissue construct.

Fabrication The process of designing, manufacturing, and assembling something.

Femoral The primary artery feeding the lower limbs.

Fluorescence The emission of light by a molecule that has absorbed energy (usually light of a lower wavelength).

Fractal The property of an object in which repeating patterns of structure occur at all scales.

Genetic predetermination A biological outcome that was determined by a genetic program.

Hematopoietic Derived from the bone marrow.

Hemodynamic forces Those physical forces associated with pressurized blood circulation.

Hierarchical A structure arranged with different levels.

Horton–Strahler nomenclature A set of rules for labelling branching orders originally developed for studying river and stream beds.

IR, near IR The region of the light spectrum with wavelengths longer than 700 nm.

Isotropic Exhibiting properties with similar values when measured in different directions. Indicates an absence of directionality in orientation.

Laminar flow Parallel layers of ordered fluid flow.

Macrovasculature That part of the entire vascular circuit comprised of conduit vessels.

Mandrel A cylindrical core around which structures are cast or fabricated.

Mechanical deformation The change in shape due to a stress field; strain.

Methyl salicylate An organic ester used to clarify tissue; winter green oil.

Microcirculation The perfusion circuit of the entire vascular tree that is comprised of the microvasculature.

Microfluidic The movement of small volumes ($\leq$ μL) of fluid through channels.

Microgravity Near gravity-free.

Microvasculature That part of the entire vascular circuit comprised of microvessels.

Microvessels The distal regions of the vascular tree consisting of small caliber vessels: arterioles, capillaries, and venules.

Minimum work/energy theory The internal energy of a system will decrease and approach a minimum value at equilibrium.

Morphology The physical appearance of vessels and cells.

Morphometry Measurement of morphological features.

Multiscaled Spanning two or more dimensional scales.

Mural cell The collection of cell types that reside in the nonendothelial layers of vessels (e.g., smooth muscle cells, pericytes).

Neovasculature A networked collection of new blood vessels.

Neovessel A newly formed, immature blood vessel.

Non-Newtonian A fluid in which the viscous stresses arising from flow is not proportional to the strain rate.

Oxygen tension The partial pressure of oxygen in a liquid.

Parenchyma Those cellular aspects of a tissue that impart specific function to the tissue.

Perfusion The distribution of blood.

Phenotype The morphological and function traits of a biological entity.

Pluronic Hydrogel.

Polymer Macromolecular complex that forms from monomers. Native polymers are naturally occurring polymers.

Postangiogenesis Those vascular processes occurring after new vessel growth leading to network and vascular maturation.

Radial stress The outward force applied to the vessel wall.

Rapid prototyping Computer-aided design in additive manufacturing.

Rasterizing Converting an image or object into a matrix of dots or spheroid.

Blood rheology The physical characteristics of blood flow.

Sacrificial casting Using a material to form a pattern or structure, which is later removed creating a patterned void.

Scaffold A 3D structure supporting construct shape and organization.

Self-assemble Assembly of disorganized components (i.e., cells) into a more ordered arrangement via local interactions between the components.

Shear rate The difference in velocity between layers of a flowing fluid.

Shear stress The viscous drag imparted by layers of a flowing fluid.

Soft lithography A method for fabricating structures via elastomeric stamps and/or molds involving photomasks.

Spheroid A collection of cells or cells in matrix forming a spherical shape.

Stochastic Involving chance or probability.

Stroma The environment of tissue that does not include the parenchyma.

Structural adaptation The structural changes that occur to a vessel wall leading to changes in diameter, length, and function.

Tissue macrophage The population of macrophages that normally reside in the tissue space (stroma).

Topology The geometric configuration of a structure.

Transgenic reporter animal An animal in which a foreign gene expressing an identifiable tag (i.e., fluorescent protein) is inserted into its genome.

Ultrasound The use of high-frequency sound waves to image structures.

Vascular cells Those cells that comprise blood vessels (e.g., ECs, mural cells).

Vascular tree The organization of vessels within a network forming a tree-like pattern: successive branching.

Vasculature The collection of vessel elements comprising a circulatory loop.

Vaso vasorum The microcirculation feeding the walls of conduit vessels.

Vessel segment A vascular component within a network. Often considered to be region of vessel between branch points.

Viscous The resistance to deformation by a fluid to stress.

Volume rendering A set of computational techniques for the 2D display of 3D image data sets.

ABBREVIATIONS

2D	Two-dimensional
3D	Three-dimensional
ASL	Arterial spin labeling
CD31	Cluster of differentiation designation for PECAM
EC	Endothelial cell
GS-1	*Griffonia simplicifolia* lectin
IR	Infrared
LDPI	Laser Doppler perfusion imaging
MEMS	Microelectromechanical systems
MRI	Magnetic resonance imaging
PECAM	Platelet endothelial cell adhesion molecule
RBC	Red blood cell, erythrocyte
μCT	Microcomputed tomography

REFERENCES

[1] LeBlanc AJ, Krishnan L, Sullivan CJ, Williams SK, Hoying JB. Microvascular repair: post-angiogenesis vascular dynamics. Microcirculation 2012;19(8):676–95.

[2] Chang CC, Nunes SS, Sibole SC, Krishnan L, Williams SK, Weiss JA, et al. Angiogenesis in a microvascular construct for transplantation depends on the method of chamber circulation. Tissue Eng Part A 2010;16(3):795–805.

[3] Rhoads RP, Johnson RM, Rathbone CR, Liu X, Temm-Grove C, Sheehan SM, et al. Satellite cell-mediated angiogenesis *in vitro* coincides with a functional hypoxia-inducible factor pathway. Am J Physiol Cell Physiol 2009;296(6):C1321–8.

[4] Nunes SS, Maijub JG, Krishnan L, Ramakrishnan VM, Clayton LR, Williams SK, et al. Generation of a functional liver tissue mimic using adipose stromal vascular fraction cell-derived vasculatures. Sci Rep 2013;3:2141.

[5] Pries AR, Reglin B, Secomb TW. Remodeling of blood vessels: responses of diameter and wall thickness to hemodynamic and metabolic stimuli. Hypertension 2005;46(4):725–31.

[6] Pries AR, Secomb TW, Gessner T, Sperandio MB, Gross JF, Gaehtgens P. Resistance to blood flow in microvessels *in vivo*. Circ Res 1994;75(5):904–15.

[7] Guillemette MD, Gauvin R, Perron C, Labbe R, Germain L, Auger FA. Tissue-engineered vascular adventitia with vasa vasorum improves graft integration and vascularization through inosculation. Tissue Eng Part A 2010;16(8):2617–26.

[8] Stenmark KR, Yeager ME, El Kasmi KC, Nozik-Grayck E, Gerasimovskaya EV, Li M, et al. The adventitia: essential regulator of vascular wall structure and function. Annu Rev Physiol 2013;75:23–47.

[9] Scotland RS, Vallance PJ, Ahluwalia A. Endogenous factors involved in regulation of tone of arterial vasa vasorum: implications for conduit vessel physiology. Cardiovasc Res 2000;46(3):403–11.

[10] Pries AR, Hopfner M, le Noble F, Dewhirst MW, Secomb TW. The shunt problem: control of functional shunting in normal and tumour vasculature. Nat Rev Cancer 2010;10(8):587–93.

[11] Pries AR, Cornelissen AJ, Sloot AA, Hinkeldey M, Dreher MR, Hopfner M, et al. Structural adaptation and heterogeneity of normal and tumor microvascular networks. PLoS Comput Biol 2009;5(5):e1000394.

[12] Stapleton PA, James ME, Goodwill AG, Frisbee JC. Obesity and vascular dysfunction. Pathophysiology 2008;15(2):79–89.

[13] Hansen-Smith FM. Capillary network patterning during angiogenesis. Clin Exp Pharmacol Physiol 2000;27(10):830–5.

[14] Fenton B, Zweifach BW. Microcirculatory model relating geometrical variation to changes in pressure and flow rate. Ann Biomed Eng 1981;9:303–21.

[15] Koller A, Dawant B, Liu A, Popel AS, Johnson PC. Quantitative analysis of arteriolar network architecture in cat sartorius muscle. Am J Physiol 1987;253(1 Pt 2):H154–64.

[16] Ellsworth ML, Liu A, Dawant B, Popel AS, Pittman RN. Analysis of vascular pattern and dimensions in arteriolar networks of the retractor muscle in young hamsters. Microvasc Res 1987;34(2):168–83.

[17] Ley K, Pries AR, Gaehtgens P. Topological structure of rat mesenteric microvessel networks. Microvasc Res 1986;32(3):315–32.

[18] Buschmann I, Pries A, Styp-Rekowska B, Hillmeister P, Loufrani L, Henrion D, et al. Pulsatile shear and Gja5 modulate arterial identity and remodeling events during flow-driven arteriogenesis. Development 2010;137(13):2187–96.

[19] Hoying JB, Utzinger U, Weiss JA. Formation of microvascular networks: role of stromal interactions directing angiogenic growth. Microcirculation 2014;21(4):278–89.

[20] Swift MR, Weinstein BM. Arterial-venous specification during development. Circ Res 2009;104(5):576–88.

[21] Pries AR, Secomb TW, Gaehtgens P. Design principles of vascular beds. Circ Res 1995;77(5):1017–23.

[22] Sullivan CJ, Hoying JB. Flow-dependent remodeling in the carotid artery of fibroblast growth factor-2 knockout mice. Arterioscler Thromb Vasc Biol 2002;22(7):1100–5.

[23] Troidl K, Schaper W. Arteriogenesis versus angiogenesis in peripheral artery disease. Diabetes Metab Res Rev 2012;28(Suppl 1):27–9.

[24] Pries AR, Secomb TW, Gaehtgens P. Structural adaptation and stability of microvascular networks: theory and simulations. Am J Physiol 1998;275(2 Pt 2):H349–60.

[25] Nunes SS, Greer KA, Stiening CM, Chen HY, Kidd KR, Schwartz MA, et al. Implanted microvessels progress through distinct neovascularization phenotypes. Microvasc Res 2010;79(1):10–20.

[26] Chiu LL, Montgomery M, Liang Y, Liu H, Radisic M. Perfusable branching microvessel bed for vascularization of engineered tissues. Proc Natl Acad Sci USA 2012;109(50):E3414–23.

[27] Chang CC, Hoying JB. Directed three-dimensional growth of microvascular cells and isolated microvessel fragments. Cell Transplant 2006;15(6):533–40.

[28] Chang CC, Krishnan L, Nunes SS, Church KH, Edgar LT, Boland ED, et al. Determinants of microvascular network topologies in implanted neovasculatures. Arterioscler Thromb Vasc Biol 2012;32(1):5–14.

[29] Qutub AA, Mac Gabhann F, Karagiannis ED, Vempati P, Popel AS. Multiscale models of angiogenesis. IEEE Eng Med Biol Mag 2009;28(2):14–31.

[30] Bauer AL, Jackson TL, Jiang Y. A cell-based model exhibiting branching and anastomosis during tumor-induced angiogenesis. Biophys J 2007;92(9):3105–21.

[31] Edgar LT, Sibole SC, Underwood CJ, Guilkey JE, Weiss JA. A computational model of *in vitro* angiogenesis based on extracellular matrix fibre orientation. Comput Methods Biomech Biomed Engin 2013;16(7):790–801.

[32] Secomb TW, Alberding JP, Hsu R, Dewhirst MW, Pries AR. Angiogenesis: an adaptive dynamic biological patterning problem. PLoS Comput Biol 2013;9(3):e1002983.

[33] Pries AR, Secomb TW. Control of blood vessel structure: insights from theoretical models. Am J Physiol Heart Circ Physiol 2005;288(3):H1010–5.

[34] Lorthois S, Cassot F. Fractal analysis of vascular networks: insights from morphogenesis. J Theor Biol 2010;262(4):614–33.

[35] Pries AR, Secomb TW. Modeling structural adaptation of microcirculation. Microcirculation 2008;15(8):753–64.

[36] Carmeliet P, Jain RK. Molecular mechanisms and clinical applications of angiogenesis. Nature 2011;473(7347):298–307.

[37] Mitsos S, Katsanos K, Koletsis E, Kagadis GC, Anastasiou N, Diamantopoulos A, et al. Therapeutic angiogenesis for myocardial ischemia revisited: basic biological concepts and focus on latest clinical trials. Angiogenesis 2011;15(1):1–22.

[38] Engelmann MG, Theiss HD, Hennig-Theiss C, Huber A, Wintersperger BJ, Werle-Ruedinger AE, et al. Autologous bone marrow stem cell mobilization induced by granulocyte colony-stimulating factor after subacute ST-segment elevation myocardial infarction undergoing late revascularization: final results from the G-CSF-STEMI (granulocyte colony-stimulating factor ST-segment elevation myocardial infarction) trial. J Am Coll Cardiol 2006;48(8):1712–21.

[39] Roberts N, Jahangiri M, Xu Q. Progenitor cells in vascular disease. J Cell Mol Med 2005;9(3):583–91.

[40] Gimble JM, Guilak F, Bunnell BA. Clinical and preclinical translation of cell-based therapies using adipose tissue-derived cells. Stem Cell Res Ther 2010;1(2):19.

[41] Williams AR, Hare JM. Mesenchymal stem cells: biology, pathophysiology, translational findings, and therapeutic implications for cardiac disease. Circ Res 2011;109(8):923–40.

[42] Rehman J, Traktuev D, Li J, Merfeld-Clauss S, Temm-Grove CJ, Bovenkerk JE, et al. Secretion of angiogenic and antiapoptotic factors by human adipose stromal cells. Circulation 2004;109(10):1292–8.

[43] Kidd KR, Nagle RB, Williams SK. Angiogenesis and neovascularization associated with extracellular matrix-modified porous implants. J Biomed Mater Res 2002;59(2):366–77.

[44] Ratner BD. Reducing capsular thickness and enhancing angiogenesis around implant drug release systems. J Control Release 2002;78(1–3):211–8.

[45] Koike N, Fukumura D, Gralla O, Au P, Schechner JS, Jain RK. Tissue engineering: creation of long-lasting blood vessels. Nature 2004;428(6979):138–9.

[46] Newman PJ. The biology of PECAM-1. J Clin Invest 1997;99(1):3–8.

[47] Chang CC, Boland ED, Williams SK, Hoying JB. Direct-write bioprinting three-dimensional biohybrid systems for future regenerative therapies. J Biomed Mater Res B Appl Biomater 2011;98(1):160–70.

[48] Shepherd BR, Hoying JB, Williams SK. Microvascular transplantation after acute myocardial infarction. Tissue Eng 2007;13(12):2871–9.

[49] Shepherd BR, Chen HY, Smith CM, Gruionu G, Williams SK, Hoying JB. Rapid perfusion and network remodeling in a microvascular construct after implantation. Arterioscler Thromb Vasc Biol 2004;24(5):898–904.

[50] Laib AM, Bartol A, Alajati A, Korff T, Weber H, Augustin HG. Spheroid-based human endothelial cell microvessel formation *in vivo*. Nat Protoc 2009;4(8):1202–15.

[51] Finkenzeller G, Graner S, Kirkpatrick CJ, Fuchs S, Stark GB. Impaired *in vivo* vasculogenic potential of endothelial progenitor cells in comparison to human umbilical vein endothelial cells in a spheroid-based implantation model. Cell Prolif 2009;42(4):498–505.

[52] Gruionu G, Hoying JB, Gruionu LG, Laughlin MH, Secomb TW. Structural adaptation increases predicted perfusion capacity after vessel obstruction in arteriolar arcade network of pig skeletal muscle. Am J Physiol Heart Circ Physiol 2005;288(6):H2778–84.

[53] Whatley BR, Li X, Zhang N, Wen X. Magnetic-directed patterning of cell spheroids. J Biomed Mater Res A 2014;102(5):1537–47.

[54] Steffens L, Wenger A, Stark GB, Finkenzeller G. *In vivo* engineering of a human vasculature for bone tissue engineering applications. J Cell Mol Med 2009;13(9B):3380–6.

[55] Chan JM, Zervantonakis IK, Rimchala T, Polacheck WJ, Whisler J, Kamm RD. Engineering of *in vitro* 3D capillary beds by self-directed angiogenic sprouting. PLoS ONE 2012;7(12):e50582.

[56] Huh D, Matthews BD, Mammoto A, Montoya-Zavala M, Hsin HY, Ingber DE. Reconstituting organ-level lung functions on a chip. Science 2010;328(5986):1662–8.

[57] Zheng Y, Chen J, Craven M, Choi NW, Totorica S, Diaz-Santana A, et al. *In vitro* microvessels for the study of angiogenesis and thrombosis. Proc Natl Acad Sci USA 2012;109(24):9342–7.

[58] Hooper RC, Hernandez KA, Boyko T, Harper A, Joyce J, Golas AR, et al. Fabrication and *in vivo* microanastomosis of vascularized tissue-engineered constructs. Tissue Eng Part A 2014;20(19–20):2711–9.

[59] Chrobak KM, Potter DR, Tien J. Formation of perfused, functional microvascular tubes *in vitro*. Microvasc Res 2006;71(3):185–96.

[60] Neumann T, Nicholson BS, Sanders JE. Tissue engineering of perfused microvessels. Microvasc Res 2003;66(1):59–67.

[61] Sekine H, Shimizu T, Sakaguchi K, Dobashi I, Wada M, Yamato M, et al. *In vitro* fabrication of functional three-dimensional tissues with perfusable blood vessels. Nat Commun 2013;4:1399.

[62] Baranski JD, Chaturvedi RR, Stevens KR, Eyckmans J, Carvalho B, Solorzano RD, et al. Geometric control of vascular networks to enhance engineered tissue integration and function. Proc Natl Acad Sci USA 2013;110(19):7586–91.

[63] Miller JS, Stevens KR, Yang MT, Baker BM, Nguyen DH, Cohen DM, et al. Rapid casting of patterned vascular networks for perfusable engineered three-dimensional tissues. Nat Mater 2012;11(9):768–74.

[64] Smith CM, Stone AL, Parkhill RL, Stewart RL, Simpkins MW, Kachurin AM, et al. Three-dimensional bioassembly tool for generating viable tissue-engineered constructs. Tissue Eng 2004;10(9–10):1566–76.

[65] Wilson WC Jr, Boland T. Cell and organ printing 1: protein and cell printers. Anat Rec A Discov Mol Cell Evol Biol 2003;272(2):491–6.

[66] Jakab K, Damon B, Neagu A, Kachurin A, Forgacs G. Three-dimensional tissue constructs built by bioprinting. Biorheology 2006;43(3–4):509–13.

[67] Xu T, Jin J, Gregory C, Hickman JJ, Boland T. Inkjet printing of viable mammalian cells. Biomaterials 2005;26(1):93–9.

[68] Jakab K, Norotte C, Damon B, Marga F, Neagu A, Besch-Williford CL, et al. Tissue engineering by self-assembly of cells printed into topologically defined structures. Tissue Eng Part A 2008;14(3):413–21.

[69] Visconti RP, Kasyanov V, Gentile C, Zhang J, Markwald RR, Mironov V. Towards organ printing: engineering an intra-organ branched vascular tree. Expert Opin Biol Ther 2010;10(3):409–20.

[70] Norotte C, Marga FS, Niklason LE, Forgacs G. Scaffold-free vascular tissue engineering using bioprinting. Biomaterials 2009;30(30):5910–7.

[71] Mironov V, Visconti RP, Kasyanov V, Forgacs G, Drake CJ, Markwald RR. Organ printing: tissue spheroids as building blocks. Biomaterials 2009;30(12):2164–74.

[72] Cui X, Boland T. Human microvasculature fabrication using thermal inkjet printing technology. Biomaterials 2009;30(31):6221–7.

[73] Fleming PA, Argraves WS, Gentile C, Neagu A, Forgacs G, Drake CJ. Fusion of uniluminal vascular spheroids: a model for assembly of blood vessels. Dev Dyn 2010;239(2):398–406.

[74] Williams SK, Touroo JS, Church KH, Hoying JB. Encapsulation of adipose stromal vascular fraction cells in alginate hydrogel spheroids using a direct-write three-dimensional printing system. Biores Open Access 2013;2(6):448–54.

[75] Leong KF, Cheah CM, Chua CK. Solid freeform fabrication of three-dimensional scaffolds for engineering replacement tissues and organs. Biomaterials 2003;24(13):2363–78.

[76] Fedorovich NE, Wijnberg HM, Dhert WJ, Alblas J. Distinct tissue formation by heterogeneous printing of osteo- and endothelial progenitor cells. Tissue Eng Part A 2011;17(15–16):2113–21.

[77] Gillis JA, Morris SF. Three-dimensional printing of perforator vascular anatomy. Plast Reconstr Surg 2014;133(1):2e–80e.

[78] Khan IS, Kelly PD, Singer RJ. Prototyping of cerebral vasculature physical models. Surg Neurol Int 2014;5:11.

[79] Webb PA. A review of rapid prototyping (RP) techniques in the medical and biomedical sector. J Med Eng Technol 2000;24(4):149–53.

[80] Jay SM, Saltzman WM. Controlled delivery of VEGF via modulation of alginate microparticle ionic crosslinking. J Control Release 2009;134(1):26–34.

[81] Orive G, De Castro M, Kong HJ, Hernandez RM, Ponce S, Mooney DJ, et al. Bioactive cell-hydrogel microcapsules for cell-based drug delivery. J Control Release 2009;135(3):203–10.

[82] Cohen DL, Malone E, Lipson H, Bonassar LJ. Direct freeform fabrication of seeded hydrogels in arbitrary geometries. Tissue Eng 2006;12(5):1325–35.

[83] Elcin YM, Dixit V, Gitnick G. Extensive *in vivo* angiogenesis following controlled release of human vascular endothelial cell growth factor: implications for tissue engineering and wound healing. Artif Organs 2001;25(7):558–65.

[84] Hunter SK, Kao JM, Wang Y, Benda JA, Rodgers VG. Promotion of neovascularization around hollow fiber bioartificial organs using biologically active substances. ASAIO J 1999;45(1):37–40.

[85] Stabenfeldt SE, Gourley M, Krishnan L, Hoying JB, Barker TH. Engineering fibrin polymers through engagement of alternative polymerization mechanisms. Biomaterials 2012;33(2):535–44.

[86] Chang WG, Andrejecsk JW, Kluger MS, Saltzman WM, Pober JS. Pericytes modulate endothelial sprouting. Cardiovasc Res 2013;100(3):492–500.

[87] Benjamin LE, Hemo I, Keshet E. A plasticity window for blood vessel remodelling is defined by pericyte coverage of the preformed endothelial network and is regulated by PDGF-B and VEGF. Development 1998;125(9):1591–8.

[88] Lindahl P, Johansson BR, Leveen P, Betsholtz C. Pericyte loss and microaneurysm formation in PDGF-B-deficient mice. Science 1997;277(5323):242–5.

[89] Aird WC. Endothelial Biomedicine. 1st ed. New York: Cambridge University Press; 2007.

[90] Davies PF. Hemodynamic shear stress and the endothelium in cardiovascular pathophysiology. Nat Clin Pract Cardiovasc Med 2009;6(1):16–26.

[91] Papaioannou TG, Stefanadis C. Vascular wall shear stress: basic principles and methods. Hellenic J Cardiol 2005;46(1):9–15.

[92] Isenberg BC, Williams CF, Tranquillo RT, Tranquillo RT. Small-diameter artificial arteries engineered *in vitro*, 20060106 DCOM-20060209(1524-4571 (Electronic)).

[93] Abaci HE, Devendra R, Smith Q, Gerecht S, Drazer G. Design and development of microbioreactors for long-term cell culture in controlled oxygen microenvironments. Biomed Microdevices 2012;14(1):145–52.

[94] Couet F, Mantovani D. A new bioreactor adapts to materials state and builds a growth model for vascular tissue engineering. Artif Organs 2012;36(4):438–45.

[95] Tosun Z, McFetridge PS. Improved recellularization of ex vivo vascular scaffolds using directed transport gradients to modulate ECM remodeling. Biotechnol Bioeng 2013;110(7):2035–45.

[96] Song L, Zhou Q, Duan P, Guo P, Li D, Xu Y, et al. Successful development of small diameter tissue-engineering vascular vessels by our novel integrally designed pulsatile perfusion-based bioreactor. PLoS ONE 2012;7(8):e42569.

[97] Unsworth BR, Lelkes PI. Growing tissues in microgravity. Nat Med 1998;4(8):901–7.

[98] Fantin A, Vieira JM, Gestri G, Denti L, Schwarz Q, Prykhozhij S, et al. Tissue macrophages act as cellular chaperones for vascular anastomosis downstream of VEGF-mediated endothelial tip cell induction. Blood 2010;116(5):829–40.

[99] Koh YJ, Koh BI, Kim H, Joo HJ, Jin HK, Jeon J, et al. Stromal vascular fraction from adipose tissue forms profound vascular network through the dynamic reassembly of blood endothelial cells. Arterioscler Thromb Vasc Biol 2011;31(5):1141–50.

[100] Cheng G, Liao S, Kit Wong H, Lacorre DA, di Tomaso E, Au P, et al. Engineered blood vessel networks connect to host vasculature via wrapping-and-tapping anastomosis. Blood 2011;118(17):4740–9.

[101] Jones EA, Le NF, Eichmann A. What determines blood vessel structure? Genetic prespecification vs. hemodynamics. Physiology (Bethesda) 2006;21:388–95.

[102] Moyon D, Pardanaud L, Yuan L, Breant C, Eichmann A. Plasticity of endothelial cells during arterial-venous differentiation in the avian embryo. Development 2001;128(17):3359–70.

[103] Nunes SS, Rekapally H, Chang CC, Hoying JB. Vessel arterial-venous plasticity in adult neovascularization. PLoS ONE 2011;6(11):e27332.

[104] Krishnan L, Underwood CJ, Maas S, Ellis BJ, Kode TC, Hoying JB, et al. Effect of mechanical boundary conditions on orientation of angiogenic microvessels. Cardiovasc Res 2008;78(2):324–32.

[105] Edgar LT, Underwood CJ, Guilkey JE, Hoying JB, Weiss JA. Extracellular matrix density regulates the rate of neovessel growth and branching in sprouting angiogenesis. PLoS ONE 2014;9(1):e85178.

[106] Edgar L, Hoying J, Utzinger U, Underwood CJ, Krishnan L, Baggett B, et al. Mechanical interaction of angiogenic microvessels with the extracellular matrix. J Biomech Eng 2014;136(2):021001.

[107] Krishnan L, Weiss JA, Wessman MD, Hoying JB. Design and application of a test system for viscoelastic characterization of collagen gels. Tissue Eng 2004;10(1–2):241–52.

[108] Underwood CJ, Edgar LT, Hoying JB, Weiss JA. Cell-generated traction forces and the resulting matrix deformation modulate microvascular alignment and growth during angiogenesis. American J. Physiol.: Heart and Circulation Physiology. 2014; 307(2): H152–64.

[109] Krishnan L, Chang CC, Nunes SS, Williams SK, Weiss JA, Hoying JB. Manipulating the microvasculature and its microenvironment. Crit Rev Biomed Eng 2013;41(2):91–123.

[110] Motoike T, Loughna S, Perens E, Roman BL, Liao W, Chau TC, et al. Universal GFP reporter for the study of vascular development. Genesis 2000;28(2):75–81.

[111] Allen J, Howell K. Microvascular imaging: techniques and opportunities for clinical physiological measurements. Physiol Meas 2014;35(7):R91–R141.

[112] Badea AF, Tamas-Szora A, Clichici S, Socaciu M, Tabaran AF, Baciut G, et al. Contrast enhanced ultrasonography (CEUS) in the characterization of tumor microcirculation. Validation of the procedure in the animal experimental model. Med Ultrason 2013;15(2):85–94.

[113] Ortiz D, Briceno JC, Cabrales P. Microhemodynamic parameters quantification from intravital microscopy videos. Physiol Measur 2014;35(3):351–67.

[114] Marshall MV, Rasmussen JC, Tan IC, Aldrich MB, Adams KE, Wang X, et al. Near-infrared fluorescence imaging in humans with indocyanine green: a review and update. Open Surg Oncol J 2010;2(2):12–25.

[115] Gao Y, Goodnough CL, Erokwu BO, Farr GW, Darrah R, Lu L, et al. Arterial spin labeling-fast imaging with steady-state free precession (ASL-FISP): a rapid and quantitative perfusion technique for high-field MRI. NMR Biomed 2014;27(8):996–1004.

[116] Taylor A, Herrmann A, Moss D, See V, Davies K, Williams SR, et al. Assessing the efficacy of nano- and micro-sized magnetic particles as contrast agents for MRI cell tracking. PLoS ONE 2014;9(6):e100259.

[117] Cardinal TR, Hoying JB. A modified fluorescent microsphere-based approach for determining resting and hyperemic blood flows in individual murine skeletal muscles. Vascul Pharmacol 2007;47(1):48–56.

[118] Leblanc AJ, Touroo JS, Hoying JB, Williams SK. Adipose stromal vascular fraction cell construct sustains coronary microvascular function after acute myocardial infarction. Am J Physiol Heart Circ Physiol 2012;302(4):H973–82.

[119] Holdsworth DW, Thornton MM. Micro-CT in small animal and specimen imaging. Trends Biotechnol. 2002;20(8):S34–9.

Chapter 20

Bioprinting of Blood Vessels

Anthony J. Melchiorri and John P. Fisher

Fischell Department of Bioengineering, University of Maryland, College Park, MD, USA

ABSTRACT

A major challenge in the development of artificial organs is vascularization. In artificial organ fabrication, it may be advantageous to design custom vasculature for adequate vascularization. One such technique is bioprinting. There is a tremendous demand for the fabrication of macroscale vascular grafts and microscale vessels. We introduce motivations for vessel printing, along with an introduction to vessel biology. Then, we discuss challenges associated with vessel bioprinting, followed by an examination of methods. For direct bioprinting, we will examine inkjet/deposition, laser-based, extrusion-based, stereolithography, and syringe-based printing. We will next review hybrid techniques and alternative printing techniques. We will then provide a discussion of postprinting considerations for vessel bioprinting. Finally, we provide a glimpse of how vessel bioprinting fits into the bioprinting arena, along with future trends.

Keywords: vascular graft; microvasculature; vascularization; endothelium; filament extrusion; spheroid deposition; spheroid fusion; stereolithography; laser printing

1 INTRODUCTION

To transport nutrients, oxygen, and waste, the human body utilizes an intricate network of blood vessels, ranging from large arteries and veins with diameters on the scale of centimeters to microvasculature on the scale of micrometers.

Such vessels are crucial to the growth of new tissue and to the sustenance and restoration of existing tissues. Due to the undeniable importance of mature and functional blood vessels in almost all tissues, there is an enormous demand for transplantable blood vessels. Transplantable vessels are necessary in a variety of clinical conditions. Annually, there are approximately 200,000 coronary artery bypass graft procedures performed in the United States [1]. Additionally, peripheral artery disease affects 8 million individuals and about 500,000 patients suffer from end-stage renal disease. Such conditions often require the transplantation of autologous blood vessels. However, these vessels may be unavailable due to disease or prior operations. In addition, options for autologous vessels do not perform equivalently. For example, in coronary artery bypass grafts, autologous transplantation of the internal mammary artery may offer better outcomes and increased patency retention rates compared to transplantation of the saphenous vein [2–4]. Synthetic vessels have been used with moderate success in large-diameter applications (>6 mm), though small-diameter applications (<6 mm) often lead to failure rates upward of 30% due primarily to complications such as restenosis and thrombosis [5]. Many of these inadequacies and unfavorable failure rates are linked to complications associated with blood–material interactions of these permanent synthetic vessels.

Essentials of 3D Biofabrication and Translation. http://dx.doi.org/10.1016/B978-0-12-800972-7.00020-7

To address these challenges, researchers have pursued tissue engineering strategies to eliminate the need for a permanent synthetic scaffold within the vasculature. The traditional processes associated with tissue engineering require the seeding and expansion of cells on a scaffold until populations are sufficient for harvest. Such a process can be slow, labor and time intensive, and clinically difficult [6–8]. Bioprinting offers a potential solution to these challenges. Printing patient- or scaffold-specific vascular grafts or blood vessels may improve outcomes in a variety of applications. For example, congenital heart diseases in pediatric patients may present vast anatomical differences between patients despite sharing a similar moniker. In single ventricle anomalies, as one case study, one of the two ventricles is of an inadequate size for normal cardiac function. These malformations in the ventricle and the surrounding anatomy may differ drastically between patients. Thus, each of these defects requires careful planning and attention from the patient's clinicians in order to restore function. Often, such defects require the use of vascular grafts, cut and adjusted to fit the patient during surgery. Using medical images to guide the design and fabrication of a custom vessel before surgery would help reduce some of the challenges associated with current congenital heart disease corrective surgeries, potentially reducing the time necessary for surgery completion.

Besides macroscale vessels, researchers have also pursued tissue-engineering strategies for the development of microvasculature networks for the vascularization of artificial organs and tissues. An enormous challenge in tissue engineering of these organs and tissues is a lack of vascularization. Vascularization is of course necessary for the transport of oxygen and other nutrients, along with the removal of waste byproducts. Without adequate vascularization, diffusion through a tissue-engineered construct may be limited to a few hundred microns. Inadequate vascularization leads to tissue necrosis and increased tissue-engineered construct morbidity [9]. In order for the fabrication of artificial organs to become a reality, researchers must first tackle the challenge of integrating a functional vascular network. Integration of a functional vascular network is a challenge faced by all manners of large-scale tissue engineering. One example is the bioprinting of patient-specific bone constructs for critical-sized bone defects [10–12]. Custom fabrication of a vessel or vessel network specifically designed to vascularize the bone construct may alleviate tissue morbidity by promoting healthy vascularization. Thus, vascular tissue engineering is not only crucial for macroscale grafts, but also for encouraging vasculogenesis and/or angiogenesis within other tissue-engineered organs or tissues.

Bioprinting of blood vessels may be advantageous both to the improvement and customization of direct cardiovascular implants, as well as the stepping stone necessary for achieving long-term clinical success in bulk tissue and organ constructs through custom vascularization networks. Blood vessel bioprinting has enabled the printing of functional vessels with capillary-sized diameters, suitable for nutrient and oxygen delivery along with waste removal, along with vascular grafts that may be used for congenital heart diseases or coronary artery bypass grafts. Researchers continue to develop a variety of printing strategies involving printing hardware, bioinks, biopapers, new materials, and software to explore the potential of blood vessel bioprinting [13–16]. These fabrication technologies are designed with the underlying goal of promoting healthy tissue growth to overcome current barriers in prosthetic blood vessel technologies, especially when it comes to the endothelium. While much research has been performed in developing materials for vascular applications, no material matches the blood compatibility of the natural endothelium [17,18]. The endothelium, consisting of a monolayer of endothelial cells (ECs), is largely responsible for vascular homeostasis. ECs release a variety of factors to help control and prevent platelet activation and inhibition, thrombogenesis, and fibrinolysis [19,20]. A healthy endothelial layer also aids in preventing smooth muscle cells (SMCs) from spreading into the inner lumen as it occurs in intimal hyperplasia, preventing complications such as restenosis [21]. Thus, functional vessel tissue layers are critical to supporting proper blood flow and maintaining vessel patency, and an understanding of blood vessel composition is necessary for effective vessel bioprinting. Because of the functional performance of the hierarchically organized tissues that make up blood vessels, it is important to understand native blood vessel composition and biology in order to achieve success in blood vessel bioprinting.

2 BLOOD VESSEL COMPOSITION

Generally, blood vessels are comprised of three layers identified, from the inner lumen outward, as the endothelium or tunica intima, the medial layer or tunica media, and the adventitia or tunic adventitia, as demonstrated in Fig. 20.1. Not all blood vessels contain these three layers. Depending on the type and size of a blood vessel, the thickness of these three layers may be drastically different. For example, capillaries mediate nutrient delivery, waste removal, and gas exchange by consisting of primarily an endothelial layer. This single layer of cells lines an inner lumen with a diameter of only 5–10 μm [22,23]. The aorta, on the other hand, is the largest vessel in the human body and consists of the tunica adventitia, tunica media, and tunica intima. Because of its size, an array of vascular networks actually provides blood circulation to the outer layers of the aorta. The vessel walls may be generally characterized by three types of cells prominent to vessel architecture and organization: ECs, SMCs, and fibroblasts [24,25]. As previously mentioned, the ECs comprise the endothelium. The medial

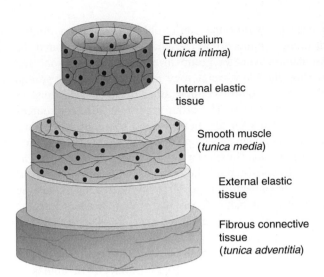

Endothelium
(*tunica intima*)

Internal elastic
tissue

Smooth muscle
(*tunica media*)

External elastic
tissue

Fibrous connective
tissue
(*tunica adventitia*)

FIGURE 20.1 **Basic architecture of blood vessels in the human body.**

layer is primarily comprised of SMCs and the adventitia consists mostly of fibroblasts. The thrombo-resistant endothelium supports healthy vascular homeostasis, providing a selectively permeable barrier between the circulating blood, vessel, and surrounding tissues. Healthy ECs also help in regulating platelets activation, adhesion, and aggregation, leukocyte adhesion, and SMC proliferation and migration [26–29]. These functions of the endothelial layer are crucial to preventing complications such as thrombosis or emboli formation and restenosis of blood vessels due to intimal hyperplasia, or the ingrowth of SMCs. SMCs can influence mechanical adaptation of the blood vessels through extracellular matrix (ECM) production, while also possessing secretory capabilities [30]. The ECM components secreted by SMCs, such as collagen, elastic fibers, elastic lamellae, and proteoglycans, aid in the maintenance of the compliance and elasticity of the blood vessel. Successful vessel bioprinting must incorporate tissue and layer organization consistent with the intended vessel's size, function, and biology.

3 CHALLENGES ASSOCIATED WITH VESSEL BIOPRINTING

Bioprinting blood vessels is not free from the challenges faced in the ongoing efforts to develop fully functional tissue-engineered vascular grafts and blood vessels. Such vessels must conform to an extensive list of requirements associated with other vascular grafts: mechanical strength must be adequate to prevent complications, such as aneurysm, and withstand hemodynamic stresses while reducing complications associated with noncompliance; biocompatibility must be adequate to ensure cell attachment and proliferation of vascular cell populations, while retaining nontoxic and nonimmunogenic inertness; materials need to

be suturable and easily handled during surgery; vessels and grafts should resist inducing thrombosis and infection; host tissue must be able to be incorporated through proper healing and tissue formation; and, for pediatric patients, grafts must allow for, and adapt to, growth of the patient [31].

Besides these requirements, bioprinting of vascular grafts and blood vessels introduces additional challenges due to limited material selection and varying printing techniques. For example, some bioprinted grafts may have insufficient mechanical properties after fabrication due to material or fabrication constraints. Many bioprinted vessel constructs rely on the use of biological or cell-infused hydrogels, which may not have the mechanical strength necessary to withstand hemodynamic forces after fabrication. It is only with long-term culturing that these constructs become strong via the ECM formation of the cultured cells. For example, a graft printed utilizing a layered filament approached yielded freshly bioprinted vessels that demonstrated a burst pressure of only 6 mmHg [32]. However, after the grafts were cultured, the vessels demonstrated burst pressures of 315 ± 81 mmHg at 3 days and 773 ± 78 mmHg at 21 days. Other materials, like polyurethanes, may be printed with adequate mechanical properties. However, these materials and the processes used to print them do not allow for the direct inclusion of vascular cells. Cells must be seeded after printing the constructs and these grafts may be limited in scale. Once a scaffold is successfully printed and the cells or tissues have matured, the fabricated vascular tissues must oftentimes be suturable to existing vessel networks to ensure immediate and effective blood perfusion. In these microvasculature networks, printing resolution must be adequate to print patent capillaries with diameters on the scale of micrometers. Even if vessels can be incorporated into an artificial organ, they must be manufactured to mimic natural vascular anatomy. Thus, vessels must be bifurcated over multiple scales to replicate the natural vascular trees that provide nutrient delivery, waste removal, and gas exchange in native organs.

Designing a proper vasculature tree for the purpose of blood vessel bioprinting involves more than just concerns over material and tissue compatibility. Researchers must also integrate computer models that effectively capture the anatomical formations of native vascular trees. Some efforts have been made in computer models though much research is still necessary to better mimic bifurcating and branching networks [14,33,34]. These designs must also be optimized so that shear stress can be maintained within the networks. While the actual printing of microvasculature is important, computer-aided design of blood vessel structures is an integral component to the success of bioprinted blood vessels. One strategy involves the use of microcomputed tomography, which is used to generate 3D tissue structures. This approach has been utilized in the reconstruction of complex capillary beds. However, there is some difficulty in this regard due to the contrast agents used in X-ray microtomography (microCT) imaging. Such agents generally do not lend themselves to producing high-resolution images of small vessels. Without accurate imaging of these vessels, it can be difficult to accurately produce a computer aided design model for printing of a capillary bed. To overcome these challenges, one group demonstrated the use of Batson's methyl methacrylate corrosion casting (BMCC) to create a vessel cast for acquiring high-resolution microCT images [33]. Animals were sacrificed and heparinized normal saline solution was perfused through the vasculature. Following heparin perfusion, the vasculature was perfused with Batson's #17 solution modified with methyl methacrylate. The polymers cured to the animal's tissue and the tissues were corroded away. The BMCC produced more accurate and more complete models compared to a standard contrast agent method of microCT imaging. While such a method may produce better capillary bed models for 3D fabrication, BMCC modeling and imaging requires that the capillary bed specimen be sacrificed for the initial model reconstruction. However, this method of imaging and subsequent model reconstruction avoids the blocking of microvascular and capillary vessels attributed to the use of contrasting metals.

Instead of designing and fabricating the vascular tree, some researchers propose simply printing microfluidic channels within organs to mimic a vascular network [13]. Such a microfluidic network may allow for fluid perfusion throughout the construct and enable gas transfer. These channels would then be able to support the survival and maturation of the artificial tissue or organ.

4 DIRECT VESSEL BIOPRINTING

Bioprinting of blood vessels has primarily consisted of three main techniques: (1) inkjet based, (2) laser based, and (3) extrusion based. Other methods of 3D printing, although not as widely used yet in vessel printing, have been investigated for vessel or graft fabrication including digital stereolithography and syringe-based deposition. Each technique possesses unique advantages and disadvantages, which must be considered depending on the intended application.

4.1 Inkjet or Drop Dispensing

Inkjet technologies have been a historically popular method of bioprinting due to the availability and adaptability of commercial inkjet printers. Bioprinting of blood vessels are no exception. For example, one group modified a basic Hewlett–Packard Deskjet 500 thermal inkjet printer to enable printing ECs and fibrin simultaneously to form microvasculature patterns [35]. The group fabricated micron-sized fibrin channels. When these channels were produced simultaneously with the ECs, ECs preferentially aligned

themselves within the fibrin channels to form confluent linings similar to an endothelial monolayer found the vasculature. Another group investigated drop-on-demand printing via inkjet technology to selectively print protein solutions to control cellular attachment on substrates as early as 2003 [36]. The group used two different sources of type 1 collagen for their printing solution: rat-tail and calf-skin collagen. These protein solutions were used in a modified Canon Bubble Jet for printing onto substrates. The substrates consisted of glass coverslips treated with type II agarose to prevent nonspecific cell attachment. Following the printing of patterns, rat SMCs were seeded onto the coverslips. The group achieved successful patterning and subsequent attachment in features designed to be 350 μm or larger. Such two-dimensional structures can be important in the fabrication of new scaffolding strategies or studies on cellular behavior. For example, researchers utilized inkjet printing to modify substrates with patterns of DNA [37]. An inkjet printer deposited complementary DNA onto DNA-polyethylene glycol (PEG)-phospholipid modified substrate surface. These printed patterns offered high-resolution renderings capable of precise and accurate immobilization of cells. Inkjet printing can also be utilized to pattern droplets of other bioactive molecules, such as peptides, to induce cell adhesion and migration. One such study utilized CGRDS for cell adhesion and CWQPPRARI for promoting cell migration [38]. Surfaces patterned with these peptides demonstrated improved endothelialization compared to unmodified surfaces by guiding endothelial cell adhesion and expansion.

These technologies, of course, are not limited to 2D patterning of vasculature structures, either, or thermal inkjet heads. Other techniques may incorporate vascular cells within hydrogel materials and subsequently deposit these hydrogel spheroids containing cells. This method may require subsequent fusion of cell-containing particles. In these methods, living cells are directly deposited as droplets instead of being seeded onto pre-existing scaffold structures as demonstrated in Fig. 20.2.

One predominant method for 3D inkjet/drop dispensing methods of bioprinting involves the use of a biopaper (composed of a biocompatible gel, such as collagen) onto which bioink (multicellular spheroids) is printed. This method also takes advantage of the self-assembly phenomena demonstrated by the multicellular spheroids as they are cultured into tissue constructs. Bioink printing can be a rapid method to produce accurate, viable tissues, though the biopaper is a critical component for predicting success of the graft. The concentration of polymer in the biopaper must be carefully tuned for proper attachment and adaptation of the cells within the bioink.

The process of dispensing vascular structures, both acellular and cellular, may introduce a number of limitations. These processes may limit print resolution due to hardware

FIGURE 20.2 **Deposition of cellular spheroids to form vessel.** Spheres are printed, spheroid-by-spheroid, by layer. After multicellular spheroids are deposited, natural fusion of the aggregates through culturing will produce a complete tissue construct.

and material constraints. The material choice in these processes is often limited by the constraints provided by a biological environment. For bioprinting of cells, materials are often restricted to hydrogels that have mechanical properties on their own that may be unsuitable for withstanding the natural hemodynamic forces in blood vessels. However, researchers are making progress in meeting these mechanical requirements. Pataky et al. demonstrated the 3D printing of alginate hydrogel bioinks into vessel-like structures with diameters of roughly 90 μm [39]. Vessel structures were formed through the stacking of alginate beads. The group demonstrated that these vessels could withstand pressures up to 100 mBar before rupturing, whereas the group estimated that relevant physiological pressures of 1–20 mBar would be necessary. Polyurethanes can also be inkjet printed with more appropriate mechanical properties, but also do not directly incorporate the printing of cells. One group developed a pH-sensitive, biodegradable polyurethane suitable for fabricating scaffolds via inkjet printing [40]. While the group did not specifically design the material for vessel printing, they demonstrated some of the steps necessary for developing a material intended for cardiovascular implants. Cytocompatibility was tested by assessing the viability of fibroblasts, and platelet adhesion of the printed scaffolds was assessed using fresh pig blood. The group was able to tune the materials by including a chain extender, N,N-bis(2-hyroxyethyl)-2-aminoethane-sulfonic acid (BES) to ensure that the tensile strength and elongation of the material could be improved and tuned for applications, such as vascular graft printing. In fact, including BES in the polyurethane formulation decreased platelet attachment compared to polyurethane scaffolds with BES. Cells must be seeded after graft fabrication instead of during the process. Particular methods of inkjet or deposition printing may introduce additional problems. For instance, bioink on biopaper printing requires removal of the biopaper from the fused tissue construct, which in itself can be quite challenging.

Inkjet printing may also introduce significant cell damage and compromise cell viability. Cells also are unable to

be printed in high densities due to the orifice diameter that may encourage cell sedimentation and aggregation [35,41]. Droplet fusion is also critical to vessel structural integrity. If fusion and shape of the droplets cannot be controlled precisely, structural integrity of the vessel may be comprised. This may prevent vessels from allowing blood perfusion or may result in more catastrophic problems, such as vessel rupture. In the design and consideration of deposition printing techniques, it is important to consider aspects of vascular cell behavior. For example, Fleming et al. demonstrated the concept of vascular spheroid fusion [42]. Vascular spheroids, consisting of an SMC outer layer and EC inner layer with a central lumen, fused to form elongated structures within collagen hydrogels. Spheroids adjacent to each other fused to form a single lumen. In these spheroids, ECs and SMCs self-assembled within combined spheroids to reform a single outer layer of SMCs and an inner layer of ECs with an empty inner lumen. Utilizing the organization of multicellular constructs may offer tremendous impact on vessel bioprinting.

To aid in the production of heterogeneous constructs, research utilizing multihead inkjet-based bioprinting has enabled the printing of multiple cell types organized within a construct. Heterogeneity may be determined by cell density or through the use of varying cell types. In these heterogeneous constructs, vascular cell types can be incorporated to promote the vascularization of the bulk artificial organ construct. For example, one group designed an inkjet drop-on-demand system that was capable of printing a pattern construct consisting of three different cell types: (1) human amniotic fluid-derived stem cells, (2) SMCs, and (3) bovine aortic ECs [43]. To provide a scaffold for these cells, an alginate–collagen mixture provided layer-by-layer substrate support. The substrate was immersed in a $CaCl_2$ cross-linking solution between each printed layer of cells to ensure each preceding layer was bonded to the next. The printed construct supported the viability and vascularization of the 3D constructs *in vivo*. In another application of inkjet printing with a bioink, researchers printed multicellular spheroids of cardiac cells and ECs onto a biopaper made of collagen and supplemented with VEGF [44]. The cell spheroids fused together over a considerable period of time, much like droplets of liquid, to aid in the self-assembly of these 3D-printed constructs. Constructs with ECs demonstrated the formation of conduits, indicative of vascularization, whereas constructs printed without ECs demonstrated no such structures.

4.2 Laser Printing

Laser-assisted technologies are another method of vessel printing developed in recent years. Barron et al. introduce biological laser printing (BioLP) as an extension of matrix-assisted pulsed laser evaporation direct-write (MAPLE DW) [45,46]. Originally, MAPLE DW was developed for the fabrication of electronics systems on a micron-scale, which Barron et al. adapted to deposit biological material while avoiding any direct interaction between the laser and the biological material. This is accomplished by the inclusion of a laser-absorbing interlayer between the laser pulse and the biological materials. Material is transferred from a fluid biolayer onto a substrate. Thus, laser bioprinters require three main components [47]: (1) a pulsed laser source, (2) a target from which the biomaterial is printed, and (3) a substrate for printed material deposition. In BioLP, cells suspended in the target biomaterial are transferred to a collector slide via laser energy. These laser pulses create bubbles and the formation of the bubbles cause shock waves that propel the cells toward the collector substrate, in the process illustrated by Fig. 20.3. This technique also offers high speed of fabrication, high efficiency, and low volume transfers. Adapting BioLP technology, Guillemot et al. developed a high-throughput workstation relevant to vascular bioprinting [48]. In their adaptation of BioLP, they utilize concepts of bubble dynamics driven droplet ejection, enabling micrometer resolution. Guillemot et al. printed ECs

encapsulated in sodium alginate solutions and were able to print consistent droplets with five to seven cells per droplet [48]. Importantly, their methods utilized a lower glycerol concentration compared to other laser printing techniques utilizing cells. Previous studies show that glycerol concentrations of 10% v/v or higher may compromise cell viability.

Laser printing and patterning of vascular cells and structures can be affected by the viscosity of the deposition material, laser energy, pulse frequency, and laser printing speed [49]. Fabrication using laser systems benefits from high resolution, which enables precise control of cell patterning. Individual cells can be printed within 5.6 ± 2.5 μm of the intended pattern [50]. Such precision has not yet been demonstrated in other bioprinting techniques for the fabrication of vessels. However, laser printing must overcome the current limitations of printing true 3D structures, the need for photocrosslinkable biomaterials, lengthy fabrication times, laser shock induced cell deformation, interactions of cells with light, and random setting of cells within the precursor solution [13].

While 2D laser printing has been largely successful, more recent work has focused on translating these technologies into the fabrication of functional, 3D structures of vascular cells. In laser-induced forward transfer (LIFT) application, Gaebel et al. laser-printed a cardiac patch utilizing human umbilical vein endothelial cell (HUVECs) and MSCs [51]. The cells were patterned to induce cardiac regeneration and their coculturing demonstrated increased vessel formation in infarcted hearts with the printed patches. Laser printing can be adapted to enable the printing of heterogeneous 3D patterns onto a basement membrane gel [45]. In one study, researchers demonstrated the use of this technique to print cocultured cells in the formation of branching/stemming structures [52]. These structures consisted of both HUVECs and human umbilical vein smooth muscle cell (HUVSMCs). HUVECs were deposited first to allow time for lumen formation. Following this period, the researchers utilized the BioLP method again to deposit HUVSMCs on and around the initial lumen structure formed by the HUVECS. The two cell populations appeared to interact to promote a vessel network; HUVSMCs limited HUVEC migration and growth beyond the designated vessel pattern, and the cells appeared to form cell–cell junctions necessary for vascular homeostasis.

4.3 Extrusion-Based Cell-Laden Hydrogel Deposition

The basic premise of extrusion-based bioprinting relies on the extrusion of continuous filaments. These filaments may consist of a combination of biomaterials and/or cells. This process combines a fluid-dispensing system for extrusion with a three-axis motor control system for printing [13]. The three-axis control system enables precise deposition of

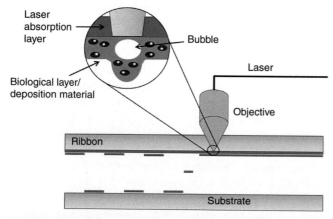

FIGURE 20.3 Schematic of general laser printing used to deposit patterned biological material containing vascular cells onto a substrate.

FIGURE 20.4 Demonstration of filaments organized via extrusion to bioprint blood vessels. Blue filaments represent acellular, support filaments that will be dissolved away. Red filaments represent cell-laden filaments that will form the tissue construct.

filaments into a 3D construct. In the filament tube printing methodology for vessels, grafts might be fabricated by using acellular filaments to provide a structural platform for the deposition of cell-containing filaments in a tube shape. To maintain the inner lumen of the cell-containing vessel, acellular tubes may also deposited within the inner lumen of the designated vessel, as shown in Fig. 20.4. Cell viability in extrusion-based system can easily be influence by nozzle geometry, dispensing pressure, material flow rate, and material concentration [53].

Bioprinted blood vessels can be fabricated more rapidly utilizing extrusion-based methodologies compared to other methodologies outlined here. In addition, the deposition of cylindrical filaments can provide more reliable mechanical integrity to vessel structures [54]. Still, this technique suffers from some disadvantages that may be inhibitive toward current vessel bioprinting. For example, materials able to be used in extrusion-based systems are limited [54]. In addition, resolution is much lower in these applications. Such a disadvantage may be inhibitive due to the microscale resolution and precision necessary for capillary-sized vessels fabricated for microvasculature networks. Similar to inkjet printing of vessels in some manners, nozzle pressure, nozzle diameter, and loaded cell density may affect shear stress experienced by cells during the printing process [55]. If shear stress is too high, cell viability suffers and may decrease the likelihood of successful viable vessel formation. The hydrogels traditionally used in these approaches are intrinsically weak. Their ability to withstand high pressure and mechanical loading may suffer, which can lead to vessel failure. The type of layer-by-layer fabrication used in filament extrusion fabrication may introduce its own host of issues. As with other modes of fabrication, there are constraints on the materials that can be used. Inherent to the nature of filament

extrusion models, resolution and minimum vessel diameters may also be markedly reduced compared to other methods of printing. Filaments extruded in a layer-by-layer fashion also may introduce seams and structural inconsistencies or artifacts that prevent high resolution and accurate rendering of the initial computer model.

Filament stacking models have been used to bioprint vessel-like structures relying on an agarose filament support and cellularized PEG hydrogel filaments [56]. The materials developed by this group enabled higher cell density suspensions. Two four-armed PEG derivatives were synthesized for their unique properties. These multiarmed PEGs possess a compact and symmetrical core. After converting the multiarmed PEGs into acrylate derivatives tetra-acrylate derivatives, the hydrogels could be used to cocrosslink thiolated hyaluronic acid and gelatin derivates, allowing the materials to be used as in extrudable hydrogel printing. These modified tetra-acrylate derivatives demonstrated equivalent or superior cell growth and proliferation support compared to polyethylene glycol-diacrylate (PEGDA)-cross-linked gels, in addition to demonstrating a higher shear storage modulus. This system was used to print NIH/3T3 cells in hydrogel macrofilaments organized in tubular constructs with internal diameters of 500 μm. Cells demonstrated viability for up to 4 weeks, supporting the use of this technique and these materials for potential use in blood vessel bioprinting.

Among the first studies to test 3D fiber deposition scaffold with heterogeneous cell populations *in vivo*, Fedorovich et al. used cell-laden extrusion fabrication to fabricate a scaffold containing regions of either endothelial progenitor cell (EPCs) or mesenchymal stem cells (MSCs) [57]. Fedorovich et al. used a 3D fiber extrusion system called the Bioscaffolder. Using this bioprinting system, the group printed heterogeneously organized constructs with osteo- and endothelial progenitor cells. As predicted, the cells differentiated heterogeneously within the construct. Perfusable blood vessels were formed in EPC-laden areas and bone tissue formed in areas laden with multipotent stromal stem cells.

Using a cell-laden hydrogel extrusion, researchers focused on the computer modeling aspect of bioprinting macroscale blood vessels, demonstrating a hybrid printing technique for a human aorta [58]. Computer algorithms were developed. Overall, the general strategy to print these macroscale vessels was to use printed cell aggregates surrounded by support hydrogels to impart mechanical strength. To tackle the challenges of various structures that researchers may wish to bioprint, two different strategies were developed and utilized: cake-like supports and zig-zag patterning. Cake-like supports are formed by bioprinting a structure that supports itself in layer-by-layer printing. Each layer is printed in a series of long cylindrical tubes dispensed by the bioprinter. Each of these cylinders is printed in a circular structure, so that the bioprinter is essentially printing cylinders made of the hoops. These cylinders support the next layer of prints by offsetting cylinder printing by the diameter of the support material. This enables each cylinder to be supported by two cylinders beneath it, suitable for hollow structures, such as an aorta. Alternatively, more complex structures can be formed using the group's so-called zig-zag patterning approach [58]. While the cellular-component of the vessel is still dispensed in filaments forming hoops, the support material is printed in straight cylinders. Each subsequent layer is dispensed at a 90° angle from the last. In addition, the hoops that make up the vascular structure are dispensed in a curved formation to form a zig-zag pattern when viewed from a transverse cross-section. Thus, each successive layer is supported by the lower layers and from the inner curves.

4.4 Stereolithography and Digital Light Processing

Stereolithography and digital light processing (DLP) utilizes a photosensitive printing resin for scaffold fabrication. The resin is irradiated, with either ultraviolet or visible light, through the bottom of a reservoir tray. One entire layer is irradiated at a time, curing thin sheets of the resin. After a layer is cured, the scaffolds are repositioned by the machine to create a void between the bottom of the reservoir and the cured sections of the scaffold. This void is refilled by resin and the next layer is cured. Layer by layer, this process is repeated until vascular scaffold fabrication is complete. This type of fabrication process is used in a variety of other applications and has not been utilized for cellular vessel bioprinting; rather, current technologies limit vessel bioprinting to acellular structures that may serve as scaffolds for vessel tissue growth.

For stereolithographic blood vessel printing, material choice and development is a crucial aspect as it encompasses several of the requirements necessary for successful vascular graft or blood vessel fabrication. Fabrication with such materials is limited in stereolithography due to the photopolymerization requirements. These materials may affect mechanical properties, biocompatibility, host integration, and prosthetic adaptation. However, to address these issues, stereolithography enables the tuning of the material resin used to produce the vessels. One group demonstrated the successful development and tuning of a urethane-based photoelastomer suitable for vascular tissue regeneration [59]. By using a urethane acrylate (UA)-based polymer, the group could take advantage of additive manufacturing via photopolymerization. Such photoelastomer mechanical properties can also be mechanically tuned via the inclusion of reactive diluents. This particular study formulates a working printing resin using 30% wt UA and 70% wt 2-hydroxethyl acrylate. Printing with such materials could be achieved with DLP additive manufacturing and yielded

fabricated scaffolds with mechanical properties similar to natural blood vessels. Researchers utilized the printed patterning approach in an early technique to create capillary networks [60]. The pattern substrates were created with optical lithography. Following pattern printing, ECs were cultured on the patterned surfaces. The cultured cells were then transferred to the ECM and subsequently formed tubular structures. These microscale tubes could be transplanted into mice and function as capillary-like networks.

4.5 Syringe-Based Bioprinting

Vessels may also be printed utilizing a syringe-based bioprinting system. A syringe pump controls the flow of cell-laden alginate solution through a coaxial nozzle system. The alginate solution cross-links when it intersects with the flow of the cross-linking solution, the two solutions meeting at the outlet of the coaxial nozzle. Vessels formed using this technique may possess mechanical properties inadequate for hemodynamic mechanical forces. For instance, Yu et al. demonstrated this technique with cartilage progenitor cells to study cell viability and functionality in vessel-like tubular channels [61]. The maximum tensile stress was 5.65 ± 1.78 kPa and the Young's modulus was 5.91 ± 1.12 kPa. With the technique, though, the group could print tubular cell-laden tubular channels with inner diameters of 135 ± 13 μm.

4.6 Choosing the Appropriate Technique

Depending on the application, each vessel bioprinting method offers distinct advantages and disadvantages. If mechanical properties after printing are crucial, stereolithography or inkjet printing of materials like polyurethane would be necessary. High resolution for vessel structures that necessitate detailed printing can be attained with lithography- and laser-based approaches. Stereolithography, spheroid deposition, filament extrusion, and syringe-based printing appear to be suitable for the fabrication of larger vascular structures, as well. Of course, stereolithography is the only printing technique that currently does not support direct inclusion of cellular components during vessel printing. Less resolution is attained with prints, especially with filament extrusion and spheroid deposition of vessel components. Examples of relevant factors in making these choices can be seen in Table 20.1.

5 HYBRID VESSEL AND GRAFT FABRICATION

Another group sought to develop a method of producing scaffold-free, perfusable vascular grafts [14]. In this procedure, cells were pelleted into multicellular spheroids and the multicellular spheroids were subsequently used as the

TABLE 20.1 Overview of Various Direct Vessel Printing Methodologies and Relevant Properties

Printing Method	Direct Cell Incorporation	Resolution (μm)	Materials
Inkjet	Yes	10–100	Cell solutions, proteins, hydrogels, polyurethanes
Spheroid deposition	Yes	100	Cell aggregates, hydrogels
Laser-based	Yes	5	Proteins, cell solutions, hydrogels
Filament extrusion	Yes	100	Hydrogels
Stereolithography	No	10	Polyurethane, photoelastomers, acrylates
Syringe-based	Yes	100	Hydrogel

bioink. The bioink pellets were printed onto previously extruded agarose rods. These agarose rods served as a mold to support the layer-by-layer deposition of the multicellular spheroids into various tubular conduit branching patterns. These spheroids fused together to form continuous vessels if left cultured over 5–7 days. Using this technique, the group demonstrated bioprinted vessels that could be printed in varying layers. For example, the inner lumen histology and immunohistochemistry confirmed that they successfully bioprinted layers of fused multicellular spheroids organized with an SMC inner layer and a fibroblast outer layer. The group further demonstrated the potential of these cells by utilizing a perfusion bioreactor to culture the conduits and demonstrate vascular maturity.

In fact, researchers have combined other methods of scaffold fabrication with bioprinting methodologies for the fabrication of vascular scaffolds. One group constructed electrospun vascular grafts made from polylactide and heparin [62]. After electrospinning, the grafts underwent fused deposition modeling to apply a single coil of polycaprolactone around the tube graft. The authors of the study claimed that the inner electrospun layer enables the incorporation of a drug delivery system with a microenvironment conducive to endothelialization. The fused deposition modeling outer layer bolsters the mechanical strength of the fabricated vessel.

Another hybrid approach utilizes the combined techniques of two-photon polymerization (2PP) and LIFT [63]. Introducing these novel approaches in a unique strategy, 2PP was used to photocrosslink acrylated PEG hydrogels. The 3D scaffolds produced by 2PP then undergo LIFT to introduce cell seeding. The group observed that the LIFT technique imparted no deleterious effects onto the deposited cells. Using 2PP, the group fabricated porous, hexagonal rings of PEGDA. Subsequently, LIFT was used to guide the seeding of SMCs and ECs.

Besides combining printing and fabrication techniques, Xu et al. recently demonstrated simultaneous printing and transfection of ECs [64]. It has been demonstrated that the printing process, specifically the application of an electric field or hydrodynamic pressures, influences cell permeability [65]. Altered cell permeability enables the introduction of macromolecules into cells. To take advantage of this phenomenon, Xu et al. incorporated green fluorescent proteins encapsulated in plasmids, which were coprinted with aortic ECs. Through a variety of experiments, they showed that plasmid concentration, cartridge model, and plasmid size influence gene transfection efficiency. Transfection was shown to be achieved simultaneously with the printing of ECs.

6 CASTING

As an alternative to direct fabrication of the scaffold or blood vessels, another popular methodology is cast fabrication of blood vessels and scaffolds. The general methodology relies on computer generation of a casting model or mold. The vessel or vascular network can be 3D printed and subsequently used as mold for casting a polymeric scaffold or used as a substrate for cellular growth and tissue formation. Thus, like many of the direct printing of vessels and vascular structures, the cast scaffolds may or may not directly incorporate living cells and tissues.

In one acellular method, Sodian et al. demonstrated the fabrication of a DLP-produced cast for a patient-specific aortic arch scaffold [66]. They demonstrated the proposed technique of obtaining an MRI for a patient and modeling the defect (in this case an isthmus stenosis). After defining the defect, they reconstructed a model of a corrected vascular structure by reconstructing a nonstenotic aorta. The reconstructed, corrected model was fabricated via stereolithography and a vascular scaffold consisting of polyglycolide and poly-4-hydroxybutyrate was cast onto the model vasculature to produce a scaffold ready for tissue engineering and suitable for implantation. This approach could potentially take advantage of a variety of materials suitable for cardiovascular applications. In addition, the cast scaffolds may or may not take advantage of cell seeding before implantation.

In a casting approach incorporating live tissue, Miller et al. printed 3D filament networks of carbohydrate glass for the casting of functional vascular networks [67]. Carbohydrate glass serves as biocompatible, sacrificial material for the fabrication of printed vascular networks. The printed networks provided a substrate for the growth of cells and development of living vascular networks, at which the point the carbohydrate glass structure was dissolved. This resulted in a perfusable, functional vascular tissue network based on the original sacrificial template model. However, in order to secure the cells within the lattice structure, cells were suspended in an ECM prepolymer that was infused into the glass networks. The glass networks were also coated with a layer of poly(D-lactide-*co*-glycolide) before casting the ECM to prevent any disruption to the ECM cross-linking process and osmotic damage to the suspended cells. With the coating in place, the cells suspended in ECM prepolymer could be conveyed throughout the structure.

7 STRATEGIES FOR INDIRECT MICROVESSEL INFILTRATION

Instead of bioprinting vessels directly, researchers have also investigated methods of bioprinting to induce vascularization of artificial organs after fabrication. Many of these constructs rely on the infiltration of microvessels into the tissue-engineered construct after implantation [68,69]. Tissue engineering constructs have been modified with materials or biofunctional molecules to induce angiogenesis, such as with vascular endothelial growth factor to induce vascularization of 3D-printed bone scaffolds [70]. However, such strategies do not provide precise control over microvascular growth and infiltration. This lack of specificity may be detrimental to 3D-printed construct success, as full vascularization of the construct cannot be guaranteed. Microvessel penetration depth is also limited, which may further core morbidity of the tissue-engineered construct. Finally, uncontrolled and passive induction of microvascular infiltration may destroy sections of the construct due to unintended or deleterious growth of the vessels.

Still, researchers have pursued a variety of strategies to incorporate vessel formation and vascularization into printing technologies. One such strategy is 3D printing of an acellular scaffold with predefined channels. By incorporating predefined channels into the overall scaffold design and fabrication, it is thought that vascular growth can be promoted to ensure adequate construct vascularization. A 2014 study demonstrates this strategy through the printing of a PLGA lattice structure scaffold [71]. After printing, the channels within the scaffold were saturated with a fibrin/collagen hydrogel containing adipose-derived stem cells (ADSCs) and subsequently exposed to growth factors to induce vascular differentiation. Another group utilized ceramic lithography to fabricate an implant with predefined channels for blood vessel implantation [72]. To further assess the potential of predefined channels and pores for vascularization of scaffolds, one research group studied the effects of pore size in 3D-printed polycaprolactone scaffolds to facilitate cell seeding and vascularized bone tissue formation

[73]. By adjusting pore size, they altered the infill density of the scaffolds. (Infill density refers to the volume of unfilled space within the graft that can accommodate cells and tissues.) The group seeded ADSC aggregates suspended in fibrinogen. Through cell seeding distribution experiments, they found that scaffolds with 40% infill density performed the best out of the groups tested. ADSCs appeared to vascularize the fibrin-filled porous structures, demonstrating both the presence of CD31+ ECs and pericyte-like cells.

8 POSTPRINTING

Many of the challenges presented in the description of various blood vessel printing strategies necessitate the maturation of the included or seeded cells into functional tissues. Many researchers assert that accelerated tissue maturation is necessary for these printed vessels to achieve the functional and biomechanical properties necessary for implantation [74]. Some of these challenges have been explored via material solutions, such as rapid polymerization and increased hydrogel stiffness. However, altering the materials in such a manner can negatively affect tissue fusion [75]. One potential innovation is that of an irrigation dripping, triple fusion bioreactor that has removable porous minitubes. This design proposes three individual circuits [76]. One is used for maintaining a wet, physiological environment around the construct, the second enables media perfusion through the branched vasculature tree, and the third enables temporal perfusion. There are a variety of bioreactors that have been developed in the fabrication of tissue-engineered vascular grafts that can simulate the biological environment necessary for tissue development [77–79]. These technologies primarily focus on the maturation of nonbioprinted tissues, but could be adapted to support structures postprinting. Continued research into platforms to support and accelerate vessel maturation is necessary for improving the applicability of bioprinted vessels. Of course, this will also necessitate the use of nondestructive techniques to monitor the maturation of the printed tissues.

To address some of these challenges, researchers are integrating other imaging and culturing technologies into the printing process to gauge and support tissue maturation within printed vessels. One group designed a system of hydrogel and cell printing utilizing a custom-tailored flow chamber for this purpose [80]. Several layers of collagen were printed and subsequently allowed time for gelation. Then, gelatin-containing HUVECs were deposited in a straight line across the collagen layers. Following the deposition of the cell/gelatin mixture, collagen was again printed in layers to fill the rest of the flow chamber. After printing, the gelatin within the cell/gelatin mixture liquefied, allowing the cells to adhere to the collagen layers that now served as the surface of the channel. Media flow through the channel could then be established. The cells demonstrated high

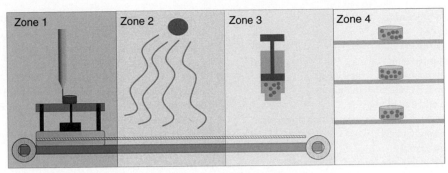

FIGURE 20.5 **Basic schematic of the *BioCell Printer*, demonstrating a self-contained process of fabrication that can be used to incorporate vascularization, from printed scaffold to postprinting culturing of deposited tissues.** Zone 1, scaffolds are printed; zone 2, scaffolds are sterilized; zone 3, cells are deposited; zone 4, constructs are cultured for tissue maturation.

viability in the patterned tubes and were used to investigate the functionality of a mesoscopic fluorescence molecular tomography of the vessel constructs, enabling observations of fluid flow and fluorescent-labeled cells with high sensitivity and accuracy. Another group documented the production methodology utilized by *BioCell Printing* [81]. The entire system is a self-contained method of 3D fabrication involving four zones, as shown in Fig. 20.5. In Zone 1, a multihead printer dispenses the scaffold, which is then sterilized in Zone 2. Cells are seeded on the scaffolds in Zone 3 using another dispenser and finally cultured in Zone 4. This self-contained methodology may reduce the risk of contamination, while providing a translatable platform for clinical application of bioprinted vessels and vasculature trees incorporated into artificial organs.

9 CONCLUSIONS AND FUTURE TRENDS

Emerging technologies in blood vessel printing will continue to provide potential solutions for the development of vascular graft applications and the vascularization of artificial organs. The continued integration of printing with blood vessel tissue engineering is still very much a work in progress but may offer enormous advantages in the form of custom-tailored vascular grafts and networks that could enable the biofabrication of more complex artificial organs. Vascularization of tissue-engineered constructs will likely remain one of the predominant challenges in tissue engineering of artificial organs for clinical applications. While much work has focused on the materials and cellular components of bioprinted vessels, there must be a focus on the process of bioprinting as a whole. For example, future work will necessitate more sophisticated vasculature modeling tools to help researchers, engineers, and clinicians design the appropriate anatomical models for integration within an artificial organ. Researchers must also consider the integral role of postprinting vessels. Current technology dictates the extended culture of cells to reach maturation appropriate for implantation. Self-contained printing and postprinting

processes like that of the *BioCell Printing* system may be an indicator of the future direction of vessel printing [81]. As always, healthy endothelialization of printed vessels has, and will, remain a primary objective of vessel bioprinting due to the endothelial layer's intimate and crucial role in the success of any blood-material contacting artificial organ or graft. Challenges aside, the bioprinting of blood vessels continues to rapidly evolve, unlocking an emerging technology that may significantly impact the cardiovascular and artificial organ fabrication aspects of tissue engineering and biofabrication in a dramatic way.

GLOSSARY

Bioink Biological components (may consist of cells and material) deposited to form the bioprinted pattern or object.

Biopaper Biocompatible gel or material upon which bioink is deposited.

Endothelial cells Cells that cover the inner lumen of native blood vessels and help regulate functions and interactions of blood and vascular tissues.

Endothelial progenitor cells Cells that circle the bloodstream and may differentiate into ECs.

Endothelialization The process of tunica intimal formation by ECs and differentiating EPCs.

Filament extrusion The process by which cell-laden or acellular materials may be deposited in fibers for scaffold fabrication.

Laser printing A printing technique utilizing laser pulses to deposit a material from a ribbon onto a substrate.

Smooth muscle cells Nonstriated muscle cells that primarily makeup the tunica media.

Spheroid fusion The process by which two multicellular aggregates spontaneously fuse to form a singular construct.

Stereolithography A fabrication technique that constructs scaffolds layer by layer utilizing a light source to cure each layer of material.

ABBREVIATIONS

2PP	Two-photon polymerization
BES	*N*,*N*-bis(2-hyroxyethyl)-2-aminoethane-sulfonic acid
BioLP	Biological laser printing
BMCC	Batson's methyl methacrylate corrosion casting

DLP	Digital light processing
EC	Endothelial cell
ECM	Extracellular matrix
EPC	Endothelial progenitor cell
HUVEC	Human umbilical vein endothelial cell
HUVSMC	Human umbilical vein smooth muscle cell
LIFT	Laser-induced forward transfer
MAPLE DW	Matrix-assisted pulsed laser evaporation direct write
microCT	X-ray microtomography
PEG	Polyethylene glycol
PEGDA	Polyethylene glycol-diacrylate
SMC	Smooth muscle cell

ACKNOWLEDGMENTS

This work was supported by the National Institute of Arthritis and Musculoskeletal and Skin Diseases of the National Institutes of Health under Award Number R01 AR061460 and a seed grant from Children's National Sheikh Zayed Institute for Pediatric Surgical Innovation and the A. James Clark School of Engineering at the University of Maryland. The content is solely the responsibility of the authors and does not necessarily represent the official views of the National Institutes of Health. This work was additionally supported by National Science Foundation Graduate Research Fellowships (to AJM), as well as National Science Foundation CBET 1264517.

REFERENCES

[1] Lloyd-Jones D, Adams RJ, Brown TM, Carnethon M, Dai S, De Simone G, et al. Heart disease and stroke statistics – 2010 update: a report from the American Heart Association. Circulation 2010;121(7):e46–e215.

[2] Motwani JG, Topol EJ. Aortocoronary saphenous vein graft disease: pathogenesis, predisposition, and prevention. Circulation 1998;97(9):916–31.

[3] Lytle BW, Loop FD, Cosgrove DM, Ratliff NB, Easley K, Taylor PC. Long-term (5 to 12 years) serial studies of internal mammary artery and saphenous vein coronary bypass grafts. J Thorac Cardiovasc Surg 1985;89(2):248–58.

[4] Desai ND, Cohen EA, Naylor CD, Fremes SE. A randomized comparison of radial-artery and saphenous-vein coronary bypass grafts. N Engl J Med 2004;2302–9.

[5] Abbott WM, Callow A, Moore W, Rutherford R, Veith F, Weinberg S. Evaluation and performance standards for arterial prostheses. J Vasc Surg 1993;17(4):746–56.

[6] Berglund JD, Galis ZS. Designer blood vessels and therapeutic revascularization. Br J Pharmacol 2003;140(4):627–36.

[7] Schmedlen RH, Elbjeirami WM, Gobin AS, West JL. Tissue engineered small-diameter vascular grafts. Clin Plast Surg 2003;30(4):507–17.

[8] Jain RK, Au P, Tam J, Duda DG, Fukumura D. Engineering vascularized tissue. Nat Biotechnol 2005;23(7):821–3.

[9] Muschler GF, Nakamoto C, Griffith LG. Engineering principles of clinical cell-based tissue engineering. J Bone Joint Surg Am 2004;86-A(7):1541–58.

[10] Zhang ZY, Teoh SH, Chong MSK, Lee ESM, Tan LG, Mattar CN, et al. Neo-vascularization and bone formation mediated by fetal mesenchymal stem cell tissue-engineered bone grafts in critical-size femoral defects. Biomaterials 2010;31(4):608–20.

[11] Askarinam A, James AW, Zara JN, Goyal R, Corselli M, Pan A, et al. Human perivascular stem cells show enhanced osteogenesis and vasculogenesis with Nel-like molecule I protein. [Internet]. Tissue Eng Part A 2013;19(11–12):1386–97.

[12] Wang L, Fan H, Zhang ZY, Lou AJ, Pei GX, Jiang S, et al. Osteogenesis and angiogenesis of tissue-engineered bone constructed by prevascularized b-tricalcium phosphate scaffold and mesenchymal stem cells. Biomaterials 2010;31(36):9452–61.

[13] Ozbolat IT, Yu Y. Bioprinting toward organ fabrication: challenges and future trends. IEEE Trans Biomed Eng 2013;60(3):691–9.

[14] Norotte C, Marga FS, Niklason LE, Forgacs G. Scaffold-free vascular tissue engineering using bioprinting. Biomaterials 2009;30(30): 5910–7.

[15] Murohara T. Printing a tissue: a new engineering strategy for cardiovascular regeneration. Arterioscler Thromb Vasc Biol 2010;30(7): 1277–8.

[16] Visconti RP, Kasyanov V, Gentile C, Zhang J, Markwald RR, Mironov V. Towards organ printing: engineering an intra-organ branched vascular tree. [Internet]. Expert Opin Biol Ther 2010;409–20.

[17] Tanzi MC. Bioactive technologies for hemocompatibility. Expert Rev Med Devices 2005;2(4):473–92.

[18] Avci-Adali M, Stoll H, Wilhelm N, Perle N, Schlensak C, Wendel HP. In vivo tissue engineering: mimicry of homing factors for self-endothelialization of blood-contacting materials. Pathobiology 2013;80(4):176–81.

[19] Rubanyi GM. The role of endothelium in cardiovascular homeostasis and diseases. J Cardiovasc Pharmacol 1993;22(Suppl. 4):S1–S14.

[20] Van Hinsbergh VWM. The endothelium: vascular control of haemostasis. Eur J Obstet Gynecol Reprod Biol 2001;95(2): 198–201.

[21] Patel SD, Waltham M, Wadoodi A, Burnand KG, Smith A. The role of endothelial cells and their progenitors in intimal hyperplasia. Ther Adv Cardiovasc Dis 2010;4(2):129–41.

[22] Potter RF, Groom AC. Capillary diameter and geometry in cardiac and skeletal muscle studied by means of corrosion casts. Microvasc Res 1983;25(1):68–84.

[23] Mathura KR, Vollebregt KC, Boer K, De Graaff JC, Ubbink DT, Ince C. Comparison of OPS imaging and conventional capillary microscopy to study the human microcirculation. J Appl Physiol 2001;91(1):74–8.

[24] Zhang WJ, Liu W, Cui L, Cao Y. Tissue engineering of blood vessel. J Cell Mol Med 2007;11(5):945–57.

[25] Sarkar S, Schmitz-Rixen T, Hamilton G, Seifalian AM. Achieving the ideal properties for vascular bypass grafts using a tissue engineered approach: a review. Med Biol Eng Comput 2007;45(4): 327–36.

[26] Furchgott RF, Zawadzki JV. The obligatory role of endothelial cells in the relaxation of arterial smooth muscle by acetylcholine. Nature 1980;288(5789):373–6.

[27] Casscells W. Migration of smooth muscle and endothelial cells. Critical events in restenosis. Circulation 1992;86(3):723–9.

[28] Cines DB, Pollak ES, Buck CA, Loscalzo J, Zimmerman GA, McEver RP, et al. Endothelial cells in physiology and in the pathophysiology of vascular disorders. Blood 1998;91(10):3527–61.

[29] Pearson JD. Endothelial cell function and thrombosis. Best Pract Res Clin Haematol 1999;12(3):329–41.

[30] Matsumoto T, Nagayama K. Tensile properties of vascular smooth muscle cells: bridging vascular and cellular biomechanics. J Biomech 2012;45(5):745–55.

[31] Kakisis JD, Liapis CD, Breuer C, Sumpio BE. Artificial blood vessel: the Holy Grail of peripheral vascular surgery. J Vasc Surg 2005;41(2):349–54.

[32] Marga F, Jakab K, Khatiwala C, Shepherd B, Dorfman S, Hubbard B, et al. Toward engineering functional organ modules by additive manufacturing. Biofabrication 2012;4(2):022001.

[33] Mondy WL, Cameron D, Timmermans J-P, De Clerck N, Sasov A, Casteleyn C, et al. Micro-CT of corrosion casts for use in the computer-aided design of microvasculature. Tissue Eng Part C Methods 2009;15(4):729–38.

[34] Mondy WL, Cameron D, Timmermans J-P, De Clerck N, Sasov A, Casteleyn C, et al. Computer-aided design of microvasculature systems for use in vascular scaffold production. Biofabrication 2009;1(3):035002.

[35] Cui X, Boland T. Human microvasculature fabrication using thermal inkjet printing technology. Biomaterials 2009;30(31):6221–7.

[36] Roth EA, Xu T, Das M, Gregory C, Hickman JJ, Boland T. Inkjet printing for high-throughput cell patterning. Biomaterials 2004;25(17):3707–15.

[37] Sakurai K, Teramura Y, Iwata H. Cells immobilized on patterns printed in DNA by an inkjet printer. Biomaterials 2011;32(14):3596–602.

[38] Boivin M-C, Chevallier P, Hoesli CA, Lagueux J, Bareille R, Rémy M, et al. Human saphenous vein endothelial cell adhesion and expansion on micropatterned polytetrafluoroethylene. J Biomed Mater Res A 2013;101(3):694–703.

[39] Pataky K, Braschler T, Negro A, Renaud P, Lutolf MP, Brugger J. Microdrop printing of hydrogel bioinks into 3D tissue-like geometries. Adv Mater 2011;391–6.

[40] Zhang C, Wen X, Vyavahare NR, Boland T. Synthesis and characterization of biodegradable elastomeric polyurethane scaffolds fabricated by the inkjet technique. Biomaterials 2008;29(28):3781–91.

[41] Xu T, Jin J, Gregory C, Hickman J, Boland T. Inkjet printing of viable mammalian cells. Biomaterials 2005;26(1):93–9.

[42] Fleming PA, Argraves WS, Gentile C, Neagu A, Forgacs G, Drake CJ. Fusion of uniluminal vascular spheroids: a model for assembly of blood vessels. Dev Dyn 2010;239(2):398–406.

[43] Xu T, Zhao W, Zhu J-M, Albanna MZ, Yoo JJ, Atala A. Complex heterogeneous tissue constructs containing multiple cell types prepared by inkjet printing technology. Biomaterials 2013;34(1):130–9.

[44] Jakab K, Norotte C, Damon B, Marga F, Neagu A, Besch-Williford CL, et al. Tissue engineering by self-assembly of cells printed into topologically defined structures. Tissue Eng Part A 2008;14(3):413–21.

[45] Barron JA, Wu P, Ladouceur HD, Ringeisen BR. Biological laser printing: a novel technique for creating heterogeneous 3-dimensional cell patterns. Biomed Microdevices 2004;6(2):139–47.

[46] Barron JA, Spargo BJ, Ringeisen BR. Biological laser printing of three dimensional cellular structures. Appl Phys A 2004; 79(4–6):1027–1030.

[47] Ringeisen BR, Othon CM, Barron JA, Young D, Spargo BJ. Jet-based methods to print living cells. Biotechnol J 2006;1(9):930–48.

[48] Guillemot F, Souquet A, Catros S, Guillotin B, Lopez J, Faucon M, et al. High-throughput laser printing of cells and biomaterials for tissue engineering. Acta Biomater 2010;6(7):2494–500.

[49] Ovsianikov A, Gruene M, Pflaum M, Koch L, Maiorana F, Wilhelmi M, et al. Laser printing of cells into 3D scaffolds. Biofabrication 2010;2(1):014104.

[50] Schiele NR, Chrisey DB, Corr DT. Gelatin-based laser direct-write technique for the precise spatial patterning of cells. Tissue Eng Part C Methods 2011;17(3):289–98.

[51] Gaebel R, Ma N, Liu J, Guan J, Koch L, Klopsch C, et al. Patterning human stem cells and endothelial cells with laser printing for cardiac regeneration. Biomaterials 2011;32(35):9218–30.

[52] Wu PK, Ringeisen BR. Development of human umbilical vein endothelial cell (HUVEC) and human umbilical vein smooth muscle cell (HUVSMC) branch/stem structures on hydrogel layers via biological laser printing (BioLP). Biofabrication 2010;2(1):014111.

[53] Nair K, Yan K, Sun W. A multilevel numerical model quantifying cell deformation in encapsulated alginate structures. J Mech Mater Struct 2007;2(6):1121–1139.

[54] Melchels FPW, Domingos MAN, Klein TJ, Malda J, Bartolo PJ, Hutmacher DW. Additive manufacturing of tissues and organs. Prog Polym Sci 2012;1079–104.

[55] Nair K, Gandhi M, Khalil S, Yan KC, Marcolongo M, Barbee K, et al. Characterization of cell viability during bioprinting processes. Biotechnol J 2009;4(8):1168–77.

[56] Skardal A, Zhang J, Prestwich GD. Bioprinting vessel-like constructs using hyaluronan hydrogels crosslinked with tetrahedral polyethylene glycol tetracrylates. Biomaterials 2010;31(24):6173–81.

[57] Fedorovich NE, Wijnberg HM, Dhert WJA, Alblas J. Distinct tissue formation by heterogeneous printing of osteo- and endothelial progenitor cells. Tissue Eng Part A 2011;17(15–16):2113–21.

[58] Kucukgul C, Ozler B, Karakas HE, Gozuacik D, Koc B. 3D Hybrid bioprinting of macrovascular structures. Procedia Eng 2013;59: 183–92.

[59] Baudis S, Pulka T, Steyrer B, Wilhelm H, Weigel G, Stampfl J, et al. 3D printing of urethane-based photoelastomers for vascular tissue regeneration. Mater Res Soc 2010;1239:3–8.

[60] Kobayashi A, Miyake H, Hattori H, Kuwana R, Hiruma Y, Nakahama K, et al. In vitro formation of capillary networks using optical lithographic techniques. Biochem Biophys Res Commun 2007;358(3): 692–7.

[61] Yu Y, Zhang Y, Martin JA, Ozbolat IT. Evaluation of cell viability and functionality in vessel-like bioprintable cell-laden tubular channels. J Biomech Eng 2013;135(9):91011.

[62] Centola M, Rainer A, Spadaccio C, De Porcellinis S, Genovese JA, Trombetta M. Combining electrospinning and fused deposition modeling for the fabrication of a hybrid vascular graft. Biofabrication 2010;2(1):014102.

[63] Ovsianikov A, Gruene M, Pflaum M, Koch L, Maiorana F, Wilhelmi M, et al. Laser printing of cells into 3D scaffolds. Biofabrication 2010;2(1):014104.

[64] Xu T, Rohozinski J, Zhao W, Moorefield EC, Atala A, Yoo JJ. Inkjet-mediated gene transfection into living cells combined with targeted delivery. Tissue Eng Part A 2009;15(1):95–101.

[65] Wells DJ. Gene therapy progress and prospects: electroporation and other physical methods. Gene Ther 2004;11(18):1363–9.

[66] Sodian R, Fu P, Lueders C, Szymanski D, Fritsche C, Gutberlet M, et al. Tissue engineering of vascular conduits: fabrication of custom-made scaffolds using rapid prototyping techniques. [Internet]. Thorac Cardiovasc Surg 2005;53(3):144–9.

[67] Miller JS, Stevens KR, Yang MT, Baker BM, Nguyen D-HT, Cohen DM, et al. Rapid casting of patterned vascular networks for perfusable engineered three-dimensional tissues. Nat Mater 2012;11(9): 768–74.

[68] Laschke MW, Vollmar B, Menger MD. Inosculation: connecting the life-sustaining pipelines. Tissue Eng Part B Rev 2009;15(4):455–65.

[69] Borselli C, Ungaro F, Oliviero O, d'Angelo I, Quaglia F, La Rotonda MI, et al. Bioactivation of collagen matrices through sustained

VEGF release from PLGA microspheres. J Biomed Mater Res A 2010;92(1):94–102.

[70] Singh S, Wu BM, Dunn JC. The enhancement of VEGF-mediated angiogenesis by polycaprolactone scaffolds with surface cross-linked heparin. Biomaterials 2011;32(8):2059–69.

[71] Zhao X, Liu L, Wang J, Xu Y, Zhang W, Khang G, et al. In vitro vascularization of a combined system based on a 3D printing technique. 2014, doi: 10.1002/term.1863.

[72] Bian W, Li D, Lian Q, Zhang W, Zhu L, Li X, et al. Design and fabrication of a novel porous implant with pre-set channels based on ceramic stereolithography for vascular implantation. Biofabrication 2011;3(3):034103.

[73] Temple JP, Hutton DL, Hung BP, Huri PY, Cook CA, Kondragunta R, et al. Engineering anatomically shaped vascularized bone grafts with hASCs and 3D-printed PCL scaffolds. J Biomed Mater Res A 2014;1–29.

[74] Mironov V, Kasyanov V, Markwald RR. Organ printing: from bioprinter to organ biofabrication line. Curr Opin Biotechnol 2011;22(5):667–73.

[75] Jakab K, Neagu A, Mironov V, Markwald RR, Forgacs G. Engineering biological structures of prescribed shape using self-assembling multicellular systems. Proc Natl Acad Sci USA 2004;101(9):2864–9.

[76] Mironov V, Kasyanov V, Drake C, Markwald RR. Organ printing: promises and challenges. Regen Med 2008;3(1):93–103.

[77] Barron V, Lyons E, Stenson-Cox C, McHugh PE, Pandit A. Bioreactors for cardiovascular cell and tissue growth: a review. Ann Biomed Eng 2003;31(9):1017–30.

[78] Couet F, Meghezi S, Mantovani D. Fetal development, mechanobiology and optimal control processes can improve vascular tissue regeneration in bioreactors: an integrative review. Med Eng Phys 2012;269–78.

[79] Sodian R, Lemke T, Fritsche C, Hoerstrup SP, Fu P, Potapov EV, et al. Tissue-engineering bioreactors: a new combined cell-seeding and perfusion system for vascular tissue engineering. Tissue Eng 2002;8(5):863–70.

[80] Zhao L, Lee VK, Yoo SS, Dai G, Intes X. The integration of 3-D cell printing and mesoscopic fluorescence molecular tomography of vascular constructs within thick hydrogel scaffolds. Biomaterials 2012;33(21):5325–32.

[81] Bartolo P, Domingos M, Gloria A, Ciurana J. BioCell Printing: integrated automated assembly system for tissue engineering constructs. CIRP Ann Manuf Technol 2011;60(1):271–4.

Chapter 21

Bioprinting of Cardiac Tissues

Daniel Y.C. Cheung, Bin Duan, and Jonathan T. Butcher
Department of Biomedical Engineering, Cornell University, Ithaca, NY, USA

Chapter Outline

ABSTRACT

Cardiovascular disease remains the leading cause of death in the world. Cardiac and heart valve tissue engineering in the past decade have made progress to recapitulate the native properties of the heart for cardiac regeneration and repair using a myriad of strategies, including concepts of biomaterials, micro- and nano-technologies, and decellularization. However, one main obstacle for tissue engineering cardiac tissue is incorporating the complex, heterogeneous micro- and macro-cardiac tissue structures for different cell types to adhere, proliferate, and ultimately repair and remodel damaged tissue. 3D bioprinting has emerged as a strategy to address this challenge by providing the technology to fabricate high-resolution scaffolds with heterogeneous biochemical and mechanical properties. This chapter gives an overview of current strategies and limitations for cardiac and heart valve tissue engineering and the advances made in 3D bioprinting for cardiac repair.

Keywords: tissue engineering; heart valve; 3D printing; bioreactor; microfabrication

1 INTRODUCTION

Cardiovascular disease (CVD) and consequent heart failure remains the leading cause of death in the world [1]. According to the World Health Organization, CVD accounted for 17.3 million deaths in 2008, representing 30% of total deaths worldwide. In the United States, CVD ranks as the highest cause of mortality and affect over 27 million Americans. Moreover, cardiovascular operations and interventional procedures increased by 28% from 2000 to 2010, and the direct and indirect costs of CVD and stroke in the United States for 2010 was estimated to be more than $300 billion, and the yearly cost for treatment of all forms of CVD will exceed $560 billion by the year 2015. Although many excellent surgical techniques and pharmacological therapies can slow the progression to end-stage disease and prolong human life, it is still challenging to prevent or reverse the progression of CVD [2]. Currently, the gold standard therapy for end-stage heart failure is heart transplantations. But the number of donor organs is far from sufficient and recipients have to rely on a constant immunosuppressive regime postoperatively, suggesting that heart transplantation cannot constitute

Essentials of 3D Biofabrication and Translation. http://dx.doi.org/10.1016/B978-0-12-800972-7.00021-9

a pragmatic therapy for heart disease. The human heart possesses a small but insufficient natural regenerative capacity [3]. The current clinical interventions for CVD, such as insertion of stents, pacemakers, and/or prosthetic valves, can temporarily recover function but are unable to regenerate tissue. This is particularly problematic in congenital defects where the repair strategy must accommodate the growth of the child. Artificial prosthetic devices cannot adapt and respond to changing biological and mechanical cues, thus exposing the patient to a variety of risks due to structural and/or hemostatic failures that are often difficult to predict. The ideal interventions would deliver therapies that restore the naturally optimized and adaptive biomechanical and biochemical behaviors of the patient's original tissue. Current approaches to achieve this include cell-based therapy, tissue engineering, or biofabrication [4–6]. Cardiac regeneration strategies have the potential to benefit people suffering from several categories of heart disease, including ischemic heart disease, primary myocardial disease, valvular heart disease, and congenital heart defects (CHDs) [7]. In this book chapter, we first briefly review the anatomy and physiology and pathology of the healthy and diseased heart. We then introduce the conventional clinical treatments for different categories of these diseases and their limitations, which motivate the pursuit of cardiac regeneration strategies. We then discuss the state-of-the-art approaches and the implementation of 3D bioprinting for cardiac tissue regeneration.

2 BRIEF OVERVIEW OF HEART ANATOMY AND PHYSIOLOGY

The human heart is a pump that is divided into four chambers: left and right ventricles (RVs), and left and right atrium (RA). The right side of the heart is responsible for pumping deoxygenated blood from the RA to the RV and through the pulmonary artery to the lungs, where fresh oxygen is transported into the blood cells. The freshly oxygenated blood is returned to the heart in the pulmonary veins and enters the left atrium (LA) then left ventricle (LV). From here, the oxygenated blood is pumped out to the rest of the body through the aorta and into smaller arteries and capillaries, where oxygen is transported from the blood and into the surrounding tissue. Then the oxygen-poor blood is returned to the RA via superior and inferior vena cava, completing the blood circulation loop.

2.1 Heart Muscle

The heart wall is a muscle made up of three distinct tissue layers: endocardium, myocardium, and epicardium. Endocardium is the innermost layer, consisting of endothelial cells covering a basement membrane that consists of extracellular matrix (ECM) proteins [8]. The outermost layer of the heart is the epicardium. It is composed of a layer of epithelial cells and a thin underneath layer of loose connective tissue that contains nerves and coronary blood vessels. The myocardium layer, the vast majority of heart tissue, lies between the endocardium and the epicardium and is composed of cardiac muscle. Myocardial tissue consists of tightly packed cardiomyocytes and fibroblasts, with dense supporting vasculature and collagen-based ECM.

There are three key functional requirements of native myocardium: (1) high density of contractile myocytes within biomechanically supporting cells and tissue matrix, (2) efficient oxygen exchange between the cells and blood, and (3) synchronous contractions orchestrated by electrical signal propagation together form a set of design requirements for engineering cardiac tissue [9]. The myocardium consumes large amounts of oxygen and cannot tolerate hypoxia due to the high density and high metabolic demand of the CMs.

2.2 Heart Valves

The heart valves are structures within the heart that ensure unidirectional flow of blood into the ventricles, the lungs, or the body. The four heart valves can be separated into two categories: the atrioventricular/inflow valves (mitral and tricuspid) and the semilunar/outflow valves (aortic and pulmonary). Alternatively, the valves can be categorized by location: the right side (tricuspid and pulmonary) and left side (mitral and aortic). The tricuspid valve is situated between the RA and ventricle while the mitral valve is located between the left atrium and ventricle. The aortic and pulmonary valves are located between the respective arteries and ventricles.

Each valve is composed of cusps (also known as leaflets) that are attached to a fibrous annulus wall (root wall), and both the leaflets and root wall are biomechanically and structurally anisotropic [10]. The root wall is a bulb-shaped fibrous structure that is highly flexible and elastic to accommodate for the high blood flow rate (~5 L/min), and the root wall is stiffer in the circumferential direction when compared to the longitudinal direction. The structure is composed of three layers similar to blood vessels, each with its own distinct cell type: intimal (endothelial cells), medial (smooth muscle cells, SMCs), and adventitial (fibroblast). Much is still unknown whether the cellular functions within the root sinus are similar to those of blood vessels.

The leaflets are the functional tissues that ensure unidirectional blood flow. During systole, high pressure from the ventricle side overcomes aortic pressure and opens the leaflets. Conversely, during diastole, the higher aortic pressure "pushes" the leaflets back toward the ventricle with sealed coaptation such that no regurgitation occurs. Leaflet tissue is composed of three distinct layers: fibrosa, spongiosa, and ventricularis [11]. The fibrosa, the thickest layer containing mainly of type-I collagen, resides on the aortic side. Its collagen fibers bundles are mainly arranged circumferentially. Elastin surrounds the collagen bundles, which allows the valves to close properly by storing energy during systole and releasing the energy back to the collagen fibers during diastole. The spongiosa is the middle layer and acts like a buffer zone by allowing shearing between the two layers. It contains high levels of glycosaminoglycans (GAGs) along with loosely connected collagen fibers and other fibrous proteins oriented in a random manner. Last, the ventricularis, located on the ventricle side, is the thinnest layer. The architecture of the components in the leaflets provides anisotropic properties with distinctive circumferential and radial responses [10]. Specifically, the circumferential direction exhibits a larger Young's modulus when compared to the radial direction. The leaflets are essentially avascular and receive nutrients via hemodynamic convection and diffusion from the root wall blood stream.

Two main cell types are found in the leaflets: valvular endothelial cells (VECs) and valvular interstitial cells (VICs). VECs line the perimeter of the leaflets and are sensitive to biomechanical stimuli, including flow-induced shear stresses, pressure-induced stretch, and cyclic strain. They also provide signal transduction to VICs, and the intercellular crosstalk between the VECs and VICs may be important for homeostasis because dysfunction of the endothelial layer has been associated with valvular pathology [12]. VICs are the most common cells in the valve but exist as a constellation of subphenotypes, including embryonic progenitor endothelial/mesenchymal cells, quiescent (qVICs), activated (aVICs), progenitor (pVICs), and osteoblastic (obVICs) VICs [13]. Additionally, there is a small percentage of SMCs in the leaflets, residing primarily in the base of the ventricularis [14].

3 PREVALENCE AND SEVERITY OF HEART DISEASE AND DEFECTS

Cardiac regeneration efforts are focused on repairing CHDs and damage to the heart caused by three main categories of heart disease: (1) ischemic heart disease, (2) primary myocardial disease (cardiomyopathy), and (3) valvular heart disease. CHDs are problems with the heart's structure that are present at birth, involving the defect(s) in interior walls of the heart, valves, or the arteries and veins [15]. These defects change the normal flow of blood through the heart and cause abnormal development and dysfunction of the heart.

The major cause of death within CVD is ischemic disease (e.g., heart attack), which represents 42% (7.3 million) of all CVD deaths. Acute myocardial infarction (MI, medical term for heart attack) is triggered by the occlusion of one of the coronary arteries supplying blood to the heart.

The lack of blood flow consequently causes ischemia (lack of oxygen) and if the blood flow is not restored quickly, CMs die within the blood-deprived myocardium. Upon MI, a vigorous inflammatory response is provoked, and dead cells are removed. Consequently, fibroblasts and endothelial cells migrate and form granulation tissue that ultimately becomes noncontracting fibrotic scars [9]. Scar formation reduces contractile function of the heart and leads to ventricle wall thinning and remodeling and ultimately leading to heart failure.

Primary myocardial disease (cardiomyopathy) is typically characterized by heart muscle that is structurally and functionally abnormal, in the absence of coronary artery disease, hypertension, valve disease, and congenital heart disease [16]. It refers to a variety of cardiac enlargement through a large number of mechanisms and frequently leads to cardiac dilation and congestive heart failure. Another major type of heart disease is coronary heart disease. The disease is caused by plaque building up inside the coronary arteries. The plaque can harden or rupture, thus narrow the coronary arteries and reduce the flow of oxygen-rich blood to the heart, resulting in heart attack and valve disease as well.

Heart valve disease is characterized by valvular stenosis (narrowing of the orifice during systole due to stiff leaflets) or valvular insufficiency (regurgitation or backflow of blood during diastole). Valvular disease can be acquired at birth (congenital heart disease) or developed later in life, with main risk factors including old age and previous heart disease. The most common congenital heart valve disease affecting 1–2% of the population is the bicuspid aortic valve (BAV) disease, which is characterized by the presence of two leaflets instead of three leaflets. The anatomy of a bicuspid valve can be categorized into a presence or absence of a raphe, which is a ridge formed between two adjacent leaflets when they are fully coated. The lack of a raphe indicates the presence of only two leaflets while the presence of a raphe indicates the fusion of two out of the three leaflets. Although many people with BAV can live normal lives, leaflets can become heavily calcified prematurely, leading to a stenotic valve [12]. Rheumatic fever and infective endocarditis, both bacterial infections, can also cause valvular disease. In rheumatic fever, an autoimmune response to the proteins released by the bacteria *Streptococcus pyogenes* causes valvular complications, including inflammation and calcification [12]. Infective endocarditis is the inflammation of the inner heart tissues, including the leaflets, which can cause severe valvular dysfunction.

While BAV is the most common CHD and affects the largest population, other congenital heart diseases affect the myocardium or valves – or a combination of the two. Ventricular septal defect (VSD) is a hole in part of the septum separating the left and RVs that leads to oxygenated blood flowing into the RV instead of the aorta. The amount of blood going into the RV varies depending on the size of the defect. Likewise, atrial septal defect could also occur. Hypoplastic left heart syndrome is characterized by a severely underdeveloped left ventricle. Tetralogy of Fallot is a combination of four defects: pulmonary valve stenosis, a large VSD, right ventricular hypertrophy, and an overriding aorta located above the VSD.

4 CURRENT CLINICAL TREATMENTS FOR HEART DISEASE

4.1 Treatments for CVD

The current status of surgical treatments for heart disease includes several mechanistic procedures such as insertion of a cardiac stent [17] and implementation of a pacemaker for cardiac synchronization in the left ventricle (left ventricular assist device). To date, whole heart transplantation remains the most efficient form of therapy, particularly in the end-stage heart disease. However, all these clinical therapeutic strategies, invasive or pharmacological approaches, can only offer a temporary and unsustainable care, rather than efficiently regenerating the diseased heart [18].

A great number of cells are lost following an MI episode (up to 25% of LV mass within a few hours post-MI), and CMs rarely divide [19]. Therefore, therapies that can induce and/or orchestrate efficient and stable regeneration of myocardial function are needed. Recently, remarkable progress has been made in myocardial regeneration therapy using cellular cardiomyoplasty, a technique that involves injecting cells into the myocardium [20]. Various cell types including somatic stem/progenitor cells (e.g., bone marrow derived mesenchymal stem cells, adipose derived mesenchymal stem cells, and cardiac derived stem cells) [21,22], embryonic stem cells [23], induced pluripotent stem cells [24], and skeletal muscle myoblasts [25]. They have their own pros and cons. Several pivotal reports provided more detailed reviews of the various cell types and the contributing mechanism, including clinical trials [26,27]. Although the direct injection method is surgically less invasive, the delivered cells have been shown to readily permeate out of the diseased myocardial tissue and have poor engraftment [4]. In addition, the viability of injected cells is low, with only 1–32% surviving after transplantation [28].

Apart from the choice of cell sources, the delivery mode and implanted niche is also very important. Myocardial tissue engineering (MTE) aims to mimic the native environment of the host cardiac tissue by implementation of biomimetic materials, bioactive molecules, oxygen supply, and various stimuli [29,30]. MTE approaches include injectable constructs (cells with hydrogels or hydrogels alone), decellularization of the native heart and repopulation with donor cells, scaffold tissue patches, and culturing of scaffold-free cell sheets [5,7,20]. Considering the complexity associated

with heart physiology and heart disease pathology, successful cardiac regeneration strategies need to replicate the myocardial microenvironment and restore myocardial functions. The native myocardium has three key features [9]: (1) high density of CMs and supporting cells, (2) synchronous contractions, and (3) efficient oxygen exchange between the cells and blood. The formation of fibrotic scar tissue is mainly due to the loss of billions of CMs, the inability to proliferate to recover, and the hostile hypoxic environment that they would have to reside within. It is thus challenging to sustain the viability and function of injected cells or engraftment and, most importantly, to integrate the engineered myocardial tissue with the host myocardium. The integration of MTE construct with the host tissue actively contributes to mechanical force generation and facilitates the electrical signals conduction and synchronous contractions [31]. Another challenge is to have sufficient vascularization and oxygen diffusion in MTE constructs. Diffusion alone can only maintain the viability of myocardiac cells less than 200 μm thick. An ideal MTE construct should promote neovascularization and organized architecture and provide a means for effective and timely connection with the host blood supply.

4.2 Treatments for Valvular Disease

4.2.1 Mechanical Heart Valves

Artificial heart valves were first introduced in the 1950s, with the first implantation performed by Dr. Charles Hufnagel in 1952 with a caged-ball valve positioned in the descending thoratic aorta [32]. Although the position did little to alleviate the patients' aortic stenosis, his surgery provided the proof-of-concept of implanting artificial structures for heart valve disease. In 1960, Dr. Dwight Harken performed the first successful heart valve implantation and paved the way for new advances in heart valve prosthetics. Since the first successful implantation, many mechanical heart valve designs have been implemented with great success, including nontilting disc valves, tilting disc valves, and bileaflet valves.

Although the invention and implementation of mechanical valves have benefited hundreds of thousands of lives, the implantation of nonliving materials in such a demanding hemodynamic environment incurs significant chronic morbidity. The large shear forces produced from the valve geometry [33] damages blood cells, decreasing oxygen supply to the body, and activates platelets toward coagulation. Patients with mechanical valves must remain on anticoagulant medication for the remainder of their lives [34]. Even with optimal anticoagulation therapy, patients, especially younger ones, with mechanical valves have a high risk of experiencing major thromboembolic or hemorrhagic stroke within their lifetime – approximately 2–5% annually and close to 100% in 20 years [35]. While mechanical valves are essentially free of structural valve degeneration, 5% of patients will undergo reoperation within 10 years and increasing to 10% by 15 years, with common problems including endocarditis, pannus formation, paravalvular leak, valve thrombosis, and nonstructural dysfunction [35].

4.2.2 Bioprosthetic Valves

An alternative to the mechanical valves is bioprosthetic heart valves. Many biotechnology companies produce commercially available valves, including Edwards Lifesciences (Carpentier–Edwards), Medtronic, St. Jude Medical, and CyroLife (SynerGraft®). The sources for these bioprosthetics vary from different animals (procine, bovine, and ovine) to human autografts. Generally, the valves are chemically treated (e.g., glutaraldehyde) not only preserve the valves but to also decrease immunogenicity. The main advantage of bioprosthetics is the improved hemodynamics when compared to mechanical valves [36]. Without any blood-shearing effects from the naturally derived valves, patients do not need life-long anticoagulant medication. In addition, bioprosthetics are currently the only type of valve that can undergo the transcatheter aortic valve replacement (TAVR)/transcatheter aortic valve implantation (TAVI) procedure, a minimally invasive surgical procedure that repairs the damaged valve by implanting a new one into the old position and "overriding" the damaged valve [37]. This valve-in-valve approach is better for high-risk patients (e.g., patients with severe aortic stenosis) who cannot undergo full open-heart surgery.

However, these valves are not as durable as mechanical valves. In general, they fail within 20 years of implantation, with general modes of failure including calcification, cusps tears, pannus, infective endocarditis, and thrombosis [38]. In younger patients, deterioration is more rapid, suggesting an age-related mechanism [39]. For older patients or those who cannot survive open-heart surgeries, the TAVI approach can be used [37]. If the initial bioprosthetic presents severe regurgitation, it can be replaced with another valve via TAVI [40].

4.2.3 Ross Procedure

The Ross procedure, also known as the pulmonary autograft, offers an interesting solution to replace diseased valves. The pulmonary valve is used to replace the diseased aortic valve, and a bioprosthetic fills in the low-pressure pulmonary position. The major advantage of the Ross procedure is the ability for the pulmonary valve to grow, remodel, and adapt in the more strenuous hemodynamic environment. Additionally, the procedure is performed on small children when commercial sizes are not available.

Similar to the other valve replacement solutions, the Ross procedure has its own set of drawbacks. First, the procedure is a two-valve procedure to repair one diseased valve. A second valve is still needed, and the lack of valve replacements is still an issue. There have also been cases reported where repeated surgeries were performed due to regurgitation and autograft dilatation, particularly for pediatric patients but a long-term randomized trial suggested adult Ross procedures could work well [41,42]. Despite some shortcomings, the Ross procedure offers promising long-term survival for adults, with high-percentage of freedom from reoperation (94.7 and 87.7% at 10 and 15 years, respectively) and freedom from moderate or severe regurgitation (94.1 and 85.6% at 10 and 15 years, respectively) [43].

Many of the pitfalls with prosthetic valve replacements are attributed to the inability to mimic the natural biochemical and hemodynamic characteristics. Bioprosthetics offers better hemodynamics, but because it is ultimately nonliving, premature – deterioration will always be a concern.

5 ENGINEERING DESIGN CRITERIA FOR TISSUE-ENGINEERED HEART VALVES

The original engineering design criteria were first articulated by Dr Dwight Harken and later extended by Sacks and others [12,44,45]. The criteria have been adapted and can be split into three categories: (1) essential, (2) desirable, and (3) other considerations (Table 21.1).

TABLE 21.1 Engineering Design Criteria for Tissue Engineered Heart Valves

Essential	Desirable	Other Considerations
• Nonobstructive • Closure of leaflets must be prompt and complete (no regurgitation) • Nonthrombogenic • Anatomically correct geometry for physiological hemodynamics • Heterogeneity – must incorporate different cell types for a synergistic remodeling and to mimic the heterogeneity of the leaflets and root wall	• Last the lifetime of the patient while withstanding millions of cycles • Adapt and grow with the patient by providing continuous remodeling processes (essential for children) • Mechanical strength of conduit to promote proper growth but also withstanding elevated blood pressure • Ease of biological functionalization • Ease of fabrication/cost	• Off the shelf versus patient specific • *In vitro* conditioning versus direct implantation upon synthesis • Demographic specific versus one size fits all

6 STATE-OF-THE-ART APPROACHES FOR CARDIAC TISSUE REGENERATION

Advanced manufacturing techniques, biomimetic materials, and combinatorial strategies have been developed and implemented to support achieving the goals of cardiac regeneration. Shimizu et al. first reported the fabrication of 3D beating constructs by stacking of four layers of neonatal rat ventricular myocyte populated cell sheets [46]. The scaffold-free technique is very versatile and suitable for transplantation and cardiac repair. Several other groups have modified this original concept by coculturing different cells within the construct [47], stack more cell sheet layers to increase the cell number, strength, and functions [48,49] (Fig. 21.1a), and to overexpress factors like vascular endothelial growth factor (VEGF) to enhance blood vessel formation [52]. Shimizu et al. stacked four-layered neonatal rat CM sheets and demonstrated the electrical communication between the sheets by connexin [53].

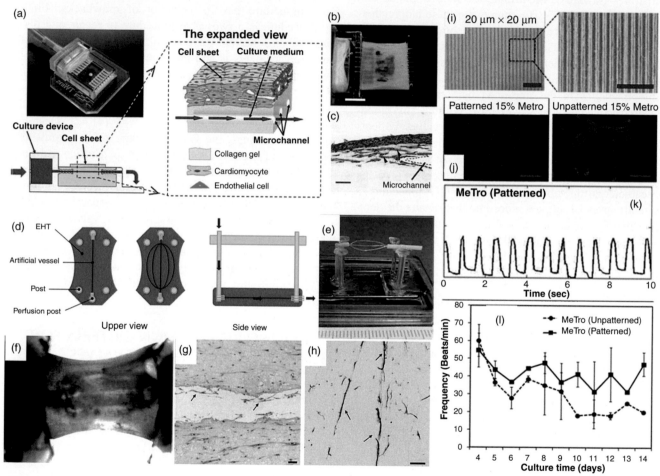

FIGURE 21.1 Advanced MTE approaches and methods. (a–c) Perfusion culture system for cell sheet technology [49]. The collagen-gel base with microchannels (Ø = 300 μm) imitates the conditions of a subcutaneous structure (Fig. 21.1a). Triple-layer cell sheets were incubated and the culture medium could diffuse into the collagen gel and provide oxygen and nutrients to the cell sheet. Fresh rat red blood cells were perfused into the bioreactor (b). When endothelial cells are included, rat blood cells spread throughout the cell sheets like real subcutaneous vessels (Scale bar, 1 cm). HE stained section of the construct containing endothelial cells shows a lot of migrating cells and a vascular formation (c). The red blood cells flowed into the newly created vascular network (Scale bar, 200 mm). (d–h) A novel approach to generate small tubular structures suitable for perfusion of engineered heart tissue [50]. Alginate fibers with aligned and branched structure were first fabricated as sacrificial vessel template (d and e). By dissolving alginate fibers with alginate lyase, microchannels were created (f). Perfusion of multivessel construct in the cell culture with methylene blue for better visualization. HE staining and immunohistochemcial staining with CD31 confirmed that longitudinal sections of perfused constructs showed the lumen and the formation of an intima-resembling endothelial cell layer (g–h). Arrows indicate endothelial cells. Scale bar 50 mm. (i–l) Highly elastic micropatterned tropoelastin hydrogels promote the CM maturation and support the synchronous beating of CM in response to electrical field stimulation [51]. CM elongation and alignment on the surface of micropatterned gels with 20 × 20 μm channels produced by micromolding (scale bar = 200 μm) (i). Cell morphology on patterned and unpatterned gels (j). Representative F-actin (red)/DAPI (blue) stained images demonstrated cell alignment on patterned gels; (k and l) Beating characterization of CM seeded on patterned and unpatterned hydrogels. CMs seeded on patterned gels displayed synchronous contractions (k). The beat frequency for unpatterned gel gradually decreased and for patterned gels, the beat frequency was remarkably consistent across the culture time with synchronized beating for at least 2 weeks (l).

The cell sheets were subcutaneously implanted and could survive for at least 1 year. They showed that the implanted CM sheet had spontaneous beating, heart tissue-like structure and neovascularization with contractile force in proportion to the host's growth. Miyagawa et al. demonstrated that the neonatal CM sheet could survive in infarcted myocardium and communicate electrically with the host myocardium, improving the cardiac performance [54]. A major disadvantage of the cell sheet technique is that it is difficult to obtain thick CM sheets. The thickness of one layer CM sheet is approximately 45 μm and it is not strong enough to repair the tissue damage. Although cell sheets can be stacked, the limiting factor is the oxygen and nutrition transportation and vascularization. When more than a four-layered graft was implanted, however, central necrosis was observed instead of the rapidly organized microvasculature because of the insufficient oxygen supply [55]. Further concerns are to determine how many CM sheets can be layered *in vivo*, how to generate autologous cell sheets, how to implant them into the damaged myocardium and how to connect the microvasculature network in the thick-layered CM sheets to the host myocardium.

Neovascularization and perfusion are undoubtedly a prerequisite for long-term survival and function of an engraftment. In order to promote the blood vessel growth, microchannels in MTE construct were generated by enzymatically dissolving embedded alginate-fibers and were populated by endothelial cells [50]. Similarly, Sakaguchi et al. generated microchannels in collagen gels through perfusion [49] (Fig. 21.1b and c). The collagen-gel base with microchannels imitates the conditions of a subcutaneous structure. Triple-layer cell sheets were incubated and the culture medium could diffuse into the collagen gel and provide oxygen and nutrients to the cell sheet. When endothelial cells are included, rat blood cells spread throughout the cell sheets like real subcutaneous vessels. Vollert et al. fabricated fibrin gels with small tubular structures by dissolving sacrificial microfibers [50,56] (Fig. 21.1d–h).

Mechanical strain is crucial for cardiac myocytes to align and to mature. The simplest form of strain is static tension. Any strategy that allows engineered constructs to perform auxotonic contractions is best suited, probably because of the mimic of normal conditions of the heart contracting against the hydrostatic pressure during the circulation. Efforts that have also focused on increasing the electrical properties of MTE constructs have included blending carbon nanofibers [57], developing new electroactive hydrogels [58], and implementing electrical stimulation [51]. A very recent study incorporated nanoelectronic arrays into synthetic 3D matrices and the resulting nanoelectronic hybrid scaffolds were used to culture, among others, CMs and SMCs [59]. The integrated capability of the nanoelectronic arrays into the nanoelectronic hybrid scaffolds allowed for real-time monitoring of the electrical activity of the CMs and the response of the CMs to drug dosage. Mihic et al. evaluated the effects of uniaxial cyclic stretch on the maturation of human embryonic stem cell-derived CMs (hESM-CMs) on gelatin-based scaffolds [60]. The cyclical stretching procedure produced a high percentage of troponin T-positive cells, which resulted in the formation of a 3D cardiac tissue. hESC-CM maturation can also be achieved through construct topology [61]. Engineered cardiac constructs usually beat spontaneously but at varying rates and with some irregularity over time. A composite method of seeding collagen sponges with cardiac cells in Matrigel have shown that pacing for 8 days improved cell orientation, tissue structure, and function [62]. Highly aligned cell aggregates were also reported to promote the CM maturation and support the synchronous beating of CMs in response to electrical field stimulation [51] (Fig. 21.1i–l). Annabi et al. developed methacrylated tropoelastin (MeTro), which is a highly elastic micropatternable hydrogel. CMs elongated and aligned on the surface of micropatterned gels with 20 × 20 μm channels and displayed synchronous contractions. The beat frequency for unpatterned gel gradually decreased and for patterned gels, the beat frequency was remarkably consistent across the culture time with synchronized beating for at least 2 weeks. These results also demonstrate the importance to accurately recapitulate the complex and compact ultrastructural organization of the myocardium from the centimeter to the nanometer level. In a recent study, CMs were cultured on electrospun fibers that were deposited in a collector with an insulating gap [63]. The electrostatic forces present in the collector drove the anisotropic alignment of CMs, promoting cell elongation and formation of native-like tissue.

7 STRATEGIES FOR HEART VALVE TISSUE ENGINEERING

Tissue engineering has emerged as a means to overcome limitations of current heart valve prosthetic treatments by providing a biocompatible scaffold for which cells can grow and remodel. There have been a variety of methods used to fabricate tissue-engineered heart valves (TEHV), including decellularizing bioprosthetics, using synthetic and biological polymers for molding, and electrospinning. A summary of selected TEHV strategies can be found in Table 21.2.

7.1 Decellularized Bioprosthetic Valves

Instead of fixing xenografts, researchers have opted to decellularize natural scaffolds with correct geometry, which reduces immunogenicity [88,89]. Decellularization is the process of eliminating cells and cellular debris on xeno- or homografts through the use of detergents and enzymes (e.g., RNases and deoxyribonucleases) while maintaining the ECM components intact and undamaged. Similar to the fixed bioprosthetic valves currently used, decellularized bioprosthetics offer similar advantages when compared to mechanical valves, including improved geometry and inherent heterogeneity of remnant

TABLE 21.2 Summary of Selected Current Strategies to Fabricate Tissue Engineered Heart Valves

Year	Scaffold Material	Synthesis Technique	Cell Source	Study	Results	References
2000	Decellularlized ovine pulmonary valve	Decellularization	Ovine myofibroblast and endothelial cells	Sheep implantation (12 weeks)	1. Showed histological restitution of tissue and endothelium 2. Correct phenotypic markers appeared 3. Histological signs of inflammation	[64]
2000	PGA/P4HB	Heat application welding technique	Autologous ovine myofibroblast and endothelial cells	Bioreactor (14 days), lamb implantation (20 weeks)	1. Mobile, functioning leaflets 2. No stenosis, thrombosis, or aneurysm 3. Showed increase FCM and DNA production, comparable to native 4. Complete polymer degradation	[65]
2000	PHO	Molding with salt-leaching	Vascular cells from ovine carotid artery and jugular vein	Lamb model (17 weeks), pulmonary site	1. No thrombus formation 2. Showed comparable GAG and collagen in ECM to native 3. Supraphysiological UTS	[66]
2002	PGA/P4HB	Heat application welding technique	Human marrow stromal cells	Bioreactor (14 days)	1. Showed cellular remodeling 2. Comparable mechanical properties	[67]
2002	PHOH/P4HB	Stereolithography	N/A	Bioreactor	1. Mild stenosis and regurgitation in all valves 2. Correct opening and closing motions	[68]
2003	Decellularized porcine pulmonary valves	Decellularization	Porcine vascular endothelial cells	Sheep implantation (6 months), pulmonary site	1. Monolayer of ECs 2. No cusp calcification 3. Ingrowth of fibroblasts	[69]
2005	Decellularized porcine and human pulmonary valves	Decellularization	N/A	In vitro study of monocytic cell migration study based on matrix residual proteins	1. Reduced migration of monocytes in decellularized human tissue 2. Decellularized porcine matrix showed significantly higher monocytic cell migration	[70]
2005	Decellularized porcine human heart valves	Decellularization	HUVEC	In vitro thromogenicity study	1. Decellularized matrix activates platelet adhesion 2. Seeding matrix with endothelial cells showed no platelet activation	[71]
2005	PGA/PLLA	Molding of melt-extruded nonwoven sheets	Ovine bone marrow derived mesenchymal stem cell	Sheep model (4–8 months), pulmonary site	1. Trivial/mild regurgitation 2. Showed ECM deposition and correct cell phenotype	[72]
2006	Decellularized ovine pulmonary valves	Decellularization	Ovine venous endothelial cells	Bioreactor conditioning	1. Nonconfluent endothelium 2. Partial to complete cell wash-off 3. Metabolically active cells	[73]
2006	Decellularized porcine pulmonary valves	Decellularization	Canine bone marrow-derived cells	Dog implantation (3 weeks), abdominal aortic and pulmonary sites	1. Positive stains for EC-like and myofibroblast-like cells 2. Survival of implanted BMCs	[74]
2006	Decellularized human pulmonary valve	Decellularization	Autologous human endothelial progenitor cells	Bioreactor conditioning (21 days), human implantation, pulmonary site	1. Showed monolayer of ECs 2. Increase valve growth rate, annulus diameter, and body surface area 1.5 years after operation 3. Trivial/mild regurgitation	[75]
2006	PCL	Electrospinning	Human fibroblast	Bioreactor	1. Poor cell penetration 2. Good cell proliferation 3. Tearing of scaffold in physiological flow	[76]

TABLE 21.2 Summary of Selected Current Strategies to Fabricate Tissue Engineered Heart Valves *(cont.)*

Year	Scaffold Material	Synthesis Technique	Cell Source	Study	Results	References
2006	Bovine type-I collagen	Mold synthesized from rapid prototyping	Human aortic VICs	Static (4 weeks)	1. Showed cell viability and proliferation on scaffold	[77]
2006	PGA/P4HB	Molding	Human vena saphena magna	Bioreactor	1. Proper opening motion 2. Suboptimal closure dynamics (stenotic) 3. Increased tissue formation 4. Increased mechanical properties	[78]
2007	Fibrin	Molding	Ovine carotid artery-derived cells	Bioreactor (12 days)	1. Increased cell attachment and alignment 2. Enhanced ECM protein deposition	[79]
2008	PCL	Electrospinning	N/A	Bioreactor	1. Full opening and coaptation 2. Slight rotation of leaflets	[80]
2009	Fibrin	Molding	Ovine carotid artery-derived cells (ECs, SMCs, fibroblasts)	Bioreactor conditioning (28 days), sheep implantation (3 months), pulmonary position	1. Showed tissue development, cell distribution, and ingrowth of blood vessels 2. Confluent EC monolayer 3. Mature collagen formation 4. Leaflet retraction (stenotic)	[81]
2013	Decellularized PGA/P4HB	Heat application welding technique, decellularization	Human vena saphena magna fibroblast	Bioreactor conditioning (4 weeks), baboon implantation (8 weeks), pulmonary position	1. Showed rapid cellular repopulation and tissue growth 2. Excellent ECM deposition 3. Leaflet retraction with mild/trivial regurgitation	[82]
2013	Polyurethane	Polyurethane spraying technique	Human fibroblast and endothelial cells	Bioreactor conditioning (5 days)	1. Showed confluent cell layer 2. Viable cells 3. Establishment of ECM	[83]
2013	Decellularized PGA/P4HB	Heat application welding technique, decellularization	Ovine vascular derived cells	Bioreactor conditioning (4 weeks), sheep implantation (6 months), pulmonary position	1. Showed cellular repopulation, tissue remodeling, and ECM deposition 2. Showed confluent EC monolayer 3. Mild/trivial regurgitation (leaflet retraction)	[84]
2014	Fibroblast sheets	Molding of self-assembled sheets	Human dermal fibroblast	Bioreactor to show feasibility	1. 8 tissue sheets 2. Uniform population of cells 3. Dense ECM with collagen fibers 4. Leaflets opened and closed	[85]
2014	Fibroblast sheets	Assembly of self-assembled sheets on Perimount stent	Human dermal fibroblast	Bioreactor to show feasibility	1. 9 tissue sheets 2. Dense ECM with uniform cell distribution 3. Similar elastic behavior compared to bovine pericardium 4. Leaflets opened and closed in bioreactor	[86]
2014	PEGdma/PLA	Electrospinning	Porcine VICs and VECs	Bioreactor conditioning	1. Similar fiber diameter, pore size, and tensile strength but higher Young's modulus when compared to native porcine 2. Observed cell adherence and spreading 3. Opening and closing of leaflets in bioreactor 4. Functionalized scaffolds	[87]

ECM proteins and fibers. If the decellularization process is performed correctly, the scaffold will have reduced immunogenicity, an advantage over traditionally fixed bioprosthetics. Early results of decellularized allografts for pulmonary valve replacement in children and young adult seemed promising, showing higher freedom from explantation when compared to homografts [90]. Currently, CryoLife's SynerGraft is the only commercial acellular bioprosthetic produced.

In addition to improved geometry and the presence of natural heterogeneity of the ECM fibers, decellularization has little effect on viscoelastic properties of the explants when compared to native counterparts. In one study, Jiao et al. compared the viscoelastic properties of fresh and/or cryopreserved ovine, baboon, and human aortic and pulmonary heart valve tissues versus their respective decellularized counterparts by using torsional wave experiments [91]. Although the sample size was small ($n = 2$–4 for each group), no significant differences in storage modulus were observed between fresh/cryopreserved and decellularized groups in each species for each tissue type. This study suggests that the remnant fibers in the decellularized scaffolds have similar properties as native valves. Also, the acellular scaffolds can be recellularized in hopes of encouraging cell repopulation and matrix remodeling, but the chemicals used in the decellularization process make cell adherence onto the scaffold difficult [92]. Ideally, autologous cells would be seeded onto coated scaffolds (e.g., fibronectin) for better clinical translation and patient integration. *In vivo* experiments showed tissue remodeling and integration [69,74], but some displayed mild or trivial regurgitation [75], inflammation [64], or cell wash-off [73]. To date, many SynerGraft valves have been implanted into humans in the pulmonary position in adults and children [93,94]. The SynerGraft showed improved freedom from moderate or severe insufficiency and displayed less stenosis when compared to cryopreserved allografts [95].

One major pitfall of decellularized/recellularized bioprosthetics is the possibility of an immune response from the remaining cells or cellular debris and calcification [96,97]. A prime example using decellularized explanted porcine valves (SynerGraft) caused deaths in 3/4 children patients due to graft degradation from severe inflammatory response and calcification [98]. The early failure of the SynerGraft helped researchers refine the quality assurance and quality control of the process. Additionally, the decellularization process is not perfect. The complete removal of DNA, RNA, and proteins is necessary, but new products formed during the decellularization process may still elicit host response [99]. The success of the procedure – the complete removal of cells while maintaining ECM – is dependent on the protocol since results have been varied [98]. Another key limitation of the decell/recell approach is the limited supply of human valves [100], but using xenogenic grafts could potentially fill the void [101], as exemplified with the SynerGraft and the Matrix P® grafts [102].

7.2 Molded Polymer Scaffolds

A large majority of scaffolds for TEHV applications mold synthetic polymers into a valve shape [103,104]. Several biocompatible polymers include poly(glycolic acid) (PGA), poly(lactic acid) (PLA), poly(lactic-*co*-glycolic acid), poly(4-hydroxybutyrate) (P4HB), poly(hyaluronic acid), and poly(vinyl alcohol) [96]. The wide range of starting raw materials provides many benefits for synthetic scaffolds for TEHV. A variety of blends of scaffolds can be produced with varying mechanical properties and degradation rates via combining, mixing, and altering different polymers at concentrations. These scaffolds are readily moldable and reproducible, making the technique compatible with different manufacturing processes. Other researchers have synthesized scaffolds from biological polymers (e.g., fibrin) or a combination of biological and synthetic polymers, all of which are inherently biocompatible so anticoagulation is not needed [76,81,105,106].

Fabricating scaffolds using a combination of synthetic and biological polymers had varying degrees of success *in vivo*. In one study, Flanagan et al. molded a fibrin-based scaffold into a valve shape, seeded the construct with autologous cells, conditioned in a bioreactor for 28 days, and implanted into sheep for 3 months [81]. The explanted valves showed tissue development, cell distribution, and native valve consistency. However, the leaflets lost coaptation due to tissue contraction. Tremblay et al. developed a new method of constructing heart valves by molding self-assembled fibroblast sheets [85]. The fabrication process took approximately 8 months, which is not clinically tenable unless the process can be expedited. However, the constructed valve demonstrated opening and closing of the leaflets during bioreactor conditioning with low pulsatile flow and showed dense ECM, mainly collagen, and uniform cell distribution. More recently, Dubé et al. constructed TEHV using self-assembled fibroblast sheets, taking only 7 weeks for the sheet to assemble and yielded promising results [86]. Although not related to aortic heart valve, Moreira et al. have developed TexMi, a textile-reinforced tubular scaffold for the mitral position with some key features, including asymmetry of leaflets and inclusion of chordae tendineae [107]. They seeded the polyethylene terephthalate (PET) mesh using fibrin as a cell carrier for ovine umbilical vein cells. After 3 weeks of conditioning, histological and IHC stains showed tissue development, aligned collagen fibers and elastin deposition, and lack of cell-mediated tissue contraction.

Hoerstrup and coworkers have combined *de novo* synthesis of TEHV with decellularization. Weber et al. decellularized tissue-engineered heart valves (dTEHV) constructed with biodegradable synthetic materials (PGA and P4HB) and human fibroblasts [82]. They showed in nonhuman primate models that compared to the decellularized native heart valves, the dTEHV showed more cellular repopulation. However, the dTEHV also displayed mild to moderate shortening of the leaflets, leading to regurgitation. In a similar study,

Driessen-Mol et al. implanted the dTEHV into sheep [84]. The study yielded similar results: cell repopulation was apparent, and leaflet shortening and regurgitation were observed.

The lack of leaflet coaptation is a common point of failure in TEHV. Baaijen's group used finite element modeling to observe the stresses imposed on and generated by the leaflets during diastole and systole [108]. Their model indicated that the imposed stresses are equal to the stresses generated by the leaflets, suggesting that the imposed stresses cannot counteract the generated stresses, leading to leaflet retraction. Additionally, material degradation became an issue when the scaffold degrades faster than cellular ECM production, leading to early structural incompetency [104,109]. Cases have also reported calcification and higher thrombotic activity in TEHV composed of synthetic polymers. Although scaffolds can be mechanically tuned, leaflets can stiffen or tear. Reliance on a single homogeneous material to construct the leaflet and root walls may therefore result in leaflets that are too stiff or too compliant root walls, which themselves are characteristic of valvular pathology (aneurysm and stenosis, respectively) [110].

7.3 Electrospun Scaffolds

Few groups have used electrospinning to fabricate heart valve scaffolds with aligned fibers to mimic the native anisotropy of the whole TEHV to promote cell growth and differentiation. Electrospinning is a technique to fabricate fine fibers from a solution by applying a high voltage to a liquid droplet. The droplet is stretched to form a thin fiber, which then travels to a grounded target, which can be a mold or a flat surface, to produce a nonwoven scaffold.

Courtney et al. first showed that electrospun fibers composed of poly(ester urethane) ureas can mimic heart valve anisotropic mechanical properties [111]. Since then, other groups have used the technology to create scaffolds and biomaterials. Masoumi et al. used directional electrospinning to fabricate a poly(glycerol sebacate):poly(caprolactone) (PGS:PCL) fibrous scaffold to emulate native ECM networks and seeded the construct with primary human VICs [112]. The biochemically active cells aligned along the scaffold and displayed various mechanical properties. Additionally, Tseng et al. took another approach by incorporating electrospun chemically modified PCL fibers into poly(ethylene glycol) diacrylic acid (PEGDA) hydrogel to improve material strength and introduce anisotropic mechanical behavior [113]. The scaffold showed anisotropic elastic moduli (3.79 ± 0.90 MPa and 0.46 ± 0.21 MPa in the parallel and perpendicular directions, respectively) and seeded cells aligned in the direction of the embedded fibers. Although there has been no study with a complete electrospun scaffold seeded with cells, creating anisotropic TEHV is promising. Del Gaudio et al. were amongst the first to use electrospinning to synthesize a whole heart valve conduit

[80]. The group electrospun PCL onto a custom-made aluminum heart valve-shaped target to fabricate a homogeneous TEHV that incorporated randomly oriented microfibers with approximately 90% porosity and observed full opening and coaptation in a pulse-duplicator, suggesting the technical feasibility of producing TEHV from electrospinning. Most recently, Hinderer et al. fabricated electrospun poly(ethylene glycol) dimethacrylate and poly(L-lactide) (PEGdma/PLA) that were seeded with VICs and VECs and showed promising results [87]. Additionally, they showed a proof of principle of functionalizing the scaffold with collagen type I and versican to mimic the heart valve during early development.

Electrospun TEHV scaffolds have shown *in vitro* feasibility, but few attempts of *in vivo* studies have been performed. Current researches have been using homogeneous materials for scaffold fabrication, but heterogeneous scaffolds could be produced by using additional syringes filled with different materials [114], but few researchers have tried fabricating heterogeneous TEHV via electrospinning. Although valved conduits have been produced, they are not anatomically shaped. Electrospinning onto a native heart valve-shaped target may help address this, but it would be significantly challenging to develop heterogeneous tissue thickness or composition until better methods to control fiber deposition is created.

8 BIOPRINTING OF MYOCARDIAL TISSUE

3D printing, also referred to as rapid prototyping or solid freeform fabrication, can fabricate 3D objects through computer-aided design and/or computer-aided manufacturing in a layer-by-layer manner. It provides precise control over both macro- and microarchitecture of the scaffolds and fulfills a customized design with a complex anatomic shape. As previously mentioned, the cardiac tissue engineering requires high density of CMs and various supporting cells, efficient oxygen exchange and vascularization, and synchronous contractions. 3D bioprinting can pattern and assemble cells and biomaterials with defined organization and spatial arrangements [115]. It also allows the production of 3D constructs with multiple cell types and more physiologically relevant microenvironments.

Several types of 3D bioprinting techniques, including extrusion, laser, and self-assembly-based bioprinting, have been used to biofabricate cardiac constructs using various cell types and biomaterials. We outline the general principles and current progress and highlight the pros and cons for different techniques.

8.1 Extrusion-Based Bioprinting

Extrusion-based bioprinting implements mechanical force driven by air pressure or motor to extrude biomaterials (normally hydrogels) through a nozzle in a controlled

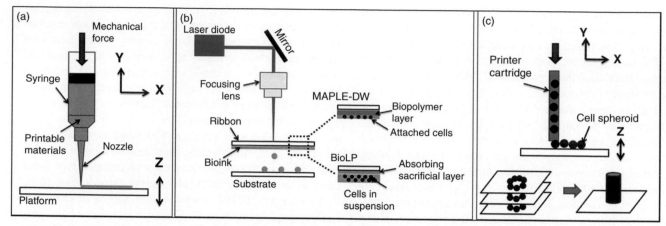

FIGURE 21.2 Schematic representation of 3D bioprinting technologies. (a) Extrusion-based bioprinting. (b) Laser-based bioprinting, including the ribbon design utilized in both MAPLE-DW and BioLP. (c) Self-assembly-based bioprinting: (bottom) scheme of bioassembly of tubular tissue construct using bioprinting of self-assembled cell spheroids.

manner to construct a 3D structure. The bioprinter usually consists of a three-axis robot that controls the movement of the syringes in the x and y directions and the movement of deposition stage in the z direction, and force system and extrusion nozzle(s) (Fig. 21.2a). The typical diameter of the nozzle is about 150–300 μm and a smaller diameter was proved to damage the cells [116]. Usually the cell-laden hydrogels need to have high viscosity to hold the shape during and after extrusion and the extruded hydrogels are further cross-linked through various physical or chemical approaches to fabricate integrated 3D constructs. Gaetani et al. extruded RGD-modified alginate with human fetal CM progenitor cells and fabricated porous scaffolds with grid structure [117]. They demonstrated that printed human fetal CM progenitor cells retained their commitment for the cardiac lineage and showed enhanced gene expression of the early cardiac transcription factors as well as the sarcomeric protein. Heterogeneous architecture and cell distributions can be achieved by using multiple cell types and biomaterials with different properties. But a common problem for extrusion-based bioprinting is the lack of mechanical integrity that matches the native myocardial tissue and allows mechanical support during *in vivo* regeneration/repair. In order to solve this problem, cell-laden hydrogels have also been extruded and combined with printing of structural synthetic polymers, which usually have much higher mechanical properties.

8.2 Laser-Based Bioprinting

Currently, the most widely used techniques to deposit biological materials and living cells are described as laser-assisted bioprinting (LAB) techniques [118]. The LAB technique requires three key components: (1) a pulsed laser source, (2) a target plate (the ribbon) usually made of quartz and coated with the bioink to be printed, and (3) a

receiving substrate that faces the ribbon (Fig. 21.2b). Based on the nature of the ribbon, the LAB approach is also categorized as matrix-assisted pulsed laser evaporation-direct write (MAPLE-DW) and biological laser printing (BioLP) [119]. For MAPLE-DW, the ribbon is coated with a biopolymer hydrogel layer, which serves as an attachment layer for cells and absorbs the laser. Volatilization at the ribbon-biopolymer interface induces cavitation, which generates a high speed (20–100 m/s) jet that transfers a small volume of the biopolymer and cells to the substrate [120]. BioLP, also known as absorbing film-assisted laser-induced forward transfer, uses a ribbon with a thin (1–100 nm) metal or metal oxide (usually Au, Ti or TiO_2) layer to absorb a high-powered laser pulse. Then, a very small volume of the bioink is injected to the substrate due to rapid thermal expansion. A range of bioink hydrogels and cell types has been used for cell patterning and various tissue regenerations.

LAB techniques are powerful to generate cell patterns with high spatial resolution and to manipulate small drop size ($\geq$ 10 μm) [121]. However, these techniques are more suitable to fabricate a 2D structure, rather than 3D constructs with clinical size and anatomical shape. In addition, the requirements of laser and ribbon preparation make LAB not as accessible or as consistent as other 3D bioprinting techniques. Therefore, LAB techniques can be used to print cells onto prefabricated scaffolds via other techniques. Gaebel et al. applied the LAB technique and patterned human umbilical vein endothelial cells (HUVEC) and human mesenchymal stem cell on a polyester urethane urea cardiac patch [122]. Patches were cultivated *in vitro* or transplanted *in vivo* to the infarcted zone of rat hearts. It was demonstrated that the LAB cell seeding pattern definitely modified growth characteristics of cocultured HUVEC and human mesenchymal stem cell leading to increased vessel formation and found significant functional improvement of infarcted hearts following transplantation.

8.3 Self-Assembly-Based Bioprinting

Another 3D bioprinting technique implements the biological self-assembly approach to fabricate tissue-engineered microtissues like spheroids containing several thousand cells. The microtissues or spheroids can be homogeneous, containing a single cell type, or heterogeneous, made from a mixture of several cell types. During 3D bioprinting, the microtissue bioinks are deposited in close spatial organization so that they fuse together to generate an organotypic structure [120] (Fig. 21.2c). Mironov et al. pioneered the bioprinting of intraorgan vascular tree by using spherical building blocks and 3D bioprinting technique [123]. Norotte et al. also demonstrated that tubular structures could be built from heterogeneous endothelial and SMC mixtures and the branching conduit was also printed using spherical units [124,125]. Self-assembly-based 3D bioprinting also has great potential for organ printing and generating engineered cardiac tissue. This approach maintains a physiological microenvironment with high cell density, cell–cell interactions, and paracrine signaling. However, the resolution of the self-assembly approach is relatively low due to the use of cell aggregates, which have the size of 300–500 μm in diameter. In addition,

the preparation of cell aggregate bioinks is normally time consuming and lacks control.

9 3D PRINTING OF TISSUE-ENGINEERED HEART VALVES

All previous TEHV strategies have failed to replicate the heterogeneity and anatomical shape of the heart valve, including valve sinuses and coronary ostia, which are two critical structures ensuring optimal hemodynamics [126]. However, 3D printing is shown to be an emerging approach for fabricating TEHV that incorporates these two key properties.

Sodian et al. were amongst the first group to utilize rapid-prototyping techniques to fabricate TEHV constructs [68,127]. The group used X-ray computed tomography (CT) to capture the complex anatomy of human pulmonary and aortic homografts and coupled this technique with stereolithography to reconstruct the valves using P4HB and polyhydroxyoctanoate (PHOH) (Fig. 21.3 a–c). The constructed heart valve was also able to open and close in a pulsatile bioreactor, further indicating that 3D printing techniques could reproduce complex structures and may be useful in custom fabrication of TEHV. In addition to printing TEHV for valve replacement, Sodian et al. also used rapid prototyping of 3D

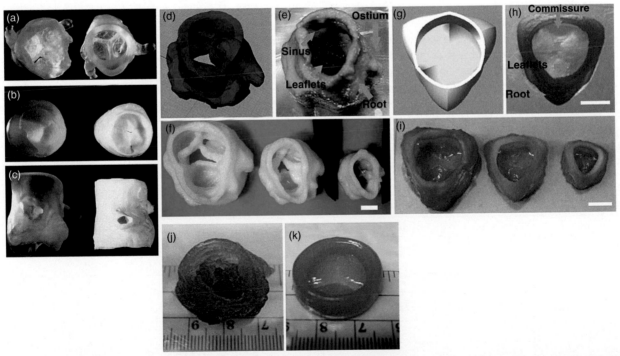

FIGURE 21.3   3D bioprinting of heart valves. Stereolithographic models of heart valves reconstructed from CT scanning [68]. (a) Plastic models of human aortic valve from the aortic side (left) and ventricular side (right). PHOH (left) and P4HB (right) heart valve reconstructions. (b) Pulmonary reconstructions from the ventricular side. (c) Aortic reconstruction, including sinus of Valsalva and coronary arteries. (d–f) Heterogeneous printing of porcine aortic valve and axisymmetric valve model [134]. 3D porcine aortic valve model (d). Printed model using 700 MW PEGDA hydrogel for the root wall and a blend of 700/8000 MW PEGDA hydrogels for the leaflet (e). Printed homogeneous (700 MW PEGDA) scaffolds at different scales (ID = 22, 17, and 12 mm) (scale bar = 1 cm) (f). (g) 3D axisymmetric model. (h) Printed heterogeneous axisymmetric model (scale bar = 1 cm). (i) Shape fidelity of differently sized heterogeneous axisymmetric model (scale bar = 1 cm). (j) Heterogeneous valve constructed from alginate and gelatin [135]. (k) Heterogeneous flat valve model using a combination of methacrylated hyaluronic acid and methacrylated gelatin [136].

printing to construct models for surgical planning, including congenital heart surgery [128], aortic valve replacement [129–132], and patients with cardiac tumors [133]. Scans of the patient heart by either computer tomography or magnetic resonance imaging were reconstructed using stereolithography or inkjet layer-by-layer deposition.

Advancing the technique further, we have first shown the feasibility of 3D printing living, heterogeneous heart valves using extrusion-based bioprinting with alginate/gelatin hydrogels (Fig. 21.3j) [135]. Duan et al. encapsulated VICs and SMCs separately into different syringes, one for each cell type. The Fab@Home 3D printer was used to print an imported scan of a porcine aortic valve. The printed structure was cross-linked in $CaCl_2$. The initial study showed high cell viability for 7 days in the conduit, but tensile strength was lacking. Phenotype (αSMA+ and vimentin+) was sustained for VIC and SMC with trends similar to native properties.

Hockaday et al. improved on the previous design by printing anatomically accurate, living, and heterogeneous valves using a combination of 700 MW and 8000 MW poly(ethylene glycol) diacrylate (PEGDA) at clinically relevant sizes (>10 mm), which was achieved by using an on-board photocrosslinking system to simultaneously print and cure the hydrogels (Fig. 21.3d–i) [134]. The print was completed in a relatively short amount of time (45 min), suggesting fast turnover rate. The moduli of the PEGDA hydrogels used were within range of the native aortic leaflet in the radial direction at physiological tensile strains but were too soft for the aortic root sinus. However, the PEGDA range of moduli matches the properties of pulmonary leaflets and root sinuses. The shape fidelity of the printed valves yielded promising results with many regions of the printed valve matching the initial scanned valve. However, the analysis showed that there were more regions overprinted than underprinted, especially when printing smaller valves using the same print paths.

Duan et al. continued to explore different materials for 3D heterogeneous printing [137]. They printed heterogeneous flat-shaped trileaflet valve conduits using a combination of methacrylated hyaluronic acid and methacrylated gelatin with human aortic VICs encapsulated (Fig. 21.3k). Altering the ratio enabled control of the polymer stiffness and viscosity. The cells were shown with high viability and matrix remodeling was shown with collagen and GAG deposition.

The Sodian and Butcher groups have shown technical feasibility of 3D printing heart valves, but some possible limitations should be addressed. First, the resolution of the printed conduit largely depends on the tip diameter and the speed at which the bioink extrudes out. However, smaller tip sizes cannot be feasibly used for bioprinting due to cell-shearing effects [116]. Until this problem is overcome, there will be a physical limitation on printing minute details. Additionally, native tissues are anisotropic, but the current studies have not been able to mimic this property, which can help with cell growth and differentiation.

10 POTENTIAL CELL SOURCE FOR 3D PRINTING

Early cell sources for TEHV were derived from donor tissues by isolating endothelial, smooth muscle, and fibroblast cells [65,138]. However, cells isolated from autograft tissue performed better than allogenic tissue. Unfortunately, harvesting autograft tissue is a sacrificial process, whereas healthy aortic valves are nonsacrificial. Therefore, stem cell sources have been investigated for use in tissue engineering heart valves because of their potential to self-renew and differentiate into multiple lineages. Weber and colleagues have tested many sources for seeding TEHV, many of which showed tissue remodeling after bioreactor conditioning [65] (Table 21.3).

TABLE 21.3 Potential Cell Sources for Heart Valve Engineering

Source	Phenotype	Supply	Ease of Isolation	Construct	References
Adipose tissue	MSC	+++	+++	TEHV	[139]
Amniotic fluid	MSC	+	+	TEHV	[140]
	EPC	+	+	TEHV	[140]
Blood	EPC	+++	+	TEHV	[141]
Bone marrow	MSC	+++	+++	Myocardium TEHV	[67,72,122]
	Mononuclear cells	+++	+++	TEHV	[142]
Chorionic villi	MSC	+	+	TEHV	[143]
Heart	CM progenitor cells	++	+	Myocardium	[117]
Umbilical cord matrix/blood	VEC	+	+	Myocardium	[122]
	MSC	+	+	TEHV	[144–146]

+++ = Larger supply, easier to isolate.

11   CONCLUSIONS AND FUTURE DIRECTION

Recent advances in 3D printing of cardiac tissue and heart valves have shown great potential toward treating CVDs. 3D printing supports fabrication of complex and customized geometries with high spatial control to produce macro- and microarchitectures while supporting multiple cell types, all of which are useful for patient-specific treatment. Several techniques have been used for cardiac tissue and heart valve printing, including extrusion-, laser-assisted-, and self-assembly-based printing, resulting in simple and complex homogeneous and heterogeneous structures that show high potential for mimicking native tissue mechanical properties and cell integration. However, even with the recent advances in technology, some persisting challenges must be addressed before clinical translation can move forward. Higher temporal and spatial resolution in micro- and macroscales need to be advanced to mimic minute structures, especially for the heart valve and vasculature for cardiac tissues. Additional biomaterials should be explored further for bioprinting, specifically addressing material properties that can be used in printing while maintaining biocompatibility and structural integrity once printed. Last, fabrication consistency must be established for accurate and precise scaffold constructions. By addressing some or all of these challenges, researchers can make more progress with *in vivo* work and eventually move forward toward the clinic.

GLOSSARY

Bioprosthetic Device engineered to replace a damaged body part that is biologically derived (e.g., from pig). These devices can also have mechanical parts, such as stents.

CVD A class of diseases that involve the heart or blood vessels. Common CVDs include blood vessel diseases, heart attack, stroke, heart failure, heart-valve problems.

Cardiomyoplasty A surgical procedure in which healthy muscle is isolated from another part of the body (usually back or abdomen) and wrapped around the heart to provide support for the failing heart. A special device similar to pacemaker is implanted to make the skeletal muscle contract.

Congenital heart disease (CHD) A category of heart diseases that occurs during fetal development and results in abnormalities in the heart after birth. Common CHD affect the heart valves, the septum between the atria and ventricles, and the myocardium.

Decellularization Process of removing cellular content within a tissue or construct. In the context of tissue engineering, the process should retain ECM composition and proteins but not factors that may induce immunogenicity.

Electrospinning Technique that uses high electrical charge to fabricate fine fibers from polymer solution. The resulting structure can have fiber alignment and variable porosity.

Heart valve calcification Heart valves may accumulate deposits of calcium with age. The deposits of calcium can accumulate on the valve's leaflets, result in stiffening of the leaflets of the valve and prevent the aortic valve from properly opening and closing.

Ross procedure An alternative heart-valve replacement procedure that replaces the diseased aortic heart valve with the pulmonary valve and places a bioprosthetic into the pulmonary position.

TAVI/TAVR A minimally invasive surgical procedure that replace the stenotic valve without removing the old valve. Instead, it delivers a replacement valve into the aortic valve's place through blood vessel. The replacement valve is delivered via one of several access methods: transfemoral (in the upper leg), transapical (through the wall of the heart), subclavian (beneath the collar bone), and direct aortic (through a minimally invasive surgical incision into the aorta).

ABBREVIATIONS

aVIC	Activated valvular interstitial cell
BAV	Bicuspid aortic valve
BioLP	Biological laser printing
CHD	Congenital heart defect/disease
CM	Cardiomyocyte
CVD	Cardiovascular disease
dTEHV	Decellularized tissue engineered heart valve
ECM	Extracellular matrix
GAG	Glycosaminoglycan
HUVEC	Human umbilical vein endothelial cell
LA	Left atrium
LAB	Laser-assisted bioprinting
LV	Left ventricle
MAPLE-DW	Matrix-assisted pulsed laser evaporation-direct write
MI	Myocardial infarction
MTE	Myocardial tissue engineering
obVIC	Osteoblastic valvular interstitial cell
P4HB	Poly(4-hydroxybutyrate)
PCL	Poly(caprolactone)
PEGDA	Poly(ethylene glycol) diacrylate
PGA	Poly(glycolic acid)
PGS	Poly(glycerol sebacate)
PHOH	Polyhydroxyoctanoate
PLA	Poly(lactic acid)
pVIC	Progenitor valvular interstitial cell
qVIC	Quiescent valvular interstitial cell
RA	Right atrium
RV	Right ventricle
SMC	Smooth muscle cell
TAVI	Transcatheter aortic valve implantation
TAVR	Transcatheter aortic valve replacement
TEHV	Tissue-engineered heart valve
VEC	Valvular endothelial cell
VIC	Valvular interstitial cell
VSD	Ventricular septal defect

REFERENCES

[1] Alwan FA. Global status report on noncommunicable diseases 2010 2011:176.

[2] Eschenhagen T, Zimmermann WH. Engineering myocardial tissue. Circ Res 2005;97:1220–31.

[3] Kikuchi K, Poss KD. Cardiac regenerative capacity and mechanisms. Annu Rev Cell Dev Biol 2012;28:719–41.

[4] Wang F, Guan JJ. Cellular cardiomyoplasty and cardiac tissue engineering for myocardial therapy. Adv Drug Deliv Rev 2010;62: 784–97.

[5] Georgiadis V, Knight RA, Jayasinghe SN, Stephanou A. Cardiac tissue engineering: renewing the arsenal for the battle against heart disease. Integr Biol 2014;6(2):111–26.

[6] Ozbolat IT, Yu Y. Bioprinting toward organ fabrication: challenges and future trends. IEEE Trans Biomed Eng 2013;60:691–9.

[7] Dunn DA, Hodge AJ, Lipke EA. Biomimetic materials design for cardiac tissue regeneration. Wiley Interdiscip Rev Nanomed Nanobiotechnol 2013;6:15–39.

[8] Bowers SLK, Baudino TA. Laying the groundwork for growth: cell–cell and cell–ECM interactions in cardiovascular development. Birth Defects Res C Embryo Today 2010;90:1–7.

[9] Vunjak-Novakovic G, Tandon N, Godier A, Maidhof R, Marsano A, Martens TP, et al. Challenges in cardiac tissue engineering. Tissue Eng Part B Rev 2010;16(2):169–87.

[10] Hasan A, Ragaert K, Swieszkowski W, Selimović Š, Paul A, Camci-Unal G, et al. Biomechanical properties of native and tissue engineered heart valve constructs. J Biomech 2014;47(9):1949–63.

[11] Misfeld M, Sievers HH. Heart valve macro- and microstructure. Philos Trans R Soc B Biol Sci 2007;362(1484):1421–36.

[12] Butcher JT, Mahler GJ, Hockaday LA. Aortic valve disease and treatment: the need for naturally engineered solutions. Adv Drug Deliv Rev 2011;63(4–5):242–68. Apr 30.

[13] Liu AC, Joag VR, Gotlieb AI. The emerging role of valve interstitial cell phenotypes in regulating heart valve pathobiology. Am J Pathol 2007;171(5):1407–18.

[14] Latif N, Sarathchandra P, Chester AH, Yacoub MH. Expression of smooth muscle cell markers and co-activators in calcified aortic valves. Eur Heart J 2014; eht547.

[15] Gelb BD. Recent advances in understanding the genetics of congenital heart defects. Curr Opin Pediatr 2013;25:561–6.

[16] Elliott P, Andersson B, Arbustini E, Bilinska Z, Cecchi F, Charron P, et al. Classification of the cardiomyopathies: a position statement from the European society of cardiology working group on myocardial and pericardial diseases. Eur Heart J 2008;29: 270–6.

[17] Wilson W, Osten M, Benson L, Horlick E. Evolving trends in interventional cardiology: endovascular options for congenital disease in adults. Can J Cardiol 2014;30:75–86.

[18] Silvestri A, Boffito M, Sartori S, Ciardelli G. Biomimetic materials and scaffolds for myocardial tissue regeneration. Macromol Biosci 2013;13:984–1019.

[19] Murry CE, Wiseman RW, Schwartz SM, Hauschka SD. Skeletal myoblast transplantation for repair of myocardial necrosis. J Clin Invest 1996;98:2512–23.

[20] Doppler SA, Deutsch MA, Lange R, Krane M. Cardiac regeneration: current therapies-future concepts. J Thorac Dis 2013 Oct;5(5): 683–97.

[21] Cashman TJ, Gouon-Evans V, Costa KD. Mesenchymal stem cells for cardiac therapy: practical challenges and potential mechanisms. Stem Cell Rev Rep 2013;9:254–65.

[22] Buikema JW, Van der Meer P, Sluijter JPG, Domian IJ. Concise review: engineering myocardial tissue: the convergence of stem cells biology and tissue engineering technology. Stem Cells 2013;31:2587–98.

[23] Dierickx P, Doevendans PA, Geijsen N, van Laake LW. Embryonic template-based generation and purification of pluripotent stem cell-derived cardiomyocytes for heart repair. J Cardiovasc Transl Res 2012;5:566–80.

[24] Christoforou N, Liau B, Chakraborty S, Chellapan M, Bursac N, Leong KW. Induced pluripotent stem cell-derived cardiac progenitors differentiate to cardiomyocytes and form biosynthetic tissues. PloS ONE 2013;8:e65963.

[25] Blumenthal B, Golsong P, Poppe A, Heilmann C, Schlensak C, Beyersdorf F, et al. Polyurethane scaffolds seeded with genetically engineered skeletal myoblasts: a promising tool to regenerate myocardial function. Artif Organs 2010;34:E46–54.

[26] Ghodsizad A, Ruhparwar A, Bordel V, Mirsaidighazi E, Klein HM, Koerner MM, et al. Clinical application of adult stem cells for therapy for cardiac disease. Cardiovasc Ther 2013;31:323–34.

[27] Sarig U, Machluf M. Engineering cell platforms for myocardial regeneration. Expert Opin Biol Ther 2011;11:1055–77.

[28] Jawad H, Ali NN, Lyon AR, Chen QZ, Harding SE, Boccaccini AR. Myocardial tissue engineering: a review. J Tissue Eng Regen Med 2007;1:327–42.

[29] Leor J, Amsalem Y, Cohen S. Cells, scaffolds, and molecules for myocardial tissue engineering. Pharmacol Ther 2005;105:151–63.

[30] Zammaretti P, Jaconi M. Cardiac tissue engineering: regeneration of the wounded heart. Curr Opin Biotechnol 2004;15:430–4.

[31] Hsiao CW, Bai MY, Chang Y, Chung MF, Lee TY, Wu CT, et al. Electrical coupling of isolated cardiomyocyte clusters grown on aligned conductive nanofibrous meshes for their synchronized beating. Biomaterials 2013;34(4):1063–72.

[32] Gott VL, Alejo DE, Cameron DE. Mechanical heart valves: 50 years of evolution. Ann Thorac Surg 2003;76(6):S2230–9.

[33] Dasi LP, Simon HA, Sucosky P, Yoganathan AP. Fluid mechanics of artificial heart valves. Clin Exp Pharmacol Physiol 2009;36(2): 225–37.

[34] Cheema FH, Hussain N, Kossar AP, Polvani G. Patents and heart valve surgery–I: mechanical valves. Recent Pat Cardiovasc Drug Discov 2013;8(1):17–34.

[35] Carpentier AF, Chikwe J, Filsoufi F. Prosthetic valve selection for middle-aged patients with aortic stenosis. Nat Rev Cardiol 2010;7(12):711–9.

[36] Bre LP, McCarthy R, Wang W. Prevention of bioprosthetic heart valve calcification: strategies and outcomes. Curr Med Chem 2013;21(22):2553–64.

[37] Leon MB, Smith CR, Mack M, Miller DC, Moses JW, Svensson LG, et al. Transcatheter aortic-valve implantation for aortic stenosis in patients who cannot undergo surgery. N Engl J Med 2010;363(17):1597–607.

[38] Siddiqui RF, Abraham JR, Butany J. Bioprosthetic heart valves: modes of failure. Histopathology 2009;55(2):135–44.

[39] Chen PC, Sager MS, Zurakowski D, Pigula FA, Baird CW, Mayer JE Jr, et al. Younger age and valve oversizing are predictors of structural valve deterioration after pulmonary valve replacement in patients with tetralogy of Fallot. J Thorac Cardiovasc Surg 2012;143(2):352–60.

[40] Diemert P, Lange P, Greif M, Seiffert M, Conradi L, Massberg S, et al. Edwards Sapien XT valve placement as treatment option for aortic regurgitation after transfemoral CoreValve implantation: a multicenter experience. Clin Res Cardiol Off J Ger Card Soc 2014;103(3):183–90.

[41] El-Hamamsy I, Eryigit Z, Stevens LM, Sarang Z, George R, Clark L, et al. Long-term outcomes after autograft versus homograft aortic root replacement in adults with aortic valve disease: a randomised controlled trial. Lancet. 2010 Aug 14;376(9740):524–31.

[42] El-Hamamsy I, Poirier NC. What is the role of the Ross procedure in today's armamentarium? Can J Cardiol 2013;29(12):1569–76.

[43] Kalfa D, Mohammadi S, Kalavrouziotis D, Kharroubi M, Doyle D, Marzouk M, et al. Long-term outcomes of the Ross procedure in adults with severe aortic stenosis: single-centre experience with 20 years of follow-up. Eur J Cardiothorac Surg 2014;47(1):159–67. ezu038.

[44] Sacks MS, Schoen FJ, Mayer JE. Bioengineering challenges for heart valve tissue engineering. Annu Rev Biomed Eng 2009; 11(1):289–313.

[45] Harken DE, Taylor WJ, Lefemine AA, Lunzer S, Low HBC, Cohen ML, et al. Aortic valve replacement with a gaged ball valve. Am J Cardiol 1962;9(2):292–9.

[46] Shimizu T, Yamato M, Isoi Y, Akutsu T, Setomaru T, Abe K, et al. Fabrication of pulsatile cardiac tissue grafts using a novel 3-dimensional cell sheet manipulation technique and temperature-responsive cell culture surfaces. Circ Res 2002;90:E40–8.

[47] Sasagawa T, Shimizu T, Sekiya S, Yamato M, Okano T. Comparison of angiogenic potential between prevascular and non-prevascular layered adipose-derived stem cell-sheets in early post-transplanted period. J Biomed Mater Res A 2014;102:358–65.

[48] Sasagawa T, Shimizu T, Sekiya S, Haraguchi Y, Yamato M, Sawa Y, et al. Design of prevascularized three-dimensional cell-dense tissues using a cell sheet stacking manipulation technology. Biomaterials 2010;31:1646–54.

[49] Sakaguchi K, Shimizu T, Horaguchi S, Sekine H, Yamato M, Umezu M, et al. In vitro engineering of vascularized tissue surrogates. Sci Rep 2013;3:1316.

[50] Yeh TS, Fang YHD, Lu CH, Chiu SC, Yeh CL, Yen TC, et al. Baculovirus-transduced, VEGF-expressing adipose-derived stem cell sheet for the treatment of myocardium infarction. Biomaterials 2014;35:174–84.

[51] Shimizu T, Sekine H, Isoi Y, Yamato M, Kikuchi A, Okano T. Long-term survival and growth of pulsatile myocardial tissue grafts engineered by the layering of cardiomyocyte sheets. Tissue Eng 2006;12(3):499–507.

[52] Miyagawa S, Sawa Y, Sakakida S, Taketani S, Kondoh H, Memon IA, et al. Tissue cardiomyoplasty using bioengineered contractile cardiomyocyte sheets to repair damaged myocardium: their integration with recipient myocardium. Transplantation 2005;80(11):1586–95.

[53] Shimizu T, Sekine H, Yang J, Isoi Y, Yamato M, Kikuchi A, et al. Polysurgery of cell sheet grafts overcomes diffusion limits to produce thick, vascularized myocardial tissues. FASEB J 2006;20(6):708–10.

[54] Vollert I, Seiffert M, Bachmair J, Sander M, Eder A, Conradi L, et al. In vitro perfusion of engineered heart tissue through endothelialized channels. Tissue Eng Part A 2014;20:854–63.

[55] Annabi N, Tsang K, Mithieux SM, Nikkhah M, Ameri A, Khademhosseini A, et al. Highly elastic micropatterned hydrogel for engineering functional cardiac tissue. Adv Funct Mater 2013;23:4950–9.

[56] Thomson KS, Korte FS, Giachelli CM, Ratner BD, Regnier M, Scatena M. Prevascularized microtemplated fibrin scaffolds for cardiac tissue engineering applications. Tissue Eng Part A 2013;19:967–77.

[57] Martins AM, Eng G, Caridade SG, Mano JF, Reis RL, Vunjak-Novakovic G. Electrically conductive chitosan/carbon scaffolds for cardiac tissue engineering. Biomacromolecules 2014;15:635–43.

[58] Cui H, Liu Y, Cheng Y, Zhang Z, Zhang P, Chen X, et al. In vitro study of electroactive tetraaniline-containing thermosensitive hydrogels for cardiac tissue engineering. Biomacromolecules 2014;15:1115–23.

[59] Tian B, Liu J, Dvir T, Jin L, Tsui JH, Qing Q, et al. Macroporous nanowire nanoelectronic scaffolds for synthetic tissues. Nat Mater 2012;11(11):986–94.

[60] Mihic A, Li J, Miyagi Y, Gagliardi M, Li SH, Zu J, et al. The effect of cyclic stretch on maturation and 3D tissue formation of human embryonic stem cell-derived cardiomyocytes. Biomaterials 2014;35:2798–808.

[61] Salick MR, Napiwocki BN, Sha J, Knight GT, Chindhy SA, Kamp TJ, et al. Micropattern width dependent sarcomere development in human ESC-derived cardiomyocytes. Biomaterials 2014;35: 4454–64.

[62] Nunes SS, Miklas JW, Liu J, Aschar-Sobbi R, Xiao Y, Zhang B, et al. Biowire: a platform for maturation of human pluripotent stem cell-derived cardiomyocytes. Nat Methods 2013;10(8):781–7.

[63] Orlova Y, Magome N, Liu L, Chen Y, Agladze K. Electrospun nanofibers as a tool for architecture control in engineered cardiac tissue. Biomaterials 2011;32(24):5615–24.

[64] Steinhoff G, Stock U, Karim N, Mertsching H, Timke A, Meliss RR, et al. Tissue engineering of pulmonary heart valves on allogenic acellular matrix conduits in vivo restoration of valve tissue. Circulation 2000;102(Suppl. 3). Iii–50–Iii–55.

[65] Hoerstrup SP, Sodian R, Daebritz S, Wang J, Bacha EA, Martin DP, et al. Functional living trileaflet heart valves grown in vitro. Circulation 2000;102(19 Suppl. 3):III44–9.

[66] Sodian R, Hoerstrup SP, Sperling JS, Daebritz S, Martin DP, Moran AM, et al. Early in vivo experience with tissue-engineered trileaflet heart valves. Circulation 2000;102(Suppl. 3). Iii–22–Iii–29.

[67] Hoerstrup SP, Kadner A, Melnitchouk S, Trojan A, Eid K, Tracy J, et al. Tissue engineering of functional trileaflet heart valves from human marrow stromal cells. Circulation 2002;106(12 Suppl. 1). I–143–I–150.

[68] Sodian R, Loebe M, Hein A, Martin DP, Hoerstrup SP, Potapov EV, et al. Application of stereolithography for scaffold fabrication for tissue engineered heart valves. ASAIO J 2002;48(1):12–6.

[69] Dohmen PM, Ozaki S, Nitsch R, Yperman J, Flameng W, Konertz W. A tissue engineered heart valve implanted in a juvenile sheep model. Med Sci Monit Int Med J Exp Clin Res 2003;9(4):BR97–BR104.

[70] Rieder E, Seebacher G, Kasimir MT, Eichmair E, Winter B, Dekan B, et al. Tissue engineering of heart valves decellularized porcine and human valve scaffolds differ importantly in residual potential to attract monocytic cells. Circulation 2005;111(21):2792–7.

[71] Kasimir M-T, Weigel G, Sharma J, Rieder E, Seebacher G, Wolner E, et al. The decellularized porcine heart valve matrix in tissue engineering: platelet adhesion and activation. Thromb Haemost 2005;94(3):562–7.

[72] Sutherland FWH, Perry TE, Yu Y, Sherwood MC, Rabkin E, Masuda Y, et al. From stem cells to viable autologous semilunar heart valve. Circulation 2005;111(21):2783–91.

[73] Lichtenberg A, Cebotari S, Tudorache I, Sturz G, Winterhalter M, Hilfiker A, et al. Flow-dependent re-endothelialization of tissue-engineered heart valves. J Heart Valve Dis 2006;15(2):287–93. discussion 293–294.

[74] Kim SS, Lim SH, Hong YS, Cho SW, Ryu JH, Chang BC, et al. Tissue engineering of heart valves in vivo using bone marrow-derived cells. Artif Organs 2006;30(7):554–7.

[75] Cebotari S, Lichtenberg A, Tudorache I, Hilfiker A, Mertsching H, Leyh R, et al. Clinical application of tissue engineered human heart valves using autologous progenitor cells. Circulation 2006;114(1 Suppl.). I–132–I–137.

[76] Lieshout MV, Peters G, Rutten M, Baaijens F. A knitted, fibrin-covered polycaprolactone scaffold for tissue engineering of the aortic valve. Tissue Eng 2006;12(3):481–7.

[77] Taylor PM, Sachlos E, Dreger SA, Chester AH, Czernuszka JT, Yacoub MH. Interaction of human valve interstitial cells with collagen matrices manufactured using rapid prototyping. Biomaterials 2006;27(13):2733–7.

[78] Mol A, Rutten MCM, Driessen NJB, Bouten CVC, Zünd G, Baaijens FPT, et al. Autologous human tissue-engineered heart valves prospects for systemic application. Circulation 2006;114 (1 Suppl). I–152–I–158.

[79] Flanagan TC, Cornelissen C, Koch S, Tschoeke B, Sachweh JS, Schmitz-Rode T, et al. The *in vitro* development of autologous fibrin-based tissue-engineered heart valves through optimised dynamic conditioning. Biomaterials 2007;28(23):3388–97.

[80] Del Gaudio C, Bianco A, Grigioni M. Electrospun bioresorbable trileaflet heart valve prosthesis for tissue engineering: *in vitro* functional assessment of a pulmonary cardiac valve design. Ann Dell-lIstituto Super Sanità 2008;44(2):178–86.

[81] Flanagan TC, Sachweh JS, Frese J, Schnöring H, Gronloh N, Koch S, et al. *In vivo* remodeling and structural characterization of fibrin-based tissue-engineered heart valves in the adult sheep model. Tissue Eng Part A 2009;15(10):2965–76.

[82] Weber B, Dijkman PE, Scherman J, Sanders B, Emmert MY, Grünenfelder J, et al. Off-the-shelf human decellularized tissue-engineered heart valves in a non-human primate model. Biomaterials 2013;34(30):7269–80.

[83] Thierfelder N, Koenig F, Bombien R, Fano C, Reichart B, Wintermantel E, et al. *In vitro* comparison of novel polyurethane aortic valves and homografts after seeding and conditioning. ASAIO J 2013;59(3):309–16.

[84] Driessen-Mol A, Emmert MY, Dijkman PE, Frese L, Sanders B, Weber B, et al. Transcatheter implantation of homologous "off-the-shelf" tissue engineered heart valves with self-repair capacity: long term functionality and rapid *in vivo* remodeling in sheep. J Am Coll Cardiol 2013;63(13):1320–9.

[85] Tremblay C, Ruel J, Bourget JM, Laterreur V, Vallières K, Tondreau MY, et al. A new construction technique for tissue-engineered heart valves using the self-assembly method. Tissue Eng Part C Methods 2014;20(11):905–15. 140228041418006.

[86] Dubé J, Bourget JM, Gauvin R, Lafrance H, Roberge CJ, Auger FA, et al. Progress in developing a living human tissue-engineered tri-leaflet heart valve assembled from tissue produced by the self-assembly approach. Acta Biomater [Internet]; 2014 [cited 2014 May 18]. Available from: http://www.sciencedirect.com/science/article/pii/S1742706114002025

[87] Hinderer S, Seifert J, Votteler M, Shen N, Rheinlaender J, Schäffer TE, et al. Engineering of a bio-functionalized hybrid off-the-shelf heart valve. Biomaterials 2014;35(7):2130–9.

[88] Gerson CJ, Elkins RC, Goldstein S, Heacox AE. Structural integrity of collagen and elastin in SynerGraft® decellularized–cryopreserved human heart valves. Cryobiology 2012;64(1):33–42.

[89] Iwai S, Torikai K, Coppin CM, Sawa Y. Minimally immunogenic decellularized porcine valve provides *in situ* recellularization as a stentless bioprosthetic valve. J Artif Organs 2007;10(1):29–35.

[90] Cebotari S, Tudorache I, Ciubotaru A, Boethig D, Sarikouch S, Goerler A, et al. Use of fresh decellularized allografts for pulmonary valve replacement may reduce the reoperation rate in children and young adults early report. Circulation 2011;124(11 Suppl. 1):S115–23.

[91] Jiao T, Clifton RJ, Converse GL, Hopkins RA. Measurements of the effects of decellularization on viscoelastic properties of tissues in ovine, baboon, and human heart valves. Tissue Eng Part A 2012;18(3–4):423–31.

[92] Masters KS, Shah DN, Walker G, Leinwand LA, Anseth KS. Designing scaffolds for valvular interstitial cells: cell adhesion and function on naturally derived materials. J Biomed Mater Res A 2004;71(1):172–80.

[93] Brown JW, Ruzmetov M, Eltayeb O, Rodefeld MD, Turrentine MW. Performance of SynerGraft decellularized pulmonary homograft in patients undergoing a Ross procedure. Ann Thorac Surg 2011;91(2):416–22. discussion 422–423.

[94] Ruzmetov M, Shah JJ, Geiss DM, Fortuna RS. Decellularized versus standard cryopreserved valve allografts for right ventricular outflow tract reconstruction: a single-institution comparison. J Thorac Cardiovasc Surg 2012;143(3):543–9.

[95] Konuma T, Devaney EJ, Bove EL, Gelehrter S, Hirsch JC, Tavakkol Z, et al. Performance of CryoValve SG decellularized pulmonary allografts compared with standard cryopreserved allografts. Ann Thorac Surg 2009;88(3):849–54. discussion 554–555.

[96] Brody S, Pandit A. Approaches to heart valve tissue engineering scaffold design. J Biomed Mater Res B Appl Biomater 2007;83B(1):16–43.

[97] Naso F, Gandaglia A, Formato M, Cigliano A, Lepedda AJ, Gerosa G, et al. Differential distribution of structural components and hydration in aortic and pulmonary heart valve conduits: impact of detergent-based cell removal. Acta Biomater 2010;6(12):4675–88.

[98] Simon P, Kasimir MT, Seebacher G, Weigel G, Ullrich R, Salzer-Muhar U, et al. Early failure of the tissue engineered porcine heart valve SYNERGRAFT® in pediatric patients. Eur J Cardiothorac Surg 2003;23(6):1002–6.

[99] Moroni F, Mirabella T. Decellularized matrices for cardiovascular tissue engineering. Am J Stem Cells 2014;3(1):1–20.

[100] Jashari R, Goffin Y, Vanderkelen A, Van Hoeck B, du Verger A, Fan Y, et al. European Homograft Bank: twenty years of cardiovascular tissue banking and collaboration with transplant coordination in Europe. Transplant Proc 2010;42(1):183–9.

[101] Bloch O, Golde P, Dohmen PM, Posner S, Konertz W, Erdbrügger W. Immune response in patients receiving a bioprosthetic heart valve: lack of response with decellularized valves. Tissue Eng Part A 2011;17(19–20):2399–405.

[102] Konertz W, Angeli E, Tarusinov G, Christ T, Kroll J, Dohmen PM, et al. Right ventricular outflow tract reconstruction with decellularized porcine xenografts in patients with congenital heart disease. J Heart Valve Dis 2011;20(3):341–7.

[103] Sewell-Loftin MK, Chun YW, Khademhosseini A, Merryman WD. EMT-inducing biomaterials for heart valve engineering: taking cues from developmental biology. J Cardiovasc Transl Res 2011;4(5):658–71.

[104] Claiborne TE, Slepian MJ, Hossainy S, Bluestein D. Polymeric trileaflet prosthetic heart valves: evolution and path to clinical reality. Expert Rev Med Devices 2012;9(6):577–94.

[105] Ahmed TAE, Dare EV, Hincke M. Fibrin: a versatile scaffold for tissue engineering applications. Tissue Eng Part B Rev 2008;14(2):199–215.

[106] Weber M, Heta E, Moreira R, Gesche VN, Schermer T, Frese J, et al. Tissue-engineered fibrin-based heart valve with a tubular leaflet design. Tissue Eng Part C Methods 2014;20(4):265–75. 131019071410009.

[107] Moreira R, Gesche VN, Hurtado-Aguilar LG, Schmitz-Rode T, Frese J, Jockenhoevel S, et al. TexMi: Development of Tissue-Engineered Textile-Reinforced Mitral Valve Prosthesis. Tissue Eng Part C Methods [Internet]; 2014 [cited 2014 Apr 27]. Available from: http://online.liebertpub.com.proxy.library.cornell.edu/doi/abs/10.1089/ten.tec.2013.0426

[108] Van Loosdregt IAEW, Argento G, Driessen-Mol A, Oomens CWJ, Baaijens FPT. Cell-mediated retraction versus hemodynamic loading – A delicate balance in tissue-engineered heart valves. J Biomech [Internet]; 2013 [cited 2014 Mar 24]. Available from: http://www.sciencedirect.com/science/article/pii/S0021929013005411

[109] Ghanbari H, Viatge H, Kidane AG, Burriesci G, Tavakoli M, Seifalian AM. Polymeric heart valves: new materials, emerging hopes. Trends Biotechnol 2009;27(6):359–67.

[110] Butcher JT, Simmons CA, Warnock JN. Mechanobiology of the aortic heart valve. J Heart Valve Dis 2008;17(1):62–73.

[111] Courtney T, Sacks MS, Stankus J, Guan J, Wagner WR. Design and analysis of tissue engineering scaffolds that mimic soft tissue mechanical anisotropy. Biomaterials 2006;27(19):3631–8.

[112] Masoumi N, Larson BL, Annabi N, Kharaziha M, Zamanian B, Shapero KS, et al. Electrospun PGS: PCL microfibers align human valvular interstitial cells and provide tunable scaffold anisotropy. Adv Healthc Mater 2014;3(6):929–39.

[113] Tseng H, Puperi DS, Kim EJ, Ayoub S, Shah JV, Cuchiara ML, et al. Anisotropic poly(ethylene glycol)/polycaprolactone (PEG/PCL) hydrogel-fiber composites for heart valve tissue engineering. Tissue Eng Part A 2014;20(19–20):2634–45.

[114] Ingavle GC, Leach JK. Advancements in Electrospinning of Polymeric Nanofibrous Scaffolds for Tissue Engineering. Tissue Eng Part B Rev [Internet]; 2013 [cited 2014 May 17]. Available from: http://online.liebertpub.com.proxy.library.cornell.edu/doi/abs/10.1089/ten.TEB.2013.0276

[115] Mironov V, Reis N, Derby B. Bioprinting: a beginning. Tissue Eng 2006;12:631–4.

[116] Chang R, Sun W. Effects of dispensing pressure and nozzle diameter on cell survival from solid freeform fabrication-based direct cell writing. Tissue Eng Part A 2008;14:41–8.

[117] Gaetani R, Doevendans PA, Metz CHG, Alblas J, Messina E, Giacomello A, et al. Cardiac tissue engineering using tissue printing technology and human cardiac progenitor cells. Biomaterials 2012;33(6):1782–90.

[118] Guillemot F, Guillotin B, Fontaine A, Ali M, Catros S, Keriquel V, et al. Laser-assisted bioprinting to deal with tissue complexity in regenerative medicine. MRS Bull 2011;36:1015–9.

[119] Schiele NR, Corr DT, Huang Y, Raof NA, Xie YB, Chrisey DB. Laser-based direct-write techniques for cell printing. Biofabrication 2010;2:032001.

[120] Ferris CJ, Gilmore KG, Wallace GG, Panhuis MIH. Biofabrication: an overview of the approaches used for printing of living cells. Appl Microbiol Biotechnol 2013;97:4243–58.

[121] Tasoglu S, Demirci U. Bioprinting for stem cell research. Trends Biotechnol 2013;31:10–9.

[122] Gaebel R, Ma N, Liu J, Guan JJ, Koch L, Klopsch C, et al. Patterning human stem cells and endothelial cells with laser printing for cardiac regeneration. Biomaterials 2011;32:9218–30.

[123] Mironov V, Visconti RP, Kasyanov V, Forgacs G, Drake CJ, Markwald RR. Organ printing: tissue spheroids as building blocks. Biomaterials 2009;30:2164–74.

[124] Jakab K, Norotte C, Damon B, Marga F, Neagu A, Besch-Williford CL, et al. Tissue engineering by self-assembly of cells printed into topologically defined structures. Tissue Eng Part A 2008;14:413–21.

[125] Norotte C, Marga FS, Niklason LE, Forgacs G. Scaffold-free vascular tissue engineering using bioprinting. Biomaterials 2009;30:5910–7.

[126] Cheng A, Dagum P, Miller DC. Aortic root dynamics and surgery: from craft to science. Philos Trans R Soc B Biol Sci 2007;362(1484):1407–19.

[127] Schaefermeier PK, Szymanski D, Weiss F, Fu P, Lueth T, Schmitz C, et al. Design and fabrication of three-dimensional scaffolds for tissue engineering of human heart valves. Eur Surg Res 2009; 42(1):49–53.

[128] Sodian R, Weber S, Markert M, Rassoulian D, Kaczmarek I, Lueth TC, et al. Stereolithographic models for surgical planning in congenital heart surgery. Ann Thorac Surg 2007;83(5):1854–7.

[129] Sodian R, Schmauss D, Markert M, Weber S, Nikolaou K, Haeberle S, et al. Three-dimensional printing creates models for surgical planning of aortic valve replacement after previous coronary bypass grafting. Ann Thorac Surg 2008;85(6):2105–8.

[130] Sodian R, Weber S, Markert M, Loeff M, Lueth T, Weis FC, et al. Pediatric cardiac transplantation: three-dimensional printing of anatomic models for surgical planning of heart transplantation in patients with univentricular heart. J Thorac Cardiovasc Surg 2008;136(4):1098–9.

[131] Sodian R, Schmauss D, Schmitz C, Bigdeli A, Haeberle S, Schmoeckel M, et al. 3-Dimensional printing of models to create custom-made devices for coil embolization of an anastomotic leak after aortic arch replacement. Ann Thorac Surg 2009;88(3):974–8.

[132] Schmauss D, Schmitz C, Bigdeli AK, Weber S, Gerber N, Beiras-Fernandez A, et al. Three-dimensional printing of models for preoperative planning and simulation of transcatheter valve replacement. Ann Thorac Surg 2012;93(2):e31–3.

[133] Schmauss D, Gerber N, Sodian R. Three-dimensional printing of models for surgical planning in patients with primary cardiac tumors. J Thorac Cardiovasc Surg 2013;145(5):1407–8.

[134] Hockaday LA, Kang KH, Colangelo NW, Cheung PYC, Duan B, Malone E, et al. Rapid 3D printing of anatomically accurate and mechanically heterogeneous aortic valve hydrogel scaffolds. Biofabrication 2012;4(3):035005.

[135] Duan B, Hockaday LA, Kang KH, Butcher JT. 3D Bioprinting of heterogeneous aortic valve conduits with alginate/gelatin hydrogels. J Biomed Mater Res A 2013;101A(5):1255–64.

[136] Duan B, Hockaday LA, Kapetanovic E, Kang KH, Butcher JT. Stiffness and adhesivity control aortic valve interstitial cell behavior within hyaluronic acid based hydrogels. Acta Biomater 2013;9(8):7640–50.

[137] Duan B, Kapetanovic E, Hockaday LA, Butcher JT. Three-dimensional printed trileaflet valve conduits using biological hydrogels and human valve interstitial cells. Acta Biomater [Internet]; 2013 [cited 2014 Feb 3]. Available from: http://www.sciencedirect.com/science/article/pii/S1742706113006016

[138] Shinoka T, Breuer CK, Tanel RE, Zund G, Miura T, Ma PX, et al. Tissue engineering heart valves: valve leaflet replacement study in a lamb model. Ann Thorac Surg 1995;60(6 Suppl.):S513–516.

[139] Colazzo F, Sarathchandra P, Smolenski RT, Chester AH, Tseng Y-T, Czernuszka JT, et al. Extracellular matrix production by adipose-derived stem cells: implications for heart valve tissue engineering. Biomaterials 2011;32(1):119–27.

[140] Schmidt D, Achermann J, Odermatt B, Genoni M, Zund G, Hoerstrup SP. Cryopreserved amniotic fluid-derived cells: a lifelong autologous fetal stem cell source for heart valve tissue engineering. J Heart Valve Dis 2008;17(4):446–55. discussion 455.

[141] Schmidt D, Dijkman PE, Driessen-Mol A, Stenger R, Mariani C, Puolakka A, et al. Minimally-invasive implantation of living tissue engineered heart valves: a comprehensive approach from autologous vascular cells to stem cells. J Am Coll Cardiol 2010;56(6):510–20.

[142] Weber B, Scherman J, Emmert MY, Gruenenfelder J, Verbeek R, Bracher M, et al. Injectable living marrow stromal cell-based autologous tissue engineered heart valves: first experiences with a one-step intervention in primates. Eur Heart J 2011;32(22):2830–40.

[143] Schmidt D, Mol A, Breymann C, Achermann J, Odermatt B, Gössi M, et al. Living autologous heart valves engineered from human prenatally harvested progenitors. Circulation 2006;114(1 Suppl.). I–125–I–131.

[144] Schmidt D, Mol A, Odermatt B, Neuenschwander S, Breymann C, Gössi M, et al. Engineering of biologically active living heart valve leaflets using human umbilical cord-derived progenitor cells. Tissue Eng 2006;12(11):3223–32.

[145] Sodian R, Lueders C, Kraemer L, Kuebler W, Shakibaei M, Reichart B, et al. Tissue engineering of autologous human heart valves using cryopreserved vascular umbilical cord cells. Ann Thorac Surg 2006;81(6):2207–16.

[146] Sodian R, Schaefermeier P, Abegg-Zips S, Kuebler WM, Shakibaei M, Daebritz S, et al. Use of human umbilical cord blood-derived progenitor cells for tissue-engineered heart valves. Ann Thorac Surg 2010;89(3):819–28.

Chapter 22

Bioprinting of Skin

Julie Marco, Anthony Atala, and James J. Yoo

Wake Forest Institute for Regenerative Medicine, Wake Forest School of Medicine, Medical Center Boulevard, Winston-Salem, NC, USA

ABSTRACT

Full-thickness skin wounds and extensive burn injuries are a major cause of morbidity and mortality. Currently, the clinical standard for wound treatment is the use of autologous split-thickness skin grafts, which is associated with donor site morbidity. Use of allografts often leads to an immune response and rejection. Other skin substitutes, synthetic and biologic, have been used, but each has advantages and disadvantages. Recently, direct application of skin cells onto wound sites has become a method to accelerate wound healing. To efficiently deliver cells onto the wound site for uniform coverage, bioprinting technology has been proposed as a delivery method. This chapter discusses the printer design and parameters specific to skin bioprinting applications for wound healing and skin repair.

Keywords: skin; bioprinting; burn; wound healing; regeneration

1 INTRODUCTION

The skin is the largest organ in the human body, and it protects the body from external insults while maintaining homeostasis. The skin's multilayer system insulates the body and regulates temperature when interacting with the outside environment [1–3]. When the skin is damaged, these functions are lost and the patient is forced to find a way to restore the wound. In fact, full-thickness skin wounds and extensive burn injuries are major causes of morbidity and mortality [4–6]. Many treatment methods have been developed that target repairing the skin; however, few have been successful with no detrimental side effects or intense scarring [7]. As an alternative to direct skin grafting, many biological

skin substitutes have been developed and applied [7–11]. Another promising approach is cell-based therapy to repair and regenerate skin tissue [12–19]. This method involves the use of cells to cover the wound defect and promote healing through regeneration. To efficiently place the skin cells uniformly onto the wound, bioprinting technology has been proposed as a delivery method [12,17,20–23]. Bioprinting of the skin seeks to deliver skin cells and components in their proper anatomical configuration to facilitate regeneration of the damaged skin while restoring function without scarring and damaging other tissues and organs of the body. The skin bioprinter is designed to rapidly deploy cells and extracellular matrices onto the wound surface to repair skin defects in situ. To accurately place cells, a scanning system was incorporated onto the printer, which determines the dimension of the wound and guides the printer to deposit cells and extracellular matrix (ECM) in layers to approximate the anatomical skin configuration. Through the precise instrument, the skin components are laid down layer by layer in order to completely restore the damaged area.

2 STRUCTURE OF THE SKIN

The skin is the largest organ in the body, covering approximately 1.8–2 m^2 of surface area in the average person [1–3]. The primary function of the skin is the protection of the body from the external environment. It serves as a barrier for prevention of water loss and regulation of temperature. The skin also protects the body from external insults, such as viral and bacterial infection, as well as physical, chemical, and UV radiation [1–3]. Therefore, the skin plays a major role in protecting and maintaining the body's homeostasis.

The skin is comprised of two main layers: the dermis and the epidermis. The dermis is the inner layer supported by the subcutaneous tissue, or the hypodermis, and contains fat that provides insulation [1,3]. The main component of the dermis is connective tissue consisting mainly of ECM, which is composed of collagen, elastin, and glycosaminoglycans. Fibroblasts, which produce collagen and elastin, are the main cell type of the dermis and comprise the remaining part of the ECM [3]. The dermis is highly vascularized and is responsible for the metabolic function of skin. Sweat glands, as well as hair follicle papilla cells, are also contained in this layer. The sweat glands play an important role in providing an exit pathway as part of the endocrine function of the skin. Hair plays a role in thermoregulation by detecting small changes in the external atmosphere. Cells in the follicles also contribute to wound healing by proliferating to help regenerate the skin [1].

The epidermis is the outermost layer of the skin and is comprised of 90–95% keratinocytes [2]. The epidermis is supported by the basal membrane, which is connected to the dermis via hemi-desmosomes, and is responsible for the renewal of the epidermis. Keratinocytes produced by the basal membrane slowly migrate to the outer layer. Another component of the epidermis is the melanocytes, which are responsible for skin pigmentation.

The epidermis can be divided into four layers: (1) stratum basale, (2) stratum spinosum, (3) stratum granulosum, and (4) stratum corneum. The stratum basale is composed of basal cells that are connected to the dermis. The stratum spinosum is similar to the stratum basale with the addition of lipid-enriched lamellar bodies. Cells in this region begin to flatten and elongate [3]. The stratum granulosum is characterized by an increase in keratin synthesis and keratin filament alignment. An increase in keratinocyte cell differentiation leads to the terminal differentiation of corneocytes that are found in the outermost layer of the epidermis, the stratum corneum, is responsible for protecting the skin from foreign substances and retaining water [1]. The brick and mortar shape of this layer makes it impervious to many substances [3] due to lamellar lipid regions that surround the corneocytes creating a highly dense protein structure that protects against absorption [1].

The melanocytes are found in the basal layer of the epidermis and contain the pigment melanin, which stimulates the proliferation of the keratinocytes, aids hair shaft elongation, and provides a barrier against UV-induced mutations [1,3]. Melanocytes also have the ability to help with temperature regulation and blood flow and can reduce thermal injury to the skin [1]. Thus, cells contained in the different skin layers aid in normal function and play a significant role in wound healing.

3 WOUND HEALING

Wound healing occurs in three stages: (1) hemostasis and inflammation, (2) proliferation, and (3) remodeling. The body benefits from a rapid restoration of the barrier function of the skin. Formation of a fibrin clot is the first step of hemostasis and inflammation. The fibrin clot has two functions: (1) to serve as a shield from foreign substances and (2) to provide a matrix through which cells can migrate. The clot also contains a variety of growth factors that aid the wound closure process [6]. The growth factors recruit neutrophils and macrophages, which initiate the inflammation phase and combat any possible contaminating bacteria that could have entered the wound site. The neutrophils also produce substances that induce chemotaxis of inflammatory cells and undergo apoptosis after a few days. The macrophages release growth factors that initiate proliferation and ECM production by the fibroblasts. This coordinated process ultimately allows the fibroblasts to migrate into the wound leading to the proliferation phase of healing.

Proliferation begins when granulation tissue is created by the fibroblast migration and production of a collagen matrix. In this phase, keratinocytes start to migrate into the wound area as evidenced by an extension of the stratum basale. The keratinocytes begin migration following the complete formation of the ECM. The keratinocytes are dependent upon the dermal layer for all metabolic support [1]. Re-epithelialization by the keratinocytes is critical to the proliferation process because it restores the barrier function to the skin. Furthermore, the wound is protected against water loss and infection. The increased rate of cell migration can lead to wound contraction and scarring, thus making the final stage of wound healing very important.

During the remodeling stage, the granulation tissue is integrated to form an organized ECM. Collagen produced during the previous phases will dictate the integrity of the wound healing. In addition, the transformation of fibroblasts into myofibroblasts via growth factors can result in contraction of the wound. Myofibroblasts resemble smooth muscle as they express actin and cause strong contractile forces [24]. Growth factors induce the formation of dendritic branches in myofibroblasts that attach to the ECM. Acid produced by the cell membrane causes the dendritic branches to shorten causing the ECM to contract and shrink. If the stress is relieved, the myofibroblasts will undergo apoptosis and produce less collagen leading to less scarring [1,24]. If stress is not relieved, a scar will form that will hinder function of the skin.

4 TREATMENT METHODS FOR SKIN REPAIR

A significant amount of research has been focused on repairing skin wounds and restoring function. Large skin wounds are a tremendous problem in both the civilian and military populations. On the battlefield, burns account for 5–20% of all causalities. Annually, 500,000 burns are treated causing healthcare expenditures to approach 2 billion dollars [25]. Patients who suffer from large-scale burn wounds or wounds that damage a large portion of the skin benefit from rapid treatments that work to stabilize and protect the wound as well as restore function to the damaged area. Current treatment methods have been designed to repair damage and restore function; however, these treatments have limitations.

4.1 Conventional Surgical Therapy

The standard treatment for many burn victims is split-thickness autologous grafts [26]. These grafts come from a donor site on the patient. The skin is removed, stretched, and then applied over the wound. This method is not suitable for patients whose burn covers a large surface area because stretching the skin over a greater surface area results in a poorer cosmetic outcome [6]. In addition, this method causes another area of the body to be damaged and is ultimately limited by availability and morbidity of donor sites [27]. One alternative to a skin graft is a skin flap, which is similar to a skin graft except for the fact that they are not completely removed from an area of the body. A flap remains connected to its original blood supply and is simply expanded to cover the wound area. This method is often used to treat patients who have had a cancerous tissue removed from the skin and adjacent skin is expanded to cover the area. While this is an effective method, it does not work for patients with larger wound areas or limited adjacent normal skin. For patients with large wound areas, allogeneic skin grafts have been obtained from cadaver skin or cultured using cell lines. This method provides more skin to be used to cover the wound. The skin can also be expanded as it is with split-thickness grafts. If taken from a cadaver, the skin has a full basal membrane that allows for the growth of keratinocytes and the restoration of the barrier function of the skin [1]; however, immune response to the donor skin is a significant problem. The immune response is caused by T-cells in the host recognizing foreign antigens that are expressed by the donor cells [28]. Patients who receive allografts require immunosuppressive drug therapy to prevent rejection [28]. The possibility of transmitting life-threatening diseases can also occur. And finally, donor cadaver skin is not often available, thus making it a less than ideal treatment option. For these reasons, other treatments have been developed to reduce immunoreactivity and infections [17].

4.2 Noncellular Therapy

Due to the limitations of using autologous tissue grafts, various types of skin substitutes have been developed.

Noncellular dermal substitutes composed of various materials have been used in the clinic [17,26,27,29–33]. Integra® (Integra LifeSciences, Plainsboro, New Jersey) is an example of a noncellular dermal substitute bilayer construct containing type I bovine collagen with chondroitin-6-sulfate and a silicone membrane that can be used to create a multilayer system [27,33]. Because Integra is an expensive treatment option and does not provide an aesthetically pleasing outcome, it is often used to prep a wound for a secondary method such as a skin graft [17,34]. A study that tested a construct similar to Integra revealed a number of problems, including development of hematomas, wrinkling of the skin, and loosening of the silicone membrane from the support layer. The membrane did provide adequate protection against water loss and infection as well as promoting healing and favorable cosmetic results [13]. Another type of artificial treatment is the Biobrane™ (Medline, Mundelein, Illinois). Like the allograft, it has a silicone membrane that serves as the epidermis while a nylon/collagen mesh creates the dermis. Unfortunately, the nylon is not biodegradable and must be removed from the wound during healing, which makes it unsuitable for large wounds. Additionally, this type of treatment has resulted in fluid build-up and tends to be used for temporary wound treatment [9,11].

4.3 Biological Therapy

Biological therapy is an alternative method that uses cell-based techniques to repair wounds and accelerate healing. For this approach, skin cells can be obtained from a tissue biopsy, culture expanded, and seeded on a matrix or applied directly on the wound [34,35]. Cultured keratinocytes from autologous or allogeneic sources have been applied to a wound to regenerate the basal membrane. While this method facilitates wound healing, it does not account for all layers of the skin, which could result in insufficient healing of the wound. Moreover, without the support of fibroblasts, culturing a structure that is capable of being transplanted becomes difficult [27].

Other types of cell-based therapy include Dermagraft®, Apligraf®, and TransCyte™. These are commercially available cellularized grafts composed of a polymer scaffold seeded with fibroblasts. The fibroblasts will produce the necessary components of the ECM to create a dermis-like structure. The scaffold is cultured in vitro and applied as needed [17]. This treatment requires 3–4 weeks to produce, therefore, availability is limited [22]. As with other treatment methods, these cellularized grafts are expensive and can be immunoreactive.

Human skin grafts have been generated from cryopreserved allogenic dermis and autologous cultured keratinocytes in an early clinical trial. Stratified autologous keratinocytes were added to a dermal bed that had been resurfaced with the cryopreserved, meshed dermis from cadaveric tissue for 4 weeks. The treatment showed that a differentiated dermis was produced along with the basement membrane between the dermis and epidermis. The keratinocytes also created a complete epidermis for patients in the study. Although the results were promising, patients experienced blistering within the graft and fragility in the grafted areas after 7 months [36].

4.4 Experimental Studies

Researchers are currently working with a variety of experimental treatments that show promise for improving wound healing. Many of these treatments utilize hydrogels in order to mimic the skin and its components. These hydrogels contain important parts of the skin, such as fibrin and collagen that are involved in wound healing. Several commercially available products are biologically inert and can be dispensed as gels or sheets [37]. Furthermore, gels that adhere to the wound and control fluid loss are necessary to promote healing [13]. It is important that the gels are not too thick that cells from the surrounding wound are not able to migrate through the gel. One study showed the possibility that an asymmetric membrane containing collagen allows for the attachment of fibroblasts and re-epithelialization – an indication of successful migration [38]. Consequently, the gel needs to be substantial enough that the gel does not begin shrinking, which can lead to contraction of the wound site.

Laser treatments, such as helium-neon and argon, have also been used in experimental phases. A study designed to evaluate healing promoted by far-infrared rays showed promising results. This type of electromagnetic wave was chosen because it had been shown to inhibit tumor growth in mice and has been used in clinical studies to treat bedsores. Side effects of this treatment involve increased skin temperature and blood flow, which makes this method risky [39].

Current wound treatments have led to many advances in the understanding of wound healing, although many of these treatments have limitations. Some of these limitations include immune response issues, significant financial costs, and the possibility of transmitting diseases or infections. Moreover, the skin that is formed via these methods does not have proper pigmentation or hair formation. In addition, most of the treatments cannot be used on large surface area burns that are difficult to stabilize. If not treated properly or promptly, the wound can undergo a lack of nutrients and infection leading to mortality for the patient [13]. In fact, patient survival is inversely proportional to the amount of time required to cover and stabilize a wound. Therefore, immediate large-size wound coverage is urgently needed.

Cell-based approaches for skin regeneration have gained much attention recently due to the prospects of accelerating wound repair of large skin defects. However, delivery methods are varied but currently involve a manually seeded matrix or a cell spray. Wounds repaired using cell spraying typically heal faster and with better cosmetic outcomes than those repaired with noncellular substitutes [40,41]. These technologies have worked well for superficial and partial thickness burns, and further advances are being attained. Another method of delivery that is currently in development is bioprinting. This method has the capability to deliver specific cells and biomaterials to predefined target sites using layer-by-layer freeform fabrication [23,42–46]. Typical freeform fabrication involves printing cells onto a scaffold. In one study amniotic-fluid-derived stem cells (AFDSCs) were delivered directly onto a skin defect using a bioprinter. Uniform delivery of these cells resulted in increased vascularization and accelerated wound healing [17]. Because of the ability to place specific cell types and biomaterials in a predefined spatial orientation, the bioprinting approach for skin repair would be a promising delivery tool for cell-based therapy.

5 SKIN BIOPRINTING

Full-thickness skin wounds and extensive burn injuries are major causes of morbidity and mortality. As an alternative to direct skin grafting, the application of skin cells onto wounds has emerged as an effective treatment modality that provides wound coverage with minimal skin grafting as cells can be expanded for larger wound areas. One of the advantages to skin bioprinting is that it has the ability to replicate the architecture of native tissues with a high degree of precision; furthermore, bioprinting can replicate the compositional complexity of the wound. Printing allows for two or more cell types to be placed in discrete locations relative to each other to allow incorporation into the printed structure, which leads to regeneration of skin with pigmentation and hair. Finally, printing is ideal for large wound areas that cover different areas of the body as composition of the tissue and geometry can be controlled.

The skin bioprinter we designed is programmed to print one layer of the skin at a time. To this end, the wound is first scanned for topography and wound area. The scan is converted to a 3D image that the printer uses to precisely map a path for the wound. Culture expanded fibroblasts and keratinocytes are suspended separately in a fibrin/collagen solution. The fibroblast solution is printed first. While it is printed, the solution is cross-linked with thrombin that is held in a separate cartridge. After the fibroblasts are printed, the keratinocytes are printed over the fibroblasts using the same method. The scan taken previously indicates the volume required for each cell type application. The scan allows for highly accurate results while the cell/gel combination mimics the normal skin composition and leads to better healing and improved cosmetic results with minimal scarring (Fig. 22.1).

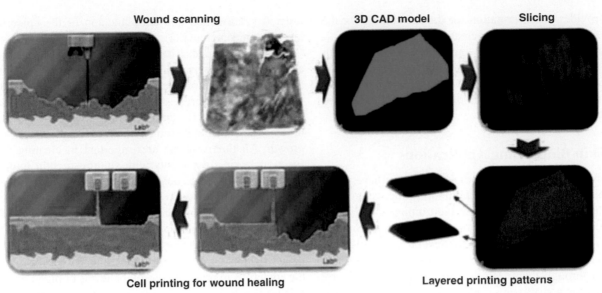

Wound scanning **3D CAD model** **Slicing**

Cell printing for wound healing **Layered printing patterns**

FIGURE 22.1 Flow diagram of the skin bioprinting procedure. The process begins with scanning of the wound and is completed with the printing of cells after a 3D image indicating depths and topography has been created using different software programs. *(Scanning images: courtesy of Lab^TV).*

5.1 Skin Bioprinter Components

The skin bioprinter we developed in our laboratory has many different components and capabilities that allow it to print exactly onto the wound without compromising sterility and the printed structure (Fig. 22.2). The printer is built into a portable frame that allows the device to be rolled over the patient and aligned properly. It can fit through doorways and has locks that prevent movement once placed over the patient. The frame encompasses all parts needed for printing including the scanner and computer and allows for easy transportation and less crowding in an operating room. The printer frame carries multiple cartridges used by the printer

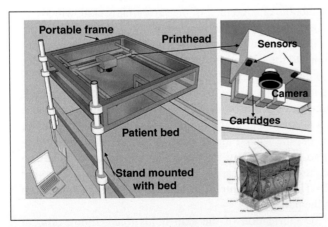

FIGURE 22.2 Schematic diagram of skin bioprinter design. The portable unit frame is designed to minimize direct contact with the patient, which increases its sterility. The enhanced view of the print head shows bioink cartridges that allow for the storage of different materials needed for printing. The red sensors allow the print head to be positioned accurately above the wound.

to allow for many different cell types and substances to be printed. The cartridges can also be filled with different cell densities and allow the operator to control the number of cells printed per wound. As mentioned previously, the keratinocytes and fibroblasts are housed separately in fibrin/collagen solutions. Thrombin is mixed separately as it is used for cross-linking purposes and is combined with the fibrin/collagen solution upon injection. Once cross-linked, a gel is created that allows the survival and migration of cells. This gel was chosen because it promotes blood clotting and contains components of the ECM needed for wound healing. The gel also produces a tight seal with the outer edges of the wound [17]. The cells are printed from the cartridge at 2 mm intervals between drops via a pressure system designed to eject the cells without killing them or changing their structure. The cartridges are controlled by the computer and are designed to withstand differences in pressure. The pressure within each cartridge is changed in order to accurately dispense the cells. A pressure-driven system allows for higher throughput of material and cells. The nozzles for the cartridges never come in direct contact with the wound bed and promote better sterility.

To accurately determine the extent of wound for accurate delivery, a scanner is attached to the frame. Sensors are applied to the wound area to serve as identification points. The operator needs only to stand above the patient in order for the scanner to work. This allows for aseptic conditions because the operator can stand a distance away and still obtain an image. Once the scan is gathered two programs are used to determine the nozzle path and the fill volume. The first program generates a 2D image that gives the overall area and nozzle path for the printer. Then, the image is transferred to a 3D image using a second program. The 3D image provides the information for the fill volume for the wound. After the 3D image is created the printer automatically prints the cells onto the wound via a print head with *XYZ* coordinates capable of moving distances as small as 100 µm. The combination of the scanner and *XYZ* coordinates direct the high precision of the printer and reduces human error (Fig. 22.1).

5.2 Skin Bioprinting Applications

Although it is still a new field, multiple groups have made significant advancements with skin bioprinting using varying methods. Due to the fact that bioprinting is still in its early stages, researchers are trying a number of approaches to optimize the cell delivery system. Several hydrogels have been tested for their ability to be printed and to evaluate cell migration through them. Furthermore, researchers are investigating the use of laser printing cells in smaller volumes compared to those required by the skin bioprinter.

One research group has found that laser-induced forward transfer, or laser printing, is successful in arranging biological materials and cells in precise patterns. Laser printing is capable of placing small volumes in a construct with a high degree of precision. In addition, higher densities of cells can be suspended in a variety of hydrogels with variable viscosities with the laser printing methodology [20]. This method is fast and reliable, and the printed cells do not differ from regular cultured cells. When mesenchymal stem cells were printed, their phenotype was not alternated, thus demonstrating their potential for bioprinting. These results demonstrated the possibility of using laser printing for the purpose of creating complex matrices, like the skin, ex vivo. Further studies are necessary to determine if constructs printed using the laser printing method can be incorporated into wound beds.

Bioprinting of AFSCs has also been tested for healing skin wounds. AFSCs are interesting because they have a high proliferation rate and immunomodulatory activity. AFSCs combined with fibrin gel were bioprinted onto skin wound. Wounds treated with AFSCs were capable of accelerating closure of full-thickness wounds and increased neovascularization. Growth factors were secreted by AFSCs that aided the wound healing process [17]. While this study with AFS cells showed some success, further studies are necessary to evaluate clinical applications.

Another research group used methods similar to the ones described previously with the addition of a collagen-based hydrogel. Multiple layers were created that contained human dermal fibroblasts as the support layer and an onlay of human keratinocytes. By controlling the cell density and location, the research group was successful in reproducing morphological features of human skin *in vivo*. Viability remained high after printing for both cell types, and uniform distribution of cells was observed in each layer; however,

ordered stratification and keratinization of the epidermal layer was incomplete [22].

Skin bioprinting has developed significantly over the recent years, and the system continues to be optimized in the field of biofabrication. Successes have been shown in the regeneration of different skin models but additional models need to be explored. In particular, the printed skin constructs have not demonstrated pigmentation, nerve endings, or hair growth. All three aspects are important in creating fully functional skin for patients with significant skin injuries. It might be possible to create these features by printing the cells associated with them in their exact location in the skin. The precision of the bioprinting system would allow the cell types to be printed in the structure needed to regenerate either the nerves or the hair follicles. The printer would also be able to place the melanocytes in their exact location in the epidermis in order to control the levels of pigmentation. If these aspects can be achieved, bioprinting would result in the complete regeneration of fully functional skin.

6 CONCLUSIONS

Large skin wounds are a tremendous problem in the civilian and military populations. Billions of dollars are spent each year on the treatment of different skin wounds. Current treatment methods, including split-thickness grafts, xenografts, and acellular dermal substitutes, have all been developed with the intent to stabilize and heal the wound. Unfortunately, these techniques do not present the best outcomes. Rejection of the skin substitute by the immune system results in the use of expensive immunosuppressants to combat the body's response to the treatment. These drugs are an inconvenience for the patient and significantly increase the treatment costs. These treatments do not provide the best cosmetic outcomes for the patients as evidenced by scarring, contraction, or stretching of the surrounding skin, which can decrease movement and function of the treated area.

Skin bioprinting technology seeks to find alternative solutions to the limitations of current treatment methods. The system may use a patient's own cells that have been culture expanded to cover the wound area. Using autologous cells eliminates the risk of an immune response. Through the use of a scanner and computer systems, the printed structure matches the wound topography and fills the volume. In doing so, contraction of the skin is limited because cells from the surrounding area do not migrate in to fill the wound area. The lack of contraction leads to an increase in function for the patient.

3D fabrication of the skin holds many future clinical applications for burns, scar contracture, and other severe wounds as well as for patients suffering from cosmetic problems such as vitiligo. Burn wounds, one of the most common and critical wounds experienced on the battlefield, demand immediate treatment. The portable bioprinter allows treatment to quickly reach these patients for the stabilization and treatment of wounds. In addition, the printer allows for other skin treatments to be developed. If skin can be successfully printed with pigmentation, patients suffering from conditions like vitiligo would have an option to remove the depigmented skin and replace with printed skin that matches their own. Skin cancer patients would also benefit from printed therapies if large areas of skin are forced to be removed due to tumors. The possibilities and applications of skin bioprinting are exciting, and with further successful studies, grafts and other artificial treatments may become obsolete to be replaced with the quick, effective, and precise printing treatment methods.

ABBREVIATION

ECM Extracellular matrix

REFERENCES

[1] Bouwstra JA, Honeywell-Nguyen PL, Gooris GS, Ponec M. Structure of the skin barrier and its modulation by vesicular formulations. Prog Lipid Res 2003;42(1):1–36.

[2] Forslind B, Engström S, Engblom J, Norlén L. A novel approach to the understanding of human skin barrier function. J Dermatol Sci 1997;14(2):115–25.

[3] Menon GK. New insights into skin structure: scratching the surface. Adv Drug Deliv Rev 2002;54(Suppl. 1):S3–17.

[4] Cherry DK, Hing E, Woodwell DA, Rechtsteiner EA. National Ambulatory Medical Care Survey: 2006 summary. Natl Health Stat Report 2008;(3):1–39.

[5] Pitts SR, Niska RW, Xu J, Burt CW. National Hospital Ambulatory Medical Care Survey: 2006 emergency department summary. Natl Health Stat Report 2008;(7):1–38.

[6] Sheridan RL, Greenhalgh D. Special problems in burns. Surg Clin North Am 2014;94(4):781–91.

[7] Rojas Y, Finnerty CC, Radhakrishnan RS, Herndon DN. Burns: an update on current pharmacotherapy. Expert Opin Pharmacother 2012;13(17):2485–94.

[8] Achora S, Muliira JK, Thanka AN. Strategies to promote healing of split thickness skin grafts: an integrative review. J Wound Ostomy Continence Nurs 2014;41(4):335–9. quiz E1-2.

[9] Nathoo R, Howe N, Cohen G. Skin substitutes: an overview of the key players in wound management. J Clin Aesthet Dermatol 2014;7(10):44–8.

[10] Nyame TT, Chiang HA, Orgill DP. Clinical applications of skin substitutes. Surg Clin North Am 2014;94(4):839–50.

[11] Tauzin H, Rolin G, Viennet C, Saas P, Humbert P, Muret P. A skin substitute based on human amniotic membrane. Cell Tissue Bank 2014;15(2):257–65.

[12] Auxenfans C, Menet V, Catherine Z, Shipkov H, Lacroix P, Bertin-Maghit M, et al. Cultured autologous keratinocytes in the treatment of large and deep burns: a retrospective study over 15 years. Burns 2015;41(1):71–9.

[13] Burke JF, Yannas IV, Quinby WC Jr, Bondoc CC, Jung WK. Successful use of a physiologically acceptable artificial skin in the treatment of extensive burn injury. Ann Surg 1981;194(4):413–28.

[14] Gu J, Liu N, Yang X, Feng Z, Qi F. Adiposed-derived stem cells seeded on PLCL/P123 eletrospun nanofibrous scaffold enhance wound healing. Biomed Mater 2014;9(9):035012.

[15] Hassan WU, Greiser U, Wang W. Role of adipose-derived stem cells in wound healing. Wound Repair Regen 2014;22(3):313–25.

[16] Ribeiro J, Pereira T, Amorim I, Caseiro AR, Lopes MA, Lima J, et al. Cell therapy with human MSCs isolated from the umbilical cord Wharton jelly associated to a PVA membrane in the treatment of chronic skin wounds. Int J Med Sci 2014;11(10):979–87.

[17] Skardal A, Mack D, Kapetanovic E, Atala A, Jackson JD, Yoo J, et al. Bioprinted amniotic fluid-derived stem cells accelerate healing of large skin wounds. Stem Cells Transl Med 2012;1(11):792–802.

[18] Teng M, Huang Y, Zhang H. Application of stems cells in wound healing – an update. Wound Repair Regen 2014;22(2):151–60.

[19] You HJ, Han SK. Cell therapy for wound healing. J Korean Med Sci 2014;29(3):311–9.

[20] Koch L, Deiwick A, Schlie S, Michael S, Gruene M, Coger V, et al. Skin tissue generation by laser cell printing. Biotechnol Bioeng 2012;109(7):1855–63.

[21] Koch L, Kuhn S, Sorg H, Gruene M, Schlie S, Gaebel R, et al. Laser printing of skin cells and human stem cells. Tissue Eng Part C Methods 2010;16(5):847–54.

[22] Lee V, Singh G, Trasatti JP, Bjornsson C, Xu X, Tran TN, et al. Design and fabrication of human skin by three-dimensional bioprinting. Tissue Eng Part C Methods 2014;20(6):473–84.

[23] Murphy SV, Atala A. 3D bioprinting of tissues and organs. Nat Biotechnol 2014;32(8):773–85.

[24] Martin P. Wound healing – aiming for perfect skin regeneration. Science 1997;276(5309):75–81.

[25] Champion HR, Bellamy RF, Roberts CP, Leppaniemi A. A profile of combat injury. J Trauma 2003;54(Suppl. 5):S13–9.

[26] Lineen E, Namias N. Biologic dressing in burns. J Craniofac Surg 2008;19(4):923–8.

[27] Atiyeh BS, Gunn SW, Hayek SN. State of the art in burn treatment. World J Surg 2005;29(2):131–48.

[28] Rosenberg AS, Singer A. Cellular basis of skin allograft rejection: an *in vivo* model of immune-mediated tissue destruction. Annu Rev Immunol 1992;10:333–58.

[29] Flasza M, Kemp P, Shering D, Qiao J, Marshall D, Bokta A, et al. Development and manufacture of an investigational human living dermal equivalent (ICX-SKN). Regen Med 2007;2(6):903–18.

[30] McGuigan FX. Skin substitutes as alternatives to autografting in a wartime trauma setting. J Am Acad Orthop Surg 2006;14(10 Spec No):S87–9.

[31] Pereira C, Gold W, Herndon D. Review paper: burn coverage technologies: current concepts and future directions. J Biomater Appl 2007;22(2):101–21.

[32] Pham C, Greenwood J, Cleland H, Woodruff P, Maddern G. Bioengineered skin substitutes for the management of burns: a systematic review. Burns 2007;33(8):946–57.

[33] Sheridan RL, Tompkins RG. Skin substitutes in burns. Burns 1999;25(2):97–103.

[34] Schlabe J, Johnen C, Schwartlander R, Moser V, Hartmann B, Gerlach JC. Isolation and culture of different epidermal and dermal cell types from human scalp suitable for the development of a therapeutical cell spray. Burns 2008;34(3):376–84.

[35] Zweifel CJ, Contaldo C, Köhler C, Jandali A, Künzi W, Giovanoli P. Initial experiences using non-cultured autologous keratinocyte

suspension for burn wound closure. J Plast Reconstr Aesthet Surg 2008;61(11):pe1–4.

[36] Langdon RC, Cuono CB, Birchall N, Madri JA, Kuklinska E, McGuire J, et al. Reconstitution of structure and cell function in human skin grafts derived from cryopreserved allogeneic dermis and autologous cultured keratinocytes. J Invest Dermatol 1988;91(5):478–85.

[37] Murphy SV, Skardal A, Atala A. Evaluation of hydrogels for bioprinting applications. J Biomed Mater Res A 2013;101(1):272–84.

[38] Chen KY, Liao WJ, Kuo SM, Tsai FJ, Chen YS, Huang CY, et al. Asymmetric chitosan membrane containing collagen I nanospheres for skin tissue engineering. Biomacromolecules 2009;10(6): 1642–9.

[39] Toyokawa H, Matsui Y, Uhara J, Tsuchiya H, Teshima S, Nakanishi H, et al. Promotive effects of far-infrared ray on full-thickness skin wound healing in rats. Exp Biol Med (Maywood) 2003;228(6): 724–9.

[40] Grant I, Warwick K, Marshall J, Green C, Martin R. The co-application of sprayed cultured autologous keratinocytes and autologous fibrin sealant in a porcine wound model. Br J Plast Surg 2002;55(3):219–27.

[41] Hartmann B, Ekkernkamp A, Johnen C, Gerlach JC, Belfekroun C, Küntscher MV. Sprayed cultured epithelial autografts for deep dermal burns of the face and neck. Ann Plast Surg 2007;58(1):70–3.

[42] Roth EA, Xu T, Das M, Gregory C, Hickman JJ, Boland T. Inkjet printing for high-throughput cell patterning. Biomaterials 2004;25(17):3707–15.

[43] Varghese D, Deshpande M, Xu T, Kesari P, Ohri S, Boland T. Advances in tissue engineering: cell printing. J Thorac Cardiovasc Surg 2005;129(2):470–2.

[44] Xu T, Gregory CA, Molnar P, Cui X, Jalota S, Bhaduri SB, et al. Viability and electrophysiology of neural cell structures generated by the inkjet printing method. Biomaterials 2006;27(19):3580–8.

[45] Xu T, Jin J, Gregory C, Hickman JJ, Boland T. Inkjet printing of viable mammalian cells. Biomaterials 2005;26(1):93–9.

[46] Xu T, Petridou S, Lee EH, Roth EA, Vyavahare NR, Hickman JJ, et al. Construction of high-density bacterial colony arrays and patterns by the ink-jet method. Biotechnol Bioeng 2004;85(1):29–33.

Chapter 23

Bioprinting of Nerve

Christopher Owens*, Francoise Marga*,**, and Gabor Forgacs*,**,†

*Department of Physics and Astronomy, University of Missouri, Columbia, MO, USA; **Modern Meadow Inc. Missouri Innovation Center, Columbia, MO, USA; †Department of Biomedical Engineering, University of Missouri, Columbia, MO, USA

Chapter Outline

ABSTRACT

The reliable tissue engineering of tubular structures, composing a significant portion of the human body, would have far reaching implications. Such structures, in the form of blood vessels or nerve conduits, could be used directly as grafts to replace respectively clogged vessels or to restore function after peripheral nerve injuries. Perhaps even more importantly they could be employed to vascularize and innervate more complex engineered organs, a still outstanding unresolved problem of tissue engineering. Here we first provide an overview on the fabrication of tubular organoids using extrusion bioprinting technology that operates with multicellular bioink units whose postprinting self-assembly, which gives rise to the ready-to-use biological structure, is based on morphogenetic principles evident in early embryonic development. We then detail the construction of nerve grafts.

Keywords: bioprinting; tissue engineering; scaffold-free; self-assembly; tissue fusion; cell sorting; tubular structure; nerve graft

1 INTRODUCTION

This chapter presents a specific example of extrusion bioprinting as applied to the biofabrication of tubular organ structures, in particular a nerve graft, without a scaffold. The major components of the technology, common to most bioprinting methods, are the "bioink," the "biopaper," and the bioprinter, which will be detailed later. The distinguishing feature of the technology is that it combines the advantages of automated deposition of discrete bioink units, such as speed, reproducibility, and reliability with the self-organizing principles of multicellular systems [1,2]. Accordingly, in our approach, the process of building the biological structure is divided into three well-separated steps. The first one implies the preparation of the bioink units, multicellular aggregates of specific shape (i.e., minitissues) whose cellular content is consistent with the composition of the desired tissue or organ. In the second step the discrete bioink units are delivered into the biopaper by the bioprinter. Here the term biopaper implies the receiving environment into which the bioink units are deposited according to a template that is consistent with the morphology of the desired structure. More specifically, in our approach, the biopaper represents a temporary support for the assembly of bioink units. It is not part of the final product, thus it is not a scaffold. In the final step the assembly of the discrete bioink units matures into the final continuous biological structure. It is this maturation step that is governed by biological self-organization. First, the self-assembling bioink particles fuse. As they represent minitissues, their fusion is an example of tissue fusion, one of the most fundamental

Essentials of 3D Biofabrication and Translation. http://dx.doi.org/10.1016/B978-0-12-800972-7.00023-2

morphogenetic processes of early embryonic development [3]. Tissue fusion might be accompanied by other morphogenetic processes, such as cell sorting [3], in the course of which the various cell types in the heterocellular bioink acquire their physiological location within the developing tissue or organ. Thus, postprinting structure formation in this technology mimics early embryonic developmental phenomena [1]. Second, the fused object is exposed to near physiological conditions in a bioreactor where it continues to mature to reach properties that eventually allow its implantation into the living organism.

2 BIOPRINTING BASED ON BIOLOGICAL SELF-ASSEMBLY

Self-assembly is the autonomous organization of components from an initial state into final pattern or structure without external intervention [4,5]. Living organisms, in particular the developing embryo, are quintessential self-organizing systems. Histogenesis and organogenesis are examples of self-assembly processes, in which, through cell–cell and cell–extracellular matrix (ECM) interactions, the developing organism and its parts gradually acquire their final shape. Ultimately, the success of engineering and fabricating functional living structures will depend on understanding the principles of cellular self-assembly and our ability to employ them. This fact is being gradu-

ally recognized across the tissue engineering community, as except for a few spectacular successes [6–8] the field has yet to present viable solutions to the growing demand for novel regenerative technologies. Thus, future biofabrication approaches (including but not restricted to the field of bioprinting) aimed at reestablishing the functionality of damaged tissues and organs will need to focus on mobilizing developmental-morphogenetic processes coupled with requirements of adult biology, in short, the body's innate regenerative capability [9–11].

We now detail the major components of our fabrication process of living structures of definite shape and functionality, collectively shown in Fig. 23.1 that utilizes developmental principles and phenomena and is based on the bioprinting of self-assembling multicellular building blocks.

2.1 The Bioink

Multicellular bioink building blocks are prepared from cell suspensions. They can be homogeneous, containing a single cell type or heterogeneous, made from a mixture of several cell types. Bioink units typically employed in our technology are either spherical or cylindrical in shape. Multiple methods have been described to prepare spherical aggregates [12–15].

Here we describe some of the methods we most frequently employ for the preparation of the spherical or cylindrical

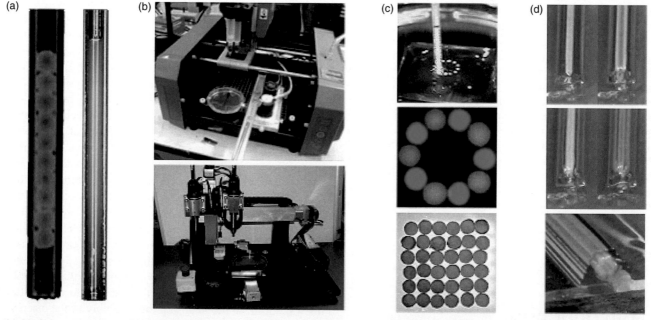

FIGURE 23.1 Components of the print-based tissue engineering technology. (a) The bioink-filled micropipette printer cartridge filled with multicellular building blocks that can be spheroidal (left) or cylindrical (right) depending on the method of preparation. (b) The bioprinter. Three-dimensional printing is achieved by displacement of the three-axis positioning system (stage in *y* and printing heads along *x* and *z* (top: Neatco, Carlisle, Canada; bottom: Organovo-Invetech, San Diego)). (c) Spheroids are delivered one by one into the hydrogel biopaper (itself printed) according to a computer script. (d) Layer-by-layer deposition of cylindrical units of biopaper (shown in blue) and multicellular cylindrical building blocks. The outcome of printing (spheroids in c multicellular cylinders in d) is a set of discrete units, which postprinting fuse to form a continuous structure. *(Figure reprinted with permission from Ref. [2].)*

FIGURE 23.2 The EHAM aggregate maker. The device is used to prepare a stamp that creates a pattern in an agarose substrate similar to an egg holder or a multiwell plate. A 2-in. diameter stainless steel disc containing a 0.2-in. deep rectangular grove is machined (top left). The xy dimensions of the groove are chosen such that it accommodates 100 wells (top right) in a 10×10 staggered array (a close-packed configuration), with 0.04″ of clearance between the wall of the grove and the closest well. The device is filled with a Sylgard silicone rubber base mixed with its proprietary hardening agent, then heated to 100°C to speed up the curing process. Once the silicone rubber cures, it is carefully peeled off from the device (middle left). The silicone "negative" of the device serves as a stamp (middle right) that is immersed into a pool of liquid agarose. Once the agarose solidifies, the silicone stamp is removed, sterilized with alcohol, and the process can be repeated as needed. The agarose wells are filled with cell solution (bottom left). Upon settling (bottom center), cells round up into spheroids (bottom right).

units. One method to form spherical bioink units uses the "egg holder aggregate maker" (Fig. 23.2; EHAM) [16]. For this device, small wells are created in an agarose mold using special plastic templates. The wells are positioned in a hexagonal pattern. The cell suspension is pipetted over the holes in agarose mold as quickly as possible and the entire assembly incubated for about 2 days. During this time cells settle down in the holes, and form spherical cell aggregates. This method is mostly used to prepare spheroids composed of a single cell type (Fig. 23.2).

In another method the cell suspension is centrifuged and the resulting pellet is transferred into a capillary micropipette. After a short incubation in medium at 37°C, cell–cell interactions are restored and the cylindrical slurry becomes sturdy enough to be extruded into liquid. The spherical building blocks are obtained by mechanically cutting uniform fragments (Fig. 23.3) that spontaneously round-up as a manifestation of tissue liquidity [17,18]. If the slurry is composed of multiple cell types, sorting and rounding will occur in parallel. As for embryonic tissues, the sorting behavior is driven by differences in tissue surface tension of the cell aggregates [3,19–23].

For the fabrication of cylindrical building blocks the slurry is extruded from the micropipette into a nonadhesive mold where it is left to mature overnight to improve the block's cohesion.

Finally, the bioink units are aspirated into the micropipette printer cartridge before deposition (Fig. 23.1a).

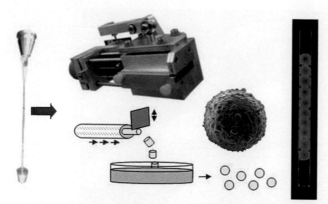

FIGURE 23.3 The aggregate cutter device. The micropipette on the left is inserted into the cutter (middle top). The cylindrical slurry inside the micropipette is advanced by a piston, which is controlled by a stepping motor at one end. As the extruded cellular material at the other end reaches the required length it is cut by a blade whose movement is coordinated with that of the piston. The resulting small cylinders with an aspect ratio close to one, round up overnight and are ready to be packaged into the micropipette printer cartridge (right). *(Figure reprinted with permission from Ref. [2].)*

2.2 Bioink Deposition and Structure Formation

The micropipette printer cartridges are inserted into the special purpose bioprinter (examples are shown in Fig. 23.1b) equipped with a single or multiple printing heads. In the case of multiple heads, the temporary support material, typically an inert biocompatible hydrogel, is printed along with the cellular material. Three-dimensional structures are built layer by layer.

In our technology, the role of the temporary support is critical. The multicellular bioink units, as they are contiguously deposited one by one, have no connections. Until such connections develop through cell–cell interactions the structure needs to be stabilized. This is achieved using a support that holds the tenuous assembly of the discrete bioink units together. For support we typically (but not exclusively) use cell-inert agarose. Specifically, the printing of discrete agarose units and bioink units is coordinated and proceeds according to a template. The template on one hand is consistent with the shape of the final desired structure and on the other hand allows to embed the bioink units into agarose (examples of templates are given later). This construction assures the safe postprinting fusion of the neighboring bioink units that, depending on the nature of the composing cells, takes 3–6 days. (As agarose is impervious to cells there is no cell migration into the support.) Once fusion is complete, the agarose is peeled of the cellular construct, which subsequently is transferred into a bioreactor for further maturation (in particular for the buildup of the reinforcing ECM). As the support structure is not part of the final product it does not constitute

a scaffold. Thus, the presented biofabrication process is an example of an entirely scaffold-free tissue engineering technology. (Note that the bioink units could in principle be prepared to contain a hydrogel, in which case the final product would contain a scaffold.)

3 BIOPRINTING TUBULAR STRUCTURES

The ultimate goal of tissue engineering is to build complete functional human organs by bioprinting or any other method. The obstacles toward this goal are however still numerous. For example, relatively thick 3D bioprinted tissue constructs will not survive without a macro- to microvascular network and their functionality will be compromised in the absence of a proper nervous system. Such networks will be indispensable during the preimplantation and postimplantation phase. After printing, large constructs will need to be perfused to nourish cells that otherwise would be farther than the 250 μm diffusion limit. Once the *in vitro* maturation of the organ is complete, upon implantation, its vascular network will be easily connected to the host vasculature and favor a rapid integration by preventing hypoxia and cellular death.

Nerves run along blood vessels *in vivo* and as transplanted tissues are getting larger, their innervation may become crucial for full functionality. During the *in vitro* maturation neural stimulation could be applied to improve organ development (for example muscle tissues need to be stimulated). The challenge is to provide a microenvironment favorable to the regrowth or ingrowth of the axons after implantation to restore the neural connections between severed peripheral nerves and the connections between the transplanted tissues into the central nervous system (see for example the strategies explored for physiological voiding of tissue-engineered bladder [24]). Creating tubular networks mimicking vasculature and neural network has to be achieved before highly vascularized and innervated fully functional engineered tissues can even be contemplated. In the last decade, research has been focused on producing straight tubular conduits to serve as vascular grafts or as guidance conduits in peripheral nerve repair. We have successfully applied our bioprinting-based tissue engineering technology to build straight small and intermediate diameter vascular and nerve grafts, as well as branching tubes – a first step toward the development of vascular or neural trees.

3.1 The Template: Vascular and Nerve Grafts

Tubular structures can be built vertically or horizontally by layer-by-layer deposition according to the schematics presented in Figs 23.4 and 23.5. In the first method, multicellular spheroids are deposited in a circular pattern and the pattern is repeated for each layer along the Z direction. After fusion of the spheroids the construct is a hollow tube (Fig. 23.4).

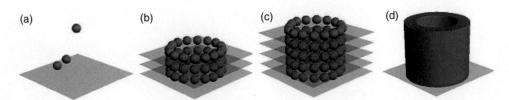

FIGURE 23.4 Schematics of building a tubular structure vertically, layer by layer, with spherical bioink units. (a) First, a sheet of biocompatible hydrogel is printed that serves as a support for the deposition of bioink building blocks. (b and c) The alternate deposition of layers of hydrogel (for example collagen) and building blocks is pursued until the desired height is reached. (d) After a few days the fusion of the building blocks and the subsequent removal of the hydrogel result in a hollow tube. *(Figure reprinted with permission from Ref. [2].)*

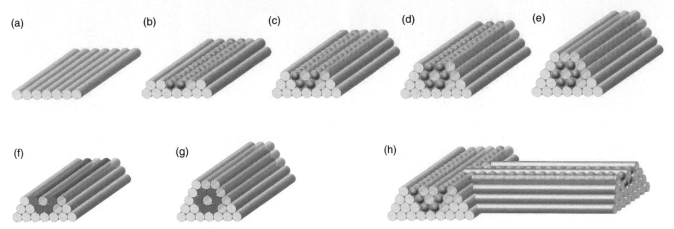

FIGURE 23.5 Possible template for a tubular structure built of agarose rods and multicellular building blocks. (a–e) Layer-by-layer deposition to build the tube with the smallest possible diameter. Printing was typically performed with units of 300 μm and 500 μm in diameter. (f–g) Same tubular configuration as in a–e, printed with cylindrical multicellular building blocks. (h) Example of deposition of a branched tubular construct shown here with spherical bioink units. *(Figure reprinted with permission from Ref. [25].)*

In the second method, the support material is shaped into long rods of similar diameter than the spherical or cylindrical cellular units. The similar dimension of both units allows stacking them, one layer at the time, as shown in Fig. 23.5. The printing along the horizontal axes facilitates the building of tubes several centimeters long, as the structure is fully supported in its longest dimension. Figure 23.5 presents the construction of a tubular configuration with the smallest lumen that can be produced with either spherical (Fig. 23.5a–e) or cylindrical (Fig. 23.5f, g) bioink units. A similar scheme can be applied to produce branching tubes (Fig. 23.5h).

Tubes of customizable geometry can easily be engineered according to a simple design template as illustrated by examples of vascular and nerve grafts presented in Fig. 23.6. Simple straight tubes of human smooth muscle cells (HUSMC) of different diameters were printed using 6 or 13 cellular units with, respectively, 1 and 7 agarose rods to create the lumen (Fig. 23.6a). After assembly, the multicellular cylinders fused within 2–4 days into the final tubular structure, and the supporting agarose rods were removed [26].

With the print-based technology, tubular structures can be built from mixtures of cells. Thus, the technology offers a unique opportunity to exploit cell–cell interactions, such as those involving endothelial cells (EC) and HUSMC, from the start of vessel formation. An endothelium is expected to form during the postprinting fusion and sorting (Fig. 23.6b). Applying flow to the forming endothelium early on will stimulate ECs to behave as they do *in vivo*. The anticipated benefits will be a more physiological ECM (both in composition and organization), better attachment of the endothelium, and faster maturation. To mimic even closer the blood vessel structure, vascular tubes displaying a double-layered wall similar to the media and adventitial layers found in the microvasculature were constructed. For this purpose we used both HUSMC and human fibroblast cylinders as building blocks according to the pattern shown in Fig. 23.6c. Smooth muscle actin staining of HUSMCs showed the sharp boundary between the HUSMC and human fibroblast layers in the engineered constructs after 3 days of fusion (Fig. 23.6c).

For nerve graft, it has been shown that a hollow guidance tube is less efficient than a structure presenting longitudinal spatial guidances for the axons to follow. A multichannel tube can be built by replacing some of the lumen-forming agarose rods with multicellular units (Fig. 23.6d).

3.2 The Support Structure

Bioprinting is an automated rapid prototyping method that allows the building of three-dimensional custom-shaped tissue and organ modules without the use of any scaffold,

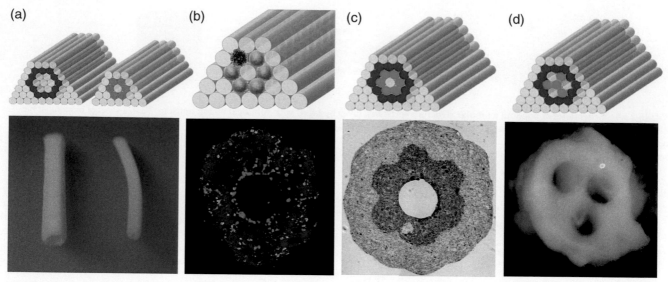

FIGURE 23.6 Design templates (top) and fused constructs (bottom). (a) Tubes of different diameter. (b) Construct built with spheroids composed of HUSMC (red) and EC (green). A cross-section of the fused tube (bottom) shows that most ECs (green) have migrated toward the lumen. (c) Double-layered vascular tube. The inner layer is made of HUSMC building blocks, the outer one of fibroblast building blocks. The transversal section (bottom) was immunostained for smooth muscle actin (brown) and shows that the building blocks fused but the two cell types remain segregated mimicking the media and adventitia of blood vessels. (d) Design template for a multichannel tube. The external wall is built with units composed of bone marrow stem cells (BMSC, red). The lumen is divided into three individual channels after fusion of the four cellular rods of heterocellular composition (BMSC and SCs) placed alternatively with agarose rods). *(Figure reprinted with permission from Ref. [2].)*

thus making the final construct fully biological, as well as structurally and functionally closer to native tissues. However, the precise spatial arrangement of the spheroid or cylindrical bioink units requires the use of support material to fill up the area void of cells and maintain the cellular units during fusion and maturation. Different materials have been employed for 3D tissue bioprinting as the technology involved. For the proof of concept, initially, the method shown in Fig. 23.4 was used. Despite some success, the speed, reproducibility and scalability of the technique were challenged by several shortcomings. The first was associated with the use of collagen as hydrogel. The collagen was deposited in sheets (~500 μm in thickness) in which the spheroid building blocks were embedded. The success of the printing depended strongly on the control of the gelation state of the different collagen layers that was essential for a smooth deposition of the building blocks. Uneven gelation compromised the spatial accuracy of the construct, in particular for constructs higher than a few layers [27]. The collagen was also remodeled by the cells and at least partially integrated into the final structure (breaching the definition of an inert support), and its removal was challenging. Another limitation arose when the printing of constructs of larger size and more complex pattern (i.e., branching tubes) were attempted. The preparation of the spheroids in large quantities (>1000 for some branched tubular structures) became time-consuming. In addition, the filling of the micropipette printer cartridge (done manually)

was cumbersome, as the spheroids had to be aspirated one by one and without a gap between them. This required a high degree of monodispersity of the spheroids that was difficult to achieve.

To circumvent these issues, we first replaced the collagen sheets with agarose rods (Fig. 23.5) and second, used cylindrical multicellular building blocks instead of the spheroids (Fig. 23.5f,g). The agarose rods were formed *in situ* and deposited rapidly and accurately by the bioprinter. Agarose is a biocompatible inert gel that cells neither invade nor rearrange. The agarose rods keep their integrity during postprinting fusion and are easy to remove. Thus, agarose can serve as a temporary support without being part of the final construct.

In principle any biocompatible, nonadhesive hydrogel (alginate, dextran, chitosan, methylcellulose, keratin-based, and silk hydrogel) could be used to build the support structure. However, agarose's physical properties made it an ideal candidate for bioprinting, as it is a low viscosity liquid at 55°C; therefore, it is easy to draw into the printer cartridge. It gels rapidly and uniformly into the capillaries by cooling in cold PBS (4°C) and is easily extruded as a smooth and regular cylindrical unit. The temperature-driven gelation process allows forming the agarose rods of precise and reproducible dimension, directly in the printer, without adding other chemicals (that could be detrimental to the cells) or the need to work in solution (i.e., calcium solution for gelation of alginate [28]).

4 BIOFABRICATION OF A NERVE GRAFT

4.1 Overview of Tissue Engineering of a Nerve Graft

A damaged or severed nerve requires the axons, the signal conducting fibers, to reinnervate distally from the injury site [29]. Because there can be thousands of axons in a nerve bundle, the statistical chance that all of the axons will fully recover from axotomy (cutting) is slim, with only 40% of cases reporting satisfactory motor recovery [30]. Thus, the goal of nerve repair surgery is to minimize lost function due to nerve injury. If the severed ends of the nerve can be reconnected without too much strain, the damage is best repaired by suturing these ends together [31]. However, if this is not possible, a nerve guide is required to bridge the gap between the two ends of the injured site to properly regrow axons and maximize the return of motor and sensory function [32]. Functional recovery rapidly dwindles with the size of the gap and the guide (>3 cm) due to the limited amount of axons reaching the specific target, leaving some patients with life-long disability and/or debilitating neuropathic pain.

Currently, the gold standard for nerve repair requiring a conduit is an autograft, a nerve segment transplanted from another part of the patient to the damaged site. This type of repair, however, implies morbidity to the donor site, the risk of multiple surgeries [31], and problems matching nerve type (sensory vs. motor) and diameter [33]. If the graft is instead taken from an organ donor, an immune response could cause complication [31,34]. Thus, efforts have focused on developing a conduit that biomimics the properties of nerve tissue, while minimizing their drawbacks. An ideal nerve graft should be permeable and biocompatible, provide a path for the growing axons, minimize the risk of immune response, and be able to accommodate neurotrophic factors either directly or through the seeding of cells that produce these factors.

Initially, non-neuronal autologous tissues have been used to fill the gap. Autologous vein grafts showed good axon regeneration and functional repair at a level similar to autologous nerve grafts in 1 cm nerve gaps for rats and in clinical application for gaps smaller than 3 cm [35]. Repair of larger gaps failed as the vein graft collapsed due to constriction by surrounding scar tissue [36]. Potential solutions included filling the vein with other tissues such as nerve slices [37] or fresh muscle [38]. Fresh muscle not only added support, but it also provided longitudinal orientation favorable for axon growth (reviewed in Ref. [39]). More sophisticated methods included the rolling of amniotic membrane [40,41] or small intestine mucosa [42] into a tubular conduit. These tissues were chosen based on their high content of ECM with hope that they would provide a favorable environment for axon growth.

The majority of grafts currently in use are constructed from natural or synthetic polymers intended to mimic the ECM. The conduits approved for clinical use by the FDA are synthesized from biodegradable polymers such as polyglycolic acid, poly-DL-lactide-caprolactone, or from the biological material collagen [42,43]. Conduits composed of artificial, biodegradable polymers have the advantage of being easier to modify and quality controlled. These polymers however are generally hydrophobic, and contrary to purely biological materials do not allow proper cell signaling to control the cellular response during repair [44].

To create a growth-permissive environment for cells, nerve guidance tubes can be filled with hydrogels (e.g., agarose, collagen, chitosan, hyaluronic acid, keratin) in association with ECM components (e.g., laminin, fibronectin, proteoglycans), neurotrophic factors (e.g., nerve growth factor, fibroblast growth factor) [45] and supporting cells such as Schwann cells (SC). For example, it has been reported that polycaprolactone conduits containing 10% laminin have a positive effect on neurite growth [46]. The disadvantage of using biological materials, however, is that they must be derived from animal sources and thus risk immune rejection and infection. To avoid such issues, peptides mimicking the binding domains of fibronectin, laminin, and collagen molecules have been used to produce new biosynthetic scaffolds with similar advantages of mixed conduits, while minimizing the risk of immune rejection [47,48].

Aside from the raw materials to produce the graft, studies have been conducted on the effectiveness of seeding different cell types in the graft. SCs are the most common type of seeded cells, and have been shown to improve nerve regeneration. Because of the limited availability of autologous SC, alternative cell types have been considered (e.g., olfactory ensheathing cells [49], hair follicle stem cells [50], bone marrow stem cells (BMSCs) [51], adipose-derived stem cells [52], and skin-derived stem cells [53]). Neural stem cells, mesenchymal stem cells (MSCs), and induced pluripotent stem cells are optimal candidates as they can differentiate into any neural cell type. In particular, MSCs have the capacity to differentiate into neural and glial cells, or glial-like cells that promote neurite extension *in vitro* [54,55]. They have been shown to secrete neurotrophic factors while trans-differentiating into glial-like cells incorporated into the growing nerve. Similarly, induced pluripotent stem cells, which are derived from skin fibroblasts, have also been shown to improve nerve recovery repairing sciatic nerve defects in rats [53,56].

An alternative approach is to use a scaffold-free method, utilizing the self-organizing and self-assembly properties of cells and tissues to develop their own ECM. In one approach, sheets of fibroblasts are cocultured with embryonic-derived neural cells, resulting in a bilayered sheet. The sheet is pinned at two points and detached so that it rolls into a cylindrical conduit. The tube thus created represents a fully biological graft with cell neurotrophic factors [57]. These conduits were tested *in vivo* and shown to restore normal conduction velocity after 28 days, suggesting

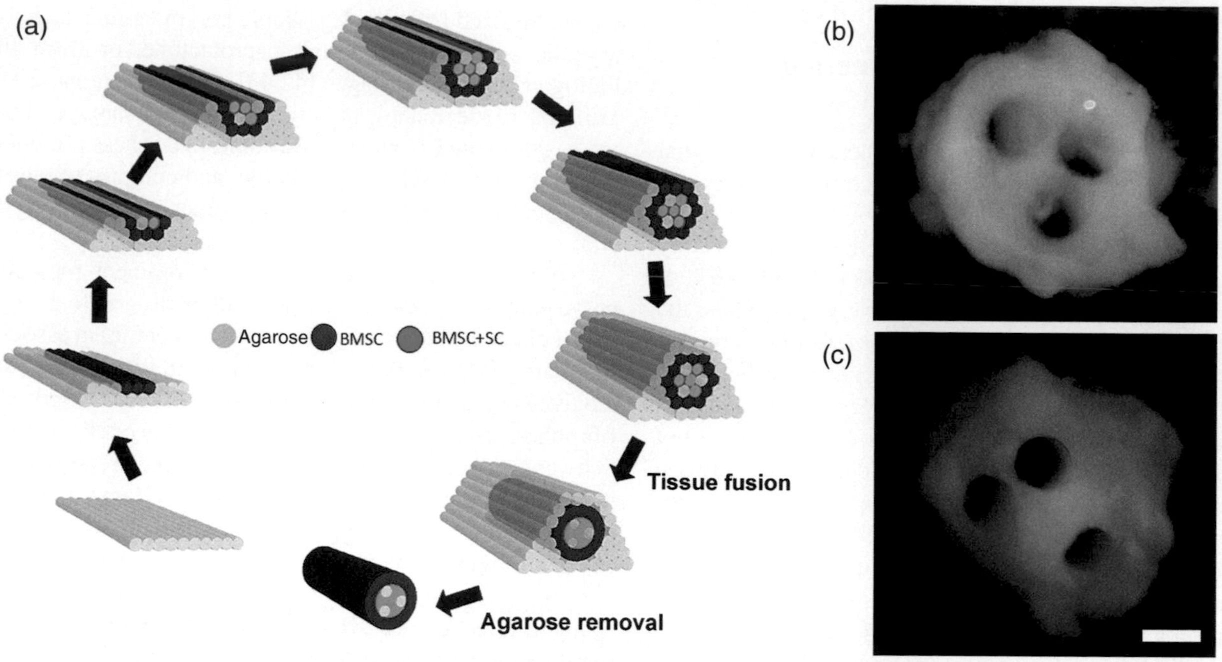

FIGURE 23.7 Building of a nerve graft by bioprinting. (a) Possible scheme for layer-by-layer deposition of cellular (red: BMSC, green: 90% BMSC + 10% SC) and agarose cylinders to build a three-lumen tube. (b) Cross-sectional view of a bioprinted nerve graft with three acellular channels. (c) Fluorescently labeled SC concentrated at the central region of the graft. Scale bar in c: 500 μm. *(Figure reprinted with permission from Ref. [25].)*

a viable strategy for nerve repair. Conduits composed of adipose-derived stem cells are also being tested with this method [52].

We applied our bioprinting-based technology, as described in Sections 2 and 3 to develop a unique alternative for repair of nerve gaps. In what follows, we describe the design and performance of the bioprinted nerve graft when used as filler for a collagen tube and implanted in a rat sciatic injury model [25].

4.2 Biofabrication of a Nerve Graft by Bioprinting

As the cell density of supporting cells is a critical factor for axons regeneration, a tube composed exclusively of SC was initially considered to establish proof of concept. However, preparation of bioink units with SC revealed that their weak cell–cell adhesion leads to cellular slurries that lose their cylindrical shape in the grooves of the agarose mold and are too fragile to be used in the printing process.

BMSCs were chosen as additional source of support cells. Their cell–cell adhesion is sufficient to form sturdy cellular cylinders either when used alone or mixed with up to 10% of SC that can serve as bioink units after a 6 h maturation. Cylindrical bioink units were printed according to a judiciously chosen template (Figs 23.6d and 23.7). BMSC form the most external layer of the tubular conduit. The lumen of the tube is partially closed by the addition of

four cellular units composed of 90% of BMSC and 10% of SC and three agarose rods of 500 μm diameter. After printing, medium was added until the construct was totally submerged, and the construct was placed in the incubator for 7 days. Structure formation occurred by the postprinting fusion of the discrete cellular units (i.e., cylinders). Agarose acted as a temporary spacer for the acellular channels and as a surrounding support structure (Fig. 23.7).

Fusion of the cellular cylinders leads to the formation of a tubular structure with three longitudinal parallel acellular channels. The channels are preferentially lined with SC that guide the growth of axons and may form the protective layer (myelin) around them (Fig. 23.7b,c). After maturation, the agarose rods are removed and the resulting multichannel construct is ready for implantation to bridge a severed nerve directly or wrapped within a collagen nerve guide. The construct is either sutured to the proximal and distal stumps or attached using fibrin glue.

4.3 Selection of the Bioink and Structure of the Graft

The design of the bioprinted nerve graft described earlier was guided by the following considerations:

1. The choice of a rat sciatic nerve injury model for implantation. The bioprinted nerve graft needed to have a diameter comparable to the one of a rat sciatic nerve (~2 mm) and be at least 1 cm long.

2. The creation of a multilumen pattern to increase the luminal surface area available for support cells to favor axons regeneration.
3. The cell type to support axon regeneration and be suitable for the preparation of the cylindrical bioink units.

The geometrical parameters of the bioprinted nerve grafts, such as wall thickness, diameter, and number of channels, can easily be controlled and modified to fit nerve diameter and length of injury. The diameter of the cylindrical units can be varied between 250 μm and 500 μm. The 500 μm units we used were based on the result obtained by Hadlock et al. [42]. These authors varied the diameter and number of channels in 2.3 mm diameter poly(lactic-*co*-glycolic acid) conduits and obtained the best regeneration using a five-channel construct, with each channel of 500 μm diameter. The number of channels in our construct was reduced to three so the final diameter of the graft was close to the diameter of a rat sciatic nerve. For repair of larger diameter nerves encountered in clinics, several of these constructs could be bundled. The bundle could be wrapped into a collagen tube (as we did for implantation in rats) or into a collagen sheet produced by cultured fibroblasts.

The cellular composition of our graft was defined by two criteria: (1) the ability of the cells to support axon growth and (2) SC were mixed with BMSC, the latter being able to produce robust cellular cylindrical units. In the right conditions, BMSC can differentiate into Schwann-like cells and support axon regeneration [49,58,59]. To bring bioprinted nerve grafts to clinical application, other supporting cells, more easily available from the patient, should be investigated such as adipose and skin derived stem cells.

Our bioprinted fully biological nerve graft circumvents some of the pitfalls causing the failure of nerve guidance conduits to promote axon regeneration, namely, the lack of longitudinally oriented structural features, low density of supporting cells such as SC, and immunological and inflammatory responses triggered by the scaffold.

4.4   Postprinting Maturation of the Graft: Implantation into an Animal Model

As described earlier, a three-channel nerve graft construct was printed using cylindrical bioink particles composed of cells known for supporting nerve regeneration (SC and BMSC). A cell-inert temporary support structure was printed along with the cellular cylinders according to the template shown in Fig. 23.7. The cellular cylinders fused over time to form a continuous tube with internal channels. Once postprinting structure formation has taken place, the support structure was removed and the construct was ready for implantation into laboratory rats [25]. The rats' left sciatic nerve was exposed and a 1 cm defect was created by tran-

section of the nerve. The resulting gap was bridged with the bioprinted nerve tube wrapped in a collagen tube (Neurogen, Integra Life Sciences, Plainboro, NJ). The free ends of the native sciatic nerve were intubulated into the nerve tube and sutured to it.

Similar procedures were followed to prepare control rats where the gap in the nerve was repaired by an autologous graft (i.e., the cut segment of the sciatic nerve inverted and reattached) or with an empty collagen nerve guide.

4.5   Testing the Functionality of the Graft

After implantation, we tested the regenerative capabilities of the graft. Among the multiple evaluation methods for nerve repair, axon counts and electrophysiology are among the most reliable methods to assess graft efficiency [60]. We have performed short- (3 weeks) [25] and long-term (40 week) studies [61] with the latter including functional evaluation of the nerve through electrophysiology.

4.5.1   Histological Evaluation: Axon Count

Three weeks after implantation, some of the bioprinted grafts were harvested, including the proximal and distal stump. Manual counting showed that about 40% of the axons present in the proximal stump had made it to the distal stump (Fig. 23.8). This fraction is similar to other studies that compared other bioengineered nerve grafts in an 8 week period [62,63].

In the long-term study, after the electrophysiology was performed (see Section 4.5.2), the grafts were excised and histological sections were taken at midgraft. To visualize the axons the Bielschowsky's staining was used in the 3 week and 40 week studies, as a standard metric of repair [60]. In the grafts harvested in the 40 week study, to estimate the total number of repaired axons, we attempted to calculate an axon density across the graft's cross-section. As the biofabricated grafts contained the three lumina, the distribution of the axons in the cross-section was inhomogeneous, which made the use of axon density unreliable. However, a qualitative comparison of the axons across each of the grafts could still be performed, as shown in Fig. 23.9. The figure displays reasonable axonal regrowth across each of the graft groups (intact control, autologous, collagen tube, biofabricated). It is also important to note that the biofabricated graft shows a higher density of axons near the lumina (Fig. 23.9e), where the cylindrical bioink units containing SC were located (Fig. 23.7). Additionally, in these initial trials, the agarose rods used to build the construct remained in the bioprinted graft during surgical implantation to prevent collapse of the lumina during the surgery. It appears the presence of agarose within the channel, prevented regrowth through the lumina, as there were no visible axons through these areas of the construct (Fig. 23.9b,e). These

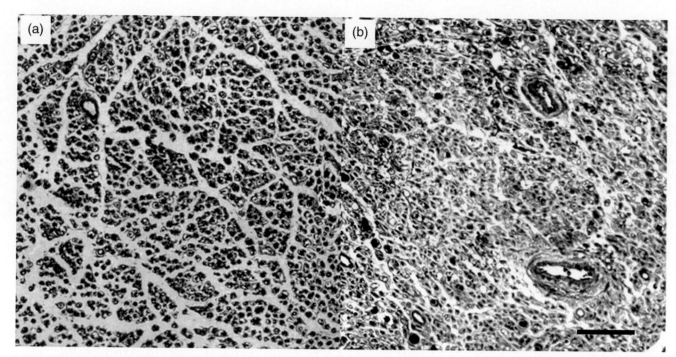

FIG. 23.8 Histological evaluation of the bioprinted nerve graft. (a) Partial view of proximal section. (b) Partial view of distal section. For global aspect, sections were stained with hematoxylin–eosin. To detect axons (black dots), Bielschowsky's staining was performed. Scale bar: 100 μm. *(Figure reprinted with permission from Ref. [25].)*

results together with the data on axon count in our earlier work [25] at both the proximal and distal ends of the conduit (Fig. 23.9) provide convincing evidence that extensive axonal regrowth has taken place across these biofabricated grafts.

4.5.2 Electrophysiological Studies

Electrophysiological studies were performed to evaluate motor and sensor recovery on rats with repaired sciatic nerves by different types of grafts; autologous graft, collagen tube, and the biofabricated construct. This 40-week study compared the data taken for the compound muscle action potential (CMAP) and the somatic pressor reflex (SPR), normalizing each experiment to the unoperated control leg for each rat.

4.5.2.1 Compound Muscle Action Potential

The CMAP measurement (recording of muscular voltage response to electrical stimulation of the repaired nerve) shown in Fig. 23.10 reveals several features of interest [64]. Latency to response (the time delay between nerve stimulation to the beginning of muscle response linked to the most conductive axons) and latency to peak (time delay between nerve stimulation and the peak voltage of the muscle response related to the average conductivity of all the axons) correlate with the speed of the nerve signal. This response characterizes a functioning nerve, in addition to describing the quality of regrowth and myelination of the axons.

Latency to peak was selected to compare the CMAP across the three different repair groups since a stimulus artifact compromised an accurate evaluation of latency to peak [65] (see Ref. [61] for details).

We quantitatively compared the latency to peak data across all three groups of grafts. As expected, the latencies typically decreased with increasing current (starting with the threshold voltage eliciting a response) and were longer for the repaired legs than for the respective controls. It is also expected that as the stimulation current increases, the repaired leg latency should approach the latency of the control leg. A two-way ANOVA showed a significant difference in the overall treatment (repair vs. control) at the threshold voltage for the biofabricated and collagen tube grafts. This difference disappeared at five times the threshold voltage for the biofabricated graft, but was still present for the collagen tube (which was present at all levels of stimulation), hinting that regarding latency our construct has superior repair to the collagen tube. The autologous group performed best, for which the ANOVA found no difference between the control and the repaired graft at any of the levels of stimulation [61].

4.5.2.2 Sensory Response

Sensory recovery was evaluated by measuring the mean arterial pressure (MAP) response to afferent stimulation of the sciatic nerve (somatic pressor reflex). The stimulating electrode was positioned on the nerve distal to the repair

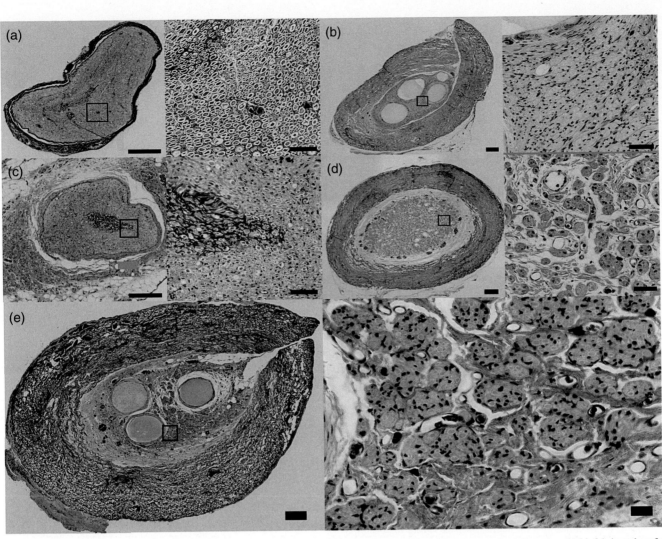

FIGURE 23.9 **Bielschowsky's staining of histological sections of the various types of grafts.** (a) Intact control sciatic nerve; (b) biofabricated graft; (c) autologous graft; and (d) collagen tube graft. In these images the left panels (showing the overall view of the cross-sectional area) and right panels (enlarged view of the boxed regions on the left panels) correspond respectively to 200× and 600× magnifications. (e) A more detailed view of one of the biofabricated grafts (left panel) and the axons across the boxed area (right panel). Axons in these images appear as black dots. All sections in these figures were taken at mid graft. Scale bars: 200 μm and 40 μm, respectively, for the left and right panels in all the images (a–e). *(Figure reprinted with permission from Ref. [61].)*

site (in the operated leg) and baseline blood pressure (prior to electrical stimulation) was noted. Then square wave pulses (1 ms duration at 20 Hz frequency) at progressively increasing current were applied and changes in MAP were recorded when the response reached a plateau between increases in current, or was averaged for the region between current changes if no plateau occurred (Fig. 23.11). Maximum blood pressure (i.e., the maximum response to stimulation) was obtained by noting the highest blood pressure achieved in the data set (Fig. 23.11).

Results in Fig. 23.11 for the sensory recovery provide further support (in addition to the histological and CMAP studies) for the excellent performance of the biofabricated graft.

4.5.3 In Vitro Physiological Model Nerve Graft

While the *in vivo* studies on the biofabricated nerve conduit gave promising results, we wished to establish a simpler method for the testing and optimization of the grafts without the need for tedious *in vivo* experiments. In 2010, Vyas et al. [66] introduced an *in vitro* model for peripheral nerve repair. Using neonatal mice, they designed and demonstrated the regrowth of peripheral nerve axons emanating from a spinal cord section into a nerve graft attached to it. In order to test the effectiveness of our bioprinted construct, we initiated a study with the objective to modify the Vyas protocol for a rat model.

The advantages of this *in vitro* method are multifold. Because survival surgery on an animal is no longer necessary, experiments are simpler, more reproducible and can

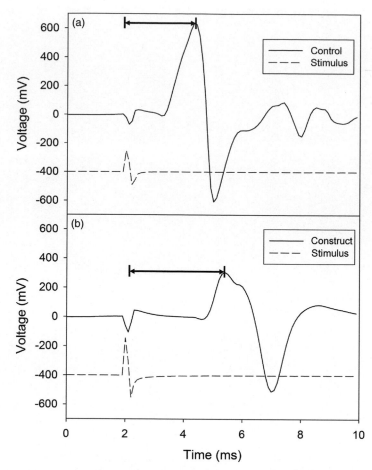

FIGURE 23.10 Electrophysiology for motor function recovery: measurement of CMAP. Examples of evoked CMAP in the control leg (a) and the leg repaired with the biofabricated graft (b) from the same animal are shown. The broken and solid lines correspond respectively to the stimulator output (not to scale) and the recorded CMAP. Here, a stimulus artifact (the slowly decaying portion of the curve following the stimulus) and response are clearly discernible in the recording. In some experiments, the latency to response can be obscured by a stimulus artifact [67], as it may overlap with the start of the CMAP. For consistency, latency to peak was evaluated in all rats in the experiments as this quantity was never obscured by the artifact. Horizontal markers denote the latency to peak measurement. *(Figure reprinted with permission from Ref. [61].)*

be conducted with less surgery training. By removing the need for extra care for the animal, and any other costs associated with more complicated surgeries, the overall costs of the studies are significantly lower. Additionally, more trials can be performed per animal, greatly increasing the efficiency of the experiment. Finally, this method allows the bioprinted construct (which on its own is too fragile at the time of implantation into an animal) to be used independently of the collagen tube.

The modified experimental setup is shown in Fig. 23.12. Six days old neonatal rats are put to sleep using carbon dioxide, and then promptly decapitated. The spinal cord is carefully removed from the rat pup, preserving the integrity of as many nerve roots as possible. The median and ulnar nerves are also harvested. A 500 μm slice of the lumbar spinal section, including ventral roots (cut to 3–4 mm length), is placed on a collagen membrane. The ventral root is then placed about 500 μm inside either a 4 mm long collagen

tube or bioprinted construct, with a 4 mm section of median or ulnar nerve placed on the opposing side in a similar fashion.

In addition to the collagen tube, different bioprinted constructs can be employed to compare the effectiveness of different ratios of the BMSC and SC mixture. To accommodate the smaller size of the nerves in such experiments we use six-cylinder single lumen constructs (1.5 mm total diameter), similar to the bioprinted tubes shown in Fig. 23.6a.

After 10 days of their assembly, the grafts are fixed and examined using histology to count the axons that grow through each of the components. The number of axons that travel through the middle of each graft (either the collagen tube or the biofabricated graft) are compared to the number of axons traveling through the ventral root of the spinal section. With this information, we can get a better idea of the short-term nerve regeneration capabilities of the

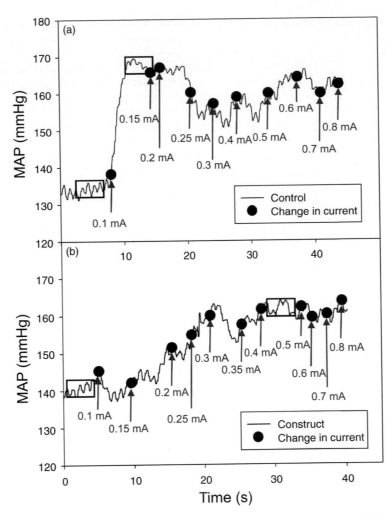

FIGURE 23.11 Somatic pressor reflex. Measurements of MAP in the control leg (a) and the leg repaired with the biofabricated graft in the same animal (b). The solid black line represents the continuous raw record of MAP. The labeled black dots denote the approximate time of the current increase with the associated value of current at that point in time. The base and maximum values of MAP were obtained by averaging values, respectively, in the first and second boxed areas. *(Figure reprinted with permission from Ref. [61].)*

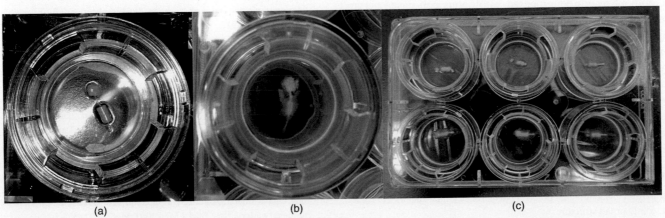

FIGURE 23.12 *In vitro* experimental arrangements. (a) A collagen tube is flanked by a fresh spinal cord section/root and an ulnar nerve. (b) Fixed bioprinted construct surrounded by neural tissue. (c) An array of fixed bioprinted constructs of varying length in the six-well plate.

biofabricated graft, and fine-tune the cellular ratios for an optimal graft.

5 CONCLUSIONS

This chapter presented an overview of tissue engineering efforts to fabricate tubular organ structures, in particular nerve grafts, by a specific application of extrusion bioprinting, We first introduced the major components of the technology, such as the bioink and the biopaper. Next, we described the general features of the scaffold-free process, from the delivery of the bioink units and the supporting biopaper to the postprinting formation of the tubular biological structure by self-assembly processes akin to those, manifest in early embryonic development. Finally, we detailed the construction of the bioengineered fully cellular nerve graft.

Peripheral nerve injuries with frequent occurrence in traffic accidents and battlefield traumas often result in debilitating conditions. However, unlike in the central nervous system, in the peripheral nervous system regeneration is possible. The regeneration process could benefit from the application of tissue engineering. Indeed, numerous attempts have recently been made to build nerve conduits of simple and complex architecture using a wide variety of biomaterials. We recognized that bioprinting could greatly facilitate such efforts and embarked on the application of our technology to the repair of damaged peripheral nerves. Specifically, we designed and printed purely biological grafts, which have been implanted into laboratory animals. The functionality of the bioprinted grafts was evaluated by histology, as well as by electrophysiological methods. The results of our proof of concept studies suggest that bioprinting represents a promising approach to the fabrication of biomimetic nerve conduits. Future work will focus on the optimization of graft geometry and cellular composition using the described *in vitro* regeneration model.

The long-term goal of our efforts is the effective development of a bioengineered autologous cell-based nerve guide by bioprinting with functionality comparable to an autologous nerve graft. The successful completion of such a project would revolutionize the current clinical practice of treatment of human peripheral nerve injuries. The benefits would be numerous, the most evident of which include reduction of valuable operative time and costs, morbidity of harvesting human nerve grafts and, ultimately, the improvement in the patients' quality of life.

GLOSSARY

Bioprinter Machine that precisely dispenses cellular material to form a 3-D organ structure.

Bioink The cellular material dispensed by the bioprinter, usually in the form of spherical or cylindrical aggregates of cells.

Biopaper An inert hydrogel, usually agarose, which provides temporary support to the cells as they self-assemble to form the organ-like structure.

Self-assembly The unaided process through which cells interact with one another to form tissues and organs. This process includes tissue fusion, cell sorting, and interactions with ECM material.

Egg holder aggregate maker (EHAM) Hexagonal array of small wells into which a cell suspension is added, which come to form a small spherical aggregate in each well.

Histogenesis The series of organized integrated processes that transforms undifferentiated cells into a tissue in the developing embryo.

Organogenesis The series of organized integrated processes that transforms an amorphous mass of differentiated cells into a complete organ in the developing embryo.

Compound muscle action potential Electrical study of muscle function evoked by the stimulation of a motor nerve.

Somatic pressor reflex Sensory response to an afferent stimulation, from the organ toward the central nervous system.

Stimulus artifact The overlap of the stimulating electrical pulse with the response to this pulse during the measurement of the compound muscle action potential.

ABBREVIATIONS

ECM	Extracellular matrix
EHAM	Egg holder aggregate maker
HUSMC	Human smooth muscle cells
EC	Endothelial cells
BMSC	Bone marrow stem cells
SC	Schwann cells
MSC	Mesenchymal stem cells
CMAP	Compound muscle action potential
MAP	Mean arterial pressure

REFERENCES

[1] Forgacs G, Newman SA. Biological physics of the developing embryo. Cambridge: Cambridge University Press; 2005.

[2] Jakab K, Norotte C, Marga F, Murphy K, Vunjak-Novakovic G, Forgacs G. Tissue engineering by self-assembly and bio-printing of living cells. Biofabrication 2010;2(2):022001.

[3] Perez-Pomares JM, Foty RA. Tissue fusion and cell sorting in embryonic development and disease: biomedical implications. BioEssays 2006;28(8):809–21.

[4] Whitesides GM, Boncheva M. Beyond molecules: self-assembly of mesoscopic and macroscopic components. Proc Natl Acad Sci USA 2002;99(8):4769–74.

[5] Whitesides GM, Grzybowski B. Self-assembly at all scales. Science 2002;295(5564):2418–21.

[6] Atala A, Bauer SB, Soker S, Yoo JJ, Retik AB. Tissue-engineered autologous bladders for patients needing cystoplasty. Lancet 2006; 367(9518):1241–6.

[7] Macchiarini P, Jungebluth P, Go T, Asnaghi MA, Rees LE, Cogan TA, et al. Clinical transplantation of a tissue-engineered airway. The Lancet 2008;372(9655):2023–30.

[8] Shin'Oka T, Imai Y, Ikada Y. Transplantation of a tissue-engineered pulmonary artery. New Engl J Med 2001;344(7):532–3.

[9] Hellman KB, Nerem RM. Advancing tissue engineering and regenerative medicine. Tissue Eng 2007;13(12):2823–4.

[10] Ingber DE, Mow VC, Butler D, Niklason L, Huard J, Mao J, et al. Tissue engineering and developmental biology: going biomimetic. Tissue Eng 2006;12(12):3265–83.

[11] Vunjak-Novakovic G, Kaplan DL. Tissue engineering: the next generation. Tissue Eng 2006;12(12):3261–3.

[12] Marga F, Neagu A, Kosztin I, Forgacs G. Developmental biology and tissue engineering. Birth Defects Res C Embryo Today 2007;81(4):320–8.

[13] Mironov V, Visconti RP, Kasyanov V, Forgacs G, Drake CJ, Markwald RR. Organ printing: tissue spheroids as building blocks. Biomaterials 2009;30(12):2164–74.

[14] Norotte C, Marga FS, Niklason LE, Forgacs G. Scaffold-free vascular tissue engineering using bioprinting. Biomaterials 2009;30(30):5910–7.

[15] Lin RZ, Chang HY. Recent advances in three-dimensional multicellular spheroid culture for biomedical research. Biotechnol J 2008;3(9–10):1172–84.

[16] McCune M, Shafiee A, Forgacs G, Kosztin I. Predictive modeling of post bioprinting structure formation. Soft Matter 2014;10:1790–800.

[17] Phillips HM, Steinberg MS. Embryonic tissues as elasticoviscous liquids. I. Rapid and slow shape changes in centrifuged cell aggregates. J Cell Sci 1978;30:1–20.

[18] Steinberg MS, Poole TJ. Liquid behavior of embryonic tissues. In: Bellairs R, Curtis ASG, Dunn G, editors. Cell behaviour. Cambridge: Cambridge University Press; 1982. p. 583–607.

[19] Foty RA, Pfleger CM, Forgacs G, Steinberg MS. Surface tensions of embryonic tissues predict their mutual envelopment behavior. Development 1996;122(5):1611–20.

[20] Jia D, Dajusta D, Foty RA. Tissue surface tensions guide *in vitro* self-assembly of rodent pancreatic islet cells. Dev Dyn 2007;236(8):2039–49.

[21] Lecuit T, Lenne PF. Cell surface mechanics and the control of cell shape, tissue patterns and morphogenesis. Nat Rev Mol Cell Biol 2007;8(8):633–44.

[22] Norotte C, Marga F, Neagu A, Kosztin A, Forgacs G. Experimental confirmation of tissue liquidity based on the exact solution of the Laplace equation. Eur Phys Lett 2008;81:46003.

[23] Steinberg MS, Takeichi M. Experimental specification of cell sorting, tissue spreading, and specific spatial patterning by quantitative differences in cadherin expression. Proc Natl Acad Sci USA 1994;91(1):206–9.

[24] Horst M, Madduri S, Gobet R, Sulser T, Milleret V, Hall H, et al. Engineering functional bladder tissues. J Tissue Eng Regen Med 2013;7:515–22.

[25] Marga F, Jakab K, Khatiwala C, Shepherd B, Dorfman S, Hubbard B, et al. Toward engineering functional organ modules by additive manufacturing. Biofabrication 2012;4(2):022001.

[26] Norotte C, Marga FS, Niklason LE, Forgacs G. Scaffold-free vascular tissue engineering using bioprinting. Biomaterials 2009;30:5910–7.

[27] Jakab K, Norotte C, Damon B, Marga F, Neagu A, Besch-Williford CL, et al. Tissue engineering by self-assembly of cells printed into topologically defined structures. Tissue Eng Part A 2008;14(3).

[28] Tan Y, Richards DJ, Trusk TC, Visconti RP, Yost MJ, Kindy MS, et al. 3D printing facilitated scaffold-free tissue unit fabrication. Biofabrication 2014;6(2):024111.

[29] Fawcett JW, Keynes RJ. Peripheral nerve regeneration. Annu Rev Neurosci 1990;13:43–60.

[30] Moneim M, Omer G. Clinical outcome following acute nerve repair. In: Omeg G, Spinner M, Beek AV, editors. Management of peripheral nerve problems. Philadelphia: Saunders; 1998. p. 414–9.

[31] Siemionow M, Brzeziki G. Current techniques and concepts in peripheral nerve repair. Int Rev Neurobiol 2009;87:141–72.

[32] Schmidt CE, Leach JB. Neural tissue engineering: strategies for repair and regeneration. Annu Rev Biomed Eng 2003;5:293–347.

[33] Wolford LM, Stevao ELL. Considerations in nerve repair. Proc (Bayl Univ Med Cent) 2003;16(2):152–6.

[34] Moore AM, Ray WZ, Chenard KE, Tung T, Mackinnon SE. Nerve allotransplantation as it pertains to composite tissue transplantation. Hand 2009;4(3):239–44.

[35] Chiu D, Lovelace R, Yu L, Wolff M, Stengel S, Middleton L, et al. Comparative electrophysiologic evaluation of nerve grafts and autogenous vein grafts as nerve conduits: an experimental study. J Reconstr Microsurg 1988;4(4):303–9.

[36] Tang JB, Gu YQ, Song YS. Repair of digital nerve defect with autogeneous vein graft during flexor tendon surgery in zone 2. J Hand Surg Br 1993;18:449–53.

[37] Tang J. Vein conduits with interposition of nerve tissue for peripheral nerve defects. J Reconstr Microsurg 1995;11(1):21–6.

[38] Battiston B, Tos P, Geuna S, Giacobini-Robecchi MG, Guglielmone R. Nerve repair by means of vein filled with muscle grafts: I. Microsurgery 2000;20:32–6.

[39] Meek M, Varejao A, Geuna S. Use of skeletal muscle tissue in peripheral nerve repair: review of the literature. Tissue Eng 2004;10:1027–36.

[40] Mohammad J, Shenaq J, Rabinovsky E, Shenaq S. Modulation of peripheral nerve regeneration: a tissue engineering approach. The role of amnion tube nerve conduit across a 1 centimeter nerve gap. Plast Reconstr Surg 2000;105(2):660–6.

[41] O'Neill A, Randolph M, Bujold K, Kochevar I, Redmond R, Winograd I. Preparation and integration of human amnion nerve conduits using a light-activated technique. Plast Reconstr Surg 2009;124(2):428–37.

[42] Hadlock T, Sundback C, Hunter D, Vacanti J, Cheney M. A new artificial nerve graft containing rolled Schwann cell monolayers. Microsurgery 2001;21(3):96–101.

[43] Meek MF, Coert JH. US Food and Drug Administration/ Conformit Europe-approved absorbable nerve conduits for clinical repair of peripheral and cranial nerves. Ann Plast Surg 2008;60:110–6.

[44] Prabhakaran MP, Venugopal JR, Chyan TT, Hai LB, Chan CK, Lim AY, et al. Electrospun biocomposite nanofibrous scaffolds for neural tissue engineering. Tissue Eng Part A 2008;14(11):1787–97.

[45] Steed M, Mukhatyar V, Valmikinathan C, Bellamkonda R. Advances in bioengineered conduits for peripheral nerve regeneration. Atlas Oral Maxillofac Surg Clin North Am 2011;19(1):119–30.

[46] Neal R, Tholpady S, Foley P, Swami N, Ogle R, Botchwey E. Alignment and composition of laminin-poly-caprolactone nanofiber blends enhance peripheral nerve regeneration. J Biomed Mater Res A 2011;100(2):406–23.

[47] Hersel U, Dahmen C, Kessler H. RGD modified polymers: biomaterials for stimulated cell adhesion and beyond. Biomaterials 2003;24(24):4385–415.

[48] Seo S, Min SK, Bae HK, Roh D, Kang HK, Roh S, et al. A laminin-2 derived peptide promotes early-stage peripheral nerve regeneration in a dual-component artificial nerve graft. J Tissue Eng Regen Med 2013;7(10):788–800.

[49] Radtke C, Kocsis JD, Vogt PM. Chapter 22: Transplantation of olfactory ensheathing cells for peripheral nerve regeneration. Int Rev Neurobiol 2009;87:405–15.

[50] Hoffman RM. The pluripotency of hair follicle stem cells. Cell Cycle 2006;5(3):232–3.

[51] Wang D, Liu XL, Zhu JK, Jiang L, Hu J, Zhang Y, et al. Bridging small-gap peripheral nerve defects using acellular nerve allograft implanted with autologous bone marrow stromal cells in primates. Brain Res 2008;1188:44–53.

[52] Adams AM, Arruda EM, Larkin LM. Use of adipose-derived stem cells to fabricate scaffoldless tissue-engineered neural conduits *in vitro*. Neuroscience 2012;201:349–56.

[53] Grimoldi N, Colleoni F, Tiberio F, Vetrano I, Cappellari A, Costa A, et al. Stem cell salvage of injured peripheral nerve. Cell Transplant 2013;24(2):213–22.

[54] Brohlin M, Mahay D, Novikov LN, Terenghi G, Wiberg M, Shawcross SG, et al. Characterisation of human mesenchymal stem cells following differentiation into Schwann cell-like cells. Neurosci Res 2009;64(1):41–9.

[55] Ladak A, Olson J, Tredget EE, Gordon T. Differentiation of mesenchymal stem cells to support peripheral nerve regeneration in a rat model. Exp Neurol 2011;228(2):242–52.

[56] Uemura S, Fujita T, Sakaguchi Y, Kumamoto E. Actions of a novel water-soluble benzodiazepine-receptor agonist JM-1232 (-) on synaptic transmission in adult rat spinal substantia gelatinosa neurons. Biochem Biophys Res Commun 2012;418(4):695–700.

[57] Baltich J, Hatch-Vallier L, Adams AM, Arruda EM, Larkin LM. Development of a scaffoldless three-dimensional engineered nerve using a nerve-fibroblast co-culture. *In Vitro* Cell Dev Biol Anim 2010;46:438–44.

[58] Krampera M, Pizzolo G, Aprili G, Franchini M. Mesenchymal stem cells for bone, cartilage, tendon and skeletal muscle repair. Bone 2006;39(4):678–83.

[59] Yamakawa T, Kakinoki R, Ikeguchi R, Nakayama K, Morimoto Y, Nakamura T. Nerve regeneration promoted in a tube with vascularity containing bone marrow-derived cells. Cell Transplant 2007;16(8):811–22.

[60] Vleggeert-Lankamp CLAM. The role of evaluation methods in the assessment of peripheral nerve regeneration through synthetic conduits: a systematic review. J Neurosurg 2007;107:1168–89.

[61] Owens CM, Marga F, Forgacs G, Heesch CM. Biofabrication and testing of a fully cellular nerve graft. Biofabrication 2013;5(4):045007.

[62] Glasby MA, Gschmeissner S, Hitchcock RJI, Huang CL-H. Regeneration of the sciatic nerve in rats: the effect of muscle basement membrane. J Bone Joint Surg 1986;68 B(5):829–33.

[63] Belkas JS, Munro CA, Shoichet MS, Midha R. Peripheral nerve regeneration through a synthetic hydrogel nerve tube. Restor Neurol Neurosci 2005;23:19–29.

[64] Mallik A, Weir AI. Nerve conduction studies: essentials and pitfalls in practice. J Neurol Neurosurg Psychiat 2005;76(Suppl. II):ii23–31.

[65] Freeman TL, Johnson E, Freeman ED, Brown DP. Nerve conduction studies (NCS). In: Cuccurullo S, editor. Physical medicine and rehabilitation board review. New York: Demos Medical Publishing; 2004.

[66] Vyas A, Li Z, Aspalter M, Feiner J, Hoke A, Zhou C, et al. An *in vitro* model of adult mammalian nerve repair. Exp Neurol 2010;(223):112–8.

[67] Simpson JA. Fact and fallacy in measurement of conduction velocity in motor nerves. J Neurol Neurosurg Psychiat 1964;27:381–5.

Chapter 24

Bioprinting: An Industrial Perspective

Kristina Roskos, Ingrid Stuiver, Steve Pentoney, and Sharon Presnell

Organovo, Inc., San Diego, CA, USA

Chapter Outline

ABSTRACT

Automated 3D fabrication technologies are being adopted in the life sciences, in particular, in the precise and reproducible engineering of tissues. Active research, development, and commercial efforts are ongoing in both the engineering of hardware and software components that comprise 3D printing or bioprinting platforms, and in the production and application of tissues generated by these platforms. 3D bioprinting brings an unprecedented ability to build complex, multicellular, compartmentalized tissues that can be used as *in vitro* models or implantable grafts. In an industry setting, many challenges exist, including the sourcing of raw materials, the development of methods that enable living cells to be dispensed without compromising their viability or function, and the need to scale biological processes sufficiently to enable successful commercialization. This chapter will review the progress in the field through an industrial lens, and discuss the considerations of bioprinters and bioprinted tissues in a commercial environment.

Keywords: bioprinting; automation; extracellular matrix

1 INTRODUCTION

Automated three-dimensional (3D) fabrication technologies have been around since the 1980s, predominantly serving the rapid prototyping industry. Additive manufacturing strategies – where objects are built up layer by layer from raw materials – have been widely adopted. Within the last decade, these technologies have evolved such that even complex objects can be manufactured on-demand with high resolution. In 2013, GE announced that in 2016 they will introduce the first 3D-printed parts in an aircraft engine platform; each CFM LEAP™ engine (produced jointly by GE and SAFRAN) will have 3D-printed fuel nozzles that are not only lighter than predecessor parts, but have intricate design features that could not be produced any other way. Like the emergence of personal computers in the 1970s, 3D printing is also making its way into the consumer market with a variety of relatively low-cost printers that can be used for anything from extravagant decoration of desserts to plastic replicas of your pet. At the time of this writing, MakerBot®

offers a consumer 3D printer that fabricates objects from polymers (the MakerBot Replicator®) for around $2500.

As the GE example demonstrates, the value of additive manufacturing extends far beyond the instrumentation platforms to the three-dimensional objects they produce, many of which have highly unique attributes and substantial commercial value. Based on data collected in 2012, the $2.2B 3D printing market included $361M of revenue generated from 3D-printed products used in the healthcare industry [1,2]. 3D-printed models are often used in simulations of difficult surgeries, so that when the actual surgery is conducted the surgical time is reduced and patient safety is increased. In the dental industry, 3D fabrication enables the digital workflow for inhouse preparation of tooth models, appliances, abutments, coping/crowns, and bridges of any size or combination. 3D printing is also used in Invisalign® plastic braces, generated as a custom-made series of appliances that gradually and progressively realign teeth. Approximately 50,000 of these appliances are 3D-printed each day [3]! Other healthcare applications for 3D-printed devices include customized braces for musculoskeletal support and components of artificial limbs and prosthetics.

During recent years, there has been a significant increase in the application of 3D additive manufacturing concepts and tools to the biomedical space. In addition to medical devices like external limb braces designed for support purposes, a significant number of implantable devices have been generated using 3D printing of biocompatible materials, including hip cups and cranial implants. The incorporation of living cells into 3D-printed structures has led to the advent of bioprinting, the subject matter of this book. While the Merriam–Webster medical dictionary does not yet define "bioprinting," it is helpful to consider bioprinting as the automated deposition of cells and/or cell-containing materials (often referred to as "bioink") in defined, spatially controlled patterns. "3D bioprinting" extends that definition to encompass the fabrication of a three-dimensional object, in which the spatially controlled placement of cells and/or materials creates distinct architectural features in the x, y, and z dimensions. Figure 24.1 provides insight into the emergence of the field of bioprinting within the 3D printing arena, through an accounting of peer-reviewed publications from 2000 to 2013.

The high heat, high shear deposition tools generally used for 3D printing plastic or metal objects are not compatible with maintaining the viability of living cells during the 3D bioprinting fabrication process. Thus, the development of instrumentation platforms that operate at physiologically relevant temperatures and under low shear conditions have driven the field of 3D bioprinting forward. Of equal importance is the development of the bioink – the cells and cell-comprising materials dispensed by the bioprinter – which can be formulated across a broad spectrum of compositions, including biomaterial/cell mixtures and 100% cellular formulations. Care and conditioning of the 3D constructs after fabrication is also vital because the value of bioprinting

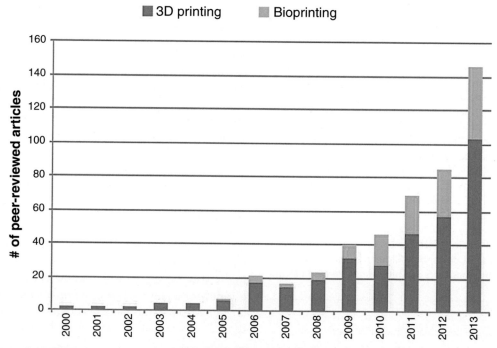

FIGURE 24.1 Peer-reviewed citations on the topics of "3D Printing" and "Bioprinting" from 2000 to 2013, sourced from PubMed database using search terms. 3D printing (and permutations); bioprinting (and permutations). Numbers reported are inclusive of both publications with primary data and review articles or publications that speculate a particular tissue could be made.

platforms will ultimately be defined by the performance of the tissues they produce.

This chapter will focus on the industrial opportunities for bioprinting technologies and applications, with a brief review of 3D printing in the healthcare industry, the transition from 3D printing to 3D bioprinting, and discussions of current and future commercial opportunities for bioprinting platforms and their unique outputs in the *in vitro* and therapeutic arenas.

2 THE ROLE OF 3D FABRICATION IN THE HEALTHCARE INDUSTRY

2.1 Transition of 3D Additive Manufacturing Concepts and Tools Into the Biological Space

2.1.1 3D Printing of Devices for External Use

Presently, physicians and dentists depend on a steady supply of medical devices such as hearing aids, musculoskeletal braces, and dental appliances that are manufactured in whole or in part by commercial 3D printers. While first-generation 3D fabrication techniques relied on subtractive processes whereby material was removed from a solid block using various methods, advanced 3D prototyping technologies utilize additive processes in which the desired part is built up, or "printed," layer by layer. Objects of virtually any shape can now be fabricated from a fairly wide range of nonbiological materials using additive technologies such as stereolithography, fused deposition modeling, selective laser sintering, laminated object manufacturing, and digital light processing. Additive manufacturing generally involves using computer-aided design software that enables an engineer to design and generate a high-resolution prototype of the desired part. As a step toward personalized medicine, these advanced 3D printing platforms enable the customization of some devices such as dental crowns [4,5] and bridges [6] so that they are a perfect fit. In many cases, replacement parts are modeled on three-dimensional maps created from scanned images of the patient's original teeth.

Three-dimensional printed prosthetics can be customized to give the patient an optimized fit, feel, and look. Many businesses currently employ 3D printing to produce custom or personalized prosthetic sockets that accurately fit the limbs of those in need. Recently, a collaborative effort between 3D Systems and California-based Ekso™ Bionics yielded a highly customized exoskeleton suit – a combination of 3D printed parts and robotics – that enabled a patient paralyzed from the waist down to walk again. Complex systems like the Ekso are being used as a component of physical rehabilitation as well, helping individuals regain strength through movement after traumatic injuries.

Additive manufacturing technologies have also led to rapid growth in the field of surgical modeling. Anatomical-ly detailed replicas of wounds, implantation sites, defects, or surgical challenges, such as the separation of conjoined twins, provide a means for surgeons to rehearse a surgery in advance [7–10]. The models serve as a functional guide for the placement of surgical tools and facilitate the development of effective strategies to access areas that are difficult to reach. For reconstructive surgeries, anatomic models can serve as templates against which implantable devices can be shaped and contoured in advance, so that on the day of surgical installation they are a perfect fit. This strategy has been used routinely in the field of orthopedics, where models of the injury and/or implantation site are used to preshape titanium plates in the reconstruction of highly contoured bones like the mandible [7,8]. As a result, surgical and anesthesia times are shortened and many risks are mitigated, usually culminating in a shorter hospital stay and improved safety for the patients.

Many businesses outsource the 3D fabrication of custom parts, while others have found it more cost effective to purchase and maintain one or more industrial 3D printers in house. Considerations for utilizing 3D printers for inhouse manufacturing of devices include cost of acquisition and maintenance of the printer, expertise and training of personnel, requirements for quality assurance of the manufactured device, and regulations regarding safe handling of incoming raw materials and any waste products generated during the manufacturing process.

2.1.2 3D Printing of Implantable Devices

There are several medical device companies that now manufacture implantable medical devices using additive manufacturing platforms. Metal implants, most often used in orthopedic repairs and reconstructions, are most frequently made from powders of titanium, titanium aluminide, and cobalt chrome using electron beam melting, a form of solid freeform fabrication that builds 3D objects by melting metal particles together with a high-powered, high-precision electron beam. Solid and porous sections of the implant can be built in the same process step, thus eliminating expensive secondary processing, and this fabrication technique produces very limited waste. As one might expect, customization of these implants can be accomplished by translating 3D images of the patient's body to the 3D printer using computer-aided design. Multiple companies manufacture orthopedic surgical implants such as hip stems and cups, femoral heads, knee replacements, shoulder and small joint components. The biocompatible materials used in fabrication have evolved to promote interaction and integration with the surrounding tissue at the site of implantation. For example, Trabecular Titanium™, a biocompatible material used in orthopedic implants, actively promotes osteoblast proliferation and differentiation [11]. Interestingly, titanium has also been shown in some studies to down-regulate inflammatory gene expression and temper the foreign body immune reaction [12].

3D-printed orthopedic products are not limited to metal compositions; a variety of medical-grade polymers have been used in the production of implantable devices, enabling product designs that consider not only shape and architecture but various biomechanical properties as well. In 2013, Connecticut-based Oxford Performance Materials received 510(k) clearance from the United States Food and Drug Administration (FDA) to begin use of the company's OsteoFab™ Patient Specific Cranial Device for skull bone replacement. The one-of-a-kind implantable devices are made from implant-grade high-performance thermoplastic poly(ether-ketone-ketone) (PEKK) polymer using selective laser sintering, guided by a CT scan or MRI to provide a perfect anatomic fit for the patient. The properties of the PEKK polymer foster ingrowth of cells into the device along the edges, providing a means for integration of the implant with the abutting host bone.

Recently, 3D printing was used to fabricate soft tissue structures from biocompatible polymers. Zopf et al. printed a sleeve made of polycaprolactone that would fit around an infant's bronchus until the tracheal tissue developed sufficiently to permit airflow to both lungs [13,14]. The investigators received emergency use approval from the FDA before implanting the device for the first time in a human. Subsequently, the team successfully tested the device in the pig model [15], which provided the data necessary to move forward with device testing in clinical trials [14].

Although still largely in the research stage, significant progress has been made in the 3D printing of scaffolds from a variety of biocompatible polymers with the goal of creating an implantable tissue, either by engineering the scaffold to recruit host cells from the site of implantation or by seeding the scaffold with living cells prior to implantation [16–18]. Additional work has been conducted that involved the 3D printing of an intricate plastic mold that creates a void space in the detailed shape of a specific target tissue [19,20] such as an ear. Hydrogels containing living cells are then dispensed or cast into the mold, yielding a three-dimensional ear-shaped gel object containing living cells that could be implanted to repair or replace the external portion of the ear. Using this approach, Reiffel et al. used the scanned image of a child's ear to create a seven-part mold that was 3D-printed from acrylonitrile butadiene styrene using a Stratasys FDM 2000 3D printer [21]. Type 1 collagen hydrogels laden with bovine auricular chondrocytes were cast into the mold to create 3D ear-shaped constructs. After 3–5 days in culture, the ears were implanted into immunocompromised rodents, where cell-comprising constructs demonstrated superiority over the acellular constructs with respect to shape retention, overall weight, and development of cartilage within 3 months of implantation [21]. Grunert and coworkers took a similar approach in the fabrication of 3D intervertebral discs, created with two cell types suspended in alginate or collagen hydrogels [22,23]. Studies performed in rats have shown that the intervertebral

discs remained viable for over 8 months *in vivo* and develop structural similarities to a native disc [22,23]. This shared vision of three-dimensional tissue engineering and the convergence of advances in biology and additive manufacturing has spawned new approaches that seek to build complex tissues with controlled, real-time incorporation of living cells during the fabrication process – *bioprinting*.

2.1.3 Bioprinting

In 2004, Thomas Boland's group at Clemson published the first in a series of seminal papers demonstrating that mammalian cells could be dispensed in specific patterns using a modified inkjet printer in the laboratory [24,25]. His team's pioneering work sparked the imaginations of tissue engineers and biologists across the globe, many of whom began to reproduce his efforts successfully and contribute to the evolution of the technology and the development of applications. Inkjet printing is based on the process of pressurizing a reservoir of print material and then using either thermal or piezoelectric forces to generate a pressure wave through the material, resulting in a small amount "jetting" out of the reservoir onto a surface. By combining liquid-state biomaterials with cells in the inkjet cartridge, thin layers of cells can be generated and patterned with fine resolution [26,27]. Initial technical challenges with inkjet deposition were around preservation of cell viability through the printing process, especially with more sensitive cell types such as primary cells or large, granular cells. More recent advancements in inkjet printing have focused on broadening the repertoire of materials that can be dispensed, and on methods that reduce the shear stress on cells, enhancing postprint viability and enabling the inkjet printing of shear-sensitive cells [28,29].

The ability to produce two-dimensional patterned cultures through the spatially controlled deposition of cells led to the desire to achieve the same kind of architectural control in larger scale, three-dimensional structures. The inkjet innovation provided a means to create geometrically defined patterns, but presented some challenges in achieving three-dimensionality in the *z* axis. The process of casting cell-containing gels into intricate molds enabled significant three-dimensionality, but had limited ability to create intricate architectural features comprising distinct cellular components. The convergence of advances in cell biology and engineering again drove the evolution of automated three-dimensional fabrication technologies that could build up tissues layer-by-layer using living human cells.

The bioprinting technologies in focus here construct 3D tissue mass with living cells incorporated at the time of manufacture. Summarized in Table 24.1, as well as throughout this book, bioprinting approaches rely on inkjet-, positive displacement/extrusion-, or laser-based methods of dispensing. Shortly after the Boland lab published their work on inkjet printing, the Forgacs lab published a series of papers describing the formation of cellular building

TABLE 24.1 Automated Cell Dispensing Technologies

Technology	Description	Examples
Inkjet	A pressurized reservoir of print material and either thermal or piezoelectric forces generate a pressure wave through the material, resulting in a small amount "jetting" out of the reservoir onto a surface	• Clemson University Lab-built custom systems: Roth et al., 2004 [24], Xu et al., 2006 [34] • Piezoelectric inkjet: Chahal et al., 2012 [29]; Parsa et al., 2010 [28] • Thermal inkjet: Xu et al., 2013 [35]
Positive displacement/extrusion bioprinting	Cellular bioink or hybrid biomaterial: cellular bioink is extruded through a size-controlled orifice for printing	• University of Missouri Lab-built custom system: Jakab et al., 2004 [30] • Lab-built custom systems: Smith et al., 2004 [36], Skardal et al., 2012 [37], Zhao et al., 2014 [38], Kolesky et al., 2014 [39] • Aerotech Inc. custom system: Xie et al., 2006 [40,41] • EnvisionTEC 3D Bio-Plotter: Fedorovich et al., 2008 [42] • Lab-modified Fab@Home system: Skardal et al. 2010 [43] • SYS + ENG "BioScaffolder": Fedorovich et al., 2011 [44], Fedorovich et al., 2011 [45,46] • Organovo, Inc. Novogen Bioprinter®: Jakab 2010 [32], Marga et al., 2012 [47] • Organovo, Inc. Novogen Bioprinter with UV Module: Bertasoni et al., 2014 [48]
Laser	Laser focuses on a transfer slide, ribbon or drum containing the print material. The focused laser releases a plug of print material on to a receiving substrate	• Biological Laser Printing (BioLP): Guillemot et al., 2010 [49] • Laser-Assisted Bioprinting (LaBP or LAB): Gruene et al., 2011 [50]; Guillotin et al., 2010 [51] • Laser-Induced Forward Transfer (LIFT): Gruene et al., 2011 [50] • Adsorbing Film Assisted-Laser Induced Forward Transfer (AFA-LIFT): Koch et al., 2013 [52] • Matrix-Assisted Pulse Laser Evaporation - Direct Write (MAPLE-DW): Koch et al., 2013 [52]

Cell-based printing approaches that have been utilized to generate patterned two-dimensional or three-dimensional objects comprising cells.

blocks they termed bioink, leveraging the natural tendencies for cells that interact with each other in close proximity to self-assemble into multicellular aggregates [30–33]. These building blocks were comprised of tens of thousands of cells each, and it was observed that when multiple cell types were included within a single building block that they would often sort out into an organized architecture relative to each other based on their physical and biological properties [32]. The automated 3D fabrication platform developed to enable the precise assembly of compartmentalized three-dimensional tissues using these purely cellular bioink building blocks became the foundation for the NovoGen® MMX bioprinting platform used by San Diego-based Organovo, Inc., to develop 3D human tissues for commercial applications in drug development and therapy. While the base technology has evolved substantially to meet the requirements for the design and manufacturing of specific tissue targets in a commercial environment, the NovoGen platform remains highly unique in its ability to construct complex three-dimensional tissues wherein all or part of the tissue is constructed from bioink that is 100% cellular.

A large number of 3D fabrication efforts rely on the deposition of polymeric biomaterials in the form of hydrogels or biomaterial filaments that contain a percentage of cells, usually in the range of 0.5–20 million cells/mL, from about 0.1% to about 25% of the total volume of material dispensed at the time of fabrication. EnvisionTEC's 3D Bioplotter™ and the SYS + ENG BioScaffolder, both of which were used by Fedorovich et al. in the fabrication of constructs designed for bone repair and replacement, dispense hydrogels, such as Pluronic F127, and alginate that contain mesenchymal stem cells at concentrations up to 10 million cells/mL [53,54]. Recent work by Skardal et al. used a custom-built 3D printer to dispense cell-laden hydrogels for skin repair, wherein the cells were suspended in a fibrin collagen gel at a concentration of 16.6 million cells/mL [37]. Lab-built systems have deployed a wide variety of hydrogel materials in fabrication, including methacrylated gelatin, methacrylated hyaluronic acid, polyethylene glycol diacrylate, collagen I, fibrinogen, and Matrigel® (see Table 24.2 for specific examples) [37–39].

TABLE 24.2 Specific Examples of 3D Fabrication Using Extrusion/Positive Displacement Platforms to Build 3D Mass with Cells or Cell-Laden Hydrogels

Extrusion Technologies	Incorporated Biomaterials/ Hydrogels	Cell Types	Cell Concentration Range
Lab-built, Forgacs et al./ University of Missouri [30,31,33]	None required	Chinese hamster ovary (CHO) cells; embryonic chick cardiomyocytes; human fibroblasts, smooth muscle cells	100% cellular
Lab-built, Stuart Williams/University of Arizona "BioAssembly Tool" (BAT) [36]	Pluronic F127	Human fibroblasts	10% cellular
	Collagen type I	Bovine aortic endothelial cells (BAECs)	$5–20 \times 10^6$ cells/mL
EnvisionTEC 3D Bio-Plotter® [42]	Alginate, Lutrol F127, PEO-PPO-PEO block copolymer, Matrigel, agarose, methylcellulose	Bone marrow stromal cells (BMSCs)	$2.5–5 \times 10^5$ cells/mL
Organovo Novogen Bioprinter [32,47]	None required	Chinese hamster ovary (CHO) cells	100% cellular
		Human aortic smooth muscle cells (HASMCs), human aortic endothelial cells (HAECs), human dermal fibroblasts (HDFb)	100% cellular
Modified Fab@Home [43]	Hyaluronic acid methacrylate (HA-MA), gelatin methacrylate (GE-MA), extracel	HepG2 C3A, NIH/3T3, Int-407	2.5×10^6 cells/mL
SYS + ENG "BioScaffolder" [44–46]	Alginate	Multipotent stromal cells (MSCs)	$5–10 \times 10^6$ cells/mL
		Chondrocytes	$3–10 \times 10^6$ cells/mL
	Matrigel	MSCs, endothelial progenitor cells (EPCs)	10^6 cells/mL
Lab-built, Aleksander Skardal/ Wake Forest Institute for Regenerative Medicine [37]	Fibrin-collagen gel	Amniotic fluid-derived stem (AFS) cells, bone marrow-derived mesenchymal stem cells (MSCs)	16.6×10^6 cells/mL
Lab-built, Wei Sun/Drexel University [38]	Gelatin, alginate, fibrinogen	HeLa	10^6 cells/mL
Lab-built, Jennifer Lewis/Harvard University [39]	Pluronic F127, gelatin methacrylate (GelMA)	Human neonatal dermal fibroblasts (HNDFs), 10T1/2	2×10^6 cells/mL
Organovo Novogen Bioprinter with UV Module [48]	Gelatin methacrylate (GelMA)	HepG2, NIH/3T3	$1–6 \times 10^6$ cells/mL

Specific examples of 3D fabrication using extrusion/positive displacement platforms to build up 3D mass with cells or cell-laden hydrogels. Multiple lab-built and commercial instrument platforms have been deployed to build 3D constructs with hydrogels, optionally containing cells at concentrations up to 16.6 million cells/mL. The NovoGen® platform enables deposition of bioink composed of 100% cells as well as a full spectrum of hydrogel: cell mixtures. Note: 100% cellular bioink is generally >100 million cells/mL, depending on cell size.

2.1.4 Mimicking Native Tissue

All tissues in the body are held together in part by interactions of cells with each other and with their surrounding environment, including the extracellular matrix (ECM) and soluble factors. Cross-sections of normal human tissues reveal the relative cellular density, tissue composition, and the spatial patterning created by alternating zones of specific cell types, ECM, and void spaces. Even in tissues like the lung, which are dominated by void spaces, the solid portions of the tissue are characterized by the presence of cells in intimate contact with each other and/or the ECM. Figure 24.2a–c shows the relative contributions of cells, tissue mass, and void space to common tissues such as liver, lung, and skin. The relative proportion of the overall tissue

mass represented by nuclear area is a reflection of the cellular density within the solid components of a given tissue. Densely cellular tissues like the liver and the epidermis are predominated by cells, whereas dense connective tissues like the dermis contain a significant amount of ECM with significantly fewer cells per millimeter square, as evidenced by the clear difference in (nuclear area):(tissue mass) ratios. Tissues predominated by void spaces, such as the lung, have substantially different (tissue mass):(void space) ratios compared to solid tissues such as the skin or liver. The relative cellularity, composition, and spatial patterning within a given tissue impact physical and functional properties of the tissue – form and function are inherently linked. Bioprinting platforms have a unique ability to build architecturally defined tissues using bioink components that approximate the cellular density of native tissues; note the similarity in cellular density, relative nuclear area, relative tissue mass area, and relative void space area between native human liver and bioprinted human liver in Fig. 24.2a–c.

Within a tissue, the intimate relationships among cells, ECM, and void spaces drive function and support homeostasis. As pathologists will appreciate, disruptions in normal tissue architecture are hallmarks of injury or disease. Histopathologic observations often provide clear insights

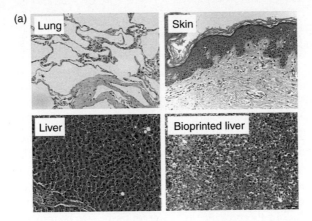

(b)

Tissue type	Relative nuclear area	Relative tissue mass area	Relative void space
Lung	0.040 ± 0.003	0.276 ± 0.024	0.683 ± 0.021
Liver	0.112 ± 0.009	0.816 ± 0.033	0.072 ± 0.031
Dermis	0.041 ± 0.009	0.630 ± 0.110	0.329 ± 0.104
Epidermis	0.220 ± 0.051	0.754 ± 0.060	0.026 ± 0.011
Bioprinted liver	0.084 ± 0.003	0.793 ± 0.035	0.122 ± 0.032

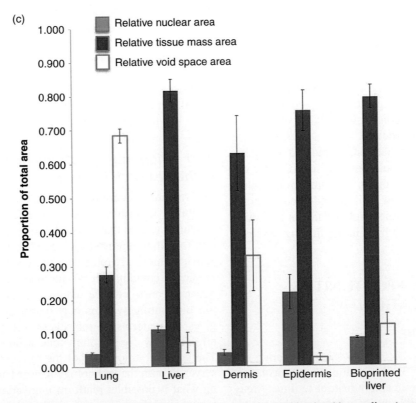

FIGURE 24.2 (a) Representative H&E sections of normal human lung, liver, and skin, and bioprinted human liver (generated with the NovoGen bioprinting platform using primary human hepatocytes, hepatic stellate cells, and endothelial cells). (b) Three sections of each tissue were processed using Image Pro Premier to calculate the nuclear area, tissue mass area, and void space area. (c) Graphical representation of relative nuclear, tissue mass, and void space areas for each tissue.

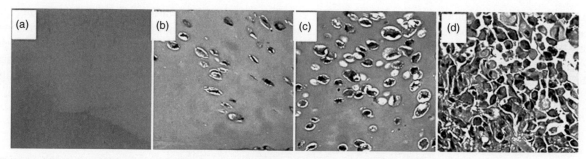

FIGURE 24.3 Bioink formulations with varying cell concentrations. (a) 100% Novogel®-3 (Organovo, Inc., San Diego, CA) with 0% cells, (b) Novogel-3 with 10% cells w/v, (c) Novogel-3 with 35% cells w/v, (d) 100% cellular bioink. Bioink was fixed and cross-sectioned immediately after deposition with the Novogen Bioprinter. Cross-sections were subjected to H&E staining and imaged (200×).

as to not only the presence of a problem, but even the root cause. There are particular applications in which the three-dimensional tissue that is created is intended to serve as an *in vitro* model of native tissue for use in disease modeling or drug testing; modeling native tissue function may depend on modeling native tissue form, including the presence of cellular compartments, intercellular interactions, and microarchitectural features such as zones of specialized ECM, void spaces, and vascular structures. Tissues intended for transplantation may have certain compartments that need to function like native tissue rapidly upon implantation, while other aspects of the tissue may be engineered to have specially designed void spaces that encourage ingrowth of cells from the host to drive tissue integration. Thus, integrated platforms that offer the greatest flexibility across the full spectrum of 3D fabrication methodologies will be key to the future. Figure 24.3 highlights the relative cellularity of bioink formulations ranging from 0% to 100% cellular material. Bioprinting provides a flexible means of producing tissues that possess specific targeted attributes of native tissue; for example, it is possible to fabricate dense, solid tissues comprised mostly of cells, or composite tissues comprised of cell:matrix blends, or tissues comprising specific void spaces that are key to native tissue function.

3D fabrication platforms that foster rapid and highly reproducible achievement of tissue-like form and function offer distinct advantages, especially in the commercial realm. The remainder of this chapter will consider commercial applications of bioprinted tissues and provide an industrial perspective on the role of 3D bioprinting in drug development and transplantation.

3 COMMERCIAL OPPORTUNITIES FOR 3D BIOFABRICATION

3.1 Instrumentation

As highlighted by Tables 24.1 and 24.2, engineering efforts aimed at advancing 3D bioprinting instrumentation are ongoing in both academic and industrial settings. Areas of focus on the engineering front include fabrication speed,

feature resolution, flexibility in dispense method, and broadening the spectrum of cells and biomaterials that a system can dispense. There is a concurrent need to continue the development of software and algorithms that enable a tissue design – whether user-defined or created by a 3D image of a target tissue – to be translated into tissue-building programs on the 3D bioprinter that take the inherent limitations of cell biology into consideration. It is unlikely that hardware and software systems originally designed for the production of intricate plastic parts will translate directly into the production of dense cellular tissues that comprise multiple cell types, all of which need to remain viable during the printing process. Instrumentation platforms engineered to dispense only polymeric materials can draw from a much broader repertoire of hardware options and fabrication strategies, including those that rely on high heat, high shear, or exposure to solvents or reagents that would destroy living cells if they were present. 3D constructs composed predominantly or entirely of polymeric materials that are generated from 3D printing platforms may be successful as implantable products, but it is unlikely that they would serve as tissue models for *in vitro* tissue modeling, as they simply lack too many key features of native tissue.

3.1.1 Bioprinting Instrumentation in the Research Environment

The research environment requires the greatest degree of flexibility in instrumentation features and specifications. Modular systems that can undergo frequent and rapid iteration to address the needs of specific applications are ideal, and it is more acceptable in the research environment to employ a combination of automated and manual steps to generate constructs. The value of the flexibility in research-use bioprinting systems may outweigh what the end-user gives up in terms of robustness, repeatability, and overall suitability for a commercial production environment. The creative atmosphere of the research environment also serves to push the technology forward and provide insights on what is possible; platform innovations that prove to be robust in performance and yield tissues with demonstrated functionality will likely transition over time to the commercial production environment. Because the practical applications of bioprinted 3D constructs nearly always involve *in vitro* culture and/or *in vivo* testing, even research-grade bioprinting instruments need to operate with basic sterility as well as protection for the operator with respect to the inherent biological, chemical, and physical safety hazards associated with the cells, biomaterials, and hardware. Cell-contacting components should be sterilizable or disposable, and the instrument itself should be able to operate in a clean environment such as a biosafety cabinet or sterile enclosure to minimize risks of contaminating the 3D constructs during fabrication. Due to the fine resolution and precise movements that are required for 3D bioprinting, engineering designs should consider the potential for vibration, heat generation, and shedding of particles or lubricants as any of these phenomena could adversely affect the outcome. Only when the basic requirements of sterility and biocompatibility are met, should practical use of tissue constructs that are produced be considered.

3.1.2 Bioprinting Instrumentation in the Production Environment

As applications for bioprinted tissues advance into commercial use, one might consider the bioprinting platform primarily as the means to manufacture a specific product or set of products. It is anticipated that some aspects of the instrumentation would be modified or customized to enhance production speed, minimize errors, reduce the operator interface requirements, and maximize reproducibility, while still operating within the required parameters of safety and sterility for the specific product(s). For example, a bioprinter designed to generate tubular structures that are 5 cm in length would likely operate under a different set of specifications than a bioprinter designed to efficiently produce arrays of microscale tissues in multiwell plate formats. For any given application, engineers must consider the requirements for throughput, physical constraints introduced by the receptacle into which the constructs will be fabricated, and the limitations imposed by the characteristics of the specific cell types and/or biomaterials to be printed. Depending on the intended use of the bioprinted constructs, the equipment used in their manufacture may fall under certain codes of federal regulation (CFRs). For example, for nonclinical laboratory studies in which the results are intended to support applications for research or marketing permits for FDA-regulated products, such as food additives, pharmaceuticals, medical devices, or biological products, compliance with CFR title 21, part 58 is required. Detailed requirements for equipment that is utilized within a regulated laboratory environment can be found through the FDA database at http://www.accessdata.fda.gov/scripts/cdrh/cfdocs/cfcfr/cfrsearch.cfm.

3.1.3 Bioprinting Instrumentation in the Clinical Setting

As bioprinted tissue products advance from the research setting into clinical testing, the production protocols and instrumentation will transition into the highly-regulated current good manufacturing practices (CGMP) arena, where validation, verification, traceability, and tight manufacturing controls must be in place at each step [55]. Equipment utilized in the production of clinical products must maintain product sterility and eliminate open procedures and manual steps wherever possible. In addition to the documentation required to commission and maintain the equipment under CGMP, there may be additional requirements to capture in real-time the 3D build program that was executed to verify that program against the documented standard operating procedure.

3.2 Bioprinted Tissues

3.2.1 As 3D Models of Mammalian Biology

Classical *in vitro* cell biology research and commonly deployed cell-based assays typically involve the use of cell lines or primary cells cultured on tissue culture treated plastic vessels in monolayer formats. Cells cultured by these traditional means are frequently sacrificed to isolate nucleic acids or proteins that are subsequently used to characterize the behavior of the cells in response to some type of treatment, and the culture media surrounding the cells is often used to detect and measure secreted proteins, metabolic activities, or the generation of specific metabolites upon exposure to a drug. Cell-based assays have been a mainstay of the drug development industry, with a growing presence in the early stages of the discovery process – target identification and mechanism of action [56]. Since the seminal work by Mina Bissell demonstrated the enhanced predictive capability of 3D tumor spheroids compared to traditional two-dimensional (2D) cultures of tumor cells [38,57–59], significant efforts have been invested in the development of "organotypic" three-dimensional cell systems across a broad spectrum of cell types and applications. Strategies to achieve three-dimensionality include multicellular spheroids and seeding of cells onto porous biomaterial scaffolds or within hydrogels (reviewed in Refs [60,61]). While both of these strategies generate three-dimensional mass, neither enables architectural control such that multiple cell types can be reproducibly patterned relative to each other within the 3D space. As highlighted in Fig. 24.3, cells seeded onto scaffolds or embedded within porous hydrogels may not develop a native tissue-like cellular density or support the intimate intercellular interactions that are required to observe certain functional or histologic attributes. Bioprinting enables the automated fabrication of microscale tissues that are tens to hundreds of cell layers thick in the *x*, *y*, and *z* axes; multiple

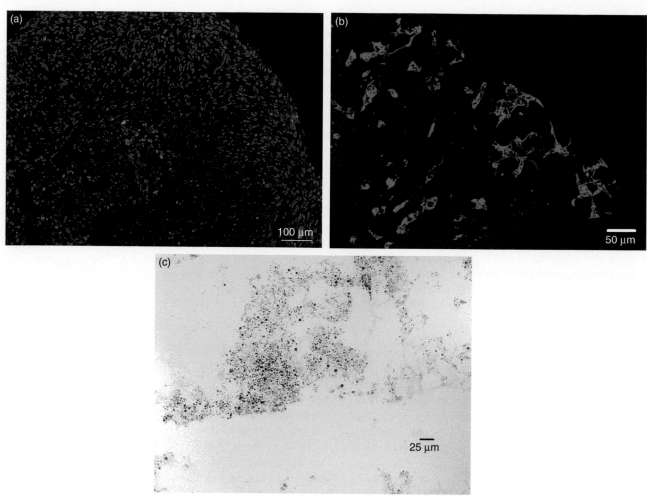

FIGURE 24.4 **Breast tissue was bioprinted using the Novogen bioprinting platform.** A complex stroma of adipose tissue and fibroblasts fully surrounds an epithelial compartment comprising human breast cancer cells, yielding a 3 mm × 3 mm × 0.75 mm solid tissue. Tissues were formalin-fixed, paraffin-embedded, and sectioned. (a) Ductal epithelial cells (E-Cadherin+; green) are evident in the core of the tissue in this cross-section. Fibroblasts within the stromal mantle are stained with TE7 (red). (b) Higher magnification images highlight organization of endothelial cells (CD31+; green) within the stroma. (c) Adipose tissue, bioprinted as a component of the stromal compartment, displays evidence of adipogenesis with the uptake of Oil Red O (red color).

bioink compositions comprising distinct tissue-specific cell types can be deposited by the bioprinter in defined patterns in all three spatial axes, so that the resulting tissue has a defined architecture with native tissue-like compartmentalization. Thus, one can contemplate tissue designs that incorporate key architectural elements of target tissues and, depending on the bioink formulation, yield tissues with dense tissue-like cellularity within a short time of fabrication. Flexible bioprinter platforms enable these microscale tissues to be manufactured directly into common arrayed culture formats for ease in handling and use in *in vitro* assays. The automation platform delivers high reproducibility from tissue to tissue during the fabrication process, which translates into biological reproducibility in the resulting tissue across a spectrum of biochemical, molecular, and histologic assays. Bioprinting is being deployed in the fabrication of a wide range of tissues in the research setting, including bone, heart, blood vessels,

skin, liver, and tumor. Figure 24.4 shows examples of bioprinted breast tumor tissue comprising a complex stroma of adipose tissue and fibroblasts fully surrounding an epithelial compartment, with endothelial cells forming microvascular structures throughout the 3 mm × 3 mm × 0.75 mm tissue [62]. Note the dense tissue-like cellularity, the compartmentalized organization, and the microarchitectural features such as intercellular junctions and CD31+ endothelial structures. Like many tumors *in vivo*, breast cancer cells embedded within the bioprinted tissue were resistant to the chemotherapeutic agent taxol, and accumulation of fluorescently labeled taxol was observed within the adipose compartment of the stroma, while tagged methotrexate penetrated the tissue fully [62]. Complex, compartmentalized tumor models may enable studies of drug distribution and effects at the cellular level within the context of a living human tissue environment.

3.2.2 As Tools for Predictive Toxicology and Pharmacology

FDA guidelines recommend that new chemical entities, under investigation as pharmaceuticals, should undergo a battery of preclinical tests to identify potential safety concerns in the clinical setting. In addition to testing the drug in animal models, which can sometimes fail to predict human outcomes due to species differences, *in vitro* tests are recommended in which cultured human cells are exposed to the drug to measure acute cytotoxic effects that may be a sign of trouble for critical organs like the heart, liver, or kidneys. The majority of *in vitro* tests performed in the drug development arena utilize two-dimensional monolayer cultures containing a single cell type, which allows detection of toxic or adverse effects on that particular cell type after a relatively short (24–72 h) exposure [63]. However, the short lifespan of these cultures limits exploration of drug responses with chronic or repeated dosing regimens. Furthermore, as the mechanisms of drug action, distribution, and toxicity are rarely monocellular, extrapolation of results from simplified *in vitro* systems to complex *in vivo* systems can be very poor. Viability and function of cells *in vitro* are often extended when the cells are grown in three-dimensional formats or in coculture systems with other tissue-relevant cell types present [64]. Thus, 3D tissue models that contain multiple human tissue-specific cell types and replicate the cellular density and architecture of native tissue significantly expand the repertoire of safety and efficacy assessments that can be conducted in the *in vitro* preclinical market. The pharmaceutical industry is anxious for the availability of reliable and standardized means to predict the human clinical response. For many complex tissue types, bioprinted human tissues may be a long sought-after answer.

Drug-induced liver injury has been the single most frequent cause of safety-related drug marketing withdrawals for more than 50 years. Many drugs have been removed from the market after approval due to unpredicted liver toxicity, including isoniazid, labetalol, trovafloxacin, tolcapone, and felbamate [65]. Other marketed drugs, like acetaminophen, have liver injury risks that are dose-dependent and can also be exacerbated by other factors such as alcohol exposure. The pattern of liver injury can vary, encompassing everything from direct injury of the hepatocytes, to cholestatic injury resulting in the accumulation of bile, to fibrosis, or steatosis. Progressive drug-induced liver injury that occurs over time and involves the interplay between multiple cell types is unlikely to be detectable in short-term, monocellular cultures. Diclofenac is a nonsteroidal anti-inflammatory drug that causes rare but serious liver toxicity in humans with a delayed onset that likely involves cumulative oxidative stress and mitochondrial damage [66]. Figure 24.5 shows bioprinted human liver tissue fabricated from spatially patterned primary hepatocytes, endothelial cells, and hepatic stellate cells, which remain viable and retain key

liver-associated functions for at least 40 days *in vitro* [67]. Treatment of these 3D human liver tissues on day 14 of culture with diclofenac, which is metabolized by cytochrome P450 2C9 (CYP450 2C9) and CYP450 3A4, shows an expected dose-dependent induction of CYP450 3A4 at lower doses and a significant increase in lactate dehydrogenase (LDH) at higher doses, signaling the accumulation of hepatocellular damage [67].

3.2.3 As Customized or Patient-Specific Disease Models

The recent advances in induced pluripotent stem cell (iPSC) technology enable the construction of complex 3D tissues with one or more cell types derived from specific patients or patient populations. This represents a significant step toward targeted drug discovery and personalized medicine. In addition to the benefits of assessing patient- or population-specific tissue with respect to drug safety and efficacy, there are potential operational benefits associated with stem cell-sourced human cells. Many highly specialized primary human cell types that are of great importance in drug discovery and development are difficult to isolate, do not expand in culture or lose key functional features after a few passages. Additionally, many of these cells are derived from human organs that were deemed unsuitable for transplantation due to unfavorable donor characteristics or poor tissue quality. Stem cell-derived sources of cells such as hepatocytes, cardiomyocytes, and renal tubular cells, could yield a more reliable and reproducible cell supply for commercial use. There are cost considerations, as the processes for iPSC generation, expansion, and subsequent differentiation are still being optimized to minimize the fully loaded cost of goods. The feasibility of 3D bioprinting human tissue comprising iPSC-derived cells has been established; Fig. 24.6 highlights histologic and functional attributes of 3D bioprinted human liver, in which the hepatocytes were derived from iPSC (Cellular Dynamics, Intl.).

3.2.4 As Therapeutic Grafts for Tissue Repair or Replacement

To date, investigations involving bioprinted tissues as implantable therapies are in the early stages of research and development. The majority of publications involving bioprinted tissues are centered on demonstrating proof of concept with a specific fabrication platform or a particular cell type. Skardal et al. used an automated instrument platform to dispense cell-laden fibrin/collagen gels into skin wounds in immunocompromised mice [37], and Fedorovich et al. have fabricated cell-laden biomaterial scaffolds using commercially available 3D fabrication instruments from EnvisionTEC and SYS + ENG, and subsequently implanted them into immunocompromised mice to study osteoid tissue formation [68]. Implanted 3D tissue constructs, regardless

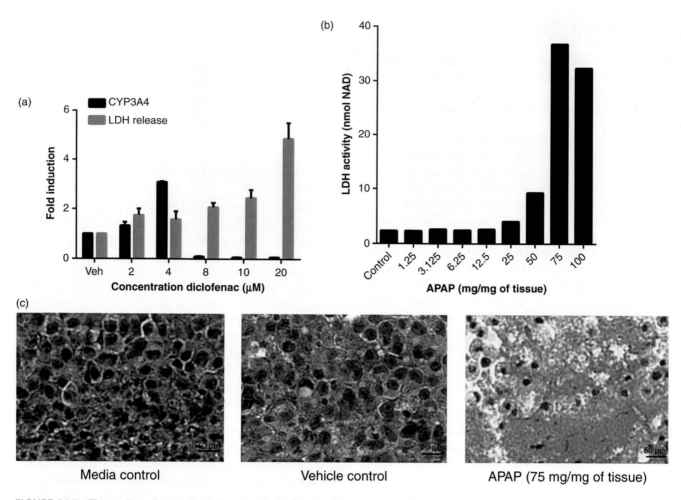

FIGURE 24.5 Human liver tissue was fabricated with the Novogen bioprinter. (a) After 14 days of fabrication, 3D bioprinted liver tissues were exposed to increasing concentrations of diclofenac for 24 h. CYP3A4 activity was induced at low concentrations of diclofenac, but lost at 2.5 mg/mL or greater. Higher concentrations of diclofenac were toxic to the tissue, as evidenced by increasing LDH activity. (b) Bioprinted liver tissues were exposed to increasing doses of acetaminophen, and toxicity response detected with LDH. (c) Bioprinted liver tissue in cross-section, maintained for 7 days in maintenance media. (d) Histologic sections of bioprinted liver tissue 24 h posttreatment with either vehicle or 75 mg acetaminophen (APAP) per mg tissue. Note the tissue damage and necrosis in the APAP-treated tissue and the correlation with LDH release signifying damage in panel B.

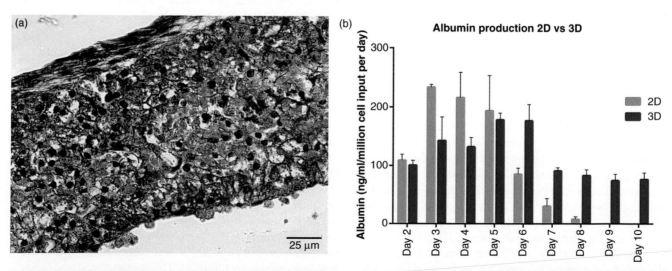

FIGURE 24.6 Human liver tissue was bioprinted on the Novogen bioprinter using iPSC-derived human hepatocytes, hepatic stellate cells, and endothelial cells. (a) Periodic acid schiff (PAS) stain of a cross-section of bioprinted liver tissue, showing dense cellularity and the presence of glycogen storage granules within the cytoplasm of the hepatocytes. (b) 3D Bioprinted liver tissues and 2D cocultures were compared daily with respect to albumin secretion (ELISA), showing loss of albumin production in 2D cultures after Day 7 and retention of albumin production in the 3D tissues at Day 10.

of how they are fabricated, undergo extensive remodeling as they interact and integrate with host tissue at the site of implantation. While 3D tissues intended for use *in vitro* need all key functional cell types present in order to yield a tissue-like response, implanted tissue constructs can recruit key cell types from the host, thus eliminating the need to isolate, culture, and incorporate those cells during the fabrication process. For example, preclinical models of engineered blood vessels, intestine, urologic tissue, and trachea have shown that endothelial or epithelial layers can form along the luminal surface of the implanted construct under proper conditions, even if they were not seeded onto the scaffold before implantation [69–72]. For example, Marga et al. implanted a fully cellular bioprinted tissue conduit comprising bone marrow stromal cells and Schwann cells into a rodent model of peripheral nerve gap, and found that host nerve cells grew through parallel channels that had been engineered into the graft at the time of fabrication [47].

The maturation state at the time of implantation is a key consideration for bioprinted tissues, with each tissue target likely possessing a unique set of requirements. For example, bioprinted tissues intended to repair or replace structural tissues, such as bone, tendon, ligament, or cartilage, will likely require a threshold biomechanical strength at the time of implantation. Bioprinted conduits will likely need to withstand the mechanical forces associated with surgical installation, and be of sufficient strength and elasticity to respond to the flow and pressure conditions at the site of implantation. Thus, postfabrication conditioning regimens are of equal importance to tissue design and fabrication strategy. Figure 24.2 highlights key attributes of a tissue – intercellular relationships and tissue organization – that drive tissue strength as well as many functional properties. The physical properties of any given tissue are determined in part by cell–cell adhesion, cell–matrix adhesion, and the composition and form of ECM within the tissue. With proper conditioning, bioprinted tissues can be stimulated to deposit matrix within the tissue over time, creating stromal compartments that mimic the native tissue in architecture and composition [47].

Additive manufacturing systems provide flexible platforms upon which purely cellular bioink, biomaterial:cellular bioink, or biomaterials alone can be leveraged to build 3D tissues that are structurally and compositionally complex. One can contemplate fully cellular tissues [47], biomaterial constructs containing cells (Table 24.2), purely biomaterial devices [73–75], or hybrids comprised of two or more of these elements. Simple tissues, presented as layered sheets or patches, or as tubular structures, will likely be the first to advance from the research arena into clinical trials. Bioprinted tissues have great potential for patient-specific customization, whether through control of shape and size to ensure a perfect fit, or incorporation of patient-tailored components (e.g. cells) to aid engraftment and postimplant

function. Although there are lay press articles suggesting that work is underway in multiple laboratories aimed at generating complex organs, such as kidney and lung, scientific publications are not yet in the public domain.

A major concern in the engineering of implantable tissues, regardless of the mode of fabrication, is the ability to scale up. Large quantities of cells will be required to generate tissues of sufficient mass to be compelling for surgical implantation, thus developing cost-effective means of generating high cell numbers while not compromising on quality is critical. Larger tissues create significant challenges in nutrient and oxygen delivery, and systems are needed to ensure cellular components of tissues survive and develop or retain key functions prior to implantation. Surgical installation of sizable tissue may require provision of a blood supply, either by wrapping the implanted construct in a well-vascularized tissue such as the omentum, or by engineering or incorporating a "vascular tree" that can be anastomosed to the host. Significant work is ongoing in both academia and industry to overcome the challenge of providing physiologically relevant vascularization of tissue-engineered constructs.

3.2.5 Key Considerations for Commercial Implementation

As any technology transitions from the research environment through development and into commercial use, success hinges on ensuring reproducibility and performance, and clearly defining and communicating how the product is used and for what purposes. The majority of efforts to date in 3D bioprinting have focused on demonstrating that a particular 3D construct can be fabricated using some form of additive manufacturing platform. As the field matures and tissue products begin to advance commercially, product development efforts will largely be focused on improving performance and reproducibility and reducing cost of goods.

3.2.5.1 Raw Materials Sourcing

For any bioprinted tissue, the cells, biomaterials, culture media, and receptacles into which the tissues are fabricated and used need to be sourced to ensure a consistent supply within a set of defined specifications. Biological components, such as living cells, are inherently variable, thus the consistency and performance of products that contain them will be best served by tightly controlling donor characteristics, culture conditions, harvesting procedures, and cryopreservation protocols. Specific lots of cells, media, growth factors, and reagents used in cell harvesting should be qualified by confirming sterility and equivalence with proven lots across a spectrum of phenotypic and functional performance characteristics of the product. Any materials that are used during the production of bioink or in the 3D fabrication process, especially those that come in contact with the finished product, should be held consistent and undergo lot testing and qualification as well. In many cases, failure of a single raw material can cause failure of an entire lot of product; for example, if the sterility of a single cell type in a 3D-bioprinted tissue was compromised, it is likely that the entire batch of composite tissues would also be contaminated. Thus, it is important to ensure quality controls are in place for all raw materials involved in 3D tissue fabrication. Polymeric or metal materials used in the fabrication of receptacles, bioreactors, or other hardware involved in the housing or conditioning of the tissues must be biocompatible, and must not leach cytotoxic chemicals into the aqueous culture media that comes in contact with the tissues.

3.2.5.2 Postfabrication Maturation and Maintenance

The creation of 3D tissues depends heavily on well-controlled bioink formulation and subsequent bioprinting process optimization. However, the ultimate value of a bioprinted tissue product is dependent on its ability to survive over time and retain the key functional and architectural features that made it valuable in the first place. The environment in which the tissue is maintained, during the first 24 h after fabrication and then through various stages of maturation and culture, plays a major role in whether or not the cells within the construct survive and function. Whereas simple monocellular systems only require optimization of media components for a single cell type, multicellular systems comprised of two or more cell types may have significantly different requirements with respect to media additives and growth factors. Furthermore, many media formulations that are optimized for cells growing in two-dimensional monolayers may prove inadequate for cells in three-dimensional structures, where factors like cell shape, receptor distribution, and the presence of secreted proteins from other cell types alter the requirements for any particular cell. Thick, three-dimensional tissues also present a challenge for diffusion of large proteins from the media into the deeper layers of the tissue; thus, there are size limitations that apply unless gaps or channels are engineered into the tissues to support media exchange over greater distances. High cell density and the presence of multiple cell types within some bioprinted tissues create the need for optimization of media formulations and feeding regimens to ensure survival and function of the tissues over time. Many tissues benefit from a dynamic environment where media is stirred or circulated and/or biomechanical forces are applied [76–78]. In bioprinted conduits comprised of smooth muscle cells and fibroblasts, controlled peristaltic flow through the lumen fosters ECM deposition and, consequently, enhances the burst strength of the tubular structures from <10 mmHg to >700 mmHg within 3 weeks of fabrication [47]. Microfluidic devices aimed at supporting living tissues *in vitro* constitute a major effort in the bioengineering space, and a review of current approaches and systems can be found in Ref. [79].

3.2.5.3 Extracting Data From 3D Tissue Systems

Bioprinted tissues are three-dimensional, generally with multiple cell types present. The majority of cell-based assay platforms are aimed at measuring molecules or proteins produced by living cells in response to some form of treatment. Nondestructive tests, such as ELISA assays that measure secreted proteins or spent media analyses that assess metabolism, are attractive because they enable dynamic data to be captured from individual tissue constructs over time. However, when 3D tissues are involved, the majority of standard cell-based assay protocols require further optimization and validation to allow for proper interpretation of results; the influx/efflux of molecules between the tissue and the media, and the distribution of molecule(s) within the multicellular tissue, are key to interpreting treatment-based outcomes. Destructive tests, such as qRT-PCR and western blot/protein analyses, can be conducted on bioprinted tissues provided the sample collection protocols are optimized and validated to ensure protein or gene targets are efficiently extracted from the tissues. In general, extraction protocols for proteins and nucleic acid that have been developed specifically for tissues (biopsy samples, for example) provide superior results to extraction protocols that have been developed for cell monolayers.

While mainstay well-based assays can be used to collect industry-standard molecular and biochemical data, 3D bioprinted tissues present a unique opportunity to study treatment outcomes within the context of native human tissue – spatially and compositionally. As demonstrated in the examples shown in Figs 24.2–24.6 of this chapter and throughout the book, many bioprinted tissues have sufficient biochemical strength and cell–cell interactions to be processed histologically and subjected to a variety of immunological or histological stains to assess cell-specific and tissue compartment-specific treatment outcomes. As nondestructive imaging technologies and cell-tracing capabilities evolve, it may be possible to observe specific effects at the cellular level within complex, multicellular 3D structures. For example, it is feasible to use custom-designed algorithms to isolate specific tissue compartments from 3D image stacks so that quantitative analyses can be conducted in a compartment-specific manner. Rose and coworkers have deployed this strategy successfully to isolate perivascular regions in the cortex of live animals, and track real-time biodistribution of injected fluorophores with subcellular resolution [80]. Together, complex heterogeneous 3D tissue models that mimic native tissue biology and next-generation tools for imaging and analysis could revolutionize the drug discovery and development industry.

4 SUMMARY

In summary, the application of additive manufacturing technologies to the field of tissue engineering provides an unprecedented ability to build complex three-dimensional tissues with precise architectural control. Although many clinical and commercial challenges remain, the field of tissue engineering is poised to evolve and provide enabling tools for the drug discovery industry and next-generation therapeutics for use in regenerative medicine.

GLOSSARY

Biodistribution A method of tracking where compounds of interest travel in an experimental animal or human subject.

Bioink Multicellular aggregates or cell-comprising hydrogels used in the fabrication of tissue via bioprinting.

Construct (noun) In the context of this book, a fabricated tissue comprised of cells or cells + biomaterial components.

Monocellular Composed of a single cell type.

Organotypic culture A complex tissue maintained outside of the body, comprised of multiple cell types, representing an organ in composition, architecture, and/or function.

ABBREVIATIONS

CFR	Code of federal regulation
CGMP	Current good manufacturing practices
ECM	Extracellular matrix
FDA	Food and Drug Administration
FDM	Fused deposition modeling
iPSC	induced pluripotent stem cells
PEKK	Poly(ether-ketone-ketone)

REFERENCES

[1] Associates W. The internet of things and CES; 2014. Available from: http://www.qtma.ca/wp-content/uploads/2014/01/The-Internet-of-Things-and-CES.pdf.

[2] Yandell K. Organs on demand. The scientist; 2013.

[3] Tenenbaum DJ. 3-D printing: wave of the future. The WHY files; 2013.

[4] Ebert J, Ozkol E, Zeichner A, Uibel K, Weiss O, Koops U, et al. Direct inkjet printing of dental prostheses made of zirconia. J Dent Res 2009;88(7):673–6.

[5] Venkatesh KV, Nandini VV. Direct metal laser sintering: a digitised metal casting technology. J Indian Prosthodont Soc 2013;13(4): 389–92.

[6] van Noort R. The future of dental devices is digital. Dent Mater 2012;28(1):3–12.

[7] Castilho M, Dias M, Gbureck U, Groll J, Fernandes P, Pires I, et al. Fabrication of computationally designed scaffolds by low temperature 3D printing. Biofabrication 2013;5(3):035012.

[8] Lambrecht JT, Berndt DC, Schumacher R, Zehnder M. Generation of three-dimensional prototype models based on cone beam computed tomography. Int J Comput Assist Radiol Surg 2009;4(2):175–80.

[9] Mick PT, Arnoldner C, Mainprize JG, Symons SP, Chen JM. Face validity study of an artificial temporal bone for simulation surgery. Otol Neurotol 2013;34(7):1305–10.

[10] Rengier F, Mehndiratta A, von Tengg-Kobligk H, Zechmann CM, Unterhinninghofen R, Kauczor HU, et al. 3D printing based on imaging data: review of medical applications. Int J Comput Assist Radiol Surg 2010;5(4):335–41.

[11] Benazzo F, Botta L, Scaffino MF, Caliogna L, Marullo M, Fusi S, et al. Trabecular titanium can induce *in vitro* osteogenic differentiation of human adipose derived stem cells without osteogenic factors. J Biomed Mater Res A 2014;102I 7:2061–71.

[12] Sollazzo V, Pezzetti F, Massari L, Palmieri A, Brunelli G, Zollino I, et al. Evaluation of gene expression in MG63 human osteoblastlike cells exposed to tantalum powder by microarray technology. Int J Periodontics Restorative Dent 2011;31(4):e17–28.

[13] Fessenden M. 3-D printed windpipe gives infant breath of life. Scientific American; 2013.

[14] Zopf DA, Hollister SJ, Nelson ME, Ohye RG, Green GE. Bioresorbable airway splint created with a three-dimensional printer. N Engl J Med 2013;368(21):2043–5.

[15] Zopf DA, Flanagan CL, Wheeler M, Hollister SJ, Green GE. Treatment of severe porcine tracheomalacia with a 3-dimensionally printed, bioresorbable, external airway splint. JAMA Otolaryngol Head Neck Surg 2014;140(1):66–71.

[16] Poldervaart MT, Gremmels H, van Deventer K, Fledderus JO, Oner FC, Verhaar MC, et al. Prolonged presence of VEGF promotes vascularization in 3D bioprinted scaffolds with defined architecture. J Control Release 2014;184:58–66.

[17] Domingos M, Intranuovo F, Russo T, De Santis R, Gloria A, Ambrosio L, et al. The first systematic analysis of 3D rapid prototyped poly(epsilon-caprolactone) scaffolds manufactured through BioCell printing: the effect of pore size and geometry on compressive mechanical behaviour and *in vitro* hMSC viability. Biofabrication 2013;5(4):045004.

[18] Inzana JA, Olvera D, Fuller SM, Kelly JP, Graeve OA, Schwarz EM, et al. 3D printing of composite calcium phosphate and collagen scaffolds for bone regeneration. Biomaterials 2014;35(13):4026–34.

[19] Park JH, Jung JW, Kang HW, Cho DW. Indirect three-dimensional printing of synthetic polymer scaffold based on thermal molding process. Biofabrication 2014;6(2):025003.

[20] Lee JY, Choi B, Wu B, Lee M. Customized biomimetic scaffolds created by indirect three-dimensional printing for tissue engineering. Biofabrication 2013;5(4):045003.

[21] Reiffel AJ, Kafka C, Hernandez KA, Popa S, Perez JL, Zhou S, et al. High-fidelity tissue engineering of patient-specific auricles for reconstruction of pediatric microtia and other auricular deformities. PLoS One 2013;8(2):e56506.

[22] Grunert P, Gebhard HH, Bowles RD, James AR, Potter HG, Macielak M, et al. Tissue-engineered intervertebral discs: MRI results and histology in the rodent spine. J Neurosurg Spine 2014;20(4):443–51.

[23] Hudson KD, Alimi M, Grunert P, Hartl R, Bonassar LJ. Recent advances in biological therapies for disc degeneration: tissue engineering of the annulus fibrosus, nucleus pulposus and whole intervertebral discs. Curr Opin Biotechnol 2013;24(5):872–9.

[24] Roth EA, Xu T, Das M, Gregory C, Hickman JJ, Boland T. Inkjet printing for high-throughput cell patterning. Biomaterials 2004;25(17):3707–15.

[25] Xu T, Jin J, Gregory C, Hickman JJ, Boland T. Inkjet printing of viable mammalian cells. Biomaterials 2005;26(1):93–9.

[26] Cui X, Boland T. Human microvasculature fabrication using thermal inkjet printing technology. Biomaterials 2009;30(31):6221–7.

[27] Xu T, Baicu C, Aho M, Zile M, Boland T. Fabrication and characterization of bio-engineered cardiac pseudo tissues. Biofabrication 2009;1(3):035001.

[28] Parsa S, Gupta M, Loizeau F, Cheung KC. Effects of surfactant and gentle agitation on inkjet dispensing of living cells. Biofabrication 2010;2(2):025003.

[29] Chahal D, Ahmadi A, Cheung KC. Improving piezoelectric cell printing accuracy and reliability through neutral buoyancy of suspensions. Biotechnol Bioeng 2012;109(11):2932–40.

[30] Jakab K, Neagu A, Mironov V, Markwald RR, Forgacs G. Engineering biological structures of prescribed shape using self-assembling multicellular systems. Proc Natl Acad Sci USA 2004;101(9): 2864–9.

[31] Jakab K, Norotte C, Damon B, Marga F, Neagu A, Besch-Williford CL, et al. Tissue engineering by self-assembly of cells printed into topologically defined structures. Tissue Eng Part A 2008;14(3):413–21.

[32] Jakab K, Norotte C, Marga F, Murphy K, Vunjak-Novakovic G, Forgacs G. Tissue engineering by self-assembly and bio-printing of living cells. Biofabrication 2010;2(2):022001.

[33] Norotte C, Marga FS, Niklason LE, Forgacs G. Scaffold-free vascular tissue engineering using bioprinting. Biomaterials 2009;30(30):5910–7.

[34] Xu T, Gregory CA, Molnar P, Cui X, Jalota S, Bhaduri SB, et al. Viability and electrophysiology of neural cell structures generated by the inkjet printing method. Biomaterials 2006;27(19):3580–8.

[35] Xu T, Zhao W, Zhu JM, Albanna MZ, Yoo JJ, Atala A. Complex heterogeneous tissue constructs containing multiple cell types prepared by inkjet printing technology. Biomaterials 2013;34(1):130–9.

[36] Smith CM, Stone AL, Parkhill RL, Stewart RL, Simpkins MW, Kachurin AM, et al. Three-dimensional bioassembly tool for generating viable tissue-engineered constructs. Tissue Eng 2004;10(9–10):1566–76.

[37] Skardal A, Mack D, Kapetanovic E, Atala A, Jackson JD, Yoo J, et al. Bioprinted amniotic fluid-derived stem cells accelerate healing of large skin wounds. Stem Cells Transl Med 2012;1(11):792–802.

[38] Zhao Y, Yao R, Ouyang L, Ding H, Zhang T, Zhang K, et al. Three-dimensional printing of Hela cells for cervical tumor model *in vitro*. Biofabrication 2014;6(3):035001.

[39] Kolesky DB, Truby RL, Gladman AS, Busbee TA, Homan KA, Lewis JA. 3D Bioprinting of vascularized, heterogeneous cell-laden tissue constructs. Adv Mater 2014;26(19):3124–30.

[40] Xie J, Marijnissen JC, Wang CH. Microparticles developed by electrohydrodynamic atomization for the local delivery of anticancer drug to treat C6 glioma *in vitro*. Biomaterials 2006;27(17):3321–32.

[41] Xie J, Lim LK, Phua Y, Hua J, Wang CH. Electrohydrodynamic atomization for biodegradable polymeric particle production. J Colloid Interface Sci 2006;302(1):103–12.

[42] Fedorovich NE, De Wijn JR, Verbout AJ, Alblas J, Dhert WJ. Three-dimensional fiber deposition of cell-laden, viable, patterned constructs for bone tissue printing. Tissue Eng Part A 2008;14(1): 127–33.

[43] Skardal A, Sarker SF, Crabbe A, Nickerson CA, Prestwich GD. The generation of 3-D tissue models based on hyaluronan hydrogel-coated microcarriers within a rotating wall vessel bioreactor. Biomaterials 2010;31(32):8426–35.

[44] Fedorovich NE, Schuurman W, Wijnberg HM, Prins HJ, van Weeren PR, Malda J, et al. Biofabrication of osteochondral tissue equivalents by printing topologically defined, cell-laden hydrogel scaffolds. Tissue Eng Part C Methods 2012;18(1):33–44.

[45] Fedorovich NE, Kuipers E, Gawlitta D, Dhert WJ, Alblas J. Scaffold porosity and oxygenation of printed hydrogel constructs affect functionality of embedded osteogenic progenitors. Tissue Eng Part A 2011;17(19–20):2473–86.

[46] Fedorovich NE, Wijnberg HM, Dhert WJ, Alblas J. Distinct tissue formation by heterogeneous printing of osteo- and endothelial progenitor cells. Tissue Eng Part A 2011;17(15–16):2113–21.

[47] Marga F, Jakab K, Khatiwala C, Shepherd B, Dorfman S, Hubbard B, et al. Toward engineering functional organ modules by additive manufacturing. Biofabrication 2012;4(2):022001.

[48] Bertassoni LE, Cardoso JC, Manoharan V, Cristino AL, Bhise NS, Araujo WA, et al. Direct-write bioprinting of cell-laden methacrylated gelatin hydrogels. Biofabrication 2014;6(2):024105.

[49] Guillemot F, Souquet A, Catros S, Guillotin B, Lopez J, Faucon M, et al. High-throughput laser printing of cells and biomaterials for tissue engineering. Acta Biomater 2010;6(7):2494–500.

[50] Gruene M, Deiwick A, Koch L, Schlie S, Unger C, Hofmann N, et al. Laser printing of stem cells for biofabrication of scaffold-free autologous grafts. Tissue Eng Part C Methods 2011;17(1):79–87.

[51] Guillotin B, Souquet A, Catros S, Duocastella M, Pippenger B, Bellance S, et al. Laser assisted bioprinting of engineered tissue with high cell density and microscale organization. Biomaterials 2010;31(28):7250–6.

[52] Koch L, Gruene M, Unger C, Chichkov B. Laser assisted cell printing. Curr Pharm Biotechnol 2013;14(1):91–7.

[53] Fedorovich NE, Alblas J, Hennink WE, Oner FC, Dhert WJ. Organ printing: the future of bone regeneration? Trends Biotechnol 2011;29(12):601–6.

[54] Fedorovich NE, Oudshoorn MH, van Geemen D, Hennink WE, Alblas J, Dhert WJ. The effect of photopolymerization on stem cells embedded in hydrogels. Biomaterials 2009;30(3):344–53.

[55] Current good manufacturing practice and investigational new drugs intended for use in clinical trials. Final rule. Fed Regist 2008; 73(136):40453–40463.

[56] Schenone M, Dancik V, Wagner BK, Clemons PA. Target identification and mechanism of action in chemical biology and drug discovery. Nat Chem Biol 2013;9(4):232–40.

[57] Fischbach C, Chen R, Matsumoto T, Schmelzle T, Brugge JS, Polverini PJ, et al. Engineering tumors with 3D scaffolds. Nat Methods 2007;4(10):855–60.

[58] Peck Y, Wang DA. Three-dimensionally engineered biomimetic tissue models for *in vitro* drug evaluation: delivery, efficacy and toxicity. Expert Opin Drug Deliv 2013;10(3):369–83.

[59] Janorkar AV, Harris LM, Murphey BS, Sowell BL. Use of three-dimensional spheroids of hepatocyte-derived reporter cells to study the effects of intracellular fat accumulation and subsequent cytokine exposure. Biotechnol Bioeng 2011;108(5):1171–80.

[60] Astashkina A, Mann B, Grainger DW. A critical evaluation of *in vitro* cell culture models for high-throughput drug screening and toxicity. Pharmacol Ther 2012;134(1):82–106.

[61] Berg L, Hsu YC, Lee JA. Consideration of the cellular microenvironment: physiologically relevant co-culture systems in drug discovery. Adv Drug Deliv Rev 2014;190–204. 69-70.

[62] King SM, Presnell SC, Nguyen DG. Development of 3D bioprinted human breast cancer for *in vitro* drug screening. American association for cancer research national conference. San Diego, CA: American Association for Cancer Research; 2014.

[63] Nelson LJ, Walker SW, Hayes PC, Plevris JN. Low-shear modelled microgravity environment maintains morphology and differentiated functionality of primary porcine hepatocyte cultures. Cells Tissues Organs 2010;192(2):125–40.

[64] Hewitt NJ, Lechon MJ, Houston JB, Hallifax D, Brown HS, Maurel P, et al. Primary hepatocytes: current understanding of the regulation of metabolic enzymes and transporter proteins, and pharmaceutical practice for the use of hepatocytes in metabolism, enzyme induction, transporter, clearance, and hepatotoxicity studies. Drug Metab Rev 2007;39(1):159–234.

[65] Temple S. The development of neural stem cells. Nature 2001;414(6859):112–7.

[66] Boelsterli UA. Diclofenac-induced liver injury: a paradigm of idiosyncratic drug toxicity. Toxicol Appl Pharmacol 2003;192(3):307–22.

[67] Robbins JB, O'Nei CM, Gorgen VA, Presnell SC, Shepherd BR. Bioprinted three dimensional (3D) human liver constructs provide a model for interrogating liver biology. American Society for Cell Biology. New Orleans, LA: Mol Biol Cell; 2013. p. 963.

[68] Fedorovich NE, Leeuwenburgh SC, van der Helm YJ, Alblas J, Dhert WJ. The osteoinductive potential of printable, cell-laden hydrogel-ceramic composites. J Biomed Mater Res A 2012;100(9):2412–20.

[69] Amoh Y, Hamada Y, Aki R, Kawahara K, Hoffman RM, Katsuoka K. Direct transplantation of uncultured hair-follicle pluripotent stem (hfPS) cells promotes the recovery of peripheral nerve injury. J Cell Biochem 2010;110(1):272–7.

[70] Jain D, Ludlow JW, Halberstadt C, Payne R, Wagner BJ, Jayo MJ, et al. An autologous smooth muscle cell (SMC) seeded PLGA-based scaffold (neo-bladder conduit) for establishing an incontinent urinary diversion. Regenerative medicine: advancing the next generation therapies. Hilton Head Island, SC; 2009.

[71] Dahl S, Smith O. Science Translational Medicine Podcast: 2 February 2011. Sci Transl Med 2011;3(68):68pc2.

[72] Jayo MJ, Rivera E, Sharp W, Wagner BJ, Ludlow JW, Jain D, Bertram TA. Regeneration of native-like mucocutaneous region at the skin-conduit junction following Neo-Urinary Conduit™ Implantation. Tissue Engineering and Regenerative Medicine International Society (TERMIS) – North America Annual Conference; December 5–8, 2010.

[73] Waran V, Devaraj P, Hari Chandran T, Muthusamy KA, Rathinam AK, Balakrishnan YK, et al. Three-dimensional anatomical accuracy of cranial models created by rapid prototyping tech-niques validated using a neuronavigation station. J Clin Neurosci 2012;19(4):574–7.

[74] Waran V, Menon R, Pancharatnam D, Rathinam AK, Balakrishnan YK, Tung TS, et al. The creation and verification of cranial models using three-dimensional rapid prototyping technology in field of transnasal sphenoid endoscopy. Am J Rhinol Allergy 2012;26(5):e132–e1326.

[75] Waran V, Pancharatnam D, Thambinayagam HC, Raman R, Rathinam AK, Balakrishnan YK, et al. The utilization of cranial models created using rapid prototyping techniques in the development of models for navigation training. J Neurol Surg A Cent Eur Neurosurg 2014;75(1):12–5.

[76] Allori AC, Davidson EH, Reformat DD, Sailon AM, Freeman J, Vaughan A, et al. Design and validation of a dynamic cell-culture system for bone biology research and exogenous tissue-engineering applications. J Tissue Eng Regen Med 2013; doi: 10.1002/term.1810. [Epub ahead of print].

[77] Sailon AM, Allori AC, Davidson EH, Reformat DD, Allen RJ, Warren SM. A novel flow-perfusion bioreactor supports 3D dynamic cell culture. J Biomed Biotechnol 2009;2009:873816.

[78] Vinci B, Duret C, Klieber S, Gerbal-Chaloin S, Sa-Cunha A, Laporte S, et al. Modular bioreactor for primary human hepatocyte culture: medium flow stimulates expression and activity of detoxification genes. Biotechnol J 2011;6(5):554–64.

[79] Tehranirokh M, Kouzani AZ, Francis PS, Kanwar JR. Microfluidic devices for cell cultivation and proliferation. Biomicrofluidics 2013;7(5):51502.

[80] Fenrich KK, Zhao EY, Wei Y, Garg A, Rose PK. Isolating specific cell and tissue compartments from 3D images for quantitative regional distribution analysis using novel computer algorithms. J Neurosci Methods 2014;226:42–56.

Subject Index